Reference data

CIBSE Guide C

CIBSE

BUTTERWORTH
HEINEMANN

Oxford   Auckland   Boston   Johannesburg   Melbourne   New Delhi

Butterworth-Heinemann
Linacre House, Jordan Hill, Oxford OX2 8DP
225 Wildwood Avenue, Woburn, MA 01801-2041
A division of Reed Educational and Professional Publishing Ltd

℞ A member of the Reed Elsevier plc group

First published 2001

Note from the Publisher

This publication is primarily intended to provide guidance to those responsible for the design,
installation, commissioning, operation and maintenance of buildings. It is not intended to be
exhaustive or definitive and it will be necessary for users of the guidance given to exercise
their own professional judgement when deciding whether to abide by or depart from it.

British Library Cataloguing in Publication Data
Reference data : CIBSE guide C
  1. Buildings — Environmental engineering
  I. Butcher, Ken   II. Chartered Institution of Building Services Engineers
  696

ISBN 0 7506 5360 4

Typeset in Great Britain by J&L Composition Ltd, Filey, North Yorkshire
Printed and Bound in Great Britain by MPG Books, Bodmin, Cornwall

Foreword

This 2001 edition of CIBSE Guide C contains significant changes from the previous 1986 edition. Although basic physical data does not change with time, the refinement of measurement and calculation techniques means that even this basic data needs to be reviewed at intervals. Additionally, revisions to practice in calculation methods, product selection and usage led to a requirement for changes in the presentation of the data. Much of the data in the 1986 edition actually dates from 1970 and was therefore ripe for review.

The changes made for the 2001 edition are summarised below:

Section 1: Properties of humid air. The data tables remain unchanged but the introduction has been updated. A critique of the data has been published in *Building Services Engineering Research and Technology*, to which interested readers are referred (reference 11 to section 1).

Section 2: Properties of water and steam. This section has been reviewed and found to need no amendment.

Section 3: Heat transfer. This section has been completely rewritten. The theoretical basis has been reviewed in the light of current knowledge and the calculation procedures updated. Reference tables have been updated to reflect current practice and needs.

Section 4: Flow of fluids in pipes and ducts. This section has been completely rewritten. Both calculation methods and reference tables have been updated. In particular, the opportunity has been taken to replace the limited and oversimplified tables of resistance coefficients with a more comprehensive and rigorous treatment.

Section 5: Fuels and combustion. This section has been reviewed and comprehensively updated to take account of changes to fuels and fuel characteristics.

Section 6: Units, standard and mathematical data. This section has been reviewed and updated. Some obsolete data has been deleted and some extra data added.

These changes have taken a significant amount of time and effort to complete and I would like to express my thanks to the volunteer authors, contributors, reviewers and CIBSE staff for their valuable contributions.

Finally, we hope that all users will find this Guide a useful and authoritative source of reference and guidance.

Paul Compton
Chairman, CIBSE Guide C

Guide C Steering Committee

P D Compton (Chairman) (Colt International Ltd)
W P Jones (consultant)
P Koch (Université Joseph Fourier, Grenoble; Coventry University)
D L Loveday (Loughborough University)
M R I Purvis (University of Portsmouth)
A C Watson (secretary)

Section task groups, principal authors, contributors and acknowledgements

Section 1: Properties of humid air

Principal author
W P Jones (consultant)

The tables of data are reprinted from the previous edition of Guide C and were prepared by a task group, see below.

Task group members
W P Jones (Chairman) (consultant)
J F Armour
B G Lawrence

Section 2: Properties of water and steam

The tables of data are reprinted from the previous edition of Guide C.

Section 3: Heat transfer

Principal authors
D L Loveday (Loughborough University)
A H Taki (De Montford University)

Contributors
H B Awbi (University of Reading)
P D Compton (Colt International Ltd)
R M Harris (Centre for Window and Cladding Technology)
M J Holmes (Ove Arup & Partners International Ltd)
B P Holownia (Loughborough University)
J Moss (Ove Arup & Partners International Ltd)
T Muneer (Napier University)
H K Versteeg (Loughborough University)

Acknowledgements
American Society of Heating, Refrigerating and Air-Conditioning Engineers Inc
International Standards Organisation
The McGraw-Hill Companies
Pearson Education Ltd

Section 4: Flow of fluids in pipes and ducts

Principal author
P Koch (Université Joseph Fourier, Grenoble; Coventry University)

Contributor
F Sprenger (Coventry University)

Acknowledgements
American Society of Heating, Refrigerating and Air-Conditioning Engineers Inc
Centre Technique des Industries Aérauliques et Thermiques
Coventry University, School of Science and the Environment
CRC Press Inc
Sheet Metal and Air Conditioning Contractors' National Association

Section 5: Fuels and combustion

Task group members/Principal authors
M R I Purvis (Chairman) (University of Portsmouth)
R Dando (Coal Research Establishment)
M Drew (BP Amoco plc)
R J Harris (Advantica Technologies Ltd)
K Mildren (University of Portsmouth)

Acknowledgements
BP Amoco plc

Section 6: Units, standard and mathematical data

Principal author
P D Compton (Chairman) (Colt International Ltd)

Co-ordinating editors

K J Butcher (CIBSE)
A C Watson (CIBSE)

CIBSE Director of Information and Policy

J Balian

Contents

1   Properties of humid air

1.1   Psychrometric data

1.1.1   Basis of calculation

The method of formulation suggested by Goff and Gratch[1,2], based on the ideal gas laws with a modification to take account of intermolecular forces, has been adopted for calculating the thermodynamic properties of moist air. This approach remains in line with current practice[3,4].

The thermodymamic properties of dry air and saturated water vapour are well established and, although more recent research work[4,5,6] has been done, the results are not significantly different from those obtained in earlier work[7,8]. Hence the thermodynamic properties of dry air and water vapour, determined by the National Bureau of Standards[7] and the National Engineering Laboratory[8], have been retained for the evaluation of the thermodynamic properties of moist air.

Since the properties of dry air and saturated water vapour are accurately known, the properties of a mixture of the two can be established for the saturated case. For the enthalpy and specific volume of moist air, at a condition other than saturated, the method is exemplified by the following equations:

$$h = h_a + \mu (h_s - h_a) / 100 \qquad (1.1)$$

$$v = v_a + \mu (v_s - v_a) / 100 \qquad (1.2)$$

where

h = specific enthalpy of moist air (kJ.kg^{-1} dry air)

h_a = specific enthalpy of dry air (kJ.kg^{-1})

μ = percentage saturation (%)

h_s = specific enthalpy of saturated moist air (kJ.kg^{-1} dry air)

v = specific volume of moist air (m^3.kg^{-1} dry air)

v_a = specific volume of dry air (m^3.kg^{-1})

v_s = specific volume of saturated moist air (m^3.kg^{-1} dry air)

The relevant specific property of moist, unsaturated air is determined by adding a proportion of the property of saturated water vapour to the same property of dry air, on a mass basis.

A consequence of this is that the humidity of moist air is expressed as percentage saturation (defined in terms of the mass of water vapour present), rather than relative humidity (defined in terms of vapour pressure). The details of the psychrometric calculations are given in references 9, 10 and 11.

1.1.2   Standards adopted

All data are tabulated for an internationally agreed standard atmospheric pressure[12] of 101.325 kPa.

The zero datum adopted by the National Engineering Laboratory[8] for the expression of the thermodynamic properties of steam is the triple point of water, +0.01°C.

The zero datum for the specific enthalpies of both dry air and liquid water has been taken here as 273.15 K (0°C).

1.1.3   Formulae used for calculations

Saturated vapour pressure over water[8]:

$$\log p_s = 30.59051 - 8.2 \log (t + 273.16)$$
$$+ 2.4804 \times 10^{-3} (t + 273.16)$$
$$- [3142.31 / (t + 273.16)] \qquad (1.3)$$

where

p_s = saturated vapour pressure over water at temperature t (kPa)

t = temperature, greater than or equal to 0°C (°C)

Saturated vapour pressure over ice[7]:

$$\log p_s = 12.5380997 - [266391 / (t + 273.15)] \quad (1.4)$$

where

p_s = saturated vapour pressure over ice at temperature t, less than 0°C (kPa)

Moisture content:

$$g_s = \frac{0.62197 f_s p_s}{101.325 - f_s p_s} \quad (1.5)$$

where

g_s = moisture content of saturated moist air (kg.kg^{-1} dry air)

f_s = dimensionless enhancement factor[1,2,3,4,11]

Percentage saturation:

$$\mu = \frac{100 \, g}{g_s} \quad (1.6)$$

where

μ = percentage saturation (%)

g = moisture content of unsaturated moist air (kg.kg^{-1} dry air)

Vapour pressure of water vapour in unsaturated moist air:

$$p_v = \frac{p_a g}{f_s(0.62197 + g)} \quad (1.7)$$

where

p_v = vapour pressure of superheated water vapour in unsaturated moist air (kPa)

p_a = atmospheric (barometric) pressure (kPa)

Relative humidity:

$$\phi = \frac{100 p_v}{p_s} \quad (1.8)$$

where

ϕ = relative humidity (%)

Wet-bulb temperature:

Knowing the value of the vapour pressure, p_v, from equation (1.7) the wet-bulb temperature is derived from the following equations by an iterative technique:

$$p_v = p_{s1} - 101.325 A(t - t'_{s1}) \quad (1.9)$$

where

p_{s1} = saturated vapour pressure at temperature t_{s1} (kPa)

A = 6.66×10^{-4} K^{-1} when $t'_{s1} \geqslant$ 0°C

A = 5.94×10^{-4} K^{-1} when $t'_{s1} <$ 0°C

t = dry-bulb temperature (°C)

t'_{s1} = sling or mechanically aspirated wet-bulb temperature (°C)

or

$$p_v = p_{sc} - 101.325 B(t - t'_{sc}) \quad (1.10)$$

where

p_{sc} = saturated vapour pressure at temperature t_{sc} (kPa)

t'_{sc} = screen wet-bulb temperature (°C)

B = 7.99×10^{-4} K^{-1} when $t_{sc} \geqslant$ 0°C

B = 7.20×10^{-4} K^{-1} when $t_{sc} <$ 0°C

Adiabatic saturation temperature:

$$t^\star = t - \frac{h_{fg}(g_{sa} - g)}{(c_{pa} + g c_{ps})} \quad (1.11)$$

where

$t^\star$ = adiabatic saturation temperature (°C)

h_{fg} = latent heat of evaporation of water at temperature t_a (kJ.kg^{-1})

g_{sa} = moisture content of saturated air at temperature t_a (kg.kg^{-1} dry air)

g = moisture content of moist air at the particular psychrometric state (kg.kg^{-1} dry air)

c_{pa} = mean specific heat capacity of dry air between t and t_a (kJ.kg^{-1}.K^{-1})

c_{ps} = mean specific heat capacity of water vapour between t and t_a (kJ.kg^{-1}.K^{-1})

In the case of the adiabatic saturation temperature above ice, h_{fg} is replaced by h_{ig}, the latent heat of fusion of water at a temperature t_a.

Dew-point:

For a particular psychrometric state, equation (1.7) is used to calculate the vapour pressure. An iterative technique is then used with equation (1.3) or (1.4) to determine the temperature for which the calculated vapour pressure is a saturated vapour pressure.

Specific volume:

$$v = \left[\frac{82.0567(273.15 + t)}{28.966(101.325 - p_v) / 101.325} \right] - [A_{aa}x_a^2 + 2A_{aw}x_a(1 - x_a) + A_{ww}(1 - x_a)^2] \quad (1.12)$$

where

v = specific volume (m^3.kg^{-1} dry air)

t = dry-bulb temperature (°C)

A_{aa} = second virial coefficient for dry air[4] (m^3.kg^{-1})

A_{aw} = interaction coefficient for moist air[4] (m^3.kg^{-1})

A_{ww} = second virial coefficient for water vapour (m^3.kg^{-1})

$$x_a = \frac{0.62197}{0.62197 + g} \quad (1.13)$$

where

$$x_a \quad = \text{mole fraction of dry air}$$

In the original work[1,2] and in more recent research[4], a third virial coefficient for water vapour, A_{www}, appears in equation (1.12) but it is complicated to calculate and its influence is insignificant. It is ignored here, without any loss of accuracy.

Equation (1.2) yields answers of adequate precision and is easier to use than equation (1.12).

Specific enthalpy:

$$h \quad = h_a + g\,h_g \qquad (1.14)$$

where

$h \quad = $ specific enthalpy of moist air (kJ.kg^{-1} dry air)

$h_a \quad = $ specific enthalpy of dry air[7] (kJ.kg^{-1})

$g \quad = $ moisture content (kg.kg^{-1} dry air)

$h_g \quad = $ specific enthalpy of water vapour at the dry-bulb temperature[8] (kJ.kg^{-1} dry air)

Equation (1.1) gives answers having the same accuracy as those obtained from equation (1.14) and is simpler to use.

1.1.4 Psychrometric properties at non-standard barometric pressures

The tabulated psychrometric data are accurate within the range of barometric pressure from 95 kPa to 105 kPa and hence are suitable for the whole of the United Kingdom. For pressures outside these limits an application of the ideal gas laws will give answers of a little less accuracy. Better answers may be obtained by the use of equations (1.15) and (1.16):

$$g_s \quad = \frac{0.624 p_s}{(p_a - 1.004 p_s)} \qquad (1.15)$$

$$v \quad = \frac{(0.287 + 0.461g)(273.15 + t)}{p_a} \qquad (1.16)$$

Corrections to specific enthalpy may be taken from Table 1.1.

Figure 1.1, which gives the relationship between height above sea level and barometric pressure, is drawn from the equation:

$$p_a = 101.325 \exp[(-9.81\,\rho a)/(101\,325)] \qquad (1.17)$$

Table 1.1 Corrections to specific enthalpy at non-standard pressures

Adiabatic saturation temperature °C	Approximate additive corrections to specific enthalpy (kJ.kg^{-1} dry air) at various barometric pressures / kPa								
	82.5	85.0	87.5	90.0	92.5	95.0	97.5	101.325	102.5
30	16.90	14.23	11.68	9.29	6.95	4.80	2.86	0	−0.82
29	15.90	13.40	11.00	8.72	6.55	4.57	2.70	0	−0.77
28	14.95	12.58	10.30	8.18	6.16	4.30	2.54	0	−0.72
27	14.00	11.78	9.65	7.67	5.80	4.05	2.40	0	−0.68
26	13.05	11.02	9.03	7.18	5.44	3.82	2.27	0	−0.64
25	12.20	10.28	8.42	6.70	5.12	3.58	2.14	0	−0.60
24	11.43	9.64	7.90	6.30	4.80	3.36	2.00	0	−0.56
23	10.68	9.03	7.40	5.88	4.43	3.15	1.86	0	−0.52
22	10.00	8.45	6.93	5.51	4.20	2.94	1.73	0	−0.48
21	9.37	7.92	6.50	5.18	3.92	2.74	1.61	0	−0.45
20	8.77	7.42	6.10	4.84	3.65	2.55	1.50	0	−0.42
19	8.22	6.95	5.70	4.53	3.43	2.39	1.40	0	−0.39
18	7.73	6.49	5.35	4.24	3.20	2.23	1.30	0	−0.37
17	7.25	6.09	5.00	3.97	3.00	2.07	1.21	0	−0.35
16	6.79	5.68	4.65	3.72	2.80	1.94	1.13	0	−0.32
15	6.33	5.32	4.34	3.48	2.62	1.82	1.07	0	−0.30
14	5.90	4.95	4.07	3.24	2.44	1.70	1.00	0	−0.28
13	5.50	4.60	3.80	3.03	2.28	1.60	0.93	0	−0.26
12	5.13	4.30	3.53	2.82	2.12	1.50	0.86	0	−0.24
11	4.78	4.04	3.28	2.62	1.97	1.40	0.80	0	−0.22
10	4.44	3.77	3.08	2.46	1.82	1.30	0.74	0	−0.20
9	4.15	3.51	2.88	2.30	1.70	1.21	0.70	0	−0.20
8	3.88	3.30	2.68	2.14	1.60	1.12	0.66	0	−0.19
7	3.62	3.08	2.51	2.00	1.50	1.06	0.62	0	−0.19
6	3.40	2.88	2.37	1.87	1.40	1.00	0.59	0	−0.18
5	3.20	2.72	2.23	1.74	1.31	0.92	0.56	0	−0.17
4	3.06	2.60	2.10	1.64	1.24	0.88	0.53	0	−0.17
3	2.92	2.47	2.02	1.59	1.19	0.84	0.50	0	−0.16
2	2.78	2.36	1.94	1.54	1.15	0.80	0.48	0	−0.16
1	2.65	2.25	1.86	1.49	1.10	0.76	0.46	0	−0.15
0	2.52	2.16	1.79	1.44	1.08	0.72	0.44	0	−0.15

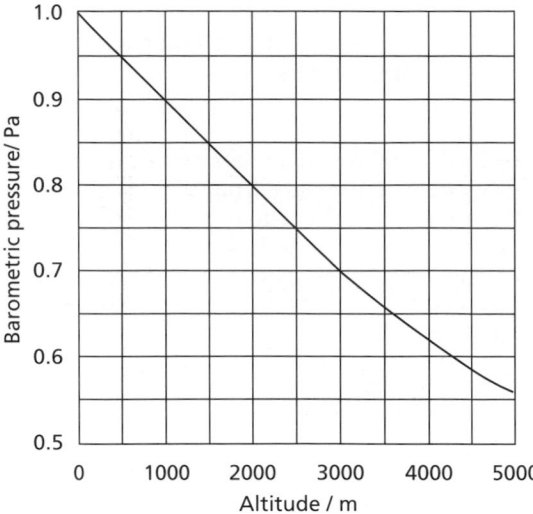

Figure 1.1 Variation of barometric pressure with altitude

where

p_a = the particular atmospheric (barometric) pressure (kPa)

ρ = density of air (kg.m^{-3})

a = altitude above sea level (m)

Alternatively, the standard relationship[12] for altitude, atmospheric pressure and temperature may be used. This is reproduced in Table 1.2.

Table 1.2 Standard atmospheric data for altitudes to 10 000 m

Altitude (m)	Temperature (°C)	Pressure (kPa)
−500	18.2	107.478
0	15.0	101.325
500	11.8	95.461
1000	8.5	89.874
2000	2.0	79.495
3000	−4.5	70.108
4000	−11.0	61.640
5000	−17.5	54.020
6000	−24.0	47.181
7000	−30.5	41.061
8000	−37.0	35.600
9000	−43.5	30.742
10000	−50.0	26.436

1.2 CIBSE psychrometric chart (−10 to +60°C)

The chart has been designed[13] and constructed using the two fundamental properties of mass (moisture content) and energy (specific enthalpy) as linear co-ordinates. Other physical properties are not then shown as linear scales[14]. The 30°C dry-bulb line has been constructed at right angles to lines of constant moisture content, which are horizontal. The scale of specific enthalpy is obliquely inclined to the vertical scale of moisture content. In this way, lines of constant dry-bulb temperature are approximately vertical,

diverging slightly on each side of the 30°C line, and the traditional appearance of the chart is preserved.

The wet-bulb values shown are those read from a sling or mechanically aspirated psychrometer and lines of percentage saturation are plotted instead of relative humidity. Within the comfort zone, there is no practical difference between percentage saturation and relative humidity. In any case, the difference diminishes as saturated or dry conditions are approached.

The psychrometric data used were taken from the tables of the properties of humid air in this section of CIBSE Guide C.

1.3 CIBSE psychrometric chart (10 to 120°C)

The psychrometric chart for 10 to 120°C has been based on the ideal gas laws. This does not give a significant difference when compared with a chart constructed using more accurate data, based on the method of Goff and Gratch[1,2]. The principles of calcuation and drawing are detailed elsewhere[14].

References

1   Goff, J A and Gratch, S, 'Thermodynamic properties of moist air', *Trans. ASHVE*, **51**, 125–164 (1945)

2   Goff, J A, 'Standardisation of thermodynamic properties of moist air', *Trans. ASHVE*, **55**, 459–484 (1949)

3   *ASHRAE Handbook* Fundamentals, Section 6 Psychrometrics (1997)

4   Hyland, R W and Wexler, A, 'Formulations for the thermodynamic properties of dry air from 173.15 K to 473.15 K and of saturated moist air from 173.15 K to 372.15 K at pressures to 5 MPa', *Trans. ASHRAE*, **89**(2A), 520–535 (1982)

5   Hyland, R W and Wexler, A 'Formulations for the thermodynamic properties of the saturated phases of H₂O from 173.15 K to 473.15 K', *Trans. ASHRAE*, **89**(2A), 500–519 (1983)

6   Stimson, H F, 'Some precise measurements of the vapour pressure of water in the range from 25°C to 100°C', *J. Res. NBS* **73A** (1969)

7   Tables of thermal properties of gases, *NBS Circular 564* (National Bureau of Standards, November 1955)

8   National Engineering Laboratory steam tables (HMSO 1964)

9   Jones, W P and Lawrence, B G, 'New psychrometric data for air' *Technical memorandum no. 11* (Polytechnic of the South Bank)

10   Some fundamental data used by building services engineers (IHVE, 1973)

11   Jones, W P, 'A review of CIBSE psychrometry', *Building Serv. Eng. Res. Technol.* **15**(4), 189–198 (1994)

12   NACA 1955, Standard atmosphere – tables and data for altitudes to 65 000 feet, NACA Report 1235:66, Washington DC

13   Jones, W P, 'The Psychrometric Chart in SI Units', *JIHVE*, **38**, 93 (1970)

14   Bull, L C, 'Design and use of the new IHVE psychrometric chart', *JIHVE*, **32**, 268 (October 1964)

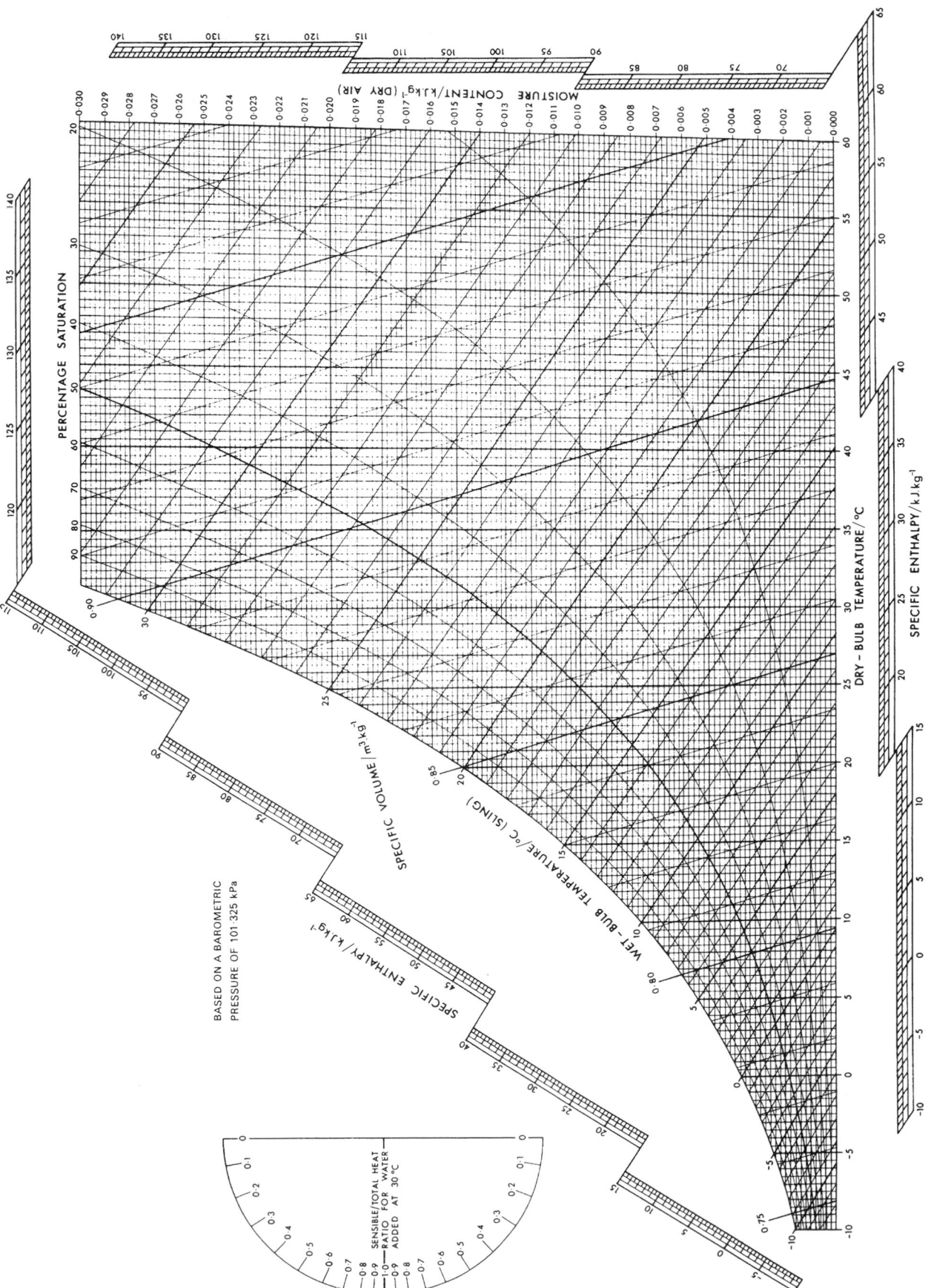

Figure 1.2 CIBSE Psychrometric chart (−10 to +60°C)

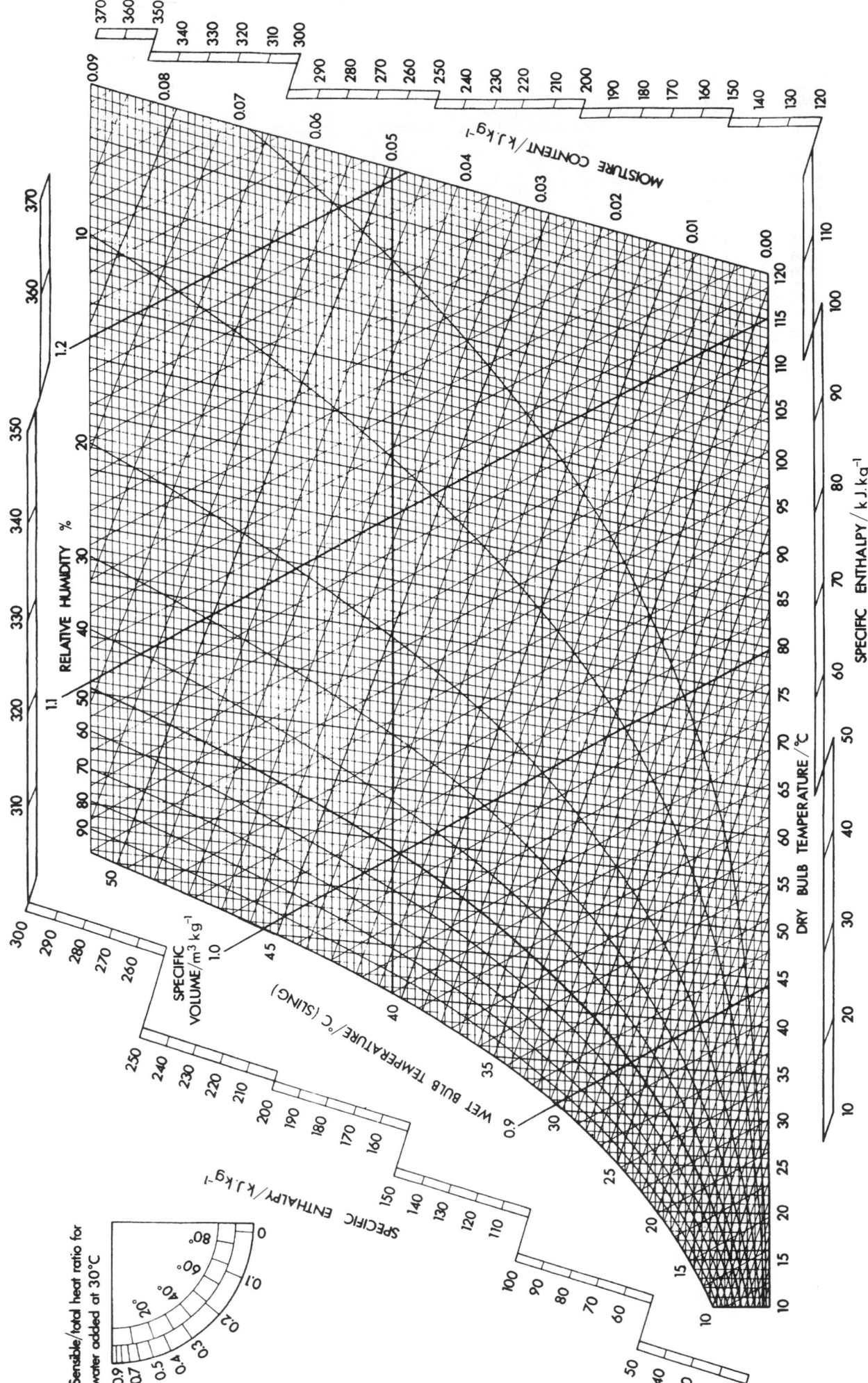

Figure 1.3 CIBSE Psychrometric chart (10 to 120°C)

−10°C DRY BULB

Percentage saturation μ/%	Relative humidity ϕ/%	Value of stated parameter per kg dry air			Vapour pressure p_s/kPa	Dew point temperature t_d/°C	Adiabatic saturation temperature $t^\star$/°C	Wet bulb temperature	
		Moisture content g/g.kg^{-1}	Specific enthalpy h/kJ.kg^{-1}	Specific volume v/m^3.kg^{-1}				Screen t'_{sc}/°C	Sling t'_{sl}/°C
100	100.00	1.607	−6.065	0.7468	0.2600	−10.0	−10.0	−10.0	−10.0
96	96.01	1.543	−6.224	0.7468	0.2496	−10.5	−10.1	−10.1	−10.1
92	92.02	1.479	−6.384	0.7467	0.2392	−10.9	−10.3	−10.2	−10.3
88	88.03	1.414	−6.544	0.7466	0.2288	−11.4	−10.4	−10.3	−10.4
84	84.03	1.350	−6.703	0.7465	0.2185	−11.9	−10.5	−10.4	−10.5
80	80.04	1.286	−6.863	0.7465	0.2081	−12.5	−10.7	−10.5	−10.6
76	76.05	1.221	−7.022	0.7464	0.1977	−13.1	−10.8	−10.7	−10.8
72	72.05	1.157	−7.182	0.7463	0.1873	−13.6	−10.9	−10.8	−10.9
70	70.05	1.125	−7.262	0.7463	0.1821	−14.0	−11.0	−10.8	−10.9
68	68.06	1.093	−7.342	0.7462	0.1769	−14.3	−11.0	−10.9	−11.0
66	66.06	1.061	−7.421	0.7462	0.1717	−14.6	−11.1	−10.9	−11.1
64	64.06	1.029	−7.501	0.7462	0.1665	−14.9	−11.2	−11.0	−11.1
62	62.06	0.996	−7.581	0.7461	0.1613	−15.3	−11.2	−11.0	−11.2
60	60.06	0.964	−7.661	0.7461	0.1562	−15.6	−11.3	−11.1	−11.3
58	58.06	0.932	−7.740	0.7460	0.1510	−16.0	−11.4	−11.1	−11.3
56	56.06	0.900	−7.820	0.7460	0.1458	−16.4	−11.4	−11.2	−11.4
54	54.06	0.868	−7.900	0.7460	0.1406	−16.8	−11.5	−11.3	−11.5
52	52.06	0.836	−7.980	0.7459	0.1354	−17.2	−11.6	−11.3	−11.5
50	50.06	0.804	−8.060	0.7459	0.1302	−17.6	−11.6	−11.4	−11.6
48	48.06	0.772	−8.139	0.7458	0.1250	−18.0	−11.7	−11.4	−11.7
46	46.06	0.739	−8.219	0.7458	0.1198	−18.5	−11.8	−11.5	−11.7
44	44.06	0.707	−8.299	0.7458	0.1146	−18.9	−11.8	−11.5	−11.8
42	42.06	0.675	−8.379	0.7457	0.1094	−19.4	−11.9	−11.6	−11.8
40	40.06	0.643	−8.459	0.7457	0.1042	−19.9	−12.0	−11.6	−11.9
38	38.06	0.611	−8.538	0.7457	0.0990	−20.5	−12.0	−11.7	−12.0
36	36.06	0.579	−8.618	0.7456	0.0938	−21.0	−12.1	−11.8	−12.0
34	34.06	0.546	−8.698	0.7456	0.0885	−21.6	−12.2	−11.8	−12.1
32	32.06	0.514	−8.778	0.7455	0.0833	−22.2	−12.3	−11.9	−12.2
30	30.05	0.482	−8.858	0.7455	0.0781	−22.9	−12.3	−11.9	−12.2
28	28.05	0.450	−8.937	0.7455	0.0729	−23.6	−12.4	−12.0	−12.3
24	24.05	0.386	−9.097	0.7454	0.0625	−25.2	−12.5	−12.1	−12.4
20	20.04	0.321	−9.256	0.7453	0.0521	−27.0	−12.7	−12.2	−12.6
16	16.03	0.257	−9.416	0.7452	0.0417	−29.2	−12.8	−12.3	−12.7
12	12.03	0.193	−9.576	0.7452	0.0313	−31.9	−12.9	−12.4	−12.8
8	8.02	0.129	−9.735	0.7451	0.0208	−35.7	−13.1	−12.6	−13.0
4	4.01	0.064	−9.895	0.7450	0.0104	−41.9	−13.2	−12.7	−13.1
0	0.00	0.000	−10.054	0.7449	0.0000	—	−13.4	−12.8	−13.2

−9.5°C DRY BULB

Percentage saturation μ/%	Relative humidity ϕ/%	Value of stated parameter per kg dry air			Vapour pressure p_s/kPa	Dew point temperature t_d/°C	Adiabatic saturation temperature $t^\star$/°C	Wet bulb temperature	
		Moisture content g/g.kg^{-1}	Specific enthalpy h/kJ.kg^{-1}	Specific volume v/m^3.kg^{-1}				Screen t'_{sc}/°C	Sling t'_{sl}/°C
100	100.00	1.680	−5.380	0.7484	0.2717	−9.5	−9.5	−9.5	−9.5
96	96.01	1.613	−5.547	0.7483	0.2609	−10.0	−9.6	−9.6	−9.6
92	92.02	1.546	−5.713	0.7482	0.2500	−10.4	−9.8	−9.7	−9.8
88	88.03	1.478	−5.880	0.7481	0.2392	−10.9	−9.9	−9.8	−9.9
84	84.04	1.411	−6.047	0.7480	0.2284	−11.5	−10.0	−9.9	−10.0
80	80.04	1.344	−6.214	0.7480	0.2175	−12.0	−10.2	−10.1	−10.1
76	76.05	1.277	−6.381	0.7479	0.2066	−12.6	−10.3	−10.2	−10.3
72	72.05	1.210	−6.548	0.7478	0.1958	−13.2	−10.4	−10.3	−10.4
70	70.06	1.176	−6.631	0.7478	0.1904	−13.5	−10.5	−10.3	−10.5
68	68.06	1.142	−6.715	0.7477	0.1849	−13.8	−10.6	−10.4	−10.5
66	66.06	1.109	−6.798	0.7477	0.1795	−14.1	−10.6	−10.5	−10.6
64	64.06	1.075	−6.882	0.7476	0.1740	−14.5	−10.7	−10.5	−10.7
62	62.06	1.042	−6.965	0.7476	0.1686	−14.8	−10.8	−10.6	−10.7
60	60.06	1.008	−7.048	0.7476	0.1632	−15.2	−10.9	−10.6	−10.8
58	58.07	0.974	−7.132	0.7475	0.1578	−15.5	−10.9	−10.7	−10.9
56	56.07	0.940	−7.215	0.7475	0.1524	−15.9	−11.0	−10.7	−10.9
54	54.07	0.907	−7.299	0.7474	0.1469	−16.3	−11.1	−10.8	−11.0
52	52.07	0.874	−7.382	0.7474	0.1415	−16.7	−11.1	−10.9	−11.1
50	50.07	0.840	−7.466	0.7474	0.1360	−17.1	−11.2	−10.9	−11.1
48	48.07	0.806	−7.549	0.7473	0.1306	−17.5	−11.3	−11.0	−11.2
46	46.07	0.773	−7.633	0.7473	0.1252	−18.0	−11.3	−11.0	−11.3
44	44.07	0.739	−7.716	0.7472	0.1197	−18.5	−11.4	−11.1	−11.3
42	42.07	0.706	−7.800	0.7472	0.1143	−19.0	−11.5	−11.2	−11.4
40	40.06	0.672	−7.883	0.7472	0.1089	−19.5	−11.6	−11.2	−11.5
38	38.06	0.638	−7.966	0.7471	0.1034	−20.0	−11.6	−11.3	−11.5
36	36.06	0.605	−8.050	0.7471	0.0980	−20.6	−11.7	−11.3	−11.6
34	34.06	0.571	−8.133	0.7470	0.0926	−21.2	−11.8	−11.4	−11.7
32	32.06	0.538	−8.217	0.7470	0.0871	−21.8	−11.8	−11.4	−11.7
30	30.06	0.504	−8.300	0.7470	0.0817	−22.5	−11.9	−11.5	−11.8
28	28.05	0.470	−8.384	0.7469	0.0762	−23.2	−12.0	−11.6	−11.9
24	24.05	0.403	−8.550	0.7468	0.0654	−24.7	−12.1	−11.7	−12.0
20	20.04	0.336	−8.717	0.7467	0.0545	−26.5	−12.3	−11.8	−12.2
16	16.04	0.269	−8.884	0.7467	0.0436	−28.7	−12.4	−11.9	−12.3
12	12.03	0.201	−9.051	0.7466	0.0327	−31.5	−12.5	−12.0	−12.4
8	8.02	0.134	−9.218	0.7465	0.0218	−35.3	−12.7	−12.1	−12.6
4	4.01	0.067	−9.385	0.7464	0.0109	−41.5	−12.8	−12.3	−12.7
0	0.00	0.000	−9.552	0.7463	0.0000	—	−13.0	−12.4	−12.8

−9°C DRY BULB

Percentage saturation μ/%	Relative humidity ϕ/%	Value of stated parameter per kg dry air			Vapour pressure p_s/kPa	Dew point temperature t_d/°C	Adiabatic saturation temperature $t^\star$/°C	Wet bulb temperature	
		Moisture content $g/g.kg^{-1}$	Specific enthalpy h/kJ.kg^{-1}	Specific volume v/m^3.kg^{-1}				Screen t'_{sc}/°C	Sling t'_{sl}/°C
100	100.00	1.755	−4.687	0.7499	0.2840	−9.0	−9.0	−9.0	−9.0
96	96.01	1.686	−4.862	0.7498	0.2726	−9.5	−9.1	−9.1	−9.1
92	92.02	1.615	−5.036	0.7497	0.2613	−9.9	−9.3	−9.2	−9.3
88	88.03	1.545	−5.210	0.7496	0.2500	−10.4	−9.4	−9.3	−9.4
84	84.04	1.475	−5.385	0.7495	0.2386	−11.0	−9.6	−9.5	−9.5
80	80.04	1.405	−5.559	0.7494	0.2273	−11.5	−9.7	−9.6	−9.7
76	76.05	1.334	−5.734	0.7494	0.2160	−12.1	−9.8	−9.7	−9.8
72	72.06	1.264	−5.908	0.7493	0.2046	−12.7	−10.0	−9.8	−9.9
70	70.06	1.229	−5.996	0.7492	0.1989	−13.0	−10.0	−9.9	−10.0
68	68.06	1.194	−6.083	0.7492	0.1933	−13.3	−10.1	−9.9	−10.1
66	66.06	1.159	−6.170	0.7492	0.1876	−13.6	−10.2	−10.0	−10.1
64	64.06	1.124	−6.257	0.7491	0.1819	−14.0	−10.3	−10.1	−10.2
62	62.07	1.089	−6.345	0.7491	0.1762	−14.3	−10.3	−10.1	−10.3
60	60.07	1.054	−6.432	0.7490	0.1706	−14.7	−10.4	−10.2	−10.4
58	58.07	1.018	−6.519	0.7490	0.1649	−15.0	−10.5	−10.2	−10.4
56	56.07	0.983	−6.606	0.7489	0.1592	−15.4	−10.5	−10.3	−10.5
54	54.07	0.948	−6.694	0.7489	0.1535	−15.8	−10.6	−10.4	−10.6
52	52.07	0.913	−6.781	0.7489	0.1479	−16.2	−10.7	−10.4	−10.6
50	50.07	0.878	−6.868	0.7488	0.1422	−16.6	−10.8	−10.5	−10.7
48	48.07	0.843	−6.955	0.7488	0.1365	−17.1	−10.8	−10.5	−10.8
46	46.07	0.808	−7.042	0.7487	0.1308	−17.5	−10.9	−10.6	−10.8
44	44.07	0.773	−7.130	0.7487	0.1251	−18.0	−11.0	−10.6	−10.9
42	42.07	0.738	−7.217	0.7487	0.1195	−18.5	−11.0	−10.7	−11.0
40	40.07	0.702	−7.304	0.7486	0.1138	−19.0	−11.1	−10.8	−11.0
38	38.07	0.667	−7.391	0.7486	0.1081	−19.5	−11.2	−10.8	−11.1
36	36.06	0.632	−7.479	0.7485	0.1024	−20.1	−11.3	−10.9	−11.2
34	34.06	0.597	−7.566	0.7485	0.0967	−20.7	−11.3	−10.9	−11.3
32	32.06	0.562	−7.653	0.7484	0.0910	−21.3	−11.4	−11.0	−11.3
30	30.06	0.527	−7.740	0.7484	0.0854	−22.0	−11.5	−11.1	−11.4
28	28.06	0.492	−7.828	0.7484	0.0797	−22.7	−11.6	−11.1	−11.5
24	24.05	0.421	−8.002	0.7483	0.0683	−24.3	−11.7	−11.3	−11.6
20	20.04	0.351	−8.177	0.7482	0.0569	−26.1	−11.9	−11.4	−11.7
16	16.04	0.281	−8.351	0.7481	0.0455	−28.3	−12.0	−11.5	−11.9
12	12.03	0.211	−8.526	0.7480	0.0342	−31.1	−12.1	−11.6	−12.0
8	8.02	0.140	−8.700	0.7479	0.0228	−34.9	−12.3	−11.7	−12.2
4	4.01	0.070	−8.875	0.7479	0.0114	−41.1	−12.4	−11.9	−12.3
0	0.00	0.000	−9.049	0.7478	0.0000	—	−12.6	−12.0	−12.5

−8.5°C DRY BULB

Percentage saturation μ/%	Relative humidity ϕ/%	Value of stated parameter per kg dry air			Vapour pressure p_s/kPa	Dew point temperature t_d/°C	Adiabatic saturation temperature $t^\star$/°C	Wet bulb temperature	
		Moisture content $g/g.kg^{-1}$	Specific enthalpy h/kJ.kg^{-1}	Specific volume v/m^3.kg^{-1}				Screen t'_{sc}/°C	Sling t'_{sl}/°C
100	100.00	1.835	−3.986	0.7514	0.2967	−8.5	−8.5	−8.5	−8.5
96	96.01	1.761	−4.169	0.7513	0.2849	−9.0	−8.6	−8.6	−8.6
92	92.02	1.688	−4.351	0.7512	0.2730	−9.4	−8.8	−8.7	−8.8
88	88.03	1.615	−4.534	0.7511	0.2612	−9.9	−8.9	−8.9	−8.9
84	84.04	1.541	−4.716	0.7510	0.2493	−10.5	−9.1	−9.0	−9.1
80	80.05	1.468	−4.898	0.7509	0.2375	−11.0	−9.2	−9.1	−9.2
76	76.05	1.394	−5.081	0.7509	0.2256	−11.6	−9.4	−9.2	−9.3
72	72.06	1.321	−5.263	0.7508	0.2138	−12.2	−9.5	−9.3	−9.5
70	70.06	1.284	−5.354	0.7507	0.2079	−12.5	−9.6	−9.4	−9.5
68	68.06	1.248	−5.446	0.7507	0.2019	−12.8	−9.7	−9.5	−9.6
66	66.07	1.211	−5.537	0.7506	0.1960	−13.2	−9.7	−9.5	−9.7
64	64.07	1.174	−5.628	0.7506	0.1901	−13.5	−9.8	−9.6	−9.8
62	62.07	1.138	−5.719	0.7506	0.1842	−13.8	−9.9	−9.7	−9.8
60	60.07	1.101	−5.810	0.7505	0.1782	−14.2	−9.9	−9.7	−9.9
58	58.07	1.064	−5.902	0.7505	0.1723	−14.6	−10.0	−9.8	−10.0
56	56.07	1.028	−5.993	0.7504	0.1664	−14.9	−10.1	−9.8	−10.0
54	54.07	0.991	−6.084	0.7504	0.1604	−15.3	−10.2	−9.9	−10.1
52	52.07	0.954	−6.175	0.7503	0.1545	−15.7	−10.2	−10.0	−10.2
50	50.07	0.917	−6.266	0.7503	0.1486	−16.2	−10.3	−10.0	−10.3
48	48.07	0.881	−6.358	0.7502	0.1426	−16.6	−10.4	−10.1	−10.3
46	46.07	0.844	−6.449	0.7502	0.1367	−17.1	−10.5	−10.1	−10.4
44	44.07	0.807	−6.540	0.7502	0.1308	−17.5	−10.5	−10.2	−10.5
42	42.07	0.771	−6.631	0.7501	0.1248	−18.0	−10.6	−10.3	−10.5
40	40.07	0.734	−6.722	0.7501	0.1189	−18.5	−10.7	−10.3	−10.6
38	38.07	0.697	−6.814	0.7500	0.1130	−19.1	−10.8	−10.4	−10.7
36	36.07	0.660	−6.905	0.7500	0.1070	−19.7	−10.8	−10.5	−10.8
34	34.07	0.624	−6.996	0.7499	0.1011	−20.3	−10.9	−10.5	−10.8
32	32.06	0.587	−7.087	0.7499	0.0951	−20.9	−11.0	−10.6	−10.9
30	30.06	0.550	−7.178	0.7498	0.0892	−21.5	−11.1	−10.6	−11.0
28	28.06	0.514	−7.270	0.7498	0.0832	−22.3	−11.1	−10.7	−11.1
24	24.05	0.440	−7.452	0.7497	0.0714	−23.8	−11.3	−10.8	−11.2
20	20.05	0.367	−7.634	0.7496	0.0595	−25.7	−11.4	−11.0	−11.3
16	16.04	0.294	−7.817	0.7495	0.0476	−27.9	−11.6	−11.1	−11.5
12	12.03	0.220	−7.999	0.7495	0.0357	−30.7	$t^\star$−11.8	−11.2	−11.6
8	8.02	0.147	−8.182	0.7494	0.0238	−34.5	−11.9	−11.3	−11.8
4	4.01	0.073	−8.364	0.7493	0.0119	−40.7	−12.1	−11.5	−11.9
0	0.00	0.000	−8.546	0.7492	0.0000	—	−12.2	−11.6	−12.1

−8°C DRY BULB

Percentage saturation μ/%	Relative humidity ϕ/%	Value of stated parameter per kg dry air			Vapour pressure p_s/kPa	Dew point temperature t_d/°C	Adiabatic saturation temperature $t^\star$/°C	Wet bulb temperature	
		Moisture content g/g.kg^{-1}	Specific enthalpy h/kJ.kg^{-1}	Specific volume v/m³.kg^{-1}				Screen t'_{sc}/°C	Sling t'_{sl}/°C
100	100.00	1.917	−3.278	0.7529	0.3100	−8.0	−8.0	−8.0	−8.0
96	96.01	1.840	−3.468	0.7528	0.2976	−8.5	−8.1	−8.1	−8.1
92	92.02	1.764	−3.659	0.7527	0.2852	−8.9	−8.3	−8.2	−8.3
88	88.03	1.687	−3.850	0.7526	0.2729	−9.5	−8.4	−8.4	−8.4
84	84.04	1.610	−4.040	0.7525	0.2605	−10.0	−8.6	−8.5	−8.6
80	80.05	1.534	−4.231	0.7524	0.2481	−10.5	−8.7	−8.6	−8.7
76	76.06	1.457	−4.421	0.7524	0.2357	−11.1	−8.9	−8.7	−8.9
72	72.06	1.380	−4.612	0.7523	0.2334	−11.7	−9.0	−8.9	−9.0
70	70.06	1.342	−4.707	0.7522	0.2172	−12.0	−9.1	−8.9	−9.1
68	68.07	1.304	−4.803	0.7522	0.2110	−12.3	−9.2	−9.0	−9.2
66	66.07	1.265	−4.898	0.7521	0.2048	−12.7	−9.3	−9.1	−9.2
64	64.07	1.227	−4.993	0.7521	0.1986	−13.0	−9.3	−9.1	−9.3
62	62.07	1.189	−5.089	0.7520	0.1924	−13.4	−9.4	−9.2	−9.4
60	60.07	1.150	−5.184	0.7520	0.1862	−13.7	−9.5	−9.3	−9.4
58	58.07	1.112	−5.279	0.7519	0.1800	−14.1	−9.6	−9.3	−9.5
56	56.08	1.074	−5.375	0.7519	0.1738	−14.5	−9.6	−9.4	−9.6
54	54.08	1.035	−5.470	0.7519	0.1676	−14.9	−9.7	−9.4	−9.7
52	52.08	0.997	−5.565	0.7518	0.1614	−15.3	−9.8	−9.5	−9.7
50	50.08	0.958	−5.661	0.7518	0.1552	−15.7	−9.9	−9.6	−9.8
48	48.08	0.920	−5.756	0.7517	0.1490	−16.1	−10.0	−9.6	−9.9
46	46.08	0.882	−5.851	0.7517	0.1428	−16.6	−10.0	−9.7	−10.0
44	44.08	0.844	−5.946	0.7516	0.1366	−17.1	−10.1	−9.8	−10.0
42	42.07	0.805	−6.042	0.7516	0.1304	−17.6	−10.2	−9.8	−10.1
40	40.07	0.767	−6.137	0.7515	0.1242	−18.1	−10.3	−9.9	−10.2
38	38.07	0.728	−6.232	0.7515	0.1180	−18.6	−10.3	−10.0	−10.3
36	36.07	0.690	−6.328	0.7514	0.1118	−19.2	−10.4	−10.0	−10.3
34	34.07	0.652	−6.423	0.7514	0.1056	−19.8	−10.5	−10.1	−10.4
32	32.07	0.614	−6.518	0.7513	0.0994	−20.4	−10.6	−10.2	−10.5
30	30.06	0.575	−6.614	0.7513	0.0932	−21.1	−10.7	−10.2	−10.6
28	28.06	0.537	−6.709	0.75113	0.0870	−21.8	−10.7	−10.3	−10.6
24	24.06	0.460	−6.900	0.7512	0.0746	−23.4	−10.9	−10.4	−10.8
20	20.05	0.383	−7.090	0.7511	0.0621	−25.2	−11.0	−10.5	−10.9
16	16.04	0.307	−7.281	0.7510	0.0497	−27.4	−11.2	−10.7	−11.1
12	12.03	0.230	−7.472	0.7509	0.0373	−30.2	−11.4	−10.8	−11.2
8	8.02	0.153	−7.662	0.7508	0.0249	−34.1	−11.5	−10.9	−11.4
4	4.01	0.077	−7.853	0.7507	0.0124	−40.4	−11.7	−11.1	−11.6
0	0.00	0.000	−8.044	0.7506	0.0000	—	−11.8	−11.2	−11.7

−7.5°C DRY BULB

Percentage saturation μ/%	Relative humidity ϕ/%	Value of stated parameter per kg dry air			Vapour pressure p_s/kPa	Dew point temperature t_d/°C	Adiabatic saturation temperature $t^\star$/°C	Wet bulb temperature	
		Moisture content g/g.kg^{-1}	Specific enthalpy h/kJ.kg^{-1}	Specific volume v/m³.kg^{-1}				Screen t'_{sc}/°C	Sling t'_{sl}/°C
100	100.00	2.003	−2.560	0.7544	0.3238	−7.5	−7.5	−7.5	−7.5
96	96.01	1.922	−2.760	0.7543	0.3108	−8.0	−7.7	−7.6	−7.6
92	92.02	1.842	−2.959	0.7542	0.2979	−8.5	−7.8	−7.8	−7.8
88	88.03	1.762	−3.158	0.7541	0.2850	−9.0	−8.0	−7.9	−7.9
84	84.04	1.682	−3.357	0.7541	0.2721	−9.5	−8.1	−8.0	−8.1
80	80.05	1.602	−3.556	0.7540	0.2592	−10.0	−8.3	−8.1	−8.2
76	76.06	1.522	−3.756	0.7539	0.2462	−10.6	−8.4	−8.3	−8.4
72	72.06	1.442	−3.955	0.7538	0.2333	−11.2	−8.6	−8.4	−8.5
70	70.07	1.402	−4.054	0.7537	0.2268	−11.5	−8.7	−8.5	−8.6
68	68.07	1.362	−4.154	0.7537	0.2204	−11.9	−8.7	−8.5	−8.7
66	66.07	1.322	−4.254	0.7536	0.2139	−12.2	−8.8	−8.6	−8.8
64	64.07	1.282	−4.353	0.7536	0.2074	−12.5	−8.9	−8.7	−8.8
62	62.08	1.242	−4.453	0.7535	0.2010	−12.9	−9.0	−8.7	−8.9
60	60.08	1.202	−4.552	0.7535	0.1945	−13.2	−9.0	−8.8	−9.0
58	58.08	1.162	−4.652	0.7534	0.1880	−13.6	−9.1	−8.9	−9.1
56	56.08	1.122	−4.752	0.7534	0.1816	−14.0	−9.2	−8.9	−9.1
54	54.08	1.081	−4.851	0.7533	0.1751	−14.4	−9.3	−9.0	−9.2
52	52.08	1.041	−4.951	0.7533	0.1686	−14.8	−9.4	−9.1	−9.3
50	50.08	1.001	−5.051	0.7532	0.1621	−15.2	−9.4	−9.1	−9.4
48	48.08	0.961	−5.150	0.7532	0.1557	−15.7	−9.5	−9.2	−9.4
46	46.08	0.921	−5.250	0.7531	0.1492	−16.1	−9.6	−9.3	−9.5
44	44.08	0.881	−5.349	0.7531	0.1427	−16.6	−9.7	−9.3	−9.6
42	42.08	0.841	−5.449	0.7530	0.1362	−17.1	−9.8	−9.4	−9.7
40	40.08	0.801	−5.549	0.7530	0.1298	−17.6	−9.8	−9.5	−9.8
38	38.08	0.761	−5.648	0.7529	0.1233	−18.2	−9.9	−9.5	−9.8
36	36.07	0.721	−5.748	0.7529	0.1168	−18.7	−10.0	−9.6	−9.9
34	34.07	0.681	−5.848	0.7528	0.1103	−19.3	−10.1	−9.7	−10.0
32	32.07	0.641	−5.947	0.7528	0.1038	−20.0	−10.2	−9.7	−10.1
30	30.07	0.601	−6.047	0.7528	0.0973	−20.6	−10.2	−9.8	−10.1
28	28.06	0.561	−6.146	0.7527	0.0909	−21.4	−10.3	−9.9	−10.2
24	24.06	0.481	−6.346	0.7526	0.0779	−22.9	−10.5	−10.0	−10.4
20	20.05	0.400	−6.545	0.7525	0.0649	−24.8	−10.6	−10.1	−10.5
16	16.04	0.320	−6.744	0.7524	0.0519	−27.0	−10.8	−10.3	−10.7
12	12.03	0.240	−6.943	0.7523	0.0390	−29.8	−11.0	−10.4	−10.9
8	8.02	0.160	−7.142	0.7522	0.0260	−33.7	−11.1	−10.5	−11.0
4	4.01	0.080	−7.341	0.7521	0.0130	−40.0	−11.3	−10.7	−11.2
0	0.00	0.000	−7.541	0.7520	0.0000	—	−11.5	−10.8	−11.3

−7°C DRY BULB

Percentage saturation μ/%	Relative humidity ϕ/%	Value of stated parameter per kg dry air			Vapour pressure p_s/kPa	Dew point temperature t_d/°C	Adiabatic saturation temperature $t\star$/°C	Wet bulb temperature	
		Moisture content g/g.kg^{-1}	Specific enthalpy h/kJ.kg^{-1}	Specific volume v/m^3.kg^{-1}				Screen t'_{sc}/°C	Sling t'_{sl}/°C
100	100.00	2.092	−1.834	0.7560	0.3381	−7.0	−7.0	−7.0	−7.0
96	96.01	2.008	−2.042	0.7559	0.3246	−7.5	−7.2	−7.1	−7.2
92	92.02	1.924	−2.250	0.7558	0.3111	−8.0	−7.3	−7.3	−7.3
88	88.04	1.841	−2.458	0.7557	0.2976	−8.5	−7.5	−7.4	−7.5
84	84.04	1.757	−2.667	0.7556	0.2842	−9.0	−7.6	−7.5	−7.6
80	80.05	1.673	−2.875	0.7555	0.2707	−9.5	−7.8	−7.7	−7.8
76	76.06	1.590	−3.083	0.7554	0.2572	−10.1	−7.9	−7.8	−7.9
72	72.07	1.506	−3.291	0.7553	0.2437	−10.7	−8.1	−7.9	−8.1
70	70.07	1.464	−3.395	0.7552	0.2369	−11.0	−8.2	−8.0	−8.1
68	68.07	1.422	−3.499	0.7552	0.2302	−11.4	−8.3	−8.1	−8.2
66	66.07	1.380	−3.603	0.7551	0.2234	−11.7	−8.3	−8.1	−8.3
64	64.08	1.339	−3.708	0.7551	0.2166	−12.0	−8.4	−8.2	−8.4
62	62.08	1.297	−3.812	0.7550	0.2099	−12.4	−8.5	−8.3	−8.5
60	60.08	1.255	−3.916	0.7550	0.2031	−12.8	−8.6	−8.3	−8.5
58	58.08	1.213	−4.020	0.7549	0.1964	−13.1	−8.7	−8.4	−8.6
56	56.08	1.171	−4.124	0.7549	0.1896	−13.5	−8.8	−8.5	−8.7
54	54.08	1.130	−4.228	0.7548	0.1828	−13.9	−8.8	−8.5	−8.8
52	52.08	1.088	−4.332	0.7548	0.1761	−14.3	−8.9	−8.6	−8.9
50	50.08	1.046	−4.436	0.7457	0.1693	−14.8	−9.0	−8.7	−8.9
48	48.08	1.004	−4.540	0.7547	0.1626	−15.2	−9.1	−8.8	−9.0
46	46.08	0.962	−4.644	0.7546	0.1558	−15.7	−9.2	−8.8	−9.1
44	44.08	0.920	−4.748	0.7546	0.1490	−16.1	−9.2	−8.9	−9.2
42	42.08	0.878	−4.852	0.7545	0.1423	−16.6	−9.3	−9.0	−9.3
40	40.08	0.837	−4.956	0.7545	0.1355	−17.2	−9.4	−9.0	−9.3
38	38.08	0.795	−5.061	0.7544	0.1287	−17.7	−9.5	−9.1	−9.4
36	36.08	0.753	−5.165	0.7544	0.1220	−18.3	−9.6	−9.2	−9.5
34	34.07	0.711	−5.269	0.7543	0.1152	−18.9	−9.7	−9.2	−9.6
32	32.07	0.669	−5.373	0.7543	0.1084	−19.5	−9.7	−9.3	−9.7
30	30.07	0.628	−5.477	0.7542	0.1017	−20.2	−9.8	−9.4	−9.7
28	28.07	0.586	−5.581	0.7542	0.0949	−20.9	−9.9	−9.4	−9.8
24	24.06	0.502	−5.789	0.7541	0.0814	−22.5	−10.1	−9.6	−10.0
20	20.05	0.418	−5.997	0.7540	0.0678	−24.3	−10.3	−9.7	−10.1
16	16.04	0.335	−6.206	0.7539	0.0542	−26.6	−10.4	−9.9	−10.3
12	12.04	0.251	−6.414	0.7538	0.0407	−29.4	−10.6	−10.0	−10.5
8	8.02	0.167	−6.622	0.7537	0.0271	−33.3	−10.8	−10.1	−10.6
4	4.01	0.084	−6.830	0.7536	0.0136	−39.6	−10.9	−10.3	−10.8
0	0.00	0.000	−7.038	0.7535	0.0000	—	−11.1	−10.4	−11.0

−6.5°C DRY BULB

Percentage saturation μ/%	Relative humidity ϕ/%	Value of stated parameter per kg dry air			Vapour pressure p_s/kPa	Dew point temperature t_d/°C	Adiabatic saturation temperature $t\star$/°C	Wet bulb temperature	
		Moisture content g/g.kg^{-1}	Specific enthalpy h/kJ.kg^{-1}	Specific volume v/m^3.kg^{-1}				Screen t'_{sc}/°C	Sling t'_{sl}/°C
100	100.00	2.184	−1.099	0.7575	0.3530	−6.5	−6.5	−6.5	−6.5
96	96.01	2.097	−1.316	0.7574	0.3390	−7.0	−6.7	−6.6	−6.7
92	92.03	2.010	−1.534	0.7573	0.3249	−7.5	−6.8	−6.8	−6.8
88	88.04	1.922	−1.751	0.7572	0.3108	−8.0	−7.0	−6.9	−7.0
84	84.05	1.835	−1.969	0.7571	0.2967	−8.5	−7.1	−7.0	−7.1
80	80.06	1.748	−2.186	0.7570	0.2826	−9.1	−7.3	−7.2	−7.3
76	76.06	1.660	−2.404	0.7569	0.2685	−9.6	−7.5	−7.3	−7.4
72	72.07	1.573	−2.621	0.7568	0.2544	−10.2	−7.6	−7.5	−7.6
70	70.07	1.529	−2.730	0.7567	0.2474	−10.6	−7.7	−7.5	−7.7
68	68.08	1.485	−2.838	0.7567	0.2403	−10.9	−7.8	−7.6	−7.8
66	66.08	1.442	−2.947	0.7566	0.2333	−11.2	−7.9	−7.7	−7.8
64	64.08	1.398	−3.056	0.7566	0.2262	−11.6	−8.0	−7.7	−7.9
62	62.08	1.354	−3.165	0.7565	0.2192	−11.9	−8.1	−7.8	−8.0
60	60.08	1.311	−3.273	0.7565	0.2121	−12.3	−8.1	−7.9	−8.1
58	58.08	1.267	−3.382	0.7564	0.2051	−12.7	−8.2	−8.0	−8.2
56	56.09	1.223	−3.491	0.7563	0.1980	−13.0	−8.3	−8.0	−8.2
54	54.09	1.180	−3.600	0.7563	0.1909	−13.4	−8.4	−8.1	−8.3
52	52.09	1.136	−3.708	0.7562	0.1839	−13.9	−8.5	−8.2	−8.4
50	50.09	1.092	−3.817	0.7562	0.1768	−14.3	−8.6	−8.2	−8.5
48	48.09	1.048	−3.926	0.7561	0.1698	−14.7	−8.6	−8.3	−8.6
46	46.09	1.005	−4.034	0.7561	0.1627	−15.2	−8.7	−8.4	−8.7
44	44.09	0.961	−4.143	0.7560	0.1556	−15.7	−8.8	−8.5	−8.7
42	42.08	0.917	−4.252	0.7560	0.1486	−16.2	−8.9	−8.5	−8.8
40	40.08	0.874	−4.361	0.7559	0.1415	−16.7	−9.0	−8.6	−8.9
38	38.08	0.830	−4.470	0.7559	0.1344	−17.2	−9.1	−8.7	−9.0
36	36.08	0.786	−4.578	0.7558	0.1274	−17.8	−9.2	−8.7	−9.1
34	34.08	0.743	−4.687	0.7558	0.1203	−18.4	−9.2	−8.8	−9.2
32	32.08	0.699	−4.796	0.7557	0.1132	−19.1	−9.3	−8.9	−9.2
30	30.07	0.655	−4.904	0.7557	0.1062	−19.7	−9.4	−9.0	−9.3
28	28.07	0.612	−5.013	0.7556	0.0991	−20.5	−9.5	−9.0	−9.4
24	24.06	0.524	−5.230	0.7555	0.0850	−22.0	−9.7	−9.2	−9.6
20	20.06	0.437	−5.448	0.7554	0.0708	−23.9	−9.9	−9.3	−9.7
16	16.05	0.350	−5.666	0.7553	0.0566	−26.1	−10.0	−9.5	−9.9
12	12.04	0.262	−5.883	0.7552	0.0425	−29.0	−10.2	−9.6	−10.1
8	8.03	0.175	−6.100	0.7551	0.0283	−32.9	−10.4	−9.8	−10.3
4	4.01	0.087	−6.318	0.7550	0.0142	−39.2	−10.6	−9.9	−10.4
0	0.00	0.000	−6.535	0.7549	0.0000	—	−10.7	−10.0	−10.6

−6°C DRY BULB

Percentage saturation μ/%	Relative humidity ϕ/%	Value of stated parameter per kg dry air			Vapour pressure p_s/kPa	Dew point temperature t_d/°C	Adiabatic saturation temperature $t^\star$/°C	Wet bulb temperature	
		Moisture content $g/g.kg^{-1}$	Specific enthalpy h/kJ.kg^{-1}	Specific volume v/m^3.kg^{-1}				Screen t'_{sc}/°C	Sling t'_{sl}/°C
100	100.00	2.281	−0.354	0.7590	0.3686	−6.0	−6.0	−6.0	−6.0
96	96.01	2.190	−0.581	0.7589	0.3539	−6.5	−6.2	−6.1	−6.2
92	92.03	2.098	−0.808	0.7588	0.3392	−7.0	−6.3	−6.3	−6.3
88	88.04	2.007	−1.035	0.7587	0.3245	−7.5	−6.5	−6.4	−6.5
84	84.05	1.916	−1.262	0.7586	0.3098	−8.0	−6.7	−6.6	−6.6
80	80.06	1.825	−1.490	0.7585	0.2950	−8.6	−6.8	−6.7	−6.8
76	76.07	1.733	−1.717	0.7584	0.2804	−9.1	−7.0	−6.9	−7.0
72	72.07	1.642	−1.944	0.7583	0.2656	−9.8	−7.2	−7.0	−7.1
70	70.08	1.597	−2.058	0.7582	0.2583	−10.1	−7.3	−7.1	−7.2
68	68.08	1.551	−2.171	0.7582	0.2509	−10.4	−7.3	−7.1	−7.3
66	66.08	1.505	2.285	0.7581	0.2436	−10.7	−7.4	−7.2	−7.4
64	64.08	1.460	−2.398	0.7581	0.2362	−11.1	−7.5	−7.3	−7.5
62	62.09	1.414	−2.512	0.7580	0.2288	−11.4	−7.6	−7.4	−7.6
60	60.09	1.368	−2.625	0.7579	0.2215	−11.8	−7.7	−7.4	−7.6
58	58.09	1.323	−2.739	0.7579	0.2141	−12.2	−7.8	−7.5	−7.7
56	56.09	1.277	−2.852	0.7578	0.2067	−12.6	−7.9	−7.6	−7.8
54	54.09	1.232	−2.966	0.7578	0.1994	−13.0	−8.0	−7.6	−7.9
52	52.09	1.186	−3.080	0.7577	0.1920	−13.4	−8.0	−7.7	−8.0
50	50.09	1.140	−3.193	0.7577	0.1846	−13.8	−8.1	−7.8	−8.1
48	48.09	1.095	−3.307	0.7576	0.1772	−14.3	−8.2	−7.9	−8.1
46	46.09	1.049	−3.420	0.7576	0.1699	−14.7	−8.3	−7.9	−8.2
44	44.09	1.004	−3.534	0.7575	0.1625	−15.2	−8.4	−8.0	−8.3
42	42.09	0.958	−3.648	0.7575	0.1551	−15.7	−8.5	−8.1	−8.4
40	40.09	0.912	−3.761	0.7574	0.1478	−16.2	−8.6	−8.2	−8.5
38	38.09	0.867	−3.875	0.7573	0.1404	−16.8	−8.7	−8.2	−8.6
36	36.08	0.821	−3.988	0.7573	0.1330	−17.4	−8.7	−8.3	−8.7
34	34.08	0.776	−4.102	0.7572	0.1256	−18.0	−8.8	−8.4	−8.7
32	32.08	0.730	−4.216	0.7572	0.1182	−18.6	−8.9	−8.5	−8.8
30	30.08	0.684	−4.329	0.7571	0.1108	−19.3	−9.0	−8.5	−8.9
28	28.07	0.639	−4.443	0.7571	0.1035	−20.0	−9.1	−8.6	−9.0
24	24.07	0.547	−4.670	0.7570	0.0887	−21.6	−9.3	−8.8	−9.2
20	20.06	0.456	−4.897	0.7568	0.0739	−23.5	−9.5	−8.9	−9.3
16	16.05	0.365	−5.124	0.7567	0.0592	−25.7	−9.6	−9.1	−9.5
12	12.04	0.274	−5.351	0.7566	0.0444	−28.6	−9.8	−9.2	−9.7
8	8.03	0.182	−5.578	0.7565	0.0296	−32.4	−10.0	−9.4	−9.9
4	4.01	0.091	−5.806	0.7564	0.0148	−38.8	−10.2	−9.5	−10.1
0	0.00	0.000	−6.033	0.7563	0.0000	—	−10.4	−9.7	−10.2

5.5°C DRY BULB

Percentage saturation μ/%	Relative humidity ϕ/%	Value of stated parameter per kg dry air			Vapour pressure p_s/kPa	Dew point temperature t_d/°C	Adiabatic saturation temperature $t^\star$/°C	Wet bulb temperature	
		Moisture content $g/g.kg^{-1}$	Specific enthalpy h/kJ.kg^{-1}	Specific volume v/m^3.kg^{-1}				Screen t'_{sc}/°C	Sling t'_{sl}/°C
100	100.00	2.381	0.400	0.7606	0.3847	−5.5	−5.5	−5.5	−5.5
96	96.01	2.286	0.164	0.7605	0.3694	−6.0	−5.7	−5.6	−5.7
92	92.03	2.191	−0.074	0.7604	0.3540	−6.5	−5.8	−5.8	−5.8
88	88.04	2.095	−0.311	0.7602	0.3387	−7.0	−6.0	−5.9	−6.0
84	84.05	2.000	−0.548	0.7601	0.3234	−7.5	−6.2	−6.1	−6.2
80	80.06	1.905	−0.785	0.7600	0.3080	−8.1	−6.4	−6.2	−6.3
76	76.07	1.810	−1.022	0.7599	0.2926	−8.7	−6.5	−6.4	−6.5
72	72.08	1.714	−1.260	0.7598	0.2773	−9.3	−6.7	−6.5	−6.7
70	70.08	1.667	−1.378	0.7597	0.2696	−9.6	−6.8	−6.6	−6.8
68	68.08	1.619	−1.497	0.7597	0.2619	−9.9	−6.9	−6.7	−6.8
66	66.08	1.572	−1.616	0.7596	0.2542	−10.3	−7.0	−6.8	−6.9
64	64.09	1.524	−1.734	0.7596	0.2466	−10.6	−7.1	−6.8	−7.0
62	62.09	1.476	−1.853	0.7595	0.2389	−11.0	−7.2	−6.9	−7.1
60	60.09	1.429	−1.972	0.7594	0.2312	−11.3	−7.2	−7.0	−7.2
58	58.09	1.381	−2.090	0.7594	0.2235	−11.7	−7.3	−7.1	−7.3
56	56.09	1.334	−2.209	0.7593	0.2158	−12.1	−7.4	−7.1	−7.4
54	54.09	1.286	−2.327	0.7593	0.2081	−12.5	−7.5	−7.2	−7.4
52	52.09	1.238	−2.446	0.7592	0.2004	−12.9	−7.6	−7.3	−7.5
50	50.09	1.191	−2.565	0.7592	0.1927	−13.3	−7.7	−7.4	−7.6
48	48.09	1.143	−2.683	0.7591	0.1850	−13.8	−7.8	−7.4	−7.7
46	46.09	1.095	−2.802	0.7590	0.1773	−14.2	−7.9	−7.5	−7.8
44	44.09	1.048	−2.920	0.7590	0.1696	−14.7	−8.0	−7.6	−7.9
42	42.09	1.000	−3.039	0.7589	0.1619	−15.2	−8.1	−7.7	−8.0
40	40.09	0.952	−3.158	0.7589	0.1542	−15.8	−8.1	−7.7	−8.1
38	38.09	0.905	−3.276	0.7588	0.1465	−16.3	−8.2	−7.8	−8.1
36	36.09	0.857	−3.395	0.7588	0.1388	−16.9	−8.3	−7.9	−8.2
34	34.09	0.810	−3.514	0.7587	0.1311	−17.5	−8.4	−8.0	−8.3
32	32.08	0.762	−3.632	0.7586	0.1234	−18.1	−8.5	−8.0	−8.4
30	30.08	0.714	−3.751	0.7586	0.1157	−18.8	−8.6	−8.1	−8.5
28	28.08	0.667	−3.869	0.7585	0.1080	−19.6	−8.7	−8.2	−8.6
24	24.07	0.572	−4.107	0.7584	0.0926	−21.2	−8.9	−8.4	−8.8
20	20.06	0.476	−4.344	0.7583	0.0772	−23.0	−9.1	−8.5	−9.0
16	16.05	0.381	−4.581	0.7582	0.0618	−25.3	−9.3	−8.7	−9.1
12	12.04	0.286	−4.818	0.7581	0.0463	−28.1	−9.5	−8.8	−9.3
8	8.03	0.190	−5.056	0.7579	0.0309	−32.0	−9.6	−9.0	−9.5
4	4.01	0.095	−5.293	0.7578	0.0154	−38.4	−9.8	−9.1	−9.7
0	0.00	0.000	−5.530	0.7577	0.0000	—	−10.0	−9.3	−9.9

−5°C DRY BULB

Percentage saturation μ/%	Relative humidity ϕ/%	Value of stated parameter per kg dry air			Vapour pressure p_s/kPa	Dew point temperature t_d/°C	Adiabatic saturation temperature t^*/°C	Wet bulb temperature	
		Moisture content g/g.kg⁻¹	Specific enthalpy h/kJ.kg⁻¹	Specific volume v/m³.kg⁻¹				Screen t'_{sc}/°C	Sling t'_{sl}/°C
100	100.00	2.486	1.166	0.7621	0.4015	−5.0	−5.0	−5.0	−5.0
96	96.02	2.386	0.918	0.7620	0.3855	−5.5	−5.2	−5.1	−5.2
92	92.03	2.287	0.670	0.7619	0.3695	−6.0	−5.4	−5.3	−5.3
88	88.04	2.187	0.423	0.7618	0.3535	−6.5	−5.5	−5.5	−5.5
84	84.05	2.088	0.175	0.7617	0.3375	−7.0	−5.7	−5.6	−5.7
80	80.06	1.988	−0.073	0.7615	0.3215	−7.6	−5.9	−5.8	−5.9
76	76.07	1.889	−0.320	0.7614	0.3054	−8.2	−6.1	−5.9	−6.0
72	72.08	1.790	−0.568	0.7613	0.2894	−8.8	−6.2	−6.1	−6.2
70	70.08	1.740	−0.692	0.7612	0.2814	−9.1	−6.3	−6.1	−6.3
68	68.09	1.690	−0.816	0.7612	0.2734	−9.4	−6.4	−6.2	−6.4
66	66.09	1.640	−0.940	0.7611	0.2654	−9.8	−6.5	−6.3	−6.5
64	64.09	1.591	−1.064	0.7611	0.2573	−10.1	−6.6	−6.4	−6.6
62	62.09	1.541	−1.188	0.7610	0.2493	−10.5	−6.7	−6.4	−6.6
60	60.09	1.491	−1.311	0.7609	0.2413	−10.8	−6.8	−6.5	−6.7
58	58.10	1.442	−1.435	0.7609	0.2333	−11.2	−6.9	−6.6	−6.8
56	56.10	1.392	−1.559	0.7608	0.2252	−11.6	−7.0	−6.7	−6.9
54	54.10	1.342	−1.683	0.7608	0.2172	−12.0	−7.1	−6.8	−7.0
52	52.10	1.292	−1.807	0.7607	0.2092	−12.4	−7.2	−6.8	−7.1
50	50.10	1.243	−1.931	0.7606	0.2012	−12.9	−7.3	−6.9	−7.2
48	48.10	1.193	−2.055	0.7606	0.1931	−13.3	−7.4	−7.0	−7.3
46	46.10	1.143	−2.178	0.7605	0.1851	−13.8	−7.4	−7.1	−7.4
44	44.10	1.094	−2.302	0.7605	0.1771	−14.3	−7.5	−7.1	−7.5
42	42.10	1.044	−2.426	0.7604	0.1690	−14.8	−7.6	−7.2	−7.5
40	40.09	0.994	−2.550	0.7603	0.1610	−15.3	−7.7	−7.3	−7.6
38	38.09	0.944	−2.674	0.7603	0.1530	−15.9	−7.8	−7.4	−7.7
36	36.09	0.895	−2.798	0.7602	0.1449	−16.4	−7.9	−7.5	−7.8
34	34.09	0.845	−2.922	0.7602	0.1369	−17.0	−8.0	−7.5	−7.9
32	32.09	0.795	−3.045	0.7601	0.1288	−17.7	−8.1	−7.6	−8.0
30	30.08	0.746	−3.169	0.7600	0.1208	−18.4	−8.2	−7.7	−8.1
28	28.08	0.696	−3.293	0.7600	0.1127	−19.1	−8.3	−7.8	−8.2
24	24.07	0.597	−3.541	0.7599	0.0966	−20.7	−8.5	−7.9	−8.4
20	20.06	0.497	−3.789	0.7597	0.0806	−22.6	−8.7	−8.1	−8.6
16	16.05	0.398	−4.036	0.7596	0.0645	−24.9	−8.9	−8.3	−8.8
12	12.04	0.298	−4.284	0.7595	0.0484	−27.7	−9.1	−8.4	−8.9
8	8.03	0.199	−4.532	0.7594	0.0322	−31.6	−9.3	−8.6	−9.1
4	4.02	0.099	−4.780	0.7593	0.0161	−38.0	−9.5	−8.8	−9.3
0	0.00	0.000	−5.027	0.7591	0.0000	—	−9.7	−8.9	−9.5

−4.5°C DRY BULB

Percentage saturation μ/%	Relative humidity ϕ/%	Value of stated parameter per kg dry air			Vapour pressure p_s/kPa	Dew point temperature t_d/°C	Adiabatic saturation temperature t^*/°C	Wet bulb temperature	
		Moisture content g/g.kg⁻¹	Specific enthalpy h/kJ.kg⁻¹	Specific volume v/m³.kg⁻¹				Screen t'_{sc}/°C	Sling t'_{sl}/°C
100	100.00	2.594	1.941	0.7637	0.4190	−4.5	−4.5	−4.5	−4.5
96	96.02	2.490	1.683	0.7636	0.4023	−5.0	−4.7	−4.7	−4.7
92	92.03	2.387	1.424	0.7635	0.3856	−5.5	−4.9	−4.8	−4.9
88	88.04	2.283	1.166	0.7633	0.3689	−6.0	−5.0	−5.0	−5.0
84	84.06	2.179	0.907	0.7632	0.3522	−6.5	−5.2	−5.1	−5.2
80	80.07	2.075	0.648	0.7631	0.3355	−7.1	−5.4	−5.3	−5.4
76	76.08	1.972	0.390	0.7629	0.3187	−7.7	−5.6	−5.4	−5.6
72	72.08	1.868	0.131	0.7628	0.3020	−8.3	−5.8	−5.6	−5.7
70	70.09	1.816	0.002	0.7628	0.2936	−8.6	−5.9	−5.7	−5.8
68	68.09	1.764	−0.128	0.7627	0.2853	−8.9	−6.0	−5.8	−5.9
66	66.09	1.712	−0.257	0.7626	0.2769	−9.3	−6.1	−5.8	−6.0
64	64.10	1.660	−0.386	0.7626	0.2685	−9.6	−6.2	−5.9	−6.1
62	62.10	1.608	−0.516	0.7625	0.2602	−10.0	−6.3	−6.0	−6.2
60	60.10	1.556	−0.645	0.7624	0.2518	−10.4	−6.3	−6.1	−6.3
58	58.10	1.505	−0.774	0.7624	0.2434	−10.7	−6.4	−6.2	−6.4
56	56.10	1.453	−0.904	0.7623	0.2350	−11.1	−6.5	−6.2	−6.5
54	54.10	1.401	−1.033	0.7623	0.2267	−11.5	−6.6	−6.3	−6.6
52	52.10	1.349	−1.162	0.7622	0.2183	−12.0	−6.7	−6.4	−6.7
50	50.10	1.297	−1.292	0.7621	0.2099	−12.4	−6.8	−6.5	−6.8
48	48.10	1.245	−1.421	0.7621	0.2015	−12.8	−6.9	−6.6	−6.8
46	46.10	1.193	−1.550	0.7620	0.1932	−13.3	−7.0	−6.6	−6.9
44	44.10	1.141	−1.680	0.7619	0.1848	−13.8	−7.1	−6.7	−7.0
42	42.10	1.090	−1.809	0.7619	0.1764	−14.3	−7.2	−6.8	−7.1
40	40.10	1.038	−1.938	0.7618	0.1680	−14.8	−7.3	−6.9	−7.2
38	38.10	0.986	−2.068	0.7618	0.1596	−15.4	−7.4	−7.0	−7.3
36	36.10	0.934	−2.197	0.7617	0.1512	−16.0	−7.5	−7.0	−7.4
34	34.09	0.882	−2.326	0.7616	0.1428	−16.6	−7.6	−7.1	−7.5
32	32.09	0.830	−2.455	0.7616	0.1344	−17.2	−7.7	−7.2	−7.6
30	30.09	0.778	−2.585	0.7615	0.1261	−17.9	−7.8	−7.3	−7.7
28	28.08	0.726	−2.714	0.7614	0.1177	−18.7	−7.9	−7.4	−7.8
24	24.08	0.623	−2.973	0.7613	0.1009	−20.3	−8.1	−7.5	−8.0
20	20.07	0.519	−3.231	0.7612	0.0841	−22.2	−8.3	−7.7	−8.2
16	16.06	0.415	−3.490	0.7611	0.0673	−24.4	−8.5	−7.9	−8.4
12	12.04	0.311	−3.749	0.7609	0.0505	−27.3	−8.7	−8.0	−8.6
8	8.03	0.208	−4.007	0.7608	0.0336	−31.2	−8.9	−8.2	−8.8
4	4.02	0.104	−4.266	0.7607	0.0168	−37.7	−9.1	−8.4	−9.0
0	0.00	0.000	−4.524	0.7606	0.0000	—	−9.3	−8.5	−9.2

−4°C DRY BULB

Percentage saturation μ/%	Relative humidity ϕ/%	Value of stated parameter per kg dry air			Vapour pressure p_s/kPa	Dew point temperature t_d/°C	Adiabatic saturation temperature $t^\star$/°C	Wet bulb temperature	
		Moisture content $g/g.kg^{-1}$	Specific enthalpy $h/kJ.kg^{-1}$	Specific volume $v/m^3.kg^{-1}$				Screen t'_{sc}/°C	Sling t'_{sl}/°C
100	100.00	2.707	2.729	0.7653	0.4371	−4.0	−4.0	−4.0	−4.0
96	96.02	2.599	2.458	0.7651	0.4197	−4.5	−4.2	−4.2	−4.2
92	92.03	2.490	2.188	0.7650	0.4023	−5.0	−4.4	−4.3	−4.4
88	88.05	2.382	1.918	0.7649	0.3849	−5.5	−4.6	−4.5	−4.5
84	84.06	2.274	1.648	0.7647	0.3674	−6.0	−4.7	−4.6	−4.7
80	80.07	2.166	1.378	0.7646	0.3500	−6.6	−4.9	−4.8	−4.9
76	76.08	2.057	1.108	0.7645	0.3326	−7.2	−5.1	−5.0	−5.1
72	72.09	1.949	0.838	0.7643	0.3151	−7.8	−5.3	−5.1	−5.3
70	70.09	1.895	0.703	0.7643	0.3064	−8.1	−5.4	−5.2	−5.4
68	68.09	1.841	0.568	0.7642	0.2977	−8.5	−5.5	−5.3	−5.5
66	66.10	1.787	0.433	0.7641	0.2889	−8.8	−5.6	−5.4	−5.6
64	64.10	1.732	0.298	0.7641	0.2802	−9.2	−5.7	−5.5	−5.7
62	62.10	1.678	0.163	0.7640	0.2715	−9.5	−5.8	−5.5	−5.7
60	60.10	1.624	0.028	0.7640	0.2627	−9.9	−5.9	−5.6	−5.8
58	58.10	1.570	−0.107	0.7639	0.2540	−10.3	−6.0	−5.7	−5.9
56	56.11	1.516	−0.242	0.7638	0.2452	−10.7	−6.1	−5.8	−6.0
54	54.11	1.462	−0.377	0.7638	0.2365	−11.1	−6.2	−5.9	−6.1
52	52.11	1.408	−0.512	0.7637	0.2278	−11.5	−6.3	−6.0	−6.2
50	50.11	1.353	−0.647	0.7636	0.2190	−11.9	−6.4	−6.0	−6.3
48	48.11	1.299	−0.782	0.7636	0.2103	−12.4	−6.5	−6.1	−6.4
46	46.11	1.245	−0.917	0.7635	0.2016	−12.8	−6.6	−6.2	−6.5
44	44.11	1.191	−1.052	0.7634	0.1928	−13.3	−6.7	−6.3	−6.6
42	42.10	1.137	−1.187	0.7634	0.1840	−13.8	−6.8	−6.4	−6.7
40	40.10	1.083	−1.322	0.7633	0.1753	−14.4	−6.9	−6.5	−6.8
38	38.10	1.029	−1.457	0.7632	0.1666	−14.9	−7.0	−6.5	−6.9
36	36.10	0.974	−1.592	0.7632	0.1578	−15.5	−7.1	−6.6	−7.0
34	34.10	0.920	−1.727	0.7631	0.1490	−16.1	−7.2	−6.7	−7.1
32	32.09	0.866	−1.862	0.7630	0.1403	−16.8	−7.3	−6.8	−7.2
30	30.09	0.812	−1.997	0.7630	0.1315	−17.5	−7.4	−6.9	−7.3
28	28.09	0.758	−2.132	0.7629	0.1228	−18.2	−7.5	−7.0	−7.4
24	24.08	0.650	−2.402	0.7628	0.1052	−19.8	−7.7	−7.1	−7.6
20	20.07	0.541	−2.672	0.7626	0.0877	−21.7	−7.9	−7.3	−7.8
16	16.06	0.433	−2.942	0.7625	0.0702	−24.0	−8.1	−7.5	−8.0
12	12.05	0.325	−3.212	0.7624	0.0526	−26.9	−8.3	−7.7	−8.2
8	8.03	0.217	−3.482	0.7622	0.0351	−30.8	−8.5	−7.8	−8.4
4	4.02	0.108	−3.752	0.7621	0.0176	−37.3	−8.8	−8.0	−8.6
0	0.00	0.000	−4.022	0.7620	0.0000	—	−9.0	−8.2	−8.8

−3.5°C DRY BULB

Percentage saturation μ/%	Relative humidity ϕ/%	Value of stated parameter per kg dry air			Vapour pressure p_s/kPa	Dew point temperature t_d/°C	Adiabatic saturation temperature $t^\star$/°C	Wet bulb temperature	
		Moisture content $g/g.kg^{-1}$	Specific enthalpy $h/kJ.kg^{-1}$	Specific volume $v/m^3.kg^{-1}$				Screen t'_{sc}/°C	Sling t'_{sl}/°C
100	100.00	2.824	3.526	0.7668	0.4560	−3.5	−3.5	−3.5	−3.5
96	96.02	2.711	3.244	0.7667	0.4378	−4.0	−3.7	−3.7	−3.7
92	92.03	2.598	2.962	0.7666	0.4197	−4.5	−3.9	−3.8	−3.9
88	88.05	2.485	2.680	0.7664	0.4015	−5.0	−4.1	−4.0	−4.1
84	84.06	2.372	2.399	0.7663	0.3833	−5.5	−4.3	−4.2	−4.2
80	80.07	2.260	2.117	0.7662	0.3651	−6.1	−4.5	−4.3	−4.4
76	76.08	2.146	1.835	0.7660	0.3469	−6.7	−4.7	−4.5	−4.6
72	72.09	2.034	1.553	0.7659	0.3287	−7.3	−4.9	−4.7	−4.8
70	70.09	1.977	1.412	0.7658	0.3196	−7.6	−5.0	−4.7	−4.9
68	68.10	1.920	1.272	0.7657	0.3105	−8.0	−5.1	−4.8	−5.0
66	66.10	1.864	1.131	0.7657	0.3014	−8.3	−5.2	−4.9	−5.1
64	64.10	1.808	0.990	0.7656	0.2923	−8.7	−5.3	−5.0	−5.2
62	62.11	1.751	0.849	0.7655	0.2832	−9.0	−5.4	−5.1	−5.3
60	60.11	1.695	0.708	0.7655	0.2741	−9.4	−5.5	−5.2	−5.4
58	58.11	1.638	0.567	0.7654	0.2650	−9.8	−5.6	−5.3	−5.5
56	56.11	1.582	0.426	0.7653	0.2559	−10.2	−5.7	−5.3	−5.6
54	54.11	1.525	0.285	0.7653	0.2468	−10.6	−5.8	−5.4	−5.7
52	52.11	1.469	0.144	0.7652	0.2376	−11.0	−5.9	−5.5	−5.8
50	50.11	1.412	0.003	0.7651	0.2285	−11.4	−6.0	−5.6	−5.9
48	48.11	1.356	−0.138	0.7650	0.2194	−11.9	−6.1	−5.7	−6.0
46	46.11	1.299	−0.278	0.7650	0.2103	−12.4	−6.2	−5.8	−6.1
44	44.11	1.243	−0.419	0.7649	0.2011	−12.9	−6.3	−5.9	−6.2
42	42.11	1.186	−0.560	0.7648	0.1920	−13.4	−6.4	−5.9	−6.3
40	40.11	1.130	−0.701	0.7648	0.1829	−13.9	−6.5	−6.0	−6.4
38	38.11	1.073	−0.842	0.7647	0.1738	−14.5	−6.6	−6.1	−6.5
36	36.10	1.017	−0.983	0.7646	0.1646	−15.1	−6.7	−6.2	−6.6
34	34.10	0.960	−1.124	0.7646	0.1555	−15.7	−6.8	−6.3	−6.7
32	32.10	0.904	−1.265	0.7645	0.1464	−16.3	−6.9	−6.4	−6.8
30	30.09	0.847	−1.406	0.7644	0.1372	−17.0	−7.0	−6.5	−6.9
28	28.09	0.791	−1.546	0.7644	0.1281	−17.8	−7.1	−6.5	−7.0
24	24.08	0.678	−1.828	0.7642	0.1098	−19.4	−7.3	−6.5	−7.2
20	20.07	0.565	−2.110	0.7641	0.0915	−21.3	−7.5	−6.5	−7.4
16	16.06	0.452	−2.392	0.7639	0.0732	−23.6	−7.7	−6.5	−7.6
12	12.05	0.339	−2.674	0.7638	0.0549	−26.5	−8.0	−6.5	−7.8
8	8.03	0.226	−2.956	0.7637	0.0366	−30.4	−8.2	−6.5	−8.0
4	4.02	0.113	−3.237	0.7635	0.0183	−36.9	−8.4	−6.5	−8.2
0	0.00	0.000	−3.519	0.7634	0.0000	—	−8.6	−6.5	−8.5

−3°C DRY BULB

Percentage saturation μ/%	Relative humidity ϕ/%	Value of stated parameter per kg dry air			Vapour pressure p_s/kPa	Dew point temperature t_d/°C	Adiabatic saturation temperature $t^\star$/°C	Wet bulb temperature	
		Moisture content g/g.kg^{-1}	Specific enthalpy h/kJ.kg^{-1}	Specific volume v/m^3.kg^{-1}				Screen t'_{sc}/°C	Sling t'_{sl}/°C
100	100.00	2.946	4.336	0.7684	0.4756	−3.0	−3.0	−3.0	−3.0
96	96.02	2.828	4.042	0.7683	0.4567	−3.5	−3.2	−3.2	−3.2
92	92.03	2.711	3.747	0.7681	0.4377	−4.0	−3.4	−3.3	−3.4
88	88.05	2.593	3.453	0.7680	0.4188	−4.5	−3.6	−3.5	−3.6
84	84.06	2.475	3.159	0.7678	0.3998	−5.0	−3.8	−3.7	−3.8
80	80.07	2.357	2.865	0.7677	0.3808	−5.6	−4.0	−3.8	−4.0
76	76.09	2.239	2.571	0.7676	0.3619	−6.2	−4.2	−4.0	−4.2
72	72.09	2.121	2.277	0.7674	0.3429	−6.8	−4.4	−4.2	−4.4
70	70.10	2.062	2.130	0.7673	0.3334	−7.2	−4.5	−4.3	−4.5
68	68.10	2.004	1.983	0.7673	0.3239	−7.5	−4.6	−4.4	−4.6
66	66.11	1.945	1.836	0.7672	0.3144	−7.8	−4.7	−4.5	−4.6
64	64.11	1.886	1.689	0.7671	0.3049	−8.2	−4.8	−4.5	−4.7
62	62.11	1.827	1.542	0.7670	0.2954	−8.6	−4.9	−4.6	−4.8
60	60.11	1.768	1.395	0.7670	0.2859	−8.9	−5.0	−4.7	−4.9
58	58.11	1.709	1.248	0.7669	0.2764	−9.3	−5.1	−4.8	−5.1
56	56.12	1.650	1.101	0.7668	0.2669	−9.7	−5.2	−4.9	−5.2
54	54.12	1.591	0.954	0.7668	0.2574	−10.1	−5.3	−5.0	−5.3
52	52.12	1.532	0.807	0.7667	0.2479	−10.5	−5.4	−5.1	−5.4
50	50.12	1.473	0.660	0.7666	0.2384	−11.0	−5.5	−5.2	−5.5
48	48.12	1.414	0.513	0.7665	0.2288	−11.4	−5.6	−5.3	−5.6
46	46.12	1.355	0.366	0.7665	0.2193	−11.9	−5.7	−5.3	−5.7
44	44.12	1.296	0.218	0.7664	0.2098	−12.4	−5.9	−5.4	−5.8
42	42.11	1.238	0.072	0.7663	0.2003	−12.9	−6.0	−5.5	−5.9
40	40.11	1.178	−0.076	0.7663	0.1908	−13.4	−6.1	−5.6	−6.0
38	38.11	1.120	−0.223	0.7662	0.1812	−14.0	−6.2	−5.7	−6.1
36	36.11	1.061	−0.370	0.7661	0.1717	−14.6	−6.3	−5.8	−6.2
34	34.11	1.002	−0.517	0.7660	0.1622	−15.2	−6.4	−5.9	−6.3
32	32.10	0.943	−0.664	0.7660	0.1527	−15.9	−6.5	−6.0	−6.4
30	30.10	0.884	−0.811	0.7659	0.1432	−16.6	−6.6	−6.0	−6.5
28	28.09	0.825	−0.958	0.7658	0.1336	−17.3	−6.7	−6.2	−6.6
24	24.09	0.707	−1.252	0.7657	0.1146	−18.9	−6.9	−6.3	−6.8
20	20.08	0.589	−1.546	0.7655	0.0955	−20.8	−7.2	−6.5	−7.0
16	16.06	0.471	−1.840	0.7654	0.0764	−23.1	−7.4	−6.7	−7.2
12	12.05	0.354	−2.134	0.7652	0.0573	−26.0	−7.6	−6.9	−7.5
8	8.03	0.236	−2.428	0.7651	0.0382	−30.0	−7.8	−7.1	−7.7
4	4.02	0.118	−2.722	0.7650	0.0191	−36.5	−8.1	−7.3	−7.9
0	0.00	0.000	−3.016	0.7648	0.0000	—	−8.3	−7.5	−8.1

−2.5°C DRY BULB

Percentage saturation μ/%	Relative humidity ϕ/%	Value of stated parameter per kg dry air			Vapour pressure p_s/kPa	Dew point temperature t_d/°C	Adiabatic saturation temperature $t^\star$/°C	Wet bulb temperature	
		Moisture content g/g.kg^{-1}	Specific enthalpy h/kJ.kg^{-1}	Specific volume v/m^3.kg^{-1}				Screen t'_{sc}/°C	Sling t'_{sl}/°C
100	100.00	3.073	5.158	0.7700	0.4960	−2.5	−2.5	−2.5	−2.5
96	96.02	2.950	4.851	0.7698	0.4762	−3.0	−2.7	−2.7	−2.7
92	92.04	2.827	4.544	0.7697	0.4565	−3.5	−2.9	−2.8	−2.9
88	88.05	2.704	4.237	0.7695	0.4367	−4.0	−3.1	−3.0	−3.1
84	84.07	2.582	3.930	0.7694	0.4170	−4.6	−3.3	−3.2	−3.3
80	80.08	2.458	3.623	0.7692	0.3972	−5.1	−3.5	−3.4	−3.5
76	76.09	2.336	3.317	0.7691	0.3774	−5.7	−3.7	−3.6	−3.7
72	72.10	2.213	3.010	0.7689	0.3576	−6.4	−3.9	−3.7	−3.9
70	70.10	2.151	2.856	0.7689	0.3477	−6.7	−4.0	−3.8	−4.0
68	68.11	2.090	2.703	0.7688	0.3378	−7.0	−4.1	−3.9	−4.1
66	66.11	2.028	2.550	0.7687	0.3279	−7.4	−4.2	−4.0	−4.2
64	64.11	1.967	2.396	0.7686	0.3180	−7.7	−4.4	−4.1	−4.3
62	62.12	1.905	2.243	0.7686	0.3081	−8.1	−4.5	−4.2	−4.4
60	60.12	1.844	2.089	0.7685	0.2982	−8.4	−4.6	−4.3	−4.5
58	58.12	1.782	1.936	0.7684	0.2883	−8.8	−4.7	−4.4	−4.6
56	56.12	1.721	1.782	0.7683	0.2783	−9.2	−4.8	−4.5	−4.7
54	54.12	1.660	1.629	0.7683	0.2684	−9.6	−4.9	−4.5	−4.8
52	52.12	1.598	1.476	0.7682	0.2585	−10.1	−5.0	−4.6	−4.9
50	50.12	1.537	1.322	0.7681	0.2486	−10.5	−5.1	−4.7	−5.0
48	48.12	1.475	1.169	0.7680	0.2387	−11.0	−5.2	−4.8	−5.1
46	46.12	1.414	1.015	0.7680	0.2288	−11.4	−5.3	−4.9	−5.2
44	44.12	1.352	0.862	0.7679	0.2188	−11.9	−5.4	−5.0	−5.3
42	42.12	1.291	0.708	0.7678	0.2089	−12.4	−5.5	−5.1	−5.4
40	40.12	1.229	0.555	0.7677	0.1990	−13.0	−5.7	−5.2	−5.6
38	38.12	1.168	0.402	0.7677	0.1890	−13.5	−5.8	−5.3	−5.7
36	36.11	1.106	0.248	0.7676	0.1791	−14.1	−5.9	−5.4	−5.8
34	34.11	1.045	0.095	0.7675	0.1692	−14.8	−6.0	−5.5	−5.9
32	32.11	0.983	−0.059	0.7674	0.1592	−15.4	−6.1	−5.6	−6.0
30	30.10	0.922	−0.212	0.7674	0.1493	−16.1	−6.2	−5.7	−6.1
28	28.10	0.860	−0.366	0.7673	0.1394	−16.9	−6.3	−5.8	−6.2
24	24.09	0.738	−0.672	0.7671	0.1195	−18.5	−6.5	−5.9	−6.4
20	20.08	0.615	−0.979	0.7670	0.0996	−20.4	−6.8	−6.1	−6.6
16	16.07	0.492	−1.286	0.7668	0.0797	−22.7	−7.0	−6.3	−6.9
12	12.05	0.369	−1.593	0.7667	0.0598	−25.6	−7.2	−6.5	−7.1
8	8.04	0.246	−1.900	0.7665	0.0399	−29.6	−7.5	−6.7	−7.3
4	4.02	0.122	−2.207	0.7664	0.0199	−36.1	−7.7	−6.9	−7.5
0	0.00	0.000	−2.514	0.7662	0.0000	—	−7.9	−7.1	−7.8

−2°C DRY BULB

Percentage saturation μ/%	Relative humidity ϕ/%	Value of stated parameter per kg dry air			Vapour pressure p_s/kPa	Dew point temperature t_d/°C	Adiabatic saturation temperature $t^\star$/°C	Wet bulb temperature	
		Moisture content $g/g.kg^{-1}$	Specific enthalpy $h/kJ.kg^{-1}$	Specific volume $v/m^3.kg^{-1}$				Screen t'_{sc}/°C	Sling t'_{sl}/°C
100	100.00	3.205	5.992	0.7716	0.5171	−2.0	−2.0	−2.0	−2.0
96	96.02	3.077	5.672	0.7714	0.4966	−2.5	−2.2	−2.2	−2.2
92	92.04	2.949	5.352	0.7713	0.4760	−3.0	−2.4	−2.4	−2.4
88	88.05	2.820	5.032	0.7711	0.4554	−3.5	−2.6	−2.5	−2.6
84	84.07	2.692	4.712	0.7710	0.4348	−4.1	−2.8	−2.7	−2.8
80	80.08	2.564	4.392	0.7708	0.4141	−4.6	−3.0	−2.9	−3.0
76	76.09	2.436	4.072	0.7706	0.3935	−5.2	−3.3	−3.1	−3.2
72	72.10	2.308	3.752	0.7705	0.3729	−5.9	−3.5	−3.3	−3.4
70	70.11	2.244	3.591	0.7704	0.3626	−6.2	−3.6	−3.4	−3.5
68	68.11	2.179	3.431	0.7703	0.3522	−6.5	−3.7	−3.4	−3.6
66	66.11	2.115	3.271	0.7703	0.3419	−6.9	−3.8	−3.5	−3.7
64	64.12	2.051	3.111	0.7702	0.3316	−7.2	−3.9	−3.6	−3.8
62	62.12	1.987	2.951	0.7701	0.3212	−7.6	−4.0	−3.7	−4.0
60	60.12	1.923	2.791	0.7700	0.3109	−8.0	−4.1	−3.8	−4.1
58	58.12	1.859	2.631	0.7699	0.3006	−8.4	−4.2	−3.9	−4.2
56	56.13	1.795	2.471	0.7699	0.2902	−8.8	−4.3	−4.0	−4.3
54	54.13	1.731	2.311	0.7698	0.2799	−9.2	−4.5	−4.1	−4.4
52	52.13	1.667	2.151	0.7697	0.2696	−9.6	−4.6	−4.2	−4.5
50	50.13	1.602	1.991	0.7696	0.2592	−10.0	−4.7	−4.3	−4.6
48	48.13	1.538	1.831	0.7695	0.2489	−10.5	−4.8	−4.4	−4.7
46	46.13	1.474	1.671	0.7695	0.2385	−11.0	−4.9	−4.5	−4.8
44	44.13	1.410	1.511	0.7694	0.2282	−11.5	−5.0	−4.6	−4.9
42	42.12	1.346	1.350	0.7693	0.2178	−12.0	−5.1	−4.7	−5.0
40	40.12	1.282	1.190	0.7692	0.2075	−12.5	−5.2	−4.8	−5.1
38	38.12	1.218	1.030	0.7692	0.1971	−13.1	−5.4	−4.9	−5.3
36	36.12	1.154	0.870	0.7691	0.1868	−13.7	−5.5	−5.0	−5.4
34	34.11	1.090	0.710	0.7690	0.1764	−14.3	−5.6	−5.0	−5.5
32	32.11	1.026	0.550	0.7689	0.1661	−15.0	−5.7	−5.1	−5.6
30	30.11	0.962	0.390	0.7688	0.1557	−15.7	−5.8	−5.2	−5.7
28	28.10	0.897	0.230	0.7688	0.1453	−16.4	−5.9	−5.4	−5.8
24	24.09	0.769	−0.090	0.7686	0.1246	−18.0	−6.2	−5.5	−6.0
20	20.08	0.641	−0.410	0.7684	0.1038	−20.0	−6.4	−5.8	−6.3
16	16.07	0.513	−0.730	0.7683	0.0831	−22.3	−6.6	−6.0	−6.5
12	12.05	0.385	−1.050	0.7681	0.0623	−25.2	−6.9	−6.1	−6.7
8	8.04	0.256	−1.371	0.7680	0.0416	−29.2	−7.1	−6.4	−7.0
4	4.02	0.128	−1.691	0.7678	0.0208	−35.7	−7.4	−6.6	−7.2
0	0.00	0.000	−2.011	0.7677	0.0000	—	−7.6	−6.7	−7.4

−1.5°C DRY BULB

Percentage saturation μ/%	Relative humidity ϕ/%	Value of stated parameter per kg dry air			Vapour pressure p_s/kPa	Dew point temperature t_d/°C	Adiabatic saturation temperature $t^\star$/°C	Wet bulb temperature	
		Moisture content $g/g.kg^{-1}$	Specific enthalpy $h/kJ.kg^{-1}$	Specific volume $v/m^3.kg^{-1}$				Screen t'_{sc}/°C	Sling t'_{sl}/°C
100	100.00	3.342	6.840	0.7732	0.5391	−1.5	−1.5	−1.5	−1.5
96	96.02	3.208	6.506	0.7730	0.5177	−2.0	−1.7	−1.7	−1.7
92	92.04	3.075	6.172	0.7729	0.4962	−2.5	−1.9	−1.9	−1.9
88	88.06	2.941	5.838	0.7727	0.4747	−3.0	−2.1	−2.1	−2.1
84	84.07	2.807	5.504	0.7725	0.4533	−3.6	−2.4	−2.2	−2.3
80	80.09	2.674	5.171	0.7724	0.4318	−4.1	−2.6	−2.4	−2.5
76	76.10	2.540	4.837	0.7722	0.4103	−4.7	−2.8	−2.6	−2.8
72	72.11	2.406	4.503	0.7720	0.3888	−5.4	−3.0	−2.8	−3.0
70	70.11	2.339	4.336	0.7720	0.3780	−5.7	−3.1	−2.9	−3.1
68	68.12	2.272	4.169	0.7719	0.3672	−6.0	−3.2	−3.0	−3.2
66	66.12	2.206	4.002	0.7718	0.3565	−6.4	−3.3	−3.1	−3.3
64	64.12	2.139	3.835	0.7717	0.3457	−6.7	−3.5	−3.2	−3.4
62	62.13	2.072	3.668	0.7716	0.3349	−7.1	−3.6	−3.3	−3.5
60	60.13	2.005	3.501	0.7715	0.3242	−7.5	−3.7	−3.4	−3.6
58	58.13	1.938	3.334	0.7715	0.3134	−7.9	−3.8	−3.5	−3.7
56	56.13	1.872	3.167	0.7714	0.3026	−8.3	−3.9	−3.6	−3.8
54	54.13	1.805	3.000	0.7713	0.2918	−8.7	−4.0	−3.7	−3.9
52	52.13	1.738	2.833	0.7712	0.2810	−9.1	−4.1	−3.8	−4.1
50	50.13	1.671	2.666	0.7711	0.2703	−9.6	−4.3	−3.9	−4.2
48	48.13	1.604	2.499	0.7710	0.2595	−10.0	−4.4	−4.0	−4.3
46	46.13	1.537	2.332	0.7710	0.2487	−10.5	−4.5	−4.1	−4.4
44	44.13	1.470	2.165	0.7709	0.2379	−11.0	−4.6	−4.2	−4.5
42	42.13	1.404	1.998	0.7708	0.2271	−11.5	−4.7	−4.3	−4.6
40	40.13	1.337	1.831	0.7707	0.2163	−12.1	−4.8	−4.4	−4.7
38	38.13	1.270	1.664	0.7706	0.2056	−12.6	−5.0	−4.5	−4.8
36	36.12	1.203	1.497	0.7706	0.1948	−13.2	−5.1	−4.5	−5.0
34	34.12	1.136	1.330	0.7705	0.1840	−13.8	−5.2	−4.6	−5.1
32	32.12	1.069	1.163	0.7704	0.1732	−14.5	−5.3	−4.7	−5.2
30	30.11	1.003	0.996	0.7703	0.1623	−15.2	−5.4	−4.8	−5.3
28	28.11	0.936	0.829	0.7702	0.1515	−16.0	−5.5	−5.0	−5.4
24	24.10	0.802	0.496	0.7701	0.1299	−17.6	−5.8	−5.2	−5.7
20	20.09	0.668	0.162	0.7699	0.1083	−19.5	−6.0	−5.4	−5.9
16	16.07	0.535	−0.172	0.7697	0.0866	−21.8	−6.3	−5.6	−6.1
12	12.06	0.401	−0.506	0.7696	0.0650	−24.8	−6.5	−5.8	−6.4
8	8.04	0.267	−0.840	0.7694	0.0433	−28.8	−6.8	−6.0	−6.6
4	4.02	0.134	−1.174	0.7692	0.0217	−35.3	−7.0	−6.2	−6.8
0	0.00	0.000	−1.508	0.7691	0.0000	—	−7.3	−6.4	−7.1

−1°C DRY BULB

Percentage saturation μ/%	Relative humidity ϕ/%	Value of stated parameter per kg dry air			Vapour pressure p_s/kPa	Dew point temperature t_d/°C	Adiabatic saturation temperature $t^\star$/°C	Wet bulb temperature	
		Moisture content $g/g.kg^{-1}$	Specific enthalpy h/kJ.kg^{-1}	Specific volume v/m³.kg^{-1}				Screen t'_{sc}/°C	Sling t'_{sl}/°C
100	100.00	3.484	7.702	0.7748	0.5620	−1.0	−1.0	−1.0	−1.0
96	96.02	3.345	7.354	0.7746	0.5396	−1.5	−1.2	−1.2	−1.2
92	92.04	3.206	7.005	0.7744	0.5172	−2.0	−1.4	−1.4	−1.4
88	88.06	3.066	6.657	0.7743	0.4949	−2.5	−1.7	−1.6	−1.6
84	84.07	2.927	6.309	0.7741	0.4725	−3.1	−1.9	−1.8	−1.9
80	80.09	2.787	5.960	0.7739	0.4501	−3.7	−2.1	−1.9	−2.1
76	76.10	2.648	5.612	0.7738	0.4277	−4.3	−2.3	−2.1	−2.3
72	72.11	2.509	5.264	0.7736	0.4052	−4.9	−2.5	−2.3	−2.5
70	70.12	2.439	5.090	0.7735	0.3940	−5.2	−2.7	−2.4	−2.6
68	68.12	2.369	4.916	0.7734	0.3828	−5.6	−2.8	−2.5	−2.7
66	66.12	2.300	4.741	0.7733	0.3716	−5.9	−2.9	−2.6	−2.8
64	64.13	2.230	4.567	0.7732	0.3604	−6.3	−3.0	−2.7	−2.9
62	62.13	2.160	4.393	0.7732	0.3492	−6.6	−3.1	−2.8	−3.1
60	60.13	2.091	4.219	0.7731	0.3379	−7.0	−3.2	−2.9	−3.2
58	58.14	2.021	4.045	0.7730	0.3267	−7.4	−3.4	−3.0	−3.3
56	56.14	1.951	3.871	0.7729	0.3155	−7.8	−3.5	−3.1	−3.4
54	54.14	1.882	3.696	0.7728	0.3042	−8.2	−3.6	−3.2	−3.5
52	52.14	1.812	3.522	0.7727	0.2930	−8.6	−3.7	−3.3	−3.6
50	50.14	1.742	3.348	0.7726	0.2818	−9.1	−3.8	−3.4	−3.7
48	48.14	1.672	3.174	0.7726	0.2705	−9.6	−3.9	−3.5	−3.9
46	46.14	1.603	3.000	0.7725	0.2593	−10.0	−4.1	−3.6	−4.0
44	44.14	1.533	2.826	0.7724	0.2480	−10.5	−4.2	−3.7	−4.1
42	42.14	1.463	2.652	0.7723	0.2368	−11.1	−4.3	−3.8	−4.2
40	40.13	1.394	2.477	0.7722	0.2255	−11.6	−4.4	−3.9	−4.3
38	38.13	1.324	2.303	0.7721	0.2143	−12.2	−4.5	−4.0	−4.4
36	36.13	1.254	2.129	0.7720	0.2030	−12.8	−4.7	−4.1	−4.6
34	34.12	1.185	1.955	0.7720	0.1918	−13.4	−4.8	−4.2	−4.7
32	32.12	1.115	1.781	0.7719	0.1805	−14.1	−4.9	−4.3	−4.8
30	30.12	1.045	1.607	0.7718	0.1692	−14.8	−5.0	−4.4	−4.9
28	28.11	0.976	1.433	0.7717	0.1580	−15.5	−5.2	−4.6	−5.0
24	24.10	0.836	1.084	0.7715	0.1354	−17.2	−5.4	−4.8	−5.3
20	20.09	0.697	0.736	0.7714	0.1129	−19.1	−5.7	−5.0	−5.5
16	16.07	0.558	0.388	0.7712	0.0903	−21.4	−5.9	−5.2	−5.8
12	12.06	0.418	0.039	0.7710	0.0678	−24.4	−6.2	−5.4	−6.0
8	8.04	0.279	−0.309	0.7708	0.0452	−28.4	−6.4	−5.6	−6.2
4	4.02	0.139	−0.657	0.7707	0.0226	−35.0	−6.7	−5.8	−6.5
0	0.00	0.000	−1.006	0.7705	0.0000	—	−6.9	−6.0	−6.7

−0.5°C DRY BULB

Percentage saturation μ/%	Relative humidity ϕ/%	Value of stated parameter per kg dry air			Vapour pressure p_s/kPa	Dew point temperature t_d/°C	Adiabatic saturation temperature $t^\star$/°C	Wet bulb temperature	
		Moisture content $g/g.kg^{-1}$	Specific enthalpy h/kJ.kg^{-1}	Specific volume v/m³.kg^{-1}				Screen t'_{sc}/°C	Sling t'_{sl}/°C
100	100.00	3.632	8.577	0.7764	0.5857	−0.5	−0.5	−0.5	−0.5
96	96.02	3.487	8.214	0.7762	0.5624	−1.0	−0.7	−0.7	−0.7
92	92.04	3.342	7.851	0.7760	0.5391	−1.5	−0.9	−0.9	−0.9
88	88.06	3.196	7.485	0.7759	0.5158	−2.0	−1.2	−1.1	−1.2
84	84.08	3.051	7.124	0.7757	0.4924	−2.6	−1.4	−1.3	−1.4
80	80.09	2.906	6.761	0.7755	0.4691	−3.2	−1.6	−1.5	−1.6
76	76.11	2.760	6.398	0.7753	0.4457	−3.8	−1.9	−1.7	−1.8
72	72.12	2.615	6.035	0.7751	0.4224	−4.4	−2.1	−1.9	−2.0
70	70.12	2.542	5.853	0.7751	0.4107	−4.7	−2.2	−2.0	−2.2
68	68.13	2.470	5.672	0.7750	0.3990	−5.1	−2.3	−2.1	−2.3
66	66.13	2.397	5.490	0.7749	0.3873	−5.4	−2.4	−2.2	−2.4
64	64.13	2.325	5.309	0.7748	0.3756	−5.8	−2.6	−2.3	−2.5
62	62.14	2.252	5.127	0.7747	0.3639	−6.1	−2.7	−2.4	−2.6
60	60.14	2.179	4.945	0.7746	0.3522	−6.5	−2.8	−2.5	−2.7
58	58.14	2.107	4.764	0.7745	0.3405	−6.9	−2.9	−2.6	−2.8
56	56.14	2.034	4.582	0.7744	0.3288	−7.3	−3.0	−2.7	−3.0
54	54.14	1.961	4.400	0.7743	0.3171	−7.7	−3.2	−2.8	−3.1
52	52.14	1.889	4.219	0.7742	0.3054	−8.2	−3.3	−2.9	−3.2
50	50.14	1.816	4.037	0.7742	0.2937	−8.6	−3.4	−3.0	−3.3
48	48.14	1.744	3.856	0.7741	0.2820	−9.1	−3.5	−3.1	−3.4
46	46.14	1.671	3.674	0.7740	0.2702	−9.6	−3.6	−3.2	−3.6
44	44.14	1.598	3.492	0.7739	0.2585	−10.1	−3.8	−3.3	−3.7
42	42.14	1.526	3.311	0.7738	0.2468	−10.6	−3.9	−3.4	−3.8
40	40.14	1.453	3.129	0.7737	0.2351	−11.1	−4.0	−3.5	−3.9
38	38.14	1.380	2.948	0.7736	0.2234	−11.7	−4.1	−3.6	−4.0
36	36.13	1.308	2.766	0.7735	0.2116	−12.3	−4.3	−3.7	−4.2
34	34.13	1.235	2.584	0.7734	0.1999	−12.9	−4.4	−3.8	−4.3
32	32.13	1.162	2.403	0.7734	0.1882	−13.6	−4.5	−3.9	−4.4
30	30.12	1.090	2.221	0.7733	0.1764	−14.3	−4.6	−4.0	−4.5
28	28.12	1.017	2.040	0.7732	0.1647	−15.1	−4.8	−4.2	−4.6
24	24.11	0.872	1.676	0.7730	0.1412	−16.7	−5.0	−4.4	−4.9
20	20.09	0.726	1.313	0.7728	0.1177	−18.7	−5.3	−4.6	−5.1
16	16.08	0.581	0.950	0.7726	0.0942	−21.0	−5.5	−4.8	−5.4
12	12.06	0.436	0.587	0.7725	0.0706	−23.9	−5.8	−5.0	−5.6
8	8.04	0.291	0.224	0.7723	0.0471	−28.0	−6.1	−5.2	−5.9
4	4.02	0.145	−0.140	0.7721	0.0236	−34.6	−6.3	−5.5	−6.2
0	0.00	0.000	−0.503	0.7719	0.0000	—	−6.6	−5.7	−6.4

0°C DRY BULB

Percentage saturation μ/%	Relative humidity ϕ/%	Value of stated parameter per kg dry air			Vapour pressure p_s/kPa	Dew point temperature t_d/°C	Adiabatic saturation temperature $t^\star$/°C	Wet bulb temperature	
		Moisture content g/g.kg^{-1}	Specific enthalpy h/kJ.kg^{-1}	Specific volume v/m³.kg^{-1}				Screen t'_{sc}/°C	Sling t'_{sl}/°C
100	100.00	3.789	9.475	0.7780	0.6108	0.0	0.0	0.0	0.0
96	96.02	3.637	9.096	0.7778	0.5865	−0.5	−0.2	−0.2	−0.2
92	92.04	3.486	8.717	0.7776	0.5622	−1.0	−0.5	−0.4	−0.4
88	88.06	3.334	8.338	0.7775	0.5379	−1.5	−0.7	−0.6	−0.7
84	84.08	3.183	7.959	0.7773	0.5136	−2.1	−0.9	−0.8	−0.9
80	80.10	3.031	7.580	0.7771	0.4892	−2.7	−1.1	−1.0	−1.1
76	76.11	2.880	7.201	0.7769	0.4649	−3.3	−1.4	−1.2	−1.3
72	72.12	2.728	6.822	0.7767	0.4405	−3.9	−1.6	−1.4	−1.6
70	70.13	2.652	6.633	0.7766	0.4283	−4.2	−1.7	−1.5	−1.7
68	68.13	2.576	6.443	0.7765	0.4161	−4.6	−1.9	−1.6	−1.8
66	66.14	2.501	6.254	0.7764	0.4039	−4.9	−2.0	−1.7	−1.9
64	64.14	2.425	6.064	0.7763	0.3918	−5.3	−2.1	−1.8	−2.0
62	62.14	2.349	5.875	0.7762	0.3796	−5.7	−2.2	−1.9	−2.2
60	60.14	2.273	5.685	0.7761	0.3674	−6.0	−2.4	−2.0	−2.3
58	58.15	2.198	5.496	0.7761	0.3552	−6.4	−2.5	−2.1	−2.4
56	56.15	2.122	5.306	0.7760	0.3429	−6.8	−2.6	−2.2	−2.5
54	54.15	2.046	5.117	0.7759	0.3307	−7.3	−2.7	−2.4	−2.6
52	52.15	1.970	4.927	0.7758	0.3185	−7.7	−2.8	−2.5	−2.8
50	50.15	1.894	4.738	0.7757	0.3063	−8.1	−3.0	−2.6	−2.9
48	48.15	1.819	4.548	0.7756	0.2941	−8.6	−3.1	−2.7	−3.0
46	46.15	1.743	4.359	0.7755	0.2819	−9.1	−3.2	−2.8	−3.1
44	44.15	1.667	4.169	0.7754	0.2696	−9.6	−3.4	−2.9	−3.3
42	42.15	1.591	3.980	0.7753	0.2574	−10.1	−3.5	−3.0	−3.4
40	40.14	1.516	3.790	0.7752	0.2452	−10.7	−3.6	−3.1	−3.5
38	38.14	1.440	3.601	0.7751	0.2330	−11.2	−3.7	−3.2	−3.6
36	36.14	1.364	3.411	0.7750	0.2207	−11.8	−3.9	−3.3	−3.8
34	34.14	1.288	3.222	0.7749	0.2085	−12.5	−4.0	−3.4	−3.9
32	32.13	1.212	3.032	0.7748	0.1962	−13.1	−4.1	−3.5	−4.0
30	30.13	1.137	2.843	0.7747	0.1840	−13.8	−4.3	−3.6	−4.1
28	28.12	1.061	2.653	0.7747	0.1718	−14.6	−4.4	−3.8	−4.3
24	24.11	0.909	2.274	0.7745	0.1473	−16.3	−4.7	−4.0	−4.5
20	20.10	0.758	1.895	0.7743	0.1228	−18.2	−4.9	−4.2	−4.8
16	16.08	0.606	1.516	0.7741	0.0982	−20.5	−5.2	−4.4	−5.0
12	12.06	0.455	1.137	0.7739	0.0737	−23.5	−5.5	−4.7	−5.3
8	8.04	0.303	0.758	0.7737	0.0491	−27.6	−5.7	−4.9	−5.5
4	4.02	0.152	0.379	0.7735	0.0246	−34.2	−6.0	−5.1	−5.8
0	0.00	0.000	0.000	0.7733	0.0000	—	−6.3	−5.3	−6.1

0.5°C DRY BULB

Percentage saturation μ/%	Relative humidity ϕ/%	Value of stated parameter per kg dry air			Vapour pressure p_s/kPa	Dew point temperature t_d/°C	Adiabatic saturation temperature $t^\star$/°C	Wet bulb temperature	
		Moisture content g/g.kg^{-1}	Specific enthalpy h/kJ.kg^{-1}	Specific volume v/m³.kg^{-1}				Screen t'_{sc}/°C	Sling t'_{sl}/°C
100	100.00	3.930	10.334	0.7796	0.6333	0.5	0.5	0.5	0.5
96	96.02	3.772	9.941	0.7794	0.6082	0.0	0.3	0.3	0.3
92	92.05	3.615	9.548	0.7792	0.5830	−0.6	0.0	0.1	0.1
88	88.07	3.458	9.154	0.7790	0.5578	−1.1	−0.2	−0.1	−0.2
84	84.08	3.301	8.761	0.7788	0.5325	−1.6	−0.5	−0.3	−0.4
80	80.10	3.144	8.368	0.7787	0.5073	−2.2	−0.7	−0.5	−0.7
76	76.11	2.986	7.975	0.7785	0.4821	−2.8	−0.9	−0.8	−0.9
72	72.13	2.829	7.581	0.7783	0.3468	−3.5	−1.2	−1.0	−1.1
70	70.13	2.751	7.385	0.7782	0.4442	−3.8	−1.3	−1.1	−1.3
68	68.14	2.672	7.188	0.7781	0.4315	−4.2	−1.4	−1.2	−1.4
66	66.14	2.594	6.992	0.7780	0.4189	−4.5	−1.6	−1.3	−1.5
64	64.14	2.515	6.795	0.7779	0.4062	−4.9	−1.7	−1.4	−1.6
62	62.15	2.436	6.598	0.7778	0.3936	−5.2	−1.8	−1.5	−1.7
60	60.15	2.358	6.402	0.7777	0.3810	−5.6	−1.9	−1.6	−1.9
58	58.15	2.279	6.205	0.7776	0.3683	−6.0	−2.1	−1.7	−2.0
56	56.15	2.200	6.008	0.7775	0.3556	−6.4	−2.2	−1.8	−2.1
54	54.16	2.122	5.812	0.7774	0.3430	−6.8	−2.3	−1.9	−2.2
52	52.16	2.043	5.615	0.7773	0.3303	−7.3	−2.4	−2.0	−2.4
50	50.16	1.965	5.418	0.7772	0.3177	−7.7	−2.6	−2.1	−2.5
48	48.16	1.886	5.222	0.7771	0.3050	−8.2	−2.7	−2.3	−2.6
46	46.16	1.808	5.025	0.7770	0.2923	−8.7	−2.8	−2.4	−2.7
44	44.15	1.729	4.829	0.7769	0.2796	−9.2	−3.0	−2.5	−2.9
42	42.15	1.650	4.632	0.7768	0.2670	−9.7	−3.1	−2.6	−3.0
40	40.15	1.572	4.435	0.7767	0.2543	−10.2	−3.2	−2.7	−3.1
38	38.15	1.493	4.239	0.7766	0.2416	−10.8	−3.3	−2.8	−3.2
36	36.14	1.415	4.042	0.7765	0.2289	−11.4	−3.5	−2.9	−3.4
34	34.14	1.336	3.846	0.7764	0.2162	−12.1	−3.6	−3.0	−3.5
32	32.14	1.258	3.649	0.7763	0.2035	−12.7	−3.7	−3.1	−3.6
30	30.13	1.179	3.452	0.7762	0.1908	−13.4	−3.9	−3.3	−3.7
28	28.13	1.100	3.256	0.7761	0.1781	−14.2	−4.0	−3.4	−3.9
24	24.11	0.943	2.862	0.7759	0.1527	−15.9	−4.3	−3.6	−4.1
20	20.10	0.786	2.469	0.7757	0.1273	−17.8	−4.6	−3.8	−4.4
16	16.08	0.629	2.076	0.7755	0.1019	−20.2	−4.8	−4.1	−4.7
12	12.07	0.471	1.683	0.7753	0.0764	−23.1	−5.1	−4.3	−4.9
8	8.05	0.314	1.290	0.7752	0.0510	−27.2	−5.4	−4.5	−5.2
4	4.02	0.157	0.896	0.7750	0.0255	−33.8	−5.7	−4.8	−5.5
0	0.00	0.000	0.503	0.7748	0.0000	—	−6.0	−5.0	−5.8

1°C DRY BULB

Percentage saturation μ/%	Relative humidity ϕ/%	Value of stated parameter per kg dry air			Vapour pressure p_s/kPa	Dew point temperature t_d/°C	Adiabatic saturation temperature $t\star$/°C	Wet bulb temperature	
		Moisture content g/g.kg^{-1}	Specific enthalpy h/kJ.kg^{-1}	Specific volume v/m^3.kg^{-1}				Screen t'_{sc}/°C	Sling t'_{sl}/°C
100	100.00	4.075	11.20	0.7812	0.6566	1.0	1.0	1.0	1.0
96	96.03	3.912	10.80	0.7810	0.6305	0.4	0.8	0.8	0.8
92	92.05	3.749	10.39	0.7808	0.6044	−0.1	0.5	0.6	0.5
88	88.07	3.586	9.98	0.7806	0.5783	−0.7	0.3	0.4	0.3
84	84.09	3.423	9.573	0.7804	0.5521	−1.2	0.1	0.2	0.1
80	80.10	3.260	9.165	0.7802	0.5260	−1.8	−0.3	−0.1	−0.2
76	76.12	3.097	8.757	0.7800	0.4998	−2.4	−0.5	−0.3	−0.5
72	72.13	2.934	8.349	0.7798	0.4736	−3.0	−0.7	−0.5	−0.7
70	70.14	2.852	8.145	0.7797	0.4605	−3.4	−0.9	−0.6	−0.8
68	68.14	2.771	7.941	0.7796	0.4474	−3.7	−1.0	−0.7	−0.9
66	66.15	2.690	7.737	0.7795	0.4343	−4.1	−1.1	−0.8	−1.1
64	64.15	2.608	7.533	0.7794	0.4212	−4.4	−1.3	−1.0	−1.2
62	62.15	2.526	7.329	0.7793	0.4081	−4.8	−1.4	−1.1	−1.3
60	60.16	2.445	7.125	0.7792	0.3950	−5.2	−1.5	−1.2	−1.4
58	58.16	2.364	6.921	0.7791	0.3819	−5.6	−1.6	−1.3	−1.6
56	56.16	2.282	6.717	0.7790	0.3688	−6.0	−1.8	−1.4	−1.7
54	54.16	2.200	6.513	0.7789	0.3556	−6.4	−1.9	−1.5	−1.8
52	52.16	2.119	6.309	0.7788	0.3425	−6.9	−2.0	−1.6	−1.9
50	50.16	2.038	6.105	0.7787	0.3294	−7.3	−2.2	−1.7	−2.1
48	48.16	1.956	5.901	0.7786	0.3162	−7.8	−2.3	−1.8	−2.2
46	46.16	1.874	5.697	0.7785	0.3031	−8.3	−2.4	−2.0	−2.3
44	44.16	1.793	5.493	0.7784	0.2900	−8.8	−2.6	−2.1	−2.5
42	42.16	1.712	5.289	0.7783	0.2768	−9.3	−2.7	−2.2	−2.6
40	40.16	1.630	5.085	0.7782	0.2637	−9.8	−2.8	−2.3	−2.7
38	38.15	1.548	4.881	0.7781	0.2505	−10.4	−3.0	−2.4	−2.8
36	36.15	1.467	4.678	0.7780	0.2374	−11.0	−3.1	−2.5	−3.0
34	34.15	1.386	4.474	0.7779	0.2242	−11.7	−3.2	−2.6	−3.1
32	32.14	1.304	4.270	0.7778	0.2110	−12.3	−3.4	−2.8	−3.2
30	30.14	1.222	4.066	0.7777	0.1979	−13.0	−3.5	−2.9	−3.4
28	28.13	1.141	3.862	0.7776	0.1847	−13.8	−3.6	−3.0	−3.5
24	24.12	0.978	3.454	0.7774	0.1584	−15.5	−3.9	−3.2	−3.8
20	20.10	0.815	3.046	0.7772	0.1320	−17.4	−4.2	−3.5	−4.0
16	16.09	0.652	2.638	0.7770	0.1056	−19.8	−4.5	−3.7	−4.3
12	12.07	0.489	2.230	0.7768	0.0792	−22.8	−4.8	−3.9	−4.6
8	8.05	0.326	1.822	0.7766	0.0528	−26.8	−5.1	−4.2	−4.9
4	4.02	0.163	1.414	0.7764	0.0264	−33.5	−5.3	−4.4	−5.1
0	0.00	0.000	1.006	0.7762	0.0000	—	−5.6	−4.7	−5.4

1.5°C DRY BULB

Percentage saturation μ/%	Relative humidity ϕ/%	Value of stated parameter per kg dry air			Vapour pressure p_s/kPa	Dew point temperature t_d/°C	Adiabatic saturation temperature $t\star$/°C	Wet bulb temperature	
		Moisture content g/g.kg^{-1}	Specific enthalpy h/kJ.kg^{-1}	Specific volume v/m^3.kg^{-1}				Screen t'_{sc}/°C	Sling t'_{sl}/°C
100	100.00	4.225	12.09	0.7828	0.6807	1.5	1.5	1.5	1.5
96	96.03	4.056	11.66	0.7826	0.6536	0.9	1.3	1.3	1.3
92	92.05	3.887	11.24	0.7824	0.6265	0.4	1.0	1.1	1.0
88	88.07	3.718	10.82	0.7822	0.5995	−0.2	0.8	0.9	0.8
84	84.09	3.549	10.40	0.7820	0.5724	−0.8	0.5	0.7	0.6
80	80.11	3.380	9.972	0.7818	0.5453	−1.4	0.3	0.4	0.3
76	76.12	3.211	9.548	0.7816	0.5181	−2.0	−1.0	0.2	0.1
72	72.14	3.042	8.125	0.7814	0.4910	−2.6	−0.3	−0.1	−0.3
70	70.14	2.958	8.914	0.7813	0.4774	−3.0	−0.4	−0.2	−0.4
68	68.15	2.873	8.702	0.7812	0.4638	−3.3	−0.6	−0.3	−0.5
66	66.15	2.789	8.491	0.7811	0.4503	−3.7	−0.7	−0.4	−0.6
64	64.15	2.704	8.279	0.7810	0.4367	−4.0	−0.8	−0.5	−0.8
62	62.16	2.620	8.068	0.7809	0.4231	−4.4	−1.0	−0.6	−0.9
60	60.16	2.535	7.856	0.7808	0.4095	−4.8	−1.1	−0.8	−1.0
58	58.16	2.451	7.644	0.7806	0.3959	−5.2	−1.2	−0.9	−1.1
56	56.17	2.366	7.438	0.7805	0.3823	−5.6	−1.4	−1.0	−1.3
54	54.17	2.282	7.221	0.7804	0.3687	−6.0	−1.5	−1.1	−1.4
52	52.17	2.197	7.010	0.7803	0.3551	−6.4	−1.6	−1.2	−1.5
50	50.17	2.113	6.798	0.7802	0.3415	−6.9	−1.8	−1.3	−1.7
48	48.17	2.028	6.587	0.7801	0.3279	−7.4	−1.9	−1.4	−1.8
46	46.17	1.944	6.375	0.7800	0.3142	−7.8	−2.0	−1.6	−1.9
44	44.17	1.859	6.163	0.7799	0.3006	−8.4	−2.2	−1.7	−2.1
42	42.16	1.775	5.952	0.7798	0.2870	−8.9	−2.3	−1.8	−2.2
40	40.16	1.690	5.740	0.7797	0.2734	−9.4	−2.4	−1.9	−2.3
38	38.16	1.606	5.529	0.7796	0.2597	−10.0	−2.6	−2.0	−2.5
36	36.16	1.521	5.317	0.7795	0.2461	−10.6	−2.7	−2.1	−2.6
34	34.15	1.436	5.106	0.7794	0.2324	−11.3	−2.9	−2.3	−2.7
32	32.15	1.352	4.894	0.7793	0.2188	−11.9	−3.0	−2.4	−2.9
30	30.14	1.268	4.682	0.7792	0.2052	−12.6	−3.1	−2.5	−3.0
28	28.14	1.183	4.471	0.7791	0.1915	−13.4	−3.3	−2.6	−3.1
24	24.12	1.014	4.048	0.7789	0.1642	−15.1	−3.6	−2.9	−3.4
20	20.11	0.845	3.625	0.7787	0.1369	−17.0	−3.8	−3.1	−3.7
16	16.09	0.676	3.202	0.7784	0.1095	−19.4	−4.1	−3.3	−4.0
12	12.07	0.507	2.778	0.7782	0.0822	−22.4	−4.4	−3.6	−4.2
8	8.05	0.338	2.355	0.7780	0.0548	−26.5	−4.7	−3.8	−4.5
4	4.03	0.169	1.932	0.7778	0.0274	−33.2	−5.0	−4.1	−4.6
0	0.00	0.000	1.509	0.7776	0.0000	—	−5.3	−4.3	−5.1

2°C DRY BULB

| Percentage saturation μ/% | Relative humidity ϕ/% | Value of stated parameter per kg dry air | | | Vapour pressure p_s/kPa | Dew point temperature t_d/°C | Adiabatic saturation temperature $t\star$/°C | Wet bulb temperature | |
		Moisture content g/g.kg⁻¹	Specific enthalpy h/kJ.kg⁻¹	Specific volume v/m³.kg⁻¹				Screen t'_{sc}/°C	Sling t'_{sl}/°C
100	100.00	4.380	12.98	0.7845	0.7055	2.0	2.0	2.0	2.0
96	96.03	4.205	12.54	0.7843	0.6774	1.4	1.8	1.8	1.8
92	92.05	4.030	12.10	0.7840	0.6494	0.8	1.5	1.6	1.5
88	88.07	3.855	11.67	0.7838	0.6213	0.2	1.3	1.4	1.3
84	84.09	3.679	11.23	0.7836	0.5933	−0.3	1.0	1.1	1.0
80	80.11	3.504	10.79	0.7834	0.5652	−0.9	0.8	0.9	0.8
76	76.13	3.329	10.35	0.7832	0.5371	−1.5	0.5	0.7	0.5
72	72.14	3.154	9.911	0.7829	0.5089	−2.2	0.3	0.5	0.3
70	70.15	3.066	9.692	0.7828	0.4949	−2.5	0.1	0.4	0.2
68	68.15	2.979	9.472	0.7827	0.4808	−2.9	−0.1	0.3	−0.1
66	66.16	2.891	9.253	0.7826	0.4667	−3.2	−0.3	0.1	−0.2
64	64.16	2.803	9.033	0.7825	0.4526	−3.6	−0.4	−0.1	−0.3
62	62.16	2.716	8.814	0.7824	0.4386	−4.0	−0.5	−0.2	−0.5
60	60.17	2.628	8.594	0.7823	0.4245	−4.3	−0.7	−0.3	−0.6
58	58.17	2.540	8.375	0.7822	0.4104	−4.7	−0.8	−0.4	−0.7
56	56.17	2.453	8.156	0.7821	0.3963	−5.2	−0.9	−0.6	−0.9
54	54.17	2.365	7.936	0.7820	0.3822	−5.6	−1.1	−0.7	−1.0
52	52.17	2.278	7.717	0.7819	0.3681	−6.0	−1.2	−0.8	−1.1
50	50.17	2.190	7.497	0.7817	0.3540	−6.5	−1.4	−0.9	−1.3
48	48.17	2.102	7.278	0.7816	0.3399	−6.9	−1.5	−1.0	−1.4
46	46.17	2.015	7.058	0.7815	0.3257	−7.4	−1.6	−1.1	−1.5
44	44.17	1.927	6.839	0.7814	0.3116	−7.9	−1.8	−1.3	−1.7
42	42.17	1.840	6.620	0.7813	0.2975	−8.5	−1.9	−1.4	−1.8
40	40.17	1.752	6.400	0.7812	0.2834	−9.0	−2.0	−1.5	−1.9
38	38.16	1.664	6.181	0.7811	0.2692	−9.6	−2.2	−1.6	−2.1
36	36.16	1.577	5.961	0.7810	0.2551	−10.2	−2.3	−1.7	−2.2
34	34.16	1.489	5.742	0.7809	0.2410	−10.9	−2.5	−1.9	−2.3
32	32.15	1.402	5.523	0.7808	0.2268	−11.5	−2.6	−2.0	−2.5
30	30.15	1.314	5.303	0.7807	0.2127	−12.2	−2.8	−2.1	−2.6
28	28.14	1.226	5.084	0.7805	0.1985	−13.0	−2.9	−2.2	−2.8
24	24.13	1.051	4.645	0.7803	0.1702	−14.7	−3.2	−2.5	−3.0
20	20.11	0.876	4.206	0.7801	0.1419	−16.7	−3.5	−2.7	−3.3
16	16.09	0.701	3.767	0.7799	0.1135	−19.0	−3.8	−3.0	−3.6
12	12.07	0.526	3.328	0.7797	0.0852	−22.0	−4.1	−3.2	−3.9
8	8.05	0.350	2.890	0.7795	0.0568	−26.1	−4.4	−3.5	−4.2
4	4.03	0.175	2.451	0.7792	0.0284	−32.8	−4.7	−3.7	−4.5
0	0.00	0.000	2.012	0.7790	0.0000	—	−5.0	−4.0	−4.8

2.5°C DRY BULB

| Percentage saturation μ/% | Relative humidity ϕ/% | Value of stated parameter per kg dry air | | | Vapour pressure p_s/kPa | Dew point temperature t_d/°C | Adiabatic saturation temperature $t\star$/°C | Wet bulb temperature | |
		Moisture content g/g.kg⁻¹	Specific enthalpy h/kJ.kg⁻¹	Specific volume v/m³.kg⁻¹				Screen t'_{sc}/°C	Sling t'_{sl}/°C
100	100.00	4.540	13.89	0.7861	0.7311	2.5	2.5	2.5	2.5
96	96.03	4.359	13.44	0.7859	0.7020	1.9	2.3	2.3	2.3
92	92.05	4.177	12.98	0.7856	0.6730	1.3	2.0	2.1	2.0
88	88.08	3.996	12.53	0.7854	0.6439	0.7	1.7	1.8	1.8
84	84.10	3.814	12.07	0.7852	0.6148	0.1	1.5	1.6	1.5
80	80.12	3.632	11.62	0.7850	0.5857	−0.5	1.2	1.4	1.3
76	76.13	3.451	11.16	0.7847	0.5566	−1.1	1.0	1.2	1.0
72	72.15	3.269	10.71	0.7845	0.5274	−1.8	0.7	0.9	0.8
70	70.15	3.178	10.48	0.7844	0.5129	−2.1	0.6	0.8	0.6
68	68.16	3.088	10.25	0.7843	0.4983	−2.4	0.5	0.7	0.5
66	66.16	2.997	10.02	0.7842	0.4837	−2.8	0.3	0.6	0.4
64	64.17	2.906	9.796	0.7841	0.4691	−3.2	0.2	0.5	0.2
62	62.17	2.815	9.568	0.7840	0.4545	−3.5	0.1	0.4	0.1
60	60.17	2.724	9.341	0.7838	0.4399	−3.9	−0.3	0.3	−0.2
58	58.18	2.633	9.113	0.7837	0.4253	−4.3	−0.4	0.1	−0.3
56	56.18	2.543	8.886	0.7836	0.4107	−4.7	−0.5	−0.1	−0.4
54	54.18	2.452	8.658	0.7835	0.3961	−5.2	−0.7	−0.3	−0.6
52	52.18	2.361	8.430	0.7834	0.3815	−5.6	−0.8	−0.4	−0.7
50	50.18	2.270	8.203	0.7833	0.3669	−6.1	−1.0	−0.5	−0.9
48	48.18	2.179	7.976	0.7832	0.3522	−6.5	−1.1	−0.6	−1.0
46	46.18	2.089	7.748	0.7830	0.3376	−7.0	−1.2	−0.7	−1.1
44	44.18	1.998	7.520	0.7829	0.3230	−7.5	−1.4	−0.9	−1.3
42	42.18	1.907	7.293	0.7829	0.3083	−8.1	−1.5	−1.0	−1.4
40	40.17	1.816	7.065	0.7827	0.2937	−8.6	−1.7	−1.1	−1.5
38	38.17	1.725	6.838	0.7826	0.2791	−9.2	−1.8	−1.2	−1.7
36	36.17	1.635	6.610	0.7825	0.2644	−9.8	−2.0	−1.4	−1.8
34	34.16	1.544	6.383	0.7824	0.2498	−10.5	−2.1	−1.5	−2.0
32	32.16	1.453	6.155	0.7823	0.2351	−11.1	−2.2	−1.6	−2.1
30	30.15	1.362	5.928	0.7821	0.2204	−11.8	−2.4	−1.7	−2.3
28	28.15	1.271	5.700	0.7820	0.2058	−12.6	−2.5	−1.9	−2.4
24	24.13	1.090	5.245	0.7818	0.1764	−14.3	−2.8	−2.1	−2.7
20	20.12	0.908	4.790	0.7816	0.1471	−16.3	−3.1	−2.4	−3.0
16	16.10	0.727	4.335	0.7813	0.1177	−18.7	−3.4	−2.6	−3.3
12	12.08	0.545	3.880	0.7811	0.0883	−21.7	−3.8	−2.9	−3.6
8	8.05	0.363	3.425	0.7809	0.0589	−25.8	−4.1	−3.1	−3.9
4	4.03	0.182	2.970	0.7807	0.0294	−32.5	−4.4	−3.4	−4.2
0	0.00	0.000	2.515	0.7804	0.0000	—	−4.7	−3.7	−4.5

3°C DRY BULB

Percentage saturation μ/%	Relative humidity ϕ/%	Value of stated parameter per kg dry air			Vapour pressure p_s/kPa	Dew point temperature t_d/°C	Adiabatic saturation temperature $t^\star$/°C	Wet bulb temperature	
		Moisture content g/g.kg^{-1}	Specific enthalpy h/kJ.kg^{-1}	Specific volume v/m³.kg^{-1}				Screen t'_{sc}/°C	Sling t'_{sl}/°C
100	100.00	4.706	14.81	0.7877	0.7575	3.0	3.0	3.0	3.0
96	96.03	4.518	14.34	0.7875	0.7274	2.4	2.7	2.8	2.8
92	92.06	4.329	13.87	0.7873	0.6973	1.8	2.5	2.6	2.5
88	88.08	4.141	13.40	0.7870	0.6672	1.2	2.2	2.3	2.2
84	84.10	3.953	12.93	0.7868	0.6371	0.6	2.0	2.1	2.0
80	80.12	3.765	12.45	0.7866	0.6069	−0.1	1.7	1.9	1.7
76	76.14	3.576	11.98	0.7863	0.5768	−0.7	1.4	1.6	1.5
72	72.15	3.388	11.51	0.7861	0.5466	−1.3	1.2	1.4	1.2
70	70.16	3.294	11.27	0.7860	0.5314	−1.7	1.0	1.3	1.1
68	68.16	3.200	11.04	0.7859	0.5164	−2.0	0.9	1.2	1.0
66	66.17	3.106	10.80	0.7857	0.5012	−2.4	0.8	1.1	0.8
64	64.17	3.012	10.57	0.7856	0.4861	−2.7	0.6	0.9	0.7
62	62.18	2.918	10.33	0.7855	0.4710	−3.1	0.5	0.8	0.6
60	60.18	2.824	10.09	0.7854	0.4559	−3.5	0.4	0.7	0.4
58	58.18	2.729	9.859	0.7853	0.4407	−3.9	0.2	0.6	0.3
56	56.18	2.635	9.623	0.7852	0.4256	−4.3	0.1	0.5	0.2
54	54.19	2.541	9.387	0.7850	0.4105	−4.7	−0.3	0.3	−0.2
52	52.19	2.447	9.151	0.7849	0.3953	−5.2	−0.4	0.2	−0.3
50	50.19	2.353	8.915	0.7848	0.3802	−5.6	−0.5	0.1	−0.5
48	48.19	2.259	8.679	0.7847	0.3650	−6.1	−0.7	−0.2	−0.6
46	46.19	2.166	8.444	0.7846	0.3499	−6.6	−0.8	−0.3	−0.7
44	44.18	2.071	8.208	0.7844	0.3347	−7.1	−1.0	−0.5	−0.9
42	42.18	1.976	7.972	0.7843	0.3195	−7.7	−1.1	−0.6	−1.0
40	40.18	1.882	7.736	0.7842	0.3044	−8.2	−1.3	−0.7	−1.2
38	38.18	1.788	7.500	0.7841	0.2892	−8.8	−1.4	−0.8	−1.3
36	36.17	1.694	7.264	0.7840	0.2740	−9.4	−1.6	−1.0	−1.4
34	34.17	1.600	7.028	0.7839	0.2588	−10.1	−1.7	−1.1	−1.6
32	32.16	1.506	6.792	0.7837	0.2436	−10.7	−1.9	−1.2	−1.7
30	30.16	1.412	6.556	0.7836	0.2284	−11.5	−2.0	−1.4	−1.9
28	28.15	1.318	6.320	0.7835	0.2132	−12.2	−2.2	−1.5	−2.0
24	24.14	1.129	5.849	0.7833	0.1828	−13.9	−2.5	−1.7	−2.3
20	20.12	0.941	5.377	0.7830	0.1524	−15.9	−2.8	−2.0	−2.6
16	16.10	0.753	4.905	0.7828	0.1220	−18.3	−3.1	−2.3	−2.9
12	12.08	0.565	4.433	0.7826	0.0915	−21.3	−3.4	−2.5	−3.2
8	8.06	0.376	3.962	0.7823	0.0610	−25.4	−3.7	−2.8	−3.5
4	4.03	0.188	3.490	0.7821	0.0305	−32.1	−4.1	−3.1	−3.8
0	0.00	0.000	3.018	0.7819	0.0000	—	−4.4	−3.3	−4.2

3.5°C DRY BULB

Percentage saturation μ/%	Relative humidity ϕ/%	Value of stated parameter per kg dry air			Vapour pressure p_s/kPa	Dew point temperature t_d/°C	Adiabatic saturation temperature $t^\star$/°C	Wet bulb temperature	
		Moisture content g/g.kg^{-1}	Specific enthalpy h/kJ.kg^{-1}	Specific volume v/m³.kg^{-1}				Screen t'_{sc}/°C	Sling t'_{sl}/°C
100	100.00	4.877	15.75	0.7894	0.7848	3.5	3.5	3.5	3.5
96	96.03	4.682	15.26	0.7891	0.7536	2.9	3.2	3.3	3.2
92	92.06	4.486	14.77	0.7889	0.7225	2.3	3.0	3.0	3.0
88	88.08	4.291	14.28	0.7887	0.6913	1.7	2.7	2.8	2.7
84	84.10	4.096	13.79	0.7884	0.6600	1.1	2.4	2.6	2.5
80	80.12	3.901	13.30	0.7882	0.6288	0.4	2.2	2.3	2.2
76	76.14	3.706	12.81	0.7879	0.5976	−0.3	1.9	2.1	1.9
72	72.16	3.511	12.32	0.7877	0.5663	−0.9	1.6	1.9	1.7
70	70.16	3.414	12.08	0.7876	0.5506	−1.2	1.5	1.7	1.5
68	68.17	3.316	11.84	0.7874	0.5350	−1.6	1.4	1.6	1.4
66	66.17	3.218	11.59	0.7873	0.5193	−1.9	1.2	1.5	1.3
64	64.18	3.121	11.35	0.7872	0.5037	−2.3	1.1	1.4	1.1
62	62.18	3.024	11.10	0.7871	0.4880	−2.7	0.9	1.3	1.0
60	60.19	2.926	10.86	0.7869	0.4723	−3.1	0.8	1.1	0.9
58	58.19	2.828	10.61	0.7868	0.4567	−3.5	0.7	1.0	0.7
56	56.19	2.731	10.37	0.7867	0.4410	−3.9	0.5	0.9	0.6
54	54.19	2.633	10.12	0.7866	0.4253	−4.3	0.4	0.8	0.5
52	52.19	2.536	9.879	0.7865	0.4096	−4.8	0.2	0.7	0.3
50	50.19	2.438	9.634	0.7663	0.3939	−5.2	0.1	0.5	0.2
48	48.19	2.341	9.390	0.7862	0.3782	−5.7	−0.3	0.4	−0.2
46	46.19	2.243	9.146	0.7861	0.3625	−6.2	−0.4	0.3	−0.3
44	44.19	2.146	8.901	0.7860	0.3468	−6.7	−0.6	0.2	−0.5
42	42.19	2.048	8.656	0.7858	0.3311	−7.2	−0.7	−0.2	−0.6
40	40.19	1.951	8.412	0.7857	0.3154	−7.8	−0.9	−0.3	−0.8
38	38.18	1.853	8.167	0.7856	0.2997	−8.4	−1.0	−0.5	−0.9
36	36.18	1.756	7.923	0.7855	0.2839	−9.0	−1.2	−0.6	−1.1
34	34.17	1.658	7.678	0.7854	0.2682	−9.6	−1.4	−0.7	−1.2
32	32.17	1.560	7.434	0.7852	0.2525	−10.3	−1.5	−0.8	−1.4
30	30.16	1.463	7.189	0.7851	0.2367	−11.1	−1.7	−1.0	−1.5
28	28.16	1.365	6.944	0.7850	0.2210	−11.8	−1.8	−1.1	−1.7
24	24.14	1.170	6.455	0.7847	0.1895	−13.5	−2.1	−1.4	−2.0
20	20.12	0.975	5.966	0.7845	0.1579	−15.5	−2.4	−1.6	−2.3
16	16.10	0.780	5.477	0.7843	0.1264	−17.9	−2.8	−1.9	−2.6
12	12.08	0.585	4.988	0.7840	0.0948	−20.9	−3.1	−2.2	−2.9
8	8.06	0.390	4.499	0.7838	0.0632	−25.0	−3.4	−2.5	−3.2
4	4.03	0.195	4.010	0.7835	0.0316	−31.8	−3.7	−2.7	−3.5
0	0.00	0.000	3.520	0.7833	0.0000	—	−4.1	−3.0	−3.9

4°C DRY BULB

Percentage saturation μ/%	Relative humidity ϕ/%	Value of stated parameter per kg dry air			Vapour pressure p_s/kPa	Dew point temperature t_d/°C	Adiabatic saturation temperature $t^\star$/°C	Wet bulb temperature	
		Moisture content g/g.kg^{-1}	Specific enthalpy h/kJ.kg^{-1}	Specific volume v/m^3.kg^{-1}				Screen t'_{sc}/°C	Sling t'_{sl}/°C
100	100.00	5.053	16.70	0.7910	0.8129	4.0	4.0	4.0	4.0
96	96.03	4.851	16.19	0.7908	0.7807	3.4	3.7	3.8	3.7
92	92.06	4.649	15.68	0.7905	0.7484	2.8	3.5	3.5	3.5
88	88.08	4.446	15.18	0.7903	0.7161	2.2	3.2	3.3	3.2
84	84.11	4.244	14.67	0.7900	0.6937	1.6	2.9	3.1	2.9
80	80.13	4.042	14.16	0.7898	0.6514	0.9	2.7	2.8	2.7
76	76.15	3.840	13.66	0.7895	0.6190	0.2	2.4	2.6	2.4
72	72.16	3.638	13.15	0.7893	0.5866	−0.5	2.1	2.3	2.1
70	70.17	3.537	12.90	0.7891	0.5704	−0.8	2.0	2.2	2.0
68	68.17	3.436	12.64	0.7890	0.5542	−1.2	1.8	2.1	1.9
66	66.18	3.335	12.39	0.7889	0.5380	−1.5	1.7	2.0	1.7
64	64.19	3.234	12.14	0.7888	0.5218	−1.9	1.5	1.8	1.6
62	62.19	3.133	11.88	0.7886	0.5056	−2.3	1.4	1.7	1.4
60	60.19	3.032	11.63	0.7885	0.4893	−2.7	1.2	1.6	1.3
58	58.20	2.931	11.38	0.7884	0.4731	−3.1	1.1	1.5	1.2
56	56.20	2.830	11.12	0.7883	0.4568	−3.5	1.0	1.3	1.0
54	54.20	2.728	10.87	0.7881	0.4406	−3.9	0.8	1.2	0.9
52	52.20	2.628	10.61	0.7880	0.4244	−4.3	0.7	1.1	0.7
50	50.20	2.526	10.36	0.7879	0.4081	−4.8	0.5	1.0	0.6
48	48.20	2.425	10.11	0.7877	0.3918	−5.3	0.4	0.8	0.5
46	46.20	2.324	9.854	0.7876	0.3756	−5.8	0.2	0.7	0.3
44	44.20	2.223	9.600	0.7875	0.3593	−6.3	0.1	0.6	0.2
42	42.20	2.122	9.347	0.7874	0.3430	−6.8	−0.4	0.4	−0.2
40	40.19	2.021	9.093	0.7872	0.3267	−7.4	−0.5	0.3	−0.4
38	38.19	1.920	8.840	0.7871	0.3105	−8.0	−0.7	0.2	−0.5
36	36.19	1.819	8.586	0.7870	0.2942	−8.6	−0.8	0.1	−0.7
34	34.18	1.718	8.333	0.7869	0.2779	−9.2	−1.0	−0.3	−0.8
32	32.18	1.617	8.080	0.7867	0.2616	−9.9	−1.1	−0.5	−1.0
30	30.17	1.516	7.826	0.7866	0.2453	−10.7	−1.3	−0.6	−1.1
28	28.16	1.415	7.573	0.7865	0.2289	−11.4	−1.5	−0.7	−1.3
24	24.15	1.213	7.066	0.7862	0.1963	−13.1	−1.8	−1.0	−1.6
20	20.13	1.011	6.559	0.7860	0.1636	−15.1	−2.1	−1.3	−1.9
16	16.11	0.808	6.052	0.7857	0.1310	−17.5	−2.4	−1.6	−2.2
12	12.09	0.606	5.545	0.7855	0.0982	−20.5	−2.8	−1.8	−2.6
8	8.06	0.404	5.038	0.7852	0.0655	−24.7	−3.1	−2.1	−2.9
4	4.03	0.202	4.531	0.7850	0.0328	−31.5	−3.4	−2.4	−3.2
0	0.00	0.000	4.024	0.7847	0.0000	—	−3.8	−2.7	−3.5

4.5°C DRY BULB

Percentage saturation μ/%	Relative humidity ϕ/%	Value of stated parameter per kg dry air			Vapour pressure p_s/kPa	Dew point temperature t_d/°C	Adiabatic saturation temperature $t^\star$/°C	Wet bulb temperature	
		Moisture content g/g.kg^{-1}	Specific enthalpy h/kJ.kg^{-1}	Specific volume v/m^3.kg^{-1}				Screen t'_{sc}/°C	Sling t'_{sl}/°C
100	100.00	5.235	17.66	0.7927	0.8420	4.5	4.5	4.5	4.5
96	96.03	5.025	17.14	0.7924	0.8086	3.9	4.2	4.3	4.2
92	92.06	4.816	16.61	0.7922	0.7751	3.3	4.0	4.0	4.0
88	88.09	4.607	16.09	0.7919	0.7417	2.7	3.7	3.8	3.7
84	84.11	4.397	15.56	0.7916	0.7082	2.1	3.4	3.5	3.4
80	80.13	4.188	15.04	0.7914	0.6747	1.4	3.1	3.3	3.2
76	76.15	3.978	14.51	0.7911	0.6412	0.7	2.8	3.0	2.9
72	72.17	3.769	13.98	0.7909	0.6076	−0.1	2.6	2.8	2.6
70	70.17	3.664	13.72	0.7907	0.5908	−0.4	2.4	2.7	2.5
68	68.18	3.560	13.46	0.7906	0.5741	−0.7	2.3	2.5	2.3
66	66.19	3.455	13.20	0.7905	0.5573	−1.1	2.1	2.4	2.2
64	64.19	3.350	12.93	0.7903	0.5405	−1.5	2.0	2.3	2.0
62	62.20	3.246	12.67	0.7902	0.5237	−1.8	1.8	2.2	1.9
60	60.20	3.141	12.41	0.7901	0.5069	−2.2	1.7	2.0	1.7
58	58.20	3.036	12.15	0.7899	0.4900	−2.6	1.5	1.9	1.6
56	56.21	2.932	11.88	0.7898	0.4732	−3.1	1.4	1.8	1.5
54	54.21	2.827	11.62	0.7897	0.4564	−3.5	1.2	1.6	1.3
52	52.21	2.722	11.36	0.7895	0.4396	−3.9	1.1	1.5	1.2
50	50.21	2.617	11.09	0.7894	0.4227	−4.4	0.9	1.4	1.0
48	48.21	2.513	10.83	0.7893	0.4059	−4.9	0.8	1.3	0.9
46	46.21	2.408	10.57	0.7891	0.3890	−5.4	0.6	1.1	0.7
44	44.21	2.303	10.31	0.7890	0.3722	−5.9	0.5	1.0	0.6
42	42.20	2.199	10.04	0.7889	0.3553	−6.4	0.3	0.9	0.4
40	40.20	2.094	9.781	0.7888	0.3385	−7.0	0.2	0.7	0.3
38	38.20	1.989	9.518	0.7886	0.3216	−7.6	−0.3	0.6	0.1
36	36.19	1.884	9.256	0.7885	0.3047	−8.2	−0.5	0.5	−0.3
34	34.19	1.780	8.993	0.7884	0.2878	−8.8	−0.6	0.3	−0.5
32	32.18	1.675	8.730	0.7882	0.2710	−9.5	−0.8	0.2	−0.6
30	30.18	1.570	8.467	0.7881	0.2541	−10.3	−0.9	0.1	−0.8
28	28.17	1.466	8.205	0.7880	0.2372	−11.0	−1.1	−0.4	−0.9
24	24.15	1.256	7.679	0.7877	0.2034	−12.7	−1.4	−0.6	−1.3
20	20.13	1.047	7.154	0.7874	0.1695	−14.7	−1.8	−0.9	−1.6
16	16.11	0.838	6.628	0.7872	0.1357	−17.1	−2.1	−1.2	−1.9
12	12.09	0.628	6.103	0.7869	0.1018	−20.2	−2.4	−1.5	−2.2
8	8.06	0.419	5.578	0.7867	0.0679	−24.3	−2.8	−1.8	−2.6
4	4.03	0.209	5.052	0.7864	0.0340	−31.1	−3.1	−2.1	−2.9
0	0.00	0.000	4.527	0.7861	0.0000	—	−3.5	−2.4	−3.2

5°C DRY BULB

Percentage saturation μ/%	Relative humidity ϕ/%	Value of stated parameter per kg dry air			Vapour pressure p_s/kPa	Dew point temperature t_d/°C	Adiabatic saturation temperature $t\star$/°C	Wet bulb temperature	
		Moisture content $g/g.kg^{-1}$	Specific enthalpy $h/kJ.kg^{-1}$	Specific volume $v/m^3.kg^{-1}$				Screen t'_{sc}/°C	Sling t'_{sl}/°C
100	100.00	5.422	18.64	0.7944	0.8719	5.0	5.0	5.0	5.0
96	96.03	5.206	18.10	0.7941	0.8373	4.4	4.7	4.8	4.7
92	92.06	4.989	17.55	0.7938	0.8027	3.8	4.4	4.5	4.5
88	88.09	4.772	17.01	0.7935	0.7681	3.2	4.2	4.3	4.2
84	84.12	4.555	16.46	0.7933	0.7334	2.5	3.9	4.0	3.9
80	80.14	4.338	15.92	0.7930	0.6987	1.9	3.6	3.8	3.6
76	76.16	4.121	15.37	0.7927	0.6640	1.2	3.3	3.5	3.3
72	72.17	3.904	14.83	0.7925	0.6293	0.4	3.0	3.2	3.1
70	70.18	3.796	14.56	0.7923	0.6119	0.0	2.9	3.1	2.9
68	68.19	3.687	14.29	0.7922	0.5945	−0.3	2.7	3.0	2.8
66	66.19	3.579	14.01	0.7920	0.5771	−0.7	2.6	2.9	2.6
64	64.20	3.470	13.74	0.7919	0.5598	−1.0	2.4	2.7	2.5
62	62.20	3.362	13.47	0.7918	0.5424	−1.4	2.3	2.6	2.3
60	60.21	3.254	13.20	0.7916	0.5250	−1.8	2.1	2.5	2.2
58	58.21	3.145	12.92	0.7915	0.5075	−2.2	2.0	2.3	2.0
56	56.21	3.037	12.65	0.7914	0.4901	−2.6	1.8	2.2	1.9
54	54.21	2.928	12.38	0.7912	0.4727	−3.1	1.7	2.1	1.7
52	52.22	2.820	12.11	0.7911	0.4553	−3.5	1.5	1.9	1.6
50	50.22	2.711	11.84	0.7910	0.4378	−4.0	1.4	1.8	1.4
48	48.22	2.603	11.56	0.7908	0.4204	−4.5	1.2	1.7	1.3
46	46.21	2.494	11.29	0.7907	0.4030	−5.0	1.1	1.5	1.1
44	44.21	2.386	11.02	0.7905	0.3855	−5.5	0.9	1.4	1.0
42	42.21	2.278	10.75	0.7904	0.3680	−6.0	0.7	1.3	0.8
40	40.21	2.169	10.47	0.7903	0.3506	−6.6	0.6	1.1	0.7
38	38.20	2.061	10.20	0.7901	0.3331	−7.2	0.4	1.0	0.5
36	36.20	1.952	9.930	0.7900	0.3156	−7.8	0.3	0.9	0.4
34	34.19	1.884	9.658	0.7899	0.2981	−8.4	0.1	0.7	0.2
32	32.19	1.735	9.385	0.7897	0.2806	−9.1	−0.4	0.6	0.1
30	30.18	1.627	9.113	0.7896	0.2632	−9.9	−0.6	0.5	−0.4
28	28.17	1.518	8.841	0.7895	0.2456	−10.6	−0.7	0.3	−0.6
24	24.16	1.301	8.296	0.7892	0.2106	−12.4	−1.1	−0.3	−0.9
20	20.14	1.084	7.752	0.7889	0.1756	−14.4	−1.4	−0.6	−1.2
16	16.12	0.868	7.208	0.7886	0.1405	−16.8	−1.8	−0.9	−1.6
12	12.09	0.651	6.663	0.7884	0.1054	−19.8	−2.1	−1.2	−1.9
8	8.06	0.434	6.119	0.7881	0.0703	−24.0	−2.5	−1.5	−2.2
4	4.03	0.217	5.574	0.7878	0.0352	−30.8	−2.8	−1.8	−2.6
0	0.00	0.000	5.030	0.7875	0.0000	—	−3.2	−2.1	−2.9

5.5°C DRY BULB

Percentage saturation μ/%	Relative humidity ϕ/%	Value of stated parameter per kg dry air			Vapour pressure p_s/kPa	Dew point temperature t_d/°C	Adiabatic saturation temperature $t\star$/°C	Wet bulb temperature	
		Moisture content $g/g.kg^{-1}$	Specific enthalpy $h/kJ.kg^{-1}$	Specific volume $v/m^3.kg^{-1}$				Screen t'_{sc}/°C	Sling t'_{sl}/°C
100	100.00	5.618	19.64	0.7960	0.9028	5.5	5.5	5.5	5.5
96	96.03	5.393	19.07	0.7958	0.8670	4.9	5.2	5.2	5.2
92	92.07	5.167	18.51	0.7955	0.8312	4.3	4.9	5.0	4.9
88	88.09	4.942	17.94	0.7952	0.7953	3.7	4.6	4.7	4.7
84	84.12	4.718	17.38	0.7949	0.7594	3.0	4.4	4.5	4.4
80	80.14	4.493	16.82	0.7916	0.7235	2.4	4.1	4.2	4.1
76	76.16	4.268	16.25	0.7943	0.6876	1.6	3.8	4.0	3.8
72	72.18	4.044	15.69	0.7941	0.6516	0.9	3.5	3.7	3.5
70	70.19	3.931	15.40	0.7939	0.6336	0.5	3.3	3.6	3.4
68	68.19	3.819	15.12	0.7938	0.6156	0.1	3.2	3.4	3.2
66	66.20	3.707	14.84	0.7936	0.5976	−0.3	3.0	3.3	3.1
64	64.21	3.594	14.56	0.7935	0.5796	−0.6	2.9	3.2	2.9
62	62.21	3.482	14.28	0.7934	0.5616	−0.1	2.7	3.0	2.8
60	60.21	3.370	13.99	0.7932	0.5436	−1.4	2.6	2.9	2.6
58	58.22	3.257	13.71	0.7931	0.5256	−1.8	2.4	2.8	2.5
56	56.22	3.145	13.43	0.7929	0.5076	−2.2	2.3	2.6	2.3
54	54.22	3.033	13.15	0.7928	0.4895	−2.7	2.1	2.5	2.2
52	52.22	2.920	12.87	0.7926	0.4715	−3.1	1.9	2.4	2.0
50	50.22	2.808	12.58	0.7925	0.4534	−3.6	1.8	2.2	1.9
48	48.22	2.696	12.30	0.7924	0.4354	−4.0	1.6	2.1	1.7
46	46.22	2.584	12.02	0.7922	0.4173	−4.5	1.5	2.0	1.6
44	44.22	2.471	11.74	0.7921	0.3992	−5.1	1.3	1.8	1.4
42	42.22	2.359	11.46	0.7919	0.3811	−5.6	1.1	1.7	1.2
40	40.21	2.246	11.17	0.7918	0.3630	−6.2	1.0	1.5	1.1
38	38.21	2.134	10.89	0.7917	0.3450	−6.8	0.8	1.4	0.9
36	36.21	2.022	10.61	0.7915	0.3269	−7.4	0.7	1.3	0.8
34	34.20	1.910	10.33	0.7914	0.3088	−8.0	0.5	1.1	0.6
32	32.19	1.797	10.05	0.7912	0.2906	−8.7	0.3	1.0	0.5
30	30.19	1.685	9.764	0.7911	0.2725	−9.5	0.2	0.8	0.3
28	28.18	1.573	9.482	0.7909	0.2544	−10.2	−0.4	0.7	0.1
24	24.16	1.348	8.918	0.7907	0.2181	−12.0	−0.7	0.4	−0.4
20	20.14	1.123	8.353	0.7904	0.1818	−14.0	−1.1	0.1	−0.9
16	16.12	0.899	7.789	0.7901	0.1455	−16.4	−1.4	−0.5	−1.2
12	12.09	0.674	7.225	0.7898	0.1092	−19.4	−1.8	−0.8	−1.6
8	8.07	0.449	6.661	0.7895	0.0728	−23.6	−2.1	−1.2	−1.9
4	4.03	0.225	6.097	0.7892	0.0364	−30.5	−2.5	−1.4	−2.3
0	0.00	0.000	5.533	0.7890	0.0000	—	−2.9	−1.8	−2.6

6°C DRY BULB

Percentage saturation μ/%	Relative humidity ϕ/%	Value of stated parameter per kg dry air			Vapour pressure p_s/kPa	Dew point temperature t_d/°C	Adiabatic saturation temperature $t\star$/°C	Wet bulb temperature	
		Moisture content g/g.kg^{-1}	Specific enthalpy h/kJ.kg^{-1}	Specific volume v/m^3.kg^{-1}				Screen t'_{sc}/°C	Sling t'_{sl}/°C
100	100.00	5.816	20.65	0.7977	0.9346	6.0	6.0	6.0	6.0
96	96.04	5.584	20.06	0.7974	0.8976	5.4	5.7	5.7	5.7
92	92.07	5.351	19.48	0.7971	0.8605	4.8	5.4	5.5	5.4
88	88.10	5.118	18.89	0.7968	0.8234	4.2	5.1	5.2	5.1
84	84.12	4.886	18.31	0.7966	0.7862	3.5	4.8	5.0	4.9
80	80.15	4.653	17.72	0.7963	0.7491	2.8	4.5	4.7	4.6
76	76.17	4.420	17.14	0.7960	0.7119	2.1	4.2	4.4	4.3
72	72.19	4.188	16.56	0.7957	0.6747	1.4	3.9	4.2	4.0
70	70.19	4.071	16.26	0.7955	0.6560	1.0	3.8	4.0	3.8
68	68.20	3.955	15.97	0.7954	0.6374	0.6	3.6	3.9	3.7
66	66.21	3.839	15.68	0.7952	0.6188	0.2	3.5	3.8	3.5
64	64.21	3.722	15.39	0.7951	0.6002	−0.2	3.3	3.6	3.4
62	62.22	3.606	15.09	0.7949	0.5815	−0.6	3.2	3.5	3.2
60	60.22	3.490	14.80	0.7948	0.5628	−1.0	3.0	3.4	3.1
58	58.23	3.373	14.51	0.7946	0.5442	−1.4	2.8	3.2	2.9
56	56.23	3.257	14.22	0.7945	0.5255	−1.8	2.7	3.1	2.8
54	54.23	3.141	13.93	0.7944	0.5068	−2.2	2.5	2.9	2.6
52	52.23	3.024	13.63	0.7942	0.4882	−2.7	2.4	2.8	2.4
50	50.23	2.907	13.34	0.7941	0.4695	−3.2	2.2	2.7	2.3
48	48.23	2.792	13.05	0.7939	0.4508	−3.6	2.0	2.5	2.1
46	46.23	2.675	12.76	0.7938	0.4321	−4.1	1.9	2.4	2.0
44	44.23	2.559	12.46	0.7936	0.4134	−4.7	1.7	2.2	1.8
42	42.23	2.443	12.17	0.7935	0.3946	−5.2	1.6	2.1	1.7
40	40.22	2.326	11.88	0.7933	0.3759	−5.8	1.4	2.0	1.5
38	38.22	2.210	11.59	0.7932	0.3572	−6.4	1.2	1.8	1.3
36	36.21	2.094	11.30	0.7930	0.3385	−7.0	1.1	1.7	1.2
34	34.21	1.978	11.00	0.7929	0.3197	−7.6	0.9	1.5	1.0
32	32.20	1.861	10.71	0.7927	0.3010	−8.3	0.7	1.4	0.8
30	30.19	30.19	10.42	0.7926	0.2822	−9.1	0.6	1.2	0.7
28	28.19	1.628	10.13	0.7924	0.2634	−9.9	0.4	1.1	0.5
24	24.17	1.396	9.542	0.7921	0.2259	−11.6	−0.4	0.8	0.2
20	20.15	1.163	8.958	0.7919	0.1883	−13.6	−0.7	0.5	−0.6
16	16.12	0.931	8.373	0.7916	0.1507	−16.0	−1.1	0.2	−0.9
12	12.10	0.698	7.789	0.7913	0.1131	−19.1	−1.5	−0.5	−1.3
8	8.07	0.465	7.205	0.7910	0.0754	−23.3	−1.8	−0.8	−1.6
4	4.04	0.233	6.620	0.7907	0.0377	−30.1	−2.2	−1.1	−2.0
0	0.00	0.000	6.036	0.7904	0.0000	—	−2.6	−1.4	−2.3

6.5°C DRY BULB

Percentage saturation μ/%	Relative humidity ϕ/%	Value of stated parameter per kg dry air			Vapour pressure p_s/kPa	Dew point temperature t_d/°C	Adiabatic saturation temperature $t\star$/°C	Wet bulb temperature	
		Moisture content g/g.kg^{-1}	Specific enthalpy h/kJ.kg^{-1}	Specific volume v/m^3.kg^{-1}				Screen t'_{sc}/°C	Sling t'_{sl}/°C
100	100.00	6.022	21.67	0.7994	0.9674	6.5	6.5	6.5	6.5
96	96.04	5.782	21.08	0.7991	0.9291	5.9	6.2	6.2	6.2
92	92.07	5.541	20.46	0.7988	0.8907	5.3	5.9	6.0	5.9
88	88.10	5.300	19.86	0.7985	0.8523	4.7	5.6	5.7	5.6
84	84.13	5.059	19.25	0.7982	0.8139	4.0	5.3	5.4	5.3
80	80.15	4.818	18.65	0.7979	0.7754	3.3	5.0	5.2	5.0
76	76.17	4.577	18.04	0.7976	0.7370	2.6	4.7	4.9	4.7
72	72.19	4.336	17.44	0.7973	0.6984	1.9	4.4	4.6	4.4
70	70.20	4.216	17.13	0.7971	0.6792	1.5	4.2	4.5	4.3
68	68.21	4.095	16.83	0.7970	0.6599	1.1	4.1	4.4	4.1
66	66.21	3.975	16.53	0.7968	0.6406	0.7	3.9	4.2	4.0
64	64.22	3.854	16.22	0.7967	0.6213	0.2	3.8	4.1	3.8
62	62.23	3.734	15.92	0.7965	0.6020	−0.2	3.6	3.9	3.7
60	60.23	3.614	15.62	0.7964	0.5827	−0.6	3.4	3.8	3.5
58	58.23	3.493	15.32	0.7962	0.5634	−1.0	3.3	3.7	3.3
56	56.24	3.373	15.01	0.7961	0.5441	−1.4	3.1	3.5	3.2
54	54.24	3.252	14.71	0.7959	0.5247	−1.8	3.0	3.4	3.0
52	52.24	3.132	14.41	0.7958	0.5054	−2.3	2.8	3.2	2.9
50	50.24	3.011	14.11	0.7956	0.4860	−2.7	2.6	3.1	2.7
48	48.24	2.891	13.80	0.7955	0.4667	−3.2	2.5	2.9	2.5
46	46.24	2.770	13.50	0.7953	0.4473	−3.7	2.3	2.8	2.4
44	44.24	2.650	13.20	0.7952	0.4280	−4.3	2.1	2.7	2.2
42	42.23	2.529	12.90	0.7950	0.4086	−4.8	2.0	2.5	2.1
40	40.23	2.409	12.59	0.7949	0.3892	−5.4	1.8	2.4	1.9
38	38.23	2.289	12.29	0.7947	0.3698	−6.0	1.6	2.2	1.7
36	36.22	2.168	11.99	0.7945	0.3504	−6.6	1.4	2.1	1.6
34	34.22	2.048	11.68	0.7944	0.3310	−7.2	1.3	1.9	1.4
32	32.21	1.927	11.38	0.7942	0.3116	−7.9	1.1	1.8	1.2
30	30.20	1.807	11.08	0.7941	0.2922	−8.7	0.9	1.6	1.1
28	28.19	1.686	10.78	0.7939	0.2728	−9.5	0.8	1.5	0.9
24	24.18	1.445	10.17	0.7936	0.2339	−11.2	0.4	1.2	0.5
20	20.15	1.204	9.566	0.7933	0.1950	−13.2	0.1	0.9	0.2
16	16.13	0.964	8.960	0.7930	0.1560	−15.6	−0.8	0.6	−0.6
12	12.10	0.723	8.355	0.7927	0.1171	−18.7	−1.2	0.3	−0.9
8	8.07	0.482	7.750	0.7924	0.0781	−22.9	−1.5	−0.5	−1.3
4	4.04	0.241	7.144	0.7921	0.0391	−29.8	−1.9	−0.8	−1.7
0	0.00	0.000	6.539	0.7918	0.0000	—	−2.3	−1.1	−2.1

7°C DRY BULB

Percentage saturation μ/%	Relative humidity ϕ/%	Value of stated parameter per kg dry air			Vapour pressure p_s/kPa	Dew point temperature t_d/°C	Adiabatic saturation temperature $t^\star$/°C	Wet bulb temperature	
		Moisture content $g/g.kg^{-1}$	Specific enthalpy h/kJ.kg^{-1}	Specific volume v/m^3.kg^{-1}				Screen t'_{sc}/°C	Sling t'_{sl}/°C
100	100.00	6.235	22.72	0.8011	1.001	7.0	7.0	7.0	7.0
96	96.04	5.986	22.09	0.8008	0.9616	6.4	6.7	6.7	6.7
92	92.07	5.736	21.46	0.8005	0.9219	5.8	6.4	6.5	6.4
88	88.10	5.487	20.84	0.8002	0.8822	5.2	6.1	6.2	6.1
84	84.13	5.238	20.21	0.7999	0.8424	4.5	5.8	5.9	5.8
80	80.16	4.988	19.58	0.7995	0.8026	3.8	5.5	5.6	5.5
76	76.18	4.739	18.95	0.7992	0.7628	3.1	5.2	5.4	5.2
72	72.20	4.489	18.33	0.7989	0.7229	2.3	4.8	5.1	4.9
70	70.21	4.365	18.01	0.7988	0.7030	2.0	4.7	5.0	4.7
68	68.22	4.240	17.70	0.7986	0.6830	1.5	4.5	4.8	4.6
66	66.22	4.115	17.39	0.7984	0.6631	1.1	4.4	4.7	4.4
64	64.23	3.990	17.07	0.7983	0.6431	0.7	4.2	4.5	4.3
62	62.23	3.866	16.76	0.7981	0.6231	0.3	4.0	4.4	4.1
60	60.24	3.741	16.45	0.7980	0.6032	−0.1	3.9	4.2	3.9
58	58.24	3.616	16.13	0.7978	0.5832	−0.6	3.7	4.1	3.8
56	56.24	3.492	15.82	0.7977	0.5632	−1.0	3.5	4.0	3.6
54	54.25	3.367	15.51	0.7975	0.5432	−1.4	3.4	3.8	3.5
52	52.25	3.242	15.19	0.7973	0.5232	−1.9	3.2	3.7	3.3
50	50.25	3.118	14.88	0.7972	0.5031	−2.3	3.0	3.5	3.1
48	48.25	2.993	14.57	0.7970	0.4831	−2.8	2.9	3.4	3.0
46	46.25	2.868	14.25	0.7969	0.4631	−3.3	2.7	3.2	2.8
44	44.24	2.744	13.94	0.7967	0.4430	−3.8	2.5	3.1	2.6
42	42.24	2.619	13.62	0.7965	0.4230	−4.4	2.4	2.9	2.5
40	40.24	2.494	13.31	0.7964	0.4029	−5.0	2.2	2.8	2.3
38	38.23	2.369	13.00	0.7962	0.3828	−5.6	2.0	2.6	2.1
36	36.23	2.245	12.68	0.7961	0.3628	−6.2	1.8	2.5	2.0
34	34.22	2.120	12.37	0.7959	0.3427	−6.8	1.7	2.3	1.8
32	32.22	1.995	12.06	0.7958	0.3226	−7.5	1.5	2.2	1.6
30	30.21	1.871	11.74	0.7956	0.3025	−8.3	1.3	2.0	1.4
28	28.20	1.746	11.43	0.7954	0.2824	−9.1	1.1	1.9	1.3
24	24.18	1.496	10.80	0.7951	0.2421	−10.8	0.8	1.6	0.9
20	20.16	1.247	10.18	0.7948	0.2018	−12.8	0.4	1.2	0.6
16	16.13	0.998	9.550	0.7945	0.1616	−15.3	−0.5	0.9	0.2
12	12.11	0.748	8.923	0.7942	0.1212	−18.3	−0.8	0.6	−0.6
8	8.07	0.499	8.296	0.7939	0.0808	−22.6	−1.2	0.3	−1.0
4	4.04	0.249	7.669	0.7935	0.0404	−29.5	−1.6	−0.5	−1.4
0	0.00	0.000	7.042	0.7932	0.0000	—	−2.0	−0.8	−1.8

7.5°C DRY BULB

Percentage saturation μ/%	Relative humidity ϕ/%	Value of stated parameter per kg dry air			Vapour pressure p_s/kPa	Dew point temperature t_d/°C	Adiabatic saturation temperature $t^\star$/°C	Wet bulb temperature	
		Moisture content $g/g.kg^{-1}$	Specific enthalpy h/kJ.kg^{-1}	Specific volume v/m^3.kg^{-1}				Screen t'_{sc}/°C	Sling t'_{sl}/°C
100	100.00	6.455	23.78	0.8028	1.036	7.5	7.5	7.5	7.5
96	96.04	6.196	23.13	0.8025	0.9951	6.9	7.2	7.2	7.2
92	92.08	5.938	22.48	0.8022	0.9541	6.3	6.9	7.0	6.9
88	88.11	5.680	21.83	0.8019	0.9130	5.7	6.6	6.7	6.6
84	84.14	5.422	21.18	0.8015	0.8718	5.0	6.3	6.4	6.3
80	80.16	5.164	20.53	0.8012	0.8306	4.3	6.0	6.1	6.0
76	76.19	4.906	19.88	0.8009	0.7894	3.6	5.6	5.8	5.7
72	72.21	4.647	19.23	0.8005	0.7482	2.8	5.3	5.6	5.3
70	70.22	4.518	18.91	0.8004	0.7276	2.4	5.1	5.4	5.2
68	68.22	4.389	18.58	0.8002	0.7069	2.0	5.0	5.3	5.0
66	66.23	4.260	18.26	0.8001	0.6863	1.6	4.8	5.1	4.9
64	64.24	4.131	17.93	0.7999	0.6656	1.2	4.7	5.0	4.7
62	62.24	4.002	17.61	0.7997	0.6449	0.8	4.5	4.8	4.5
60	60.25	3.873	17.28	0.7996	0.6243	0.3	4.3	4.7	4.4
58	58.25	3.744	16.96	0.7994	0.6036	−0.1	4.1	4.5	4.2
56	56.25	3.615	16.63	0.7992	0.5829	−0.6	4.0	4.4	4.0
54	54.25	3.486	16.31	0.7991	0.5622	−1.0	3.8	4.2	3.9
52	52.26	3.356	15.99	0.7989	0.5415	−1.4	3.6	4.1	3.7
50	50.26	3.227	15.66	0.7987	0.5208	−1.9	3.5	3.9	3.5
48	48.26	3.098	15.34	0.7986	0.5000	−2.4	3.3	3.8	3.4
46	46.26	2.969	15.01	0.7984	0.4793	−2.9	3.1	3.6	3.2
44	44.25	2.840	14.69	0.7983	0.4585	−3.4	2.9	3.5	3.0
42	42.25	2.711	14.36	0.7981	0.4378	−4.0	2.8	3.3	2.9
40	40.25	2.582	14.04	0.7979	0.4170	−4.6	3.6	3.2	2.7
38	38.24	2.453	13.71	0.7978	0.3963	−5.2	2.4	3.0	2.5
36	36.24	2.324	13.39	0.7976	0.3755	−5.8	2.2	2.9	2.3
34	34.23	2.195	13.06	0.7974	0.3547	−6.4	2.1	2.7	2.2
32	32.22	2.066	12.74	0.7973	0.3339	−7.1	1.9	2.6	2.0
30	30.22	1.936	12.41	0.7971	0.3131	−7.9	1.7	2.4	1.8
28	28.21	1.807	12.09	0.7969	0.2923	−8.7	1.5	2.2	1.6
24	24.19	1.549	11.44	0.7966	0.2506	−10.4	1.1	1.9	1.3
20	20.16	1.291	10.79	0.7963	0.2089	−12.4	0.8	1.6	0.9
16	16.14	1.033	10.14	0.7960	0.1672	−14.9	0.4	1.3	0.6
12	12.11	0.775	9.493	0.7956	0.1255	−18.0	−0.5	1.0	0.2
8	8.08	0.516	8.843	0.7953	0.0837	−22.2	−0.9	0.6	−0.7
4	4.04	0.258	8.194	0.7950	0.0419	−29.1	−1.3	0.3	−1.1
0	0.00	0.000	7.545	0.7946	0.0000	—	−1.7	−0.5	−1.5

8°C DRY BULB

Percentage saturation μ/%	Relative humidity ϕ/%	Value of stated parameter per kg dry air			Vapour pressure p_s/kPa	Dew point temperature t_d/°C	Adiabatic saturation temperature $t\star$/°C	Wet bulb temperature	
		Moisture content g/g.kg^{-1}	Specific enthalpy h/kJ.kg^{-1}	Specific volume v/m³.kg^{-1}				Screen t'_{sc}/°C	Sling t'_{sl}/°C
100	100.00	6.681	24.86	0.8046	1.072	8.0	8.0	8.0	8.0
96	96.04	6.414	24.18	0.8042	1.030	7.4	7.7	7.7	7.7
92	92.08	6.146	23.51	0.8039	0.9872	6.8	7.4	7.4	7.4
88	88.11	5.879	22.84	0.8035	0.9447	6.2	7.1	7.2	7.1
84	84.14	5.612	22.17	0.8032	0.9021	5.5	6.7	6.9	6.8
80	80.17	5.345	21.49	0.8029	0.8595	4.8	6.4	6.6	6.4
76	76.19	5.078	20.82	0.8025	0.8169	4.1	6.1	6.3	6.1
72	72.21	4.810	20.15	0.8022	0.7742	3.3	5.8	6.0	5.8
70	70.22	4.677	19.81	0.8020	0.7529	2.9	5.6	5.9	5.6
68	68.23	4.543	19.48	0.8018	0.7315	2.5	5.4	5.7	5.5
66	66.24	4.909	19.14	0.8017	0.7102	2.1	5.3	5.6	5.3
64	64.24	4.276	18.80	0.8015	0.6888	1.7	5.1	5.4	5.2
62	62.25	4.142	18.47	0.8013	0.6674	1.2	4.9	5.3	5.0
60	60.25	4.009	18.13	0.8012	0.6460	0.8	4.8	5.1	4.8
58	58.26	3.875	17.80	0.8010	0.6246	0.3	4.6	5.0	4.6
56	56.26	3.741	17.46	0.8008	0.6032	−0.1	4.4	4.8	4.5
54	54.26	3.608	17.12	0.8007	0.5818	−0.6	4.2	4.7	4.3
52	52.27	3.474	16.79	0.8005	0.5604	−1.0	4.1	4.5	4.1
50	50.27	3.340	16.45	0.8003	0.5389	−1.5	3.9	4.4	4.0
48	48.27	3.207	16.12	0.8001	0.5175	−2.0	3.7	4.2	3.8
46	46.26	3.073	15.78	0.8000	0.4960	−2.5	3.5	4.0	3.6
44	44.26	2.940	15.44	0.7998	0.4745	−3.0	3.4	3.9	3.4
42	42.26	2.806	15.11	0.7996	0.4531	−3.6	3.2	3.7	3.3
40	40.26	2.672	14.77	0.7995	0.4316	−4.2	3.0	3.6	3.1
38	38.25	2.539	14.43	0.7993	0.4101	−4.8	2.8	3.4	2.9
36	36.25	2.405	14.10	0.7991	0.3886	−5.4	2.6	3.3	2.7
34	34.24	2.272	13.76	0.7990	0.3671	−6.0	2.4	3.1	2.6
32	32.23	2.138	13.43	0.7988	0.3456	−6.7	2.3	2.9	2.4
30	30.22	2.004	13.09	0.7986	0.3240	−7.5	2.1	2.8	2.2
28	28.21	1.871	12.75	0.7984	0.3025	−8.3	1.9	2.6	2.0
24	24.19	1.603	12.08	0.7981	0.2594	−10.0	1.5	2.3	1.6
20	20.17	1.336	11.41	0.7978	0.2162	−12.1	1.1	2.0	1.3
16	16.14	1.069	10.74	0.7974	0.1731	−14.5	0.7	1.6	0.9
12	12.11	0.8017	10.06	0.7971	0.1299	−17.6	0.3	1.3	0.5
8	8.08	0.5345	9.392	0.7967	0.0866	−21.9	−0.6	1.0	0.1
4	4.04	0.2672	8.720	0.7964	0.0433	−28.8	−1.0	0.6	−0.8
0	0.00	0.0000	8.048	0.7961	0.0000	—	−1.5	0.3	−1.2

8.5°C DRY BULB

Percentage saturation μ/%	Relative humidity ϕ/%	Value of stated parameter per kg dry air			Vapour pressure p_s/kPa	Dew point temperature t_d/°C	Adiabatic saturation temperature $t\star$/°C	Wet bulb temperature	
		Moisture content g/g.kg^{-1}	Specific enthalpy h/kJ.kg^{-1}	Specific volume v/m³.kg^{-1}				Screen t'_{sc}/°C	Sling t'_{sl}/°C
100	100.00	6.914	25.95	0.8063	1.109	8.5	8.5	8.5	8.5
96	96.04	6.638	25.26	0.8059	1.065	7.9	8.2	8.2	8.2
92	92.08	6.361	24.56	0.8056	1.021	7.3	7.9	7.9	7.9
88	88.12	6.085	23.86	0.8052	0.9774	6.6	7.5	7.6	7.6
84	84.15	5.808	23.17	0.8049	0.9333	6.0	7.2	7.4	7.2
80	80.18	5.532	22.47	0.8045	0.8893	5.3	6.9	7.1	6.9
76	76.20	5.255	21.78	0.8042	0.8452	4.6	6.6	6.8	6.6
72	72.22	4.978	21.08	0.8038	0.8011	3.8	6.2	6.5	6.3
70	70.23	4.840	20.73	0.8037	0.7790	3.4	6.1	6.3	6.1
68	68.24	4.702	20.38	0.8035	0.7569	3.0	5.9	6.2	5.9
66	66.25	4.564	20.04	0.8033	0.7348	2.6	5.7	6.0	5.8
64	64.25	4.425	19.69	0.8031	0.7127	2.1	5.5	5.9	5.6
62	62.26	4.287	19.34	0.8029	0.6906	1.7	5.4	5.7	5.4
60	60.26	4.149	18.99	0.8028	0.6684	1.2	5.2	5.6	5.3
58	58.27	4.010	18.64	0.8026	0.6463	0.8	5.0	5.4	5.1
56	56.27	3.872	18.29	0.8024	0.6241	0.3	4.8	5.3	4.9
54	54.27	3.734	17.95	0.8022	0.6020	−0.2	4.7	5.1	4.7
52	52.27	3.596	17.60	0.8021	0.5798	−0.6	4.5	4.9	4.6
50	50.27	3.457	17.25	0.8019	0.5576	−1.1	4.3	4.8	4.4
48	48.27	3.319	16.90	0.8017	0.5354	−1.6	4.1	4.6	4.2
46	46.27	3.181	16.55	0.8015	0.5132	−2.1	3.9	4.5	4.0
44	44.27	3.042	16.21	0.8014	0.4910	−2.6	3.8	4.3	3.9
42	42.27	2.904	15.86	0.8012	0.4688	−3.2	3.6	4.1	3.7
40	40.26	2.766	15.51	0.8010	0.4466	−3.7	3.4	4.0	3.5
38	38.26	2.627	15.16	0.8008	0.4244	−4.3	3.2	3.8	3.3
36	36.25	2.489	14.81	0.8007	0.4021	−5.0	3.0	3.7	3.1
34	34.25	2.351	14.47	0.8005	0.3799	−5.6	2.8	3.5	2.9
32	32.24	2.213	14.12	0.8003	0.3576	−6.4	2.6	3.3	2.8
30	30.23	2.074	13.77	0.8001	0.3353	−7.1	2.4	3.2	2.6
28	28.22	1.936	13.42	0.8000	0.3130	−7.9	2.3	3.0	2.4
24	24.20	1.659	12.73	0.7996	0.2684	−9.6	1.9	2.7	2.0
20	20.18	1.383	12.03	0.7993	0.2238	−11.7	1.5	2.3	1.6
16	16.15	1.106	11.33	0.7989	0.1791	−14.1	1.1	2.0	1.2
12	12.12	0.831	10.64	0.7985	0.1344	−17.2	0.7	1.7	0.9
8	8.08	0.553	9.943	0.7982	0.0896	−21.5	0.3	1.3	0.5
4	4.04	0.278	9.247	0.7978	0.0448	−28.5	−0.7	1.0	0.1
0	0.00	0.000	8.551	0.7975	0.0000	—	−1.2	0.6	−0.9

9°C DRY BULB

Percentage saturation μ/%	Relative humidity ϕ/%	Value of stated parameter per kg dry air			Vapour pressure p_s/kPa	Dew point temperature t_d/°C	Adiabatic saturation temperature $t^\star$/°C	Wet bulb temperature	
		Moisture content $g/g.kg^{-1}$	Specific enthalpy $h/kJ.kg^{-1}$	Specific volume $v/m^3.kg^{-1}$				Screen t'_{sc}/°C	Sling t'_{sl}/°C
100	100.00	7.155	27.07	0.8080	1.147	9.0	9.0	9.0	9.0
96	96.04	6.869	26.35	0.8077	1.102	8.4	8.7	8.7	8.7
92	92.08	6.583	25.63	0.8073	1.056	7.8	8.4	8.4	8.4
88	88.12	6.296	24.90	0.8069	1.011	7.1	8.0	8.1	8.0
84	84.15	6.010	24.18	0.8066	0.9655	6.5	7.7	7.8	7.7
80	80.18	5.724	23.46	0.8062	0.9200	5.8	7.4	7.5	7.4
76	76.21	5.438	22.74	0.8058	0.8744	5.0	7.0	7.2	7.1
72	72.23	5.152	22.02	0.8055	0.8287	4.3	6.7	6.9	6.7
70	70.24	5.008	21.66	0.8053	0.8059	3.9	6.5	6.8	6.6
68	68.25	4.865	21.30	0.8051	0.7830	3.5	6.3	6.6	6.4
66	66.25	4.722	20.94	0.8049	0.7602	3.0	6.2	6.5	6.2
64	64.26	4.579	20.58	0.8048	0.7373	2.6	6.0	6.3	6.0
62	62.27	4.436	20.22	0.8046	0.7144	2.2	5.8	6.2	5.9
60	60.27	4.293	19.86	0.8044	0.6915	1.7	5.6	6.0	5.7
58	58.28	4.150	19.50	0.8042	0.6686	1.3	5.4	5.8	5.5
56	56.28	4.007	19.14	0.8040	0.6457	0.8	5.3	5.7	5.3
54	54.28	3.864	18.78	0.8038	0.6228	0.3	5.1	5.5	5.2
52	52.28	3.721	18.42	0.8037	0.5999	−0.2	4.9	5.4	5.0
50	50.28	3.578	18.06	0.8035	0.5769	−0.7	4.7	5.2	4.8
48	48.28	3.434	17.70	0.8033	0.5540	−1.2	4.5	5.0	4.6
46	46.28	3.291	17.34	0.8031	0.5310	−1.7	4.4	4.9	4.4
44	44.28	3.148	16.98	0.8029	0.5080	−2.2	4.2	4.7	4.3
42	42.28	3.005	16.62	0.8027	0.4851	−2.8	4.0	4.6	4.1
40	40.27	2.862	16.26	0.8026	0.4621	−3.3	3.8	4.4	3.9
38	38.27	2.719	15.90	0.8024	0.4391	−3.9	3.6	4.2	3.7
36	36.26	2.576	15.54	0.8022	0.4160	−4.6	3.4	4.1	3.5
34	34.26	2.433	15.18	0.8020	0.3930	−5.3	3.2	3.9	3.3
32	32.25	2.290	14.82	0.8018	0.3700	−6.0	3.0	3.7	3.1
30	30.24	2.146	14.46	0.8016	0.3470	−6.7	2.8	3.6	3.0
28	28.23	2.003	14.10	0.8015	0.3239	−7.5	2.6	3.4	2.8
24	24.21	1.717	13.38	0.8011	0.2778	−9.3	2.2	3.0	2.4
20	20.18	1.431	12.66	0.8007	0.2316	−11.3	1.8	2.7	2.0
16	16.15	1.145	11.94	0.8004	0.1853	−13.8	1.4	2.4	1.6
12	12.12	0.857	11.22	0.8000	0.1391	−16.9	1.0	2.0	1.2
8	8.08	0.572	10.50	0.7996	0.0928	−21.1	0.6	1.6	0.8
4	4.04	0.286	9.774	0.7993	0.0464	−28.1	0.2	1.3	0.4
0	0.00	0.000	9.054	0.7989	0.0000	—	−0.9	0.9	−0.6

9.5°C DRY BULB

Percentage saturation μ/%	Relative humidity ϕ/%	Value of stated parameter per kg dry air			Vapour pressure p_s/kPa	Dew point temperature t_d/°C	Adiabatic saturation temperature $t^\star$/°C	Wet bulb temperature	
		Moisture content $g/g.kg^{-1}$	Specific enthalpy $h/kJ.kg^{-1}$	Specific volume $v/m^3.kg^{-1}$				Screen t'_{sc}/°C	Sling t'_{sl}/°C
100	100.00	7.403	28.20	0.8098	1.188	9.5	9.5	9.5	9.5
96	96.05	7.107	27.45	0.8094	1.141	8.9	9.2	9.2	9.2
92	92.09	6.811	26.71	0.8090	1.094	8.3	8.8	8.9	8.9
88	88.12	6.515	25.96	0.8087	1.047	7.6	8.5	8.6	8.5
84	84.16	6.219	25.22	0.8083	0.9987	7.0	8.2	8.3	8.2
80	80.19	5.923	24.47	0.8079	0.9516	6.3	7.8	8.0	7.9
76	76.21	5.626	23.73	0.8075	0.9044	5.5	7.5	7.7	7.5
72	72.24	5.330	22.98	0.8071	0.8572	4.8	7.1	7.4	7.2
70	70.25	5.182	22.61	0.8070	0.8336	4.4	7.0	7.2	7.0
68	68.26	5.034	22.23	0.8068	0.8100	3.9	6.8	7.1	6.8
66	66.26	4.886	21.86	0.8066	0.7863	3.5	6.6	6.9	6.7
64	64.27	4.738	21.49	0.8064	0.7627	3.1	6.4	6.8	6.5
62	62.28	4.590	21.12	0.8062	0.7390	2.7	6.2	6.6	6.3
60	60.28	4.442	20.74	0.8060	0.7154	2.2	6.1	6.4	6.1
58	58.29	4.294	20.37	0.8058	0.6917	1.7	5.9	6.3	5.9
56	56.29	4.146	20.00	0.8056	0.6680	1.2	5.7	6.1	5.8
54	54.29	3.998	19.62	0.8054	0.6443	0.7	5.5	6.0	5.6
52	52.29	3.850	19.25	0.8053	0.6206	0.2	5.3	5.8	5.4
50	50.29	3.702	18.88	0.8051	0.5968	−0.3	5.1	5.6	5.2
48	48.29	3.554	18.51	0.8049	0.5731	−0.8	4.9	5.5	5.0
46	46.29	3.405	18.13	0.8047	0.5493	−1.3	4.8	5.3	4.9
44	44.29	3.257	17.76	0.8045	0.5256	−1.8	4.6	5.1	4.7
42	42.29	3.109	17.39	0.8043	0.5018	−2.4	4.4	5.0	4.5
40	40.28	2.961	17.01	0.8041	0.4780	−2.9	4.2	4.8	4.3
38	38.28	2.813	16.64	0.8039	0.4542	−3.5	4.0	4.6	4.1
36	36.27	2.665	16.27	0.8037	0.4304	−4.2	3.8	4.5	3.9
34	34.26	2.517	15.90	0.8035	0.4066	−4.9	3.6	4.3	3.7
32	32.26	2.369	15.52	0.8034	0.3828	−5.6	3.4	4.1	3.5
30	30.25	2.221	15.15	0.8032	0.3589	−6.3	3.2	3.9	3.3
28	28.24	2.073	14.78	0.8030	0.3351	−7.1	3.0	3.8	3.1
24	24.22	1.777	14.03	0.8026	0.2874	−8.9	2.6	3.4	2.7
20	20.19	1.481	13.29	0.8022	0.2396	−10.9	2.2	3.1	2.3
16	16.16	1.184	12.54	0.8018	0.1918	−13.4	1.8	2.7	1.9
12	12.12	0.888	11.79	0.8015	0.1439	−16.5	1.3	2.3	1.5
8	8.09	0.592	11.05	0.8011	0.0960	−20.8	0.9	2.0	1.1
4	4.05	0.296	10.30	0.8007	0.0480	−27.8	0.5	1.6	0.7
0	0.00	0.000	9.557	0.8003	0.0000	—	−0.6	1.2	0.3

10°C DRY BULB

Percentage saturation μ/%	Relative humidity ϕ/%	Value of stated parameter per kg dry air			Vapour pressure p_s/kPa	Dew point temperature t_d/°C	Adiabatic saturation temperature $t^\star$/°C	Wet bulb temperature	
		Moisture content g/g.kg^{-1}	Specific enthalpy h/kJ.kg^{-1}	Specific volume v/m^3.kg^{-1}				Screen t'_{sc}/°C	Sling t'_{sl}/°C
100	100.00	7.659	29.35	0.8116	1.227	10.0	10.0	10.0	10.0
96	96.05	7.352	28.58	0.8112	1.180	9.4	9.7	9.7	9.7
92	92.09	7.046	27.81	0.8108	1.130	8.8	9.3	9.4	9.3
88	88.13	6.740	27.04	0.8104	1.081	8.1	9.0	9.1	9.0
84	84.16	6.433	26.27	0.8100	1.034	7.5	8.7	8.8	8.7
80	80.19	6.127	25.50	0.8096	0.9841	6.7	8.3	8.5	8.3
76	76.22	5.821	24.72	0.8092	0.9354	6.0	8.0	8.2	8.0
72	72.24	5.514	23.95	0.8088	0.8866	5.2	7.6	7.8	7.6
70	70.26	5.361	23.57	0.8086	0.8622	4.8	7.4	7.7	7.5
68	68.26	5.208	23.18	0.8084	0.8377	4.4	7.2	7.5	7.3
66	66.27	5.055	22.79	0.8082	0.8133	4.0	7.1	7.4	7.1
64	64.28	4.902	22.41	0.8080	0.7888	3.6	6.9	7.2	6.9
62	62.29	4.748	22.02	0.8078	0.7644	3.1	6.7	7.0	6.7
60	60.29	4.595	21.64	0.8076	0.7398	2.7	6.5	6.9	6.6
58	58.30	4.442	21.25	0.8074	0.7154	2.2	6.3	6.7	6.4
56	56.30	4.289	20.86	0.8072	0.6909	1.7	6.1	6.5	6.2
54	54.30	4.136	20.48	0.8070	0.6664	1.2	5.9	6.4	6.0
52	52.30	3.983	20.09	0.8069	0.6419	0.7	5.7	6.2	5.8
50	50.30	3.829	19.71	0.8067	0.6173	0.1	5.6	6.0	5.6
48	48.30	3.676	19.32	0.8065	0.5928	−0.4	5.4	5.9	5.5
46	46.30	3.523	18.94	0.8063	0.5682	−0.9	5.2	5.7	5.3
44	44.30	3.370	18.55	0.8061	0.5436	−1.4	5.0	5.5	5.1
42	42.30	3.217	18.16	0.8059	0.5190	−2.0	4.8	5.4	4.9
40	40.29	3.064	17.78	0.8057	0.4945	−2.5	4.6	5.2	4.7
38	38.29	2.910	17.39	0.8055	0.4698	−3.1	4.4	5.0	4.5
36	36.28	2.757	17.00	0.8053	0.4452	−3.8	4.2	4.8	4.3
34	34.27	2.604	16.62	0.8051	0.4206	−4.5	4.0	4.7	4.1
32	32.27	2.451	16.23	0.8049	0.3960	−5.2	3.8	4.5	3.9
30	30.26	2.298	15.85	0.8047	0.3713	−5.9	3.6	4.3	3.7
28	28.25	2.144	15.46	0.8045	0.3466	−6.7	3.4	4.1	3.5
24	24.22	1.838	14.69	0.8041	0.2973	−8.5	3.0	3.8	3.1
20	20.20	1.532	13.92	0.8037	0.2478	−10.5	2.5	3.4	2.7
16	16.16	1.225	13.15	0.8033	0.1984	−13.0	2.1	3.1	2.3
12	12.13	0.919	12.38	0.8029	0.1488	−16.1	1.7	2.7	1.9
8	8.09	0.613	11.60	0.8025	0.0993	−20.4	1.2	2.3	1.4
4	4.05	0.306	10.83	0.8021	0.0497	−27.4	0.8	1.9	1.0
0	0.00	0.000	10.06	0.8018	0.0000	—	0.3	1.6	0.6

10.5°C DRY BULB

Percentage saturation μ/%	Relative humidity ϕ/%	Value of stated parameter per kg dry air			Vapour pressure p_s/kPa	Dew point temperature t_d/°C	Adiabatic saturation temperature $t^\star$/°C	Wet bulb temperature	
		Moisture content g/g.kg^{-1}	Specific enthalpy h/kJ.kg^{-1}	Specific volume v/m^3.kg^{-1}				Screen t'_{sc}/°C	Sling t'_{sl}/°C
100	100.00	7.922	30.53	0.8133	1.269	10.5	10.5	10.5	10.5
96	96.05	7.606	29.73	0.8129	1.219	9.9	10.2	10.2	10.2
92	92.09	7.289	28.93	0.8125	1.169	9.3	9.8	9.9	9.8
88	88.13	6.972	28.13	0.8121	1.118	8.6	9.5	9.6	9.5
84	84.17	6.655	27.33	0.8117	1.068	7.9	9.1	9.3	9.2
80	80.20	6.338	26.54	0.8113	1.018	7.2	8.8	8.9	8.8
76	76.23	6.021	25.74	0.8109	0.9673	6.5	8.4	8.6	8.5
72	72.25	5.704	24.94	0.8105	0.9168	5.7	8.1	8.3	8.1
70	70.26	5.546	24.54	0.8103	0.8916	5.3	7.9	8.1	7.9
68	68.27	5.387	24.14	0.8101	0.8663	4.9	7.7	8.0	7.7
66	66.28	5.229	23.74	0.8099	0.8410	4.5	7.5	7.8	7.6
64	64.29	5.070	23.34	0.8097	0.8158	4.0	7.3	7.6	7.4
62	62.30	4.912	22.94	0.8095	0.7905	3.6	7.1	7.5	7.2
60	60.30	4.754	22.54	0.8093	0.7652	3.1	6.9	7.3	7.0
58	58.31	4.595	22.14	0.8091	0.7398	2.7	6.7	7.1	6.8
56	56.31	4.437	21.74	0.8089	0.7145	2.2	6.6	7.0	6.6
54	54.31	4.278	21.34	0.8087	0.6892	1.7	6.4	6.8	6.4
52	52.31	4.120	20.95	0.8085	0.6638	1.2	6.2	6.6	6.2
50	50.31	3.961	20.55	0.8083	0.6384	0.6	6.0	6.5	6.1
48	48.31	3.803	20.15	0.8080	0.6131	0.1	5.8	6.3	5.9
46	46.31	3.644	19.75	0.8078	0.5877	−0.5	5.6	6.1	5.7
44	44.31	3.486	19.35	0.8076	0.5623	−1.0	5.4	5.9	5.5
42	42.31	3.328	18.95	0.8074	0.5368	−1.6	5.2	5.8	5.3
40	40.30	3.169	18.55	0.8072	0.5114	−2.1	5.0	5.6	5.1
38	38.30	3.011	18.15	0.8070	0.4860	−2.7	4.8	5.4	4.9
36	36.29	2.852	17.75	0.8068	0.4605	−3.4	4.6	5.2	4.7
34	34.28	2.694	17.35	0.8066	0.4350	−4.1	4.4	5.1	4.5
32	32.27	2.535	16.95	0.8064	0.4095	−4.8	4.2	4.9	4.3
30	30.27	2.377	16.55	0.8062	0.3840	−5.5	3.9	4.7	4.1
28	28.25	2.218	16.15	0.8060	0.3585	−6.3	3.7	4.5	3.9
24	24.23	1.901	15.35	0.8056	0.3075	−8.1	3.3	4.1	3.5
20	20.20	1.584	14.56	0.8052	0.2560	−10.2	2.9	3.8	3.0
16	16.17	1.268	13.76	0.8048	0.2052	−12.6	2.4	3.4	2.6
12	12.13	0.951	12.96	0.8044	0.1540	−15.8	2.0	3.0	2.2
8	8.09	0.634	12.16	0.8040	0.1027	−20.1	1.6	2.6	1.8
4	4.05	0.317	11.36	0.8036	0.0514	−27.1	1.1	2.3	1.3
0	0.00	0.000	10.56	0.8032	0.0000	—	0.6	1.9	0.9

11°C DRY BULB

Percentage saturation μ/%	Relative humidity ϕ/%	Value of stated parameter per kg dry air			Vapour pressure p_s/kPa	Dew point temperature t_d/°C	Adiabatic saturation temperature $t^\star$/°C	Wet bulb temperature	
		Moisture content g/g.kg^{-1}	Specific enthalpy h/kJ.kg^{-1}	Specific volume v/m^3.kg^{-1}				Screen t'_{sc}/°C	Sling t'_{sl}/°C
100	100.00	8.194	31.72	0.8151	1.312	11.0	11.0	11.0	11.0
96	96.05	7.866	30.90	0.8147	1.260	10.4	10.7	10.7	10.7
92	92.09	7.539	30.07	0.8143	1.208	9.8	10.3	10.4	10.3
88	88.14	7.211	29.25	0.8139	1.156	9.1	10.0	10.1	10.0
84	84.17	6.883	28.42	0.8134	1.104	8.4	9.6	9.7	9.6
80	80.21	6.555	27.59	0.8130	1.052	7.7	9.2	9.4	9.3
76	76.24	6.228	26.77	0.8126	1.000	7.0	8.9	9.1	8.9
72	72.26	5.900	25.94	0.8122	0.9480	6.2	8.5	8.8	8.6
70	70.27	5.736	25.53	0.8120	0.9219	5.8	8.3	8.6	8.4
68	68.28	5.572	25.11	0.8117	0.8959	5.4	8.1	8.4	8.2
66	66.29	5.408	24.70	0.8115	0.8696	5.0	7.9	8.3	8.0
64	64.30	5.244	24.29	0.8113	0.8435	4.5	7.8	8.1	7.8
62	62.31	5.080	23.87	0.8111	0.8174	4.1	7.6	7.9	7.6
60	60.31	4.916	23.46	0.8109	0.7912	3.6	7.4	7.8	7.4
58	58.32	4.753	23.05	0.8107	0.7650	3.1	7.2	7.6	7.2
56	56.32	4.589	22.63	0.8105	0.7388	2.6	7.0	7.4	7.1
54	54.32	4.425	22.22	0.8103	0.7126	2.1	6.8	7.2	6.9
52	52.32	4.261	21.81	0.8101	0.6864	1.6	6.6	7.1	6.7
50	50.33	4.097	21.40	0.8099	0.6602	1.1	6.4	6.9	6.5
48	48.32	3.933	20.98	0.8096	0.6339	0.5	6.2	6.7	6.3
46	46.32	3.769	20.57	0.8094	0.6077	−0.1	6.0	6.5	6.1
44	44.32	3.605	20.16	0.8092	0.5814	−0.6	5.8	6.4	5.9
42	42.32	3.442	19.74	0.8090	0.5551	−1.1	5.6	6.2	5.7
40	40.31	3.278	19.33	0.8088	0.5288	−1.7	5.4	6.0	5.5
38	38.31	3.114	18.92	0.8086	0.5025	−2.3	5.2	5.8	5.3
36	36.30	2.950	18.50	0.8084	0.4762	−3.0	5.0	5.6	5.1
34	34.29	2.786	18.09	0.8082	0.4499	−3.7	4.7	5.4	4.9
32	32.28	2.622	17.68	0.8080	0.4235	−4.4	4.5	5.3	4.7
30	30.27	2.458	17.26	0.8077	0.3972	−5.1	4.3	5.1	4.5
28	28.26	2.294	16.85	0.8075	0.3708	−5.9	4.1	4.9	4.2
24	24.24	1.967	16.02	0.8071	0.3180	−7.7	3.7	4.5	3.8
20	20.21	1.639	15.20	0.8067	0.2651	−9.8	3.2	4.1	3.4
16	16.18	1.311	14.37	0.8063	0.2122	−12.3	2.8	3.8	3.0
12	12.14	0.983	13.54	0.8059	0.1592	−15.4	2.3	3.4	2.5
8	8.10	0.655	12.72	0.8054	0.1062	−19.7	1.9	3.0	2.1
4	4.05	0.328	11.89	0.8050	0.0531	−26.8	1.4	2.6	1.6
0	0.00	0.000	11.06	0.8046	0.0000	—	0.9	2.2	1.2

11.5°C DRY BULB

Percentage saturation μ/%	Relative humidity ϕ/%	Value of stated parameter per kg dry air			Vapour pressure p_s/kPa	Dew point temperature t_d/°C	Adiabatic saturation temperature $t^\star$/°C	Wet bulb temperature	
		Moisture content g/g.kg^{-1}	Specific enthalpy h/kJ.kg^{-1}	Specific volume v/m^3.kg^{-1}				Screen t'_{sc}/°C	Sling t'_{sl}/°C
100	100.00	8.474	32.94	0.8169	1.356	11.5	11.5	11.5	11.5
96	96.05	8.135	32.09	0.8165	1.303	10.9	11.2	11.2	11.2
92	92.10	7.796	31.23	0.8160	1.249	10.3	10.8	10.9	10.8
88	88.14	7.457	30.38	0.8156	1.195	9.6	10.4	10.5	10.5
84	84.18	7.118	29.52	0.8152	1.142	8.9	10.1	10.2	10.1
80	80.21	6.779	28.67	0.8147	1.088	8.2	9.7	9.9	9.7
76	76.24	6.440	27.81	0.8143	1.034	7.5	9.3	9.6	9.4
72	72.27	6.102	26.96	0.8139	0.9800	6.7	9.0	9.2	9.0
70	70.28	5.932	26.53	0.8136	0.9531	6.3	8.8	9.1	8.8
68	68.29	5.762	26.10	0.8134	0.9261	5.9	8.6	8.9	8.6
66	66.30	5.593	25.67	0.8132	0.8991	5.4	8.4	8.7	8.4
64	64.31	5.424	25.25	0.8130	0.8721	5.0	8.2	8.5	8.3
62	62.32	5.254	24.82	0.8128	0.8451	4.6	8.0	8.4	8.1
60	60.32	5.085	24.39	0.8126	0.8180	4.1	7.8	8.2	7.9
58	58.33	4.915	23.96	0.8123	0.7910	3.6	7.6	8.0	7.7
56	56.33	4.746	23.54	0.8121	0.7639	3.1	7.4	7.8	7.5
54	54.33	4.576	23.11	0.8119	0.7368	2.6	7.2	7.7	7.3
52	52.34	4.407	22.68	0.8117	0.7097	2.1	7.0	7.5	7.1
50	50.34	4.237	22.26	0.8115	0.6826	1.5	6.8	7.3	6.9
48	48.34	4.068	21.83	0.8112	0.6555	1.0	6.6	7.1	6.7
46	46.33	3.898	21.40	0.8110	0.6283	0.4	6.4	6.9	6.5
44	44.33	3.729	20.97	0.8108	0.6012	−0.2	6.2	6.8	6.3
42	42.33	3.559	20.55	0.8106	0.5740	−0.7	6.0	6.6	6.1
40	40.32	3.390	20.12	0.8104	0.5468	−1.3	5.8	6.4	5.9
38	38.32	3.220	19.69	0.8102	0.5196	−1.9	5.6	6.2	5.7
36	36.31	3.051	19.26	0.8099	0.4924	−2.6	5.3	6.0	5.5
34	34.30	2.881	18.84	0.8097	0.4652	−3.3	5.1	5.8	5.2
32	32.29	2.712	18.41	0.8095	0.4379	−4.0	4.9	5.6	5.0
30	30.28	2.542	17.98	0.8093	0.4107	−4.7	4.7	5.5	4.8
28	28.27	2.373	17.55	0.8091	0.3834	−5.5	4.5	5.3	4.6
24	24.25	2.034	16.70	0.8086	0.3288	−7.3	4.0	4.9	4.2
20	20.22	1.695	15.84	0.8082	0.2742	−9.4	3.6	4.5	3.7
16	16.18	1.356	14.99	0.8078	0.2194	−11.9	3.1	4.1	3.3
12	12.14	1.017	14.13	0.8073	0.1647	−15.1	2.7	3.7	2.8
8	8.10	0.678	13.28	0.8069	0.1098	−19.4	2.2	3.3	2.4
4	4.05	0.339	12.42	0.8064	0.0549	−26.4	1.7	2.9	1.9
0	0.00	0.000	11.57	0.8060	0.0000	—	1.2	2.5	1.4

12°C DRY BULB

Percentage saturation μ/%	Relative humidity ϕ/%	Value of stated parameter per kg dry air			Vapour pressure p_s/kPa	Dew point temperature t_d/°C	Adiabatic saturation temperature $t^\star$/°C	Wet bulb temperature	
		Moisture content g/g.kg^{-1}	Specific enthalpy h/kJ.kg^{-1}	Specific volume v/m^3.kg^{-1}				Screen t'_{sc}/°C	Sling t'_{sl}/°C
100	100.00	8.763	34.18	0.8187	1.402	12.0	12.0	12.0	12.0
96	96.05	8.412	33.30	0.8183	1.346	11.4	11.6	11.7	11.7
92	92.10	8.062	32.41	0.8178	1.291	10.8	11.3	11.4	11.3
88	88.15	7.711	31.53	0.8174	1.236	10.1	10.8	11.0	10.9
84	84.19	7.361	30.64	0.8169	1.180	9.4	10.6	10.7	10.6
80	80.22	7.010	29.76	0.8165	1.124	8.7	10.2	10.4	10.2
76	76.25	6.660	28.87	0.8160	1.069	8.0	9.8	10.0	9.8
72	72.28	6.309	27.99	0.8156	1.013	7.2	9.4	9.7	9.5
70	70.29	6.134	27.55	0.8153	0.9852	6.8	9.2	9.5	9.3
68	68.30	5.959	27.11	0.8151	0.9574	6.3	9.0	9.3	9.1
66	66.31	5.784	26.66	0.8149	0.9294	5.9	8.8	9.2	8.9
64	64.32	5.608	26.22	0.8147	0.9015	5.5	8.6	9.0	8.7
62	62.33	5.433	25.78	0.8144	0.8736	5.0	8.4	8.8	8.5
60	60.33	5.258	25.34	0.8142	0.8457	4.6	8.2	8.6	8.3
58	58.34	5.082	24.89	0.8140	0.8177	4.1	8.0	8.4	8.1
56	56.34	4.907	24.45	0.8138	0.7897	3.6	7.8	8.3	7.9
54	54.35	4.732	24.01	0.8135	0.7617	3.1	7.6	8.1	7.7
52	52.35	4.557	23.57	0.8133	0.7337	2.6	7.4	7.9	7.5
50	50.35	4.382	23.13	0.8131	0.7057	2.0	7.2	7.7	7.3
48	48.35	4.206	22.68	0.8129	0.6777	1.4	7.0	7.5	7.1
46	46.35	4.031	22.24	0.8126	0.6496	0.9	6.8	7.4	6.9
44	44.34	3.856	21.80	0.8124	0.6215	0.2	6.6	7.2	6.7
42	42.34	3.680	21.36	0.8122	0.5934	−0.3	6.4	7.0	6.5
40	40.34	3.505	20.92	0.8120	0.5654	−0.9	6.2	6.8	6.3
38	38.33	3.330	20.47	0.8117	0.5372	−1.5	5.9	6.6	6.1
36	36.32	3.155	20.03	0.8115	0.5091	−2.2	5.7	6.4	5.8
34	34.31	2.979	19.59	0.8113	0.4810	−2.9	5.5	6.2	5.6
32	32.30	2.804	19.15	0.8110	0.4528	−3.6	5.3	5.0	5.4
30	30.29	2.629	18.70	0.8108	0.4246	−4.3	5.1	5.8	5.2
28	28.28	2.454	18.26	0.8106	0.3964	−5.2	4.8	5.6	5.0
24	24.26	2.103	17.38	0.8101	0.3400	−6.9	4.4	5.2	4.5
20	20.22	1.753	16.49	0.8097	0.2835	−9.0	3.9	4.8	4.1
16	16.19	1.402	15.61	0.8092	0.2269	−11.5	3.5	4.4	3.6
12	12.15	1.052	14.72	0.8088	0.1703	−14.7	3.0	4.0	3.2
8	8.10	0.701	13.84	0.8083	0.1136	−19.0	2.5	3.6	2.7
4	4.05	0.350	12.96	0.8079	0.0568	−26.1	2.0	3.2	2.2
0	0.00	0.000	12.07	0.8074	0.0000	—	1.5	2.8	1.7

12.5°C DRY BULB

Percentage saturation μ/%	Relative humidity ϕ/%	Value of stated parameter per kg dry air			Vapour pressure p_s/kPa	Dew point temperature t_d/°C	Adiabatic saturation temperature $t^\star$/°C	Wet bulb temperature	
		Moisture content g/g.kg^{-1}	Specific enthalpy h/kJ.kg^{-1}	Specific volume v/m^3.kg^{-1}				Screen t'_{sc}/°C	Sling t'_{sl}/°C
100	100.00	9.060	35.44	0.8205	1.449	12.5	12.5	12.5	12.5
96	96.06	8.698	34.53	0.8201	1.391	11.9	12.1	12.2	12.1
92	92.11	8.336	33.61	0.8196	1.334	11.3	11.8	11.8	11.8
88	88.15	7.973	32.70	0.8191	1.277	10.6	11.4	11.5	11.4
84	84.19	7.611	31.78	0.8187	1.220	9.9	11.0	11.2	11.1
80	80.23	7.248	30.87	0.8182	1.162	9.2	10.7	10.8	10.7
76	76.26	6.886	29.95	0.8177	1.105	8.4	10.3	10.5	10.3
72	72.29	6.524	29.04	0.8173	1.047	7.7	9.9	10.1	9.9
70	70.30	6.342	28.58	0.8170	1.018	7.2	9.7	10.0	9.7
68	68.31	6.161	28.12	0.8168	0.9895	6.8	9.5	9.8	9.5
66	66.32	5.980	27.67	0.8166	0.9607	6.4	9.3	9.6	9.3
64	64.33	5.799	27.21	0.8163	0.9319	6.0	9.1	9.4	9.1
62	62.34	5.618	26.75	0.8161	0.9030	5.5	8.9	9.2	8.9
60	60.35	5.436	26.30	0.8159	0.8741	5.0	8.7	9.1	8.7
58	58.35	5.255	25.84	0.8156	0.8452	4.6	8.5	8.9	8.5
56	56.36	5.074	25.38	0.8154	0.8163	4.1	8.3	8.7	8.3
54	54.36	4.893	24.92	0.8152	0.7874	3.5	8.1	8.5	8.1
52	52.36	4.711	24.47	0.8149	0.7584	3.0	7.8	8.3	7.9
50	50.36	4.530	24.01	0.8147	0.7295	2.5	7.6	8.1	7.7
48	48.36	4.349	23.55	0.8145	0.7005	1.9	7.4	8.0	7.5
46	46.36	4.168	23.09	0.8142	0.6715	1.3	7.2	7.8	7.3
44	44.36	3.987	22.64	0.8140	0.6425	0.7	7.0	7.6	7.1
42	42.35	3.805	22.18	0.8138	0.6135	0.1	6.8	7.4	6.9
40	40.35	3.624	21.72	0.8135	0.5844	−0.5	6.6	7.2	6.7
38	38.34	3.443	21.26	0.8133	0.5554	−1.1	6.3	7.0	6.4
36	36.33	3.262	20.81	0.8131	0.5263	−1.8	6.1	6.8	6.2
34	34.32	3.080	20.35	0.8128	0.4972	−2.5	5.9	6.6	6.0
32	32.31	2.899	19.89	0.8126	0.4681	−3.2	5.7	6.4	5.8
30	30.30	2.718	19.44	0.8124	0.4390	−4.0	5.4	6.2	5.6
28	28.29	2.537	18.98	0.8121	0.4098	−4.8	5.2	6.0	5.3
24	24.26	2.174	18.06	0.8117	0.3515	−6.6	4.7	5.6	4.9
20	20.23	1.812	17.15	0.8112	0.2931	−8.6	4.3	5.2	4.4
16	16.19	1.450	16.23	0.8107	0.2346	−11.2	3.8	4.8	4.0
12	12.15	1.087	15.32	0.8103	0.1760	−14.3	3.3	4.4	3.5
8	8.11	0.724	14.40	0.8098	0.1174	−18.7	2.8	3.9	3.0
4	4.06	0.362	13.49	0.8093	0.0587	−25.8	2.3	3.5	2.5
0	0.00	0.000	12.58	0.8089	0.0000	—	1.8	3.1	2.0

13°C DRY BULB

Percentage saturation μ/%	Relative humidity ϕ/%	Value of stated parameter per kg dry air			Vapour pressure p_s/kPa	Dew point temperature t_d/°C	Adiabatic saturation temperature $t\star$/°C	Wet bulb temperature	
		Moisture content g/g.kg^{-1}	Specific enthalpy h/kJ.kg^{-1}	Specific volume v/m^3.kg^{-1}				Screen t'_{sc}/°C	Sling t'_{sl}/°C
100	100.00	9.367	36.73	0.8224	1.497	13.0	13.0	13.0	13.0
96	96.06	8.992	35.78	0.8219	1.438	12.4	12.6	12.7	12.6
92	92.11	8.618	34.84	0.8214	1.379	11.7	12.3	12.3	12.3
88	88.16	8.243	33.89	0.8209	1.320	11.1	11.9	12.0	11.9
84	84.20	7.868	32.94	0.8204	1.260	10.4	11.5	11.6	11.5
80	80.24	7.494	32.00	0.8200	1.201	9.7	11.1	11.3	11.2
76	76.27	7.119	31.05	0.8195	1.142	8.9	10.7	11.0	10.8
72	72.30	6.744	30.11	0.8190	1.082	8.1	10.3	10.6	10.4
70	70.31	6.557	29.63	0.8188	1.052	7.7	10.1	10.4	10.2
68	68.33	6.370	29.16	0.8185	1.023	7.3	9.9	10.2	10.0
66	66.34	6.182	28.69	0.8183	0.9929	6.9	9.7	10.1	9.8
64	64.34	5.995	28.21	0.8180	0.9631	6.4	9.5	9.9	9.6
62	62.35	5.808	27.74	0.8178	0.9333	6.0	9.3	9.7	9.4
60	60.36	5.620	27.27	0.8175	0.9034	5.5	9.1	9.5	9.2
58	58.36	5.433	26.79	0.8173	0.8736	5.0	8.9	9.3	9.0
56	56.37	5.245	26.32	0.8171	0.8437	4.5	8.7	9.1	8.8
54	54.37	5.058	25.85	0.8168	0.8138	4.0	8.5	8.9	8.6
52	52.37	4.871	25.38	0.8166	0.7839	3.5	8.3	8.7	8.3
50	50.37	0.4683	24.90	0.8163	0.7540	2.9	8.1	8.6	8.1
48	48.37	4.496	24.43	0.8161	0.7240	2.4	7.8	8.4	7.9
46	46.37	4.309	23.96	0.8158	0.6941	1.8	7.6	8.2	7.7
44	44.37	4.121	23.48	0.8156	0.6641	1.2	7.4	8.0	7.5
42	42.36	3.934	23.01	0.8154	0.6341	0.5	7.2	7.8	7.3
40	40.36	3.747	22.54	0.8151	0.6041	−0.1	6.9	7.6	7.1
38	38.35	3.559	22.06	0.8149	0.5740	−0.7	6.7	7.4	6.8
36	36.34	3.372	21.59	0.8146	0.5440	−1.4	6.5	7.2	6.6
34	34.34	3.185	21.12	0.8144	0.5139	−2.1	6.3	7.0	6.4
32	32.33	2.997	20.65	0.8141	0.4838	−2.8	6.0	6.8	6.2
30	30.31	2.810	20.17	0.8139	0.4537	−3.6	5.8	6.6	5.9
28	28.30	2.623	19.70	0.8137	0.4236	−4.4	5.6	6.4	5.7
24	24.27	2.248	18.75	0.8132	0.3633	−6.2	5.1	6.0	5.2
20	20.24	1.873	17.81	0.8127	0.3029	−8.3	4.6	5.6	4.8
16	16.20	1.499	16.86	0.8122	0.2425	−10.8	4.1	5.1	4.3
12	12.16	1.124	15.92	0.8117	0.1820	−14.0	3.6	4.7	3.8
8	8.11	0.7493	14.97	0.8112	0.1214	−18.3	3.1	4.3	3.3
4	4.06	0.3747	14.02	0.8108	0.0607	−25.5	2.6	3.8	2.8
0	0.00	0.000	13.08	0.8103	0.0000	—	2.1	3.4	2.3

13.5°C DRY BULB

Percentage saturation μ/%	Relative humidity ϕ/%	Value of stated parameter per kg dry air			Vapour pressure p_s/kPa	Dew point temperature t_d/°C	Adiabatic saturation temperature $t\star$/°C	Wet bulb temperature	
		Moisture content g/g.kg^{-1}	Specific enthalpy h/kJ.kg^{-1}	Specific volume v/m^3.kg^{-1}				Screen t'_{sc}/°C	Sling t'_{sl}/°C
100	100.00	9.683	38.04	0.8242	1.546	13.5	13.5	13.5	13.5
96	96.06	9.295	37.06	0.8237	1.486	12.9	13.1	13.2	13.1
92	92.11	8.908	36.08	0.8232	1.424	12.2	12.8	12.8	12.8
88	88.16	8.521	35.10	0.8227	1.363	11.6	12.4	12.5	12.4
84	84.21	8.133	34.12	0.8222	1.302	10.8	12.0	12.1	12.0
80	80.25	7.746	33.14	0.8217	1.241	10.2	11.6	11.8	11.6
76	76.28	7.359	32.17	0.8212	1.180	9.4	11.2	11.4	11.2
72	72.31	6.972	31.19	0.8207	1.118	8.6	10.8	11.1	10.8
70	70.32	6.778	30.70	0.8205	1.088	8.2	10.6	10.9	10.6
68	68.34	6.584	30.21	0.8202	1.057	7.8	10.4	10.7	10.4
66	66.35	6.390	29.72	0.8200	1.026	7.4	10.2	10.5	10.2
64	64.36	6.197	29.23	0.8197	0.9952	6.9	10.0	10.3	10.0
62	62.36	6.003	28.74	0.8195	0.9644	6.5	9.8	10.1	9.8
60	60.37	5.810	28.25	0.8192	0.9336	6.0	9.6	9.9	9.6
58	58.38	5.616	27.76	0.8190	0.9028	5.5	9.3	9.7	9.4
56	56.38	5.422	27.28	0.8187	0.8719	5.0	9.1	9.6	9.2
54	54.38	5.229	26.79	0.8185	0.8410	4.5	8.9	9.4	9.0
52	52.39	5.036	26.30	0.8182	0.8101	4.0	8.7	9.2	8.8
50	50.39	4.841	25.81	0.8180	0.7792	3.4	8.5	9.0	8.6
48	48.39	4.648	25.32	0.8177	0.7482	2.8	8.2	8.8	8.3
46	46.38	4.454	24.83	0.8175	0.7173	2.2	8.0	8.6	8.1
44	44.38	4.260	24.34	0.8172	0.6863	1.6	7.8	8.4	7.9
42	42.38	4.067	23.85	0.8170	0.6553	1.0	7.6	8.2	7.7
40	40.37	3.873	23.36	0.8167	0.6243	0.3	7.3	8.0	7.4
38	38.36	3.679	22.87	0.8165	0.5933	−0.3	7.1	7.8	7.2
36	36.36	3.486	22.38	0.8162	0.5622	−1.0	6.9	7.6	7.0
34	34.35	3.292	21.90	0.8160	0.5312	−1.7	6.6	7.4	6.8
32	32.34	3.098	21.41	0.8157	0.5000	−2.4	6.4	7.2	6.5
30	30.33	2.905	20.92	0.8155	0.4690	−3.2	6.2	7.0	6.3
28	28.31	2.711	20.43	0.8152	0.4378	−4.0	5.9	6.8	6.1
24	24.28	2.324	19.45	0.8147	0.3755	−5.8	5.4	6.3	5.6
20	20.25	1.936	18.47	0.8142	0.3131	−7.9	4.9	5.9	5.1
16	16.21	1.549	17.49	0.8137	0.2506	−10.4	4.4	5.5	4.6
12	12.16	1.162	16.52	0.8132	0.1881	−13.6	3.9	5.0	4.1
8	8.11	0.7746	15.54	0.8127	0.1255	−18.0	3.4	4.6	3.6
4	4.06	0.3873	14.56	0.8122	0.0628	−25.1	2.9	4.1	3.1
0	0.00	0.000	13.58	0.8117	0.0000	—	2.4	3.7	2.6

14°C DRY BULB

Percentage saturation μ/%	Relative humidity ϕ/%	Value of stated parameter per kg dry air			Vapour pressure p_s/kPa	Dew point temperature t_d/°C	Adiabatic saturation temperature $t^\star$/°C	Wet bulb temperature	
		Moisture content g/g.kg^{-1}	Specific enthalpy h/kJ.kg^{-1}	Specific volume v/m^3.kg^{-1}				Screen t'_{sc}/°C	Sling t'_{sl}/°C
100	100.00	10.010	39.37	0.8261	1.598	14.0	14.0	14.0	14.0
96	96.06	9.608	38.36	0.8256	1.535	13.4	13.6	13.7	13.6
92	92.12	9.207	37.37	0.8251	1.472	12.7	13.2	13.3	13.3
88	88.17	8.807	35.34	0.8245	1.409	12.1	12.9	13.0	12.9
84	84.21	8.407	35.32	0.8240	1.345	11.4	12.5	12.6	12.5
80	80.26	8.006	34.31	0.8235	1.282	10.7	12.1	12.2	12.1
76	76.29	7.606	33.30	0.8230	1.219	9.9	11.7	11.9	11.7
72	72.32	7.206	32.29	0.8225	1.155	9.1	11.3	11.5	11.3
70	70.34	7.005	31.78	0.8222	1.124	8.7	11.0	11.3	11.1
68	68.35	6.805	31.28	0.8219	1.092	8.3	10.8	11.1	10.9
66	66.36	6.605	30.77	0.8217	1.060	7.8	10.6	10.9	10.7
64	64.37	6.405	30.27	0.8214	1.028	7.4	10.4	10.8	10.5
62	62.38	6.205	29.76	0.8212	0.9965	6.9	10.2	10.6	10.3
60	60.38	6.005	29.26	0.8209	0.9647	6.5	10.0	10.4	10.0
58	58.39	5.804	28.75	0.8206	0.9328	6.0	9.8	10.2	9.8
56	56.39	5.604	28.24	0.8204	0.9009	5.5	9.6	10.0	9.6
54	54.40	5.404	27.74	0.8201	0.8690	5.0	9.3	9.8	9.4
52	52.40	5.204	27.23	0.8199	0.8371	4.4	9.1	9.6	9.2
50	50.40	5.004	26.73	0.8196	0.8052	3.9	8.9	9.4	9.0
48	48.40	4.804	26.22	0.8193	0.7732	3.3	8.7	9.2	8.7
46	46.40	4.604	25.72	0.8191	0.7412	2.7	8.4	9.0	8.5
44	44.39	4.403	25.21	0.8188	0.7092	2.1	8.2	8.8	8.3
42	42.39	4.203	24.70	0.8186	0.6772	1.4	8.0	8.6	8.1
40	40.38	4.003	24.20	0.8183	0.6452	0.8	7.7	8.4	7.8
38	38.38	3.803	23.69	0.8180	0.6131	0.1	7.5	8.2	7.6
36	36.37	3.603	23.17	0.8178	0.5810	−0.6	7.3	8.0	7.4
34	34.36	3.403	22.68	0.8175	0.5489	−1.3	7.0	7.8	7.1
32	32.35	3.202	22.18	0.8173	0.5168	−2.0	6.8	7.5	6.9
30	30.34	3.002	21.67	0.8170	0.4846	−2.8	6.5	7.3	6.7
28	28.32	2.802	21.16	0.8167	0.4525	−3.6	6.3	7.1	6.4
24	24.29	2.402	20.15	0.8162	0.3881	−5.4	5.8	6.7	5.9
20	20.26	2.002	19.14	0.8157	0.3236	−7.5	5.3	6.3	5.5
16	16.22	1.601	18.13	0.8152	0.2590	−10.0	4.8	5.8	5.0
12	12.17	1.201	17.12	0.8147	0.1944	−13.2	4.3	5.4	4.4
8	8.12	0.8006	16.11	0.8142	0.1297	−17.6	3.7	4.9	3.9
4	4.06	0.4003	15.10	0.8136	0.0649	−24.8	3.2	4.4	3.4
0	0.00	0.0000	14.08	0.8131	0.0000	—	2.6	4.0	2.9

14.5°C DRY BULB

Percentage saturation μ/%	Relative humidity ϕ/%	Value of stated parameter per kg dry air			Vapour pressure p_s/kPa	Dew point temperature t_d/°C	Adiabatic saturation temperature $t^\star$/°C	Wet bulb temperature	
		Moisture content g/g.kg^{-1}	Specific enthalpy h/kJ.kg^{-1}	Specific volume v/m^3.kg^{-1}				Screen t'_{sc}/°C	Sling t'_{sl}/°C
100	100.00	10.340	40.73	0.8280	1.650	14.5	14.5	14.5	14.5
96	96.06	9.929	39.68	0.8274	1.585	13.9	14.1	14.2	14.1
92	92.12	9.515	38.64	0.8269	1.520	13.2	13.7	13.8	13.7
88	88.17	9.102	37.59	0.8264	1.455	12.6	13.3	13.4	13.4
84	84.22	8.688	36.55	0.8258	1.390	11.9	12.9	13.1	13.0
80	80.26	8.274	35.50	0.8253	1.324	11.1	12.5	12.7	12.6
76	76.30	7.860	34.45	0.8248	1.259	10.4	12.1	12.3	12.2
72	72.33	7.447	33.41	0.8242	1.194	9.6	11.7	12.0	11.8
70	70.35	7.240	32.89	0.8239	1.161	9.2	11.5	11.8	11.5
68	68.36	7.033	32.36	0.8237	1.128	8.7	11.3	11.6	11.3
66	66.37	6.826	31.84	0.8234	1.095	8.3	11.1	11.4	11.1
64	64.38	6.619	31.32	0.8231	1.062	7.9	10.9	11.2	10.9
62	62.39	6.412	30.79	0.8229	1.030	7.4	10.6	11.0	10.7
60	60.40	6.206	30.27	0.8226	0.9966	6.9	10.4	10.8	10.5
58	58.40	5.999	29.75	0.8223	0.9937	6.4	10.2	10.6	10.3
56	56.41	5.792	29.23	0.8221	0.9308	5.9	10.0	10.4	10.0
54	54.41	5.585	28.70	0.8218	0.8978	5.4	9.8	10.2	9.8
52	52.41	5.378	28.18	0.8215	0.8649	4.9	9.5	10.0	9.6
50	50.41	5.171	27.66	0.8213	0.8329	4.3	9.3	9.8	9.4
48	48.41	4.964	27.13	0.8210	0.7989	3.8	9.1	9.6	9.2
46	46.41	4.758	26.61	0.8207	0.7658	3.2	8.8	9.4	8.9
44	44.41	4.551	26.09	0.8204	0.7328	2.5	8.6	9.2	8.7
42	42.40	4.344	25.57	0.8202	0.6997	1.9	8.4	9.0	8.5
40	40.40	4.137	25.04	0.8199	0.6666	1.2	8.1	8.8	8.2
38	38.39	3.930	24.52	0.8196	0.6335	0.5	7.9	8.6	8.0
36	36.38	3.723	24.00	0.8194	0.6003	−0.2	7.6	8.4	7.8
34	34.37	3.516	23.47	0.8191	0.5672	−0.9	7.4	8.1	7.5
32	32.36	3.310	22.95	0.8188	0.5340	−1.6	7.2	7.9	7.3
30	30.35	3.103	22.43	0.8186	0.5008	−2.4	6.9	7.7	7.0
28	28.33	2.896	21.91	0.8183	0.4675	−3.2	6.7	7.5	6.8
24	24.30	2.482	20.86	0.8178	0.4010	−5.0	6.1	7.0	6.3
20	20.27	2.068	19.81	0.8172	0.3344	−7.1	5.6	6.6	5.8
16	16.22	1.655	18.77	0.8167	0.2677	−9.7	5.1	6.1	5.3
12	12.18	1.241	17.72	0.8161	0.2009	−12.9	4.6	5.7	4.8
8	8.12	0.827	16.68	0.8156	0.1340	−17.3	4.0	5.2	4.2
4	4.06	0.413	15.63	0.8151	0.0670	−24.5	3.5	4.7	3.7
0	0.00	0.000	14.59	0.8145	0.0000	—	2.9	4.3	3.2

15°C DRY BULB

Percentage saturation μ/%	Relative humidity ϕ/%	Moisture content g/g.kg^{-1}	Specific enthalpy h/kJ.kg^{-1}	Specific volume v/m^3.kg^{-1}	Vapour pressure p_s/kPa	Dew point temperature t_d/°C	Adiabatic saturation temperature $t^\star$/°C	Screen t'_{sc}/°C	Sling t'_{sl}/°C
100	100.00	10.697	42.11	0.8299	1.704	15.0	15.0	15.0	15.0
96	96.06	10.260	41.03	0.8293	1.637	14.4	14.6	14.6	14.6
92	92.13	9.833	39.95	0.8288	1.570	13.7	14.2	14.3	14.2
88	88.18	9.405	38.87	0.8282	1.503	13.1	13.8	13.9	13.8
84	84.23	8.978	37.79	0.8276	1.435	12.4	13.4	13.6	13.4
80	80.27	8.550	36.71	0.8271	1.368	11.6	13.0	13.2	13.0
76	76.31	8.123	35.63	0.8265	1.301	10.9	12.6	12.8	12.6
72	72.34	7.695	34.55	0.8260	1.233	10.1	12.2	12.4	12.2
70	70.36	7.481	34.01	0.8257	1.199	9.7	12.0	12.2	12.0
68	68.37	7.268	33.47	0.8254	1.165	9.2	11.7	12.0	11.8
66	66.38	7.054	32.93	0.8251	1.131	8.8	11.5	11.8	11.6
64	64.39	6.840	32.28	0.8249	1.097	8.3	11.3	11.6	11.4
62	62.40	6.626	31.84	0.8246	1.064	7.9	11.1	11.4	11.1
60	60.41	6.413	31.30	0.8243	1.030	7.4	10.9	11.2	10.9
58	58.42	6.199	30.76	0.8240	0.9956	6.9	10.6	11.0	10.7
56	56.42	5.985	30.22	0.8237	0.9616	6.4	10.4	10.8	10.5
54	54.42	5.771	29.68	0.8235	0.9275	5.9	10.2	10.6	10.3
52	52.43	5.558	29.14	0.8232	0.8935	5.4	9.9	10.4	10.0
50	50.43	5.344	28.60	0.8229	0.8594	4.8	9.7	10.2	9.8
48	48.43	5.130	28.06	0.8226	0.8253	4.2	9.5	10.0	9.6
46	46.42	4.916	27.52	0.8224	0.7912	3.6	9.2	9.8	9.3
44	44.42	4.703	26.98	0.8221	0.7570	3.0	9.0	9.6	9.1
42	42.42	4.489	26.44	0.8218	0.7229	2.3	8.8	9.4	8.9
40	40.41	4.275	25.90	0.8215	0.6887	1.7	8.5	9.2	8.6
38	38.40	4.061	25.36	0.8212	0.6545	1.0	8.3	9.0	8.4
36	36.39	3.848	24.82	0.8210	0.6202	0.2	8.0	8.7	8.1
34	34.38	3.634	24.28	0.8207	0.5860	−0.5	7.8	8.5	7.9
32	32.37	3.420	23.74	0.8204	0.5517	−1.2	7.5	8.3	7.7
30	30.36	3.206	23.20	0.8201	0.5174	−2.0	7.3	8.1	7.4
28	28.35	2.993	22.66	0.8198	0.4831	−2.8	7.0	7.9	7.2
24	24.31	2.565	21.58	0.8193	0.4143	−4.6	6.5	7.4	6.7
20	20.27	2.138	20.49	0.8187	0.3455	−6.7	6.0	6.9	6.1
16	16.23	1.710	19.41	0.8182	0.2766	−9.3	5.4	6.5	5.6
12	12.18	1.282	18.33	0.8176	0.2076	−12.5	4.9	6.0	5.1
8	8.13	0.855	17.25	0.8171	0.1385	−16.9	4.3	5.5	4.5
4	4.07	0.427	16.17	0.8165	0.0692	−24.1	3.8	5.0	4.0
0	0.00	0.000	15.09	0.8159	0.0000	—	3.2	4.6	3.4

15.5°C DRY BULB

Percentage saturation μ/%	Relative humidity ϕ/%	Moisture content g/g.kg^{-1}	Specific enthalpy h/kJ.kg^{-1}	Specific volume v/m^3.kg^{-1}	Vapour pressure p_s/kPa	Dew point temperature t_d/°C	Adiabatic saturation temperature $t^\star$/°C	Screen t'_{sc}/°C	Sling t'_{sl}/°C
100	100.00	11.04	43.52	0.8318	1.760	15.5	15.5	15.5	15.5
96	96.07	10.60	42.41	0.8312	1.691	14.9	15.1	15.1	15.1
92	92.13	10.16	41.29	0.8306	1.621	14.2	14.7	14.8	14.7
88	88.19	9.718	40.17	0.8300	1.552	13.6	14.3	14.4	14.3
84	84.24	9.276	39.06	0.8295	1.482	12.9	13.9	14.0	13.9
80	80.28	8.834	37.94	0.8289	1.413	12.1	13.5	13.7	13.5
76	76.32	8.393	36.82	0.8283	1.343	11.4	13.1	13.3	13.1
72	72.36	7.951	35.70	0.8277	1.273	10.6	12.6	12.9	12.7
70	70.37	7.730	35.14	0.8275	1.238	10.1	12.4	12.7	12.5
68	68.38	7.509	34.59	0.8272	1.204	9.7	12.2	12.5	12.2
66	66.40	7.288	34.03	0.8269	1.168	9.3	12.0	12.3	12.0
64	64.41	7.068	33.47	0.8266	1.134	8.8	11.7	12.1	11.8
62	62.42	6.847	32.91	0.8263	1.098	8.4	11.5	11.9	11.6
60	60.42	6.626	32.35	0.8260	1.063	7.9	11.3	11.7	11.4
58	58.43	6.405	31.79	0.8257	1.028	7.4	11.1	11.5	11.1
56	56.44	6.184	31.24	0.8254	0.9932	6.9	10.8	11.3	10.9
54	54.44	5.963	30.68	0.8252	0.9581	6.4	10.6	11.1	10.7
52	52.44	5.742	30.12	0.8249	0.9229	5.8	10.4	10.9	10.4
50	50.44	5.522	29.56	0.8246	0.8877	5.3	10.1	10.6	10.2
48	48.44	5.301	29.00	0.8243	0.8525	4.7	9.9	10.4	10.0
46	46.44	5.080	28.44	0.8240	0.8173	4.1	9.6	10.2	9.7
44	44.44	4.859	27.88	0.8237	0.7820	3.4	9.4	10.0	9.5
42	42.43	4.638	27.32	0.8234	0.7467	2.8	9.2	9.8	9.3
40	40.42	4.417	26.77	0.8231	0.7114	2.1	8.9	9.6	9.0
38	38.42	4.196	26.21	0.8228	0.6761	1.4	8.7	9.3	8.8
36	36.41	3.976	25.65	0.8226	0.6407	0.7	8.4	9.1	8.5
34	34.40	3.755	25.09	0.8223	0.6053	−0.1	8.2	8.9	8.3
32	32.39	3.534	24.53	0.8220	0.5699	−0.8	7.9	8.7	8.0
30	30.37	3.313	23.97	0.8217	0.5345	−1.6	7.6	8.4	7.8
28	28.36	3.092	23.41	0.8214	0.4990	−2.4	7.4	8.2	7.5
24	24.32	2.650	22.30	0.8208	0.4280	−4.2	6.8	7.8	7.0
20	20.28	2.209	21.18	0.8203	0.3570	−6.4	6.3	7.3	6.5
16	16.24	1.767	20.06	0.8197	0.2858	−8.9	5.8	6.8	5.9
12	12.19	1.325	18.94	0.8191	0.2145	−12.2	5.2	6.3	5.4
8	8.13	0.883	17.83	0.8185	0.1431	−16.6	4.6	5.8	4.8
4	4.07	0.442	16.71	0.8179	0.0716	−23.8	4.0	5.3	4.3
0	0.00	0.000	15.59	0.8174	0.0000	—	3.5	4.8	3.7

16°C DRY BULB

Percentage saturation μ/%	Relative humidity ϕ/%	Value of stated parameter per kg dry air			Vapour pressure p_s/kPa	Dew point temperature t_d/°C	Adiabatic saturation temperature $t^\star$/°C	Wet bulb temperature	
		Moisture content $g/g.kg^{-1}$	Specific enthalpy h/kJ.kg^{-1}	Specific volume v/m^3.kg^{-1}				Screen t'_{sc}/°C	Sling t'_{sl}/°C
100	100.00	11.41	44.96	0.8337	1.817	16.0	16.0	16.0	16.0
96	96.07	10.95	43.81	0.8331	1.746	15.4	15.6	15.6	15.6
92	92.13	10.50	42.65	0.8325	1.674	14.7	15.2	15.3	15.2
88	88.19	10.04	41.50	0.8319	1.603	14.0	14.8	14.9	14.8
84	84.24	9.583	40.34	0.8313	1.531	13.3	14.4	14.5	14.4
80	80.29	9.127	39.19	0.8307	1.459	12.6	14.0	14.1	14.0
76	76.33	8.671	38.04	0.8301	1.387	11.8	13.5	13.7	13.6
72	72.37	8.214	36.88	0.8295	1.315	11.0	13.1	13.3	13.1
70	70.38	7.986	36.30	0.8292	1.279	10.6	12.9	13.1	12.9
68	68.40	7.758	35.73	0.8289	1.243	10.2	12.6	12.9	12.7
66	66.41	7.530	35.15	0.8286	1.207	9.7	12.4	12.7	12.5
64	64.42	7.302	34.57	0.8283	1.171	9.3	12.2	12.5	12.2
62	62.43	7.074	33.99	0.8280	1.134	8.8	12.0	12.3	12.0
60	60.44	6.845	33.42	0.8277	1.098	8.4	11.7	12.1	11.8
58	58.45	6.617	32.84	0.8274	1.062	7.9	11.5	11.9	11.6
56	56.45	6.389	32.26	0.8271	1.026	7.4	11.3	11.7	11.3
54	54.45	6.161	31.68	0.8268	0.9895	6.8	11.0	11.5	11.1
52	52.46	5.933	31.11	0.8265	0.9532	6.3	10.8	11.3	10.9
50	50.46	5.704	30.53	0.8262	0.9169	5.7	10.5	11.1	10.6
48	48.46	5.476	29.95	0.8259	0.8805	5.1	10.3	10.8	10.4
46	46.45	5.248	29.37	0.8257	0.8441	4.5	10.1	10.6	10.1
44	44.45	5.020	28.80	0.8254	0.8077	3.9	9.8	10.4	9.9
42	42.45	4.792	28.22	0.8251	0.7713	3.3	9.6	10.2	9.7
40	40.44	4.564	27.64	0.8248	0.7348	2.6	9.3	10.0	9.4
38	38.43	4.335	27.07	0.8245	0.6983	1.9	9.0	9.7	9.2
36	36.42	4.107	26.49	0.8242	0.6618	1.1	8.8	9.5	8.9
34	34.41	3.879	25.91	0.8239	0.6253	0.3	8.5	9.3	8.7
32	32.40	3.651	25.33	0.8236	0.5887	−0.4	8.3	9.1	8.4
30	30.38	3.423	24.76	0.8233	0.5521	−1.2	8.0	8.8	8.1
28	28.37	3.194	24.17	0.8230	0.5155	−2.0	7.7	8.6	7.9
24	24.33	2.738	23.02	0.8224	0.4422	−3.9	7.2	8.1	7.4
20	20.29	2.282	21.87	0.8218	0.3688	−6.0	6.6	7.6	6.8
16	16.25	1.825	20.71	0.8212	0.2952	−8.6	6.1	7.2	6.3
12	12.19	1.369	19.56	0.8206	0.2216	−11.8	5.5	6.7	5.7
8	8.13	0.912	18.40	0.8200	0.1478	−16.2	4.9	6.2	5.1
4	4.07	0.456	17.25	0.8194	0.0740	−23.5	4.3	5.6	4.6
0	0.00	0.000	16.10	0.8188	0.0000	—	3.7	5.1	4.0

16.5°C DRY BULB

Percentage saturation μ/%	Relative humidity ϕ/%	Value of stated parameter per kg dry air			Vapour pressure p_s/kPa	Dew point temperature t_d/°C	Adiabatic saturation temperature $t^\star$/°C	Wet bulb temperature	
		Moisture content $g/g.kg^{-1}$	Specific enthalpy h/kJ.kg^{-1}	Specific volume v/m^3.kg^{-1}				Screen t'_{sc}/°C	Sling t'_{sl}/°C
100	100.00	11.79	46.43	0.8356	1.876	16.5	16.5	16.5	16.5
96	96.07	11.31	45.24	0.8350	1.802	15.9	16.1	16.1	16.1
92	92.14	10.84	44.04	0.8344	1.728	15.2	15.7	15.8	15.7
88	88.20	10.37	42.85	0.8338	1.655	14.5	15.3	15.4	15.3
84	84.25	9.900	41.66	0.8332	1.581	13.8	14.9	15.0	14.9
80	80.30	9.428	40.46	0.8326	1.506	13.1	14.4	14.6	14.4
76	76.34	8.957	39.27	0.8319	1.432	12.3	14.0	14.2	14.0
72	72.38	8.486	38.08	0.8313	1.358	11.5	13.5	13.8	13.6
70	70.40	8.250	37.48	0.8310	1.321	11.1	13.3	13.6	13.4
68	68.41	8.014	36.88	0.8307	1.283	10.7	13.1	13.4	13.1
66	66.42	7.778	36.29	0.8304	1.246	10.2	12.9	13.2	12.9
64	64.43	7.543	35.69	0.8301	1.209	9.8	12.6	13.0	12.7
62	62.44	7.307	35.09	0.8298	1.171	9.3	12.4	12.8	12.5
60	60.45	7.071	34.50	0.8295	1.134	8.8	12.2	12.6	12.2
58	58.46	6.836	33.90	0.8292	1.097	8.3	11.9	12.3	12.0
56	56.47	6.600	33.30	0.8289	1.059	7.8	11.7	12.1	11.8
54	54.47	6.364	32.71	0.8285	1.022	7.3	11.4	11.9	11.5
52	52.47	6.128	32.11	0.8282	0.9844	6.8	11.2	11.7	11.3
50	50.47	5.893	31.51	0.8279	0.9469	6.2	11.0	11.5	11.0
48	48.47	5.657	30.92	0.8276	0.9093	5.6	10.7	11.3	10.8
46	46.47	5.421	30.32	0.8273	0.8718	5.0	10.5	11.0	10.6
44	44.47	5.186	29.72	0.8270	0.8342	4.4	10.2	10.8	10.4
42	42.46	4.950	29.13	0.8267	0.7965	3.7	10.0	10.6	10.1
40	40.45	4.714	28.53	0.8264	0.7589	3.0	9.7	10.4	9.8
38	38.45	4.478	27.93	0.8261	0.7212	2.3	9.4	10.1	9.5
36	36.44	4.243	27.34	0.8258	0.6835	1.6	9.2	9.9	9.3
34	34.42	4.007	26.74	0.8255	0.6458	0.8	8.9	9.7	9.0
32	32.41	3.771	26.14	0.8251	0.6080	−0.1	8.6	9.4	8.8
30	30.40	3.536	25.55	0.8248	0.5702	−0.8	8.4	9.2	8.5
28	28.38	3.300	24.95	0.8245	0.5324	−1.7	8.1	9.0	8.2
24	24.35	2.828	23.76	0.8239	0.4567	−3.5	7.5	8.5	7.7
20	20.30	2.357	22.56	0.8233	0.3809	−5.6	7.0	8.0	7.2
16	16.25	1.886	21.37	0.8227	0.3049	−8.2	6.4	7.5	6.6
12	12.20	1.414	20.18	0.8221	0.2289	−11.4	5.8	7.0	6.0
8	8.14	0.943	18.98	0.8214	0.1527	−15.9	5.2	6.5	5.4
4	4.07	0.471	17.79	0.8208	0.0764	−23.1	4.6	5.9	4.8
0	0.00	0.000	16.60	0.8202	0.0000	—	4.0	5.4	4.2

17°C DRY BULB

Percentage saturation μ/%	Relative humidity ϕ/%	Value of stated parameter per kg dry air			Vapour pressure p_s/kPa	Dew point temperature t_d/°C	Adiabatic saturation temperature $t\star$/°C	Wet bulb temperature	
		Moisture content g/g.kg^{-1}	Specific enthalpy h/kJ.kg^{-1}	Specific volume v/m^3.kg^{-1}				Screen t'_{sc}/°C	Sling t'_{sl}/°C
100	100.00	12.17	47.93	0.8376	1.936	17.0	17.0	17.0	17.0
96	96.07	11.69	46.69	0.8370	1.860	16.4	16.6	16.6	16.6
92	92.14	11.20	45.46	0.8363	1.784	15.7	16.2	16.2	16.2
88	88.20	10.71	44.23	0.8357	1.708	15.0	15.8	15.9	15.8
84	84.26	10.23	42.99	0.8350	1.632	14.3	15.3	15.5	15.4
80	80.31	9.739	41.76	0.8344	1.555	13.6	14.9	15.1	14.9
76	76.35	9.252	40.53	0.8338	1.479	12.8	14.5	14.7	14.5
72	72.39	8.765	39.30	0.8331	1.402	12.0	14.0	14.3	14.0
70	70.41	8.521	38.68	0.8328	1.363	11.6	13.8	14.0	13.8
68	68.42	8.278	38.06	0.8325	1.325	11.2	13.5	13.8	13.6
66	66.44	8.034	37.45	0.8322	1.287	10.7	13.3	13.6	13.4
64	64.45	7.791	36.83	0.8318	1.248	10.3	13.1	13.4	13.1
62	62.46	7.547	36.21	0.8315	1.210	9.8	12.8	13.2	12.9
60	60.47	7.304	35.60	0.8312	1.171	9.3	12.6	13.0	12.7
58	58.47	7.060	34.98	0.8309	1.132	8.8	12.4	12.8	12.4
56	56.48	6.817	34.36	0.8306	1.094	8.3	12.1	12.6	12.2
54	54.48	6.574	33.75	0.8303	1.055	7.8	11.9	12.3	11.9
52	52.49	6.330	33.13	0.8299	1.016	7.2	11.6	12.1	11.7
50	50.49	6.087	32.51	0.8296	0.9777	6.7	11.4	11.9	11.5
48	48.49	5.843	31.90	0.8293	0.9390	6.1	11.1	11.7	11.2
46	46.49	5.600	31.28	0.8290	0.9002	5.5	10.9	11.4	11.0
44	44.48	5.356	30.66	0.8287	0.8614	4.8	10.6	11.2	10.7
42	42.48	5.113	30.05	0.8283	0.8225	4.2	10.3	11.0	10.4
40	40.47	4.869	29.43	0.8280	0.7837	3.5	10.1	10.7	10.2
38	38.46	4.626	28.81	0.8277	0.7448	2.8	9.8	10.5	9.9
36	36.45	4.382	28.20	0.8274	0.7058	2.0	9.6	10.3	9.7
34	34.44	4.139	27.58	0.8271	0.6669	1.2	9.3	10.0	9.4
32	32.43	3.896	26.97	0.8267	0.6279	0.4	9.0	9.8	9.1
30	30.41	3.652	26.35	0.8264	0.5889	−0.4	8.7	9.6	8.9
28	28.39	3.409	25.73	0.8261	0.5498	−1.3	8.5	9.3	8.6
24	24.36	2.922	24.50	0.8255	0.4717	−3.1	7.9	8.8	8.0
20	20.31	2.435	23.27	0.8248	0.3934	−5.2	7.3	8.3	7.5
16	16.26	1.948	22.03	0.8242	0.3149	−7.8	6.7	7.8	6.9
12	12.21	1.461	20.80	0.8235	0.2364	−11.1	6.1	7.3	6.3
8	8.14	0.974	19.57	0.8229	0.1577	−15.5	5.5	6.8	5.7
4	4.08	0.487	18.33	0.8223	0.0789	−22.8	4.9	6.2	5.1
0	0.00	0.000	17.10	0.8216	0.0000	—	4.3	5.7	4.5

17.5°C DRY BULB

Percentage saturation μ/%	Relative humidity ϕ/%	Value of stated parameter per kg dry air			Vapour pressure p_s/kPa	Dew point temperature t_d/°C	Adiabatic saturation temperature $t\star$/°C	Wet bulb temperature	
		Moisture content g/g.kg^{-1}	Specific enthalpy h/kJ.kg^{-1}	Specific volume v/m^3.kg^{-1}				Screen t'_{sc}/°C	Sling t'_{sl}/°C
100	100.00	12.57	49.45	0.8396	1.999	17.5	17.5	17.5	17.5
96	96.08	12.07	48.18	0.8389	1.920	16.9	17.1	17.1	17.1
92	92.15	11.57	46.90	0.8382	1.842	16.2	16.7	16.7	16.7
88	88.21	11.06	45.63	0.8376	1.763	15.5	16.2	16.3	16.3
84	84.27	10.56	44.36	0.8369	1.684	14.8	15.8	15.9	15.8
80	80.32	10.06	43.08	0.8363	1.605	14.1	15.4	15.5	15.4
76	76.37	9.555	41.81	0.8356	1.526	13.3	14.9	15.1	14.9
72	72.40	9.052	40.53	0.8349	1.447	12.5	14.5	14.7	14.5
70	70.42	8.801	39.90	0.8346	1.408	12.1	14.2	14.5	14.5
68	68.44	8.549	39.26	0.8343	1.368	11.6	14.0	14.3	14.0
66	66.45	8.298	38.62	0.8340	1.328	11.2	13.8	14.1	13.8
64	64.46	8.046	37.99	0.8336	1.288	10.7	13.5	13.9	13.6
62	62.47	7.795	37.35	0.8333	1.249	10.3	13.3	13.6	13.3
60	60.48	7.544	36.71	0.8330	1.209	9.8	13.0	13.4	13.1
58	58.49	7.292	36.08	0.8326	1.169	9.3	12.8	13.2	12.9
56	56.50	7.041	35.44	0.8323	1.129	8.8	12.5	13.0	12.6
54	54.50	6.789	34.80	0.8320	1.089	8.2	12.3	12.8	12.4
52	52.50	6.538	34.16	0.8316	1.049	7.7	12.0	12.5	12.1
50	50.50	6.286	33.53	0.8313	1.009	7.1	11.8	12.3	11.9
48	48.50	6.035	32.89	0.8310	0.9695	6.5	11.5	12.1	11.6
46	46.50	5.783	32.25	0.8307	0.9294	5.9	11.3	11.8	11.4
44	44.50	5.532	31.62	0.8303	0.8894	5.3	11.0	11.6	11.1
42	42.49	5.280	30.98	0.8300	0.8493	4.6	10.7	11.4	10.8
40	40.48	5.029	30.34	0.8297	0.8092	3.9	10.5	11.1	10.6
38	38.48	4.778	29.71	0.8293	0.7690	3.2	10.2	10.9	10.3
36	36.46	4.526	29.07	0.8290	0.7288	2.5	9.9	10.7	10.1
34	34.45	4.275	28.43	0.8287	0.6886	1.7	9.7	10.4	9.8
32	32.44	4.023	27.80	0.8283	0.6484	0.8	9.4	10.2	9.5
30	30.42	3.772	27.16	0.8280	0.6081	0.0	9.1	9.9	9.2
28	28.41	3.520	26.52	0.8277	0.5678	−0.9	8.8	9.7	9.0
24	24.37	3.017	25.25	0.8270	0.4871	−2.7	8.2	9.2	8.4
20	20.32	2.514	23.97	0.8264	0.4062	−4.9	7.7	8.7	7.8
16	16.27	2.012	22.70	0.8257	0.3252	−7.4	7.1	8.1	7.2
12	12.21	1.509	21.43	0.8250	0.2441	−10.7	6.4	7.6	6.6
8	8.15	1.006	20.15	0.8244	0.1629	−15.2	5.8	7.1	6.0
4	4.08	0.503	18.88	0.8237	0.0815	−22.5	5.2	6.5	5.4
0	0.00	0.000	17.60	0.8230	0.0000	—	4.5	6.0	4.8

18°C DRY BULB

Percentage saturation μ/%	Relative humidity ϕ/%	Value of stated parameter per kg dry air			Vapour pressure p_s/kPa	Dew point temperature t_d/°C	Adiabatic saturation temperature $t^\star$/°C	Wet bulb temperature	
		Moisture content g/g.kg⁻¹	Specific enthalpy h/kJ.kg⁻¹	Specific volume v/m³.kg⁻¹				Screen t'_{sc}/°C	Sling t'_{sl}/°C
100	100.00	12.98	51.01	0.8416	2.063	18.0	18.0	18.0	18.0
96	96.08	12.46	49.69	0.8409	1.982	17.4	17.6	17.6	17.6
92	92.15	11.94	48.38	0.8402	1.901	16.7	17.2	17.2	17.2
88	88.22	11.43	47.06	0.8395	1.820	16.0	16.7	16.8	16.7
84	84.28	10.91	45.74	0.8388	1.738	15.3	16.3	16.4	16.3
80	80.33	10.39	44.43	0.8381	1.657	14.6	15.8	16.0	15.9
76	76.38	9.868	43.11	0.8375	1.576	13.8	15.4	15.6	15.4
72	72.42	9.348	41.80	0.8368	1.494	13.0	14.9	15.2	15.0
70	70.44	9.088	41.14	0.8364	1.453	12.5	14.7	15.0	14.7
68	68.45	8.829	40.48	0.8361	1.412	12.1	14.4	14.7	14.5
66	66.47	8.569	39.82	0.8357	1.371	11.7	14.2	14.5	14.3
64	64.48	8.309	39.16	0.8354	1.330	11.2	14.0	14.3	14.0
62	62.49	8.050	38.51	0.8351	1.289	10.7	13.7	14.1	13.8
60	60.50	7.790	37.85	0.8347	1.248	10.3	13.5	13.9	13.5
58	58.51	7.530	37.19	0.8344	1.207	9.8	13.2	13.6	13.3
56	56.51	7.271	36.53	0.8340	1.166	9.2	13.0	13.4	13.0
54	54.52	7.011	35.87	0.8337	1.125	8.7	12.7	13.2	12.8
52	52.52	6.751	35.22	0.8334	1.083	8.2	12.5	13.0	12.5
50	50.52	6.492	34.56	0.8330	1.042	7.6	12.2	12.7	12.3
48	48.52	6.232	33.90	0.8327	1.001	7.0	11.9	12.5	12.0
46	46.52	5.972	33.24	0.8323	0.9595	6.4	11.7	12.3	11.8
44	44.51	5.713	32.58	0.8320	0.9182	5.7	11.4	12.0	11.5
42	42.51	5.453	31.93	0.8316	0.8768	5.1	11.1	11.8	11.2
40	40.50	5.193	31.27	0.8313	0.8354	4.4	10.9	11.5	11.0
38	38.49	4.934	30.61	0.8310	0.7940	3.7	10.6	11.3	10.7
36	36.48	4.674	29.95	0.8306	0.7525	2.9	10.3	11.0	10.4
34	34.47	4.414	29.29	0.8303	0.7110	2.1	10.0	10.8	10.2
32	32.45	4.155	28.64	0.8299	0.6694	1.3	9.7	10.5	9.9
30	30.44	3.895	27.98	0.8296	0.6278	0.4	9.5	10.3	9.6
28	28.42	3.635	27.32	0.8293	0.5862	−0.5	9.2	10.0	9.3
24	24.38	3.116	26.00	0.8286	0.5029	−2.5	8.6	9.5	8.7
20	20.33	2.597	24.69	0.8279	0.4194	−4.5	8.0	9.0	8.2
16	16.28	2.077	23.37	0.8272	0.3358	−7.1	7.4	8.5	7.6
12	12.22	1.558	22.06	0.8265	0.2521	−10.3	6.7	7.9	6.9
8	8.15	1.039	20.74	0.8258	0.1682	−14.8	6.1	7.4	6.3
4	4.08	0.519	19.42	0.8252	0.0842	−22.1	5.5	6.8	5.7
0	0.00	0.000	18.11	0.8245	0.0000	—	4.8	6.3	5.0

18.5°C DRY BULB

Percentage saturation μ/%	Relative humidity ϕ/%	Value of stated parameter per kg dry air			Vapour pressure p_s/kPa	Dew point temperature t_d/°C	Adiabatic saturation temperature $t^\star$/°C	Wet bulb temperature	
		Moisture content g/g.kg⁻¹	Specific enthalpy h/kJ.kg⁻¹	Specific volume v/m³.kg⁻¹				Screen t'_{sc}/°C	Sling t'_{sl}/°C
100	100.00	13.41	52.59	0.8436	2.128	18.5	18.5	18.5	18.5
96	96.08	12.87	51.24	0.8429	2.045	17.9	18.1	18.1	18.1
92	92.16	12.33	49.88	0.8421	1.962	17.2	17.6	17.7	17.7
88	88.22	11.80	48.52	0.8414	1.878	16.5	17.2	17.3	17.2
84	84.29	11.26	47.16	0.8407	1.794	15.8	16.8	16.9	16.8
80	80.34	10.73	45.80	0.8400	1.710	15.1	16.3	16.5	16.3
76	76.39	10.19	44.44	0.8393	1.626	14.3	15.8	16.1	15.9
72	72.43	9.653	43.08	0.8386	1.542	13.5	15.4	15.6	15.4
70	70.45	9.384	42.40	0.8383	1.500	13.0	15.1	15.4	15.2
68	68.47	9.116	41.72	0.8379	1.457	12.6	14.9	15.2	14.9
66	66.48	8.848	41.04	0.8376	1.415	12.1	14.6	15.0	14.7
64	64.49	8.580	40.36	0.8372	1.373	11.7	14.4	14.7	14.5
62	62.50	8.312	39.68	0.8368	1.330	11.2	14.2	14.5	14.2
60	60.51	8.044	39.00	0.8365	1.288	10.7	13.9	14.3	14.0
58	58.52	7.776	38.32	0.8361	1.246	10.2	13.7	14.1	13.7
56	56.53	7.508	37.64	0.8358	1.203	9.7	13.4	13.8	13.5
54	54.53	7.240	36.96	0.8354	1.161	9.2	13.1	13.6	13.2
52	52.54	6.971	36.28	0.8351	1.118	8.6	12.9	13.4	13.0
50	50.54	6.703	35.60	0.8347	1.076	8.0	12.6	13.1	12.7
48	48.54	6.435	34.92	0.8344	1.033	7.5	12.3	12.9	12.4
46	46.53	6.167	34.24	0.8340	0.9904	6.8	12.1	12.7	12.2
44	44.53	5.899	33.56	0.8337	0.9478	6.2	11.8	12.4	11.9
42	42.52	5.631	32.88	0.8333	0.9051	5.5	11.5	12.2	11.6
40	40.52	5.363	32.20	0.8330	0.8624	4.8	11.3	11.9	11.4
38	38.51	5.094	31.52	0.8326	0.8196	4.1	11.0	11.7	11.1
36	36.50	4.826	30.84	0.8323	0.7278	3.4	10.7	11.4	10.8
34	34.48	4.558	30.16	0.8319	0.7340	2.6	10.4	11.2	10.5
32	32.47	4.290	29.49	0.8315	0.6911	1.7	10.1	10.9	10.2
30	30.45	4.022	28.81	0.8312	0.6482	0.8	9.8	10.7	10.0
28	28.43	3.754	28.13	0.8308	0.6052	−0.1	9.5	10.4	9.7
24	24.39	3.218	26.77	0.8301	0.5192	−2.0	8.9	9.9	9.1
20	20.34	2.681	25.41	0.8294	0.4330	−4.1	8.3	9.3	8.5
16	16.29	2.145	24.05	0.8287	0.3477	−6.7	7.7	8.8	7.9
12	12.23	1.609	22.69	0.8280	0.2603	−10.0	7.1	8.2	7.3
8	8.16	1.072	21.33	0.8273	0.1747	−14.5	6.4	7.7	6.6
4	4.08	0.536	19.97	0.8266	0.0869	−21.8	5.7	7.1	6.0
0	0.00	0.000	18.61	0.8259	0.0000	—	5.0	6.5	5.3

19°C DRY BULB

Percentage saturation μ/%	Relative humidity ϕ/%	Value of stated parameter per kg dry air			Vapour pressure p_s/kPa	Dew point temperature t_d/°C	Adiabatic saturation temperature $t\star$/°C	Wet bulb temperature	
		Moisture content $g/g.kg^{-1}$	Specific enthalpy $h/kJ.kg^{-1}$	Specific volume $v/m^3.kg^{-1}$				Screen t'_{sc}/°C	Sling t'_{sl}/°C
100	100.00	13.84	54.21	0.8456	2.196	19.0	19.0	19.0	19.0
96	96.08	13.29	52.81	0.8449	2.110	18.4	18.6	18.6	18.6
92	92.16	12.74	51.41	0.8441	2.024	17.7	18.1	18.2	18.1
88	88.23	12.18	50.00	0.8434	1.938	17.0	17.7	17.8	17.7
84	84.30	11.63	48.60	0.8427	1.851	16.3	17.2	17.4	17.3
80	80.35	11.07	47.19	0.8419	1.765	15.5	16.8	17.0	16.8
76	76.40	10.52	45.79	0.8412	1.678	14.8	16.3	16.5	16.3
72	72.44	9.966	44.39	0.8405	1.591	13.9	15.8	16.1	15.9
70	70.46	9.689	43.68	0.8401	1.547	13.5	15.6	15.9	15.6
68	68.48	9.412	42.98	0.8397	1.504	13.1	15.3	15.6	15.4
66	66.50	9.136	42.28	0.8394	1.460	12.6	15.1	15.4	15.1
64	64.51	8.859	41.58	0.8390	1.417	12.2	14.8	15.2	14.9
62	62.52	8.582	40.88	0.8386	1.373	11.7	14.6	15.0	14.7
60	60.53	8.305	40.17	0.8383	1.329	11.2	14.3	14.7	14.4
58	58.54	8.028	39.47	0.8379	1.286	10.7	14.1	14.5	14.1
56	56.55	7.751	38.77	0.8375	1.242	10.2	13.8	14.3	13.9
54	54.55	7.475	38.07	0.8372	1.198	9.6	13.6	14.0	13.6
52	52.55	7.198	37.37	0.8368	1.154	9.1	13.3	13.8	13.4
50	50.55	6.921	36.66	0.8365	1.110	8.5	13.0	13.5	13.1
48	48.55	6.644	35.96	0.8361	1.066	7.9	12.8	13.3	12.8
46	46.55	6.367	35.26	0.8357	1.022	7.3	12.5	13.1	12.6
44	44.55	6.090	34.56	0.8354	0.9782	6.7	12.2	12.8	12.3
42	42.54	5.814	33.86	0.8350	0.9342	6.0	11.9	12.6	12.0
40	40.53	5.537	33.15	0.8346	0.8902	5.3	11.6	12.3	11.8
38	38.52	5.260	32.45	0.8343	0.8460	4.6	11.4	12.1	11.5
36	36.51	4.983	31.75	0.8339	0.8018	3.8	11.1	11.8	11.2
34	34.50	4.706	31.05	0.8335	0.7576	3.0	10.8	11.6	10.9
32	32.48	4.429	30.35	0.8332	0.7134	2.2	10.5	11.3	10.6
30	30.47	4.153	29.64	0.8328	0.6691	1.3	10.2	11.0	10.3
28	28.45	3.876	28.94	0.8324	0.6248	0.3	9.9	10.8	10.0
24	24.41	3.322	27.54	0.8317	0.5360	−1.6	9.3	10.2	9.4
20	20.36	2.768	26.13	0.8310	0.4470	−3.7	8.7	9.7	8.8
16	16.30	2.215	24.73	0.8302	0.3580	−6.3	8.0	9.1	8.2
12	12.24	1.661	23.33	0.8295	0.2687	−9.6	7.4	8.6	7.6
8	8.16	1.107	21.92	0.8288	0.1792	−14.1	6.7	8.0	6.9
4	4.09	0.554	20.52	0.8280	0.0897	−21.5	6.0	7.4	6.2
0	0.00	0.000	19.11	0.8273	0.0000	—	5.3	6.8	5.6

19.5°C DRY BULB

Percentage saturation μ/%	Relative humidity ϕ/%	Value of stated parameter per kg dry air			Vapour pressure p_s/kPa	Dew point temperature t_d/°C	Adiabatic saturation temperature $t\star$/°C	Wet bulb temperature	
		Moisture content $g/g.kg^{-1}$	Specific enthalpy $h/kJ.kg^{-1}$	Specific volume $v/m^3.kg^{-1}$				Screen t'_{sc}/°C	Sling t'_{sl}/°C
100	100.00	14.29	55.87	0.8476	2.266	19.5	19.5	19.5	19.5
96	96.09	13.72	54.42	0.8469	2.177	18.9	19.1	19.1	19.1
92	92.17	13.15	52.97	0.8461	2.088	18.2	18.6	18.7	18.6
88	88.24	12.58	51.52	0.8454	1.999	17.5	18.2	18.3	18.2
84	84.30	12.00	50.07	0.8446	1.910	16.8	17.7	17.9	17.7
80	80.36	11.43	48.62	0.8438	1.821	16.0	17.3	17.4	17.3
76	76.41	10.86	47.17	0.8431	1.731	15.2	16.8	17.0	16.8
72	72.46	10.29	45.72	0.8423	1.642	14.4	16.3	16.5	16.3
70	70.48	10.00	44.99	0.8420	1.597	14.0	16.0	16.3	16.1
68	68.50	9.717	44.27	0.8416	1.552	13.6	15.8	16.1	15.8
66	66.51	9.431	43.54	0.8412	1.507	13.1	15.5	15.9	15.6
64	64.53	9.146	42.82	0.8408	1.462	12.6	15.3	15.6	15.3
62	62.54	8.860	42.09	0.8405	1.417	12.2	15.0	15.4	15.1
60	60.55	8.574	41.37	0.8401	1.372	11.7	14.8	15.2	14.8
58	58.56	8.288	40.64	0.8397	1.327	11.2	14.5	14.9	14.6
56	56.56	8.002	39.92	0.8393	1.282	10.6	14.3	14.7	14.3
54	54.57	7.716	39.19	0.8389	1.236	10.1	14.0	14.5	14.1
52	52.57	7.431	38.47	0.8386	1.191	9.6	13.7	14.2	13.8
50	50.57	7.145	37.74	0.8382	1.146	9.0	13.4	14.0	13.5
48	48.57	6.859	37.02	0.8378	1.100	8.4	13.2	13.7	13.3
46	46.57	6.573	36.29	0.8374	1.055	7.8	12.9	13.5	13.0
44	44.56	6.288	35.57	0.8370	1.010	7.1	12.6	13.2	12.7
42	42.56	6.002	34.84	0.8367	0.9642	6.5	12.3	13.0	12.4
40	40.55	5.716	34.12	0.8363	0.9187	5.8	12.0	12.7	12.1
38	38.54	5.430	33.39	0.8359	0.8732	5.0	11.7	12.5	11.9
36	36.53	5.144	32.67	0.8355	0.8276	4.3	11.5	12.2	11.6
34	34.52	4.859	31.94	0.8352	0.7820	3.4	11.2	11.9	11.3
32	32.50	4.573	31.22	0.8348	0.7363	2.6	10.9	11.7	11.0
30	30.48	4.287	30.49	0.8344	0.6906	1.7	10.6	11.4	10.7
28	28.46	4.001	29.77	0.8340	0.6448	0.7	10.2	11.1	10.4
24	24.42	3.430	28.32	0.8333	0.5532	−1.2	9.6	10.6	9.8
20	20.37	2.858	26.87	0.8325	0.4614	−3.4	9.0	10.0	9.2
16	16.31	2.286	25.42	0.8318	0.3695	−6.0	8.3	9.5	8.5
12	12.24	1.715	23.97	0.8310	0.2774	−9.3	7.7	8.9	7.9
8	8.17	1.143	22.52	0.8302	0.1851	−13.8	7.0	8.3	7.2
4	4.09	0.572	21.07	0.8295	0.0926	−21.2	6.3	7.7	6.5
0	0.00	0.000	19.62	0.8287	0.0000	—	5.6	7.1	5.8

20°C DRY BULB

Percentage saturation μ/%	Relative humidity ϕ/%	Value of stated parameter per kg dry air			Vapour pressure p_s/kPa	Dew point temperature t_d/°C	Adiabatic saturation temperature $t\star$/°C	Wet bulb temperature	
		Moisture content $g/g.kg^{-1}$	Specific enthalpy $h/kJ.kg^{-1}$	Specific volume $v/m^3.kg^{-1}$				Screen t'_{sc}/°C	Sling t'_{sl}/°C
100	100.00	14.75	57.55	0.8497	2.337	20.0	20.0	20.0	20.0
96	96.09	14.16	56.05	0.8489	2.246	19.4	19.6	19.6	19.6
92	92.17	13.57	54.56	0.8481	2.154	18.7	19.1	19.2	19.1
88	88.25	12.98	53.06	0.8473	2.062	18.0	18.7	18.8	18.7
84	84.31	12.39	51.56	0.8466	1.970	17.3	18.2	18.3	18.2
80	80.37	11.80	50.06	0.8458	1.878	16.5	17.7	17.9	17.7
76	76.43	11.21	48.57	0.8450	1.786	15.7	17.2	17.5	17.3
72	72.47	10.62	47.07	0.8442	1.694	14.9	16.7	17.0	16.8
70	70.49	10.33	46.32	0.8438	1.647	14.5	16.5	16.8	16.5
68	68.51	10.03	45.57	0.8434	1.601	14.0	16.2	16.5	16.3
66	66.53	9.736	44.82	0.8431	1.555	13.6	16.0	16.3	16.0
64	64.54	9.441	44.08	0.8427	1.508	13.1	15.7	16.1	15.8
62	62.55	9.146	43.33	0.8423	1.462	12.6	15.5	15.8	15.5
60	60.56	8.851	42.58	0.8419	1.415	12.1	15.2	15.6	15.3
58	58.57	8.556	41.83	0.8415	1.369	11.6	14.9	15.4	15.0
56	56.58	8.260	41.08	0.8411	1.322	11.1	14.7	15.1	14.7
54	54.59	7.966	40.33	0.8407	1.276	10.6	14.4	14.9	14.5
52	52.59	7.670	39.58	0.8403	1.229	10.0	14.1	14.6	14.2
50	50.59	7.376	38.84	0.8399	1.182	9.4	13.9	14.4	13.9
48	48.59	7.080	38.09	0.8395	1.136	8.8	13.6	14.1	13.7
46	46.59	6.785	37.34	0.8391	1.089	8.2	13.3	13.9	13.4
44	44.58	6.490	36.59	0.8388	1.042	7.6	13.0	13.6	13.1
42	42.58	6.195	35.84	0.8384	0.9945	6.9	12.7	13.4	12.8
40	40.57	5.900	35.09	0.8380	0.9480	6.2	12.4	13.1	12.5
38	38.56	5.605	34.34	0.8376	0.9011	5.5	12.1	12.8	12.2
36	36.55	5.310	33.60	0.8372	0.8541	4.7	11.8	12.6	12.0
34	34.53	5.015	32.85	0.8368	0.8070	3.9	11.5	12.3	11.7
32	32.52	4.720	32.10	0.8364	0.7600	3.0	11.2	12.0	11.4
30	30.50	4.425	31.35	0.8360	0.7127	2.1	10.9	11.8	11.1
28	28.48	4.130	30.60	0.8356	0.6656	1.2	10.6	11.5	10.7
24	24.43	3.540	29.10	0.8348	0.5710	−0.8	10.0	10.9	10.1
20	20.38	2.950	27.61	0.8341	0.4763	−3.0	9.3	10.4	9.5
16	16.32	2.360	26.11	0.8333	0.3814	−5.6	8.6	9.8	8.8
12	12.25	1.770	24.61	0.8325	0.2863	−8.9	8.0	9.2	8.2
8	8.17	1.180	23.11	0.8317	0.1910	−13.4	7.3	8.6	7.5
4	4.09	0.590	21.62	0.8309	0.0956	−20.8	6.5	8.0	6.8
0	0.00	0.000	20.11	0.8301	0.0000	—	5.8	7.3	6.1

20.5°C DRY BULB

Percentage saturation μ/%	Relative humidity ϕ/%	Value of stated parameter per kg dry air			Vapour pressure p_s/kPa	Dew point temperature t_d/°C	Adiabatic saturation temperature $t\star$/°C	Wet bulb temperature	
		Moisture content $g/g.kg^{-1}$	Specific enthalpy $h/kJ.kg^{-1}$	Specific volume $v/m^3.kg^{-1}$				Screen t'_{sc}/°C	Sling t'_{sl}/°C
100	100.00	15.22	59.27	0.8518	2.410	20.5	20.5	20.5	20.5
96	96.09	14.62	57.73	0.8510	2.316	19.9	20.1	20.1	20.1
92	92.18	14.01	56.18	0.8502	2.222	19.2	19.6	19.7	19.6
88	88.25	13.49	54.63	0.8493	2.127	18.5	19.1	19.2	19.2
84	84.32	12.79	53.09	0.8485	2.032	17.8	18.7	18.8	18.7
80	80.39	12.18	51.54	0.8477	1.938	17.0	18.2	18.4	18.2
76	76.44	11.57	50.00	0.8469	1.842	16.2	17.7	17.9	17.7
72	72.49	10.96	48.45	0.8461	1.747	15.4	17.2	17.5	17.2
70	70.51	10.66	47.68	0.8457	1.699	15.0	17.0	17.2	17.0
68	68.53	10.35	46.90	0.8453	1.652	14.5	16.7	17.0	16.7
66	66.54	10.05	46.13	0.8449	1.604	14.1	16.4	16.8	16.5
64	64.56	9.744	45.36	0.8445	1.556	13.6	16.2	16.5	16.2
62	62.57	9.440	44.59	0.8441	1.508	13.1	15.9	16.3	16.0
60	60.58	9.135	43.81	0.8437	1.460	12.6	15.6	16.0	15.7
58	58.59	8.831	43.04	0.8433	1.412	12.1	15.4	15.8	15.4
56	56.60	8.526	42.27	0.8429	1.364	11.6	15.1	15.5	15.2
54	54.60	8.222	41.49	0.8425	1.316	11.0	14.8	15.3	14.9
52	52.61	7.917	40.72	0.8421	1.268	10.5	14.6	15.0	14.6
50	50.61	7.613	39.95	0.8417	1.220	9.9	14.3	14.8	14.4
48	48.61	7.308	39.18	0.8413	1.172	9.3	14.0	14.5	14.1
46	46.61	7.004	38.40	0.8409	1.123	8.7	13.7	14.3	13.8
44	44.60	6.699	37.63	0.8405	1.075	8.0	13.4	14.0	13.5
42	42.59	6.395	36.86	0.8401	1.027	7.4	13.1	13.8	13.2
40	40.59	6.090	36.08	0.8397	0.9782	6.7	12.8	13.5	12.9
38	38.58	5.786	35.31	0.8393	0.9298	5.9	12.5	13.2	12.6
36	36.56	5.481	34.54	0.8388	0.8813	5.2	12.2	13.0	12.3
34	34.55	5.177	33.76	0.8384	0.8327	4.3	11.9	12.7	12.0
32	32.53	4.872	32.99	0.8380	0.7841	3.5	11.6	12.4	11.7
30	30.51	4.568	32.22	0.8376	0.7355	2.6	11.3	12.1	11.4
28	28.49	4.263	31.44	0.8372	0.6868	1.6	11.0	11.8	11.1
24	24.45	3.654	29.90	0.8364	0.5892	−0.4	10.3	11.3	10.5
20	20.39	3.045	28.35	0.8356	0.4915	−2.6	9.6	10.7	9.8
16	16.33	2.436	26.81	0.8348	0.3936	−5.2	9.0	10.1	9.2
12	12.26	1.827	25.26	0.8340	0.2955	−8.5	8.3	9.5	8.5
8	8.18	1.218	23.71	0.8332	0.1972	−13.1	7.5	8.9	7.8
4	4.09	0.609	22.17	0.8324	0.0986	−20.5	6.8	8.2	7.1
0	0.00	0.000	20.62	0.8316	0.0000	—	6.1	7.6	6.3

21°C DRY BULB

Percentage saturation μ/%	Relative humidity ϕ/%	Value of stated parameter per kg dry air			Vapour pressure p_s/kPa	Dew point temperature t_d/°C	Adiabatic saturation temperature $t^\star$/°C	Wet bulb temperature	
		Moisture content g/g.kg^{-1}	Specific enthalpy h/kJ.kg^{-1}	Specific volume v/m^3.kg^{-1}				Screen t'_{sc}/°C	Sling t'_{sl}/°C
100	100.00	15.71	61.03	0.8539	2.486	21.0	21.0	21.0	21.0
96	96.09	15.09	59.43	0.8530	2.389	20.4	20.6	20.6	20.6
92	92.18	14.46	57.84	0.8522	2.291	19.7	20.1	20.2	20.1
88	88.26	13.83	56.24	0.8514	2.194	19.0	19.6	19.7	19.6
84	84.33	13.20	54.64	0.8505	2.096	18.3	19.2	19.3	19.2
80	80.40	12.57	53.05	0.8497	1.998	17.5	18.7	18.8	18.7
76	76.45	11.94	51.45	0.8489	1.900	16.7	18.2	18.4	18.2
72	72.50	11.31	49.86	0.8480	1.802	15.9	17.7	17.9	17.7
70	70.52	11.00	49.06	0.8476	1.753	15.4	17.4	17.7	17.5
68	68.54	10.69	48.26	0.8472	1.704	15.0	17.1	17.4	17.2
66	66.56	10.37	47.46	0.8468	1.654	14.5	16.9	17.2	16.9
64	64.58	10.06	46.66	0.8464	1.605	14.1	16.6	17.0	16.7
62	62.59	9.742	45.87	0.8459	1.556	13.6	16.4	16.7	16.4
60	60.60	9.428	45.07	0.8455	1.506	13.1	16.1	16.5	16.1
58	58.61	9.114	44.27	0.8451	1.457	12.6	15.8	16.2	15.9
56	56.62	8.800	43.47	0.8447	1.407	12.1	15.5	16.0	15.6
54	54.62	8.485	42.67	0.8443	1.358	11.5	15.3	15.7	15.3
52	52.63	8.171	41.88	0.8439	1.308	11.0	15.0	15.5	15.0
50	50.63	7.857	41.08	0.8434	1.258	10.4	14.7	15.2	14.8
48	48.63	7.542	40.28	0.8430	1.209	9.8	14.4	14.9	14.5
46	46.62	7.228	39.48	0.8426	1.159	9.1	14.1	14.7	14.2
44	44.62	6.914	38.68	0.8422	1.109	8.5	13.8	14.4	13.9
42	42.61	6.600	37.89	0.8418	1.059	7.8	13.5	14.2	13.6
40	40.60	6.285	37.09	0.8413	1.009	7.1	13.2	13.9	13.3
38	38.59	5.971	36.29	0.8409	0.9593	6.4	12.9	13.6	13.0
36	36.58	5.657	35.49	0.8405	0.9093	5.6	12.6	13.3	12.7
34	34.57	5.343	34.69	0.8401	0.8592	4.8	12.3	13.1	12.4
32	32.55	5.028	33.89	0.8397	0.8091	3.9	12.0	12.8	12.1
30	30.53	4.714	33.10	0.8393	0.7589	3.0	11.6	12.5	11.8
28	28.51	4.400	32.30	0.8388	0.7087	2.1	11.3	12.2	11.5
24	24.46	3.771	30.70	0.8380	0.6080	−0.1	10.7	11.6	10.8
20	20.40	3.143	29.11	0.8372	0.5072	−2.2	10.0	11.0	10.1
16	16.34	2.514	27.51	0.8363	0.4062	−4.9	9.3	10.4	9.5
12	12.27	1.886	25.91	0.8355	0.3049	−8.2	8.6	9.8	8.8
8	8.19	1.257	24.32	0.8347	0.2035	−12.7	7.8	9.2	8.1
4	4.10	0.628	22.72	0.8338	0.1018	−20.2	7.1	8.5	7.3
0	0.00	0.000	21.13	0.8330	0.0000	—	6.3	7.9	6.6

21.5°C DRY BULB

Percentage saturation μ/%	Relative humidity ϕ/%	Value of stated parameter per kg dry air			Vapour pressure p_s/kPa	Dew point temperature t_d/°C	Adiabatic saturation temperature $t^\star$/°C	Wet bulb temperature	
		Moisture content g/g.kg^{-1}	Specific enthalpy h/kJ.kg^{-1}	Specific volume v/m^3.kg^{-1}				Screen t'_{sc}/°C	Sling t'_{sl}/°C
100	100.00	16.22	62.82	0.8560	2.563	21.5	21.5	21.5	21.5
96	96.10	15.57	61.17	0.8551	2.463	20.9	21.0	21.1	21.1
92	92.19	14.92	59.53	0.8543	2.363	20.2	20.6	20.6	20.6
88	88.27	14.27	57.88	0.8534	2.262	19.5	20.1	20.2	20.1
84	84.34	13.62	56.23	0.8525	2.162	18.7	19.6	19.8	19.6
80	80.41	12.97	54.58	0.8517	2.061	18.0	19.1	19.3	19.2
76	76.47	12.32	52.94	0.8508	1.960	17.2	18.6	18.8	18.7
72	72.52	11.68	51.29	0.8500	1.859	16.4	18.1	18.4	18.2
70	70.54	11.35	50.46	0.8495	1.808	15.9	17.9	18.1	17.9
68	68.56	11.03	49.64	0.8491	1.757	15.5	17.6	17.9	17.6
66	66.58	10.70	48.82	0.8487	1.706	15.0	17.3	17.7	17.4
64	64.59	10.38	47.99	0.8482	1.656	14.6	17.1	17.4	17.1
62	62.61	10.05	47.17	0.8478	1.605	14.1	16.8	17.2	16.9
60	60.62	9.730	46.34	0.8474	1.554	13.6	16.5	16.9	16.6
58	58.63	9.405	45.52	0.8469	1.503	13.1	16.2	16.7	16.3
56	56.64	9.081	44.70	0.8465	1.452	12.5	16.0	16.4	16.0
54	54.64	8.756	43.87	0.8461	1.401	12.0	15.7	16.1	15.8
52	52.65	8.432	43.05	0.8456	1.349	11.4	15.4	15.9	15.5
50	50.65	8.108	42.23	0.8452	1.298	10.8	15.1	15.6	15.2
48	48.65	7.784	41.40	0.8448	1.247	10.2	14.8	15.4	14.9
46	46.64	7.459	40.58	0.8443	1.196	9.6	14.5	15.1	14.6
44	44.64	7.135	39.75	0.8439	1.144	9.0	14.2	14.8	14.3
42	42.63	6.811	38.93	0.8435	1.093	8.3	13.9	14.5	14.0
40	40.62	6.486	38.11	0.8431	1.041	7.6	13.6	14.3	13.7
38	38.61	6.162	37.28	0.8426	0.9897	6.8	13.3	14.0	13.4
36	36.60	5.838	36.46	0.8422	0.9381	6.1	13.0	13.7	13.1
34	34.58	5.513	35.64	0.8418	0.8864	5.2	12.7	13.4	12.8
32	32.57	5.189	34.81	0.8413	0.8347	4.4	12.3	13.1	12.5
30	30.55	4.865	33.99	0.8409	0.7829	3.5	12.0	12.9	12.1
28	28.53	4.540	33.16	0.8405	0.7311	2.5	11.7	12.6	11.8
24	24.48	3.892	31.52	0.8396	0.6273	0.4	11.0	12.0	11.2
20	20.42	3.243	29.87	0.8387	0.5233	−1.9	10.3	11.4	10.5
16	16.35	2.594	28.22	0.8379	0.4191	−4.5	9.6	10.7	9.8
12	12.28	1.946	26.57	0.8370	0.3146	−7.8	8.9	10.1	9.1
8	8.19	1.297	24.92	0.8361	0.2100	−12.4	8.1	9.5	8.3
4	4.10	0.648	23.28	0.8353	0.1051	−19.8	7.3	8.8	7.6
0	0.00	0.000	21.63	0.8344	0.0000	—	6.6	8.1	6.8

22°C DRY BULB

Percentage saturation μ/%	Relative humidity ϕ/%	Value of stated parameter per kg dry air			Vapour pressure p_s/kPa	Dew point temperature t_d/°C	Adiabatic saturation temperature $t\star$/°C	Wet bulb temperature	
		Moisture content $g/g.kg^{-1}$	Specific enthalpy $h/kJ.kg^{-1}$	Specific volume $v/m^3.kg^{-1}$				Screen t'_{sc}/°C	Sling t'_{sl}/°C
100	100.00	16.73	64.65	0.8581	2.643	22.0	22.0	22.0	22.0
96	96.10	16.06	62.95	0.8572	2.540	21.3	21.5	21.6	21.5
92	92.19	15.39	61.25	0.8564	2.436	20.7	21.1	21.1	21.1
88	88.28	14.72	59.55	0.8555	2.333	20.0	20.6	20.7	20.6
84	84.36	14.06	57.85	0.8546	2.229	19.2	20.1	20.2	20.1
80	80.42	13.39	56.15	0.8537	2.125	18.5	19.6	19.8	19.6
76	76.48	12.72	54.45	0.8528	2.021	17.7	19.1	19.3	19.1
72	72.53	12.05	52.75	0.8519	1.917	16.8	18.6	18.8	18.6
70	70.56	11.71	51.90	0.8515	1.865	16.4	18.3	18.6	18.4
68	68.58	11.38	51.05	0.8510	1.812	16.0	18.1	18.3	18.1
66	66.60	11.04	50.20	0.8506	1.760	15.5	17.8	18.1	17.8
64	64.61	10.71	49.35	0.8501	1.707	15.0	17.5	17.9	17.6
62	62.63	10.37	48.50	0.8497	1.655	14.5	17.2	17.6	17.3
60	60.64	10.04	47.64	0.8492	1.602	14.0	17.0	17.3	17.0
58	58.65	9.705	46.79	0.8488	1.550	13.5	16.7	17.1	16.7
56	56.66	9.370	45.94	0.8483	1.497	13.0	16.4	16.8	16.5
54	54.66	9.036	45.09	0.8479	1.445	12.5	16.1	16.6	16.2
52	52.67	8.701	44.24	0.8474	1.392	11.9	15.8	16.3	15.9
50	50.67	8.366	43.39	0.8470	1.339	11.3	15.5	16.0	15.6
48	48.67	8.032	42.54	0.8465	1.286	10.7	15.2	15.8	15.3
46	46.66	7.697	41.69	0.8461	1.233	10.1	14.9	15.5	15.0
44	44.66	7.362	40.84	0.8457	1.180	9.4	14.6	15.2	14.7
42	42.65	7.028	39.99	0.8452	1.127	8.7	14.3	14.9	14.4
40	40.64	6.693	39.14	0.8448	1.074	8.0	14.0	14.7	14.1
38	38.63	6.358	38.29	0.8443	1.021	7.3	13.7	14.4	13.8
36	36.62	6.024	37.44	0.8439	0.9677	6.5	13.4	14.1	13.5
34	34.60	5.689	36.59	0.8434	0.9144	5.7	13.0	13.8	13.2
32	32.58	5.354	35.74	0.8430	0.8611	4.8	12.7	13.5	12.8
30	30.56	5.020	34.89	0.8425	0.8077	3.9	12.4	13.2	12.5
28	28.54	4.685	34.04	0.8421	0.7542	2.9	12.0	12.9	12.2
24	24.49	4.016	32.34	0.8412	0.6472	0.8	11.3	12.3	11.5
20	20.43	3.346	30.64	0.8403	0.5399	−1.5	10.6	11.7	10.8
16	16.36	2.677	28.94	0.8394	0.4324	−4.1	9.9	11.1	10.1
12	12.28	2.008	27.24	0.8385	0.3246	−7.5	9.2	10.4	9.4
8	8.20	1.339	25.53	0.8376	0.2166	−12.0	8.4	9.8	8.6
4	4.10	0.669	23.83	0.8367	0.1084	−19.5	7.6	9.1	7.9
0	0.00	0.000	22.13	0.8358	0.0000	—	6.8	8.4	7.1

22.5°C DRY BULB

Percentage saturation μ/%	Relative humidity ϕ/%	Value of stated parameter per kg dry air			Vapour pressure p_s/kPa	Dew point temperature t_d/°C	Adiabatic saturation temperature $t\star$/°C	Wet bulb temperature	
		Moisture content $g/g.kg^{-1}$	Specific enthalpy $h/kJ.kg^{-1}$	Specific volume $v/m^3.kg^{-1}$				Screen t'_{sc}/°C	Sling t'_{sl}/°C
100	100.00	17.26	66.52	0.8603	2.724	22.5	22.5	22.5	22.5
96	96.10	16.57	64.77	0.8594	2.618	21.8	22.0	22.1	22.0
92	92.20	15.88	63.01	0.8585	2.512	21.2	21.6	21.6	21.6
88	88.29	15.19	61.26	0.8575	2.405	20.5	21.1	21.2	21.1
84	84.37	14.50	59.50	0.8566	2.298	19.7	20.6	20.7	20.6
80	80.44	13.81	57.75	0.8557	2.191	19.0	20.1	20.3	20.1
76	76.50	13.12	55.99	0.8548	2.084	18.2	19.6	19.8	19.6
72	72.55	12.43	54.23	0.8539	1.976	17.3	19.0	19.3	19.1
70	70.58	12.08	53.36	0.8534	1.923	16.9	18.8	19.0	18.8
68	68.60	11.74	52.48	0.8529	1.869	16.4	18.5	18.8	18.6
66	66.62	11.39	51.60	0.8525	1.815	16.0	18.2	18.5	18.3
64	64.63	11.05	50.72	0.8520	1.761	15.5	18.0	18.3	18.0
62	62.65	10.70	49.85	0.8516	1.707	15.0	17.7	18.0	17.7
60	60.66	10.36	48.97	0.8511	1.653	14.5	17.4	17.8	17.5
58	58.67	10.01	48.09	0.8506	1.598	14.0	17.1	17.5	17.2
56	56.68	9.668	47.21	0.8502	1.544	13.5	16.8	17.3	16.9
54	54.68	9.322	46.33	0.8497	1.490	12.9	16.5	17.0	16.6
52	52.69	8.977	45.46	0.8492	1.435	12.4	16.2	16.7	16.3
50	50.69	8.632	44.58	0.8488	1.381	11.8	15.9	16.5	16.0
48	48.69	8.287	43.70	0.8483	1.326	11.2	15.6	16.2	15.7
46	46.69	7.941	42.82	0.8479	1.272	10.5	15.3	15.9	15.4
44	44.68	7.596	41.95	0.8474	1.217	9.9	15.0	15.6	15.1
42	42.67	7.251	41.07	0.8469	1.163	9.2	14.7	15.3	14.8
40	40.66	6.906	40.19	0.8465	1.108	8.5	14.4	15.1	14.5
38	38.65	6.560	39.31	0.8460	1.053	7.7	14.1	14.8	14.2
36	36.64	6.215	38.43	0.8456	0.9981	7.0	13.7	14.5	13.8
34	34.62	5.870	37.56	0.8451	0.9432	6.1	13.4	14.2	13.5
32	32.60	5.524	36.68	0.8446	0.8882	5.3	13.1	13.9	13.2
30	30.58	5.179	35.80	0.8442	0.8331	4.3	12.7	13.6	12.9
28	28.56	4.834	34.92	0.8437	0.7780	3.4	12.4	13.3	12.5
24	24.51	4.143	33.17	0.8428	0.6676	1.2	11.7	12.7	11.8
20	20.44	3.453	31.41	0.8419	0.5570	−1.1	11.0	12.0	11.1
16	16.37	2.762	29.66	0.8409	0.4461	−3.8	10.2	11.4	10.4
12	12.29	2.072	27.90	0.8400	0.3349	−7.1	9.5	10.7	9.7
8	8.20	1.381	26.15	0.8391	0.2235	−11.7	8.7	10.1	8.9
4	4.11	0.690	24.39	0.8382	0.1119	−19.2	7.9	9.4	8.1
0	0.00	0.000	22.64	0.8372	0.0000	—	7.0	8.7	7.3

23°C DRY BULB

Percentage saturation μ/%	Relative humidity ϕ/%	Value of stated parameter per kg dry air			Vapour pressure p_s/kPa	Dew point temperature t_d/°C	Adiabatic saturation temperature $t^\star$/°C	Wet bulb temperature	
		Moisture content $g/g.kg^{-1}$	Specific enthalpy $h/kJ.kg^{-1}$	Specific volume $v/m^3.kg^{-1}$				Screen t'_{sc}/°C	Sling t'_{sl}/°C
100	100.00	17.81	68.43	0.8625	2.808	23.0	23.0	23.0	23.0
96	96.11	17.10	66.62	0.8615	2.699	22.3	22.5	22.6	22.5
92	92.21	16.39	64.81	0.8606	2.589	21.7	22.1	22.1	22.1
88	88.30	15.67	63.00	0.8596	2.479	21.0	21.6	21.7	21.6
84	84.38	14.96	61.18	0.8587	2.369	20.2	21.1	21.2	21.1
80	80.45	14.25	59.37	0.8577	2.259	19.5	20.6	20.7	20.6
76	76.51	13.54	57.56	0.8568	2.149	18.6	20.0	20.2	20.1
72	72.57	12.82	55.75	0.8558	2.038	17.8	19.5	19.7	19.5
70	70.59	12.47	54.84	0.8554	1.982	17.4	19.2	19.5	19.3
68	68.61	12.11	53.94	0.8549	1.927	16.9	19.0	19.2	19.0
66	66.63	11.76	53.03	0.8544	1.871	16.5	18.7	19.0	18.7
64	64.65	11.40	52.13	0.8539	1.815	16.0	18.4	18.7	18.5
62	62.67	11.04	51.22	0.8535	1.760	15.5	18.1	18.5	18.2
60	60.68	10.69	50.31	0.8530	1.704	15.0	17.8	18.2	17.9
58	58.69	10.33	49.41	0.8525	1.648	14.5	17.5	18.0	17.6
56	56.70	9.974	48.50	0.8520	1.592	13.9	17.2	17.7	17.3
54	54.70	9.618	47.60	0.8515	1.536	13.4	17.0	17.4	17.0
52	52.71	9.262	46.69	0.8511	1.480	12.8	16.7	17.1	16.7
50	50.71	8.905	45.79	0.8506	1.424	12.2	16.3	16.9	16.4
48	48.71	8.549	44.88	0.8501	1.368	11.6	16.0	16.6	16.1
46	46.71	8.193	43.97	0.8496	1.312	11.0	15.7	16.3	15.8
44	44.70	7.837	43.07	0.8392	1.255	10.3	15.4	16.0	15.5
42	42.69	7.480	42.16	0.8487	1.199	9.7	15.1	15.7	15.2
40	40.68	7.124	41.26	0.8482	1.142	8.9	14.8	15.4	14.9
38	38.67	6.768	40.35	0.8477	1.086	8.2	14.4	15.2	14.6
36	36.66	6.412	39.44	0.8473	1.029	7.4	14.1	14.9	14.2
34	34.64	6.056	38.54	0.8468	0.9727	6.6	13.8	14.6	13.9
32	32.62	5.699	37.63	0.8463	0.9160	5.7	13.4	14.3	13.6
30	30.60	5.343	36.73	0.8458	0.8593	4.8	13.1	13.9	13.2
28	28.58	4.987	35.82	0.8453	0.8024	3.8	12.7	13.6	12.9
24	24.52	4.275	34.01	0.8444	0.6886	1.7	12.0	13.0	12.2
20	20.46	3.562	32.20	0.8434	0.5745	−0.7	11.3	12.4	11.5
16	16.39	2.850	30.39	0.8425	0.4601	−3.4	10.5	11.7	10.7
12	12.30	2.137	28.57	0.8415	0.3455	−6.8	9.8	11.0	10.0
8	8.21	1.425	26.76	0.8406	0.2306	−11.3	9.0	10.3	9.2
4	4.11	0.712	24.95	0.8396	0.1154	−18.9	8.1	9.6	8.4
0	0.00	0.000	23.14	0.8387	0.0000	—	7.3	8.9	7.6

23.5°C DRY BULB

Percentage saturation μ/%	Relative humidity ϕ/%	Value of stated parameter per kg dry air			Vapour pressure p_s/kPa	Dew point temperature t_d/°C	Adiabatic saturation temperature $t^\star$/°C	Wet bulb temperature	
		Moisture content $g/g.kg^{-1}$	Specific enthalpy $h/kJ.kg^{-1}$	Specific volume $v/m^3.kg^{-1}$				Screen t'_{sc}/°C	Sling t'_{sl}/°C
100	100.00	18.37	70.38	0.8467	2.894	23.5	23.5	23.5	23.5
96	96.11	17.64	68.51	0.8637	2.782	22.8	23.0	23.1	23.0
92	92.21	16.90	66.64	0.8627	2.669	22.2	22.5	22.6	22.6
88	88.31	16.17	64.77	0.8618	2.556	21.5	22.0	22.1	22.1
84	84.39	15.43	62.90	0.8608	2.442	20.7	21.5	21.7	21.6
80	80.46	14.70	61.03	0.8598	2.329	19.9	21.0	21.2	21.1
76	76.53	13.96	59.16	0.8588	2.215	19.1	20.5	20.7	20.5
72	72.59	13.23	57.29	0.8578	2.101	18.3	20.0	20.2	20.0
70	70.61	12.86	56.36	0.8573	2.044	17.9	19.7	20.0	19.7
68	68.63	12.49	55.42	0.8568	1.986	17.4	19.4	19.7	19.5
66	66.65	12.13	54.49	0.8563	1.929	16.9	19.1	19.4	19.2
64	64.67	11.76	53.56	0.8559	1.872	16.5	18.8	19.2	18.9
62	62.69	11.39	52.62	0.8554	1.814	16.0	18.6	18.9	18.6
60	60.70	11.02	51.69	0.8549	1.757	15.5	18.3	18.7	18.3
58	58.71	10.66	50.75	0.8544	1.699	15.0	18.0	18.4	18.0
56	56.72	10.29	49.82	0.8539	1.642	14.4	17.7	18.1	17.7
54	54.73	9.921	48.88	0.8534	1.584	13.9	17.4	17.8	17.4
52	52.73	9.554	47.95	0.8529	1.526	13.3	17.1	17.6	17.1
50	50.73	9.187	47.01	0.8524	1.468	12.7	16.8	17.3	16.8
48	48.73	8.819	46.08	0.8519	1.410	12.1	16.5	17.0	16.5
46	46.73	8.452	45.14	0.8514	1.352	11.5	16.1	16.7	16.2
44	44.72	8.804	44.21	0.8509	1.294	10.8	15.8	16.4	15.9
42	42.72	7.717	43.27	0.8504	1.236	10.1	15.5	16.1	15.6
40	40.71	7.349	42.34	0.8499	1.178	9.4	15.2	15.8	15.3
38	38.69	6.982	41.40	0.8495	1.120	8.6	14.8	15.5	14.9
36	36.68	6.614	40.47	0.8490	1.062	7.9	14.5	15.2	14.6
34	34.66	6.247	39.53	0.8485	1.003	7.0	14.1	14.9	14.3
32	32.64	5.879	38.60	0.8480	0.9447	6.2	13.8	14.6	13.9
30	30.62	5.512	37.66	0.8475	0.8862	5.2	13.4	14.3	13.6
28	28.59	5.144	36.73	0.8470	0.8276	4.3	13.1	14.0	13.2
24	24.54	4.410	34.86	0.8460	0.7102	2.1	12.4	13.4	12.5
20	20.47	3.675	32.99	0.8450	0.5925	−0.4	11.6	12.7	11.8
16	16.40	2.940	31.12	0.8440	0.4746	−3.0	10.8	12.0	11.0
12	12.31	2.205	29.25	0.8430	0.3564	−6.4	10.1	11.3	10.3
8	8.22	1.470	27.38	0.8421	0.2379	−11.0	9.2	10.6	9.5
4	4.11	0.735	25.51	0.8411	0.1191	−18.5	8.4	9.9	8.7
0	0.00	0.000	23.64	0.8401	0.0000	—	7.5	9.2	7.8

24°C DRY BULB

Percentage saturation μ/%	Relative humidity ϕ/%	Value of stated parameter per kg dry air			Vapour pressure p_s/kPa	Dew point temperature t_d/°C	Adiabatic saturation temperature $t^\star$/°C	Wet bulb temperature	
		Moisture content g/g.kg^{-1}	Specific enthalpy h/kJ.kg^{-1}	Specific volume v/m^3.kg^{-1}				Screen t'_{sc}/°C	Sling t'_{sl}/°C
100	100.00	18.95	72.37	0.8669	2.983	24.0	24.0	24.0	24.0
96	96.11	18.19	70.44	0.8659	2.867	23.3	23.5	23.6	23.5
92	92.22	17.44	68.52	0.8649	2.751	22.7	23.0	23.1	23.0
88	88.31	16.68	66.59	0.8639	2.634	21.9	22.5	22.6	22.5
84	84.40	15.92	64.66	0.8629	2.517	21.2	22.0	22.2	22.0
80	80.48	15.16	62.73	0.8619	2.400	20.4	21.5	21.7	21.5
76	76.55	14.40	60.80	0.8608	2.283	19.6	21.0	21.2	21.0
72	72.60	13.64	58.87	0.8598	2.165	18.8	20.4	20.7	20.5
70	70.63	13.27	57.90	0.8593	2.107	18.3	20.1	20.4	20.2
68	68.65	12.89	56.94	0.8588	2.048	17.9	19.9	20.2	19.9
66	66.67	12.51	55.98	0.8583	1.989	17.4	19.6	19.9	19.6
64	64.69	12.13	55.01	0.8578	1.930	16.9	19.3	19.6	19.3
62	62.71	11.75	54.05	0.8573	1.870	16.5	19.0	19.4	19.1
60	60.72	11.37	53.08	0.8568	1.811	15.9	18.7	19.1	18.8
58	58.73	10.99	52.12	0.8563	1.752	15.4	18.4	18.8	18.5
56	56.74	10.61	51.15	0.8558	1.692	14.9	18.1	18.5	18.2
54	54.75	10.23	50.19	0.8553	1.633	14.3	17.8	18.3	17.9
52	52.75	9.855	49.22	0.8547	1.573	13.8	17.5	18.0	17.6
50	50.75	9.476	48.26	0.8542	1.514	13.2	17.2	17.7	17.3
48	48.75	9.097	47.29	0.8537	1.454	12.6	16.9	17.4	16.9
46	46.75	8.718	46.33	0.8532	1.394	11.9	16.5	17.1	16.6
44	44.75	8.339	45.37	0.8527	1.335	11.3	16.2	16.8	16.3
42	42.74	7.960	44.40	0.8522	1.275	10.6	15.9	16.5	16.0
40	40.73	7.580	43.44	0.8517	1.215	9.8	15.6	16.2	15.7
38	38.71	7.202	42.47	0.8512	1.155	9.1	15.2	15.9	15.3
36	36.70	6.822	41.51	0.8507	1.095	8.3	14.9	15.6	15.0
34	34.68	6.444	40.54	0.8502	1.034	7.5	14.5	15.3	14.6
32	32.66	6.064	39.58	0.8497	0.9741	6.6	14.2	15.0	14.3
30	30.64	5.685	38.61	0.8491	0.9138	5.7	13.8	14.7	13.9
28	28.61	5.306	37.65	0.8486	0.8534	4.7	13.4	14.4	13.6
24	24.56	4.548	35.72	0.8476	0.7324	2.5	12.7	13.7	12.9
20	20.49	3.790	33.79	0.8466	0.6111	−0.0	11.9	13.0	12.1
16	16.41	3.032	31.86	0.8456	0.4894	−2.7	11.2	12.3	11.4
12	12.32	2.274	29.93	0.8446	0.3675	−6.0	10.4	11.6	10.6
8	8.22	1.516	28.00	0.8435	0.2453	−10.7	9.5	10.9	9.8
4	4.12	0.758	26.07	0.8425	0.1228	−18.2	8.7	10.2	8.9
0	0.00	0.000	24.14	0.8415	0.0000	—	7.8	9.4	8.1

24.5°C DRY BULB

Percentage saturation μ/%	Relative humidity ϕ/%	Value of stated parameter per kg dry air			Vapour pressure p_s/kPa	Dew point temperature t_d/°C	Adiabatic saturation temperature $t^\star$/°C	Wet bulb temperature	
		Moisture content g/g.kg^{-1}	Specific enthalpy h/kJ.kg^{-1}	Specific volume v/m^3.kg^{-1}				Screen t'_{sc}/°C	Sling t'_{sl}/°C
100	100.00	19.55	74.41	0.8692	3.073	24.5	24.5	24.5	24.5
96	96.12	18.76	72.42	0.8682	2.954	23.8	24.0	24.0	24.0
92	92.23	17.98	70.43	0.8671	2.834	23.2	23.5	23.6	23.5
88	88.32	17.20	68.44	0.8661	2.714	22.4	23.0	23.1	23.0
84	84.41	16.42	66.45	0.8650	2.594	21.7	22.5	22.6	22.5
80	80.49	15.64	64.46	0.8640	2.474	20.9	22.0	22.1	22.0
76	76.56	14.86	62.47	0.8629	2.353	20.1	21.4	21.6	21.5
72	72.62	14.07	60.48	0.8619	2.232	19.3	20.9	21.1	20.9
70	70.65	13.68	59.48	0.8613	2.171	18.8	20.6	20.9	20.6
68	68.67	13.29	58.48	0.8608	2.111	18.4	20.3	20.6	20.4
66	66.69	12.90	57.49	0.8603	2.050	17.9	20.0	20.3	20.1
64	64.71	12.51	56.49	0.8598	1.989	17.4	19.7	20.1	19.8
62	62.73	12.12	55.50	0.8592	1.928	16.9	19.4	19.8	19.5
60	60.74	11.73	54.50	0.8587	1.867	16.4	19.1	19.5	19.2
58	58.76	11.34	53.51	0.8582	1.806	15.9	18.8	19.3	18.9
56	56.77	10.95	52.51	0.8577	1.745	15.4	18.5	19.0	18.6
54	54.77	10.56	51.52	0.8571	1.658	14.8	18.2	18.7	18.3
52	52.78	10.16	50.52	0.8566	1.622	14.2	17.9	18.4	18.0
50	50.78	9.773	49.53	0.8561	1.561	13.6	17.6	18.1	17.7
48	48.78	9.382	48.53	0.8556	1.499	13.0	17.3	17.8	17.4
46	46.77	8.991	47.54	0.8550	1.439	12.4	16.9	17.5	17.0
44	44.77	8.600	46.54	0.8545	1.376	11.7	16.6	17.2	16.7
42	42.76	8.209	45.55	0.8540	1.314	11.0	16.3	16.9	16.4
40	40.75	7.818	44.55	0.8535	1.252	10.3	15.9	16.6	16.0
38	38.74	7.428	43.56	0.8529	1.190	9.5	15.6	16.3	15.7
36	36.72	7.037	42.56	0.8524	1.129	8.8	15.2	16.0	15.4
34	34.70	6.646	41.57	0.8519	1.067	7.9	14.9	15.7	15.0
32	32.68	6.255	40.57	0.8513	1.004	7.0	14.5	15.4	14.7
30	30.66	5.864	39.58	0.8508	0.9422	6.1	14.2	15.0	14.3
28	28.63	5.473	38.58	0.8503	0.8800	5.1	13.8	14.7	13.9
24	24.57	4.691	36.59	0.8492	0.7552	3.0	13.0	14.0	13.2
20	20.50	3.909	34.60	0.8482	0.6301	0.4	12.3	13.4	12.4
16	16.42	3.127	32.61	0.8471	0.5047	−2.3	11.5	12.7	11.7
12	12.33	2.346	30.62	0.8461	0.3790	−5.7	10.6	11.9	10.9
8	8.23	1.564	28.63	0.8450	0.2530	−10.3	9.8	11.2	10.0
4	4.12	0.782	26.64	0.8440	0.1267	−17.9	8.9	10.4	9.2
0	0.00	0.000	24.65	0.8429	0.0000	—	8.0	9.7	8.3

25°C DRY BULB

Percentage saturation μ/%	Relative humidity ϕ/%	Value of stated parameter per kg dry air			Vapour pressure p_s/kPa	Dew point temperature t_d/°C	Adiabatic saturation temperature $t\star$/°C	Wet bulb temperature	
		Moisture content g/g.kg^{-1}	Specific enthalpy h/kJ.kg^{-1}	Specific volume v/m^3.kg^{-1}				Screen t'_{sc}/°C	Sling t'_{sl}/°C
100	100.00	20.16	76.49	0.8715	3.166	25.0	25.0	25.0	25.0
96	96.12	19.35	74.43	0.8704	3.044	24.3	24.5	24.5	24.5
92	92.23	18.55	72.38	0.8693	2.921	23.7	24.0	24.1	24.0
88	88.33	17.74	70.33	0.8682	2.797	22.9	23.5	23.6	23.5
84	84.43	16.93	68.27	0.8672	2.673	22.2	23.0	23.1	23.0
80	80.51	16.13	66.22	0.8661	2.549	21.4	22.4	22.6	22.5
76	76.58	15.32	64.17	0.8650	2.425	20.6	21.9	22.1	21.9
72	72.64	14.51	62.11	0.8639	2.300	19.7	21.3	21.6	21.4
70	70.67	14.11	61.09	0.8634	2.238	19.3	21.1	21.3	21.1
68	68.69	13.71	60.06	0.8628	2.175	18.8	20.8	21.1	20.8
66	66.72	13.30	59.03	0.8623	2.113	18.4	20.5	20.8	20.5
64	64.74	12.90	58.01	0.8617	2.050	17.9	20.2	20.5	20.2
62	62.75	12.50	56.98	0.8612	1.987	17.4	19.9	20.2	19.9
60	60.77	12.10	55.95	0.8606	1.924	16.9	19.6	20.0	19.6
58	58.78	11.69	54.93	0.8601	1.861	16.4	19.3	19.7	19.3
56	56.79	11.29	53.90	0.8596	1.798	15.8	19.0	19.4	19.0
54	54.80	10.89	52.87	0.8590	1.735	15.3	18.7	19.1	18.7
52	52.80	10.48	51.85	0.8585	1.672	14.7	18.3	18.8	18.4
50	50.80	10.08	50.82	0.8579	1.609	14.1	18.0	18.5	18.1
48	48.80	9.676	49.79	0.8574	1.545	13.5	17.7	18.2	17.8
46	46.80	9.273	48.76	0.8568	1.482	12.8	17.4	17.9	17.4
44	44.79	8.870	47.74	0.8563	1.418	12.2	17.0	17.6	17.1
42	42.78	8.466	46.71	0.8558	1.355	11.5	16.7	17.3	16.8
40	40.77	8.063	45.69	0.8552	1.291	10.8	16.3	17.0	16.4
38	38.76	7.660	44.66	0.8547	1.227	10.0	16.0	16.7	16.1
36	36.74	7.257	43.63	0.8541	1.163	9.2	15.6	16.4	15.7
34	34.72	6.854	42.60	0.8536	1.100	8.4	15.3	16.1	15.4
32	32.70	6.450	41.58	0.8530	1.036	7.5	14.9	15.7	15.0
30	30.68	6.047	40.55	0.8525	0.9714	6.6	14.5	15.4	14.7
28	28.65	5.644	39.52	0.8520	0.9072	5.6	14.2	15.1	14.3
24	24.59	4.838	37.47	0.8509	0.7786	3.4	13.4	14.4	13.5
20	20.52	4.032	35.42	0.8498	0.6497	0.9	12.6	13.7	12.8
16	16.44	3.225	33.36	0.8487	0.5204	−1.9	11.8	13.0	12.0
12	12.34	2.419	31.31	0.8476	0.3908	−5.3	10.9	12.2	11.2
8	8.24	1.613	29.26	0.8465	0.2609	−10.0	10.1	11.5	10.3
4	4.12	0.806	27.20	0.8454	0.1306	−17.5	9.2	10.7	9.4
0	0.00	0.000	25.15	0.8443	0.0000	—	8.2	9.9	8.5

25.5°C DRY BULB

Percentage saturation μ/%	Relative humidity ϕ/%	Value of stated parameter per kg dry air			Vapour pressure p_s/kPa	Dew point temperature t_d/°C	Adiabatic saturation temperature $t\star$/°C	Wet bulb temperature	
		Moisture content g/g.kg^{-1}	Specific enthalpy h/kJ.kg^{-1}	Specific volume v/m^3.kg^{-1}				Screen t'_{sc}/°C	Sling t'_{sl}/°C
100	100.00	20.79	78.61	0.8738	3.262	25.5	25.5	25.5	25.5
96	96.12	20.00	76.49	0.8727	3.136	24.8	25.0	25.0	25.0
92	92.24	19.12	74.37	0.8716	3.009	24.1	24.5	24.6	24.5
88	88.34	18.29	72.26	0.8704	2.882	23.4	24.0	24.1	24.0
84	84.44	17.46	70.14	0.8693	2.754	22.7	23.5	23.6	23.5
80	80.52	16.63	68.02	0.8682	2.627	21.9	22.9	23.1	22.9
76	76.60	15.80	65.90	0.8671	2.499	21.1	22.4	22.6	22.4
72	72.66	14.97	63.78	0.8660	2.370	20.2	21.8	22.0	21.8
70	70.69	14.55	62.72	0.8654	2.306	19.8	21.5	21.8	21.6
68	68.71	14.14	61.66	0.8648	2.242	19.3	21.2	21.5	21.3
66	66.74	13.72	60.61	0.8643	2.177	18.9	20.9	21.2	21.0
64	64.76	13.30	59.55	0.8637	2.112	18.4	20.6	21.0	20.7
62	62.78	12.89	58.49	0.8632	2.048	17.9	20.3	20.7	20.4
60	60.79	12.47	57.43	0.8626	1.983	17.4	20.0	20.4	20.1
58	58.80	12.06	56.37	0.8620	1.918	16.9	19.7	20.1	19.8
56	56.81	11.64	55.31	0.8615	1.853	16.3	19.4	19.8	19.5
54	54.82	11.22	54.25	0.8609	1.788	15.7	19.1	19.5	19.2
52	52.82	10.81	53.19	0.8604	1.723	15.2	18.8	19.2	18.8
50	50.83	10.39	52.13	0.8598	1.658	14.6	18.4	18.9	18.5
48	48.83	9.978	51.07	0.8592	1.593	14.0	18.1	18.6	18.2
46	46.82	9.562	50.01	0.8587	1.527	13.3	17.8	18.3	17.9
44	44.82	9.146	48.95	0.8581	1.462	12.6	17.4	18.0	17.5
42	42.81	8.730	47.90	0.8576	1.396	11.9	17.1	17.7	17.2
40	40.80	8.315	46.84	0.8570	1.331	11.2	16.7	17.4	16.8
38	38.78	7.899	45.78	0.8564	1.265	10.5	16.4	17.1	16.5
36	36.77	7.483	44.72	0.8559	1.199	9.7	16.0	16.8	16.1
34	34.75	7.068	43.66	0.8553	1.133	8.8	15.6	16.4	15.8
32	32.72	6.652	42.60	0.8548	1.067	7.9	15.3	16.1	15.4
30	30.70	6.236	41.54	0.8542	1.001	7.0	14.9	15.8	15.0
28	28.67	5.820	40.48	0.8536	0.9353	6.0	14.5	15.4	14.7
24	24.61	4.989	38.36	0.8525	0.8027	3.8	13.7	14.7	13.9
20	20.53	4.157	36.25	0.8514	0.6698	1.3	12.9	14.0	13.1
16	16.45	3.326	34.13	0.8503	0.5366	−1.6	12.1	13.3	12.3
12	12.35	2.494	32.01	0.8491	0.4030	−5.0	11.2	12.5	11.4
8	8.25	1.663	29.89	0.8480	0.2690	−9.6	10.3	11.8	10.6
4	4.13	0.831	27.77	0.8469	0.1347	−17.2	9.4	11.0	9.7
0	0.00	0.000	25.65	0.8458	0.0000	—	8.5	10.2	8.8

26°C DRY BULB

Percentage saturation μ/%	Relative humidity ϕ/%	Value of stated parameter per kg dry air			Vapour pressure p_s/kPa	Dew point temperature t_d/°C	Adiabatic saturation temperature $t^\star$/°C	Wet bulb temperature	
		Moisture content g/g.kg^{-1}	Specific enthalpy h/kJ.kg^{-1}	Specific volume v/m^3.kg^{-1}				Screen t'_{sc}/°C	Sling t'_{sl}/°C
100	100.00	21.43	80.78	0.8761	3.360	26.0	26.0	26.0	26.0
96	96.13	20.58	78.60	0.8750	3.230	25.3	25.5	25.5	25.5
92	92.25	19.72	76.41	0.8738	3.100	24.6	25.0	25.1	25.0
88	88.35	18.86	74.23	0.8727	2.969	23.9	24.5	24.6	24.5
84	84.45	18.00	72.04	0.8715	2.838	23.2	23.9	24.1	24.0
80	80.54	17.15	69.86	0.8704	2.706	22.4	23.4	23.6	23.4
76	76.62	16.29	67.67	0.8692	2.574	21.6	22.8	23.0	22.9
72	72.68	15.43	65.49	0.8680	2.442	20.7	22.3	22.5	22.3
70	70.71	15.00	64.39	0.8675	2.376	20.3	22.0	22.2	22.0
68	68.74	14.58	63.30	0.8669	2.310	19.8	21.7	22.0	21.7
66	66.76	14.15	62.21	0.8663	2.243	19.3	21.4	21.7	21.4
64	64.78	13.72	61.12	0.8657	2.177	18.9	21.1	21.4	21.1
62	62.80	13.29	60.02	0.8652	2.110	18.4	20.8	21.1	20.8
60	60.81	12.86	58.93	0.8646	2.043	17.9	20.5	20.8	20.5
58	58.83	12.43	57.84	0.8640	1.977	17.3	20.1	20.6	20.2
56	56.84	12.00	56.75	0.8634	1.910	16.8	19.8	20.3	19.9
54	54.85	11.57	55.65	0.8628	1.843	16.2	19.5	20.0	19.6
52	52.85	11.15	54.56	0.8623	1.776	15.6	19.2	19.7	19.3
50	50.85	10.72	53.47	0.8617	1.709	15.0	18.8	19.4	18.9
48	48.85	10.29	52.38	0.8611	1.641	14.4	18.5	19.1	18.6
46	46.85	9.860	51.28	0.8605	1.574	13.8	18.2	18.7	18.3
44	44.84	9.431	50.19	0.8599	1.507	13.1	17.8	18.4	17.9
42	42.83	9.002	49.10	0.8594	1.439	12.4	17.5	18.1	17.6
40	40.82	8.574	48.01	0.8588	1.372	11.7	17.1	17.8	17.2
38	38.81	8.145	46.91	0.8582	1.304	10.9	16.8	17.5	16.9
36	36.79	7.716	45.82	0.8576	1.236	10.1	16.4	17.1	16.5
34	34.77	7.288	44.73	0.8570	1.168	9.3	16.0	16.8	16.1
32	32.75	6.859	43.64	0.8565	1.100	8.4	15.6	16.5	15.8
30	30.72	6.430	42.54	0.8559	1.032	7.4	15.3	16.1	15.4
28	28.69	6.002	41.45	0.8553	0.9641	6.4	14.9	15.8	15.0
24	24.63	5.144	39.27	0.8541	0.8275	4.3	14.1	15.1	14.2
20	20.55	4.287	37.08	0.8530	0.6905	1.7	13.2	14.3	13.4
16	16.46	3.429	34.90	0.8518	0.5532	−1.2	12.4	13.6	12.6
12	12.36	2.572	32.71	0.8507	0.4155	−4.6	11.5	12.8	11.7
8	8.25	1.715	30.53	0.8495	0.2774	−9.3	10.6	12.1	10.9
4	4.13	0.857	28.34	0.8483	0.1389	−16.9	9.7	11.2	9.9
0	0.00	0.000	26.16	0.8472	0.0000	—	8.7	10.4	9.0

26.5°C DRY BULB

Percentage saturation μ/%	Relative humidity ϕ/%	Value of stated parameter per kg dry air			Vapour pressure p_s/kPa	Dew point temperature t_d/°C	Adiabatic saturation temperature $t^\star$/°C	Wet bulb temperature	
		Moisture content g/g.kg^{-1}	Specific enthalpy h/kJ.kg^{-1}	Specific volume v/m^3.kg^{-1}				Screen t'_{sc}/°C	Sling t'_{sl}/°C
100	100.00	22.10	83.00	0.8785	3.461	26.5	26.5	26.5	26.5
96	96.13	21.22	80.75	0.8773	3.327	25.8	26.0	26.0	26.0
92	92.26	20.33	78.49	0.8761	3.193	25.1	25.5	25.5	25.5
88	88.37	19.45	76.24	0.8749	3.058	24.4	25.0	25.0	25.0
84	84.47	18.56	73.99	0.8737	2.923	23.7	24.4	24.5	24.4
80	80.56	17.68	71.73	0.8725	2.788	22.9	23.9	24.0	23.9
76	76.63	16.80	69.48	0.8713	2.652	22.1	23.3	23.5	23.3
72	72.70	15.91	67.23	0.8702	2.516	21.2	22.7	23.0	22.8
70	70.73	15.47	66.10	0.8696	2.448	20.8	22.4	22.7	22.5
68	68.76	15.03	64.97	0.8690	2.380	20.3	22.1	22.4	22.2
66	66.78	14.59	63.84	0.8684	2.311	19.8	21.8	22.1	21.9
64	64.81	14.14	62.72	0.8678	2.243	19.3	21.5	21.9	21.6
62	62.82	13.70	61.59	0.8672	2.174	18.8	21.2	21.6	21.3
60	60.84	13.26	60.46	0.8666	2.106	18.3	20.9	21.3	21.0
58	58.85	12.82	59.34	0.8660	2.037	17.8	20.6	21.0	20.6
56	56.86	12.38	58.21	0.8654	1.968	17.3	20.3	20.7	20.3
54	54.87	11.93	57.08	0.8648	1.899	16.7	19.9	20.4	20.0
52	52.88	11.49	55.96	0.8642	1.830	16.1	19.6	20.1	19.7
50	50.88	11.05	54.83	0.8636	1.761	15.5	19.3	19.8	19.3
48	48.88	10.61	53.70	0.8630	1.692	14.9	18.9	19.5	19.0
46	46.87	10.18	52.58	0.8624	1.622	14.2	18.6	19.2	18.7
44	44.87	9.723	51.45	0.8618	1.553	13.6	18.2	18.8	18.3
42	42.86	9.281	50.32	0.8612	1.483	12.9	17.9	18.5	18.0
40	40.85	8.840	49.20	0.8606	1.414	12.1	17.5	18.2	17.6
38	38.83	8.400	48.07	0.8600	1.344	11.4	17.1	17.9	17.3
36	36.81	7.956	46.94	0.8594	1.274	10.6	16.8	17.5	16.9
34	34.79	7.514	45.82	0.8588	1.204	9.7	16.4	17.2	16.5
32	32.77	7.072	44.69	0.8582	1.134	8.8	16.0	16.8	16.1
30	30.74	6.630	43.56	0.8576	1.064	7.9	15.6	16.5	15.8
28	28.71	6.188	42.44	0.8570	0.9937	6.9	15.2	16.1	15.4
24	24.65	5.304	40.18	0.8558	0.8530	4.7	14.4	15.4	14.6
20	20.57	4.420	37.93	0.8546	0.7118	2.1	13.6	14.7	13.8
16	16.48	3.536	35.67	0.8534	0.5703	−0.8	12.7	13.9	12.9
12	12.38	2.652	33.42	0.8522	0.4283	−4.2	11.8	13.1	12.0
8	8.26	1.768	31.17	0.8510	0.2860	−8.9	10.9	12.3	11.1
4	4.14	0.884	28.91	0.8498	0.1432	−16.6	9.9	11.5	10.2
0	0.00	0.000	26.66	0.8486	0.0000	—	8.9	10.7	9.2

27°C DRY BULB

Percentage saturation μ/%	Relative humidity ϕ/%	Value of stated parameter per kg dry air			Vapour pressure p_s/kPa	Dew point temperature t_d/°C	Adiabatic saturation temperature $t\star$/°C	Wet bulb temperature	
		Moisture content g/g.kg^{-1}	Specific enthalpy h/kJ.kg^{-1}	Specific volume v/m^3.kg^{-1}				Screen t'_{sc}/°C	Sling t'_{sl}/°C
100	100.00	22.78	85.29	0.8809	3.564	27.0	27.0	27.0	27.0
96	96.14	21.87	82.94	0.8797	3.426	26.3	26.5	26.5	26.5
92	92.26	20.96	80.62	0.8784	3.288	25.6	26.0	26.0	26.0
88	88.38	20.05	78.30	0.8772	3.150	24.9	25.4	25.5	25.5
84	84.48	19.14	75.97	0.8760	3.011	24.2	24.9	25.0	24.9
80	80.57	18.23	73.65	0.8747	2.872	23.4	24.3	24.5	24.4
76	76.66	17.32	71.32	0.8735	2.732	22.5	23.8	24.0	23.8
72	72.72	16.40	69.00	0.8723	2.592	21.7	23.2	23.4	23.2
70	70.76	15.95	67.84	0.8717	2.522	21.2	22.9	23.1	22.9
68	68.78	15.49	66.67	0.8710	2.452	20.8	22.6	22.9	22.6
66	66.81	15.04	65.51	0.8704	2.381	20.3	22.3	22.6	22.3
64	64.83	14.58	64.35	0.8698	2.311	19.8	22.0	22.3	22.0
62	62.85	14.13	63.19	0.8692	2.240	19.3	21.7	22.0	21.7
60	60.87	13.67	62.03	0.8686	2.169	18.8	21.3	21.7	21.4
58	58.88	13.21	60.86	0.8680	2.099	18.3	21.0	21.4	21.1
56	56.89	12.76	59.70	0.8673	2.028	17.7	20.7	21.1	20.8
54	54.90	12.30	58.54	0.8667	1.957	17.2	20.4	20.8	20.4
52	52.90	11.85	57.38	0.8661	1.886	16.6	20.0	20.5	20.1
50	50.91	11.39	56.22	0.8655	1.814	16.0	19.7	20.2	19.8
48	48.90	10.94	55.05	0.8649	1.743	15.3	19.3	19.9	19.4
46	46.90	10.48	53.89	0.8642	1.672	14.7	19.0	19.6	19.1
44	44.89	10.02	52.73	0.8636	1.600	14.0	18.6	19.2	18.7
42	42.88	9.569	51.57	0.8630	1.528	13.3	18.3	18.9	18.4
40	40.87	9.113	50.41	0.8624	1.459	12.6	17.9	18.6	18.0
38	38.86	8.658	49.24	0.8618	1.385	11.8	17.5	18.2	17.6
36	36.84	8.202	48.08	0.8612	1.313	11.0	17.2	17.9	17.3
34	34.82	7.746	46.92	0.8605	1.241	10.2	16.8	17.6	16.9
32	32.79	7.291	45.76	0.8599	1.169	9.3	16.4	17.2	16.5
30	30.77	6.835	44.60	0.8593	1.097	8.3	16.0	16.9	16.1
28	28.74	6.379	43.43	0.8587	1.024	7.3	15.6	16.5	15.7
24	24.67	5.468	41.11	0.8574	0.8792	5.1	14.8	15.8	14.9
20	20.59	4.557	38.78	0.8562	0.7337	2.6	13.9	15.0	14.1
16	16.49	3.645	36.46	0.8550	0.5878	−0.5	13.0	14.2	13.2
12	12.39	2.734	34.14	0.8537	0.4415	−3.9	12.1	13.4	12.3
8	8.27	1.823	31.81	0.8525	0.2948	−8.6	11.2	12.6	11.4
4	4.14	0.911	29.49	0.8513	0.1476	−16.2	10.2	11.8	10.4
0	0.00	0.000	27.16	0.8500	0.0000	—	9.2	10.9	9.5

27.5°C DRY BULB

Percentage saturation μ/%	Relative humidity ϕ/%	Value of stated parameter per kg dry air			Vapour pressure p_s/kPa	Dew point temperature t_d/°C	Adiabatic saturation temperature $t\star$/°C	Wet bulb temperature	
		Moisture content g/g.kg^{-1}	Specific enthalpy h/kJ.kg^{-1}	Specific volume v/m^3.kg^{-1}				Screen t'_{sc}/°C	Sling t'_{sl}/°C
100	100.00	23.49	87.59	0.8833	3.670	27.5	27.5	27.5	27.5
96	96.14	22.55	85.19	0.8821	3.528	26.8	27.0	27.0	27.0
92	92.27	21.61	82.79	0.8808	3.386	26.1	26.5	26.5	26.5
88	88.39	20.67	80.40	0.8795	3.244	25.4	25.9	26.0	25.9
84	84.50	19.73	78.00	0.8782	3.101	24.7	25.4	25.5	25.4
80	80.59	18.79	75.60	0.8770	2.958	23.9	24.8	25.0	24.8
76	76.68	17.85	73.21	0.8757	2.814	23.0	24.2	24.4	24.3
72	72.75	16.91	70.81	0.8744	2.670	22.2	23.7	23.9	23.7
70	70.78	16.44	69.61	0.8738	2.598	21.7	23.3	23.6	23.4
68	68.81	15.97	68.41	0.8731	2.525	21.3	23.0	23.3	23.1
66	66.83	15.50	67.21	0.8725	2.453	20.8	22.7	23.0	22.8
64	64.86	15.03	66.02	0.8719	2.380	20.3	22.4	22.7	22.5
62	62.88	14.56	64.82	0.8712	2.308	19.8	22.1	22.5	22.2
60	60.89	14.09	63.62	0.8706	2.235	19.3	21.8	22.2	21.8
58	58.91	13.62	62.42	0.8700	2.162	18.7	21.5	21.9	21.5
56	56.92	13.15	61.22	0.8693	2.089	18.2	21.1	21.5	21.2
54	54.93	12.68	60.02	0.8687	2.016	17.6	20.8	21.2	20.9
52	52.93	12.21	58.82	0.8680	1.943	17.0	20.4	20.9	20.5
50	50.93	11.74	57.63	0.8674	1.869	16.4	20.1	20.6	20.2
48	48.93	11.27	56.43	0.8668	1.796	15.8	19.7	20.3	19.8
46	46.93	10.80	55.23	0.8661	1.722	15.2	19.4	20.0	19.5
44	44.92	10.33	54.03	0.8655	1.649	14.5	19.0	19.6	19.1
42	42.91	9.864	52.83	0.8649	1.575	13.8	18.7	19.3	18.8
40	40.90	9.395	51.63	0.8642	1.501	13.0	18.3	19.0	18.4
38	38.88	8.925	50.44	0.8636	1.427	12.3	17.9	18.6	18.0
36	36.86	8.455	49.24	0.8629	1.353	11.5	17.5	18.3	17.6
34	34.84	7.986	48.04	0.8623	1.279	10.6	17.1	17.9	17.3
32	32.82	7.516	46.84	0.8617	1.204	9.7	16.7	17.6	16.9
30	30.79	7.046	45.64	0.8610	1.130	8.8	16.3	17.2	16.5
28	28.76	6.576	44.44	0.8604	1.055	7.8	15.9	16.8	16.1
24	24.69	5.637	42.05	0.8591	0.9060	5.6	15.1	16.1	15.3
20	20.60	4.697	39.65	0.8578	0.7562	3.0	14.2	15.3	14.4
16	16.51	3.758	37.25	0.8566	0.6059	−0.1	13.3	14.5	13.5
12	12.40	2.818	34.86	0.8553	0.4551	−3.5	12.4°C	13.7	12.6
8	8.28	1.879	32.46	0.8540	0.3039	−8.2	11.4	12.9	11.7
4	4.15	0.939	30.06	0.8527	0.1522	−15.9	10.4	12.0	10.7
0	0.00	0.000	27.67	0.8514	0.0000	—	9.4	11.1	9.7

28°C DRY BULB

Percentage saturation μ/%	Relative humidity ϕ/%	Value of stated parameter per kg dry air			Vapour pressure p_s/kPa	Dew point temperature t_d/°C	Adiabatic saturation temperature $t^\star$/°C	Wet bulb temperature	
		Moisture content g/g.kg^{-1}	Specific enthalpy h/kJ.kg^{-1}	Specific volume v/m^3.kg^{-1}				Screen t'_{sc}/°C	Sling t'_{sl}/°C
100	100.00	24.21	89.96	0.8858	3.779	28.0	28.0	28.0	28.0
96	96.14	23.24	87.49	0.8845	3.633	27.3	27.5	27.5	27.5
92	92.28	22.27	85.01	0.8832	3.487	26.6	27.0	27.0	27.0
88	88.40	21.30	82.54	0.8818	3.340	25.9	26.4	26.5	26.4
84	84.51	20.34	80.07	0.8805	3.194	25.1	25.9	26.0	25.9
80	80.61	19.37	77.60	0.8792	3.046	24.4	25.3	25.5	25.3
76	76.70	18.40	75.13	0.8779	2.898	23.5	24.7	24.9	24.7
72	72.77	17.43	72.66	0.8766	2.750	22.7	24.1	24.3	24.2
70	70.80	16.95	71.42	0.8759	2.675	22.2	23.8	24.1	23.8
68	68.83	16.46	70.18	0.8753	2.601	21.7	23.5	23.8	23.5
66	66.86	15.98	68.95	0.8746	2.526	21.3	23.2	23.5	23.2
64	64.88	15.49	67.71	0.8740	2.452	20.8	22.9	23.2	22.9
62	62.90	15.01	66.48	0.8733	2.377	20.3	22.5	22.9	22.6
60	60.92	14.53	65.24	0.8726	2.302	19.8	22.2	22.6	22.3
58	58.94	14.04	64.01	0.8720	2.227	19.2	21.9	22.3	22.0
56	56.95	13.56	62.77	0.8713	2.152	18.7	21.6	22.0	21.6
54	54.95	13.07	61.53	0.8707	2.077	18.1	21.2	21.7	21.3
52	52.96	12.59	60.30	0.8700	2.001	17.5	20.9	21.3	20.9
50	50.96	12.10	59.06	0.8693	1.926	16.9	20.5	21.0	20.6
48	48.96	11.62	57.83	0.8687	1.850	16.3	20.2	20.7	20.2
46	46.96	11.14	56.59	0.8680	1.774	15.6	19.8	20.4	19.9
44	44.95	10.65	55.36	0.8674	1.699	14.9	19.4	20.0	19.5
42	42.94	10.17	54.12	0.8667	1.624	14.2	19.1	19.7	19.2
40	40.93	9.684	52.88	0.8661	1.547	13.5	18.7	19.4	18.8
38	38.91	9.200	51.65	0.8654	1.470	12.7	18.3	19.0	18.4
36	36.89	8.716	50.41	0.8647	1.394	11.9	17.9	18.7	18.0
34	34.87	8.231	49.18	0.8641	1.318	11.1	17.5	18.3	17.6
32	32.84	7.747	47.94	0.8634	1.241	10.2	17.1	17.9	17.2
30	30.81	7.263	46.70	0.8628	1.164	9.2	16.7	17.6	16.8
28	28.78	6.779	45.47	0.8621	1.088	8.2	16.3	17.2	16.4
24	24.71	5.810	43.00	0.8608	0.9337	6.0	15.4	16.4	15.6
20	20.62	4.842	40.53	0.8595	0.7793	3.4	14.6	15.7	14.7
16	16.52	3.874	38.06	0.8581	0.6244	0.3	13.6	14.9	13.8
12	12.41	2.905	35.58	0.8568	0.4690	−3.2	12.7	14.0	12.9
8	8.29	1.937	33.11	0.8555	0.3132	−7.9	11.7	13.2	11.9
4	4.15	0.968	30.64	0.8542	0.1568	−15.6	10.7	12.3	10.9
0	0.00	0.000	28.17	0.8529	0.0000	—	9.6	11.4	9.9

28.5°C DRY BULB

Percentage saturation μ/%	Relative humidity ϕ/%	Value of stated parameter per kg dry air			Vapour pressure p_s/kPa	Dew point temperature t_d/°C	Adiabatic saturation temperature $t^\star$/°C	Wet bulb temperature	
		Moisture content g/g.kg^{-1}	Specific enthalpy h/kJ.kg^{-1}	Specific volume v/m^3.kg^{-1}				Screen t'_{sc}/°C	Sling t'_{sl}/°C
100	100.00	24.95	92.38	0.8883	3.890	28.5	28.5	28.5	28.5
96	96.15	23.96	89.83	0.8869	3.740	27.8	28.0	28.0	28.0
92	92.29	22.96	87.28	0.8856	3.590	27.1	27.4	27.5	27.5
88	88.41	21.96	84.74	0.8842	3.439	26.4	26.9	27.0	26.9
84	84.53	20.96	82.19	0.8828	3.288	25.6	26.3	26.5	26.4
80	80.63	19.96	79.64	0.8815	3.137	24.8	25.8	25.9	25.8
76	76.72	18.96	77.09	0.8801	2.985	24.0	25.2	25.4	25.2
72	72.79	17.97	74.54	0.8788	2.832	23.1	24.6	24.8	24.6
70	70.83	17.47	73.27	0.8781	2.755	22.7	24.3	24.5	24.3
68	68.86	16.97	71.99	0.8774	2.679	22.2	24.0	24.2	24.0
66	66.89	16.47	70.72	0.8767	2.602	21.7	23.6	23.9	23.7
64	64.91	15.97	69.45	0.8761	2.525	21.3	23.3	23.6	23.4
62	62.93	15.47	68.17	0.8754	2.448	20.8	23.0	23.3	23.0
60	60.95	14.97	66.90	0.8747	2.371	20.2	22.7	23.0	22.7
58	58.96	14.47	65.62	0.8740	2.294	19.7	22.3	22.7	22.4
56	56.98	13.97	64.35	0.8733	2.216	19.1	22.0	22.4	22.1
54	54.98	13.47	63.07	0.8727	2.139	18.6	21.6	22.1	21.7
52	52.99	12.98	61.80	0.8720	2.061	18.0	21.3	21.8	21.4
50	50.99	12.48	60.53	0.8713	1.984	17.4	20.9	21.4	21.0
48	48.99	11.98	59.25	0.8706	1.906	16.7	20.6	21.1	20.7
46	46.99	11.48	57.98	0.8699	1.828	16.1	20.2	20.8	20.3
44	44.98	10.98	56.70	0.8693	1.750	15.4	19.8	20.4	19.9
42	42.97	10.48	55.43	0.8686	1.672	14.7	19.5	20.1	19.6
40	40.96	9.981	54.16	0.8679	1.593	14.0	19.1	19.8	19.2
38	38.94	9.482	52.88	0.8672	1.515	13.2	18.7	19.4	18.8
36	36.92	8.983	51.61	0.8665	1.436	12.4	18.3	19.0	18.4
34	34.90	8.484	50.33	0.8659	1.358	11.5	17.9	18.7	18.0
32	32.87	7.985	49.06	0.8652	1.279	10.6	17.5	18.3	17.6
30	30.84	7.486	47.78	0.8645	1.200	9.7	17.1	17.9	17.2
28	28.81	6.987	46.51	0.8638	1.121	8.7	16.6	17.6	16.8
24	24.73	5.989	43.96	0.8625	0.9621	6.4	15.8	16.8	15.9
20	20.64	4.991	41.41	0.8611	0.8030	3.8	14.9	16.0	15.1
16	16.54	3.992	38.87	0.8597	0.6434	0.7	13.9	15.2	14.1
12	12.42	2.994	36.32	0.8584	0.4834	−2.8	13.0	14.3	13.2
8	8.30	1.996	33.77	0.8570	0.3228	−7.5	12.0	13.4	12.2
4	4.16	0.998	31.22	0.8556	0.1616	−15.3	10.9	12.5	11.2
0	0.00	0.000	28.67	0.8543	0.0000	—	9.8	11.6	10.1

29°C DRY BULB

Percentage saturation μ/%	Relative humidity ϕ/%	Value of stated parameter per kg dry air			Vapour pressure p_s/kPa	Dew point temperature t_d/°C	Adiabatic saturation temperature $t^\star$/°C	Wet bulb temperature	
		Moisture content $g/g.kg^{-1}$	Specific enthalpy h/kJ.kg^{-1}	Specific volume v/m^3.kg^{-1}				Screen t'_{sc}/°C	Sling t'_{sl}/°C
100	100.00	25.72	94.86	0.8908	4.005	29.0	29.0	29.0	29.0
96	96.15	24.69	92.23	0.8894	3.850	28.3	28.5	28.5	28.5
92	92.30	23.66	89.60	0.8880	3.696	27.6	27.9	28.0	27.9
88	88.43	22.63	86.98	0.8866	3.541	26.9	27.4	27.5	27.4
84	84.54	21.60	84.35	0.8852	3.386	26.1	26.8	26.9	26.8
80	80.65	20.57	81.72	0.8838	3.230	25.3	26.2	26.4	26.3
76	76.74	19.55	79.09	0.8824	3.073	24.5	25.6	25.8	25.7
72	72.82	18.52	76.47	0.8810	2.916	23.6	25.0	25.3	25.1
70	70.85	18.00	75.15	0.8803	2.837	23.2	24.7	25.0	24.8
68	68.88	17.49	73.84	0.8796	2.759	22.7	24.4	24.7	24.5
66	66.91	16.97	72.53	0.8789	2.680	22.2	24.1	24.4	24.1
64	64.94	16.46	71.21	0.8782	2.600	21.7	23.8	24.1	23.8
62	62.96	15.94	69.90	0.8775	2.521	21.2	23.4	23.8	23.5
60	60.98	15.43	68.58	0.8768	2.442	20.7	23.1	23.5	23.2
58	58.99	14.92	67.27	0.8761	2.362	20.2	22.8	23.2	22.8
56	57.01	14.40	65.96	0.8754	2.283	19.6	22.4	22.8	22.5
54	55.01	13.89	64.64	0.8747	2.203	19.1	22.1	22.5	22.1
52	53.02	13.37	63.33	0.8740	2.123	18.5	21.7	22.2	21.8
50	51.02	12.86	62.02	0.8733	2.043	17.8	21.4	21.9	21.4
48	49.02	12.34	60.70	0.8726	1.963	17.2	21.0	21.5	21.1
46	47.02	11.83	59.39	0.8719	1.883	16.6	20.6	21.2	20.7
44	45.01	11.32	58.08	0.8712	1.802	15.9	20.2	20.8	20.3
42	43.00	10.80	56.76	0.8705	1.722	15.2	19.9	20.5	20.0
40	40.98	10.29	55.45	0.8698	1.641	14.4	19.5	20.1	19.6
38	38.97	9.773	54.13	0.8691	1.560	13.6	19.1	19.8	19.2
36	36.95	9.258	52.82	0.8684	1.480	12.8	18.7	19.4	18.8
34	34.92	8.744	51.51	0.8677	1.398	12.0	18.3	19.1	18.4
32	32.90	8.230	50.19	0.8670	1.317	11.1	17.9	18.7	18.0
30	30.86	7.715	48.88	0.8662	1.236	10.1	17.4	18.3	17.6
28	28.83	7.201	47.57	0.8655	1.155	9.1	17.0	17.9	17.1
24	24.75	6.172	44.94	0.8641	0.9912	6.9	16.1	17.1	16.3
20	20.66	5.144	42.31	0.8627	0.8274	4.3	15.2	16.3	15.4
16	16.56	4.115	39.68	0.8613	0.6630	1.1	14.3	15.5	14.5
12	12.44	3.086	37.06	0.8599	0.4981	−2.4	13.3	14.6	13.5
8	8.31	2.057	34.43	0.8585	0.3326	−7.2	12.2	13.7	12.5
4	4.16	1.029	31.80	0.8571	0.1666	−14.9	11.2	12.8	11.4
0	0.00	0.000	29.18	0.8557	0.0000	—	10.1	11.9	10.4

29.5°C DRY BULB

Percentage saturation μ/%	Relative humidity ϕ/%	Value of stated parameter per kg dry air			Vapour pressure p_s/kPa	Dew point temperature t_d/°C	Adiabatic saturation temperature $t^\star$/°C	Wet bulb temperature	
		Moisture content $g/g.kg^{-1}$	Specific enthalpy h/kJ.kg^{-1}	Specific volume v/m^3.kg^{-1}				Screen t'_{sc}/°C	Sling t'_{sl}/°C
100	100.00	26.50	97.39	0.8933	4.122	29.5	29.5	29.5	29.5
96	96.16	25.44	94.68	0.8919	3.963	28.8	29.0	29.0	29.0
92	92.30	24.38	91.97	0.8905	3.805	28.1	28.4	28.5	28.4
88	88.44	23.32	89.27	0.8890	3.645	27.4	27.9	28.0	27.9
84	84.56	22.26	86.56	0.8876	3.485	26.6	27.3	27.4	27.3
80	80.67	21.20	83.85	0.8861	3.325	25.8	26.7	26.9	26.7
76	76.76	20.14	81.14	0.8847	3.164	25.0	26.1	26.3	26.1
72	72.84	19.08	78.43	0.8832	3.002	24.1	25.5	25.7	25.5
70	70.88	18.55	77.08	0.8825	2.922	23.7	25.2	25.4	25.2
68	68.91	18.02	75.72	0.8818	2.840	23.2	24.9	25.1	24.9
66	66.94	17.49	74.37	0.8810	2.759	22.7	24.5	24.8	24.6
64	64.97	16.96	73.01	0.8803	2.678	22.2	24.2	24.5	24.3
62	62.99	16.43	71.66	0.8796	2.596	21.7	23.9	24.2	23.9
60	61.01	15.90	70.31	0.8789	2.515	21.2	23.5	23.9	23.6
58	59.02	15.37	68.95	0.8782	2.433	20.7	23.2	23.6	23.3
56	57.04	14.84	67.60	0.8774	2.351	20.1	22.9	23.3	22.9
54	55.04	14.31	66.24	0.8767	2.269	19.5	22.5	22.9	22.6
52	53.05	13.78	64.89	0.8760	2.187	18.9	22.1	22.6	22.2
50	51.05	13.25	63.54	0.8753	2.104	18.3	21.8	22.3	21.9
48	49.05	12.72	62.18	0.8745	2.022	17.7	21.4	21.9	21.5
46	47.05	12.19	60.83	0.8738	1.939	17.0	21.0	21.6	21.1
44	45.04	11.66	59.47	0.8731	1.856	16.3	20.7	21.3	20.7
42	43.03	11.13	58.12	0.8724	1.774	15.6	20.3	20.9	20.4
40	41.01	10.60	56.76	0.8716	1.691	14.9	19.9	20.5	20.0
38	39.00	10.07	55.41	0.8709	1.607	14.1	19.5	20.2	19.6
36	36.98	9.541	54.06	0.8702	1.524	13.3	19.1	19.8	19.2
34	34.95	9.011	52.70	0.8695	1.441	12.4	18.6	19.4	18.8
32	32.92	8.481	51.35	0.8687	1.357	11.5	18.2	19.1	18.4
30	30.89	7.951	49.99	0.8680	1.273	10.6	17.8	18.7	17.9
28	28.86	7.421	48.64	0.8673	1.189	9.5	17.4	18.3	17.5
24	24.78	6.361	45.93	0.8658	1.021	7.3	16.5	17.5	16.6
20	20.68	5.301	43.22	0.8644	0.8524	4.7	15.5	16.6	15.7
16	16.57	4.240	40.51	0.8629	0.6831	1.6	14.6	15.8	14.8
12	12.45	3.180	37.80	0.8615	0.5132	−2.1	13.6	14.9	13.8
8	8.32	2.120	35.10	0.8600	0.3427	−6.8	12.5	14.0	12.8
4	4.16	1.060	32.39	0.8586	0.1717	−14.6	11.4	13.1	11.7
0	0.00	0.000	29.68	0.8571	0.0000	—	10.3	12.1	10.6

30°C DRY BULB

| Percentage saturation μ/% | Relative humidity ϕ/% | Value of stated parameter per kg dry air | | | Vapour pressure p_s/kPa | Dew point temperature t_d/°C | Adiabatic saturation temperature $t^\star$/°C | Wet bulb temperature | |
		Moisture content g/g.kg⁻¹	Specific enthalpy h/kJ.kg⁻¹	Specific volume v/m³.kg⁻¹				Screen t'_{sc}/°C	Sling t'_{sl}/°C
100	100.00	27.31	99.98	0.8959	4.242	30.0	30.0	30.0	30.0
96	96.16	26.22	97.19	0.8944	4.079	29.3	29.5	29.5	29.5
92	92.31	25.13	94.40	0.8929	3.916	28.6	28.9	29.0	28.9
88	88.45	24.03	91.61	0.8915	3.752	27.9	28.4	28.4	28.4
84	84.57	22.94	88.81	0.8900	3.588	27.1	27.8	27.9	27.8
80	80.69	21.85	86.02	0.8885	3.423	26.3	27.2	27.3	27.2
76	76.78	20.76	83.23	0.8870	3.257	25.5	26.6	26.8	26.6
72	72.87	19.66	80.44	0.8855	3.091	24.6	26.0	26.2	26.0
70	70.91	19.12	79.04	0.8847	3.008	24.1	25.6	25.9	25.7
68	68.94	18.57	77.65	0.8840	2.924	23.7	25.3	25.6	25.4
66	66.97	18.02	76.25	0.8832	2.841	23.2	25.0	25.3	25.0
64	65.00	17.48	74.85	0.8825	2.757	22.7	24.7	25.0	24.7
62	63.02	16.93	73.46	0.8817	2.673	22.2	24.3	24.7	24.4
60	61.04	16.39	72.06	0.8810	2.589	21.7	24.0	24.4	24.0
58	59.05	15.84	70.67	0.8802	2.505	21.1	23.6	24.0	23.7
56	57.07	15.29	69.27	0.8795	2.421	20.6	23.3	23.7	23.4
54	55.08	14.75	67.87	0.8788	2.336	20.0	22.9	23.4	23.0
52	53.08	14.20	66.48	0.8780	2.252	19.4	22.6	23.0	22.6
50	51.08	13.66	65.08	0.8773	2.167	18.8	22.2	22.7	22.3
48	49.08	13.11	63.69	0.8765	2.092	18.1	21.8	22.4	21.9
46	47.08	12.56	62.29	0.8758	1.997	17.5	21.4	22.0	21.5
44	45.07	12.02	60.89	0.8750	1.912	16.8	21.1	21.7	21.2
42	43.06	11.47	59.50	0.8743	1.827	16.1	20.7	21.3	20.8
40	41.05	10.92	58.10	0.8735	1.741	15.3	20.3	20.9	20.4
38	39.03	10.38	56.71	0.8728	1.656	14.6	19.9	20.6	20.0
36	37.00	9.832	55.31	0.8720	1.570	13.7	19.4	20.2	19.6
34	34.98	9.286	53.91	0.8713	1.484	12.9	19.0	19.8	19.1
32	32.95	8.739	52.52	0.8705	1.398	12.0	18.6	19.4	18.7
30	30.92	8.193	51.12	0.8698	1.312	11.0	18.2	19.0	18.3
28	28.88	7.647	49.73	0.8690	1.225	10.0	17.7	18.6	17.9
24	24.80	6.554	46.93	0.8675	1.052	7.7	16.8	17.8	17.0
20	20.70	5.462	44.14	0.8660	0.8782	5.1	15.9	17.0	16.0
16	16.59	4.370	41.35	0.8645	0.7038	2.0	14.9	16.1	15.1
12	12.47	3.277	38.56	0.8630	0.5288	−1.7	13.8	15.2	14.1
8	8.32	2.185	35.77	0.8615	0.3531	−6.5	12.8	14.3	13.0
4	4.17	1.092	32.97	0.8600	0.1769	−14.3	11.7	13.3	11.9
0	0.00	0.000	30.18	0.8585	0.0000	—	10.5	12.3	10.8

30.5°C DRY BULB

| Percentage saturation μ/% | Relative humidity ϕ/% | Value of stated parameter per kg dry air | | | Vapour pressure p_s/kPa | Dew point temperature t_d/°C | Adiabatic saturation temperature $t^\star$/°C | Wet bulb temperature | |
		Moisture content g/g.kg⁻¹	Specific enthalpy h/kJ.kg⁻¹	Specific volume v/m³.kg⁻¹				Screen t'_{sc}/°C	Sling t'_{sl}/°C
100	100.00	28.14	102.63	0.8985	4.365	30.5	30.5	30.5	30.5
96	96.17	27.02	99.75	0.8970	4.198	29.8	30.0	30.0	30.0
92	92.31	25.89	96.88	0.8955	4.030	29.1	29.4	29.5	29.4
88	88.46	24.76	94.00	0.8939	3.862	28.4	28.8	28.9	28.9
84	84.59	23.64	91.12	0.8924	3.693	27.6	28.3	28.4	28.3
80	80.71	22.51	88.24	0.8908	3.523	26.8	27.7	27.8	27.7
76	76.81	21.39	85.36	0.8893	3.353	26.0	27.1	27.2	27.1
72	72.89	20.26	82.49	0.8878	3.182	25.1	26.4	26.7	26.5
70	70.93	19.70	81.05	0.8870	3.096	24.6	26.1	26.4	26.1
68	68.97	19.14	79.61	0.8862	3.011	24.2	25.8	26.1	25.8
66	67.00	18.57	78.17	0.8854	2.925	23.7	25.5	25.7	25.5
64	65.03	18.01	76.73	0.8847	2.839	23.2	25.1	25.4	25.2
62	63.05	17.45	75.29	0.8839	2.752	22.7	24.8	25.1	24.8
60	61.07	16.88	73.85	0.8831	2.666	22.1	24.4	24.8	24.5
58	59.09	16.32	72.41	0.8824	2.579	21.6	24.1	24.5	24.1
56	57.10	15.76	70.98	0.8816	2.493	21.0	23.7	24.1	23.8
54	55.11	15.20	69.54	0.8808	2.406	20.5	23.4	23.8	23.4
52	53.11	14.63	68.10	0.8800	2.319	19.9	23.0	23.5	23.1
50	51.12	14.07	66.66	0.8793	2.231	19.3	22.6	23.1	22.7
48	49.12	13.51	65.22	0.8785	2.144	18.6	22.2	22.8	22.3
46	47.11	12.94	63.78	0.8777	2.057	18.0	21.9	22.4	21.9
44	45.10	12.38	62.34	0.8770	1.969	17.3	21.5	22.1	21.6
42	43.09	11.82	60.90	0.8762	1.881	16.5	21.1	21.7	21.2
40	41.08	11.26	59.46	0.8754	1.793	15.8	20.7	21.3	20.8
38	39.06	10.69	58.02	0.8746	1.705	15.0	20.2	21.0	20.4
36	37.04	10.13	56.59	0.8739	1.617	14.2	19.8	20.6	19.9
34	35.01	9.568	55.15	0.8731	1.528	13.3	19.4	20.2	19.5
32	32.98	9.005	53.71	0.8723	1.435	12.4	19.0	19.8	19.1
30	30.95	8.442	52.27	0.8716	1.351	11.4	18.5	19.4	18.7
28	28.91	7.879	50.83	0.8708	1.262	10.4	18.1	19.0	18.2
24	24.82	6.754	47.95	0.8692	1.084	8.2	17.1	18.2	17.3
20	20.72	5.628	45.07	0.8677	0.9046	5.5	16.2	17.3	16.4
16	16.61	4.502	42.40	0.8661	0.7250	2.4	15.2	16.4	15.4
12	12.48	3.377	39.32	0.8646	0.5448	−1.4	14.1	15.5	14.4
8	8.33	2.251	36.44	0.8631	0.3638	−6.2	13.0	14.5	13.3
4	4.17	1.125	33.56	0.8615	0.1823	−14.0	11.9	13.6	12.2
0	0.00	0.000	30.68	0.8600	0.0000	—	10.7	12.5	11.0

31°C DRY BULB

Percentage saturation μ/%	Relative humidity ϕ/%	Value of stated parameter per kg dry air			Vapour pressure p_s/kPa	Dew point temperature t_d/°C	Adiabatic saturation temperature $t^\star$/°C	Wet bulb temperature	
		Moisture content g/g.kg^{-1}	Specific enthalpy h/kJ.kg^{-1}	Specific volume v/m^3.kg^{-1}				Screen t'_{sc}/°C	Sling t'_{sl}/°C
100	100.00	29.00	105.3	0.9012	4.492	31.0	31.0	31.0	31.0
96	96.17	27.83	102.4	0.8996	4.320	30.3	30.5	30.5	30.5
92	92.33	26.67	99.41	0.8980	4.147	29.6	29.9	30.0	29.9
88	88.48	25.51	96.44	0.8964	3.974	28.9	29.3	29.4	29.3
84	84.61	24.35	93.48	0.8948	3.800	28.1	28.7	28.9	28.8
80	80.73	23.19	90.51	0.8932	3.626	27.3	28.1	28.3	28.2
76	76.83	22.04	87.55	0.8917	3.451	26.5	27.5	27.7	27.6
72	72.92	20.88	84.58	0.8901	3.275	25.6	26.9	27.1	26.9
70	70.96	20.30	83.10	0.8893	3.187	25.1	26.6	26.8	26.6
68	69.00	19.72	81.61	0.8885	3.099	24.6	26.2	26.5	26.3
66	67.03	19.14	80.13	0.8877	3.011	24.2	25.9	26.2	26.0
64	65.06	18.56	78.65	0.8869	2.922	23.7	25.6	25.9	25.6
62	63.08	17.98	77.16	0.8861	2.833	23.1	25.2	25.6	25.3
60	61.10	17.40	75.68	0.8853	2.744	22.6	24.9	25.2	24.9
58	59.12	16.82	74.20	0.8845	2.655	22.1	24.5	24.9	24.6
56	57.13	16.24	72.71	0.8837	2.566	21.5	24.2	24.6	24.2
54	55.14	15.66	71.23	0.8829	2.477	20.9	23.8	24.2	23.9
52	53.15	15.08	69.75	0.8821	2.387	20.3	23.4	23.9	23.5
50	51.15	14.50	68.27	0.8813	2.298	19.7	23.0	23.5	23.1
48	49.15	13.92	66.78	0.8805	2.208	19.1	22.7	23.2	22.7
46	47.15	13.34	65.30	0.8797	2.118	18.4	22.3	22.8	22.4
44	45.14	12.76	63.82	0.8789	2.027	17.7	21.9	22.5	22.0
42	43.12	12.18	62.33	0.8781	1.937	17.0	21.5	22.1	21.6
40	41.11	11.60	60.85	0.8773	1.846	16.3	21.1	21.7	21.2
38	39.09	11.02	59.37	0.8765	1.756	15.5	20.6	21.3	20.7
36	37.07	10.44	57.88	0.8757	1.665	14.6	20.2	21.0	20.3
34	35.04	9.858	56.40	0.8749	1.574	13.8	19.8	20.6	19.9
32	33.01	9.278	54.92	0.8741	1.483	12.9	19.3	20.2	19.5
30	30.97	8.698	53.45	0.8733	1.391	11.9	18.9	19.8	19.0
28	28.94	8.118	51.95	0.8725	1.300	10.9	18.4	19.3	18.6
24	24.85	6.958	48.99	0.8710	1.116	8.6	17.5	18.5	17.6
20	20.75	5.799	46.02	0.8694	0.9318	6.0	16.5	17.6	16.7
16	16.63	4.639	43.05	0.8678	0.7468	2.8	15.5	16.7	15.7
12	12.49	3.479	40.09	0.8662	0.5612	-1.0	14.4	15.8	14.6
8	8.34	2.320	37.12	0.8646	0.3748	-5.8	13.3	14.8	13.6
4	4.18	1.160	34.15	0.8630	0.1878	-13.6	12.1	13.8	12.4
0	0.00	0.000	31.19	0.8614	0.0000	—	10.9	12.8	11.2

31.5°C DRY BULB

Percentage saturation μ/%	Relative humidity ϕ/%	Value of stated parameter per kg dry air			Vapour pressure p_s/kPa	Dew point temperature t_d/°C	Adiabatic saturation temperature $t^\star$/°C	Wet bulb temperature	
		Moisture content g/g.kg^{-1}	Specific enthalpy h/kJ.kg^{-1}	Specific volume v/m^3.kg^{-1}				Screen t'_{sc}/°C	Sling t'_{sl}/°C
100	100.00	29.87	108.1	0.9039	4.621	31.5	31.5	31.5	31.5
96	96.18	28.68	105.0	0.9022	4.444	30.8	31.0	31.0	31.0
92	92.34	27.48	102.0	0.9006	4.267	30.1	30.4	30.4	30.4
88	88.49	26.29	98.94	0.8990	4.089	29.4	29.8	29.9	29.8
84	84.63	25.09	95.89	0.8973	3.911	28.6	29.2	29.3	29.2
80	80.75	23.90	92.83	0.8957	3.731	27.8	28.6	28.8	28.6
76	76.85	22.70	89.77	0.8940	3.552	26.9	28.0	28.2	28.0
72	72.95	21.51	86.72	0.8924	3.371	26.1	27.4	27.6	27.4
70	70.99	20.91	85.19	0.8916	3.280	25.6	27.0	27.3	27.1
68	69.02	20.31	83.66	0.8908	3.190	25.1	26.7	27.0	26.7
66	67.06	19.71	82.13	0.8899	3.099	24.6	26.4	26.6	26.4
64	65.09	19.12	80.60	0.8891	3.008	24.1	26.0	26.3	26.1
62	63.11	18.52	79.07	0.8883	2.917	23.6	25.7	26.0	25.7
60	61.13	17.92	77.55	0.8875	2.825	23.1	25.3	25.7	25.4
58	59.15	17.32	76.02	0.8867	2.734	22.6	25.0	25.3	25.0
56	57.17	16.73	74.49	0.8858	2.642	22.0	24.6	25.0	24.7
54	55.18	16.13	72.96	0.8850	2.550	21.4	24.2	24.7	24.3
52	53.18	15.53	71.43	0.8842	2.458	20.8	23.8	24.3	23.9
50	51.19	14.94	69.90	0.8834	2.365	20.2	23.5	24.0	23.5
48	49.18	14.34	68.38	0.8825	2.273	19.6	23.1	23.6	23.2
46	47.18	13.74	66.85	0.8817	2.180	18.9	22.7	23.2	22.8
44	45.17	13.14	65.32	0.8809	2.087	18.2	22.3	22.9	22.4
42	43.16	12.55	63.79	0.8801	1.994	17.5	21.9	22.5	22.0
40	41.14	11.95	62.26	0.8793	1.901	16.7	21.5	22.1	21.6
38	39.12	11.35	60.73	0.8784	1.808	15.9	21.0	21.7	21.1
36	37.10	10.75	59.20	0.8776	1.714	15.1	20.6	21.3	20.7
34	35.07	10.16	57.68	0.8768	1.621	14.2	20.2	20.9	20.3
32	33.04	9.559	56.15	0.8760	1.527	13.3	19.7	20.5	19.8
30	31.00	8.961	54.62	0.8751	1.433	12.3	19.3	20.1	19.4
28	28.96	8.364	53.09	0.8743	1.338	11.3	18.8	19.7	18.9
24	24.87	7.169	50.03	0.8727	1.150	9.0	17.8	18.8	18.0
20	20.77	5.974	46.98	0.8710	0.9597	6.4	16.8	18.0	17.0
16	16.65	4.780	43.92	0.8694	0.7692	3.2	15.8	17.0	16.0
12	12.51	3.584	40.86	0.8677	0.5780	-0.7	14.7	16.1	14.9
8	8.36	2.390	37.81	0.8661	0.3861	-5.5	13.6	15.1	13.8
4	4.19	1.194	34.75	0.8644	0.1934	-13.3	12.4	14.1	12.7
0	0.00	0.000	31.69	0.8628	0.0000	—	11.1	13.0	11.5

32°C DRY BULB

Percentage saturation μ/%	Relative humidity ϕ/%	Value of stated parameter per kg dry air			Vapour pressure p_s/kPa	Dew point temperature t_d/°C	Adiabatic saturation temperature $t^\star$/°C	Wet bulb temperature	
		Moisture content g/g.kg⁻¹	Specific enthalpy h/kJ.kg⁻¹	Specific volume v/m³.kg⁻¹				Screen t'_{sc}/°C	Sling t'_{sl}/°C
100	100.00	30.77	111.0	0.9066	4.754	32.0	32.0	32.0	32.0
96	96.18	29.54	107.8	0.9049	4.572	31.3	31.4	31.5	31.5
92	92.35	28.31	104.6	0.9032	4.390	30.6	30.9	30.9	30.9
88	88.50	27.08	101.5	0.9015	4.207	29.9	30.3	30.4	30.3
84	84.64	25.85	98.35	0.8998	4.024	29.1	29.7	29.8	29.7
80	80.77	24.62	95.20	0.8981	3.840	28.3	29.1	29.2	29.1
76	76.88	23.39	92.05	0.8965	3.655	27.4	28.5	28.7	28.5
72	72.97	22.16	88.90	0.8948	3.469	26.5	27.8	28.0	27.9
70	71.02	21.54	87.33	0.8939	3.376	26.1	27.5	27.7	27.5
68	69.05	20.92	85.75	0.8931	3.283	25.6	27.2	27.4	27.2
66	67.09	20.31	84.17	0.8922	3.189	25.1	26.8	27.1	26.9
64	65.12	19.69	82.60	0.8914	3.096	24.6	26.5	26.8	26.5
62	63.15	19.08	81.02	0.8905	3.002	24.1	26.1	26.5	26.2
60	61.17	18.46	79.45	0.8897	2.908	23.6	25.8	26.1	25.8
58	59.19	17.85	77.87	0.8888	2.814	23.0	25.4	25.8	25.5
56	57.20	17.23	76.30	0.8880	2.719	22.5	25.0	25.4	25.1
54	55.21	16.62	74.72	0.8871	2.625	21.9	24.7	25.1	24.7
52	53.22	16.00	73.15	0.8863	2.530	21.3	24.3	24.7	24.4
50	51.22	15.39	71.57	0.8854	2.435	20.7	23.9	24.4	24.0
48	49.22	14.77	70.00	0.8846	2.340	20.0	23.5	24.0	23.6
46	47.21	14.15	68.42	0.8837	2.245	19.3	23.1	23.7	23.2
44	45.21	13.54	66.85	0.8829	2.149	18.7	22.7	23.3	22.8
42	43.19	12.92	65.27	0.8820	2.053	17.9	22.3	22.9	22.4
40	41.18	12.31	63.70	0.8812	1.957	17.2	21.9	22.5	22.0
38	39.16	11.69	62.12	0.8804	1.861	16.4	21.4	22.1	21.5
36	37.13	11.08	60.55	0.8795	1.765	15.5	21.0	21.7	21.1
34	35.10	10.46	58.97	0.8787	1.669	14.7	20.5	21.3	20.7
32	33.07	9.847	57.40	0.8778	1.572	13.8	20.1	20.9	20.2
30	31.03	9.232	55.82	0.8770	1.475	12.8	19.6	20.5	19.8
28	28.99	8.616	54.25	0.8761	1.378	11.7	19.1	20.1	19.3
24	24.90	7.385	51.10	0.8744	1.183	9.5	18.2	19.2	18.3
20	20.79	6.154	47.95	0.8727	0.9884	6.8	17.2	18.3	17.3
16	16.67	4.924	44.80	0.8710	0.7923	3.6	16.1	17.3	16.3
12	12.52	3.693	41.65	0.8693	0.5954	−0.3	15.0	16.4	15.2
8	8.37	2.462	38.50	0.8676	0.3977	−5.1	13.8	15.4	14.1
4	4.19	1.231	35.35	0.8659	0.1992	−13.0	12.6	14.3	12.9
0	0.00	0.000	32.20	0.8642	0.0000	—	11.4	13.2	11.7

32.5°C DRY BULB

Percentage saturation μ/%	Relative humidity ϕ/%	Value of stated parameter per kg dry air			Vapour pressure p_s/kPa	Dew point temperature t_d/°C	Adiabatic saturation temperature $t^\star$/°C	Wet bulb temperature	
		Moisture content g/g.kg⁻¹	Specific enthalpy h/kJ.kg⁻¹	Specific volume v/m³.kg⁻¹				Screen t'_{sc}/°C	Sling t'_{sl}/°C
100	100.00	31.70	113.9	0.9094	4.890	32.5	32.5	32.5	32.5
96	96.19	30.43	110.6	0.9076	4.703	31.8	31.9	32.0	32.0
92	92.36	29.16	107.4	0.9059	4.516	31.1	31.4	31.4	31.4
88	88.52	27.89	104.1	0.9041	4.329	30.4	30.8	30.9	30.8
84	84.66	26.63	100.9	0.9024	4.140	29.6	30.2	30.3	30.2
80	80.79	25.36	97.62	0.9006	3.951	28.8	29.6	29.7	29.6
76	76.90	24.09	94.38	0.8989	3.761	27.9	28.9	29.1	29.0
72	73.00	22.82	91.13	0.8972	3.570	27.0	28.3	28.5	28.3
70	71.05	22.19	89.51	0.8963	3.474	26.6	28.0	28.2	28.0
68	69.08	21.55	87.88	0.8954	3.378	26.1	27.6	27.9	27.7
66	67.12	20.92	86.26	0.8945	3.282	25.6	27.3	27.6	27.3
64	65.15	20.29	84.64	0.8937	3.186	25.1	26.9	27.2	27.0
62	63.18	19.65	83.02	0.8928	3.089	24.6	26.6	26.9	26.6
60	61.20	19.02	81.39	0.8919	2.993	24.1	26.2	26.6	26.3
58	59.22	18.38	79.77	0.8910	2.896	23.5	25.8	26.2	25.9
56	57.24	17.75	78.15	0.8902	2.799	22.9	25.5	25.9	25.5
54	55.25	17.12	76.52	0.8893	2.702	22.4	25.1	25.5	25.2
52	53.25	16.48	74.90	0.8884	2.604	21.8	24.7	25.2	24.8
50	51.26	15.85	73.28	0.8875	2.506	21.1	24.3	24.8	24.4
48	49.26	15.22	71.65	0.8867	2.409	20.5	23.9	24.4	24.0
46	47.25	14.58	70.03	0.8858	2.311	19.8	23.5	24.1	23.6
44	45.24	13.95	68.41	0.8849	2.213	19.1	23.1	23.7	23.2
42	43.23	13.31	66.78	0.8840	2.114	18.4	22.7	23.3	22.8
40	41.21	12.68	65.16	0.8832	2.015	17.6	22.2	22.9	22.4
38	39.19	12.05	63.54	0.8823	1.916	16.8	21.8	22.5	21.9
36	37.16	11.41	61.92	0.8814	1.817	16.0	21.4	22.1	21.5
34	35.14	10.78	60.29	0.8805	1.718	15.1	20.9	21.7	21.0
32	33.10	10.14	58.67	0.8797	1.619	14.2	20.5	21.3	20.6
30	31.06	9.509	57.05	0.8788	1.519	13.2	20.0	20.8	20.1
28	29.02	8.875	55.42	0.8779	1.419	12.2	19.5	20.4	19.6
24	24.93	7.608	52.18	0.8762	1.219	9.9	18.5	19.5	18.7
20	20.81	6.340	48.93	0.8744	1.018	7.2	17.5	18.6	17.7
16	16.69	5.072	45.68	0.8727	0.8159	4.1	16.4	17.7	16.6
12	12.54	3.804	42.44	0.8709	0.6132	0.1	15.3	16.7	15.5
8	8.38	2.536	39.19	0.8691	0.4096	−4.8	14.1	15.6	14.4
4	4.20	1.268	35.94	0.8674	0.4096	−12.6	12.9	14.6	13.1
0	0.00	0.000	32.70	0.8656	0.0000	—	11.6	13.5	11.9

33°C DRY BULB

Percentage saturation μ/%	Relative humidity ϕ/%	Value of stated parameter per kg dry air			Vapour pressure p_s/kPa	Dew point temperature t_d/°C	Adiabatic saturation temperature $t^\star$/°C	Wet bulb temperature	
		Moisture content g/g.kg^{-1}	Specific enthalpy h/kJ.kg^{-1}	Specific volume v/m³.kg^{-1}				Screen t'_{sc}/°C	Sling t'_{sl}/°C
100	100.00	32.65	116.8	0.9122	5.029	33.0	33.0	33.0	33.0
96	96.19	31.34	113.5	0.9104	4.838	32.3	32.4	32.5	32.4
92	92.37	30.04	110.1	0.9086	4.646	31.6	31.9	31.9	31.9
88	88.53	28.73	106.8	0.9068	4.453	30.8	31.3	31.4	31.3
84	84.68	27.43	103.4	0.9050	4.259	30.1	30.7	30.8	30.7
80	80.81	26.12	100.1	0.9032	4.064	29.3	30.1	30.2	30.1
76	76.93	24.81	96.76	0.9014	3.869	28.4	29.4	29.6	29.4
72	73.03	23.51	93.41	0.8996	3.673	27.5	28.8	29.0	28.8
70	71.08	22.85	91.74	0.8987	3.575	27.1	28.4	28.7	28.5
68	69.12	22.20	90.06	0.8978	3.476	26.6	28.1	28.3	28.1
66	67.15	21.55	88.39	0.8969	3.377	26.1	27.7	28.0	27.8
64	65.18	20.90	86.72	0.8960	3.278	25.6	27.4	27.7	27.4
62	63.21	20.24	85.05	0.8951	3.179	25.1	27.0	27.3	27.1
60	61.24	19.59	83.38	0.8942	3.080	24.5	26.7	27.0	26.7
58	59.26	18.94	81.70	0.8933	2.980	24.0	26.3	26.7	26.3
56	57.27	18.28	80.03	0.8924	2.880	23.4	25.9	26.3	26.0
54	55.28	17.63	78.36	0.8915	2.780	22.8	25.5	26.0	25.6
52	53.29	16.98	76.69	0.8906	2.680	22.2	25.1	25.6	25.2
50	51.29	16.32	75.01	0.8896	2.580	21.6	24.7	25.2	24.8
48	49.29	15.67	73.34	0.8887	2.479	21.0	24.3	24.9	24.4
46	47.29	15.02	71.67	0.8878	2.378	20.3	23.9	24.5	24.0
44	45.28	14.37	70.00	0.8869	2.277	19.6	23.5	24.1	23.6
42	43.26	13.71	68.32	0.8860	2.176	18.9	23.1	23.7	23.2
40	41.25	13.06	66.65	0.8851	2.074	18.1	22.6	23.3	22.8
38	39.23	12.41	64.98	0.8842	1.973	17.3	22.2	22.9	22.3
36	37.20	11.75	63.31	0.8833	1.871	16.5	21.8	22.5	21.9
34	35.17	11.10	61.63	0.8824	1.769	15.6	21.3	22.1	21.4
32	33.13	10.45	59.96	0.8815	1.666	14.7	20.8	21.6	21.0
30	31.10	9.795	58.29	0.8806	1.564	13.7	20.3	21.2	20.5
28	29.05	9.142	56.62	0.8797	1.461	12.6	19.9	20.8	20.0
24	24.95	7.836	53.27	0.8779	1.255	10.3	18.9	19.9	19.0
20	20.84	6.530	49.93	0.8761	1.048	7.7	17.8	18.9	18.0
16	16.71	5.224	46.58	0.8743	0.8402	4.5	16.7	18.0	16.9
12	12.56	3.918	43.24	0.8725	0.6315	0.5	15.6	17.0	15.8
8	8.39	2.612	39.89	0.8707	0.4219	−4.4	14.4	15.9	14.6
4	4.20	1.306	36.55	0.8689	0.2114	−12.3	13.1	14.8	13.4
0	0.00	0.000	32.20	0.8671	0.0000	—	11.8	13.7	12.1

33.5°C DRY BULB

Percentage saturation μ/%	Relative humidity ϕ/%	Value of stated parameter per kg dry air			Vapour pressure p_s/kPa	Dew point temperature t_d/°C	Adiabatic saturation temperature $t^\star$/°C	Wet bulb temperature	
		Moisture content g/g.kg^{-1}	Specific enthalpy h/kJ.kg^{-1}	Specific volume v/m³.kg^{-1}				Screen t'_{sc}/°C	Sling t'_{sl}/°C
100	100.00	33.63	119.9	0.9150	5.172	33.5	33.5	33.5	33.5
96	96.20	32.28	116.4	0.9132	4.976	32.8	32.9	33.0	32.9
92	92.38	30.94	113.0	0.9113	4.778	32.1	32.4	32.4	32.4
88	88.55	29.59	109.5	0.9095	4.580	31.3	31.8	31.8	31.8
84	84.70	28.25	106.1	0.9076	4.381	30.6	31.2	31.3	31.2
80	80.84	26.90	102.6	0.9057	4.181	29.7	30.5	30.7	30.6
76	76.96	25.56	99.19	0.9039	3.980	28.9	29.9	30.1	29.9
72	73.06	24.21	95.74	0.9020	3.779	28.0	29.2	29.4	29.3
70	71.11	23.54	94.02	0.9011	3.678	27.5	28.9	29.1	28.9
68	69.15	22.87	92.29	0.9002	3.576	27.1	28.5	28.8	28.6
66	67.19	22.19	90.57	0.8992	3.475	26.6	28.2	28.5	28.2
64	65.22	21.52	88.85	0.8983	3.373	26.1	27.8	28.1	27.9
62	63.25	20.85	87.12	0.8974	3.271	25.5	27.5	27.8	27.5
60	61.27	20.18	85.40	0.8964	3.169	25.0	27.1	27.4	27.2
58	59.29	19.50	83.68	0.8955	3.067	24.5	26.7	27.1	26.8
56	57.31	18.83	81.95	0.8946	2.964	23.9	26.3	26.7	26.4
54	55.32	18.16	80.23	0.8936	2.861	23.3	26.0	26.4	26.0
52	53.33	17.49	78.51	0.8927	2.758	22.7	25.6	26.0	25.6
50	51.33	16.81	76.78	0.8918	2.655	22.1	25.2	25.6	25.2
48	49.33	16.14	75.06	0.8909	2.551	21.4	24.8	25.3	24.8
46	47.33	15.47	73.34	0.8899	2.448	20.8	24.3	24.9	24.4
44	45.32	14.80	71.61	0.8890	2.344	20.0	23.9	24.5	24.0
42	43.30	14.12	69.89	0.8881	2.240	19.3	23.5	24.1	23.6
40	41.28	13.45	68.17	0.8871	2.135	18.6	23.0	23.7	23.2
38	39.26	12.78	66.45	0.8862	2.031	17.8	22.6	23.3	22.7
36	37.23	12.11	64.72	0.8853	1.926	16.9	22.1	22.9	22.3
34	35.20	11.43	63.00	0.8843	1.821	16.0	21.7	22.4	21.8
32	33.17	10.76	61.28	0.8834	1.716	15.1	21.2	22.0	21.3
30	31.13	10.09	59.55	0.8825	1.610	14.1	20.7	21.6	20.9
28	29.08	9.416	57.83	0.8815	1.504	13.1	20.2	21.1	20.4
24	24.98	8.071	54.38	0.8797	1.292	10.8	19.2	20.2	19.4
20	20.86	6.726	50.94	0.8778	1.079	8.1	18.1	19.3	18.3
16	16.73	5.380	47.49	0.8759	0.8652	4.9	17.0	18.3	17.2
12	12.57	4.035	44.04	0.8741	0.6503	0.9	15.9	17.2	16.1
8	8.40	2.690	40.60	0.8722	0.4345	−4.1	14.6	16.2	14.9
4	4.21	1.345	37.15	0.8703	0.2177	−12.0	13.3	15.1	13.6
0	0.00	0.000	33.70	0.8685	0.0000	—	12.0	13.9	12.3

34°C DRY BULB

Percentage saturation μ/%	Relative humidity ϕ/%	Value of stated parameter per kg dry air			Vapour pressure p_s/kPa	Dew point temperature t_d/°C	Adiabatic saturation temperature $t\star$/°C	Wet bulb temperature	
		Moisture content g/g.kg^{-1}	Specific enthalpy h/kJ.kg^{-1}	Specific volume v/m^3.kg^{-1}				Screen t'_{sc}/°C	Sling t'_{sl}/°C
100	100.00	34.63	123.0	0.9179	5.319	34.0	34.0	34.0	34.0
96	96.20	33.25	119.4	0.9160	5.117	33.3	33.4	33.5	33.4
92	92.39	31.86	115.9	0.9141	4.914	32.6	32.9	32.9	32.9
88	88.56	30.48	112.3	0.9122	4.710	31.8	32.3	32.3	32.3
84	84.72	29.09	108.8	0.9102	4.506	31.1	31.6	31.8	31.7
80	80.86	27.71	105.2	0.9083	4.301	30.2	31.0	31.2	31.0
76	76.98	26.32	101.7	0.9064	4.094	29.4	30.4	30.5	30.4
72	73.09	24.94	98.12	0.9045	3.887	28.5	29.7	29.9	29.7
70	71.14	24.24	96.34	0.9035	3.784	28.0	29.3	29.6	29.4
68	69.18	23.55	94.57	0.9026	3.679	27.5	29.0	29.3	29.0
66	67.22	22.86	92.79	0.9016	3.575	27.1	28.6	28.9	28.7
64	65.25	22.17	91.02	0.9007	3.470	26.5	28.3	28.6	28.3
62	63.28	21.47	89.24	0.8997	3.366	26.0	27.9	28.2	28.0
60	61.31	20.78	87.47	0.8987	3.261	25.5	27.5	27.9	27.6
58	59.33	20.09	85.69	0.8978	3.155	24.9	27.2	27.5	27.2
56	57.35	19.39	83.92	0.8968	3.050	24.4	26.8	27.2	26.8
54	55.36	18.70	82.14	0.8959	2.944	23.8	26.4	26.8	26.5
52	53.37	18.01	80.37	0.8949	2.838	23.2	26.0	26.4	26.1
50	51.37	17.32	78.59	0.8939	2.732	22.5	25.6	26.1	25.7
48	49.37	16.62	76.82	0.8930	2.626	21.9	25.2	25.7	25.3
46	47.36	15.93	75.04	0.8920	2.519	21.2	24.8	25.3	24.8
44	45.35	15.24	73.27	0.8911	2.412	20.5	24.3	24.9	24.4
42	43.34	14.55	71.49	0.8901	2.305	19.8	23.9	24.5	24.0
40	41.32	13.85	69.71	0.8891	2.198	19.0	23.4	24.1	23.5
38	39.30	13.16	67.94	0.8882	2.090	18.2	23.0	23.7	23.1
36	37.27	12.47	66.16	0.8872	1.982	17.4	22.5	23.3	22.6
34	35.24	11.78	64.39	0.8863	1.874	16.5	22.1	22.8	22.2
32	33.20	11.08	62.61	0.8853	1.766	15.6	21.6	22.4	21.7
30	31.16	10.39	60.84	0.8843	1.657	14.6	21.1	21.9	21.2
28	29.12	9.697	59.06	0.8834	1.549	13.5	20.6	21.5	20.7
24	25.01	8.312	55.51	0.8814	1.330	11.2	19.5	20.6	19.7
20	20.89	6.927	51.96	0.8795	1.111	8.5	18.5	19.6	18.6
16	16.75	5.541	48.41	0.8776	0.8918	5.3	17.3	18.6	17.5
12	12.59	4.156	44.86	0.8757	0.6705	1.3	16.1	17.5	16.4
8	8.41	2.771	41.31	0.8737	0.4473	−3.7	14.9	16.4	15.1
4	4.22	1.385	37.76	0.8718	0.2241	−11.7	13.6	15.3	13.9
0	0.00	0.000	34.21	0.8699	0.0000	—	12.2	14.1	12.5

34.5°C DRY BULB

Percentage saturation μ/%	Relative humidity ϕ/%	Value of stated parameter per kg dry air			Vapour pressure p_s/kPa	Dew point temperature t_d/°C	Adiabatic saturation temperature $t\star$/°C	Wet bulb temperature	
		Moisture content g/g.kg^{-1}	Specific enthalpy h/kJ.kg^{-1}	Specific volume v/m^3.kg^{-1}				Screen t'_{sc}/°C	Sling t'_{sl}/°C
100	100.00	35.67	126.2	0.9208	5.468	34.5	34.5	34.5	34.5
96	96.21	34.24	122.5	0.9189	5.261	33.8	33.9	34.0	33.9
92	92.40	32.81	118.8	0.9169	5.053	33.1	33.3	33.4	33.4
88	88.58	31.39	115.2	0.9149	4.844	32.3	32.7	32.8	32.8
84	84.74	29.96	111.5	0.9129	4.634	31.5	32.1	32.2	32.1
80	80.89	28.53	107.9	0.9110	4.423	30.7	31.5	31.6	31.5
76	77.01	27.11	104.2	0.9090	4.211	29.9	30.8	31.0	30.9
72	73.12	25.68	100.6	0.9070	3.999	29.0	30.2	30.4	30.2
70	71.17	24.97	98.72	0.9060	3.892	28.5	29.8	30.0	29.9
68	69.21	24.25	96.89	0.9050	3.785	28.0	29.5	29.7	29.5
66	67.25	23.54	95.07	0.9040	3.678	27.5	29.1	29.4	29.1
64	65.29	22.83	93.24	0.9030	3.570	27.0	28.7	29.0	28.8
62	63.32	22.11	91.41	0.9021	3.463	26.5	28.4	28.7	28.4
60	61.35	21.40	89.58	0.9011	3.355	26.0	28.0	28.3	28.1
58	59.37	20.69	87.75	0.9001	3.247	25.4	27.6	28.0	27.7
56	57.39	19.97	85.92	0.8991	3.138	24.8	27.2	27.6	27.3
54	55.40	19.26	84.09	0.8981	3.029	24.3	26.8	27.2	26.9
52	53.41	18.55	82.26	0.8971	2.920	23.7	26.4	26.9	26.5
50	51.41	17.83	80.43	0.8961	2.811	23.0	26.0	26.5	26.1
48	49.41	17.12	78.61	0.8951	2.702	22.4	25.6	26.1	25.7
46	47.40	16.41	76.78	0.8941	2.592	21.7	25.2	25.7	25.3
44	45.39	15.69	74.95	0.8931	2.482	21.0	24.7	25.3	24.8
42	43.38	14.98	73.12	0.8921	2.372	20.2	24.3	24.9	24.4
40	41.36	14.27	71.29	0.8912	2.262	19.5	23.8	24.5	24.0
38	39.34	13.55	69.46	0.8902	2.151	18.7	23.4	24.1	23.5
36	37.31	12.84	67.63	0.8892	2.040	17.8	22.9	23.6	23.0
34	35.28	12.13	65.80	0.8882	1.929	16.9	22.4	23.2	22.6
32	33.24	11.41	63.97	0.8872	1.818	16.0	22.0	22.8	22.1
30	31.20	10.70	62.15	0.8862	1.706	15.0	21.5	22.3	21.6
28	29.15	9.986	60.32	0.8852	1.594	14.0	20.9	21.8	21.1
24	25.04	8.560	56.66	0.8832	1.369	11.6	19.9	20.9	20.1
20	20.92	7.133	53.00	0.8812	1.144	9.0	18.8	19.9	19.0
16	16.77	5.707	49.34	0.8793	0.9181	5.7	17.6	18.9	17.8
12	12.61	4.280	45.69	0.8773	0.6894	1.7	16.4	17.8	16.7
8	8.42	2.853	42.03	0.8753	0.4606	−3.4	15.2	16.7	15.4
4	4.22	1.427	38.37	0.8733	0.2308	−11.3	13.8	15.5	14.1
0	0.00	0.000	34.71	0.713	0.0000	—	12.4	14.3	12.7

35°C DRY BULB

Percentage saturation μ/%	Relative humidity ϕ/%	Value of stated parameter per kg dry air			Vapour pressure p_s/kPa	Dew point temperature t_d/°C	Adiabatic saturation temperature $t^\star$/°C	Wet bulb temperature	
		Moisture content g/g.kg^{-1}	Specific enthalpy h/kJ.kg^{-1}	Specific volume v/m^3.kg^{-1}				Screen t'_{sc}/°C	Sling t'_{sl}/°C
100	100.00	36.73	129.4	0.9238	5.622	35.0	35.0	35.0	35.0
96	96.21	35.26	125.6	0.9218	5.409	34.3	34.4	34.5	34.4
92	92.41	33.79	121.9	0.9197	5.196	33.6	33.8	33.9	33.8
88	88.60	32.32	118.1	0.9177	4.981	32.8	33.2	33.3	33.2
84	84.76	30.85	114.3	0.9157	4.765	32.0	32.6	32.7	32.6
80	80.91	29.38	110.6	0.9136	4.549	31.2	32.0	32.1	32.0
76	77.04	27.91	106.8	0.9116	4.331	30.4	31.3	31.5	31.3
72	73.15	26.44	103.0	0.9095	4.113	29.5	30.6	30.8	30.7
70	71.20	25.71	101.2	0.9085	4.003	29.0	30.3	30.5	30.3
68	69.25	24.97	99.27	0.9075	3.893	28.5	29.9	30.2	30.0
66	67.29	24.24	97.39	0.9065	3.783	28.0	29.6	29.8	29.6
64	65.33	23.51	95.50	0.9055	3.673	27.5	29.2	29.5	29.2
62	63.36	22.77	93.62	0.9044	3.562	27.0	28.8	29.1	28.9
60	61.38	22.04	91.74	0.9034	3.451	26.5	28.4	28.8	28.5
58	59.41	21.30	89.85	0.9024	3.340	25.9	28.1	28.4	28.1
56	57.42	20.57	87.97	0.9014	3.228	25.3	27.7	28.1	27.7
54	55.44	19.83	86.08	0.9004	3.117	24.7	27.3	27.7	27.3
52	53.45	19.10	84.20	0.8993	3.005	24.1	26.9	27.3	26.9
50	51.45	18.36	82.32	0.8983	2.893	23.5	26.4	26.9	26.5
48	49.45	17.63	80.43	0.8973	2.780	22.8	26.0	26.5	26.1
46	47.45	16.89	78.55	0.8963	2.667	22.2	25.6	26.1	25.7
44	45.44	16.16	76.66	0.8952	2.554	21.4	25.1	25.7	25.2
42	43.42	15.43	74.78	0.8942	2.441	20.7	24.7	25.3	24.8
40	41.40	14.69	72.90	0.8932	2.328	19.9	24.2	24.9	24.4
38	39.38	13.96	71.01	0.8922	2.214	19.1	23.8	24.5	23.9
36	37.35	13.22	69.13	0.8912	2.100	18.3	23.3	24.0	23.4
34	35.31	12.49	67.24	0.8901	1.985	17.4	22.8	23.6	22.9
32	33.27	11.75	65.36	0.8891	1.871	16.5	22.3	23.1	22.5
30	31.23	11.02	63.48	0.8881	1.756	15.5	21.8	22.7	22.0
28	29.18	10.28	61.59	0.8871	1.641	14.4	21.3	22.2	21.4
24	25.07	8.815	57.82	0.8850	1.410	12.1	20.2	21.2	20.4
20	20.94	7.346	54.06	0.8830	1.177	9.4	19.1	20.2	19.3
16	16.79	5.876	50.29	0.8809	0.9442	6.1	17.9	19.2	18.1
12	12.63	4.407	46.52	0.8789	0.7098	2.1	16.7	18.1	16.9
8	8.44	2.938	42.75	0.8768	0.4743	−3.0	15.4	17.0	15.7
4	4.23	1.469	38.98	0.8748	0.2377	−11.0	14.0	15.8	14.3
0	0.00	0.000	35.22	0.8727	0.0000	—	12.6	14.6	12.9

35.5°C DRY BULB

Percentage saturation μ/%	Relative humidity ϕ/%	Value of stated parameter per kg dry air			Vapour pressure p_s/kPa	Dew point temperature t_d/°C	Adiabatic saturation temperature $t^\star$/°C	Wet bulb temperature	
		Moisture content g/g.kg^{-1}	Specific enthalpy h/kJ.kg^{-1}	Specific volume v/m^3.kg^{-1}				Screen t'_{sc}/°C	Sling t'_{sl}/°C
100	100.00	37.82	132.8	0.9268	5.779	35.5	35.5	35.5	35.5
96	96.22	36.31	128.9	0.9247	5.561	34.8	34.9	34.9	34.9
92	92.43	34.79	125.0	0.9226	5.342	34.1	34.3	34.4	34.3
88	88.61	33.28	121.1	0.9205	5.121	33.3	33.7	33.8	33.7
84	84.78	31.77	117.2	0.9184	4.900	32.5	33.1	33.2	33.1
80	80.94	30.26	113.3	0.9163	4.678	31.7	32.4	32.6	32.5
76	77.07	28.74	109.5	0.9142	4.454	30.9	31.8	32.0	31.8
72	73.19	27.23	105.6	0.9121	4.230	29.9	31.1	31.3	31.1
70	71.24	26.47	103.6	0.9111	4.117	29.5	30.7	31.0	30.8
68	69.28	25.72	101.7	0.9100	4.004	29.0	30.4	30.6	30.4
66	67.33	24.96	99.76	0.9090	3.891	28.5	30.0	30.3	30.1
64	65.36	24.20	97.82	0.9079	3.778	28.0	29.6	29.9	29.7
62	63.40	23.45	95.88	0.9068	3.664	27.5	29.3	29.6	29.3
60	61.42	22.69	93.94	0.9058	3.560	26.9	28.9	29.2	28.9
58	59.45	21.93	92.00	0.9047	3.436	26.4	28.5	28.9	28.6
56	57.47	21.18	90.06	0.9037	3.321	25.8	28.1	28.5	28.2
54	55.48	20.42	88.12	0.9026	3.206	25.2	27.7	28.1	27.8
52	53.49	19.67	86.18	0.9016	3.091	24.6	27.3	27.7	27.4
50	51.49	18.91	84.23	0.9005	2.976	24.0	26.9	27.3	26.9
48	49.49	18.15	82.29	0.8995	2.860	23.3	26.4	26.9	26.5
46	47.49	17.40	80.35	0.8984	2.744	22.6	26.0	26.5	26.1
44	45.48	16.64	78.41	0.8974	2.628	21.9	25.6	26.1	25.7
42	43.46	15.88	76.47	0.8963	2.512	21.2	25.1	25.7	25.2
40	41.44	15.13	74.53	0.8953	2.395	20.4	24.6	25.3	24.8
38	39.42	14.37	72.59	0.8942	2.278	19.6	24.2	24.9	24.3
36	37.39	13.62	70.65	0.8932	2.161	18.7	23.7	24.4	23.8
34	35.35	12.86	68.71	0.8921	2.043	17.8	23.2	24.0	23.3
32	33.31	12.10	66.77	0.8910	1.925	16.9	22.7	23.5	22.8
30	31.27	11.35	64.83	0.8900	1.807	15.9	22.2	23.0	22.3
28	29.22	10.59	62.89	0.8889	1.689	14.9	21.7	22.6	21.8
24	25.10	9.076	59.01	0.8868	1.451	12.5	20.6	21.6	20.7
20	20.97	7.564	55.13	0.8847	1.212	9.8	19.4	20.6	19.6
16	16.82	6.051	51.24	0.8826	0.9719	6.6	18.3	19.5	18.5
12	12.64	4.538	47.36	0.8805	0.7307	2.5	17.0	18.4	17.2
8	8.45	3.026	43.48	0.8784	0.4883	−2.7	15.7	17.2	15.9
4	4.24	1.513	39.60	0.8763	0.2448	−10.7	14.3	16.0	14.6
0	0.00	0.000	35.72	0.8742	0.0000	—	12.8	14.8	13.1

36°C DRY BULB

Percentage saturation μ/%	Relative humidity ϕ/%	Value of stated parameter per kg dry air			Vapour pressure p_s/kPa	Dew point temperature t_d/°C	Adiabatic saturation temperature $t\star$/°C	Wet bulb temperature	
		Moisture content g/g.kg^{-1}	Specific enthalpy h/kJ.kg^{-1}	Specific volume v/m^3.kg^{-1}				Screen t'_{sc}/°C	Sling t'_{sl}/°C
100	100.00	38.94	136.2	0.9299	5.940	36.0	36.0	36.0	36.0
96	96.23	37.38	132.2	0.9277	5.716	35.3	35.4	35.4	35.4
92	92.44	35.82	128.2	0.9256	5.491	34.6	34.8	34.9	34.8
88	88.63	34.27	124.2	0.9234	5.265	33.8	34.2	34.3	34.2
84	84.81	32.71	120.2	0.9212	5.038	33.0	33.6	33.7	33.6
80	80.96	31.15	116.2	0.9191	4.810	32.2	32.9	33.1	33.0
76	77.10	29.59	112.2	0.9169	4.580	31.3	32.3	32.4	32.3
72	73.22	28.04	108.2	0.9147	4.350	30.4	31.6	31.8	31.6
70	71.27	27.26	106.2	0.9136	4.234	30.0	31.2	31.4	31.2
68	69.32	26.48	104.2	0.9125	4.118	29.5	30.8	31.1	30.9
66	67.36	25.70	102.2	0.9115	4.002	29.0	30.5	30.5	30.5
64	65.40	24.92	100.2	0.9104	3.885	28.5	30.1	30.4	30.2
62	63.44	24.14	98.19	0.9093	3.768	28.0	29.7	30.0	29.8
60	61.46	23.59	96.19	0.9082	3.651	27.4	29.3	29.7	29.4
58	59.49	22.36	94.19	0.9071	3.534	26.9	28.9	29.3	29.0
56	57.51	21.81	92.19	0.9060	3.416	26.3	28.5	28.9	28.6
54	55.52	21.03	90.19	0.9049	3.298	25.7	28.1	28.5	28.2
52	53.53	20.25	88.19	0.9039	3.180	25.1	27.7	28.2	27.8
50	51.54	19.47	86.19	0.9028	3.061	24.4	27.3	27.8	27.4
48	49.53	18.69	84.20	0.9017	2.943	23.8	26.9	27.4	26.9
46	47.53	17.91	82.20	0.9006	2.823	23.1	26.4	27.0	26.5
44	45.52	17.13	80.20	0.8995	2.704	22.4	26.0	26.5	26.1
42	43.50	16.35	78.20	0.8984	2.584	21.6	25.5	26.1	25.6
40	41.48	15.58	76.20	0.8973	2.464	20.9	25.1	25.7	25.2
38	39.46	14.80	74.20	0.8963	2.344	20.0	24.6	25.3	24.7
36	37.43	14.02	72.20	0.8952	2.223	19.2	24.1	24.8	24.2
34	35.39	13.24	70.20	0.8941	2.102	18.3	23.6	24.3	23.7
32	33.35	12.46	68.20	0.8930	1.981	17.4	23.1	23.9	23.2
30	31.30	11.68	66.21	0.8919	1.860	16.4	22.6	23.4	22.7
28	29.25	10.90	64.21	0.8908	1.738	15.3	22.0	22.9	22.2
24	25.14	9.346	60.21	0.8886	1.493	13.0	20.9	21.9	21.1
20	21.00	7.788	56.21	0.8865	1.248	10.2	19.8	20.9	20.0
16	16.84	6.230	52.21	0.8843	1.000	7.0	18.6	19.8	18.8
12	12.66	4.673	48.22	0.8821	0.7522	2.9	17.3	18.7	17.5
8	8.46	3.115	44.22	0.8799	0.5027	−2.3	15.9	17.5	16.2
4	4.24	1.558	40.22	0.8778	0.2520	−10.4	14.5	16.3	14.8
0	0.00	0.000	36.22	0.8756	0.0000	—	13.0	15.0	13.3

36.5°C DRY BULB

Percentage saturation μ/%	Relative humidity ϕ/%	Value of stated parameter per kg dry air			Vapour pressure p_s/kPa	Dew point temperature t_d/°C	Adiabatic saturation temperature $t\star$/°C	Wet bulb temperature	
		Moisture content g/g.kg^{-1}	Specific enthalpy h/kJ.kg^{-1}	Specific volume v/m^3.kg^{-1}				Screen t'_{sc}/°C	Sling t'_{sl}/°C
100	100.00	40.09	139.7	0.9330	6.106	36.5	36.5	36.5	36.5
96	96.23	38.49	135.5	0.9308	5.876	35.8	35.9	35.9	35.9
92	92.45	36.88	131.4	0.9285	5.645	35.1	35.3	35.4	35.3
88	88.65	35.28	127.3	0.9263	5.412	34.3	34.7	34.8	34.7
84	84.83	33.68	123.2	0.9241	5.179	33.5	34.1	34.2	34.1
80	80.99	32.07	119.1	0.9218	4.945	32.7	33.4	33.5	33.4
76	77.13	30.47	115.0	0.9196	4.709	31.8	32.7	32.9	32.8
72	73.25	28.87	110.8	0.9174	4.472	30.9	32.0	32.2	32.1
70	71.31	28.06	108.8	0.9162	4.354	30.5	31.7	31.9	31.7
68	69.36	27.26	106.7	0.9151	4.235	30.0	31.3	31.6	31.4
66	67.40	26.46	104.7	0.9140	4.115	29.5	30.9	31.2	31.0
64	65.44	25.66	102.6	0.9129	3.996	29.0	30.6	30.8	30.6
62	63.48	24.86	100.6	0.9118	3.876	28.4	30.2	30.5	30.2
60	61.51	24.06	98.49	0.9106	3.755	27.9	29.8	30.1	29.8
58	59.53	23.25	96.43	0.9095	3.635	27.3	29.4	29.7	29.5
56	57.55	22.45	94.37	0.9084	3.514	26.8	29.0	29.4	29.1
54	55.56	21.65	92.31	0.9073	3.392	26.2	28.6	29.0	28.6
52	53.57	20.85	90.25	0.9062	3.271	25.5	28.2	28.6	28.2
50	51.58	20.05	88.19	0.9050	3.149	24.9	27.7	28.2	27.8
48	49.58	19.24	86.14	0.9039	3.027	24.2	27.3	27.8	27.4
46	47.57	18.44	84.08	0.9028	2.905	23.6	26.8	27.4	26.9
44	45.56	17.64	82.02	0.9017	2.782	22.8	26.4	27.0	26.5
42	43.55	16.84	79.96	0.9006	2.659	22.1	25.9	26.5	26.0
40	41.53	16.04	77.90	0.8994	2.535	21.3	25.5	26.1	25.6
38	39.50	15.24	75.84	0.8983	3.412	20.5	25.0	25.6	25.1
36	37.47	14.43	73.78	0.8972	2.288	19.7	24.5	25.2	24.6
34	35.43	13.63	71.72	0.8961	2.163	18.8	24.0	24.7	24.1
32	33.39	12.83	69.67	0.8950	2.039	17.8	23.5	24.3	23.6
30	31.34	12.03	67.61	0.8938	1.914	16.8	22.9	23.8	23.1
28	29.29	11.23	65.55	0.8927	1.788	15.8	22.4	23.3	22.5
24	25.17	9.622	61.43	0.8905	1.537	13.4	21.3	22.3	21.4
20	21.03	8.018	57.31	0.8882	1.284	10.7	20.1	21.2	20.3
16	16.87	6.415	53.20	0.8860	1.030	7.4	18.9	20.1	19.1
12	12.68	4.811	49.08	0.8837	0.7743	3.3	17.6	19.0	17.8
8	8.48	3.207	44.96	0.8815	0.5175	−2.0	16.2	17.8	16.5
4	4.25	1.604	40.84	0.8792	0.2594	−10.0	14.7	16.5	15.0
0	0.00	0.000	36.73	0.8770	0.0000	—	13.2	15.2	13.5

37°C DRY BULB

Percentage saturation μ/%	Relative humidity ϕ/%	Value of stated parameter per kg dry air			Vapour pressure p_s/kPa	Dew point temperature t_d/°C	Adiabatic saturation temperature $t\star$/°C	Wet bulb temperature	
		Moisture content g/g.kg^{-1}	Specific enthalpy h/kJ.kg^{-1}	Specific volume v/m³.kg^{-1}				Screen t'_{sc}/°C	Sling t'_{sl}/°C
100	100.00	41.28	143.2	0.9362	6.274	37.0	37.0	37.0	37.0
96	96.24	39.62	139.0	0.9339	6.038	36.3	36.4	36.4	36.4
92	92.46	37.97	134.8	0.9316	5.802	35.6	35.8	35.9	35.8
88	88.67	36.32	130.5	0.9293	5.563	34.8	35.2	35.3	35.2
84	84.85	34.67	126.3	0.9270	5.324	34.0	34.5	34.6	34.6
80	81.02	33.02	122.0	0.9246	5.083	33.2	33.9	34.0	33.9
76	77.16	31.37	117.8	0.9223	4.842	32.3	33.2	33.4	33.2
72	73.29	29.72	113.6	0.9200	4.598	31.4	32.5	32.7	32.5
70	71.34	28.89	111.4	0.9189	4.476	30.9	32.1	32.4	32.2
68	69.40	28.07	109.3	0.9177	4.354	30.5	31.8	32.0	31.8
66	67.44	27.24	107.2	0.9166	4.232	30.0	31.4	31.7	31.4
64	65.48	26.42	105.1	0.9154	4.109	29.4	31.0	31.3	31.1
62	63.52	25.59	103.0	0.9143	3.985	28.9	30.6	30.9	30.7
60	61.55	24.77	100.8	0.9131	3.862	28.4	30.2	30.6	30.3
58	59.57	23.94	98.72	0.9120	3.738	27.8	29.8	30.2	29.9
56	57.59	23.11	96.60	0.9108	3.614	27.2	29.4	29.8	29.5
54	55.61	22.29	94.48	0.9096	3.489	26.6	29.0	29.4	29.1
52	53.62	21.46	92.36	0.9085	3.364	26.0	28.6	29.0	28.7
50	51.62	20.64	90.24	0.9073	3.239	25.4	28.2	28.6	28.2
48	49.62	19.81	88.12	0.9062	3.114	24.7	27.7	28.2	27.8
46	47.62	18.99	86.00	0.9050	2.988	24.0	27.3	27.8	27.4
44	45.61	18.16	83.88	0.9039	2.862	23.3	26.8	27.4	26.9
42	43.59	17.34	81.75	0.9027	2.735	22.6	26.3	26.9	26.4
40	41.57	16.51	79.63	0.9016	2.608	21.8	25.9	26.5	26.0
38	39.54	15.68	77.51	0.9004	2.481	21.0	25.4	26.0	25.5
36	37.51	14.86	75.39	0.8992	2.354	20.1	24.9	25.6	25.0
34	35.47	14.03	73.27	0.8981	2.226	19.2	24.4	25.1	24.5
32	33.43	13.21	71.15	0.8969	2.098	18.3	23.8	24.6	24.0
30	31.38	12.38	69.03	0.8958	1.969	17.3	23.3	24.1	23.4
28	29.33	11.56	66.91	0.8946	1.840	16.2	22.8	23.7	22.9
24	25.20	9.906	62.67	0.8923	1.581	13.8	21.6	22.6	21.8
20	21.06	8.255	58.43	0.8900	1.321	11.1	20.4	21.6	20.6
16	16.89	6.604	54.19	0.8877	1.060	7.8	19.2	20.4	19.4
12	12.70	4.953	49.95	0.8854	0.7970	3.7	17.9	19.3	18.1
8	8.49	3.302	45.71	0.8830	0.5327	−1.6	16.5	18.0	16.7
4	4.26	1.651	41.47	0.8807	0.2671	−9.7	15.0	16.8	15.3
0	0.00	0.000	37.23	0.8784	0.0000	—	13.4	15.4	13.7

37.5°C DRY BULB

Percentage saturation μ/%	Relative humidity ϕ/%	Value of stated parameter per kg dry air			Vapour pressure p_s/kPa	Dew point temperature t_d/°C	Adiabatic saturation temperature $t\star$/°C	Wet bulb temperature	
		Moisture content g/g.kg^{-1}	Specific enthalpy h/kJ.kg^{-1}	Specific volume v/m³.kg^{-1}				Screen t'_{sc}/°C	Sling t'_{sl}/°C
100	100.00	42.49	146.9	0.9394	6.447	37.5	37.5	37.5	37.5
96	96.25	40.79	142.5	0.9370	6.205	36.8	36.9	36.9	36.9
92	92.48	39.09	138.2	0.9346	5.962	36.1	36.3	36.3	36.3
88	88.69	37.39	133.8	0.9323	5.718	35.3	35.7	35.8	35.7
84	84.87	35.69	129.4	0.9299	5.472	34.5	35.0	35.1	35.0
80	81.04	33.99	125.1	0.9275	5.225	33.7	34.4	34.5	34.4
76	77.19	32.29	120.7	0.9251	4.977	32.8	33.7	33.8	33.7
72	73.32	30.59	116.3	0.9227	4.727	31.9	33.0	33.2	33.0
70	71.38	29.74	114.2	0.9216	4.602	31.4	32.6	32.8	32.6
68	69.43	28.89	112.0	0.9204	4.477	30.9	32.2	32.5	32.3
66	67.48	28.04	109.8	0.9192	4.351	30.4	31.9	32.1	31.9
64	65.52	27.20	107.6	0.9180	4.225	29.9	31.5	31.8	31.5
62	63.56	26.35	105.4	0.9168	4.098	29.4	31.1	31.4	31.1
60	61.59	25.50	103.2	0.9156	3.971	28.9	30.7	31.0	30.7
58	59.62	24.65	101.1	0.9144	3.844	28.3	30.3	30.6	30.3
56	57.64	23.80	98.87	0.9132	3.716	27.7	29.9	30.2	29.9
54	55.65	22.95	96.69	0.9120	3.588	27.1	29.4	29.9	29.5
52	53.66	22.10	94.50	0.9108	3.460	26.5	29.0	29.5	29.1
50	51.67	21.25	92.32	0.9097	3.331	25.9	28.6	29.0	28.7
48	49.67	20.40	90.14	0.9085	3.202	25.2	28.1	28.6	28.2
46	47.66	19.55	87.95	0.9073	3.073	24.5	27.7	28.2	27.8
44	45.65	18.70	85.77	0.9061	2.943	23.8	27.2	27.8	27.3
42	43.64	17.85	83.59	0.9049	2.813	23.0	26.7	27.3	26.8
40	41.61	17.00	81.40	0.9037	2.683	22.2	26.3	26.9	26.4
38	39.59	16.15	79.22	0.9025	2.552	21.4	25.8	26.4	25.9
36	37.55	15.30	77.04	0.9013	2.421	20.6	25.3	26.0	25.4
34	35.52	14.45	74.85	0.9001	2.290	19.7	24.7	25.5	24.9
32	33.47	13.60	72.67	0.8989	2.158	18.7	24.2	25.0	24.3
30	31.42	12.75	70.49	0.8977	2.026	17.7	23.7	24.5	23.8
28	29.37	11.90	68.30	0.8965	1.893	16.6	23.1	24.0	23.3
24	25.24	10.20	63.93	0.8942	1.627	14.3	22.0	23.0	22.1
20	21.09	8.498	59.57	0.8918	1.360	11.5	20.8	21.9	20.9
16	16.92	6.799	55.20	0.8894	1.091	8.3	19.5	20.7	19.7
12	12.72	5.099	50.83	0.8870	0.8202	4.1	18.1	19.6	18.4
8	8.50	3.399	46.47	0.8846	0.5483	−1.3	16.7	18.3	17.0
4	4.26	1.700	42.10	0.8822	0.2749	−9.4	15.2	17.0	15.5
0	0.00	0.000	37.73	0.8798	0.0000	—	13.6	15.6	13.9

38°C DRY BULB

Percentage saturation μ/%	Relative humidity ϕ/%	Value of stated parameter per kg dry air			Vapour pressure p_s/kPa	Dew point temperature t_d/°C	Adiabatic saturation temperature t^*/°C	Wet bulb temperature	
		Moisture content g/g.kg^{-1}	Specific enthalpy h/kJ.kg^{-1}	Specific volume v/m^3.kg^{-1}				Screen t'_{sc}/°C	Sling t'_{sl}/°C
100	100.00	43.74	150.7	0.9427	6.624	38.0	38.0	38.0	38.0
96	96.25	41.99	146.2	0.9402	6.376	37.3	37.4	37.4	37.4
92	92.49	40.24	141.7	0.9378	6.127	36.6	36.8	36.8	36.8
88	88.70	38.49	137.2	0.9353	5.876	35.8	36.2	36.2	36.2
84	84.90	36.74	132.7	0.9329	5.624	35.0	35.5	35.6	35.5
80	81.07	34.99	128.2	0.9304	5.371	34.2	34.8	35.0	34.9
76	77.23	33.24	123.7	0.9279	5.116	33.3	34.2	34.3	34.2
72	73.36	31.49	119.2	0.9255	4.860	32.4	33.4	33.6	33.5
70	71.42	30.62	116.9	0.9243	4.731	31.9	33.1	33.3	33.1
68	69.47	29.74	114.7	0.9230	4.602	31.4	32.7	32.9	32.7
66	67.52	28.87	112.4	0.9218	4.473	30.9	32.3	32.6	32.4
64	65.57	28.00	110.2	0.9206	4.343	30.4	31.9	32.2	32.0
62	63.60	27.12	107.9	0.9194	4.213	29.9	31.5	31.8	31.6
60	61.64	26.25	105.7	0.9181	4.083	29.3	31.1	31.5	31.2
58	59.66	25.37	103.4	0.9169	3.952	28.8	30.7	31.1	30.8
56	57.68	24.50	101.2	0.9157	3.821	28.2	30.3	30.7	30.4
54	55.70	23.62	98.95	0.9145	3.690	27.6	29.9	30.3	30.0
52	53.71	22.75	96.70	0.9132	3.558	27.0	29.5	29.9	29.5
50	51.72	21.87	94.45	0.9120	3.426	26.3	29.0	29.5	29.1
48	49.72	21.00	92.20	0.9108	3.293	25.7	28.6	29.1	28.7
46	47.71	20.12	89.95	0.9095	3.161	25.0	28.1	28.6	28.2
44	45.70	19.25	87.70	0.9083	3.027	24.2	27.6	28.2	27.7
42	43.68	18.37	85.46	0.9071	2.894	23.5	27.2	27.7	27.3
40	41.66	17.50	83.21	0.9059	2.760	22.7	26.7	27.3	26.8
38	39.63	16.62	80.96	0.9046	2.625	21.9	26.2	26.8	26.3
36	37.60	15.75	78.71	0.9034	2.491	21.0	25.7	26.4	25.8
34	35.56	14.87	76.46	0.9022	2.356	20.1	25.1	25.9	25.3
32	33.51	14.00	74.21	0.9009	2.220	19.2	24.6	25.4	24.7
30	31.46	13.12	71.96	0.8997	2.084	18.2	24.0	24.9	24.2
28	29.41	12.25	69.72	0.8985	1.948	17.1	23.5	24.4	23.6
24	25.27	10.50	65.22	0.8960	1.674	14.7	22.3	23.3	22.5
20	21.12	8.748	60.72	0.8936	1.399	12.0	21.1	22.2	21.3
16	16.94	6.999	56.22	0.8911	1.122	8.7	19.8	21.1	20.0
12	12.74	5.249	51.73	0.8886	0.8442	4.5	18.4	19.8	18.7
8	8.52	3.499	47.23	0.8862	0.5644	−0.9	17.0	18.6	17.2
4	4.27	1.750	42.73	0.8837	0.2830	−9.0	15.4	17.2	15.7
0	0.00	0.000	38.24	0.8813	0.0000	—	13.8	15.8	14.1

38.5°C DRY BULB

Percentage saturation μ/%	Relative humidity ϕ/%	Value of stated parameter per kg dry air			Vapour pressure p_s/kPa	Dew point temperature t_d/°C	Adiabatic saturation temperature t^*/°C	Wet bulb temperature	
		Moisture content g/g.kg^{-1}	Specific enthalpy h/kJ.kg^{-1}	Specific volume v/m^3.kg^{-1}				Screen t'_{sc}/°C	Sling t'_{sl}/°C
100	100.00	45.03	154.5	0.9460	6.806	38.5	38.5	38.5	38.5
96	96.26	43.23	149.9	0.9434	6.551	37.8	37.9	37.9	37.9
92	92.50	41.42	145.2	0.9409	6.295	37.1	37.3	37.3	37.3
88	88.72	39.62	140.6	0.9384	6.038	36.3	36.7	36.7	36.7
84	84.92	37.82	136.0	0.9359	5.780	35.5	36.0	36.1	36.0
80	81.10	36.02	131.4	0.9333	5.520	34.7	35.3	35.5	35.4
76	77.26	34.22	126.7	0.9308	5.258	33.8	34.6	34.8	34.7
72	73.40	32.42	122.1	0.9283	4.995	32.9	33.9	34.1	33.9
70	71.46	31.52	119.8	0.9270	4.863	32.4	33.5	33.8	33.6
68	69.52	30.62	117.5	0.9258	4.731	31.9	33.2	33.4	33.2
66	67.57	29.72	115.2	0.9245	4.598	31.4	32.8	33.0	32.8
64	65.61	28.82	112.8	0.9232	4.465	30.9	32.4	32.7	32.4
62	63.65	27.92	110.5	0.9220	4.332	30.4	32.0	32.3	32.0
60	61.68	27.02	108.2	0.9207	4.198	29.8	31.6	31.9	31.6
58	59.71	26.12	105.9	0.9194	4.064	29.3	31.2	31.5	31.2
56	57.73	25.22	103.6	0.9182	3.929	28.7	30.8	31.1	30.8
54	55.75	24.31	101.3	0.9169	3.794	28.1	30.3	30.7	30.4
52	53.76	23.41	98.94	0.9156	3.659	27.4	29.9	30.3	30.0
50	51.76	22.51	96.62	0.9144	3.523	26.8	29.4	29.9	29.5
48	49.76	21.61	94.31	0.9131	3.387	26.1	29.0	29.5	29.1
46	47.76	20.71	91.99	0.9118	3.250	25.4	28.5	29.0	28.6
44	45.75	19.81	89.68	0.9106	3.113	24.7	28.1	28.6	28.1
42	43.73	18.91	87.36	0.9093	2.976	24.0	27.6	28.2	27.7
40	41.71	18.01	85.05	0.9080	2.838	23.2	27.1	27.7	27.2
38	39.68	17.11	82.73	0.9068	2.700	22.4	26.6	27.2	26.7
36	37.64	16.21	80.42	0.9055	2.562	21.5	26.1	26.8	26.2
34	35.60	15.31	78.10	0.9042	2.423	20.6	25.5	26.3	25.6
32	33.56	14.41	75.79	0.9030	2.284	19.6	25.0	25.8	25.1
30	31.50	13.51	73.47	0.9017	2.144	18.6	24.4	25.3	24.6
28	29.45	12.61	71.15	0.9004	2.004	17.5	23.9	24.7	24.0
24	25.31	10.81	66.52	0.8979	1.723	15.2	22.7	23.7	22.8
20	21.15	9.005	61.89	0.8954	1.440	12.4	21.4	22.5	21.6
16	16.97	7.204	57.26	0.8928	1.155	9.1	20.1	21.4	20.3
12	12.77	5.403	52.63	0.8903	0.8688	4.9	18.7	20.1	18.9
8	8.53	3.602	48.00	0.8878	0.5809	−0.6	17.2	18.8	17.5
4	4.28	1.801	43.37	0.8852	0.2913	−8.7	15.7	17.5	16.0
0	0.00	0.000	38.74	0.8827	0.0000	—	14.0	16.0	14.3

39°C DRY BULB

Percentage saturation μ/%	Relative humidity ϕ/%	Moisture content g/g.kg⁻¹	Specific enthalpy h/kJ.kg⁻¹	Specific volume v/m³.kg⁻¹	Vapour pressure p_s/kPa	Dew point temperature t_d/°C	Adiabatic saturation temperature $t\star$/°C	Wet bulb Screen t'_{sc}/°C	Wet bulb Sling t'_{sl}/°C
100	100.00	46.35	158.4	0.9494	6.991	39.0	39.0	39.0	39.0
96	96.27	44.49	153.7	0.9467	6.730	38.3	38.4	38.4	38.4
92	92.52	42.64	148.9	0.9441	6.468	37.6	37.8	37.8	37.8
88	88.74	40.79	144.1	0.9415	6.204	36.8	37.1	37.2	37.2
84	84.95	38.93	139.4	0.9389	5.939	36.0	36.5	36.6	36.5
80	81.13	37.08	134.6	0.9363	5.672	35.2	35.8	35.9	35.8
76	77.30	35.22	129.8	0.9337	5.404	34.3	35.1	35.3	35.1
72	73.44	33.37	125.1	0.9311	5.134	33.4	34.4	34.6	34.4
70	71.50	32.44	122.7	0.9298	4.999	32.9	34.0	34.2	34.0
68	69.56	31.52	120.3	0.9285	4.863	32.4	33.6	33.9	33.7
66	67.61	30.59	117.9	0.9272	4.727	31.9	33.2	33.5	33.3
64	65.65	29.66	115.3	0.9259	4.590	31.4	32.8	33.1	32.9
62	63.69	28.74	113.1	0.9246	4.453	30.8	32.4	32.7	32.5
60	61.73	27.81	110.8	0.9233	4.316	30.3	32.0	32.4	32.1
58	59.76	26.88	108.4	0.9220	4.178	29.7	31.6	32.0	31.7
56	57.78	25.95	106.0	0.9207	4.040	29.2	31.2	31.6	31.3
54	55.80	25.03	103.6	0.9194	3.901	28.5	30.8	31.2	30.8
52	53.81	24.10	101.2	0.9181	3.762	27.9	30.3	30.8	30.4
50	51.81	23.17	98.84	0.9168	3.622	27.3	29.9	30.3	30.0
48	49.81	22.25	94.46	0.9155	3.483	26.6	29.4	29.9	29.5
46	47.81	21.32	94.08	0.9142	3.342	25.9	29.0	29.5	29.0
44	45.80	20.39	91.69	0.9129	3.202	25.2	28.5	29.0	28.6
42	43.78	19.47	89.31	0.9116	3.061	24.4	28.0	28.6	28.1
40	41.76	18.54	86.92	0.9102	2.919	23.6	27.5	28.1	27.6
38	39.73	17.61	84.54	0.9089	2.777	22.8	27.0	27.6	27.1
36	37.69	16.68	82.16	0.9076	2.635	22.0	26.4	27.1	26.6
34	35.65	15.76	79.77	0.9063	2.492	21.0	25.9	26.7	26.0
32	33.60	14.83	77.39	0.9050	2.349	20.1	25.4	26.1	25.5
30	31.55	13.90	75.00	0.9037	2.206	19.1	24.8	25.6	24.9
28	29.49	12.98	72.62	0.9024	2.062	18.0	24.2	25.1	24.4
24	25.35	11.12	67.85	0.8998	1.772	15.6	23.0	24.0	23.2
20	21.19	9.269	63.08	0.8972	1.481	12.8	21.8	22.9	21.9
16	17.00	7.416	58.32	0.8946	1.188	9.5	20.4	21.7	20.6
12	12.79	5.562	53.55	0.8919	0.8940	5.4	19.0	20.4	19.2
8	8.55	3.708	48.78	0.8893	0.5978	−0.3	17.5	19.1	17.8
4	4.29	1.854	44.01	0.8867	0.2998	−8.4	15.9	17.7	16.2
0	0.00	0.000	39.24	0.8841	0.0000	—	14.2	16.2	14.5

39.5°C DRY BULB

Percentage saturation μ/%	Relative humidity ϕ/%	Moisture content g/g.kg⁻¹	Specific enthalpy h/kJ.kg⁻¹	Specific volume v/m³.kg⁻¹	Vapour pressure p_s/kPa	Dew point temperature t_d/°C	Adiabatic saturation temperature $t\star$/°C	Wet bulb Screen t'_{sc}/°C	Wet bulb Sling t'_{sl}/°C
100	100.00	47.70	162.5	0.9528	7.181	39.5	39.5	39.5	39.5
96	96.27	45.80	157.6	0.9501	6.914	38.8	38.9	38.9	38.9
92	92.53	43.89	152.7	0.9474	6.645	38.1	38.3	38.3	38.3
88	88.76	41.98	147.8	0.9447	6.374	37.3	37.6	37.7	37.6
84	84.98	40.07	142.8	0.9420	6.102	36.5	37.0	37.1	37.0
80	81.16	38.16	137.9	0.9394	5.829	35.7	36.3	35.4	36.3
76	77.33	36.25	133.0	0.9367	5.553	34.8	35.6	35.7	35.6
72	73.48	34.35	128.1	0.9340	5.276	33.9	34.9	35.0	34.9
70	71.54	33.39	125.7	0.9326	5.138	33.4	34.5	34.7	34.5
68	69.60	32.44	123.2	0.9313	4.948	32.9	34.1	34.3	34.1
66	67.65	31.48	120.8	0.9300	4.858	32.4	33.7	34.0	33.8
64	65.70	30.53	118.3	0.9286	4.718	31.9	33.3	33.6	33.4
62	63.74	29.58	115.8	0.9273	4.577	31.3	32.9	33.2	33.0
60	61.78	28.62	113.4	0.9259	4.436	30.8	32.5	32.8	32.6
58	59.81	27.67	110.9	0.9246	4.295	30.2	32.1	32.4	32.1
56	57.83	26.71	108.5	0.9232	4.153	29.6	31.6	32.0	31.7
54	55.85	25.76	106.0	0.9219	4.010	29.0	31.2	31.6	31.3
52	53.86	24.81	103.6	0.9205	3.868	28.4	30.8	31.2	30.8
50	51.86	23.85	101.1	0.9192	3.724	27.8	30.3	30.8	30.4
48	49.87	22.90	98.66	0.9179	3.581	27.1	29.9	30.3	29.9
46	47.86	21.94	96.20	0.9165	3.437	26.4	29.4	29.9	29.5
44	45.85	20.99	93.75	0.9152	3.292	25.7	28.9	29.4	29.0
42	43.83	20.04	91.30	0.9138	3.147	24.9	28.4	29.0	28.5
40	41.80	19.08	88.84	0.9125	3.002	24.1	27.9	28.5	28.0
38	39.77	18.13	86.39	0.9111	2.856	23.3	27.4	28.0	27.5
36	37.74	17.17	83.93	0.9098	2.710	22.4	26.8	27.5	27.0
34	35.70	16.22	81.48	0.9084	2.563	21.5	26.3	27.0	26.4
32	33.65	15.27	79.02	0.9071	2.416	20.5	25.7	26.5	25.9
30	31.59	14.31	76.57	0.9057	2.269	19.5	25.2	26.0	25.3
28	29.53	13.36	74.11	0.9044	2.121	18.4	24.6	25.5	24.7
24	25.39	11.45	69.20	0.9017	1.823	16.1	23.4	24.4	23.5
20	21.22	9.541	64.29	0.8990	1.524	13.3	22.1	23.2	22.3
16	17.03	7.632	59.38	0.8963	1.223	9.9	20.7	22.0	20.9
12	12.81	5.724	54.47	0.8936	0.9199	5.8	19.3	20.7	19.5
8	8.57	3.816	49.56	0.8909	0.6152	0.1	17.8	19.4	18.0
4	4.30	1.908	44.66	0.8882	0.3085	−8.1	16.1	17.9	16.4
0	0.00	0.000	39.74	0.8855	0.0000	—	14.4	16.4	14.7

40°C DRY BULB

Percentage saturation μ/%	Relative humidity ϕ/%	Value of stated parameter per kg dry air			Vapour pressure p_s/kPa	Dew point temperature t_d/°C	Adiabatic saturation temperature $t^\star$/°C	Wet bulb temperature	
		Moisture content g/g.kg^{-1}	Specific enthalpy h/kJ.kg^{-1}	Specific volume v/m^3.kg^{-1}				Screen t'_{sc}/°C	Sling t'_{sl}/°C
100	100.00	49.10	166.6	0.9563	7.375	40.0	40.0	40.0	40.0
96	96.28	47.13	161.6	0.9535	7.101	39.3	39.4	39.4	39.4
92	92.54	45.17	156.5	0.9507	6.826	38.6	38.8	38.8	38.8
88	88.78	43.21	151.5	0.9480	6.548	37.8	38.1	38.2	38.1
84	85.00	41.24	146.4	0.9452	6.269	37.0	37.5	37.6	37.5
80	81.20	39.28	141.3	0.9424	5.989	36.1	36.8	36.9	36.8
76	77.37	37.31	136.3	0.9397	5.706	35.3	36.1	36.2	36.1
72	73.52	35.35	131.2	0.9369	5.422	34.3	35.3	35.5	35.4
70	71.58	34.37	128.7	0.9355	5.280	33.9	34.9	35.2	35.0
68	69.64	33.39	126.2	0.9341	5.137	33.4	34.6	34.8	34.6
66	67.70	32.40	123.7	0.9327	4.993	32.9	34.2	34.4	34.2
64	65.75	31.42	121.1	0.9314	4.849	32.4	33.8	34.0	33.8
62	63.79	30.44	118.6	0.9300	4.705	31.8	33.4	33.7	33.4
60	61.83	29.46	116.1	0.9286	4.560	31.3	32.9	33.3	33.0
58	59.86	28.48	113.5	0.9272	4.415	30.7	32.5	32.9	32.6
56	57.88	27.49	111.0	0.9258	4.269	30.1	32.1	32.5	32.2
54	55.90	26.51	108.5	0.9244	4.123	29.5	31.7	32.0	31.7
52	53.91	25.53	106.0	0.9230	3.976	28.9	31.2	31.6	31.3
50	51.92	24.55	103.4	0.9217	3.829	28.2	30.7	31.2	30.8
48	49.92	23.57	100.9	0.9203	3.682	27.6	30.3	30.8	30.4
46	47.91	22.58	98.38	0.9189	3.534	26.9	29.8	30.3	29.9
44	45.90	21.60	95.85	0.9175	3.385	26.1	29.3	29.9	29.4
42	43.88	20.62	93.32	0.9161	3.236	25.4	28.8	29.4	28.9
40	41.86	19.64	90.80	0.9147	3.087	24.6	28.3	28.9	28.4
38	39.82	18.66	88.27	0.9133	2.937	23.7	27.8	28.4	27.9
36	37.79	17.67	85.74	0.9119	2.787	22.9	27.2	27.9	27.4
34	35.74	16.69	83.21	0.9106	2.636	22.0	26.7	27.4	26.8
32	33.69	15.71	80.69	0.9092	2.485	21.0	26.1	26.9	26.3
30	31.64	14.73	78.16	0.9078	2.333	20.0	25.6	26.4	25.7
28	29.57	13.75	75.63	0.9064	2.181	18.9	25.0	25.8	25.1
24	25.43	11.78	70.58	0.9036	1.875	16.5	23.7	24.7	23.9
20	21.26	9.819	65.52	0.9008	1.568	13.7	22.4	23.5	22.6
16	17.06	7.856	60.47	0.8981	1.258	10.4	21.0	22.3	21.2
12	12.83	5.892	55.41	0.8953	0.9466	6.2	19.6	21.0	19.8
8	8.58	3.928	50.36	0.8925	0.6330	0.5	18.0	19.6	18.3
4	4.31	1.964	45.30	0.8897	0.3175	−7.7	16.3	18.2	16.6
0	0.00	0.000	40.25	0.8869	0.0000	—	14.6	16.6	14.9

41°C DRY BULB

Percentage saturation μ/%	Relative humidity ϕ/%	Value of stated parameter per kg dry air			Vapour pressure p_s/kPa	Dew point temperature t_d/°C	Adiabatic saturation temperature $t^\star$/°C	Wet bulb temperature	
		Moisture content g/g.kg^{-1}	Specific enthalpy h/kJ.kg^{-1}	Specific volume v/m^3.kg^{-1}				Screen t'_{sc}/°C	Sling t'_{sl}/°C
100	100.00	52.00	175.2	0.9634	7.778	41.0	41.0	41.0	41.0
96	96.30	49.92	169.8	0.9605	7.490	40.3	40.3	40.4	40.4
92	92.57	47.84	164.5	0.9576	7.200	39.5	39.8	39.8	39.8
88	88.83	45.76	159.1	0.9546	6.909	38.8	39.1	39.2	39.1
84	85.06	43.68	153.8	0.9517	6.616	38.0	38.4	38.5	38.5
80	81.26	41.60	148.4	0.9487	6.320	37.1	37.7	37.9	37.8
76	77.44	39.52	143.0	0.9458	6.023	36.3	37.0	37.2	37.0
72	73.60	37.44	137.7	0.9429	5.725	35.3	36.3	36.5	36.3
70	71.67	36.40	135.0	0.9414	5.575	34.8	35.9	36.1	35.9
68	69.74	35.36	132.3	0.9399	5.424	34.4	35.5	35.7	35.5
66	67.79	34.32	129.7	0.9384	5.273	33.8	35.1	35.3	35.1
64	65.84	33.28	127.0	0.9370	5.121	33.3	34.7	35.0	34.7
62	63.89	32.24	124.3	0.9355	4.969	32.8	34.3	34.6	34.3
60	61.93	31.20	121.6	0.9340	4.817	32.2	33.9	34.2	33.9
58	59.96	30.16	118.9	0.9326	4.664	31.7	33.4	33.8	33.5
56	57.99	29.12	116.3	0.9311	4.510	31.1	33.0	33.4	33.1
54	56.01	28.08	113.6	0.9296	4.356	30.5	32.5	32.9	32.6
52	54.02	27.04	110.9	0.9281	4.201	29.8	32.1	32.5	32.2
50	52.03	26.00	108.2	0.9267	4.046	29.2	31.6	32.1	31.7
48	50.03	24.96	105.5	0.9252	3.891	28.5	31.1	31.6	31.2
46	48.02	23.92	102.9	0.9237	3.735	37.8	30.7	31.2	30.7
44	46.01	22.88	100.2	0.9222	3.578	27.1	30.2	30.7	30.3
42	43.99	21.84	97.51	0.9208	3.421	26.3	29.6	30.2	29.7
40	41.96	20.80	94.83	0.9193	3.264	25.5	29.1	29.7	29.2
38	39.93	19.76	92.15	0.9178	3.106	24.7	28.6	29.2	28.7
36	37.89	18.72	89.47	0.9163	2.947	23.8	28.0	28.7	28.2
34	35.84	17.68	86.79	0.9149	2.788	22.9	27.5	28.2	27.6
32	33.79	16.64	84.12	0.9134	2.628	21.9	26.9	27.7	27.0
30	31.73	15.60	81.44	0.9119	2.468	20.9	26.3	27.1	26.4
28	29.66	14.56	78.76	0.9104	2.307	19.8	25.7	26.6	25.8
24	25.51	12.48	73.40	0.9075	1.984	17.4	24.4	25.4	24.6
20	21.33	10.40	68.04	0.9045	1.659	14.6	23.1	24.2	23.3
16	17.12	8.320	62.69	0.9016	1.332	11.2	21.7	22.9	21.9
12	12.88	6.240	57.33	0.8986	1.002	7.0	20.2	21.6	20.4
8	8.62	4.160	51.97	0.8957	0.6702	1.3	18.5	20.1	18.8
4	4.32	2.080	46.61	0.8927	0.3362	−7.1	16.8	18.6	17.1
0	0.00	0.000	41.26	0.8898	0.0000	—	14.9	17.0	15.3

42°C DRY BULB

Percentage saturation μ/%	Relative humidity ϕ/%	Value of stated parameter per kg dry air			Vapour pressure p_s/kPa	Dew point temperature t_d/°C	Adiabatic saturation temperature $t^\star$/°C	Wet bulb temperature	
		Moisture content $g/g.kg^{-1}$	Specific enthalpy $h/kJ.kg^{-1}$	Specific volume $v/m^3.kg^{-1}$				Screen t'_{sc}/°C	Sling t'_{sl}/°C
100	100.00	55.07	184.2	0.9709	8.199	42.0	42.0	42.0	42.0
96	96.31	52.87	178.5	0.9677	7.897	41.3	41.3	41.4	41.4
92	92.61	50.66	172.8	0.9646	7.593	40.5	40.7	40.8	40.8
88	88.87	48.46	167.2	0.9615	7.287	39.8	40.0	40.2	40.1
84	85.11	46.26	161.5	0.9584	6.978	39.0	39.4	39.5	39.4
80	81.33	44.06	155.8	0.9553	6.668	38.1	38.7	38.8	38.7
76	77.52	41.85	150.1	0.9521	6.356	37.2	38.0	38.1	38.0
72	73.69	39.65	144.5	0.9490	6.042	36.3	37.2	37.4	37.3
70	71.76	38.55	141.6	0.9474	5.884	35.8	36.8	37.0	36.9
68	69.83	37.45	138.8	0.9459	5.726	35.3	36.4	36.7	36.5
66	67.89	36.35	135.9	0.9443	5.566	34.8	36.0	36.3	36.1
64	65.95	35.24	133.1	0.9428	5.407	34.3	35.6	35.9	35.7
62	63.99	34.14	130.3	0.9412	5.247	33.8	35.2	35.5	35.3
60	62.03	33.04	127.4	0.9396	5.086	33.2	34.8	35.1	34.8
58	60.07	31.94	124.6	0.9381	4.925	32.6	34.3	34.7	34.4
56	58.10	30.84	121.7	0.9365	4.763	32.0	33.9	34.2	34.0
54	56.12	29.74	118.9	0.9349	4.601	31.4	33.4	33.8	33.5
52	54.13	28.64	116.1	0.9334	4.438	30.8	33.0	33.4	33.0
50	52.14	27.53	113.2	0.9318	4.275	30.1	32.5	32.9	32.6
48	50.14	26.43	110.4	0.9302	4.111	29.5	32.0	32.5	32.1
46	48.13	25.33	107.6	0.9287	3.946	28.7	31.5	32.0	31.6
44	46.12	24.23	104.7	0.9271	3.781	28.0	31.0	31.5	31.1
42	44.10	23.13	101.9	0.9255	3.616	27.2	30.5	31.0	30.6
40	42.07	22.03	99.03	0.9240	3.449	26.4	29.9	30.5	30.1
38	40.04	20.93	96.20	0.9224	3.283	25.6	29.4	30.0	29.5
36	38.00	19.82	93.36	0.9208	3.115	24.7	28.8	29.5	29.0
34	35.95	18.72	90.52	0.9193	2.947	23.8	28.3	29.0	28.4
32	33.89	17.62	87.68	0.9177	2.779	22.8	27.7	28.4	27.8
30	31.83	16.52	84.84	0.9161	2.610	21.8	27.1	27.9	27.2
28	29.76	15.42	82.00	0.9146	2.440	20.7	26.4	27.3	26.6
24	25.60	13.22	76.33	0.9114	2.099	18.3	25.1	26.1	25.3
20	21.41	11.01	70.65	0.9083	1.755	15.5	23.8	24.9	23.9
16	17.18	8.811	64.97	0.9052	1.409	12.1	22.3	23.5	22.5
12	12.93	6.608	59.40	0.9020	1.060	7.8	20.7	22.2	21.0
8	8.65	4.406	53.62	0.8989	0.7095	2.1	19.1	20.7	19.3
4	4.34	2.203	47.94	0.8957	0.3560	−6.4	17.2	19.1	17.6
0	0.00	0.000	42.26	0.8926	0.0000	—	15.3	17.4	15.7

43°C DRY BULB

Percentage saturation μ/%	Relative humidity ϕ/%	Value of stated parameter per kg dry air			Vapour pressure p_s/kPa	Dew point temperature t_d/°C	Adiabatic saturation temperature $t^\star$/°C	Wet bulb temperature	
		Moisture content $g/g.kg^{-1}$	Specific enthalpy $h/kJ.kg^{-1}$	Specific volume $v/m^3.kg^{-1}$				Screen t'_{sc}/°C	Sling t'_{sl}/°C
100	100.00	58.30	193.7	0.9786	8.640	43.0	43.0	43.0	43.0
96	96.33	55.98	187.6	0.9752	8.322	42.3	42.3	42.4	42.4
92	92.64	53.64	181.6	0.9719	8.004	41.5	41.6	41.8	41.7
88	88.92	51.31	175.6	0.9686	7.682	40.8	40.9	41.1	41.1
84	85.18	48.98	169.6	0.9653	7.359	40.0	40.2	40.5	40.4
80	81.40	46.65	163.6	0.9620	7.033	39.1	39.7	39.8	39.7
76	77.61	44.31	157.6	0.9587	6.705	38.2	38.9	39.1	39.0
72	73.78	41.98	151.6	0.9553	6.375	37.3	38.2	38.4	38.2
70	71.86	40.82	148.5	0.9537	6.209	36.8	37.8	38.0	37.8
68	69.93	39.65	145.5	0.9520	6.042	36.3	37.4	37.6	37.4
66	68.00	38.48	142.5	0.9504	5.875	35.8	37.0	37.2	37.0
64	66.05	37.32	139.5	0.9487	5.707	35.3	36.5	36.8	36.6
62	64.10	36.15	136.5	0.9470	5.538	34.7	36.1	36.4	36.2
60	62.15	34.98	133.5	0.9454	5.369	34.2	35.7	36.0	35.7
58	60.18	33.82	130.5	0.9437	5.200	33.6	35.2	35.6	35.3
56	58.21	32.65	127.5	0.9421	5.029	33.0	34.8	35.1	34.9
54	56.24	31.49	124.5	0.9404	4.858	32.4	34.3	34.7	34.4
52	54.25	30.32	121.5	0.9387	4.687	31.7	33.9	34.3	33.9
50	52.26	29.15	118.5	0.9371	4.515	31.1	33.4	33.8	33.5
48	50.26	27.99	115.5	0.9354	4.342	30.4	32.9	33.3	33.0
46	48.25	26.82	112.4	0.9337	4.169	29.7	32.4	32.9	32.5
44	46.24	25.66	109.4	0.9321	3.995	29.0	31.9	32.4	31.9
42	44.22	24.49	106.4	0.9304	3.820	28.2	31.3	31.9	31.4
40	42.19	23.32	103.4	0.9288	3.645	27.4	30.8	31.4	30.9
38	40.15	22.16	100.4	0.9271	3.469	26.5	30.2	30.8	30.3
36	38.11	20.99	97.40	0.9254	3.292	25.7	29.6	30.3	29.8
34	36.06	19.82	94.40	0.9238	3.115	24.7	29.1	29.8	29.2
32	34.00	18.66	91.39	0.9221	2.937	23.7	28.5	29.2	28.6
30	31.93	17.49	88.39	0.9204	2.759	22.7	27.8	28.6	28.0
28	29.86	16.33	85.38	0.9188	2.580	21.6	27.2	28.1	27.3
24	25.69	13.99	79.36	0.9154	2.219	19.2	25.9	26.8	26.0
20	21.49	11.66	73.34	0.9121	1.856	16.3	24.4	25.5	24.6
16	17.25	9.329	67.33	0.9088	1.491	12.9	22.9	24.2	23.1
12	12.99	6.997	61.31	0.9055	1.122	8.7	21.3	22.7	21.5
8	8.69	4.665	55.30	0.9021	0.7509	2.9	19.6	21.2	19.8
4	4.36	2.332	49.29	0.8988	0.3769	−5.7	17.7	19.6	18.3
0	0.00	0.000	43.27	0.8954	0.0000	—	15.7	17.8	16.0

44°C DRY BULB

Percentage saturation μ/%	Relative humidity ϕ/%	Value of stated parameter per kg dry air			Vapour pressure p_s/kPa	Dew point temperature t_d/°C	Adiabatic saturation temperature $t^\star$/°C	Wet bulb temperature	
		Moisture content $g/g.kg^{-1}$	Specific enthalpy h/kJ.kg^{-1}	Specific volume v/m^3.kg^{-1}				Screen t'_{sc}/°C	Sling t'_{sl}/°C
100	100.00	61.73	203.6	0.9865	9.100	44.0	44.0	44.0	44.0
96	96.35	59.26	197.2	0.9830	8.768	43.3	43.3	43.4	43.4
92	92.67	56.79	190.9	0.9795	8.434	42.5	42.6	42.8	42.7
88	88.97	54.32	184.5	0.9760	8.097	41.8	41.9	42.1	42.1
84	85.24	51.85	178.1	0.9725	7.757	40.9	41.2	41.4	41.4
80	81.48	49.38	171.7	0.9689	7.415	40.1	40.4	40.8	40.7
76	77.69	46.91	165.4	0.9654	7.071	39.2	39.9	40.0	39.9
72	73.88	44.45	159.0	0.9619	6.723	38.3	39.1	39.3	39.2
70	71.96	43.21	155.8	0.9601	6.549	37.8	38.7	38.9	38.8
68	70.04	41.98	152.6	0.9584	6.374	37.3	38.3	38.5	38.4
66	68.11	40.74	149.4	0.9566	6.198	36.8	37.9	38.1	37.9
64	66.17	39.51	146.2	0.9548	6.022	36.2	37.5	37.7	37.5
62	64.22	38.27	143.1	0.9531	5.844	35.7	37.0	37.3	37.1
60	62.27	37.04	139.9	0.9513	5.666	35.1	36.6	36.9	36.7
58	60.30	35.80	136.7	0.9495	5.488	34.6	36.2	36.5	36.2
56	58.34	34.57	133.5	0.9478	5.309	34.0	35.7	36.0	35.8
54	56.36	33.33	130.3	0.9460	5.129	33.3	35.2	35.6	35.3
52	54.38	32.10	127.1	0.9443	4.949	32.7	34.7	35.1	34.8
50	52.38	30.87	123.9	0.9425	4.767	32.0	34.3	34.7	34.3
48	50.38	29.63	120.8	0.9407	4.585	31.4	33.8	34.2	33.8
46	48.38	28.40	117.6	0.9390	4.403	30.6	33.2	33.7	33.3
44	46.36	27.16	114.4	0.9372	4.219	29.9	32.7	33.2	32.8
42	44.34	25.93	111.2	0.9354	4.035	29.1	32.2	32.7	32.3
40	42.31	24.69	108.0	0.9337	3.851	28.3	31.6	32.2	31.7
38	40.27	23.46	104.8	0.9319	3.665	27.5	31.0	31.7	31.2
36	38.23	22.22	101.6	0.9301	3.479	26.6	30.5	31.1	30.6
34	36.17	20.99	98.45	0.9284	3.292	25.7	29.9	30.6	30.0
32	34.11	19.75	95.26	0.9266	3.104	24.7	29.2	30.0	29.4
30	32.04	18.52	92.07	0.9248	2.916	23.6	28.6	29.4	28.7
28	29.97	17.28	88.89	0.9231	2.727	22.5	28.0	28.8	28.1
24	25.78	14.82	82.51	0.9195	2.347	20.1	26.6	27.5	26.7
20	21.57	12.35	76.14	0.9160	1.963	17.2	25.1	26.2	25.3
16	17.33	9.877	69.77	0.9125	1.577	13.8	23.6	24.8	23.8
12	13.05	7.408	63.40	0.9089	1.187	9.5	21.9	23.3	22.1
8	8.73	4.938	57.02	0.9054	0.7946	3.7	20.1	21.7	20.4
4	4.38	2.469	50.65	0.9018	0.3989	−5.1	18.1	20.0	18.5
0	0.00	0.000	44.28	0.8983	0.0000	—	16.0	18.2	16.4

45°C DRY BULB

Percentage saturation μ/%	Relative humidity ϕ/%	Value of stated parameter per kg dry air			Vapour pressure p_s/kPa	Dew point temperature t_d/°C	Adiabatic saturation temperature $t^\star$/°C	Wet bulb temperature	
		Moisture content $g/g.kg^{-1}$	Specific enthalpy h/kJ.kg^{-1}	Specific volume v/m^3.kg^{-1}				Screen t'_{sc}/°C	Sling t'_{sl}/°C
100	100.00	65.34	214.1	0.9948	9.582	45.0	45.0	45.0	45.0
96	96.37	62.73	207.3	0.9911	9.234	44.3	44.3	44.4	44.4
92	92.71	60.12	200.6	0.9874	8.884	43.5	43.6	43.8	43.7
88	89.02	57.50	193.8	0.9836	8.530	42.8	42.9	43.1	43.0
84	85.31	54.89	187.0	0.9799	8.174	41.9	42.2	42.4	42.4
80	81.56	52.28	180.3	0.9761	7.815	41.1	41.4	41.7	41.6
76	77.79	49.66	173.6	0.9724	7.454	40.2	40.6	41.0	40.9
72	73.98	47.05	166.8	0.9687	7.089	39.3	40.0	40.2	40.1
70	72.07	45.74	163.4	0.9668	6.906	38.8	39.7	39.9	39.7
68	70.15	44.44	160.0	0.9649	6.722	38.3	39.3	39.5	39.3
66	68.22	43.13	156.7	0.9630	6.537	37.8	38.8	39.1	38.9
64	66.28	41.82	153.3	0.9612	6.352	37.2	38.4	38.7	38.5
62	64.34	40.51	149.9	0.9593	6.165	36.7	38.0	38.2	38.0
60	62.39	39.21	146.5	0.9574	5.978	36.1	37.5	37.8	37.6
58	60.43	37.90	143.2	0.9556	5.791	35.5	37.1	37.4	37.1
56	58.46	36.59	139.8	0.9537	5.602	34.9	36.6	36.9	36.7
54	56.49	35.29	136.4	0.9518	5.413	34.3	36.1	36.5	36.2
52	54.51	33.98	133.0	0.9499	5.223	33.7	35.6	36.0	35.7
50	52.52	32.67	129.7	0.9481	5.032	33.0	35.1	35.6	35.2
48	50.52	31.37	126.3	0.9462	4.841	32.3	34.6	35.1	34.7
46	48.51	30.06	122.9	0.9443	4.648	31.6	34.1	34.6	34.2
44	46.49	28.75	119.5	0.9424	4.455	30.9	33.6	34.1	33.7
42	44.47	27.45	116.2	0.9406	4.261	30.1	33.0	33.6	33.1
40	42.44	26.14	112.8	0.9387	4.067	29.3	32.5	33.0	32.6
38	40.40	24.83	109.4	0.9368	3.871	28.4	31.9	32.5	32.0
36	38.35	23.52	106.0	0.9349	3.675	27.5	31.3	31.9	31.4
34	36.30	22.22	102.7	0.9331	3.478	26.6	30.7	31.4	30.8
32	34.23	20.91	99.29	0.9312	3.280	25.6	30.0	30.8	30.2
30	32.16	19.60	95.91	0.9293	3.082	24.5	29.4	30.2	29.5
28	30.08	18.30	92.54	0.9274	2.882	23.4	28.7	29.5	28.9
24	25.89	15.68	85.79	0.9237	2.481	21.0	27.3	28.3	27.5
20	21.66	13.07	79.04	0.9199	2.076	18.1	25.8	26.9	26.0
16	17.40	10.46	72.29	0.9162	1.668	14.7	24.2	25.4	24.4
12	13.11	7.842	65.54	0.9124	1.256	10.3	22.5	23.9	22.7
8	8.77	5.228	58.79	0.9086	0.8417	4.5	20.6	22.2	20.9
4	4.41	2.614	52.04	0.9049	0.4221	−4.4	18.6	20.5	18.9
0	0.00	0.000	45.28	0.9011	0.0000	—	16.4	18.6	16.8

46°C DRY BULB

Percentage saturation μ/%	Relative humidity ϕ/%	Value of stated parameter per kg dry air			Vapour pressure p_s/kPa	Dew point temperature t_d/°C	Adiabatic saturation temperature $t\star$/°C	Wet bulb temperature	
		Moisture content g/g.kg^{-1}	Specific enthalpy h/kJ.kg^{-1}	Specific volume v/m^3.kg^{-1}				Screen t'_{sc}/°C	Sling t'_{sl}/°C
100	100.00	69.17	225.0	1.003	10.09	46.0	46.0	46.0	46.0
96	96.39	66.40	217.9	0.9995	9.722	45.3	45.3	45.4	45.4
92	92.75	63.63	210.8	0.9955	9.354	44.5	44.6	44.7	44.7
88	89.08	60.87	203.6	0.9915	8.984	43.8	43.9	44.1	44.0
84	85.37	58.10	196.4	0.9876	8.611	42.9	43.1	43.4	43.3
80	81.64	55.33	189.3	0.9836	8.235	42.1	42.3	42.7	42.6
76	77.88	52.57	182.1	0.9796	7.855	41.2	41.5	42.0	41.9
72	74.09	49.80	175.0	0.9757	7.473	40.2	40.7	41.2	41.1
70	72.18	48.42	171.4	0.9737	7.280	39.8	40.2	40.8	40.7
68	70.27	47.03	167.8	0.9717	7.087	39.3	40.0	40.4	40.3
66	68.34	45.65	164.3	0.9697	6.893	38.7	39.8	40.0	39.8
64	66.41	44.27	160.7	0.9677	6.698	38.2	39.3	39.6	39.4
62	64.47	42.88	157.1	0.9657	6.502	37.7	38.9	39.2	39.0
60	62.52	41.50	153.5	0.9637	6.306	37.1	38.5	38.7	38.5
58	60.56	40.12	150.0	0.9618	6.108	36.5	38.0	38.3	38.1
56	58.60	38.73	146.4	0.9598	5.910	35.9	37.5	37.8	37.6
54	56.63	37.35	142.8	0.9578	5.711	35.3	37.0	37.4	37.1
52	54.64	35.97	139.2	0.9558	5.511	34.6	36.5	36.9	36.6
50	52.65	34.58	135.7	0.9538	5.311	34.0	36.0	36.4	36.1
48	50.66	33.20	132.1	0.9518	5.109	33.3	35.5	35.9	35.6
46	48.65	31.82	128.5	0.9498	4.907	32.6	35.0	35.4	35.1
44	46.63	30.43	124.9	0.9478	4.703	31.8	34.4	34.9	34.5
42	44.61	29.05	121.4	0.9458	4.499	31.0	33.9	34.4	34.0
40	42.58	27.67	117.8	0.9438	4.294	30.2	33.3	33.9	33.4
38	40.53	26.28	114.2	0.9419	4.088	29.4	32.7	33.3	32.8
36	38.48	24.90	110.6	0.9399	3.882	28.5	32.1	32.7	32.2
34	36.43	23.52	107.1	0.9379	3.674	27.5	31.5	32.2	31.6
32	34.36	22.13	103.5	0.9359	3.465	26.5	30.8	31.6	31.0
30	32.28	20.75	99.92	0.9339	3.256	25.5	30.2	30.9	30.3
28	30.19	19.37	96.34	0.9319	3.045	24.3	29.5	30.3	29.6
24	25.99	16.60	89.19	0.9279	2.622	21.9	28.0	29.0	28.2
20	21.76	13.83	82.04	0.9239	2.194	19.0	26.5	27.6	26.7
16	17.48	11.07	74.89	0.9199	1.763	15.5	24.8	26.1	25.1
12	13.17	8.300	67.74	0.9159	1.328	11.2	23.1	24.5	23.3
8	8.82	5.533	60.59	0.9120	0.8895	5.3	21.1	22.8	21.4
4	4.43	2.767	53.44	0.9080	0.4467	−3.7	19.0	20.9	19.4
0	0.00	0.000	46.29	0.9040	0.0000	—	16.7	19.0	17.1

47°C DRY BULB

Percentage saturation μ/%	Relative humidity ϕ/%	Value of stated parameter per kg dry air			Vapour pressure p_s/kPa	Dew point temperature t_d/°C	Adiabatic saturation temperature $t\star$/°C	Wet bulb temperature	
		Moisture content g/g.kg^{-1}	Specific enthalpy h/kJ.kg^{-1}	Specific volume v/m^3.kg^{-1}				Screen t'_{sc}/°C	Sling t'_{sl}/°C
100	100.00	73.20	236.6	1.012	10.61	47.0	47.0	47.0	47.0
96	96.41	70.27	229.0	1.008	10.23	46.3	46.3	46.4	46.4
92	92.78	67.35	221.5	1.004	9.847	45.5	45.6	45.7	45.7
88	89.13	64.42	213.9	0.9998	9.459	44.7	44.9	45.1	45.0
84	85.45	61.49	206.3	0.9956	9.068	43.9	44.1	44.4	44.3
80	81.73	58.56	198.8	0.9913	8.674	43.1	43.3	43.7	43.6
76	77.98	55.63	191.2	0.9871	8.276	42.2	42.5	42.9	42.8
72	74.20	52.71	183.6	0.9829	7.875	41.2	41.6	42.2	42.0
70	72.30	51.24	179.8	0.9808	7.673	40.7	41.2	41.8	41.6
68	70.39	49.78	176.0	0.9787	7.470	40.2	40.7	41.4	41.2
66	68.47	48.31	172.2	0.9766	7.266	39.7	40.3	40.9	40.8
64	66.54	46.85	168.5	0.9745	7.061	39.2	40.0	40.5	40.3
62	64.60	45.39	164.7	0.9724	6.856	38.6	39.8	40.1	39.9
60	62.66	43.92	160.9	0.9703	6.649	38.1	39.4	39.7	39.4
58	60.70	42.46	157.1	0.9681	6.442	37.5	38.9	39.2	39.0
56	58.74	40.99	153.3	0.9660	6.234	36.9	38.4	38.8	38.5
54	56.77	39.53	149.5	0.9639	6.025	36.3	37.9	38.3	38.0
52	54.79	38.07	145.7	0.9618	5.814	35.6	37.4	37.8	37.5
50	52.80	36.60	142.0	0.9597	5.603	34.9	36.9	37.3	37.0
48	50.80	35.14	138.2	0.9576	5.391	34.2	36.4	36.8	36.5
46	48.79	33.67	134.4	0.9555	5.178	33.5	35.9	36.3	35.9
44	46.78	32.21	130.6	0.9534	4.964	32.8	35.3	35.8	35.4
42	44.75	30.75	126.8	0.9513	4.749	32.0	34.7	35.3	34.8
40	42.72	29.28	123.0	0.9491	4.534	31.2	34.1	34.7	34.3
38	40.68	27.82	119.2	0.9470	4.317	30.3	33.5	34.1	33.7
36	38.62	26.35	115.4	0.9449	4.099	29.4	32.9	33.6	33.0
34	36.56	24.89	111.7	0.9428	3.880	28.5	32.3	33.0	32.4
32	34.49	23.43	107.9	0.9407	3.660	27.5	31.6	32.3	31.8
30	32.41	21.96	104.1	0.9386	3.439	26.4	30.9	31.7	31.1
28	30.32	20.50	100.3	0.9365	3.217	25.3	30.2	31.1	30.4
24	26.11	17.57	92.74	0.9322	2.770	22.8	28.8	29.7	28.9
20	21.86	14.64	85.16	0.9280	2.319	19.9	27.2	28.3	27.4
16	17.57	11.71	77.59	0.9238	1.864	16.4	25.5	26.7	25.7
12	13.24	8.784	70.02	0.9195	1.405	12.0	23.7	25.1	23.9
8	8.87	5.856	62.44	0.9153	0.9409	6.1	21.7	23.3	21.9
4	4.45	2.928	54.87	0.9110	0.4727	−3.1	19.5	21.4	19.8
0	0.00	0.000	47.30	0.9068	0.0000	—	17.1	19.3	17.5

48°C DRY BULB

Percentage saturation μ/%	Relative humidity ϕ/%	Value of stated parameter per kg dry air			Vapour pressure p_s/kPa	Dew point temperature t_d/°C	Adiabatic saturation temperature $t^\star$/°C	Wet bulb temperature	
		Moisture content g/g.kg^{-1}	Specific enthalpy h/kJ.kg^{-1}	Specific volume v/m³.kg^{-1}				Screen t'_{sc}/°C	Sling t'_{sl}/°C
100	100.00	77.47	248.8	1.022	11.16	48.0	48.0	48.0	48.0
96	96.43	74.37	240.8	1.017	10.76	47.3	47.4	47.4	47.4
92	92.83	71.27	232.8	1.013	10.36	46.5	46.7	46.7	46.7
88	89.19	68.17	224.7	1.008	9.956	45.7	46.0	46.1	46.0
84	85.52	65.08	216.7	1.004	9.546	44.9	45.3	45.4	45.3
80	81.82	61.98	208.7	0.9994	9.133	44.1	44.5	44.6	44.6
76	78.09	58.88	200.7	0.9949	8.716	43.2	43.8	43.9	43.8
72	74.32	55.78	192.7	0.9904	8.295	42.2	42.9	43.1	43.0
70	72.42	54.23	188.7	0.9882	8.084	41.7	42.5	42.7	42.6
68	70.51	52.68	184.6	0.9860	7.871	41.2	42.1	42.3	42.2
66	68.60	51.13	180.6	0.9837	7.657	40.7	41.7	41.9	41.7
64	66.68	49.58	176.6	0.9815	7.442	40.2	41.2	41.5	41.3
62	64.74	48.03	172.6	0.9792	7.227	39.6	40.8	41.0	40.8
60	62.80	46.48	168.6	0.9770	7.010	39.0	40.3	40.6	40.4
58	60.85	44.93	164.6	0.9748	6.792	38.5	39.8	40.1	39.9
56	58.89	43.38	160.6	0.9725	6.573	37.9	39.4	39.7	39.4
54	56.92	41.83	156.6	0.9703	6.353	37.2	38.9	39.2	38.9
52	54.94	40.28	152.6	0.9680	6.133	36.6	38.3	38.7	38.4
50	52.95	38.74	148.6	0.9658	5.911	35.9	37.8	38.2	37.9
48	50.95	37.19	144.5	0.9636	5.688	35.2	37.3	37.7	37.4
46	48.95	35.64	140.5	0.9613	5.464	34.5	36.7	37.2	36.8
44	46.93	34.09	136.5	0.9591	5.239	33.7	36.2	36.7	36.3
42	44.90	32.54	132.5	0.9568	5.012	32.9	35.6	36.1	35.7
40	42.87	30.99	128.5	0.9546	4.785	32.1	35.0	35.5	35.1
38	40.82	29.44	124.5	0.9523	4.557	31.3	34.4	35.0	34.5
36	38.77	27.89	120.5	0.9501	4.327	30.3	33.8	34.4	33.9
34	36.70	26.34	116.5	0.9479	4.097	29.4	33.1	33.8	33.2
32	34.63	24.79	112.5	0.9456	3.865	28.4	32.4	33.1	32.6
30	32.54	23.24	108.5	0.9434	3.632	27.3	31.7	32.5	31.9
28	30.45	21.69	104.4	0.9411	3.398	26.2	31.0	31.8	31.2
24	26.22	18.59	96.42	0.9366	2.927	23.7	29.5	30.4	29.7
20	21.96	15.49	88.40	0.9321	2.451	20.8	27.9	29.0	28.1
16	17.66	12.40	80.38	0.9276	1.971	17.3	26.1	27.4	26.4
12	13.31	9.296	72.36	0.9231	1.485	12.9	24.3	25.7	24.5
8	8.92	6.198	64.35	0.9186	0.9952	6.9	22.2	23.8	22.5
4	4.48	3.099	56.33	0.9141	0.5000	−2.4	19.9	21.8	20.2
0	0.00	0.000	48.31	0.9096	0.0000	—	17.4	19.7	17.8

49°C DRY BULB

Percentage saturation μ/%	Relative humidity ϕ/%	Value of stated parameter per kg dry air			Vapour pressure p_s/kPa	Dew point temperature t_d/°C	Adiabatic saturation temperature $t^\star$/°C	Wet bulb temperature	
		Moisture content g/g.kg^{-1}	Specific enthalpy h/kJ.kg^{-1}	Specific volume v/m³.kg^{-1}				Screen t'_{sc}/°C	Sling t'_{sl}/°C
100	100.00	81.98	261.6	1.0315	11.74	49.0	49.0	49.0	49.0
96	96.45	78.70	253.1	1.0267	11.32	48.3	48.3	48.4	48.4
92	92.87	75.43	244.6	1.0220	10.90	47.5	47.7	47.7	47.7
88	89.25	72.15	236.2	1.0172	10.48	46.7	47.0	47.0	47.0
84	85.60	68.87	227.7	1.0125	10.05	45.9	46.3	46.3	46.3
80	81.92	65.59	219.2	1.0077	9.614	45.1	45.5	45.6	45.5
76	78.20	62.31	210.7	1.0030	9.177	44.2	44.7	44.9	44.8
72	74.44	59.03	202.2	0.9982	8.736	43.2	43.9	44.1	44.0
70	72.55	57.39	197.9	0.9959	8.515	42.7	43.5	43.7	43.5
68	70.65	55.75	193.7	0.9935	8.291	42.2	43.1	43.3	43.1
66	68.74	54.11	189.4	0.9911	8.067	41.7	42.6	42.8	42.7
64	66.82	52.47	185.2	0.9887	7.842	41.2	42.2	42.4	42.2
62	64.89	50.83	181.0	0.9864	7.616	40.6	41.7	42.0	41.8
60	62.95	49.19	176.7	0.9840	7.388	40.0	41.2	41.5	41.3
58	61.00	47.55	172.5	0.9816	7.159	39.4	40.8	41.1	40.8
56	59.04	45.91	168.2	0.9792	6.930	38.8	40.3	40.6	40.3
54	57.08	44.27	164.0	0.9768	6.699	38.2	39.8	40.1	39.8
52	55.10	42.63	159.7	0.9745	6.467	37.6	39.3	39.6	39.3
50	53.11	40.99	155.5	0.9721	6.233	36.9	38.7	39.1	38.8
48	51.12	39.35	151.2	0.9697	5.999	36.2	38.2	38.6	38.3
46	49.11	37.71	147.0	0.9673	5.764	35.4	37.6	38.1	37.7
44	47.09	36.07	142.7	0.9649	5.527	34.7	37.1	37.5	37.2
42	45.06	34.43	138.5	0.9626	5.289	33.9	36.5	37.0	36.6
40	43.03	32.79	134.2	0.9602	5.050	33.1	35.9	36.4	36.0
38	40.98	31.15	130.0	0.9578	4.809	32.2	35.2	35.8	35.3
36	38.92	29.51	125.7	0.9554	4.568	31.3	34.6	35.2	34.7
34	36.85	27.88	121.5	0.9530	4.325	30.3	33.9	34.6	34.1
32	34.77	26.24	117.3	0.9507	4.081	29.3	33.2	33.9	33.4
30	32.68	24.60	113.0	0.9483	3.836	28.3	32.5	33.3	32.7
28	30.58	22.96	108.8	0.9459	3.589	27.1	31.8	32.6	31.9
24	26.35	19.68	100.3	0.9411	3.092	24.6	30.3	31.2	30.4
20	22.07	16.40	91.78	0.9364	2.590	21.7	28.6	29.6	28.8
16	17.75	13.12	83.78	0.9316	2.083	18.2	26.8	28.0	27.0
12	13.38	9.838	74.79	0.9268	1.571	13.7	24.9	26.3	25.1
8	8.97	6.559	66.30	0.9220	1.053	7.7	22.7	24.4	23.0
4	4.51	3.279	57.81	0.9173	0.5290	−1.7	20.4	22.3	20.7
0	0.00	0.000	49.31	0.9125	0.0000	—	17.8	20.0	18.1

50°C DRY BULB

Percentage saturation μ/%	Relative humidity ϕ/%	Value of stated parameter per kg dry air			Vapour pressure p_s/kPa	Dew point temperature t_d/°C	Adiabatic saturation temperature $t^\star$/°C	Wet bulb temperature	
		Moisture content g/g.kg^{-1}	Specific enthalpy h/kJ.kg^{-1}	Specific volume v/m^3.kg^{-1}				Screen t'_{sc}/°C	Sling t'_{sl}/°C
100	100.00	86.76	275.2	1.042	12.34	50.0	50.0	50.0	50.0
96	96.47	83.29	266.2	1.037	11.90	49.3	49.3	49.4	49.4
92	92.91	79.82	257.2	1.032	11.46	48.5	48.7	48.7	48.7
88	89.32	76.35	248.2	1.026	11.02	47.7	48.0	48.0	48.0
84	85.69	72.88	239.2	1.022	10.57	46.9	47.2	47.3	47.3
80	82.02	69.41	230.2	1.016	10.12	46.1	46.5	46.6	46.5
76	78.31	65.94	221.2	1.011	9.660	45.2	45.7	45.8	45.7
72	74.57	62.47	212.2	1.006	9.198	44.2	44.9	45.0	44.9
70	72.68	60.73	207.7	1.004	8.966	43.7	44.4	44.6	44.5
68	70.79	59.00	203.2	1.001	8.732	43.2	44.0	44.2	44.1
66	68.88	57.26	198.7	0.9988	8.497	42.7	43.6	43.8	43.6
64	66.97	55.53	194.2	0.9963	8.261	42.1	43.1	43.4	43.2
62	65.04	53.79	189.7	0.9937	8.023	41.6	42.7	42.9	42.7
60	63.11	52.06	185.2	0.9912	7.785	41.0	42.2	42.5	42.3
58	61.16	50.32	180.7	0.9887	7.545	40.4	41.7	42.0	41.8
56	59.21	48.58	176.2	0.9862	7.304	39.8	41.2	41.5	41.3
54	57.24	46.85	171.7	0.9836	7.061	39.2	40.7	41.0	40.8
52	55.27	45.11	167.2	0.9811	6.817	38.5	40.2	40.5	40.3
50	53.28	43.38	162.7	0.9786	6.572	37.9	39.6	40.0	39.7
48	51.28	41.64	158.2	0.9761	6.326	37.2	39.1	39.5	39.2
46	49.28	39.91	153.7	0.9735	6.079	36.4	38.5	39.0	38.6
44	47.26	38.17	149.2	0.9710	5.830	35.7	37.9	38.4	38.0
42	45.23	36.44	144.8	0.9685	5.580	34.9	37.3	37.8	37.4
40	43.19	34.70	140.3	0.9660	5.328	34.0	36.7	37.2	36.8
38	41.14	32.97	135.8	0.9634	5.075	33.2	36.1	36.7	36.2
36	39.08	31.23	131.3	0.9609	4.821	32.2	35.4	36.0	35.6
34	37.01	29.50	126.8	0.9584	4.565	31.3	34.8	35.4	34.9
32	34.93	27.76	122.3	0.9558	4.308	30.3	34.1	34.7	34.2
30	32.83	26.03	117.8	0.9533	4.050	29.2	33.3	34.1	33.5
28	30.73	24.29	113.3	0.9508	3.790	28.1	32.6	33.4	32.7
24	26.48	20.82	104.3	0.9457	3.266	25.5	31.0	31.9	31.2
20	22.19	17.35	95.29	0.9407	2.737	22.6	29.3	30.3	29.5
16	17.85	13.88	86.29	0.9356	2.202	19.0	27.5	28.7	27.7
12	13.46	10.41	77.30	0.9305	1.660	14.6	25.5	26.9	25.7
8	9.02	6.941	68.31	0.9255	1.113	8.6	23.3	24.9	23.5
4	4.54	3.470	59.31	0.9204	0.5597	−1.0	20.8	22.7	21.1
0	0.00	0.000	50.32	0.9153	0.0000	—	18.1	20.4	18.5

51°C DRY BULB

Percentage saturation μ/%	Relative humidity ϕ/%	Value of stated parameter per kg dry air			Vapour pressure p_s/kPa	Dew point temperature t_d/°C	Adiabatic saturation temperature $t^\star$/°C	Wet bulb temperature	
		Moisture content g/g.kg^{-1}	Specific enthalpy h/kJ.kg^{-1}	Specific volume v/m^3.kg^{-1}				Screen t'_{sc}/°C	Sling t'_{sl}/°C
100	100.00	91.81	289.4	1.052	12.96	51.0	51.0	51.0	51.0
96	96.50	88.14	279.9	1.047	12.51	50.3	50.3	50.4	50.3
92	92.96	84.46	270.4	1.042	12.05	49.5	49.7	49.7	49.7
88	89.38	80.79	260.8	1.036	11.58	48.7	49.0	49.0	49.0
84	85.77	77.12	251.3	1.031	11.12	47.9	48.2	48.3	48.2
80	82.12	73.45	241.8	1.026	10.64	47.1	47.5	47.6	47.5
76	78.43	69.78	232.3	1.020	10.16	46.2	46.7	46.8	46.7
72	74.70	66.10	222.8	1.015	9.682	45.2	45.8	46.0	45.9
70	72.83	64.27	218.0	1.012	9.439	44.7	45.4	45.6	45.5
68	70.94	62.43	213.2	1.009	9.194	44.2	45.0	45.2	45.0
66	69.04	60.59	208.5	1.007	8.948	43.7	44.5	44.7	44.6
64	67.13	58.76	203.7	1.004	8.700	43.1	44.1	44.3	44.1
62	65.20	56.92	198.9	1.001	8.451	42.6	43.6	43.8	43.7
60	63.27	55.09	194.2	0.9987	8.201	42.0	43.1	43.4	43.2
58	61.33	53.25	189.4	0.9960	7.949	41.4	42.6	42.9	42.7
56	59.38	51.41	184.7	0.9933	7.696	40.8	42.1	42.4	42.2
54	57.42	49.58	179.9	0.9907	7.442	40.2	41.6	41.9	41.7
52	55.44	47.74	175.1	0.9880	7.186	39.5	41.1	41.4	41.2
50	53.46	45.90	170.4	0.9853	6.929	38.8	40.5	40.9	40.6
48	51.46	44.07	165.6	0.9826	6.670	38.1	40.0	40.4	40.1
46	49.46	42.23	160.8	0.9799	6.410	37.4	39.4	39.8	39.5
44	47.44	40.40	156.1	0.9773	6.148	36.6	38.8	39.3	38.9
42	45.41	38.56	151.3	0.9746	5.885	35.8	38.2	38.7	38.3
40	43.37	36.72	146.6	0.9719	5.621	35.0	37.6	38.1	37.7
38	41.32	34.89	141.8	0.9692	5.355	34.1	36.9	37.5	37.1
36	39.25	33.05	137.0	0.9665	5.087	33.2	36.3	36.9	36.4
34	37.18	31.22	132.3	0.9639	4.818	32.2	35.6	36.2	35.7
32	35.09	29.38	127.5	0.9612	4.548	31.2	34.9	35.6	35.0
30	32.99	27.54	122.8	0.9585	4.276	30.1	34.1	34.9	34.3
28	30.88	25.71	118.0	0.9558	4.002	29.0	33.4	34.2	33.5
24	26.62	22.03	108.5	0.9504	3.450	26.4	31.8	32.7	31.9
20	22.31	18.36	98.94	0.9451	2.892	23.5	30.0	31.1	30.2
16	17.95	14.69	89.42	0.9397	2.327	19.9	28.1	29.3	28.3
12	13.54	11.02	79.90	0.9343	1.755	15.5	26.1	27.5	26.3
8	9.08	7.345	70.37	0.9289	1.177	9.4	23.8	25.4	24.1
4	4.57	3.672	60.85	0.9235	0.5920	−0.4	21.3	23.2	21.6
0	0.00	0.000	51.33	0.9182	0.0000	—	18.4	20.8	18.8

52°C DRY BULB

Percentage saturation μ/%	Relative humidity ϕ/%	Value of stated parameter per kg dry air			Vapour pressure p_s/kPa	Dew point temperature t_d/°C	Adiabatic saturation temperature $t\star$/°C	Wet bulb temperature	
		Moisture content $g/g.kg^{-1}$	Specific enthalpy $h/kJ.kg^{-1}$	Specific volume $v/m^3.kg^{-1}$				Screen t'_{sc}/°C	Sling t'_{sl}/°C
100	100.00	97.16	304.5	1.063	13.61	52.0	52.0	52.0	52.0
96	96.52	93.27	294.4	1.058	13.14	51.3	51.3	51.4	51.3
92	93.01	89.38	284.3	1.052	12.66	50.5	50.7	50.7	50.7
88	89.46	85.50	274.2	1.046	12.18	49.7	49.9	50.0	50.0
84	85.86	81.61	264.1	1.041	11.69	48.9	49.2	49.3	49.2
80	82.23	77.72	254.0	1.035	11.19	48.1	48.4	48.5	48.5
76	78.56	73.84	243.9	1.029	10.69	47.2	47.6	47.8	47.7
72	74.84	69.95	233.9	1.024	10.19	46.2	46.8	47.0	46.9
70	72.97	68.01	228.8	1.021	9.934	45.7	46.4	46.6	46.4
68	71.09	66.07	223.8	1.018	9.677	45.2	45.9	46.1	46.0
66	69.19	64.12	218.7	1.015	9.420	44.7	45.5	45.7	45.5
64	67.29	62.18	213.7	1.012	9.160	44.1	45.0	45.3	45.1
62	65.37	60.24	208.7	1.009	8.899	43.6	44.6	44.8	44.6
60	63.45	58.29	203.6	1.007	8.637	43.0	44.1	44.3	44.1
58	61.51	56.35	198.6	1.004	8.373	42.4	43.6	43.9	43.6
56	59.56	54.41	193.5	1.001	8.108	41.8	43.1	43.4	43.1
54	57.60	52.46	188.5	0.9980	7.841	41.2	42.5	42.9	42.6
52	55.63	50.52	183.4	0.9951	7.573	40.5	42.0	42.4	42.1
50	53.64	48.58	178.4	0.9923	7.303	39.8	41.5	41.8	41.6
48	51.65	46.63	173.4	0.9894	7.031	39.1	40.9	41.3	41.0
46	49.64	44.69	168.3	0.9866	6.758	38.4	40.3	40.7	40.4
44	47.62	42.75	163.3	0.9837	6.483	37.6	39.7	40.2	39.8
42	45.59	40.81	158.2	0.9809	6.207	36.8	39.1	39.6	39.2
40	43.55	38.86	153.2	0.9780	5.929	36.0	38.5	39.0	38.6
38	41.50	36.92	148.1	0.9752	5.649	35.1	37.8	38.4	37.9
36	39.43	34.98	143.1	0.9723	5.368	34.2	37.1	37.7	37.3
34	37.35	33.03	138.1	0.9695	5.085	33.2	36.4	37.1	36.6
32	35.26	31.09	133.0	0.9667	4.800	32.2	35.7	36.4	35.8
30	33.16	29.15	128.0	0.9638	4.513	31.1	35.0	35.7	35.1
28	31.04	27.20	122.9	0.9610	4.225	29.9	34.2	34.9	34.3
24	26.77	23.32	112.8	0.9553	3.644	27.4	32.5	33.4	32.7
20	22.44	19.43	102.8	0.9495	3.056	24.4	30.8	31.8	30.9
16	18.06	15.54	92.68	0.9438	2.459	20.8	28.8	30.0	29.0
12	13.63	11.66	82.59	0.9381	1.856	16.3	26.7	28.1	26.9
8	9.14	7.772	72.51	0.9324	1.245	10.2	24.3	26.0	24.6
4	4.60	3.886	62.42	0.9267	0.6263	0.3	21.7	23.7	22.0
0	0.00	0.000	52.34	0.9210	0.0000	—	18.8	21.1	19.2

53°C DRY BULB

Percentage saturation μ/%	Relative humidity ϕ/%	Value of stated parameter per kg dry air			Vapour pressure p_s/kPa	Dew point temperature t_d/°C	Adiabatic saturation temperature $t\star$/°C	Wet bulb temperature	
		Moisture content $g/g.kg^{-1}$	Specific enthalpy $h/kJ.kg^{-1}$	Specific volume $v/m^3.kg^{-1}$				Screen t'_{sc}/°C	Sling t'_{sl}/°C
100	100.00	102.87	320.3	1.075	14.29	53.0	53.0	53.0	53.0
96	96.55	98.70	309.7	1.069	13.80	52.3	52.3	52.4	52.3
92	93.06	94.59	299.0	1.063	13.30	51.5	51.7	51.7	51.7
88	89.53	90.48	288.3	1.057	12.80	50.7	50.9	51.0	51.0
84	85.96	86.37	277.6	1.051	12.29	49.9	50.2	50.3	50.2
80	82.35	82.25	266.9	1.045	11.77	49.1	49.4	49.5	49.5
76	78.69	78.14	256.3	1.039	11.25	48.2	48.6	48.8	48.7
72	74.99	74.03	245.6	1.033	10.72	47.2	47.8	47.9	47.8
70	73.13	71.97	240.2	1.030	10.45	46.7	47.4	47.5	47.4
68	71.25	69.92	234.9	1.027	10.18	46.2	46.9	47.1	47.0
66	69.36	67.86	229.6	1.024	9.914	45.7	46.5	46.7	46.5
64	67.46	65.80	224.2	1.021	9.642	45.1	46.0	46.2	46.0
62	65.55	63.75	218.9	1.018	9.369	44.6	45.5	45.7	45.6
60	63.63	61.69	213.5	1.015	9.095	44.0	45.0	45.3	45.1
58	61.69	59.63	208.2	1.012	8.818	43.4	44.5	44.8	44.6
56	59.75	57.58	202.9	1.009	8.540	42.8	44.0	44.3	44.1
54	57.79	55.52	197.5	1.006	8.260	42.1	43.5	43.8	43.6
52	55.82	53.46	192.2	1.003	7.979	41.5	42.9	43.3	43.0
50	53.84	51.41	186.8	0.9995	7.695	40.8	42.4	42.7	42.5
48	51.85	49.35	181.5	0.9965	7.410	40.1	41.8	42.2	41.9
46	49.84	47.30	176.2	0.9934	7.124	39.3	41.2	41.6	41.3
44	47.82	45.24	170.8	0.9904	6.835	38.6	40.6	41.1	40.7
42	45.79	43.18	165.5	0.9874	6.545	37.8	40.0	40.5	40.1
40	43.74	41.13	160.1	0.9844	6.252	36.9	39.3	39.9	39.5
38	41.69	39.07	154.8	0.9814	5.958	36.1	38.7	39.2	38.8
36	39.62	37.01	149.5	0.9783	5.663	35.1	38.0	38.6	38.1
34	37.53	34.96	144.1	0.9753	5.365	34.2	37.3	37.9	37.4
32	35.44	32.90	138.8	0.9723	5.065	33.1	36.5	37.2	36.7
30	33.33	30.85	133.4	0.9693	4.764	32.0	35.8	36.5	35.9
28	31.21	28.79	128.1	0.9662	4.460	30.9	35.0	35.7	35.1
24	26.92	24.68	117.4	0.9602	3.848	28.3	33.3	34.2	33.5
20	22.58	20.56	106.7	0.9541	3.227	25.3	31.5	32.5	31.7
16	18.18	16.45	96.06	0.9481	2.599	21.7	29.5	30.7	29.7
12	13.72	12.34	85.38	0.9420	1.962	17.2	27.3	28.7	27.5
8	9.21	8.225	74.70	0.9360	1.316	11.1	24.9	26.5	25.2
4	4.64	4.112	64.02	0.9299	0.6625	1.1	22.2	24.1	22.5
0	0.00	0.000	53.34	0.9238	0.0000	—	19.1	21.4	19.5

54°C DRY BULB

Percentage saturation μ/%	Relative humidity ϕ/%	Value of stated parameter per kg dry air			Vapour pressure p_s/kPa	Dew point temperature t_d/°C	Adiabatic saturation temperature $t\star$/°C	Wet bulb temperature	
		Moisture content g/g.kg^{-1}	Specific enthalpy h/kJ.kg^{-1}	Specific volume v/m³.kg^{-1}				Screen t'_{sc}/°C	Sling t'_{sl}/°C
100	100.00	108.8	337.1	1.0869	15.00	54.0	54.0	54.0	54.0
96	96.58	104.5	325.8	1.0805	14.49	53.3	53.3	53.4	53.3
92	93.11	100.1	314.5	1.0741	13.97	52.5	52.6	52.7	52.7
88	89.61	95.76	303.2	1.0678	13.44	51.7	51.9	52.0	52.0
84	86.06	91.40	291.9	1.0614	12.91	50.9	51.2	51.3	51.2
80	82.47	87.05	280.6	1.0550	12.37	50.1	50.4	50.5	50.4
76	78.83	82.70	269.2	1.0486	11.83	49.2	49.6	49.7	49.6
72	75.15	78.35	257.9	1.0422	11.27	48.2	48.8	48.9	48.8
70	73.29	76.17	252.3	1.0390	10.99	47.7	48.3	48.5	48.4
68	71.42	73.99	246.6	1.0358	10.71	47.2	47.9	48.1	47.9
66	69.54	71.82	241.0	1.0326	10.43	46.7	47.4	47.6	47.5
64	67.64	69.64	235.3	1.0294	10.16	46.1	47.0	47.2	47.0
62	65.74	67.47	229.7	1.0262	9.862	45.6	46.5	46.7	46.5
60	63.82	65.29	224.0	1.0230	9.574	45.0	46.0	46.2	46.0
58	61.89	63.11	218.4	1.0198	9.285	44.4	45.5	45.7	45.5
56	59.95	60.94	212.7	1.0166	8.993	43.8	45.0	45.2	45.0
54	57.99	58.76	207.0	1.0134	8.700	43.1	44.4	44.7	44.5
52	56.02	56.58	201.4	1.0102	8.405	42.5	43.9	44.2	44.0
50	54.04	54.41	195.7	1.0070	8.108	41.8	43.3	43.7	43.4
48	52.05	52.23	190.1	1.0038	7.809	41.1	42.7	43.1	42.8
46	50.05	50.06	184.4	1.0006	7.508	40.3	42.1	42.5	42.2
44	48.03	47.88	178.8	0.9974	7.205	39.6	41.5	42.0	41.6
42	45.99	45.70	173.1	0.9942	6.900	38.8	40.9	41.4	41.0
40	43.95	43.53	167.5	0.9909	6.593	37.9	40.2	40.7	40.4
38	41.89	41.35	161.8	0.9877	6.284	37.0	39.6	40.1	39.7
36	39.82	39.17	156.1	0.9845	5.973	36.1	38.9	39.4	39.0
34	37.73	37.00	150.5	0.9813	5.660	35.1	38.1	38.7	38.3
32	35.63	34.82	144.8	0.9781	5.345	34.1	37.4	38.0	37.5
30	33.51	32.64	139.2	0.9749	5.028	33.0	36.6	37.3	36.8
28	31.38	30.47	133.5	0.9717	4.708	31.8	35.8	36.5	36.0
24	27.08	26.12	122.2	0.9653	4.063	29.3	34.1	34.9	34.3
20	22.72	21.76	110.9	0.9588	3.409	26.2	32.2	33.2	32.4
16	18.30	17.41	99.59	0.9524	2.746	22.6	30.2	31.3	30.4
12	13.82	13.06	88.28	0.9460	2.074	18.1	27.9	29.3	28.2
8	9.28	8.705	76.97	0.9396	1.392	11.9	25.4	27.0	25.7
4	4.67	4.353	65.66	0.9331	0.7009	1.9	22.6	24.6	22.9
0	0.00	0.000	54.35	0.9267	0.0000	—	19.4	21.8	19.8

55°C DRY BULB

Percentage saturation μ/%	Relative humidity ϕ/%	Value of stated parameter per kg dry air			Vapour pressure p_s/kPa	Dew point temperature t_d/°C	Adiabatic saturation temperature $t\star$/°C	Wet bulb temperature	
		Moisture content g/g.kg^{-1}	Specific enthalpy h/kJ.kg^{-1}	Specific volume v/m³.kg^{-1}				Screen t'_{sc}/°C	Sling t'_{sl}/°C
100	100.00	115.2	354.8	1.100	15.74	55.0	55.0	55.0	55.0
96	96.60	110.6	342.9	1.093	15.21	54.3	54.3	54.4	54.3
92	93.17	106.0	330.9	1.086	14.67	53.5	53.6	53.7	53.7
88	89.69	101.4	318.9	1.079	14.12	52.7	52.9	53.0	52.9
84	86.16	96.74	306.9	1.072	13.56	51.9	52.2	52.3	52.2
80	82.59	92.14	294.9	1.066	13.00	51.1	51.4	51.5	51.4
76	78.97	87.53	283.0	1.059	12.43	50.2	50.6	50.7	50.6
72	75.31	82.92	271.0	1.052	11.85	49.2	49.7	49.9	49.8
70	73.46	80.62	265.0	1.049	11.56	48.7	49.3	49.5	49.4
68	71.60	78.32	259.0	1.045	11.2	48.2	48.9	49.0	48.9
66	69.72	76.01	253.0	1.042	10.97	47.7	48.4	48.6	48.4
64	67.83	73.71	247.0	1.038	10.68	47.1	47.9	48.1	48.0
62	65.93	71.41	241.0	1.035	10.38	46.6	47.4	47.7	47.5
60	64.02	69.10	235.0	1.032	10.08	46.0	46.9	47.2	47.0
58	62.09	66.80	229.1	1.028	9.774	45.4	46.4	46.7	46.5
56	60.16	64.50	223.1	1.025	9.479	44.8	45.9	46.2	46.0
54	58.20	62.19	217.1	1.022	9.161	44.1	45.4	45.7	45.4
52	56.24	59.89	211.1	1.018	8.852	43.5	44.8	45.1	44.9
50	54.26	57.59	205.1	1.015	8.541	42.8	44.2	44.6	44.3
48	52.27	55.28	199.1	1.011	8.237	42.1	43.7	44.0	43.8
46	50.26	52.98	193.1	1.008	7.911	41.3	43.1	43.5	43.2
44	48.24	50.68	187.1	1.004	7.593	40.5	42.4	42.9	42.5
42	46.21	48.37	181.1	1.001	7.273	39.7	41.8	42.2	41.9
40	44.16	46.07	175.2	0.9977	6.951	38.9	41.1	41.6	41.2
38	42.10	43.76	169.2	0.9943	6.636	38.0	40.4	41.0	40.6
36	40.02	41.46	163.2	0.9909	6.300	37.1	39.7	40.3	39.9
34	37.93	39.16	157.2	0.9875	5.971	36.1	39.0	39.6	39.1
32	35.83	36.86	151.2	0.9841	5.640	35.1	38.2	38.9	38.4
30	33.71	34.55	145.2	0.9807	5.305	34.0	37.4	38.1	37.6
28	31.57	32.25	139.2	0.9773	4.979	32.8	36.6	37.4	36.8
24	27.25	27.64	127.2	0.9705	4.290	30.2	34.9	35.7	35.0
20	22.88	23.03	115.3	0.9637	3.600	27.2	33.0	34.0	33.2
16	18.43	18.43	103.3	0.9568	2.901	23.5	30.9	32.0	31.1
12	13.93	13.82	91.30	0.9500	2.192	19.0	28.6	29.9	28.8
8	9.35	9.214	79.32	0.9432	1.472	12.7	26.0	27.6	26.3
4	4.71	4.607	67.34	0.9363	0.7416	2.7	23.1	25.0	23.4
0	0.00	0.000	55.36	0.9295	0.0000	—	19.7	22.1	20.1

56°C DRY BULB

Percentage saturation μ/%	Relative humidity ϕ/%	Value of stated parameter per kg dry air			Vapour pressure p_s/kPa	Dew point temperature t_d/°C	Adiabatic saturation temperature $t\star$/°C	Wet bulb temperature	
		Moisture content g/g.kg⁻¹	Specific enthalpy h/kJ.kg⁻¹	Specific volume v/m³.kg⁻¹				Screen t'_{sc}/°C	Sling t'_{sl}/°C
100	100.00	121.9	373.6	1.113	16.51	56.0	56.0	56.0	56.0
96	96.63	117.0	360.9	1.106	15.96	55.3	55.3	55.4	55.3
92	93.23	112.2	348.2	1.099	15.39	54.5	54.6	54.7	54.7
88	89.77	107.3	335.5	1.091	14.82	53.7	53.9	54.0	53.9
84	86.27	102.4	322.8	1.084	14.24	52.9	53.2	53.2	53.2
80	82.72	97.53	310.2	1.077	13.66	52.1	52.4	52.5	52.4
76	79.12	92.65	297.5	1.070	13.06	51.2	51.6	51.7	51.6
72	75.48	87.78	284.8	1.062	12.46	50.2	50.7	50.9	50.8
70	73.64	85.34	278.4	1.059	12.16	49.7	50.3	50.4	50.3
68	71.78	82.90	272.1	1.055	11.85	49.2	49.8	50.0	49.9
66	69.91	80.46	265.7	1.052	11 54	48.7	49.1	49.6	49.4
64	68.03	78.03	259.4	1.048	11.23	48.1	48.9	49.1	49.0
62	66.14	75.59	253.0	1.044	10.92	47.6	48.4	48.6	48.5
60	64.23	73.15	246.7	1.041	10.60	47.0	47.9	48.1	48.0
58	62.31	70.71	240.4	1.037	10.29	46.4	47.4	47.6	47.5
56	60.37	68.27	234.0	1.034	9.968	45.8	46.9	47.1	46.9
54	58.42	65.83	227.7	1.030	9.646	45.1	46.3	46.6	46.4
52	56.46	63.40	221.3	1.026	9.322	44.5	45.8	46.1	45.8
50	54.49	60.96	215.0	1.023	8.996	43.8	45.2	45.5	45.3
48	52.49	58.52	208.6	1.019	8.667	43.1	44.6	45.0	44.7
46	50.49	56.08	202.3	1.016	8.336	42.3	44.0	44.4	44.1
44	48.47	53.64	195.9	1.012	8.003	41.5	43.4	43.8	43.5
42	46.44	51.20	189.6	1.008	7.667	40.7	42.7	43.1	42.8
40	44.39	48.77	183.3	1.005	7.329	39.9	42.0	42.5	42.1
38	42.32	46.33	176.9	1.001	6.988	39.0	41.3	41.8	41.5
36	40.24	43.89	170.6	0.9976	6.644	38.1	40.6	41.2	40.7
34	38.15	41.45	164.2	0.9939	6.298	37.1	39.9	40.5	40.0
32	36.04	39.01	157.9	0.9903	5.950	36.0	39.1	39.7	39.2
30	33.91	36.57	151.5	0.9867	5.599	34.9	38.3	39.0	38.4
28	31.77	34.14	145.2	0.9831	5.245	33.8	37.5	38.2	37.6
24	27.44	29.26	132.5	0.9759	4.530	31.1	35.7	36.5	35.8
20	23.04	24.38	119.9	0.9686	3.804	28.1	33.7	34.7	33.9
16	18.57	19.51	107.1	0.9614	3.066	24.5	31.6	32.7	31.8
12	14.04	14.63	94.43	0.9541	2.317	19.9	29.2	30.6	29.4
8	9.43	9.753	81.74	0.9469	1.557	13.6	26.5	28.1	26.8
4	4.75	4.877	69.06	0.9396	0.7847	3.5	23.5	25.5	23.8
0	0.00	0.000	56.37	0.9323	0.0000	—	20.0	22.5	20.4

57°C DRY BULB

Percentage saturation μ/%	Relative humidity ϕ/%	Value of stated parameter per kg dry air			Vapour pressure p_s/kPa	Dew point temperature t_d/°C	Adiabatic saturation temperature $t\star$/°C	Wet bulb temperature	
		Moisture content g/g.kg⁻¹	Specific enthalpy h/kJ.kg⁻¹	Specific volume v/m³.kg⁻¹				Screen t'_{sc}/°C	Sling t'_{sl}/°C
100	100.00	129.1	393.4	1.127	17.31	57.0	57.0	57.0	57.0
96	96.66	123.9	380.0	1.119	16.74	56.3	56.3	56.3	56.3
92	93.29	118.7	366.6	1.112	16.15	55.5	55.6	55.7	55.7
88	89.86	113.6	353.1	1.104	15.56	54.8	54.9	55.0	54.9
84	86.38	108.4	339.7	1.096	14.96	53.9	54.2	54.2	54.2
80	82.86	103.3	326.2	1.089	14.34	53.1	53.4	53.5	53.4
76	79.28	98.09	312.8	1.081	13.73	52.2	52.6	52.7	52.6
72	75.65	92.93	299.3	1.073	13.10	51.2	51.7	51.9	51.8
70	73.82	90.35	292.6	1.070	12.78	50.7	51.3	51.4	51.3
68	71.97	87.77	285.9	1.066	12.46	50.2	50.8	51.0	50.9
66	70.11	85.19	279.2	1.062	12.14	49.7	50.3	50.5	50.4
64	68.24	82.60	272.5	1.058	11.81	49.1	49.9	50.1	49.9
62	66.35	80.02	265.7	1.054	11.49	48.6	49.4	49.6	49.4
60	64.45	77.44	259.0	1.050	11.16	48.0	48.9	49.1	48.9
58	62.53	74.86	252.3	1.047	10.83	47.4	48.4	48.6	48.4
56	60.60	72.28	245.6	1.043	10.49	46.8	47.8	48.1	47.9
54	58.66	69.70	238.9	1.039	10.16	46.1	47.3	47.6	47.4
52	56.70	67.12	232.1	1.035	9.816	45.5	46.7	47.0	46.8
50	54.72	64.53	225.4	1.031	9.474	44.8	46.1	46.5	46.2
48	52.73	61.95	218.7	1.027	9.130	44.1	45.5	45.9	45.6
46	50.73	59.37	212.0	1.024	8.783	43.3	44.9	45.3	45.0
44	48.71	56.79	205.2	1.020	8.433	42.5	44.3	44.7	44.4
42	46.67	54.21	198.5	1.016	8.080	41.7	43.6	44.1	43.7
40	44.62	51.63	191.8	1.012	7.725	40.9	42.9	43.4	43.1
38	42.56	49.05	185.1	1.008	7.368	40.0	42.2	42.7	42.4
36	40.47	46.46	178.4	1.004	7.007	39.0	41.5	42.0	41.6
34	38.37	43.88	171.6	1.001	6.644	38.1	40.7	41.3	40.9
32	36.26	41.30	164.9	0.9967	6.277	37.0	40.0	40.6	40.1
30	34.13	38.72	158.2	0.9929	5.908	35.9	39.1	39.8	39.3
28	31.98	36.14	151.5	0.9891	5.536	34.7	38.3	39.0	38.5
24	27.63	30.98	138.0	0.9814	4.783	32.1	36.5	37.3	36.7
20	23.21	25.81	124.6	0.9737	4.018	29.1	34.5	35.4	34.7
16	18.72	20.65	111.1	0.9660	3.240	25.4	32.3	33.4	32.5
12	14.15	15.49	97.70	0.9583	2.450	20.8	29.9	31.2	30.1
8	9.51	10.33	84.26	0.9506	1.647	14.5	27.1	28.7	27.4
4	4.80	5.163	70.82	0.9429	0.8303	4.3	24.0	25.9	24.3
0	0.00	0.000	57.37	0.9352	0.0000	—	20.3	22.8	20.7

58°C DRY BULB

Percentage saturation μ/%	Relative humidity ϕ/%	Value of stated parameter per kg dry air			Vapour pressure p_s/kPa	Dew point temperature t_d/°C	Adiabatic saturation temperature $t^\star$/°C	Wet bulb temperature	
		Moisture content g/g.kg^{-1}	Specific enthalpy h/kJ.kg^{-1}	Specific volume v/m³.kg^{-1}				Screen t'_{sc}/°C	Sling t'_{sl}/°C
100	100.00	136.70	414.5	1.141	18.15	58.0	58.0	58.0	58.0
96	96.70	131.20	400.2	1.134	17.55	57.3	57.3	57.3	57.3
92	93.35	125.70	386.0	1.125	16.94	56.5	56.6	56.7	56.7
88	89.95	120.30	371.7	1.117	16.32	55.8	55.9	56.0	55.9
84	86.50	114.80	357.5	1.109	15.70	54.9	55.2	55.2	55.2
80	83.00	109.30	343.3	1.101	15.06	54.1	54.4	54.5	54.4
76	79.45	103.70	329.0	1.093	14.42	53.2	53.6	53.7	53.6
72	75.84	98.40	314.8	1.085	13.76	52.2	52.7	52.8	52.7
70	74.02	95.67	307.6	1.081	13.43	51.7	52.3	52.4	52.3
68	72.18	92.93	300.5	1.077	13.10	51.2	51.8	52.0	51.8
66	70.32	90.20	293.4	1.073	12.76	50.7	51.3	51.5	51.4
64	68.46	87.47	286.3	1.068	12.42	50.1	50.8	51.0	50.9
62	66.58	84.73	279.2	1.064	12.08	49.6	50.3	50.6	50.4
60	64.68	82.00	272.0	1.060	11.74	49.0	49.8	50.1	49.9
58	62.77	79.27	264.9	1.056	11.39	48.4	49.3	49.6	49.4
56	60.84	76.53	257.8	1.052	11.04	47.8	48.8	49.1	48.9
54	58.90	73.80	250.7	1.048	10.69	47.1	48.2	48.5	48.3
52	56.94	71.07	243.6	1.044	10.33	46.5	47.7	48.0	47.8
50	54.97	68.33	236.4	1.040	9.976	45.8	47.1	47.4	47.2
48	52.98	65.60	229.3	1.036	9.615	45.1	46.5	46.8	46.6
46	50.98	62.87	222.2	1.032	9.252	44.3	45.9	46.2	46.0
44	48.96	60.13	215.1	1.028	8.885	43.5	45.2	45.6	45.3
42	46.92	57.40	207.9	1.024	8.516	42.7	44.5	45.0	44.7
40	44.87	54.67	200.8	1.020	8.143	41.9	43.9	44.3	44.0
38	42.80	51.93	193.7	1.016	7.767	41.0	43.1	43.6	43.3
36	40.72	49.20	186.6	1.012	7.389	40.0	42.4	42.9	42.5
34	38.61	46.47	179.5	1.008	7.007	39.0	41.6	42.2	41.8
32	36.49	43.73	172.3	1.003	6.622	38.0	40.8	41.4	41.0
30	34.35	41.00	165.2	0.9993	6.234	36.9	40.0	40.7	40.2
28	32.20	38.27	158.1	0.9952	5.843	35.7	39.1	39.8	39.3
24	27.83	32.80	143.8	0.9871	5.050	33.1	37.3	38.1	37.5
20	23.39	27.33	129.6	0.9789	4.244	30.0	35.3	36.2	35.5
16	18.87	21.87	115.4	0.9707	3.424	26.3	33.0	34.1	33.2
12	14.28	16.40	101.1	0.9626	2.590	21.7	30.5	31.8	30.8
8	9.60	10.93	86.87	0.9544	1.742	15.3	27.7	29.3	28.0
4	4.84	5.467	72.62	0.9462	0.8788	5.1	24.4	26.4	24.8
0	0.00	0.000	58.38	0.9380	0.0000	—	20.6	23.1	21.1

59°C DRY BULB

Percentage saturation μ/%	Relative humidity ϕ/%	Value of stated parameter per kg dry air			Vapour pressure p_s/kPa	Dew point temperature t_d/°C	Adiabatic saturation temperature $t^\star$/°C	Wet bulb temperature	
		Moisture content g/g.kg^{-1}	Specific enthalpy h/kJ.kg^{-1}	Specific volume v/m³.kg^{-1}				Screen t'_{sc}/°C	Sling t'_{sl}/°C
100	100.00	144.7	436.8	1.157	19.02	59.0	59.0	59.0	59.0
96	96.73	138.9	421.7	1.149	18.39	58.3	58.3	58.3	58.3
92	93.41	133.2	406.6	1.140	17.76	57.5	57.6	57.7	57.7
88	90.05	127.4	391.5	1.131	17.12	56.8	56.9	57.0	56.9
84	86.62	121.6	376.4	1.123	16.47	56.0	56.2	56.2	56.2
80	83.15	115.8	361.3	1.114	15.81	55.1	55.4	55.5	55.4
76	79.62	110.0	346.2	1.105	15.14	54.2	54.6	54.7	54.6
72	76.03	104.2	331.1	1.097	14.46	53.2	53.7	53.8	53.7
70	74.22	101.3	323.6	1.092	14.11	52.7	53.2	53.4	53.3
68	72.39	98.42	316.0	1.088	13.77	52.2	52.8	52.9	52.8
66	70.54	95.53	308.5	1.084	13.41	51.7	52.3	52.5	52.4
64	68.69	92.63	300.9	1.080	13.06	51.2	51.8	52.0	51.9
62	66.81	89.74	293.4	1.075	12.70	50.6	51.3	51.5	51.4
60	64.92	86.84	285.8	1.071	12.35	50.0	50.8	51.0	50.9
58	63.02	83.95	278.3	1.066	11.98	49.4	50.3	50.5	50.4
56	61.09	81.05	270.7	1.062	11.62	48.8	49.8	50.0	49.8
54	59.16	78.16	263.2	1.058	11.25	48.2	49.2	49.5	49.3
52	57.20	75.26	255.6	1.054	10.88	47.5	48.6	48.9	48.7
50	55.23	72.37	248.1	1.049	10.50	46.8	48.0	48.4	48.1
48	53.25	69.47	240.5	1.045	10.13	46.1	47.4	47.8	47.5
46	51.25	66.58	233.0	1.041	9.745	45.3	46.8	47.2	46.9
44	49.23	63.68	225.4	1.036	9.361	44.5	46.1	46.5	46.3
42	47.19	60.79	217.9	1.032	8.973	43.7	45.5	45.9	45.6
40	45.13	57.90	210.3	1.028	8.583	42.9	44.8	45.2	44.9
38	43.06	55.00	202.8	1.023	8.189	42.0	44.1	44.5	44.2
36	40.97	52.11	195.2	1.019	7.791	41.0	43.3	43.8	43.4
34	38.86	49.21	187.7	1.015	7.390	40.0	42.5	43.1	42.7
32	36.74	46.32	180.2	1.010	6.986	39.0	41.7	42.3	41.9
30	34.59	43.42	172.6	1.006	6.578	37.9	40.9	41.5	41.0
28	32.43	40.53	165.1	1.002	6.167	36.7	40.0	40.7	40.2
24	28.04	34.74	150.0	0.9930	5.332	34.0	38.1	38.9	38.3
20	23.58	28.95	134.9	0.9843	4.484	31.0	36.0	37.0	36.2
16	19.03	23.16	119.8	0.9756	3.619	27.3	33.7	34.8	34.0
12	14.41	17.37	104.7	0.9669	2.739	22.6	31.2	32.5	31.4
8	9.69	11.58	89.58	0.9583	1.843	16.2	28.3	29.8	28.5
4	4.89	5.790	74.48	0.9496	0.9302	5.9	24.9	26.9	25.2
0	0.00	0.000	59.39	0.9409	0.0000	—	20.9	23.4	21.4

60°C DRY BULB

Percentage saturation μ/%	Relative humidity ϕ/%	Value of stated parameter per kg dry air			Vapour pressure p_s/kPa	Dew point temperature t_d/°C	Adiabatic saturation temperature $t\star$/°C	Wet bulb temperature	
		Moisture content g/g.kg^{-1}	Specific enthalpy h/kJ.kg^{-1}	Specific volume v/m^3.kg^{-1}				Screen t'_{sc}/°C	Sling t'_{sl}/°C
100	100.00	153.3	460.4	1.173	19.92	60.0	60.0	60.0	60.0
96	96.77	147.2	444.4	1.164	19.28	59.3	59.3	59.3	59.3
92	93.48	141.1	428.4	1.155	18.62	58.6	58.6	58.7	58.7
88	90.14	134.9	412.4	1.146	17.96	57.8	57.9	58.0	57.9
84	86.75	128.8	396.4	1.137	17.28	57.0	57.2	57.2	57.2
80	83.30	122.7	380.4	1.128	16.59	56.1	56.4	56.5	56.4
76	79.80	116.5	364.4	1.118	15.90	55.2	55.6	55.7	55.6
72	76.23	110.4	348.4	1.109	15.19	54.3	54.7	54.8	54.7
70	74.43	107.3	340.4	1.105	14.83	53.8	54.2	54.4	54.3
68	72.61	104.3	332.4	1.100	14.46	53.2	53.8	53.9	53.8
66	70.78	101.2	324.4	1.096	14.10	52.7	53.3	53.5	53.4
64	68.92	98.14	316.4	1.091	13.73	52.2	52.8	53.0	52.9
62	67.06	95.06	308.4	1.086	13.36	51.6	52.3	52.5	52.4
60	65.17	91.99	300.4	1.082	12.98	51.0	51.8	52.0	51.9
58	63.28	88.93	292.4	1.077	12.60	50.4	51.3	51.5	51.4
56	61.36	85.86	284.4	1.073	12.22	49.8	50.7	51.0	50.8
54	59.43	82.79	276.4	1.068	11.84	49.2	50.2	50.4	50.3
52	57.48	79.73	268.4	1.063	11.45	48.5	49.6	49.9	49.7
50	55.51	76.66	260.4	1.059	11.06	47.8	49.0	49.3	49.1
48	53.53	73.60	252.4	1.054	10.66	47.1	48.4	48.7	48.5
46	51.52	70.53	244.4	1.050	10.26	46.3	47.8	48.1	47.9
44	49.50	67.46	236.4	1.045	9.861	45.6	47.1	47.5	47.2
42	47.47	64.40	228.4	1.041	9.455	44.7	46.4	46.8	46.5
40	45.41	61.33	220.4	1.036	9.046	43.9	45.7	46.2	45.8
38	43.34	58.26	212.4	1.031	8.632	43.0	45.0	45.5	45.1
36	41.24	55.20	204.4	1.027	8.215	42.0	44.2	44.7	44.4
34	39.13	52.13	196.4	1.022	7.795	41.0	43.4	44.0	43.6
32	37.00	49.06	188.4	1.018	7.370	40.0	42.6	43.2	42.8
30	34.84	46.00	180.4	1.013	6.941	38.9	41.8	42.4	41.9
28	32.67	42.93	172.4	1.008	6.508	37.7	40.9	41.5	41.0
24	28.27	36.80	156.4	0.9991	5.631	35.0	38.9	39.7	39.1
20	23.78	30.66	140.4	0.9899	4.737	31.9	36.8	37.8	37.0
16	19.21	24.53	124.4	0.9806	3.826	28.2	34.5	35.6	34.7
12	14.54	18.40	108.4	0.9714	2.897	23.5	31.8	33.1	32.1
8	9.79	12.27	92.40	0.9622	1.950	17.1	28.8	30.4	29.1
4	4.94	6.133	76.40	0.9529	0.9848	6.8	25.3	27.3	25.7
0	0.00	0.000	60.40	0.9437	0.0000	—	21.2	23.7	21.7

2   Properties of water and steam

2.1   Introduction

The data presented in Tables 2.1, 2.2, and 2.3 are based upon the tables prepared by the National Engineering Laboratory[1], augmented as necessary by further values obtained by interpolation from the tables of Mayhew and Rogers[2].

Other data have been published more recently, e.g. *UK Steam Tables in SI Units*[3]. These tables differ slightly from the NEL Steam Tables. For building services applications, these variations are not significant. However, for critical applications the latest internationally agreed tables should be consulted.

The units and symbols used are as follows:

t = temperature (°C)

t_s = temperature (saturation) (°C)

p = absolute pressure (kPa)

p_s = absolute saturation pressure (kPa)

ρ = density (specific mass) (kg.m^{-3})

v = specific volume (m^3.kg^{-1})

c_f = specific heat capacity (kJ.kg^{-1}.K^{-1})

h_f = specific enthalpy (saturated liquid) (kJ.kg^{-1})

h_g = specific enthalpy (saturated vapour) (kJ.kg^{-1})

h_{fg} = specific latent heat of evaporation (kJ.kg^{-1})

μ_f = dynamic viscosity (saturated liquid) (μPa.s)

$(Pr)_f$ = Prandtl Number (saturated liquid)

$(Pr)_g$ = Prandtl Number (saturated vapour)

Table 2.1 lists values of saturation temperature, specific volume, specific enthalpies of saturated liquid and vapour and the specific latent heat of evaporation at round values of absolute pressure.

The values quoted were in all cases derived from NEL Table 2, except for Prandtl Numbers, which were obtained by Lagrangian interpolation from Mayhew and Rogers, page 10.

Table 2.2 lists values of the saturation vapour pressure, specific heat capacity, dynamic viscosity, density (specific mass) and of the specific enthalpy and Prandtl Numbers of the saturated liquid and vapour at round values of temperature.

The values quoted were derived as follows:

—   Vapour pressure, density (as reciprocal of volume) and specific enthalpy from NEL Table 1.

—   Specific heat capacity, dynamic viscosity and Prandtl Numbers (all by Lagrangian interpolation) from Mayhew and Rogers, page 10.

Table 2.3 lists values of specific enthalpy for superheated steam at round values of absolute pressure for a restricted range of final temperatures.

The values quoted were in all cases derived from NEL Table 3.

References

1   *Steam Tables 1964* (Edinburgh: National Engineering Laboratory/HMSO) (1964)

2   Mayhew, Y R and Rogers, G F C, *Thermodynamics and Transport Properties of Fluids* (Oxford: Blackwell)

3   *UK Steam Table in SI Units* (Oxford: Butterwoth-Heinemann) (1970)

Table 2.1 Properties of saturated steam

Absolute pressure p/kPa	Temperature $t_\mathrm{s}/°\mathrm{C}$	Specific enthalpy			Specific volume $v/(\mathrm{m^3.kg^{-1}})$	Specific heat capacity of vapour $c_\mathrm{f}/(\mathrm{kJ.kg^{-1}.K^{-1}})$	Prandtl number $(Pr)_\mathrm{g}$	Absolute pressure p/kPa
		In saturated liquid $h_\mathrm{f}/(\mathrm{kJ.kg^{-1}})$	Latent heat of evaporation $h_\mathrm{fg}/(\mathrm{kJ.kg^{-1}})$	In saturated vapour $h_\mathrm{g}/(\mathrm{kJ.kg^{-1}})$				
1	6.98	29.3	2484.3	2513.6	129.205	1.86	1.03	1
2	17.51	73.5	2459.5	2533.0	67.010	1.87	1.03	2
4	28.98	121.4	2432.4	2553.9	34.805	1.88	1.02	4
6	36.18	151.5	2415.3	2566.8	23.742	1.88	1.02	6
8	41.53	173.9	2402.5	2576.4	18.104	1.89	1.02	8
10	45.83	191.8	2392.2	2584.1	14.673	1.89	1.03	10
20	60.09	251.5	2357.7	2609.1	7.648	1.91	1.03	20
30	69.13	289.3	2335.4	2624.8	5.228	1.93	1.03	30
40	75.89	317.7	2318.6	2636.3	3.992	1.94	1.03	40
50	81.35	340.6	2304.9	2645.4	3.239	1.95	1.04	50
60	85.95	359.9	2293.2	2653.1	2.731	1.96	1.04	60
70	89.96	376.8	2282.9	2659.7	2.364	1.97	1.04	70
80	83.51	391.7	2273.7	2665.4	2.087	1.98	1.04	80
90	96.71	405.2	2265.4	2670.6	1.869	1.99	1.04	90
100	99.63	417.5	2257.7	2675.2	1.694	2.01	1.04	100
110	102.32	428.8	2250.6	2679.5	1.549	2.02	1.05	110
120	104.81	439.3	2244.0	2683.3	1.428	2.03	1.05	120
130	107.13	449.2	2237.8	2686.9	1.325	2.04	1.05	130
140	109.32	458.4	2231.9	2690.3	1.237	2.05	1.05	140
150	111.37	367.5	2226.3	2693.4	1.159	2.05	1.05	150
160	113.32	475.4	2221.0	2696.4	1.091	2.06	1.06	160
170	115.17	483.2	2215.9	2699.1	1.031	2.07	1.06	170
180	116.93	490.7	2211.1	2701.8	0.977	2.07	1.06	180
190	118.62	497.9	2206.4	2704.2	0.929	2.08	1.06	190
200	120.23	504.7	2201.9	2706.6	0.886	2.09	1.06	200
210	121.78	511.3	2197.6	2708.9	0.846	2.10	1.06	210
220	123.27	517.6	2193.4	2711.0	0.810	2.11	1.07	220
230	124.71	523.7	2189.3	2713.1	0.777	2.12	1.07	230
240	126.09	529.6	2185.4	2715.0	0.747	2.13	1.07	240
250	127.43	535.4	2181.6	2716.9	0.719	2.13	1.07	250
260	128.73	540.9	2177.8	2718.7	0.693	2.14	1.07	260
270	129.99	546.2	2174.2	2720.5	0.669	2.15	1.08	270
280	131.21	551.5	2170.7	2722.2	0.646	2.16	1.08	280
290	132.39	556.5	2167.3	2723.8	0.625	2.16	1.08	290
300	133.54	561.4	2163.9	2725.4	0.606	2.17	1.08	300
310	134.66	566.2	2160.6	2726.9	0.587	2.18	1.09	310
320	135.76	570.9	2157.4	2728.4	0.570	2.19	1.09	320
330	136.82	575.5	2154.3	2729.8	0.554	2.20	1.09	330
340	137.86	579.9	2151.2	2731.1	0.539	2.20	1.09	340
350	138.88	584.3	2148.2	2732.5	0.524	2.21	1.10	350
360	139.87	588.5	2145.2	2733.8	0.510	2.21	1.10	360
370	140.84	592.7	2142.3	2735.0	0.497	2.22	1.10	370
380	141.79	596.8	2139.5	2736.3	0.485	2.22	1.10	380
390	142.72	600.8	2136.7	2737.5	0.473	2.23	1.11	390
400	143.63	604.7	2133.9	2738.6	0.462	2.24	1.11	400
410	144.52	608.5	2131.2	2739.8	0.452	2.24	1.11	410
420	145.39	612.3	2128.6	2740.9	0.442	2.25	1.11	420
430	146.25	616.0	2125.9	2741.9	0.432	2.26	1.11	430
440	147.09	619.6	2123.4	2743.0	0.423	2.27	1.12	440
450	147.92	623.2	2120.8	2744.0	0.414	2.28	1.12	450
460	148.73	626.7	2118.2	2745.0	0.405	2.28	1.12	460
470	149.53	630.1	2115.8	2746.0	0.397	2.29	1.13	470
480	150.31	633.5	2113.4	2746.9	0.389	2.29	1.13	480
490	151.09	636.8	2111.0	2747.8	0.382	2.30	1.13	490
500	151.85	640.1	2108.6	2748.7	0.375	2.31	1.13	500

Table 2.1 (*continued*)

Absolute pressure p / kPa	Temperature t_s / °C	Specific enthalpy			Specific volume v / (m³.kg⁻¹)	Specific heat capacity of vapour c_f / (kJ.kg⁻¹.K⁻¹)	Prandtl number $(Pr)_g$	Absolute pressure p / kPa
		In saturated liquid h_f / (kJ.kg⁻¹)	Latent heat of evaporation h_{fg} / (kJ.kg⁻¹)	In saturated vapour h_g / (kJ.kg⁻¹)				
520	153.33	646.5	2104.0	2750.5	0.361	2.32	1.13	520
540	154.77	652.8	2099.4	2752.2	0.349	2.33	1.14	540
560	156.16	658.8	2095.0	2753.8	0.337	2.34	1.14	560
580	157.52	664.7	2090.7	2755.4	0.326	2.35	1.14	580
600	158.84	670.4	2086.4	2756.8	0.316	2.37	1.14	600
620	160.12	676.0	2082.3	2758.3	0.306	2.38	1.15	620
640	161.38	681.5	2078.2	2759.9	0.297	2.39	1.15	640
660	162.60	686.8	2074.2	2761.0	0.288	2.40	1.16	660
680	163.79	692.0	2070.3	2762.3	0.280	2.41	1.16	680
700	164.96	697.1	2066.4	2763.5	0.273	2.43	1.16	700
720	166.10	702.0	2062.7	2764.7	0.266	2.44	1.17	720
740	167.21	706.9	2058.9	2765.8	0.259	2.45	1.17	740
760	168.30	711.7	2055.3	2767.0	0.252	2.47	1.17	760
780	169.37	716.4	2051.7	2768.0	0.246	2.48	1.18	780
800	170.41	720.9	2048.2	2769.1	0.240	2.49	1.18	800
820	171.44	725.4	2044.7	2770.1	0.235	2.51	1.18	820
840	172.45	729.8	2041.2	2771.1	0.229	2.52	1.19	840
860	173.43	734.2	2037.8	2772.0	0.224	2.53	1.19	860
880	174.40	738.4	2034.5	2772.9	0.220	2.55	1.20	880
900	175.36	742.6	2031.2	2773.8	0.215	2.56	1.21	900
920	176.29	746.8	2028.0	2774.7	0.210	2.57	1.21	920
940	177.21	750.8	2024.7	2775.6	0.206	2.58	1.21	940
960	178.12	754.8	2021.6	2776.4	0.202	2.60	1.22	960
980	179.01	758.7	2018.4	2777.2	0.198	2.61	1.22	980
1000	179.88	762.6	2015.3	2777.9	0.194	2.62	1.23	1000
1100	184.06	781.1	2000.4	2781.5	0.177	2.67	1.25	1100
1200	187.96	798.4	1986.2	2784.6	0.163	2.71	1.26	1200
1300	191.60	814.7	1972.6	2787.3	0.151	2.76	1.27	1300
1400	195.04	830.0	1959.6	2789.7	0.141	2.81	1.29	1400
1500	198.28	844.6	1947.1	2791.8	0.132	2.86	1.30	1500
1600	201.37	858.5	1935.1	2793.6	0.124	2.91	1.31	1600
1700	204.30	871.8	1923.4	2795.2	0.117	2.96	1.33	1700
1800	207.10	884.5	1912.1	2796.6	0.110	3.01	1.35	1800
1900	209.79	896.8	1901.1	2797.8	0.105	3.07	1.36	1900
2000	212.37	908.6	1890.4	2798.9	0.100	3.13	1.37	2000
2200	217.24	930.0	1869.7	2800.6	0.091	3.19	1.39	2200
2400	221.78	951.9	1850.0	2801.9	0.083	3.25	1.41	2400
2600	226.03	971.7	1831.0	2802.7	0.077	3.36	1.43	2600
2800	230.04	990.5	1812.7	2903.2	0.071	3.45	1.44	2800
3000	233.84	1008.3	1795.0	2803.4	0.067	3.53	1.46	3000
3200	237.44	1025.4	1777.9	2803.3	0.062	3.60	1.49	3200
3400	240.88	1041.8	1761.2	2803.0	0.059	3.68	1.53	3400
3600	244.16	1057.6	1744.9	2802.4	0.055	3.77	1.57	3600
3800	247.31	1072.7	1728.9	2802.1	0.052	3.86	1.59	3800
4000	250.33	1087.4	1713.4	2801.7	0.050	3.94	1.60	4000

Table 2.2 Properties of water at saturation

Temperature t_s/ °C	Absolute vapour pressure p_s/ kPa	Specific heat capacity c_r/ (kJ.kg^{-1}.K^{-1})	Dynamic viscosity μ_f/(μPa.s)	Density ρ/ (kg.m^{-3})	Specific enthalpy of liquid h_f/ (kJ.kg^{-1})	Prandtl number $(Pr)_g$	Temperature t_s/ °C
0.01	0.61	4.2100	1782	999.8	0.00	13.61	0.01
1	0.66	4.2096	1724	999.8	4.17	13.11	1
2	0.71	4.2088	1669	999.9	8.39	12.64	2
3	0.76	4.2075	1616	999.9	12.60	12.18	3
4	0.81	4.2059	1565	999.9	16.80	11.75	4
5	0.87	4.2040	1517	999.9	21.01	11.33	5
6	0.93	4.2019	1471	999.9	25.21	10.93	6
7	1.00	4.1997	1427	999.9	29.41	10.55	7
8	1.07	4.1974	1385	999.8	33.61	10.19	8
9	1.15	4.1952	1344	999.7	37.81	9.85	9
10	1.23	4.1930	1306	999.7	42.00	9.52	10
11	1.31	4.1913	1269	999.6	46.19	9.21	11
12	1.40	4.1897	1234	999.4	50.38	8.92	12
13	1.50	4.1883	1201	999.3	54.57	8.64	13
14	1.60	4.1871	1169	999.2	58.75	8.37	14
15	1.70	4.1860	1138	999.0	62.94	8.12	15
16	1.82	4.1852	1108	998.9	67.13	7.88	16
17	1.94	4.1845	1080	998.7	71.31	7.64	17
18	2.06	4.1839	1053	998.6	75.49	7.42	18
19	2.20	4.1834	1027	998.4	79.68	7.20	19
20	2.34	4.1830	1002	998.2	83.86	7.00	20
21	2.49	4.1826	978	997.9	88.04	6.81	21
22	2.64	4.1821	955	997.7	92.23	6.62	22
23	2.81	4.1817	932	997.5	96.41	6.44	23
24	2.98	4.1814	911	997.2	100.59	6.27	24
25	3.17	4.1810	890	997.0	104.77	6.11	25
26	3.36	4.1806	870	996.7	108.95	5.95	26
27	3.56	4.1801	851	996.5	113.13	5.80	27
28	3.78	4.1797	833	996.2	117.31	5.66	28
29	4.00	4.1794	815	995.9	121.49	5.52	29
30	4.24	4.1790	798	995.6	125.67	5.39	30
31	4.49	4.1787	781	995.3	129.85	5.26	31
32	4.75	4.1784	765	995.0	134.03	5.14	32
33	5.01	4.1782	749	994.6	138.20	5.02	33
34	5.32	4.1781	734	994.3	142.38	4.91	34
35	5.62	4.1780	719	994.0	146.56	4.80	35
36	5.94	4.1781	705	993.6	150.74	4.69	36
37	6.27	4.1782	692	993.3	154.92	4.59	37
38	6.62	4.1784	678	993.0	159.09	4.49	38
39	6.99	4.1787	666	992.6	163.27	4.39	39
40	7.38	4.1790	653	992.2	167.45	4.30	40
41	7.78	4.1794	641	991.8	171.63	4.21	41
42	8.20	4.1798	629	991.4	175.81	4.13	42
43	8.64	4.1802	618	991.0	179.99	4.05	43
44	9.10	4.1806	606	990.6	184.17	3.97	44
45	9.58	4.1810	596	990.2	188.35	3.89	45
46	10.09	4.1812	586	989.8	192.53	3.81	46
47	10.61	4.1815	576	989.3	196.71	3.74	47
48	11.16	4.1817	566	988.9	200.90	3.67	48
49	11.74	4.1818	556	988.5	205.08	3.60	49
50	12.33	4.1820	547	988.0	209.26	3.53	50
51	12.96	4.1822	538	987.6	213.44	3.47	51
52	13.61	4.1823	529	987.2	217.62	3.40	52
53	14.29	4.1825	521	986.7	221.81	3.35	53
54	15.00	4.1828	512	986.2	225.99	3.29	54

Table 2.2 (*continued*)

Temperature $t_s/\,^\circ C$	Absolute vapour pressure $p_s/\,kPa$	Specific heat capacity $c_r/\,(kJ.kg^{-1}.K^{-1})$	Dynamic viscosity $\mu_f/(\mu Pa.s)$	Density $\rho/\,(kg.m^{-3})$	Specific enthalpy of liquid $h_f/\,(kJ.kg^{-1})$	Prandtl number $(Pr)_g$	Temperature $t_s/\,^\circ C$
55	15.74	4.1830	504	985.7	230.17	3.23	55
56	16.51	4.1833	496	985.2	234.35	3.17	56
57	17.31	4.1837	489	984.7	238.54	3.12	57
58	18.15	4.1841	481	984.3	242.72	3.06	58
59	19.02	4.1845	474	983.7	246.81	3.01	59
60	19.92	4.1850	467	983.2	251.09	2.96	60
61	20.86	4.1856	460	982.7	255.27	2.91	61
62	21.84	4.1861	453	982.1	259.46	2.87	62
63	22.85	4.1867	447	981.6	263.65	2.82	63
64	23.91	4.1874	440	981.1	267.83	2.78	64
65	25.01	4.1880	434	980.5	272.02	2.74	65
66	26.15	4.1886	428	979.9	276.21	2.70	66
67	27.33	4.1892	422	979.4	280.40	2.66	67
68	28.56	4.1898	416	978.9	284.59	2.62	68
69	29.84	4.1904	410	978.3	288.78	2.58	69
70	31.16	4.1910	404	977.7	292.97	2.54	70
71	32.53	4.1916	399	977.1	297.16	2.50	71
72	33.96	4.1921	394	976.6	301.35	2.47	72
73	35.43	4.1927	388	976.0	305.54	2.43	73
74	36.96	4.1934	383	975.4	309.74	2.39	74
75	38.55	4.1940	378	974.9	313.93	2.36	75
76	40.19	4.1947	373	974.3	318.12	2.33	76
77	41.89	4.1955	369	973.6	322.32	2.30	77
78	43.65	4.1962	364	973.1	326.52	2.27	78
79	45.47	4.1971	359	972.5	330.71	2.24	79
80	47.36	4.1980	355	971.8	334.91	2.21	80
81	49.31	4.1990	350	971.2	339.11	2.18	81
82	51.33	4.1999	346	970.6	343.31	2.15	82
83	53.42	4.2009	342	969.9	347.51	2.12	83
84	55.57	4.2020	338	969.3	351.71	2.10	84
85	57.80	4.2030	334	968.6	355.91	2.07	85
86	60.11	4.2040	330	968.0	360.11	2.05	86
87	62.49	4.2050	326	967.3	364.32	2.02	87
88	64.95	4.2060	322	966.7	368.52	2.00	88
89	67.49	4.2070	318	966.0	372.73	1.97	89
90	70.11	4.2080	314	965.3	376.94	1.95	90
91	72.82	4.2090	311	964.7	381.15	1.93	91
92	75.61	4.2099	307	964.0	385.36	1.90	92
93	78.49	4.2109	304	963.3	389.57	1.88	93
94	81.46	4.2120	300	962.7	393.78	1.85	94
95	84.53	4.2130	297	961.9	397.99	1.83	95
96	87.69	4.2141	294	961.2	402.30	1.81	96
97	90.94	4.2153	291	960.5	406.42	1.79	97
98	94.30	4.2165	287	959.8	410.63	1.77	98
99	97.76	4.2177	284	959.1	414.84	1.76	99
100	101.33	4.2190	281	958.3	419.06	1.74	100
102	108.78	4.2217	275	956.9	427.50	1.70	102
104	116.68	4.2246	269	955.5	435.95	1.67	104
106	125.04	4.2274	264	954.0	444.40	1.63	106
108	133.90	4.2302	259	952.6	452.86	1.60	108
110	143.26	4.2330	253	951.0	461.32	1.57	110
112	153.16	4.2357	248	949.5	469.79	1.54	112
114	163.61	4.2386	243	948.0	478.26	1.51	114
116	174.64	4.2414	239	946.3	486.74	1.48	116
118	186.28	4.2445	234	944.7	495.20	1.45	118

Table 2.2 (*continued*)

Temperature t_s/ °C	Absolute vapour pressure p_s/ kPa	Specific heat capacity c_r/ (kJ.kg^{-1}.K^{-1})	Dynamic viscosity μ_f/(μPa.s)	Density ρ/ (kg.m^{-3})	Specific enthalpy of liquid h_f/ (kJ.kg^{-1})	Prandtl number $(Pr)_g$	Temperature t_s/ °C
120	198.53	4.2480	230	943.1	503.70	1.42	120
122	211.44	4.2527	226	941.5	512.20	1.40	122
124	225.03	4.2576	223	939.9	520.70	1.38	124
126	239.32	4.2621	219	938.3	529.20	1.36	126
128	254.34	4.2661	216	936.5	537.80	1.34	128
130	270.12	4.2700	212	934.8	546.30	1.32	130
132	286.68	4.2740	209	933.1	554.90	1.30	132
134	304.05	4.2780	206	931.4	563.40	1.28	134
136	322.27	4.2820	202	929.6	572.00	1.26	136
138	341.36	4.2860	199	927.9	580.50	1.25	138
140	361.36	4.2900	196	926.1	589.10	1.23	140
142	382.28	4.2934	193	924.3	597.70	1.21	142
144	404.18	4.2975	191	922.5	606.30	1.20	144
146	427.07	4.3029	189	920.6	614.90	1.18	146
148	450.99	4.3102	187	918.8	623.50	1.17	148
150	475.97	4.3200	185	916.9	632.20	1.17	150
152	502.05	4.3250	183	915.1	640.80	1.16	152
154	529.26	4.3320	181	913.2	649.40	1.15	154
156	557.64	4.3380	178	911.2	658.10	1.13	156
158	587.23	4.3440	176	909.3	666.80	1.12	158
160	618.05	4.3500	174	907.4	675.50	1.11	160
162	650.14	4.3557	172	905.4	684.20	1.10	162
164	683.55	4.3614	170	903.4	692.90	1.09	164
166	718.31	4.3674	167	901.5	701.60	1.07	166
168	754.45	4.3735	165	899.4	710.40	1.06	168
170	792.03	4.3800	163	897.3	719.1	1.05	170
172	831.07	4.3875	161	895.3	727.9	1.04	172
174	871.61	4.3954	159	893.3	736.7	1.03	174
176	913.71	4.4034	157	891.2	745.5	1.02	176
178	957.39	4.4117	155	889.1	754.3	1.01	178
180	1002.7	4.4200	153	886.9	763.1	1.00	180
182	1049.7	4.4277	151	884.8	772.0	0.99	182
184	1098.4	4.4354	150	882.6	780.8	0.99	184
186	1148.9	4.4434	148	880.4	789.7	0.98	186
188	1201.1	4.4515	146	879.7	798.6	0.98	188
190	1255.2	4.4600	145	876.0	807.5	0.97	190
192	1311.2	4.4695	144	873.8	816.4	0.96	192
194	1369.2	4.4794	142	781.5	825.4	0.96	194
196	1429.1	4.4894	141	869.3	834.4	0.95	196
198	1491.0	4.4997	139	867.0	843.4	0.95	198
200	1555.1	4.5100	138	864.7	852.4	0.94	200
205	1724.5	4.5337	134	858.8	875.0	0.92	205
210	1908.0	4.5600	131	852.8	897.7	0.91	210
215	2106.3	4.5937	128	846.6	920.6	0.90	215
220	2320.1	4.6300	125	840.3	943.7	0.90	220
225	2550.4	4.6644	122	833.9	966.9	0.90	225
230	2797.0	4.7000	120	827.3	990.3	0.89	230
235	3063.5	4.7387	118	820.6	1013.8	0.89	235
240	3348.0	4.7800	115	813.6	1037.6	0.88	240
245	3652.4	4.8231	112	806.5	1061.6	0.87	245
250	3977.6	4.8700	110	799.2	1085.8	0.87	250

Table 2.3 Enthalpy of superheated steam

Absolute pressure p / kPa	Saturation temperature t_s / °C	Enthalpy of superheated steam / kJ.kg^{-1} for stated final steam temperature / °C															
		100	120	140	160	180	200	220	240	260	280	300	320	340	360	380	400
100	99.6	2676	2717	2757	2797	2836	2876	2915	2955	2995	3034	3075	3115	3155	3196	3237	3278
120	104.8		2715	2755	2796	2835	2875	2914	2954	2994	3034	3074	3114	3155	3196	3237	3278
140	109.3		2713	2754	2794	2834	2874	2914	2953	2993	3033	3074	3114	3155	3195	3236	3278
160	113.3		2711	2752	2793	2833	2873	2913	2953	2993	3033	3073	3114	3154	3195	3236	3277
180	116.9		2708	2751	2792	2832	2872	2912	2952	2992	3032	3073	3113	3154	3195	3236	3277
200	120.2			2749	2790	2831	2871	2911	2951	2992	3032	3072	3113	3153	3194	3235	3277
220	123.3			2747	2789	2830	2870	2911	2951	2991	3031	3072	3112	3153	3194	3235	3276
240	126.1			2745	2787	2829	2869	2910	2950	2990	3031	3071	3112	3153	3194	3235	3276
260	128.7			2743	2786	2827	2868	2909	2949	2990	3030	3071	3111	3152	3193	3234	3276
280	131.2			2742	2785	2826	2867	2908	2949	2989	3030	3070	3111	3152	3193	3234	3275
300	133.5			2740	2783	2825	2866	2907	2948	2988	3029	3070	3110	3151	3192	3234	3275
340	137.9			2736	2780	2823	2865	2906	2947	2987	3028	3069	3110	3151	3192	3233	3274
380	141.8				2777	2820	2863	2904	2945	2986	3027	3068	3109	3150	3191	3232	3274
420	145.4				2774	2818	2861	2902	2944	2985	3026	3067	3108	3149	3190	3232	3273
460	148.7				2771	2816	2859	2901	2942	2984	3025	3066	3107	3148	3189	3231	3273
500	151.8				2768	2813	2857	2899	2941	2982	3024	3065	3106	3147	3189	3230	3272
600	158.8				2759	2807	2851	2895	2937	2979	3021	3062	3104	3145	3187	3229	3270
700	165.0					2800	2846	2890	2933	2976	3018	3060	3102	3143	3185	3227	3269
800	170.4					2793	2840	2886	2930	2973	3015	3057	3099	3141	3183	3225	3267
900	175.4					2785	2835	2881	2926	2969	3012	3055	3097	3139	3181	3223	3266
1000	179.9					2778	2829	2876	2922	2966	3009	3052	3095	3137	3179	3222	3264
1200	188.0						2817	2867	2914	2959	3003	3047	3090	3133	3176	3218	3261
1400	195.0						2803	2856	2905	2952	2997	3042	3085	3129	3172	3215	3257
1600	201.4							2846	2897	2945	2991	3036	3081	3124	3168	3211	3254
1800	207.1							2834	2888	2937	2985	3031	3076	3120	3164	3207	3251
2000	212.4							2822	2878	2930	2978	3025	3071	3116	3160	3204	3248
2500	223.9								2853	2909	2961	3011	3058	3105	3150	3195	3239
3000	233.8								2825	2888	2943	2995	3045	3093	3140	3185	3231
3500	242.5									2864	2925	2980	3031	3081	3129	3176	3222
4000	250.3									2838	2904	2963	3017	3069	3118	3166	3214
4500	257.4									2809	2883	2946	3003	3056	3107	3156	3205
5000	263.9										2859	2927	2988	3043	3096	3146	3196
6000	275.6										2806	2887	2955	3016	3072	3126	3177
7000	285.8											2841	2919	2987	3048	3104	3158
8000	295.0											2787	2880	2955	3021	3082	3139
9000	303.3												2835	2921	2993	3058	3118
10000	311.0												2784	2884	2964	3033	3097

3 Heat transfer

3.1   Introduction

This section of the CIBSE Guide is concerned with heat transfer. It is divided into a theoretical part and a practical part. The theoretical part provides a number of basic equations and discussion for convection, conduction and radiation together with a brief review of mass transfer. This part is not exhaustive and for a detailed treatment of the subject, the references cited at the end of the section should be consulted.

The practical part is intended for reference in dealing with common heat transfer problems related to the built environment. This is structured as: external environment; internal environment; human body. This part concludes with a treatment of components and equipment. For particular products, the manufacturer's data should be consulted. Example calculations are provided where appropriate to aid understanding and application.

In view of the number of equations, the notation is given at the start so as to provide a point of reference for the section.

3.1.1   Notation

A = area (m^2)

A_c = cross-sectional area (m^2)

A_i or $_j$ = area for surface i or j (m^2)

A_s = heated or cooled surface area (m^2)

A_{si} = inside surface area (m^2)

A_{so} = outside surface area (m^2)

A_1 = area of surface 1, etc. (m^2)

C = a constant (specified in text)

C_h = rate of heat transfer by convection at a human body/unit area ($W.m^{-2}$)

$C_{(min,max)}$ = fluid heat capacity rates (smaller and greater, respectively) ($W.K^{-1}$)

D = characteristic plate dimension (m)

E = emissivity factor (dimensionless)

E_{bi} = emissive power for surface i ($W.m^{-2}$)

F = view factor (or form factor or angle factor)

F_{ij} = view factor of surface j with respect to surface i

F_{p-N} = mean view factor between a person and a room surface (see Figures 3.3–3.7)

F_r = radiation exchange factor for two surfaces (dimensionless)

F_{12} = view factor of surface 2 with respect to surface 1

H = height of vertical rectangular enclosure (m)

H_1 = latent heat of water vapour (= 2450 at 20°C) ($kJ.kg^{-1}$)

I = intensity of solar radiation ($W.m^{-2}$)

I_{clo} = thermal resistance of clothing (clo)

$\mathcal{J}_i$ or $_j$ = radiosity for surface i or j ($W.m^{-2}$)

L = length of cylinder, pipe, etc. (m)

M = mass flow rate ($kg.s^{-1}$)

N = number of surfaces within the enclosure

NTU = number of exchanger heat transfer units

P = perimeter (m)

Q = heat exchange per unit length ($W.m^{-1}$)

R = thermal resistance (m².K.W⁻¹)

R_a = thermal resistance of air gap (m².K.W⁻¹)

R_e = thermal resistance of earth (m².K.W⁻¹)

R_n = thermal resistance of insulation (m².K.W⁻¹)

R_{se} = external surface resistance (m².K.W⁻¹)

R_1 = thermal resistance of element 1, etc. (m².K.W⁻¹)

T = absolute temperature (K)

T_a = absolute temperature of air (K)

T_{fs} = absolute temperature of fictitious surface (given by an area and emissivity-weighted average of other surfaces) (K)

T_i = absolute surface temperature for surface i (K)

T_r = absolute mean radiant temperature (K)

T_{rs} = absolute mean radiant temperature of sky (= 253 + t_a) (K)

T_s = absolute temperature of heat emitting/absorbing surface (K)

T_{sw} = absolute temperature of water surface (K)

T_1 = absolute temperature of surface 1, etc. (K)

U = overall thermal transmittance (W.m⁻².K⁻¹)

U_c = thermal transmittance of clean surfaces (W.m⁻².K⁻¹)

U_t = thermal transmittance of water tank (W.m⁻².K⁻¹)

V = volume flow rate (m³.s⁻¹)

V_s = surface wind speed (m.s⁻¹)

W = rate of water evaporation (kg.m⁻².s⁻¹)

Z = heat capacity rate ratio

a = length (m)

a_1 = length of side 1, etc. (m)

b = width (m)

c = specific heat capacity (kJ.kg⁻¹.K⁻¹)

c_p = specific heat capacity at constant pressure (kJ.kg⁻¹.K⁻¹)

d = diameter (m)

d_h = hydraulic mean (equivalent) diameter (m)

d_i = inside diameter (m)

d_{ic} = inside diameter of casing (m)

d_{i1} = inside diameter of element 1, etc. (m)

d_o = outside diameter (m)

d_{on} = outside diameter of insulation (m)

d_{op} = outside diameter of pipe (m)

d_{o1} = outside diameter of element 1, etc. (m)

f = a function (specified in text)

f_{cl} = clothing area factor (ratio of clothed to unclothed body surface area)

f_r = friction factor

g = acceleration due to gravity (= 9.81) (m.s⁻²)

h = heat transfer coefficient (W.m⁻².K⁻¹)

h_c = convective heat transfer coefficient (W.m⁻².K⁻¹)

h_{ci} = inside surface convective heat transfer coefficient (W.m⁻².K⁻¹)

h_{cn} = convective heat transfer coefficient for natural convection at internal surfaces (W.m⁻².K⁻¹)

h_{co} = outside surface convective heat transfer coefficient (W.m⁻².K⁻¹)

h_r = radiative heat transfer coefficient (W.m⁻².K⁻¹)

h_{si} = inside surface heat transfer coefficient (or film coefficient) (W.m⁻².K⁻¹)

h_{so} = outside surface heat transfer coefficient (or film coefficient) (W.m⁻².K⁻¹)

k = thermal conductivity (W.m⁻¹.K⁻¹)

k_e = earth thermal conductivity (W.m⁻¹.K⁻¹)

k_n = thermal conductivity of insulation (W.m⁻¹.K⁻¹)

k_w = thermal conductivity of wall (W.m⁻¹.K⁻¹)

k_{zi} = inside surface thermal conductivity of fouling substance (W.m⁻¹.K⁻¹)

k_{zo} = outside surface thermal conductivity of fouling substance (W.m⁻¹.K⁻¹)

l = thickness (m)

l_n = thickness of insulation (m)

l_w = wall thickness (m)

l_{zi} = inside fouling substance thickness (m)

l_{zo} = outside fouling substance thickness (m)

l_1 = thickness of element 1, etc. (m)

m = burial depth (m)

n = an index (specified in text)

p_{sw} = saturated vapour pressure at water surface (kPa)

p_v = vapour pressure of moisture in air remote from water surface (kPa)

q = actual rate of heat transfer (for a heat exchanger) (W)

q_c = rate of convective heat transfer by heated or cooled surface (W)

q_i = net rate of heat transfer by radiation to surface i (W)

q_{max} = maximum possible rate of heat transfer (for a heat exchanger) (W)

q_r = rate of radiant heat transfer by heated or cooled surface (W)

q_t = total rate of heat transfer by heated or cooled surface (W)

r = radius (m)

r_i = inside radius (m)

r_o = outside radius (m)

r_1 = radius of surface 1, etc. (m)

t = Celsius temperature (°C)

t_a = air temperature (°C)

t_c = casing temperature (°C)

t_{cfi} = cold fluid inlet temperature (°C)

t_{cfo} = cold fluid outlet temperature (°C)

t_{cl} = clothing surface temperature (°C)

t_d = temperature at down stream end of pipe or duct section (°C)

t_e = ground temperature (°C)

t_f = free stream fluid temperature (°C)

t_{fs} = area weighted average fictitious surface temperature (°C)

t_{hfi} = hot fluid inlet temperature (°C)

t_{hfo} = hot fluid outlet temperature (°C)

t_{hi} = shell inlet temperature (°C)

t_{ho} = shell outlet temperature (°C)

t_i = inside temperature (°C)

t_j = surface temperature of each of the surfaces comprising the fictitious surface (°C)

t_m = mean of the inlet and outlet fluid temperatures in a tube (°C)

t_n = insulation temperature (°C)

t_o = outside temperature (°C)

t_s = surface temperature (°C)

t_{si} = inside surface temperature (°C)

t_{so} = outside surface temperature (°C)

t_{s1} = temperature of surface 1, etc. (°C)

t_{ti} = tube inlet temperature (°C)

t_{to} = tube outlet temperature (°C)

t_u = temperature at upstream end of pipe or duct section (°C)

t_w = water temperature (°C)

t_{sw} = water surface temperature (°C)

v = velocity (m.s^{-1})

v_a = air velocity at water surface (m.s^{-1})

v_r = room air speed (m.s^{-1})

Δ = difference in; change in

Δt = temperature difference (K)

Δt_{ca} = difference between clothing and air temperature (K)

Δt_l = logarithmic mean temperature difference (K)

Δt_m = change in temperature per unit length (K.m^{-1})

Δt_{tg} = greatest terminal temperature difference (K)

Δt_{sa} = surface to air temperature difference (K)

Δt_{ts} = smallest terminal temperature difference (K)

Σ = sum of

Φ = heat exchange (W)

γ = coefficient of cubical expansion (K^{-1})

ε = emissivity

ε_i = emissivity of surface i

ε_w = emissivity of water

ε_1 = emissivity of surface 1, etc.

η = heat exchanger effectiveness

μ = dynamic viscosity (Pa.s)

v = kinematic viscosity (m^2.s^{-1})

ξ = layer thickness (m)

ρ = density (kg.m^{-3})

σ = Stefan-Boltzmann constant ($= 5.67 \times 10^{-8}$) (W.m^{-2}.K^{-4})

ϕ = heat exchange per unit area (W.m^{-2})

ϕ_c = convective heat exchange per unit area (W.m^{-2})

ϕ_{cd} = conductive heat exchange per unit area (W.m^{-2})

ϕ_e = evaporative heat exchange per unit area (W.m^{-2})

ϕ_r = radiative heat exchange per unit area (W.m^{-2})

θ = angle (degrees)

Dimensionless numbers

(Nu) = Nusselt number = hD/k

(Re) = Reynolds number = $\rho v D/\mu$

(Pr) = Prandtl number = $c_p \mu/k$

(Gr) = Grashof number = $\gamma g \rho^2 D^3 \Delta t/\mu^2$

(Ra) = Rayleigh number = $(Gr)(Pr)$

3.2 Heat transfer principles

The mechanisms of heat transfer are convection, conduction, radiation and mass transfer. Theoretical and experimental work has led to the development of many equations that express the magnitude of heat transfer by these modes. Some of the commonly used equations are given in this section of the CIBSE Guide, but a wider range of heat transfer equations, especially those for particular applications, may be found in some of the many references given at the end of the section.

3.2.1 Convection

Convection is a mode of heat transfer between a moving fluid and a solid, liquid or gas. Convection can be free or forced. Free convection is the movement of a fluid primarily due to buoyancy forces. Forced convection is the movement of a fluid due primarily to external means, i.e. a mechanical pump or a pressure difference induced by a fan, for example. Many complications can arise due to the following reasons:

(a)   The flow can be laminar (having smooth and orderly streamlines) or turbulent (having irregularly interwoven streamlines) or separated (having a reversed flow along the fluid boundary, as with cylinders in cross flow with vortex shedding). There is a transitional state between laminar and turbulent flow.

(b)   There is a boundary layer between the fluid and the other medium.

(c)   At the base of the boundary layer the fluid is stationary with respect to the other medium. This means that there is a thin layer in which only conduction and mass transfer can take place.

(d)   For heat to be transferred within any substance, a temperature gradient must exist. In a fluid, therefore, there are density gradients which cause buoyancy forces.

The remainder of this subsection is therefore structured as follows. Empirical correlations (in the form of dimensionless numbers) are presented for the conditions of free and forced convection and for a range of common situations. These situations are classified as being either external flows (over surfaces) or internal flows (inside enclosures). In both cases the flow regimes can be either laminar or turbulent. For each situation, guidance is given regarding its particular characteristics and the use of the appropriate equation.

The following equations can be used for heating or cooling applications provided that no phase change (e.g. condensation or evaporation) occurs since, for most practical purposes, the theoretical differences between heating and cooling are usually negligible. The equations are valid (strictly) only for ideal fluids, i.e. with non-variable property values, and for incompressible flow. Table 3.1 gives typical values of the convection coefficients for various situations.

If the fluid property values change significantly over the temperature range investigated, account can be taken of them by using values at some mean temperature[1].

Table 3.1 Typical convection coefficients h_c for different modes of convection (reproduced from *Engineering Thermodynamics, Work and Heat Transfer*[2] by GFC Rogers and YR Mayhew (1992) by permission of Pearson Education Ltd)

Mode	Coefficient / $W.m^{-2}.K^{-1}$
Forced convection	
Gases and dry vapours	10 to 10^3
Liquids	10^2 to 10^4
Liquid metals	5×10^3 to 4×10^4
Free convection	
Gases and dry vapours	0.5 to 10^3
Liquids	50 to 3×10^3
Condensation	
Filmwise	5×10^2 to 3×10^4
Dropwise	2×10^4 to 5×10^5
Boiling	5×10^2 to 2×10^4

Free convection over surfaces

Table 3.2 presents correlations for the average (Nu) number for free convection over various geometries for a range of (Ra) numbers. The appropriate characteristic length is also given. The flow regime is determined by calculating the (Ra) number for a particular situation; note that the power of the (Ra) number is usually 0.25 for laminar flow and 0.33 for turbulent flow. All fluid properties are evaluated at the average of the surface temperature (t_s) and the free stream fluid temperature (t_f).

Free convection inside enclosures

Table 3.3 presents correlations for the average (Nu) number for free convection within enclosures of various geometries for a range of (Ra) numbers. Here, the term 'enclosures' refers to cavities such as those found in walls, double glazed windows or solar collectors etc. The characteristic length is the distance between the hot and cold surfaces. All fluid properties are evaluated at the average of the hot and cold surface temperatures.

Forced convection over surfaces

Flat plates:
For laminar flow over an entire horizontal flat surface, the average (Nu) number, based on plate length D, over the entire plate is given by:

$$(Nu)_D = 0.664 \, (Re)_D^{1/2} \, (Pr)^{1/3} \qquad (3.27)$$

$$\text{for } (Pr) \geq 0.6 \text{ and } (Re)_D < 5 \times 10^5$$

For turbulent flow over an entire horizontal flat surface, the average (Nu) number, based on plate length D, over the entire plate is given by:

$$(Nu)_D = 0.037 \, (Re)_D^{4/5} \, (Pr)^{1/3} \qquad (3.28)$$

$$\text{for } 0.6 \leq (Pr) \leq 60 \text{ and } 5 \times 10^5 \leq (Re)_D \leq 10^7$$

Equations (3.27) and (3.28) are restricted to the condition of uniform surface temperature over a smooth plate[3]. However, for other conditions, including uniform heat flux over a plate, unheated sections, combinations of laminar and turbulent flows, entry region effects and evaluation of local (Nu) numbers, the reader should consult references 3 and 4. Note that fluid properties are evaluated at the average of the surface temperature (t_s) and the free stream fluid temperature (t_f).

Cross flow

Table 3.4 presents correlations for the average (Nu) number for forced convection over circular and non-circular smooth cylinders and surfaces in cross flow for a range of (Re) numbers. The characteristic length is also shown. All fluid properties are evaluated at the average of the surface temperature (t_s) and the free stream fluid temperature (t_f).

Forced convection inside enclosures (tubes)

Table 3.5 presents correlations for the average (Nu) number for forced convection, fully developed, laminar flow inside tubes of circular and non-circular cross-sections. Laminar flow in tubes occurs for $(Re)_D < 2300$. Values of (Nu) numbers are given for the following thermal conditions at the surface of the tube.

Table 3.2 Empirical correlations for the average Nusselt number for free convection over surfaces (reproduced from *Introduction to Thermodynamics and Heat Transfer* by YA Cengel (1997) by permission of The McGraw-Hill Companies)

Geometry	Characteristic length	Range of (Ra)	Nusselt number (Nu)	
Vertical plate	D	$10^4 - 10^9$	$(Nu) = 0.59\,(Ra)^{1/4}$	(3.1)
		$10^9 - 10^{13}$	$(Nu) = 0.1\,(Ra)^{1/3}$	(3.2)
		Entire range	$(Nu) = \left\{ 0.825 + \dfrac{0.387(Ra)^{1/6}}{\left[1 + (0.492/(Pr))^{9/16}\right]^{8/27}} \right\}^2$	(3.3)
			(complex but more accurate)	
Inclined plate	D		Use vertical plate equations as a first degree of approximation	
			Replace g with $g \cos\theta$ in the formula for (Gr), see page 3-3, for $(Ra) < 10^9$	
Horizontal plate (surface area A and perimeter P)	A/P			
(a) Upper surface of a hot plate or lower surface of a cold plate		$10^4 - 10^7$	$(Nu) = 0.54\,(Ra)^{1/4}$	(3.4)
		$10^7 - 10^{11}$	$(Nu) = 0.15\,(Ra)^{1/3}$	(3.5)
(b) Lower surface of a hot plate or upper surface of a cold plate		$10^5 \quad 10^{11}$	$(Nu) = 0.27\,(Ra)^{1/4}$	(3.6)
Vertical cylinder	D		A vertical cylinder can be treated as a vertical plate when: $d \geq \dfrac{35D}{(Gr)^{1/4}}$	(3.7)
Horizontal cylinder	d	$10^5 - 10^{12}$	$(Nu) = \left\{ 0.6 + \dfrac{0.387(Ra)^{1/6}}{\left[1 + (0.559/(Pr))^{9/16}\right]^{8/27}} \right\}^2$	(3.8)
Sphere	$\pi d/2$	$(Ra) \leq 10^{11}$ $(Pr) \geq 0.7$	$(Nu) = 2 + \dfrac{0.589(Ra)^{1/4}}{\left[1 + (0.469/(Pr))^{9/16}\right]^{4/9}}$	(3.9)

Table 3.3 Empirical correlations for the average Nusselt number for free convection in enclosures (the characteristic length D is as indicated on the respective diagram) (reproduced from *Introduction to Thermodynamics and Heat Transfer* by YA Cengel (1997) by permission of The McGraw-Hill Companies)

Geometry	Fluid	H/D	Range of (Pr)	Range of (Ra)	Nusselt number (Nu)	
Vertical rectangular or cylindrical enclosure	Gas or liquid	—	—	<2000	$(Nu) = 1$	(3.10)
	Gas	$11 - 42$	$0.5 - 2$	$2 \times 10^3 - 2 \times 10^5$	$(Nu) = 0.197(Ra)^{1/4}\left(\dfrac{H}{D}\right)^{-1/9}$	(3.11)
		$11 - 42$	$0.5 - 2$	$2 \times 10^5 - 10^7$	$(Nu) = 0.073(Ra)^{1/3}\left(\dfrac{H}{D}\right)^{-1/9}$	(3.12)
	Liquid	$10 - 40$	$1 - 20\,000$	$10^4 - 10^7$	$(Nu) = 0.42(Pr)^{0.012}(Ra)^{1/4}\left(\dfrac{H}{D}\right)^{-0.3}$	(3.13)
		$1 - 40$	$1 - 20$	$10^6 - 10^9$	$(Nu) = 0.046(Ra)^{1/3}$	(3.14)
Inclined rectangular enclosure					Use the correlations for vertical enclosures as a first degree approximation for $\theta \leq 20^\circ$ by replacing g with $g\cos\theta$ in the formula for (Ra), see page 3-3	
Horizontal rectangular enclosure (hot surface at the top)	Gas or liquid	—	—	—	$(Nu) = 1$	(3.15)
Horizontal rectangular enclosure (hot surface at the bottom)	Gas or liquid	—	—	< 1700	$(Nu) = 1$	(3.16)
	Gas	—	$0.5 - 2$	$1.7 \times 10^3 - 7 \times 10^3$	$(Nu) = 0.059\,(Ra)^{0.4}$	(3.17)
		—	$0.5 - 2$	$7 \times 10^3 - 3.2 \times 10^5$	$(Nu) = 0.212\,(Ra)^{1/4}$	(3.18)
		—	$0.5 - 2$	$>3.2 \times 10^5$	$(Nu) = 0.061\,(Ra)^{1/3}$	(3.19)
	Liquid	—	$1 - 5000$	$1.7 \times 10^3 - 6 \times 10^3$	$(Nu) = 0.012\,(Ra)^{0.6}$	(3.20)
		—	$1 - 5000$	$6 \times 10^3 - 3.7 \times 10^4$	$(Nu) = 0.375\,(Ra)^{0.2}$	(3.21)
		—	$1 - 20$	$3.7 \times 10^4 - 10^8$	$(Nu) = 0.13\,(Ra)^{0.3}$	(3.22)
		—	$1 - 20$	$>10^8$	$(Nu) = 0.057\,(Ra)^{1/3}$	(3.23)
Concentric horizontal cylinders	Gas or liquid	—	$1 - 5000$	$6.3 \times 10^3 - 10^6$	$(Nu) = 0.11\,(Ra)^{0.29}$	(3.24)
	liquid	—	$1 - 5000$	$10^6 - 10^8$	$(Nu) = 0.40\,(Ra)^{0.20}$	(3.25)
Concentric spheres	Gas or liquid	—	$0.7 - 4000$	$10^2 - 10^9$	$(Nu) = 0.228\,(Ra)^{0.226}$	(3.26)

Table 3.4 Empirical correlations for the average Nusselt number for forced convection over circular and non-circular cylinders in cross flow (reproduced from *Introduction to Thermodynamics and Heat Transfer* by YA Cengel (1997) by permission of The McGraw-Hill Companies)

Cross section	Fluid	Range of (Re)	Nusselt number (Nu)	
Circle 	Gas or liquid	0.4 − 4 4 − 40 40 − 4000 4000 − 40 000 40 000 − 400 000	$(Nu) = 0.989\,(Re)^{0.33}\,(Pr)^{1/3}$ $(Nu) = 0.911\,(Re)^{0.385}\,(Pr)^{1/3}$ $(Nu) = 0.683\,(Re)^{0.466}\,(Pr)^{1/3}$ $(Nu) = 0.193\,(Re)^{0.618}\,(Pr)^{1/3}$ $(Nu) = 0.027\,(Re)^{0.805}\,(Pr)^{1/3}$	(3.29) (3.30) (3.31) (3.32) (3.33)
Square 	Gas	5000 − 100 000	$(Nu) = 0.102\,(Re)^{0.675}\,(Pr)^{1/3}$	(3.34)
Square (tilted 45°) 	Gas	5000 − 100 000	$(Nu) = 0.246\,(Re)^{0.588}\,(Pr)^{1/3}$	(3.35)
Hexagon 	Gas	5000 − 100 000	$(Nu) = 0.153\,(Re)^{0.638}\,(Pr)^{1/3}$	(3.36)
Hexagon (tilted 45°) 	Gas	5000 − 19 500 19 500 − 100 000	$(Nu) = 0.160\,(Re)^{0.638}\,(Pr)^{1/3}$ $(Nu) = 0.0385\,(Re)^{0.782}\,(Pr)^{1/3}$	(3.37) (3.38)
Vertical plate 	Gas	4000 − 15 000	$(Nu) = 0.228\,(Re)^{0.731}\,(Pr)^{1/3}$	(3.39)
Ellipse 	Gas	2500 − 15 000	$(Nu) = 0.248\,(Re)^{0.612}\,(Pr)^{1/3}$	(3.40)

(a)   Constant surface temperature: a typical situation for which this condition is applicable would be when a phase change process takes place at the outer surface of the tube (such as condensation or evaporation).

(b)   Constant heat flux: a typical example would be when the tube is subjected to uniform heating such as from an electric element.

For turbulent flow within circular and non-circular tubes and ducts, the following correlation can be used:

$$(Nu)_D = 0.023 \, (Re)_D^{4/5} \, (Pr)^n \qquad (3.41)$$

where $n = 0.4$ for heating $(t_s > t_m)$ and 0.3 for cooling $(t_s < t_m)$, where t_m is the mean fluid temperature, evaluated as the average of the inlet and outlet fluid temperatures of the tube:

$$t_m = (t_{ti} + t_{to}) / 2 \qquad (3.42)$$

Equation (3.41) is known as the Dittus-Boelter equation, and has been confirmed by experiment[4] to be valid for the range $0.7 \leq (Pr) \leq 160$, and for $(Re)_D > 10000$.

Table 3.5 Nusselt numbers for fully developed laminar flow in tubes of various cross sections (hydraulic diameter $d_h = 4A_f/P$) (reproduced from *Introduction to Thermodynamics and Heat Transfer* by YA Cengel (1997) by permission of The McGraw-Hill Companies)

Cross section of tube	a/b or $\theta°$	Nusselt number (Nu)	
		t_s = const.	q_t = const.
Circle	—	3.66	4.36
Hexagon	—	3.35	4.00
Square	—	2.98	3.61
Rectangle	1	2.98	3.61
	2	3.39	4.12
	3	3.96	4.79
	4	4.44	5.33
	6	5.14	6.05
	8	5.60	6.49
	∞	7.54	8.24
Ellipse	1	3.66	4.36
	2	3.74	4.56
	4	3.79	4.88
	8	3.72	5.09
	16	3.65	5.18
Triangle	10°	1.61	2.45
	30°	2.26	2.91
	60°	2.47	3.11
	90°	2.34	2.98
	120°	2.00	2.68

Flow in tubes is considered to be turbulent for $(Re)_D >$ 4000 and to be transitional for the range $2300 \leq (Re)_D \leq 4000$.

Another expression has been proposed[7] to cover the range of $(Re)_D$ for the transitional and turbulent regimes:

$$(Nu)_D = \frac{(f_r/8)((Re)_D - 1000)(Pr)}{1 + 12.7(f_r/8)^{1/2}((Pr)^{2/3} - 1)} \quad (3.43)$$

where, for smooth tubes, the friction factor f_r is given by:

$$f_r = (0.79 \ln(Re)_D - 1.64)^{-2} \quad (3.44)$$

Equation (3.43) is valid for the range $0.5 < (Pr) < 2000$, and $2300 < (Re)_D < 5 \times 10^6$.

3.2.2 Conduction

Conduction is the transfer of heat within substances from positions of higher temperature to positions of lower temperature. Within all substances, except metals, conduction is primarily due to molecular movements although internal radiation can be significant. Within metals, most heat is transferred by free electrons[2].

The following equations assume that heat is transferred steadily, one dimensionally, and that materials are homogeneous and isotropic. These equations can be used for heating or cooling applications and are based on Fourier's Law:

$$\Phi = -kA \frac{dt}{dx} \quad (3.45)$$

In practice, it is usually necessary to consider heat transfer by conduction in two or three dimensions in order to obtain the temperature distribution throughout the medium of interest. To do this, it is necessary to solve the Fourier conduction equation in its two- or three-dimensional form. This can be achieved either analytically (using the method of separation of variables), graphically (using the flux plotting method), or numerically (using the finite difference technique). For details of these methods refer to reference 4.

Non-steady state or transient heat transfer occurs in situations where the boundary conditions are varying with respect to time. In these instances, the methods stated above can be used to solve the transient or non-steady state conduction equation in one, two or three dimensions as required[4]. References 8 and 9 also contain methods for solving such heat transfer problems. The availability of powerful computers together with software for simulating the dynamic thermal behaviour of buildings and systems has meant that the numerical method is the most commonly used for producing practical solutions.

Section 5 of CIBSE Guide A (1999) describes a number of procedures that can be used for structural problems.

Complications arise when heat is conducted through porous materials; this is because they are rarely dry, so that there are transient periods at the end and beginning of heat transfer during which moisture is also transferred. Eventually, equilibrium states are reached, though some-

times only after many months. Section 3 of CIBSE Guide A (1999) deals with the problem of moisture content in more detail.

Conduction through flat structures

For single layer flat structures and for pipes and curved structures of large radii, the heat transfer rate is given by:

$$\Phi = \frac{k(t_{so} - t_{si})A}{l} \quad (3.46)$$

For multi-layered structures or pipes having large radii:

$$\Phi = \frac{A}{R}(t_{so} - t_{si}) \quad (3.47)$$

where:

$$R = \frac{l_1}{k_1} + \frac{l_2}{k_2} + \cdots \quad (3.48)$$

These equations are valid for steady state conditions, i.e. when the rate of heat transfer is constant with respect to time, and in the direction of the temperature gradient. The criterion which describes whether the radii of curved structures are large is:

$$\frac{r + \xi}{r} \approx 1 \quad (3.49)$$

Table 3.6 gives percentage errors introduced to the value of the heat transfer rate by using the above formulae for various ratios of actual structure radii.

Conduction through cylindrical structures

For a single element:

$$\Phi = \frac{2\pi kL(t_{so} - t_{si})}{\ln\left(\dfrac{d_o}{d_i}\right)} \quad (3.50)$$

For multi-layered structures:

$$\Phi = \frac{2\pi L(t_{so} - t_{si})}{R_1 + R_2 + \cdots} \quad (3.51)$$

where:

$$R_1 = \frac{1}{k_1} \ln\left(\frac{d_{o1}}{d_{i1}}\right), \text{ etc.} \quad (3.52)$$

Table 3.6 Error in approximating a cylinder to a flat surface

$\dfrac{r + \xi}{r}$	Error %
1.2	10
1.1	5
1.05	3
1.025	1.3
1.0125	0.8

Note: r is radius of cylinder
ξ is thickness of layer

Heat flow through structures by conduction and convection

The effects of convection from the surfaces of structures can be included in the above formulae by the addition of surface convective heat transfer coefficients. For multi-layered flat structures or structures with large radii of curvature with convective heat transfer on both exposed surfaces, the heat transfer rate is given by:

$$\Phi = \frac{A(t_i - t_o)}{R_1 + R_2 + \cdots + \dfrac{1}{h_{ci}} + \dfrac{1}{h_{co}}} \quad (3.53)$$

where:

$$R_1 = \frac{l_1}{k_1}, \text{ etc.} \quad (3.54)$$

For multi-layered cylindrical structures:

$$\Phi = \frac{2\pi L(t_i - t_o)}{R_1 + R_2 + \cdots + \dfrac{2}{h_{ci}d_i} + \dfrac{2}{h_{co}d_o}} \quad (3.55)$$

where:

$$R_1 = \frac{1}{k_1} \ln\left(\frac{d_{o1}}{d_{i1}}\right), \text{ etc.} \quad (3.56)$$

For structures having continuous air spaces, the resistance of the air space, which takes account of radiative, convective, and conductive heat transfer can also be included in the above equations, see Section 3 of CIBSE Guide A (1999).

3.2.3   Radiation

Radiation, in the context of heating, is the emission, from a source, of electromagnetic waves having wavelengths between those of visible light and those of radio waves. Radiation heat transfer is governed by the Stefan-Boltzmann law for black bodies which can be written:

$$\phi = \sigma T^4 \quad (3.57)$$

A black body is defined as one which absorbs totally all radiation falling onto its surface. A grey body has a surface which absorbs all wavelengths equally but does not absorb all the radiation. Emissivity is defined as the ratio of the total emissive power of a body to the total emissive power of a black body at the same temperature. Some values of emissivity are given in Table 3.7.

Equations for black body radiation are simple, but in practice black body radiation does not often occur. Most of the following equations assume that the reflective, emissive and absorptive properties of the surfaces remain constant with wavelength, i.e. grey body radiation is assumed, though these equations are also valid for black body radiation. Unless otherwise stated, all equations require that reflections are non-specular, i.e. have no directional properties.

For problems involving surfaces which are so placed that a significant proportion of the radiation emitted by one surface does not fall upon the other surface, or escapes from between the surfaces, or for problems involving enclosures with varying surface temperatures, then the following equations can be modified. These terms will include view factors (otherwise known as form factors or angle factors), which are functions of the geometry of the surfaces only, and are a measure of the proportion of the field of view of one surface which is occupied by the other surface. Table 3.8 and Figure 3.2 contain view factors for some of the more common geometries. References 10, 11, and 12 contain further work and examples of view factors. A comprehensive catalogue of view factor relations may be found in reference 13.

It can be shown that the relation between the view factors for two surfaces is:

$$F_{12}A_1 = F_{21}A_2 \quad (3.58)$$

and that the heat exchange between two black body finite areas is given by:

$$\Phi = A_1\sigma F_{12}(T_1^4 - T_2^4) \quad (3.59)$$
$$\Phi = A_2\sigma F_{21}(T_1^4 - T_2^4) \quad (3.60)$$

Radiation between parallel flat surfaces

For surfaces so placed that negligible radiation escapes from between them, the heat exchange for grey body radiation is given by[2]:

$$\Phi = \sigma\varepsilon_{12}A_1(T_1^4 - T_2^4) \quad (3.61)$$

where:

$$\varepsilon_{12} = \frac{\varepsilon_1\varepsilon_2}{\varepsilon_1 + \varepsilon_2 - \varepsilon_1\varepsilon_2} \quad (3.62)$$

for:

$$A_1 = A_2$$

Radiation between concentric curved surfaces

The following equation is valid for grey body radiation with or without a specular reflection, assuming negligible edge losses:

$$\phi = \sigma\varepsilon_{12}(T_1^4 - T_2^4) \quad (3.63)$$

where:

$$\varepsilon_{12} = \frac{1}{\varepsilon_1} + \frac{A_1}{A_2}\left(\frac{1}{\varepsilon_2} - 1\right) \quad (3.64)$$

for:

$$A_1 < A_2$$

Note that:

$$\Phi = \phi A_1 \quad (3.65)$$

Radiation between small surfaces well separated

For surfaces which are small compared with their distance apart, the amount of radiation reflected back to the radiating

Table 3.7(a) Absorptivity and emissivity: impermeable materials

Material	Condition (where known)	Absorptivity	Emissivity
Aluminium	Polished	0.10–0.40	0.03–0.06
	Dull/rough polish	0.40–0.65	0.18–0.30
	Anodised	—	0.72
Aluminium surfaced roofing		—	0.216
Asphalt	Newly laid	0.91–0.93	—
	Weathered	0.82–0.89	—
	Block	0.85–0.98	0.90–0.98
Asphalt pavement		0.852–0.928	—
Bitumen/felt roofing		0.86–0.89	0.91
Bitumen pavement		0.86–0.89	0.90–0.98
Brass	Polished	0.30–0.50	0.03–0.05
	Dull	0.40–0.065	0.20–0.30
	Anodised	—	0.59–0.61
Bronze		0.34	—
Copper	Polished	0.18–0.50	0.02–0.05
	Dull	0.40–0.065	0.20–0.30
	Anodised	0.64	0.60
Glass	Normal	★	0.88
	Hemispherical	★	0.84
Iron	Unoxidised	—	0.05
	Bright/polished	0.40–0.65	0.20–0.377
	Oxidised	—	0.736–0.74
	Red rusted	—	0.61–0.65
	Heavily rusted	0.737	0.85–0.94
Iron, cast	Unoxidised/polished	—	0.21–0.24
	Oxidised	—	0.64–0.78
	Strongly oxidised	—	0.95
Iron, galvanised	New	0.64–0.66	0.22–0.28
	Old/very dirty	0.89–0.92	0.89
Lead	Unoxidised	—	0.05–0.075
	Old/oxidised	0.77–0.79	0.28–0.281
Rubber	Hard/glossy	—	0.945
	Grey/rough	—	0.859
Steel	Unoxidised/polished/stainless	0.20	0.074–0.097
	Oxidised	0.20	0.79–0.82
Tin	Highly polished/unoxidised	0.10–0.40	0.043–0.084
Paint			
— aluminium		0.30–0.55	0.27–0.67
— zinc		0.30	0.95
Polyvinylchloride (pvc)		—	0.90–0.92
Tile	Light colour	0.3–0.5	0.85–0.95
Varnish		—	0.80–0.98
Zinc	Polished	0.55	0.045–0.053
	Oxidised	0.05	0.11–0.25

★ See manufacturers' data

surface from the other surface is negligible. The heat exchange for grey bodies is given by:

$$\Phi = \sigma \varepsilon_1 \varepsilon_2 F_{12} A_1 (T_1^4 - T_2^4) \qquad (3.66)$$

Radiation between an enclosure and a contained surface

For a comparatively small area within an enclosure, it is possible to assume that the enclosure behaves like a black body. This is because negligible radiation from the enclosed body will be reflected back to it from the enclosure. Hence $F_{12} = 1$ and $\varepsilon_2 = 1$ so:

$$\Phi = \sigma \varepsilon_1 A_1 (T_1^4 - T_2^4) \qquad (3.67)$$

Surface 1 has the smaller area and this equation is valid for grey body radiation.

Equivalent radiative heat transfer coefficient

Very often it is useful to have a heat transfer coefficient for radiation heat exchange such that:

$$\Phi = f h_r A_1 (t_{s_1} - t_{s_2}) \qquad (3.68)$$

The function $f = f(\varepsilon_{12}, F_{12})$ is extremely complex unless either the radiation is black body (in which case $\varepsilon_{12} = 1$) or the radiation is independent of the geometry (in which case $F_{12} = F_{21} = 1$).

Table 3.7(b) Absorptivity and emissivity: inorganic, porous materials

Material	Condition (where known)	Absorptivity	Emissivity
Asbestos:			
— board		—	0.96
— paper		—	0.93–0.94
— cloth		—	0.90
— cement	New	0.61	0.95–0.96
	Very dirty	0.83	0.95–0.96
Brick	Glazed/light	0.25–0.36	0.85–0.95
	Light	0.36–0.62	0.85–0.95
	Dark	0.63–0.89	0.85–0.95
Cement mortar, screed		0.73	0.93
Clay tiles	Red, brown	0.60–0.69	0.85–0.95
	Purple/dark	0.81–0.82	0.85–0.95
Concrete		0.65–0.80	0.85–0.95
— tile		0.65–0.80	0.85–0.95
— block		0.56–0.69	0.94
Plaster		0.30–0.50	0.91
Stone:			
— granite (red)		0.55	0.90–0.93
— limestone		0.33–0.53	0.90–0.93
— marble		0.44–0.592	0.90–0.93
— quartz		—	0.90
— sandstone		0.54–0.76	0.90–0.93
— slate		0.79–0.93	0.85–0.98

Hence, for black body radiation, equation (3.68) can be simplified to:

$$\Phi = F_{12} h_r A_1 (t_{s_1} - t_{s_2}) \qquad (3.69)$$

or, for grey body radiation (with $F_{12} = 1$):

$$\Phi = \varepsilon_{12} h_r A_1 (t_{s_1} - t_{s_2}) \qquad (3.70)$$

It can be shown that:

$$h_r = \sigma(T_1 + T_2)(T_1^2 + T_2^2) \qquad (3.71)$$

Table 3.9 gives values of h_r for various values of surface temperature. When $T_1 = T_2$ then:

$$h_r = 4\sigma T^3 \qquad (3.72)$$

3.2.4 Mass transfer

Mass transfer can be described as the diffusion of one substance into another. Most commonly, a liquid vaporises into, or condenses from, a gas with an accompanying variation of vapour concentration, i.e. partial pressure, within the gas. This variation occurs within a boundary layer.

In buildings and building services systems, the most common examples of mass transfer are condensation, humidification and evaporation. These examples can all take place within air conditioning equipment and cooling towers but the first and last examples often occur within structures such as during the drying out period of buildings and condensation on walls and windows. For these examples, the previous heat transfer equations are not applicable.

Calculations for most mass transfer problems are quite complex although generally the equations follow the form of the convection heat transfer equations.

Table 3.7(c) Absorptivity and emissivity: hygroscopic materials

Material	Condition (where known)	Absorptivity	Emissivity
Paper	—		0.091–0.94
— white, bond		0.25–0.28	—
Cloth:			
— cotton, black		0.67–0.98	—
— cotton, deep blue		0.82–0.83	—
— cotton, red		0.562	—
— wool, black		0.75–0.88	—
— felt, black		0.775–0.861	—
— fabric (unspecified)		—	0.89–0.92
Wood:			
— beach		—	0.94
— oak		—	0.89–0.90
— spruce		—	0.82
— walnut		—	0.83

A detailed treatment of mass transfer is beyond the scope of this section. For a general introduction to mass transfer the reader is referred to references 1, 4, 14 and 15. Reference 2 contains data and procedures involving mixtures (as well as most other aspects of thermodynamics) and reference 16 has a section that covers the simple aspects of water sprays.

3.3 Heat transfer practice

3.3.1 External environment

At the external surfaces of buildings, heat transfer takes place via the processes of convection and radiation. While radiation heat loss is a function of surface temperature and emissivity, convection heat loss is more complex, being a function of a number of variables such as wind speed and direction, flow regime and surface roughness. This makes difficult the accurate determination of heat transfer at a building surface.

Radiative exchange at external surfaces

The longwave radiation exchange at the external surface of a building is dependent upon the difference between the incoming longwave radiation from the external environment (sky, other buildings and the ground) and the outgoing longwave radiation emitted from the surface of the building in question. Normally, calculation of the radiative exchange is simplified by the use of linear expressions, rather than retaining the fourth power temperature term. This linearisation is produced by the introduction of a longwave radiation heat transfer coefficient, h_r. In order to determine a value for h_r, certain approximations are necessary. One approximation is the assumption that the external environment radiates as a black body at external air temperature; the other approximation is that the surrounding external environment can be assigned a view factor of unity. This leads to a value for the radiation component Eh_r of 4.0 W.m^{-2}.K^{-1}[17], where E, the emissivity factor, is the product of view factor and emissivity of the building surface.

Convective exchange at external surfaces

For design purposes, the exposure of buildings has been classified into three categories based on near-surface wind

Table 3.8 View factors

Configuration	View factor
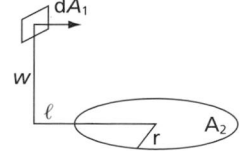	$$F_{12} = \frac{w}{2l}\left\{\frac{\left(\frac{w}{l}\right)^2 + \left(\frac{r}{l}\right)^2 + 1}{\left[\left(\left(\frac{w}{l}\right)^2 + \left(\frac{r}{l}\right)^2 + 1\right)^2 - 4\left(\frac{r}{l}\right)^2\right]^{0.5}} - 1\right\}$$
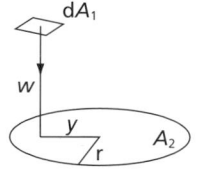	$$F_{12} = \frac{1}{2}\left\{1 - \frac{\left(\frac{w}{y}\right)^2 + \left(\frac{r}{y}\right)^2 + 1}{\left[\left(\left(\frac{w}{y}\right)^2 + \left(\frac{r}{y}\right)^2 + 1\right)^2 - 4\left(\frac{r}{y}\right)^2\right]^{0.5}}\right\}$$
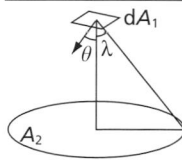	$$F_{12} = \sin^2 \lambda \cos \theta$$
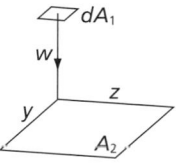	$$F_{12} = \frac{1}{2\pi}\left\{\left[\frac{y}{(w^2 + y^2)^{0.5}}\arctan\frac{z}{(w^2 + y^2)^{0.5}}\right] + \left[\frac{z}{(w^2 + z^2)^{0.5}}\arctan\frac{y}{(w^2 + z^2)^{0.5}}\right]\right\} = F_{\mathrm{p}}$$
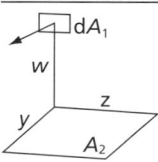	$$F_{12} = \frac{1}{2\pi}\left\{\arctan\frac{z}{w} - \left[\frac{w}{(w^2 + y^2)^{0.5}}\arctan\frac{z}{(w^2 + y^2)^{0.5}}\right]\right\} = F_{\mathrm{N}}$$
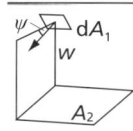	$$F_{12} = F_{\mathrm{p}}\cos\psi + F_{\mathrm{N}}\sin\psi$$
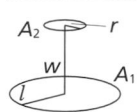	$$F_{12} = \frac{1}{2}\left\{\frac{r^2 + w^2 + l^2}{l^2} - \left[\left(\frac{r^2 + w^2 + l^2}{l^2}\right)^2 - \frac{4r^2}{l^2}\right]^{0.5}\right\}$$
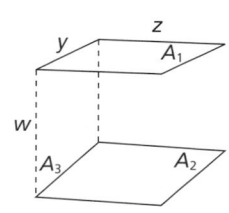	$$F_{12} = \frac{2}{\pi yz}\left\{\left[y(w^2 + z^2)^{0.5}\arctan\frac{y}{(w^2 + z^2)^{0.5}}\right] + \left[z(w^2 + y^2)^{0.5}\arctan\frac{z}{(w^2 + y^2)^{0.5}}\right]\right.$$ $$\left. - \left[wy\arctan\frac{y}{w}\right] - \left[wz\arctan\frac{z}{w}\right] - \left[\frac{w^2}{2}\ln\frac{w^2(w^2 + y^2 + z^2)}{(y^2 + w^2)(w^2 + z^2)}\right]\right\}$$ $$F_{32} = \frac{1}{\pi}\left\{\left[\arctan\frac{y}{w}\right] + \left[\frac{z}{w}\arctan\frac{y}{z}\right] - \left[\frac{(w^2 + z^2)^{0.5}}{w}\arctan\frac{y}{(w^2 + z^2)^{0.5}}\right]\right.$$ $$+ \left[\frac{w}{4y}\ln\frac{w^2(w^2 + y^2 + z^2)}{(w^2 + z^2)(w^2 + y^2)}\right] + \left[\frac{z^2}{4wy}\ln\frac{z^2(w^2 + y^2 + z^2)}{(y^2 + z^2)(w^2 + z^2)}\right]$$ $$\left. - \left[\frac{y}{4w}\ln\frac{y^2(w^2 + y^2 + z^2)}{(y^2 + z^2)(w^2 + y^2)}\right]\right\}$$
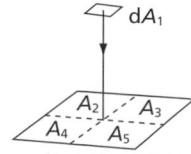	$$F_{1(2+3+4+5)} = F_{12} + F_{13} + F_{14} + F_{15}$$

Notes: (i) The arrowhead indicates the direction of the normal to the radiating element.

(ii) In some configurations a particular distance has been taken as unity, since view factors are independent of the scale.

Table 3.9 Values of radiative heat transfer coefficient h_r

T_1 (°C)	Values of h_r / W.m⁻².K⁻¹ for given values of T_2(°C)										
	−40	−20	0	5	10	15	20	25	30	40	50
−10	3.5	3.9	4.4	4.5	4.6	4.8	4.9	5.0	5.2	5.5	5.8
−5	3.6	4.0	4.5	4.6	4.8	4.9	5.0	5.2	5.3	5.6	5.9
0	3.7	4.1	4.6	4.7	4.9	5.0	5.2	5.3	5.4	5.7	6.0
5	3.8	4.3	4.7	4.9	5.0	5.1	5.3	5.4	5.6	5.9	6.2
10	3.9	4.4	4.9	5.0	5.1	5.3	5.4	5.6	5.7	6.0	6.3
15	4.1	4.5	5.0	5.1	5.3	5.4	5.6	5.7	5.9	6.2	6.5
20	4.2	4.6	5.2	5.3	5.4	5.6	5.7	5.9	6.0	6.3	6.6
25	4.3	4.8	5.3	5.4	5.6	5.7	5.9	6.0	6.2	6.5	6.8
30	4.4	4.9	5.4	5.6	5.7	5.9	6.0	6.2	6.3	6.6	7.0
40	4.7	5.2	5.7	5.9	6.0	6.2	6.3	6.5	6.6	7.0	7.3
50	5.0	5.5	6.0	6.2	6.3	6.5	6.6	6.8	7.0	7.3	7.6
60	5.3	5.8	6.4	6.5	6.7	6.8	7.0	7.1	7.3	7.7	8.0
70	5.6	6.1	6.7	6.9	7.0	7.2	7.3	7.5	7.7	8.0	8.4
80	5.9	6.5	7.1	7.2	7.4	7.5	7.7	7.9	8.1	8.4	8.8
90	6.3	6.8	7.4	7.6	7.8	7.9	8.1	8.3	8.4	8.8	9.2

speeds: 'sheltered', 'normal' and 'severe' exposure (see Section 3 of CIBSE Guide A (1999)). These exposure levels correspond to wind speeds at roof surfaces of 1.0, 3.0 and 9.0 m.s⁻¹, respectively; at wall surfaces, the corresponding near-surface wind speeds are 0.7, 2.0 and 6.0 m.s⁻¹, respectively.

Wind speed at the surface V_s and the convection coefficient h_c at the external surface have to date been correlated using the following expression derived from wind tunnel studies[18]:

$$h_c = 5.8 + 4.1V_s \qquad (3.73)$$

Equation (3.73) has been assumed to be valid for all external wind speeds.

Use of this correlation with the wind speeds given above leads to the design values of external surface resistance, R_{se}, based on the following relation:

$$R_{se} = 1/(h_c + Eh_r) \qquad (3.74)$$

These values for R_{se} are given in Table 3.10 of CIBSE Guide A (1999), for the three levels of exposure. In recent years, however, a number of experiments to measure h_c at the external surface of actual buildings have been made. These results are discussed below.

Full-scale external measurements

Reference 19 presents a review of full-scale measurements for h_c at building external surfaces, and has suggested the following combined correlation for use by designers for the case of 'normal' exposure:

$$h_c = 16.7V_s^{0.5} \qquad (3.75)$$

for h_c in W.m⁻².K⁻¹ and V_s in m.s⁻¹ up to a surface wind speed of 3.5 m.s⁻¹. Substituting a value for V_s of 2.0 m.s⁻¹ leads to a value for h_c of 23.6 W.m⁻².K⁻¹ and a corresponding modified value of the external surface resistance R_{se} for normal exposure of 0.04 m².K.W⁻¹.

3.3.2　　Internal environment

Internal surface convection coefficients

There have been several studies of internal surface convection coefficients[20,21,22,23], showing that a range of variation exists in the measured values. Reference 20 presents correlations that give typical values for convective heat transfer coefficients at heated wall, floor and ceiling surfaces of a room-sized test enclosure; the correlations were expressed in terms of a hydraulic diameter d_h, and a temperature difference Δt_{sa}:

$$h_{cn} = \frac{1.823}{d_h^{0.121}} (\Delta t_{sa})^{0.293} \qquad (3.76)$$

for a heated wall;

$$h_{cn} = \frac{2.175}{d_h^{0.076}} (\Delta t_{sa})^{0.308} \qquad (3.77)$$

for a heated floor; and

$$h_{cn} = \frac{0.704}{d_h^{0.601}} (\Delta t_{sa})^{0.133} \qquad (3.78)$$

for a heated ceiling.

Here, Δt_{sa} was the difference between the surface and air temperatures (measured at 0.1 m from the surface for wall and floor and at the centre of the enclosure for the case of the ceiling); Δt_{sa} values ranged between 5 and 25 K. In these expressions, the value for d_h was evaluated as $4A_s/P$.

Note that values for the internal surface resistance R_{si} can be estimated from a knowledge of appropriate convective and radiative heat transfer coefficients. Values are given in Table 3.9 of CIBSE Guide A (1999).

Radiation exchange between internal surfaces

For a multi-surface enclosure of N surfaces, each with an area $A_1, A_2, \ldots, A_N$ and with emissivities of $\varepsilon_1, \varepsilon_2, \ldots, \varepsilon_N$, respectively, the following equations can be used, based on radiosity formulation methods (see, for example, any standard heat transfer text), to determine the net rate of heat exchange by radiation to each surface:

$$q_i = \frac{\varepsilon_i A_i (E_{bi} - \mathcal{J}_i)}{(1 - \varepsilon_i)} \quad (3.79)$$

and

$$\frac{\varepsilon_i A_i (E_{bi} - \mathcal{J}_i)}{(1 - \varepsilon_i)} = \sum_{j=1}^{N} (\mathcal{J}_i - \mathcal{J}_j)(A_i F_{ij}) \quad (3.80)$$

For N surfaces, equation (3.80) results in a set of N linear equations with unknowns $\mathcal{J}_1, \mathcal{J}_2, \ldots, \mathcal{J}_N$. For the surface '$i$' at a known surface temperature T_i, knowledge of the value for $\mathcal{J}_i$ permits calculation of the net rate of heat transfer by radiation, q_i, to the surface 'i' by use of equation (3.79).

If the surface 'i' is considered to be a black body of known temperature T_i, then its radiosity $\mathcal{J}_i$ is equal to its emissive power E_{bi}, that is:

$$E_{bi} = \mathcal{J}_i = \sigma T_i^4 \quad (3.81)$$

and the net rate of heat transfer by radiation to the black surface 'i' can be determined from the following expression:

$$q_i = \sum_{j=1}^{N} (\mathcal{J}_i - \mathcal{J}_j)(A_i F_{ij}) \quad (3.82)$$

To calculate the rate of radiation exchange between surfaces within the multi-surface enclosure, the following rules of view factor algebra must be applied:

(a) reciprocity rule:

$$A_i F_{ij} = A_j F_{ji} \quad (3.83)$$

(b) summation rule:

$$\sum_{j=1}^{N} F_{ij} = 1 \quad (3.84)$$

Values of view factors for some of the more commonly encountered geometries are given in Table 3.8 or Figure 3.2. From a knowledge of view factors and known values of radiosities, for all surfaces, the net rate of heat transfer by radiation to each surface can be determined using equation (3.79).

Example 3.1

A cuboidal room is 6 m long, 4 m wide and 3 m high. It contains a chilled ceiling of surface temperature 18°C, and a floor of surface temperature 22°C; each of its four walls is at a surface temperature of 25°C. Determine the net rate of heat transfer by radiation to the ceiling and to the floor, assuming that the walls act as a black body surface, and that the emissivities of the ceiling and the floor are 0.9 and 0.85, respectively.

Solution

The ceiling is treated as surface 1, the floor as surface 2, and all the four walls can be treated as a single surface, surface 3. Figure 3.1 illustrates the arrangement:

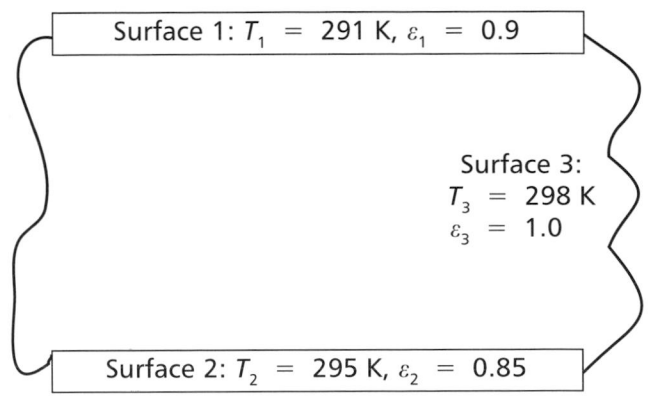

Surface 1: $T_1 = 291$ K, $\varepsilon_1 = 0.9$

Surface 3: $T_3 = 298$ K $\varepsilon_3 = 1.0$

Surface 2: $T_2 = 295$ K, $\varepsilon_2 = 0.85$

Figure 3.1 Illustration of room as a 3-surface enclosure

Using equation (3.80), calculate the radiosities $\mathcal{J}_1$ and $\mathcal{J}_2$ for surfaces 1 and 2, respectively:

For surface 1 ($i = 1$):

$$\frac{\varepsilon_1 A_1}{(1 - \varepsilon_1)}(E_{b1} - \mathcal{J}_1) = \sum_{j=1}^{3} (\mathcal{J}_1 - \mathcal{J}_j)(A_1 F_{1j}) \quad (3.85)$$

but:

$$\sum_{j=1}^{3} (\mathcal{J}_1 - \mathcal{J}_j)(A_1 F_{ij}) = (\mathcal{J}_1 - \mathcal{J}_1)(A_1 F_{11})$$
$$+ (\mathcal{J}_1 - \mathcal{J}_2)(A_1 F_{12})$$
$$+ (\mathcal{J}_1 - \mathcal{J}_3)(A_1 F_{13}) \quad (3.86)$$

therefore:

$$\frac{\varepsilon_1 A_1}{(1 - \varepsilon_1)}(E_{b1} - \mathcal{J}_1) = (\mathcal{J}_1 - \mathcal{J}_2)(A_1 F_{12})$$
$$+ (\mathcal{J}_1 - \mathcal{J}_3)(A_1 F_{13}) \quad (3.87)$$

Similarly, for surface 2 ($i = 2$):

$$\frac{\varepsilon_2 A_2}{(1 - \varepsilon_2)}(E_{b2} - \mathcal{J}_2) = (\mathcal{J}_2 - \mathcal{J}_1)(A_2 F_{21})$$
$$+ (\mathcal{J}_2 - \mathcal{J}_3)(A_2 F_{23}) \quad (3.88)$$

For surface 3, since this is assumed to be a black body surface, then from equation (3.81):

$$\mathcal{J}_3 = \sigma T_3^4 \quad (3.89)$$

The view factors are then evaluated as follows.

F_{12} (parallel plates) is 0.34, found from Table 3.8, or from a suitable chart of view factors for various geometries, as given in Figure 3.2[24].

From equation (3.84):

$$F_{11} + F_{12} + F_{13} = 1 \quad (3.90)$$

But $F_{11} = 0$ for a flat surface, so:

$$F_{13} = 1 - F_{12} \quad (3.91)$$

Therefore $F_{13} = 1 - 0.34 = 0.66$.

Similarly, $F_{23} = 0.66$.

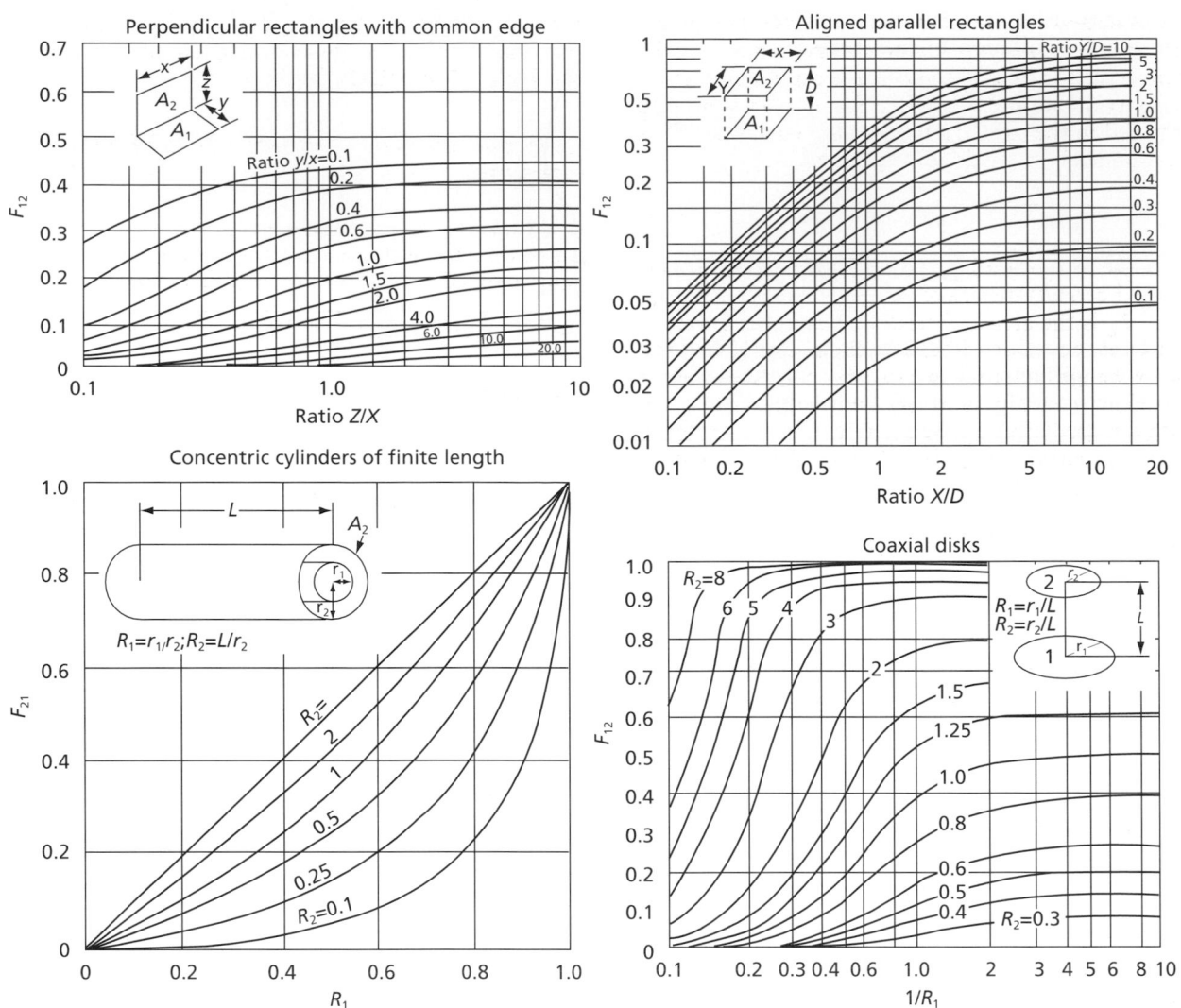

Figure 3.2 Radiation view factor for various geometries (reproduced from ASHRAE Handbook: *Fundamentals* ©1997 by permission of the American Society of Heating, Refrigerating and Air-Conditioning Engineers Inc., 1791 Tullie Circle, NE, Alberta, GA 30329, USA (www.ashrae.org))

The value for F_{21} can be found from equation (3.83):

$$A_1 F_{12} = A_2 F_{21} \qquad (3.92)$$

In this case, $A_1 = A_2$, and so $F_{21} = F_{12} = 0.34$

Equations (3.87) and (3.88) can now be solved (using matrix inversion, for example) for $\mathcal{J}_1$ and $\mathcal{J}_2$, since all other values for terms are now known. In situations where there are a large number of 'grey' surfaces, necessitating the calculation of $\mathcal{J}$-values, a solution can be obtained from matrix inversion or from the use of a suitable computer program.

Solution gives:

$$\mathcal{J}_1 = 410.06 \ \text{W.m}^{-2}$$

$$\mathcal{J}_2 = 430.17 \ \text{W.m}^{-2}$$

$$\mathcal{J}_3 = 447.14 \ \text{W.m}^{-2}$$

Finally, the values for the rates of heat transfer by radiation to each surface, q_1, q_2 and q_3 can now be evaluated from equations (3.79) and (3.80), giving:

$$q_1 = \frac{\varepsilon_1 A_1}{(1 - \varepsilon_1)}(E_{bi} - \mathcal{J}_1) \qquad (3.93)$$

$$= \frac{0.9 \times 6 \times 4}{(1 - 0.9)}(5.67 \times 10^{-8} \times 291^4 - 410.06)$$

$$= -749.9 \ \text{W}$$

Similarly, $q_2 = -103.3$ W, and $q_3 = 854.1$ W.

Thus, the ceiling removes radiant heat at a rate of 749.9 W, while the floor removes 103.3 W.

To check the results, the total rate of heat removal by the ceiling plus the floor should equal the total rate of heat supplied by the four walls, that is:

$$749.9 + 103.3 = 853.2 \ \text{W}$$

The agreement is within rounding error.

Simplified technique for rooms

The multi-surface enclosure can be simplified by treating it as a two-surface approximation[25]. Here, the radiant exchange in the room is modelled by assuming that one surface (the heated or cooled surface) radiates to a single fictitious surface (made up of the other surfaces in the room, which are considered to be at a similar temperature and emissivity to one another). View factors do not need to be determined in the case of a two-surface enclosure. This simplification leads to the following equation:

$$q_r = \sigma F_r A_s (T_s^4 - T_{fs}^4) \tag{3.94}$$

where q_r is the rate of radiant heat transfer exchanged by the heated (or cooled) surface of area A_s and of temperature T_s (K), and T_{fs} is the area and emissivity-weighted average temperature of the other surfaces in the room (K); F_r is a radiation exchange factor for two surfaces and takes a value of 0.87 for most rooms in which the emissivities of the surfaces can be regarded as approximately 0.9 (see reference 26). For this situation, reference 21 showed that equation (3.94) can be expressed as:

$$q_r = 5 \times 10^{-8} A_s [(t_s + 273)^4 - (t_{fs} + 273)^4] \tag{3.95}$$

where t_s is the temperature in °C of the heated or cooled surface; the term t_{fs} is the area-weighted average surface temperature in °C for the fictitious surface, given by:

$$t_{fs} = \left(\sum_{j=1}^{N} (A_j t_j) / \sum_{j=1}^{N} A_j ; i \neq j \right) \tag{3.96}$$

(that is, when the emissivities of those surfaces comprising the fictitious surface are nearly equal). Here, t_i is the temperature in °C of a surface making up the fictitious surface.

Equation (3.95) can be used by designers to estimate the rate of heat removal by radiation by a heated or cooled surface as a function of the temperature of the surface and of t_{fs}.

Example 3.2

Estimate the rate of heat removal by radiation by the chilled ceiling of example 3.1 for the same room conditions, but using the simplified technique for rooms.

Solution

Calculate t_{fs} for the four walls and floor using equation (3.96):

$$t_{fs} = \frac{[((2 \times 3 \times 6) + (2 \times 3 \times 4)) \times 25] + [(6 \times 4) \times 22]}{(2 \times 3 \times 6) + (2 \times 3 \times 4) + (6 \times 4)} = 24.14°C$$

The rate of heat removal by radiation q_r by the chilled ceiling can now be estimated using equation (3.95):

$$q_r = 5 \times 10^{-8} \times (6 \times 4)[(18 + 273)^4 - (24.14 + 273)^4]$$

$$q_r = -749.6 \text{ W}$$

In addition, a radiative heat transfer coefficient h_r can be determined using the following linear expression:

$$q_r = h_r A_s (t_s - t_{fs}) \tag{3.97}$$

For the ceiling of example 3.2, the value for h_r is 5.1 W.m^{-2}.K^{-1}; this is a typical value for indoor conditions (see Table 3.7 of CIBSE Guide A (1999)).

Combined convective and radiative heat transfer in enclosures

The total rate of heat exchange at a surface in an enclosure can be determined by adding the convective and radiative components discussed in the previous sections:

$$q_t = q_c + q_r \tag{3.98}$$

where q_c is given by:

$$q_c = h_c A_s (t_s - t_a) \tag{3.99}$$

Values for h_c are dependent upon conditions, but could be evaluated, as h_{cn}, for example, from equations (3.76), (3.77) or (3.78), as appropriate. The term q_r can be evaluated from equation (3.95) when the simplified technique is adopted; alternatively, q_i replaces q_r for that surface in equation (3.98), where q_i is evaluated from equation (3.79).

3.3.3 Human body heat transfer

Occupied enclosures

The presence of a human body in an enclosure can be treated as the addition of a further surface which exchanges heat by convection and by radiation with the surroundings. Such a treatment can be of relevance to the estimation of room thermal loads (see Section 6 of CIBSE Guide A (1999)), and to the determination of human thermal comfort (see Section 3 of CIBSE Guide A (1999)). Here, correlations for natural (free) and for forced convection for a human body are given, together with view factors for use in estimation of radiant exchange.

Human heat exchange by convection

The rate of heat transfer per unit area by convection C_h between the human body surface and its surroundings is given by:

$$C_h = f_{cl} h_c (t_{cl} - t_a) \tag{3.100}$$

where f_{cl} is the clothing area factor, given by reference 27 as:

$$f_{cl} \approx 1 + 0.15 I_{clo} \tag{3.101}$$

where I_{clo} is the thermal resistance of clothing, in units of clo.

Values for h_c for natural or for forced convection can be determined from a suitable correlation equation. Table 3.10 (after reference 28) summarises some of the existing correlations available to date.

Human heat exchange by radiation

This can be treated using the radiosity technique as described above in the section entitled *Radiation exchange*

Table 3.10 Human body convection coefficients

$h_c/\text{W.m}^{-2}.\text{K}^{-1}$	Reference	Remarks
Natural convection:		
3.0	Nishi and Gagge (1977)[29]	Still body in still air
$2.38\Delta t_{ca}^{0.25}$	Neilsen and Pedersen (1952)[30]	Used in Fanger's equation[27]
4.0	Rapp (1973)[31]	Recommended (by McIntyre)[28] for sedentary people
Forced convection:		
$8.3\,v_r^{0.6}$	Mitchell (1974)[32]	Best average
$12.1\,\sqrt{v_r}$	Winslow *et al.* (1937)[33]	Used in Fanger's equation[27]
$8.3\,\sqrt{v_r}$	Kerslake (1972)[34]	Recommended (by McIntyre)[28]

between internal surfaces. Here, the human body is regarded as an additional surface for radiant exchange. Human body view factors for use in the technique are given in reference 35, and are reproduced here with permission (Figures 3.3–3.7).

Example 3.3 (Adapted with permission from reference 27)

Determine the view factors between a seated person (of known position but unknown orientation) and the surrounding surfaces of the room as shown in Figure 3.7 (upper figure).

Solution

Each surface of the room is divided into rectangles as shown in Figure 3.7 (lower figures). The view factors pertaining to each rectangle can then be determined graphically from Figures 3.3 and 3.4, or analytically from the equations in Table 3.11. All the results are given in Table 3.12. As a check on the calculations, the sum of all the view factors between the person and the surrounding surfaces should equal unity.

Note that once the view factors have been determined, the rate of heat exchange by radiation between a person and the room surroundings can be calculated using the radiosity approach as illustrated in example 3.1. Here, the person can be regarded as an additional 'surface' within the enclosure.

Total human heat exchange with surroundings

The preceding sections entitled 'Human Heat Exchange by Convection' and 'Human Heat Exchange by Radiation'

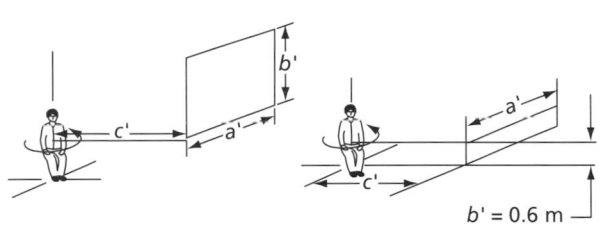

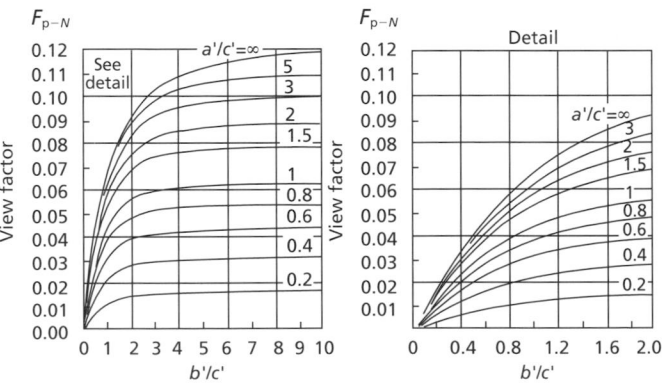

Example

$a' = 4$ m; $b' = 3$ m; $c' = 5$ m;
$b'/c' = 0.6$; $a'/c' = 0.8$; $F_{p-N} = 0.029$

Figure 3.3 Mean value of view factor between a seated person and a vertical rectangle (above or below his centre) when the person is rotated around a vertical axis. (To be used when the location but not the orientation of the person is known.) (Reproduced from ISO 7726: 1998 by permission of the International Organisation for Standardisation, ISO. This standard may be obtained from any member body or direct from the Central Secretariat, ISO, Case postal 56, 1211 Geneva 20, Switzerland. Copyright remains with the ISO)

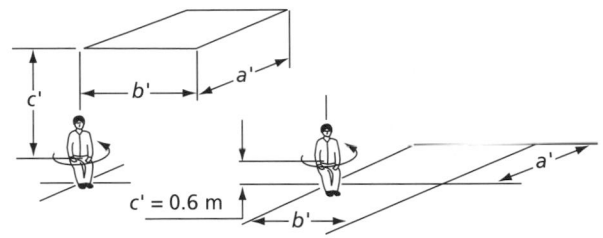

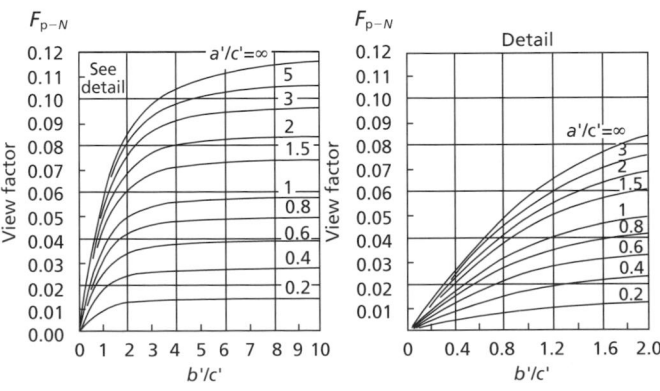

Example

$a' = 3$ m; $b' = 6$ m; $c' = 2$ m;
$b'/c' = 3.0$; $a'/c' = 1.5$; $F_{p-N} = 0.067$

Figure 3.4 Mean value of view factor between a seated person and a horizontal rectangle (on the ceiling or floor) when the person is rotated around a vertical axis. (To be used when the location but not the orientation of the person is known.) (Reproduced from ISO 7726: 1998 by permission of the International Organisation for Standardisation, ISO. This standard may be obtained from any member body or direct from the Central Secretariat, ISO, Case postal 56, 1211 Geneva 20, Switzerland. Copyright remains with the ISO)

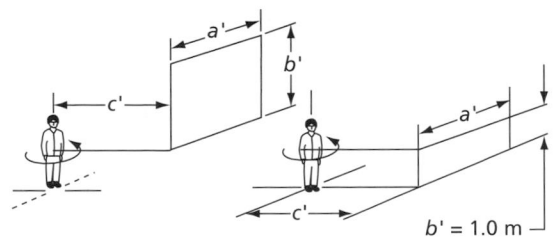

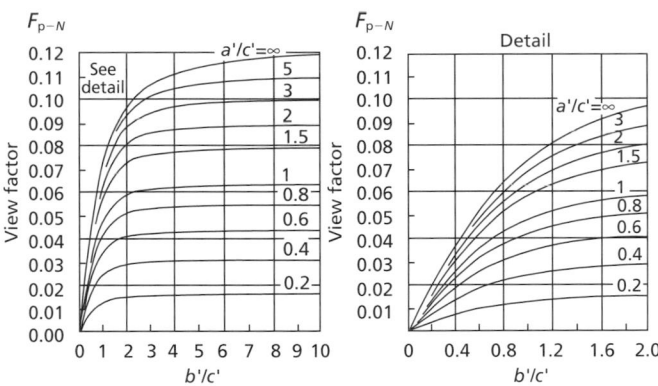

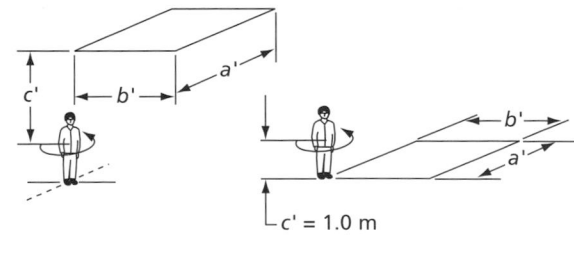

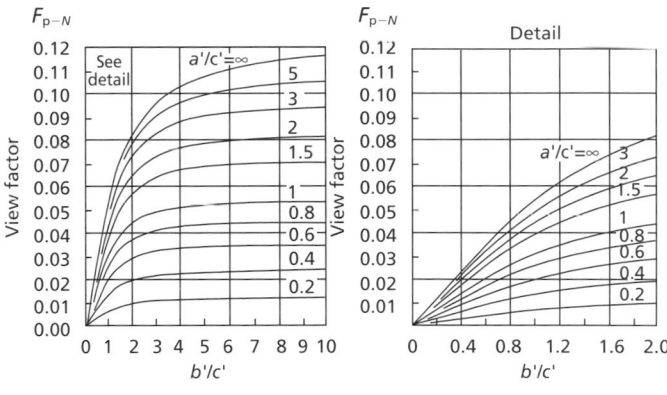

Example

$a' = 4.5$ m; $b' = 2.0$ m; $c' = 3.0$ m;
$b'/c' = 0.67$; $a'/c' = 1.5$; $F_{p-N} = 0.047$

Figure 3.5 Mean value of view factor between a standing person and a vertical rectangle (above or below his centre) when the person is rotated around a vertical axis. (To be used when the location but not the orientation of the person is known.) (Reproduced from ISO 7726: 1998 by permission of the International Organisation for Standardisation, ISO. This standard may be obtained from any member body or direct from the Central Secretariat, ISO, Case postal 56, 1211 Geneva 20, Switzerland. Copyright remains with the ISO)

Example

$a' = 1.0$ m; $b' = 15$ m; $c' = 1.5$ m;
$b'/c' = 10$; $a'/c' = 0.67$; $F_{p-N} = 0.039$

Figure 3.6 Mean value of view factor between a standing person and a horizontal rectangle (on the ceiling or floor) when the person is rotated around a vertical axis. (To be used when the location but not the orientation of the person is known.) (Reproduced from ISO 7726: 1998 by permission of the International Organisation for Standardisation, ISO. This standard may be obtained from any member body or direct from the Central Secretariat, ISO, Case postal 56, 1211 Geneva 20, Switzerland. Copyright remains with the ISO)

Table 3.11 Equations for calculation of view factors

View factor $= F'_{max} (1 - e^{-(a'/c')/\tau}) (1 - e^{-(b'/c')/\beta})$
where $\tau = A' + B' (a'/c')$
$\beta = C' + D' (b'/c') + E' (a'/c')$

	F'_{max}	A'	B'	C'	D'	E'
SEATED PERSON, Figure 3.3 Vertical surfaces: Wall, Window	0.118	1.216	0.169	0.717	0.087	0.052
SEATED PERSON, Figure 3.4 Horizontal surfaces: Floor, Ceiling	0.116	1.396	0.130	0.951	0.080	0.055
STANDING PERSON, Figure 3.5 Vertical surfaces: Wall, Window	0.120	1.242	0.167	0.616	0.082	0.051
STANDING PERSON, Figure 3.6 Horizontal surfaces: Floor, Ceiling	0.116	1.595	0.128	1.226	0.046	0.044

deal with heat transfer between a human body and its surroundings by convection and by radiation only. However, the total heat exchange includes not only convective and radiative components, but also heat exchange by evaporation, respiration and conduction. For a treatment of these, refer to Section 1 of CIBSE Guide A (1999) or to references 24 or 36.

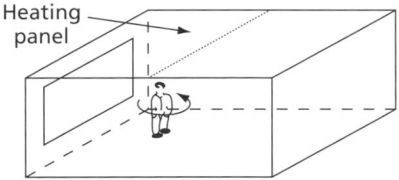

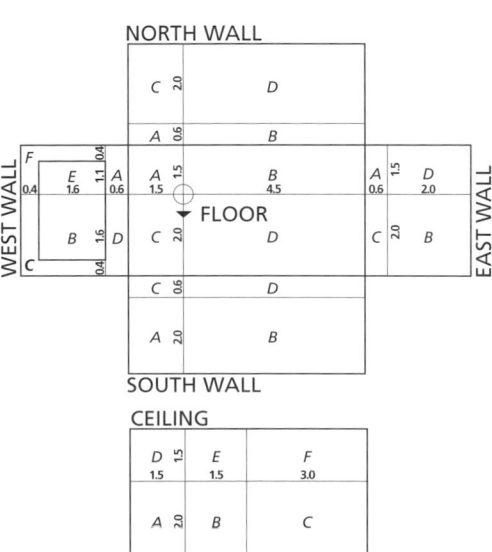

Figure 3.7 Room dimensions and person placement for example 3.3 (after reference 27)

Table 3.12 Calculation of view factors for example 3.3 (after reference 27)

			$(b'/c', a'/c')$	
North wall	$F_{\text{P-}ABCD} = F_{\text{P-}A} + F_{\text{P-}B} + F_{\text{P-}C} + F_{\text{P-}D}$			
	$F_{\text{P-}A}$　(Fig 3.4)		$(b'/c', a'/c')\ =$	(0.40, 1.0): 0.024
	$+F_{\text{P-}B}$			(0.40, 3.0): 0.033
	$+F_{\text{P-}C}$			(1.3 , 1.0): 0.033
	$+F_{\text{P-}D}$			(1.3 , 3.0): 0.072　0.179
East wall	$F_{\text{P-}ABCD} = F_{\text{P-}A} + F_{\text{P-}B} + F_{\text{P-}C} + F_{\text{P-}D}$			
	$F_{\text{P-}A}$　(Fig. 3.4)		$(b'/c', a'/c')\ =$	(0.13, 0.33):0.004
	$+F_{\text{P-}B}$			(0.44, 0.44):0.015
	$+F_{\text{P-}C}$			(0.13, 0.44):0.005
	$+F_{\text{P-}D}$			(0.44, 0.33):0.011　0.035
South wall	$F_{\text{P-}ABCD} = F_{\text{P-}A} + F_{\text{P-}B} + F_{\text{P-}C} + F_{\text{P-}D}$			
	$F_{\text{P-}A}$　(Fig. 3.4)		$(b'/c', a'/c')\ =$	(1.0 , 0.75):0.037
	$+F_{\text{P-}B}$			(1.0 , 2.3):0.060
	$+F_{\text{P-}C}$			(0.30, 0.75):0.015
	$+F_{\text{P-}D}$			(0.30, 2.3):0.024　0.136
West wall	Window: $F_{\text{P-}BE} = F_{\text{P-}B} + F_{\text{P-}E}$			
	$F_{\text{P-}B}$　(Fig. 3.4)		$(b'/c', a'/c')\ =$	(1.1 , 1.1): 0.049
	$+F_{\text{P-}E}$			(1.1 , 0.73):0.038　0.087
West wall	Rest of the wall: $F_{\text{P-}ACDF} = F_{\text{P-}A} + F_{\text{P-}C}\ F_{\text{P-}D} + F_{\text{P-}F}$			
	$\qquad = F_{\text{P-}A} + F_{\text{P-}BC} - F_{\text{P-}B} + F_{\text{P-}D} + F_{\text{P-}EF} - F_{\text{P-}E}$			
	$\qquad = F_{\text{P-}A} + F_{\text{P-}BC} + F_{\text{P-}D} + F_{\text{P-}EF} - F_{\text{P-}BE}$			
	$F_{\text{P-}A}$　(Fig. 3.4)		$(b'/c', a'/c')\ =$	(0.40, 1.0): 0.024
	$+F_{\text{P-}BC}$			(1.3 , 1.3): 0.057
	$+F_{\text{P-}D}$			(0.40, 1.3): 0.026
	$+F_{\text{P-}EF}$			(1.3 , 1.0): 0.050
	$-F_{\text{P-}BE}$			0.087　0.070
Floor	$F_{\text{P-}ABCD} = F_{\text{P-}A} + F_{\text{P-}B} + F_{\text{P-}C} + F_{\text{P-}D}$			
	$F_{\text{P-}A}$　(Fig. 3.5)		$(b'/c', a'/c')\ =$	(2.5 , 2.5): 0.078
	$+F_{\text{P-}B}$			(2.5 , 7.5): 0.090
	$+F_{\text{P-}C}$			(3.3 , 2.5): 0.082
	$+F_{\text{P-}D}$			(3.3 , 7.5): 0.095　0.345
Ceiling	Heating panel: $F_{\text{P-}ABDE} = F_{\text{P-}A} + F_{\text{P-}B} + F_{\text{P-}D} + F_{\text{P-}E}$			
	$F_{\text{P-}A}$　(Fig. 3.5)		$(b'/c', a'/c')\ =$	(1.0 , 0.75):0.030
	$+F_{\text{P-}B}$			(1.0 , 0.75):0.030
	$+F_{\text{P-}D}$			(0.75, 0.75):0.025
	$+F_{\text{P-}E}$			(0.75, 0.75):0.025　0.110
Ceiling	Rest of the ceiling: $F_{\text{P-}CF} = F_{\text{P-}C} + F_{\text{P-}F}$			
	$\qquad = F_{\text{P-}BC} - F_{\text{P-}B} + F_{\text{P-}EF} - F_{\text{P-}E}$			
	$F_{\text{P-}BC}$　(Fig. 3.5)		$(b'/c', a'/c')\ =$	(1.0 , 2.3): 0.051
	$-F_{\text{P-}B}$			(1.0 , 0.75):0.030
	$+F_{\text{P-}EF}$			(0.75, 2.3): 0.042
	$-F_{\text{P-}E}$			(0.75, 0.75):0.025　0.038
	$F_{\text{person-all surfaces}}$			1.000

3.3.4　　Equipment and components

Plane surfaces

Tables 3.13 and 3.14 give the heat emission/absorption from freely exposed plane surfaces to air. They have been prepared from the following equations.

Radiation (from equation (3.61)):

$$\phi_r = 5.67 \times 10^{-8}\varepsilon\ (T_s^4 - T_r^4) \qquad (3.102)$$

Convection (laminar flow from equation (3.6)):

$$\phi_c = 0.64\ \frac{(t_s - t_f)^{1.25}}{D^{0.25}} \qquad (3.103)$$

Convection (turbulent flow from equation (3.5)):

$$\phi_c = 1.7(t_s - t_f)^{1.33} \qquad (3.104)$$

Hence, for room applications and taking $D = 1$ m, equations (3.103) and (3.104) can be put in the general form:

$$\phi_c = C(T_s - T_a)^n \qquad (3.105)$$

The values of C and n for particular arrangements are given in Table 3.15.

The radiation and convection emissions are shown separately in Tables 3.13 and 3.14 since there may be significant differences between the mean radiant temperature of the enclosure and the air temperature within it. Absorption by the surfaces is shown by negative values of emission.

The radiation equation stated above (3.102) is strictly true only for a panel of comparatively small area relative to the rest of the room area.

In practice, the convection emission applies to draught-free conditions and appreciable increases in heat transfer

Table 3.13 Heat emission from plane surfaces by radiation (based on equation (3.102)). (Values above 1000 rounded to nearest 10, values above 10 000 rounded to nearest 100)

Heat emission for stated surface emissivity and enclosure mean radiant temperature (°C) / W.m^{-2}

Surface temp. °C	Surface emissivity 0.3							Surface emissivity 0.6							Surface emissivity 0.9						
	10	12.5	15	17.5	20	22.5	25	10	12.5	15	17.5	20	22.5	25	10	12.5	15	17.5	20	22.5	25
5	-7.5	-11	-15	-20	-24	-28	-33	-15	-23	-31	-39	-48	-56	-65	-23	-34	-46	-59	-71	-84	-98
10	0	-3.9	-7.9	-12	-16	-21	-25	0	-7.8	-16	-24	-33	-41	-50	0	-12	-24	-36	-49	-62	-75
15	7.9	4.0	0	-4.1	-8.3	-13	-17	16	8.0	0	-8.2	-17	-25	-34	24	12	0	-12	-25	-38	-51
20	16	12	8.3	4.2	0	-4.3	-8.8	33	25	17	8.4	0	-8.7	-18	49	37	25	13	0	-13	-26
30	34	30	26	22	18	14	9.2	69	61	53	44	36	27	18	103	91	79	67	54	41	28
40	54	50	46	42	38	34	29	108	100	92	84	76	67	58	162	151	139	126	114	101	87
50	76	72	68	64	60	55	51	152	144	136	128	120	111	102	228	216	204	192	179	166	153
60	100	96	92	88	84	79	75	200	192	184	176	168	159	150	300	288	276	264	251	238	225
70	126	122	118	114	110	106	101	253	245	237	229	220	211	203	379	367	355	343	330	317	304
80	155	151	147	143	139	134	130	310	302	294	286	278	269	260	465	453	441	429	416	403	390
90	186	182	178	174	170	166	161	372	365	357	348	340	331	322	559	547	535	523	510	497	484
100	220	216	212	208	204	200	195	440	432	424	416	408	399	390	660	649	637	624	612	599	585
120	297	293	289	285	280	276	272	593	586	577	569	561	552	543	890	878	866	854	841	828	815
140	386	382	378	374	370	365	361	772	764	756	747	739	730	721	1160	1150	1130	1120	1110	1100	1080
160	489	485	481	477	473	468	464	978	970	962	954	945	936	928	1470	1450	1440	1430	1420	1400	1390
180	607	603	599	595	591	587	582	1210	1210	1200	1190	1180	1170	1160	1820	1810	1800	1790	1770	1760	1750
200	742	738	734	730	726	722	717	1480	1480	1470	1460	1450	1440	1430	2230	2220	2200	2190	2180	2170	2150
220	896	892	888	884	879	875	871	1790	1780	1780	1770	1760	1750	1740	2690	2680	2660	2650	2640	2630	2610
240	1070	1070	1060	1060	1050	1050	1040	2140	2130	2120	2110	2110	2100	2090	3210	3200	3180	3170	3160	3150	3130
260	1250	1260	1260	1250	1250	1240	1240	2530	2520	2510	2500	2490	2490	2480	3790	3780	3770	3760	3740	3730	3720
280	1480	1480	1470	1470	1470	1460	1460	2960	2960	2950	2940	2930	2920	2910	4440	4430	4420	4410	4400	4380	4370
300	1720	1720	1720	1710	1710	1700	1700	3450	3440	3430	3430	3420	3410	3400	5170	5160	5150	5140	5120	5110	5100
320	1990	1990	1990	1980	1980	1970	1970	3990	3980	3970	3960	3960	3950	3940	5980	5970	5960	5950	5930	5920	5910
340	2290	2290	2280	2280	2280	2270	2270	4590	4580	4570	4560	4550	4540	4540	6880	6870	6850	6840	6830	6820	6800
360	2620	2620	2610	2610	2610	2600	2600	5240	5240	5230	5220	5210	5200	5190	7870	7850	7840	7830	7820	7800	7790
380	2980	2980	2980	2970	2970	2960	2960	5970	5960	5950	5940	5930	5930	5920	8950	8940	8930	8920	8900	8890	8880
400	3380	3380	3370	3370	3360	3360	3360	6760	6750	6740	6740	6730	6720	6710	10100	10100	10100	10100	10100	10100	10100

Table 3.14 Heat emission from plane surfaces by free convection (based on equation (3.105) and Table 3.15). (Values above 1000 rounded to nearest 10, values above 10 000 rounded to nearest 100)

Heat emission / W·m⁻²

Surface temp. °C	Horizontal looking down — Air temperature (°C)							Vertical — Air temperature (°C)							Horizontal looking up — Air temperature (°C)						
	10	12.5	15	17.5	20	22.5	25	10	12.5	15	17.5	20	22.5	25	10	12.5	15	17.5	20	22.5	25
5	−14	−2.5	−36	−49	−62	−77	−91	−12	−20	−30	−40	−51	−63	−75	−4.8	−7.9	−11	−15	−19	−23	−27
10	0	−5.8	−14	−25	−36	−49	−62	0	−4.7	−12	−20	−30	−40	−51	0	−2.0	−4.8	−7.9	−11	−15	−19
15	4.8	2.0	0	−5.8	−14	−25	−36	12	4.7	0	−4.7	−12	−20	−30	14	5.8	0	−2.0	−4.8	−7.9	−11
20	11	7.9	4.8	2.0	0	−5.8	−14	30	20	12	4.7	0	−4.7	−12	36	25	14	5.8	0	−2.0	−4.8
30	27	23	19	15	11	7.9	4.8	75	63	51	40	30	20	12	91	77	62	49	36	25	14
40	45	40	36	31	27	23	19	129	115	101	88	75	63	51	157	140	123	107	91	77	62
50	64	59	54	50	45	40	36	189	174	158	144	129	115	101	230	211	192	174	157	140	123
60	85	80	75	69	64	59	54	255	238	221	205	189	174	158	309	289	269	249	230	211	192
70	107	101	96	90	85	80	75	324	307	289	272	255	238	221	394	372	351	330	309	289	269
80	130	124	118	112	107	101	96	398	379	361	342	324	307	289	484	461	438	416	394	372	351
90	153	147	141	135	130	124	118	476	456	436	417	398	379	361	578	554	530	507	484	461	438
100	177	171	165	159	153	147	141	556	536	516	495	476	456	436	675	651	626	602	578	554	530
120	228	222	215	209	202	196	190	726	705	683	661	640	619	598	882	856	829	803	777	751	726
140	281	274	267	261	254	248	241	907	884	861	838	816	793	771	1100	1070	1050	1020	990	963	936
160	336	329	322	315	308	301	295	1100	1070	1050	1020	1000	977	954	1330	1300	1270	1240	1220	1190	1160
180	393	386	378	371	364	357	350	1300	1270	1250	1220	1200	1170	1150	1570	1540	1510	1480	1450	1420	1390
200	451	444	437	429	422	415	407	1500	1480	1450	1420	1400	1370	1350	1820	1790	1760	1730	1700	1670	1640
220	512	504	496	489	481	474	466	1720	1690	1660	1640	1610	1580	1560	2080	2050	2020	1990	1950	1920	1890
240	573	565	558	550	542	535	527	1940	1910	1880	1850	1830	1800	1770	2350	2320	2280	2250	2220	2180	2150
260	636	628	620	612	605	597	589	2160	2140	2110	2080	2050	2020	1990	2630	2590	2560	2520	2490	2460	2420
280	700	692	684	676	668	660	652	2400	2370	2340	2310	2280	2250	2220	2910	2880	2840	2800	2770	2730	2700
300	766	758	749	741	733	725	717	2640	2610	2580	2550	2520	2490	2460	3200	3170	3130	3090	3060	3020	2980
320	832	824	816	807	799	791	782	2880	2850	2820	2790	2760	2730	2700	3500	3460	3420	3390	3350	3310	3280
340	900	892	883	875	866	858	849	3130	3100	3070	3040	3010	2970	2940	3800	3760	3730	3690	3650	3610	3570
360	969	960	952	943	934	926	917	3390	3350	3320	3290	3260	3230	3190	4110	4070	4030	4000	3960	3920	3880
380	1040	1030	1020	1010	1000	995	986	3650	3610	3580	3550	3520	3480	3450	4430	4390	4350	4310	4270	4230	4190
400	1110	1100	1090	1080	1070	1060	1060	3910	3880	3840	3810	3780	3740	3710	4750	4710	4670	4630	4590	4550	4510

Table 3.15 Values of coefficients in equation (3.105)

Situation	C	n
Warm or cold vertical planes	1.4	1.33
Warm horizontal planes facing up	1.7	1.33
Cold horizontal planes facing down	1.7	1.33
Warm horizontal planes facing down	0.64	1.25
Cold horizontal planes facing up	0.64	1.25

Table 3.16 Effect of air velocity on convective heat transfer from plane surfaces

Velocity (m.s^{-1})	Multiplying factor
0	1
0.5	1.3
1.0	1.7
2.0	2.4
3.0	3.1

occur if air movement is present[10]. Table 3.16 gives some indication of the effect of air velocity on heat transfer.

Bare pipes

The heat emissions per metre run of horizontal steel and copper pipes are given in Tables 3.17 and 3.18. Absorption by the surfaces is shown by negative values of emission. These tables have been prepared using equation (3.102) and for convection:

$$\phi_c = \frac{k}{d_{op}} (Nu)\Delta t \qquad (3.106)$$

For horizontal pipes in air, free convection, laminar flow, and $(Gr) < 10^8$, the following equation can be used:

$$(Nu) = 0.53[(Gr)(Pr)]^{0.25} \qquad (3.107)$$

Hence, equation (3.105) becomes:

$$\phi_c = 1.35 \left[\frac{T_s - T_a}{d_{op}} \right]^{0.25} (T_s - T_a) \qquad (3.108)$$

The emission per metre run may then be found from:

$$Q = \pi d_{op}(\phi_r + \phi_c) \qquad (3.109)$$

Tables 3.17 and 3.18 assume that both the ambient temperature and the mean radiant temperature are equal to 20°C. It is also assumed that the external surface of the pipe is at the same temperature as the fluid contained within it.

In practice, for a given temperature difference, a change in air temperature or mean radiant temperature of ±5°C introduces an error of about 2% while a change of ±10°C gives an error of about 5%.

Table 3.17 Heat emission from single horizontal steel pipes ($\varepsilon = 0.95$) freely exposed in surroundings at 20°C

Temp. difference between surface and surroundings (°C)	* 15	20	25	32	40	50	65	80	100	125	150	200	250	300	350	400
	† 21.3	26.9	33.7	42.4	48.3	60.3	76.1	88.9	114.3	139.7	168.3	219.1	273.0	323.9	355.6	406.4
−15	−12	−15	−18	−22	−24	−29	−36	−41	−52	−62	−71	−92	−112	−130	−142	−160
−10	−7.7	−9.4	−11	−14	−16	−19	−23	−27	−33	−40	−46	−60	−73	−85	−92	−104
−5	−3.5	−4.3	−5.3	−6.5	−7.3	−8.9	−11	−13	−16	−19	−22	−28	−35	−40	−44	−50
0	0	0	0	0	0	0	0	0	0	0	0	0	0	0	0	0
5	3.6	4.5	5.5	6.7	7.5	9.1	11	13	16	19	23	29	36	42	46	52
10	8.1	9.8	12	15	16	20	25	28	35	42	49	63	77	90	98	111
15	13	16	19	23	26	32	39	45	56	67	78	100	123	143	156	176
20	18	22	27	33	37	45	55	63	78	94	109	140	171	199	217	245
25	24	29	35	43	48	58	71	81	102	122	141	182	221	258	281	317
30	29	36	44	53	60	72	88	101	126	151	175	226	275	320	349	393
35	35	43	53	64	72	87	106	122	152	182	211	272	331	385	419	473
40	42	51	62	75	84	102	125	143	179	214	248	319	389	453	493	556
45	48	59	71	87	98	118	145	166	207	247	287	369	449	524	570	642
50	55	67	81	99	111	135	165	189	236	281	327	421	512	597	649	732
55	62	75	92	112	125	152	186	213	266	317	368	474	577	673	732	825
60	69	84	102	125	140	169	207	238	297	354	411	529	644	751	817	921
65	77	93	114	138	155	188	230	263	329	392	456	586	714	832	905	1020
70	84	103	125	152	170	206	253	290	362	432	502	646	786	916	997	1120
75	92	112	137	167	186	226	277	317	396	473	549	706	860	1000	1090	1230
80	100	122	149	181	203	246	301	345	431	515	598	769	937	1090	1190	1340
100	135	164	200	244	273	331	406	466	582	695	808	1040	1270	1480	1610	1820
120	173	211	257	314	352	427	523	600	750	897	1040	1340	1640	1910	2080	2350
140	215	262	320	391	438	532	653	750	937	1120	1310	1680	2050	2400	2610	2950
160	261	318	389	476	534	648	796	915	1140	1370	1600	2060	2520	2940	3200	3620
180	311	380	465	569	638	776	954	1100	1370	1650	1920	2480	3030	3540	3860	4360
200	366	447	547	670	753	916	1130	1300	1620	1950	2270	2940	3600	4200	4580	5180
220	425	520	637	781	878	1070	1320	1510	1900	2280	2660	3450	4220	4930	5380	6090
240	490	600	735	902	1010	1240	1520	1750	2200	2650	3090	4000	4900	5740	6260	7080
260	560	686	842	1030	1160	1420	1750	2010	2530	3040	3550	4610	5650	6620	7220	8180
280	635	780	957	1180	1320	1620	2000	2300	2890	3480	4060	5280	6470	7590	8280	9380
300	717	881	1080	1330	1500	1830	2260	2610	3280	3950	4620	6010	7370	8650	9440	10 700
320	806	990	1220	1500	1690	2060	2550	2940	3710	4470	5230	6800	8350	9800	10 700	12 100

*These pipe sizes are to BS1387: 1967 and BS 3600: 1973 [37,38] †These outside diameters are to BS3600: 1997 [39]

Table 3.18(a) Heat emission from single horizontal copper pipes ($\varepsilon = 0.5$) freely exposed in surroundings at 20°C

Temp. difference between surface and surroundings (°C) ★	Heat emission/W.m^{-1} for stated pipe nominal size★ (mm)														
	8	10	12	15	18	22	28	35	42	54	67	76.1	108	133	159
−15	−4.3	−5.2	−6.0	−7.2	−8.4	−9.9	−12	−14	−17	−21	−25	−28	−37	−44	−51
−10	−2.7	−3.2	−3.8	−4.5	−5.2	−6.2	−7.5	−9.1	−11	−13	−16	−17	−23	−28	−33
−5	−1.2	−1.4	−1.7	−2.0	−2.3	−2.8	−3.4	−4.1	−4.8	−5.9	−7.1	−7.9	−11	−13	−15
0	0	0	0	0	0	0	0	0	0	0	0	0	0	0	0
5	1.2	1.5	1.7	2.0	2.4	2.8	3.5	4.2	4.9	6.0	7.2	8.1	11	13	15
10	2.8	3.3	3.9	4.6	5.4	6.4	7.8	9.4	11	14	16	18	24	29	34
15	4.5	5.4	6.3	7.5	8.8	10	13	15	18	22	26	29	39	47	55
20	6.4	7.7	8.9	11	12	15	18	21	25	31	37	41	55	66	77
25	8.4	10	12	14	16	19	23	28	33	40	48	54	72	86	100
30	10	13	15	17	20	24	29	35	41	50	60	67	90	107	125
35	13	15	18	21	24	29	35	42	49	61	72	81	109	130	151
40	15	18	21	25	29	34	42	50	58	72	85	95	128	153	178
45	17	21	24	29	33	39	48	58	67	83	99	110	148	177	205
50	20	24	27	33	38	45	55	66	77	94	112	126	169	201	234
55	22	27	31	37	43	51	62	74	86	106	127	142	190	227	264
60	25	30	34	41	48	56	69	83	96	119	141	158	212	253	294
65	27	33	38	46	53	62	76	92	107	131	156	175	235	280	326
70	30	36	42	50	58	69	84	101	117	144	172	192	258	308	358
75	33	39	46	55	64	75	91	110	128	158	188	210	282	337	392
80	36	43	50	60	69	81	99	120	139	171	204	229	307	366	426
100	48	57	66	80	92	109	133	160	186	230	274	306	412	492	572
120	61	73	84	101	118	139	170	204	238	293	350	392	527	630	733
140	74	89	104	125	145	171	209	252	294	363	433	485	653	781	910
160	89	107	125	150	174	206	252	304	354	438	523	586	790	945	1100
180	105	126	147	177	206	244	298	360	419	519	620	695	938	1120	1310
200	122	147	171	206	240	284	347	420	490	606	726	814	1100	1320	1540
220	140	169	196	237	276	327	400	484	565	701	839	941	1270	1530	1780
240	160	192	224	270	315	373	457	553	646	802	961	1080	1460	1760	2050
260	180	217	253	305	356	422	518	627	734	911	1090	1230	1660	2000	2340
280	202	243	284	342	400	475	583	707	827	1030	1230	1390	1880	2270	2650
300	225	271	316	382	447	531	653	792	927	1150	1390	1560	2120	2550	2990
320	249	301	351	425	497	590	727	882	1030	1290	1550	1740	2370	2860	3350

★ These pipe sizes are to BS2871: Part 1: 1971[40]

Heat emission/absorption under site conditions may vary from the tabulated data since:

(a) Draught-free surroundings may not occur in practice. Table 3.19 gives some indication of the effect on heat transfer.

(b) The actual surface emissivity may differ from those used in preparing the tables. Emissivities for other materials and surface finishes are given in Table 3.7.

(c) The British Standards[37,39,40] permit tolerances in pipe diameters from the mean values used in preparing the tables.

Considerations (b) and (c) may lead to the actual heat emission of a given pipe varying by a further 10% from the tabulated figure.

All these aspects should be taken into consideration in system design and an appropriate allowance made according to circumstances.

Vertical pipes

Pipes set in a vertical position have a heat emission/absorption which differs from that arising from horizontal fixing due to the variation in the thickness of the boundary layer of air about the pipe surface. Correction factors quoted in Table 3.20 are for use in conjunction with the data listed in Tables 3.17 and 3.18(a) and (b).

Multiple banks of pipes

Where horizontal pipes are arranged one above another at close pitch, the heat emission is reduced overall owing to a cumulative interference with the convection output. Table 3.21 lists correction factors which illustrate the reduction in single pipe emission.

Effect of proximity of walls

Where pipes are installed near to cold external walls, the emission by radiation is likely to increase and that by convection to remain unchanged. With internal walls, there may be a reduction in radiation and a slight increase in convection. These variations are, however, probably appreciably smaller than variations in emission due to draughts, etc.

Finned surfaces

Owing to the wide variation in geometrical arrangement and fin efficiency, it is impractical to give any formulae for heat emission by forced or free convection from such surfaces. References should be made to manufacturers' catalogues and to other test data, etc.[1,41,42,43,44,45].

Table 3.18(b) Heat emission from single horizontal copper pipes ($\varepsilon = 0.95$) freely exposed in surroundings at 20°C

Temp. difference between surface and surroundings (°C) ★	Heat emission /W.m^{-1} for stated pipe nominal size★ (mm)														
	8	10	12	15	18	22	28	35	42	54	67	76.1	108	133	159
−15	−5.2	−6.3	−7.4	−8.9	−10	−12	−15	−18	−22	−27	−32	−36	−49	−59	−69
−10	−3.3	−4.0	−4.7	−5.7	−6.6	−7.9	−9.7	−12	−14	−17	−21	−23	−32	−38	−45
−5	−1.5	−1.8	−2.1	−2.6	−3.1	−3.6	−4.5	−5.5	−6.4	−8.0	−9.7	−11	−15	−18	−21
0	0	0	0	0	0	0	0	0	0	0	0	0	0	0	0
5	1.5	1.9	2.2	2.7	3.1	3.7	4.6	5.6	6.6	8.3	10	11	15	19	22
10	3.5	4.2	4.9	5.9	6.9	8.3	10	12	15	18	22	25	34	41	48
15	5.6	6.7	7.9	9.5	11	13	16	20	23	29	35	39	53	64	76
20	7.8	9.4	11	13	16	19	23	28	32	40	49	55	75	90	105
25	10	12	14	17	20	24	30	36	42	53	63	71	97	117	137
30	13	15	18	22	25	30	37	45	53	66	79	89	120	145	170
35	15	19	22	26	31	36	45	54	63	79	95	107	145	175	204
40	18	22	25	31	36	43	53	64	75	93	112	125	171	205	240
45	21	25	29	36	42	49	61	74	86	107	129	145	197	237	278
50	24	29	34	41	47	56	69	84	98	122	147	165	225	270	317
55	27	32	38	46	53	63	78	95	111	138	166	186	253	305	357
60	30	36	42	51	60	71	87	106	124	154	185	208	283	340	399
65	33	40	47	57	66	78	97	117	137	171	205	230	313	377	442
70	37	44	51	62	73	86	106	129	151	188	225	253	345	415	486
75	40	48	56	68	79	94	116	141	165	205	247	277	377	454	532
80	43	52	61	74	86	103	126	153	179	223	268	302	411	494	579
100	58	70	82	99	116	138	170	206	242	301	362	407	554	668	783
120	75	90	105	127	149	177	218	265	311	387	466	525	715	862	1010
140	92	112	131	158	185	220	272	330	387	483	581	655	893	1080	1260
160	112	135	158	192	224	267	330	401	471	588	709	798	1090	1320	1550
180	133	161	188	228	267	319	394	479	563	704	848	956	1310	1580	1860
200	155	188	221	268	314	375	464	565	664	830	1000	1130	1550	1870	2200
220	180	218	256	311	365	436	539	658	774	969	1170	1320	1810	2190	2580
240	207	251	294	358	420	502	622	759	893	1120	1350	1530	2100	2540	2990
260	235	286	336	409	480	574	712	869	1020	1280	1550	1750	2410	2920	3440
280	266	324	380	463	545	652	809	989	1170	1460	1770	2000	2750	3340	3930
300	299	364	428	522	615	736	914	1120	1320	1660	2010	2270	3130	3790	4470
320	335	408	480	586	690	827	1030	1260	1480	1870	2260	2560	3530	4290	5060

★ These pipe sizes are to BS2871: Part 1: 1971[40]

Table 3.19 Effect of air velocity on heat transfer

Emissivity of surfaces	Correction factor to be applied to Tables 3.17 and 3.18 for stated air velocity (m.s^{-1})			
	0	0.5	1.0	2.0
0.5 (dull metal)	1.0	1.05	1.15	1.25
0.95	1.0	1.04	1.12	1.20

Table 3.20 Correction factors for heat emission/absorption from vertical pipes

Nominal pipe size (mm)	Correction factor for Tables 3.17 and 3.18
8	0.72
10	0.74
15	0.76
20	0.79
25	0.82
32	0.84
40	0.86
50	0.88
65	0.90
80	0.92
100	0.95
125	0.97
150	0.99
200	1.03
250	1.05
300	1.07

Table 3.21 Correction factors for multiple banks of pipes (horizontal, one above another at close pitch)

Number of pipes in bank	Emission from each pipe as fraction of theoretical single pipe value
2	0.95
4	0.85
6	0.75
8	0.65

Insulated pipes

If the external surface temperature of insulation concentric to a pipe is known, then the theoretical heat emission may be calculated in the same way as for an exposed pipe. In practice, however, it is the surface temperature of the pipe itself which can be more readily ascertained so that an equation relating the temperature difference between this and ambient air is more useful. The main variables are the thermal conductivity of the insulation material, its thickness and the nature of the final surface finish. The data is, therefore, given per unit area of pipe.

The heat exchange from insulated pipes is given by:

$$\phi = U(t_s - t_a) \tag{3.110}$$

or more conveniently, per metre run of pipe, by:

$$Q = \pi d_{op} U(t_s - t_a) \tag{3.111}$$

where the overall thermal transmittance is given by:

$$U = \frac{1}{R_n + \dfrac{d_{op}}{h_{so}d_{on}}} \qquad (3.112)$$

and where the thermal resistance of the insulation is given by:

$$R_n = \frac{d_{op}}{2k_n} \ln\left(\frac{d_{on}}{d_{op}}\right) \qquad (3.113)$$

The surface temperature of the exterior of the insulation is a function of the value assumed for the surface heat transfer coefficient (see Table 3.22) and may be calculated from:

$$t_n = \frac{\phi d_{op}}{h_{so}d_{on}} + t_a \qquad (3.114)$$

$$= \frac{Q}{\pi h_{so}d_{on}} + t_a \qquad (3.115)$$

Outside surface heat transfer coefficient h_{so}

Values for the outside heat transfer coefficient appropriate to normal finishes are recommended in BS5422[46] where they are defined as giving 'reasonable approximations to the surface temperature of insulation fully exposed in still air and not influenced by other external sources of heat, including sunshine'. Table 3.22 gives these values and also those for moving air.

If the average wind speed is unknown, it is recommended[46] that the following values be assumed:

sheltered situations $1\ \mathrm{m.s^{-1}}$

normal situations $3\ \mathrm{m.s^{-1}}$

exposed situations $10\ \mathrm{m.s^{-1}}$

Thermal conductivity of insulation λ_n

Thermal conductivities of commonly used insulating materials are shown in Figure 3.8. These values are derived from Section 3 of CIBSE Guide A (1999) and reference 47. The insulating effect of a hard setting finish is insignificant and may be neglected in the calculations.

Tabulated data

Table 3.23 gives heat emission or absorption per degree temperature difference between the pipe surface and the

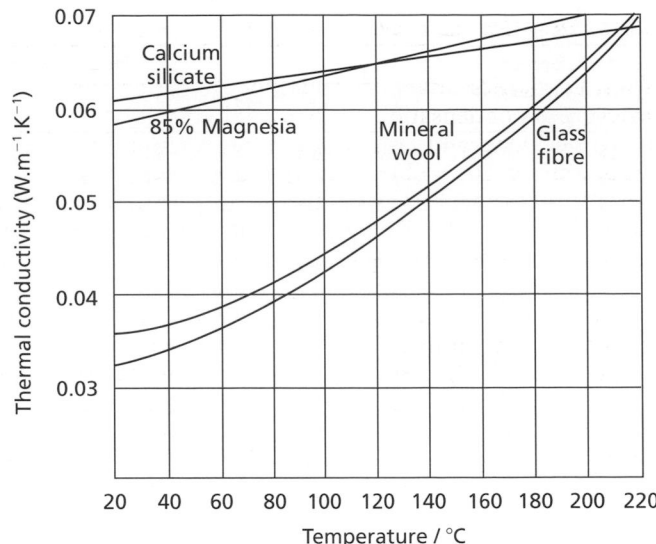

Figure 3.8 Thermal conductivities of insulating materials

ambient air using equation (3.111); here, thermal conductivities of 0.025, 0.040, 0.055 and 0.070 $\mathrm{W.m^{-1}.K^{-1}}$ are selected as examples covering the range of values encountered in thermal heating practice, together with a value of 10 $\mathrm{W.m^{-2}.K^{-1}}$ for the surface coefficient. From a practical point of view, heat transfer under site conditions may vary from the tabulated data for the following reasons:

(a)   Draught-free surroundings may not occur in practice; Table 3.19 may be used to assess the effect of air velocity on heat transfer.

(b)   For brighter surface finishes than those tabulated, the heat emission will be reduced, the amount of the reduction varying according to the surface area, i.e. the pipe size and the insulation thickness. With a bright metallic finish, e.g. aluminium, the reduction in emission will be about 8 per cent. For dull metallic finishes the reduction will be about 4 per cent. Such reductions in heat emissions lead to an increase in outside surface temperature of insulation and sheathing.

(c)   Damp insulation will also have an important effect on emission and can increase losses up to fivefold on installations which have the appearance of being satisfactory.

The corrections given for vertical pipes, multiple banks of pipes and the effect of the proximity of walls in the section on bare pipes are applicable to insulated pipes.

Table 3.22 Outside surface heat transfer coefficients h_{so} for insulated surfaces at various wind speeds

Wind speed / $\mathrm{m.s^{-1}}$	Surface coefficient / $\mathrm{W.m^{-2}.K^{-1}}$ by surface type		
	High emissivity ($\varepsilon \geq 0.9$)	Medium emissivity ($0.2 < \varepsilon < 0.9$)	Low emissivity ($\varepsilon < 0.2$)
Still air	10.0	8.0	5.7
1	13.5	11.5	9.0
2	16.5	14.5	12.5
3	20.0	18.0	15.5
5	26.0	24.0	22.0
10	40.0	38.0	36.0

Thicknesses of insulation

Minimum thicknesses of insulation for the prevention of freezing or condensation are given in BS5422[46] for a range of applications and conditions; these include: refrigeration, chilled and cold water supplies in industrial and commercial applications, central heating, air conditioning and direct hot water supply installations in both non-domestic and domestic applications.

In some circumstances it is important to relate the thickness of insulation to the financial cost involved. This can be addressed by introducing the concept of the 'economic thickness of insulation'. According to BS5422[46], the 'economic thickness' is defined as the thickness of insulation that gives a minimum total cost over a chosen evaluation period.

The costs to be considered are:

(a)   the cost of heat lost from the insulated surfaces during the evaluation period;

(b)   the cost of the insulation system during the evaluation period.

Methods of calculating these costs and hence the economic thicknesses of insulation for the applications stated above are given in BS5422[46], to which the reader is referred.

Temperature changes in insulated pipes

In a piping system, the heat gains or losses of the fluid can be significant, especially when passing through an untreated space. The temperature change in the fluid passing through a pipe is given by:

$$\Delta t = t_u - t_d = \frac{t_u - t_a}{0.5 + f} \quad (3.116)$$

where:

$$f = \frac{Mc_p \times 10^3}{\pi L d_{op} U} \quad (3.117)$$

For water, $c_p = 4.19$ kJ.kg^{-1}.K^{-1} and the loss per metre run is approximately:

$$\Delta t_m = \frac{U(t_u - t_a)d_{op}}{1330M} \quad (3.118)$$

Equation (3.118) is illustrated in Figure. 3.9 for various values of overall transmittance. The U values for various thermal conductivities, thicknesses of insulation and pipe sizes may be found by dividing the values in Table 3.23 by πd_{op} or by using equation (3.112).

Buried pipes

The heat emission from underground piping, whether buried in ducts, pressure tight casings, insulating materials *in situ* or laid directly in the earth, varies from that of the insulated pipe exposed freely to ambient air; this is the result of the additional insulating effect of the air gap within the duct or outer pipe, where present, and that of the earth cover. Equation (3.111) may be adapted to each of these cases with sufficient accuracy for practical purposes, bearing in mind the thermal resistance of the air gap is not greatly significant and that of the earth cover will vary dependent upon its wetness. BS4508[48] describes methods for the determination of heat losses.

The heat loss per metre run of buried pipe is given by the following expression:

$$Q = \pi d_{op} U(t_s - t_e) \quad (3.119)$$

where the overall thermal transmittance is given by:

$$U = \frac{1}{R_n + R_a + R_e} \quad (3.120)$$

The thermal resistance of the insulation is given by:

$$R_n = \frac{d_{op}}{2k_n} \ln\left(\frac{d_{on}}{d_{op}}\right) \quad (3.121)$$

Table 3.23 Heat emission or absorption from insulated pipes

Nominal pipe size (mm)	Heat emission or absorption from insulated pipework / W.m^{-1}.°C^{-1} temperature difference for given values of insulation thermal conductivity / W.m^{-1}.K^{-1}																			
	Thickness of insulation (mm)																			
	0.025					0.040					0.055					0.070				
	12.5	19	25	38	50	12.5	19	25	38	50	12.5	19	25	38	50	12.5	19	25	38	50
15	0.18	0.14	0.12	0.10	0.09	0.27	0.22	0.19	0.16	0.14	0.34	0.29	0.25	0.21	0.19	0.41	0.35	0.31	0.27	0.24
20	0.21	0.16	0.14	0.11	0.10	0.31	0.25	0.22	0.18	0.16	0.40	0.33	0.29	0.24	0.21	0.47	0.40	0.36	0.30	0.26
25	0.25	0.19	0.16	0.13	0.11	0.36	0.29	0.25	0.20	0.18	0.47	0.38	0.33	0.27	0.24	0.56	0.46	0.41	0.34	0.30
32	0.29	0.22	0.19	0.15	0.13	0.43	0.34	0.29	0.23	0.20	0.55	0.45	0.39	0.31	0.27	0.66	0.54	0.47	0.38	0.34
40	0.32	0.25	0.21	0.16	0.14	0.48	0.37	0.32	0.25	0.21	0.61	0.49	0.42	0.33	0.29	0.72	0.59	0.52	0.42	0.36
50	0.39	0.29	0.24	0.18	0.16	0.57	0.44	0.37	0.29	0.24	0.73	0.58	0.49	0.39	0.33	0.86	0.70	0.60	0.48	0.41
65	0.47	0.35	0.29	0.22	0.18	0.69	0.55	0.44	0.34	0.28	0.88	0.69	0.58	0.45	0.38	1.04	0.83	0.71	0.56	0.48
80	0.54	0.40	0.33	0.24	0.20	0.79	0.60	0.50	0.38	0.32	1.0	0.78	0.66	0.50	0.43	1.19	0.94	0.80	0.63	0.53
100	0.67	0.49	0.40	0.29	0.24	0.98	0.74	0.61	0.45	0.38	1.25	0.96	0.80	0.61	0.51	1.47	1.16	0.98	0.75	0.63
125	0.81	0.58	0.47	0.34	0.28	1.18	0.88	0.72	0.53	0.44	1.49	1.14	0.95	0.71	0.59	1.76	1.38	1.16	0.88	0.73
150	0.96	0.69	0.55	0.40	0.32	1.37	1.02	0.83	0.61	0.50	1.74	1.32	1.09	0.81	0.67	2.05	1.59	1.33	1.01	0.84
200	1.22	0.88	0.70	0.50	0.40	1.78	1.32	1.07	0.77	0.63	2.26	1.70	1.40	1.03	0.84	2.66	2.05	1.71	1.27	1.05
250	1.50	1.07	0.86	0.60	0.48	2.19	1.61	1.30	0.94	0.75	2.77	2.09	1.71	1.25	1.01	3.27	2.51	2.08	1.54	1.26
300	1.77	1.26	1.00	0.70	0.56	2.58	1.89	1.52	1.09	0.87	3.26	2.44	2.00	1.45	1.17	3.84	2.94	2.48	1.79	1.46

Note: The pipes sizes are to BS1387 and BS3600[37, 38]. It is assumed that the outside surface of the insulation has been painted, is in still air at 20°C and $h_{so} = 10$ W.m^{-2}.K^{-1}

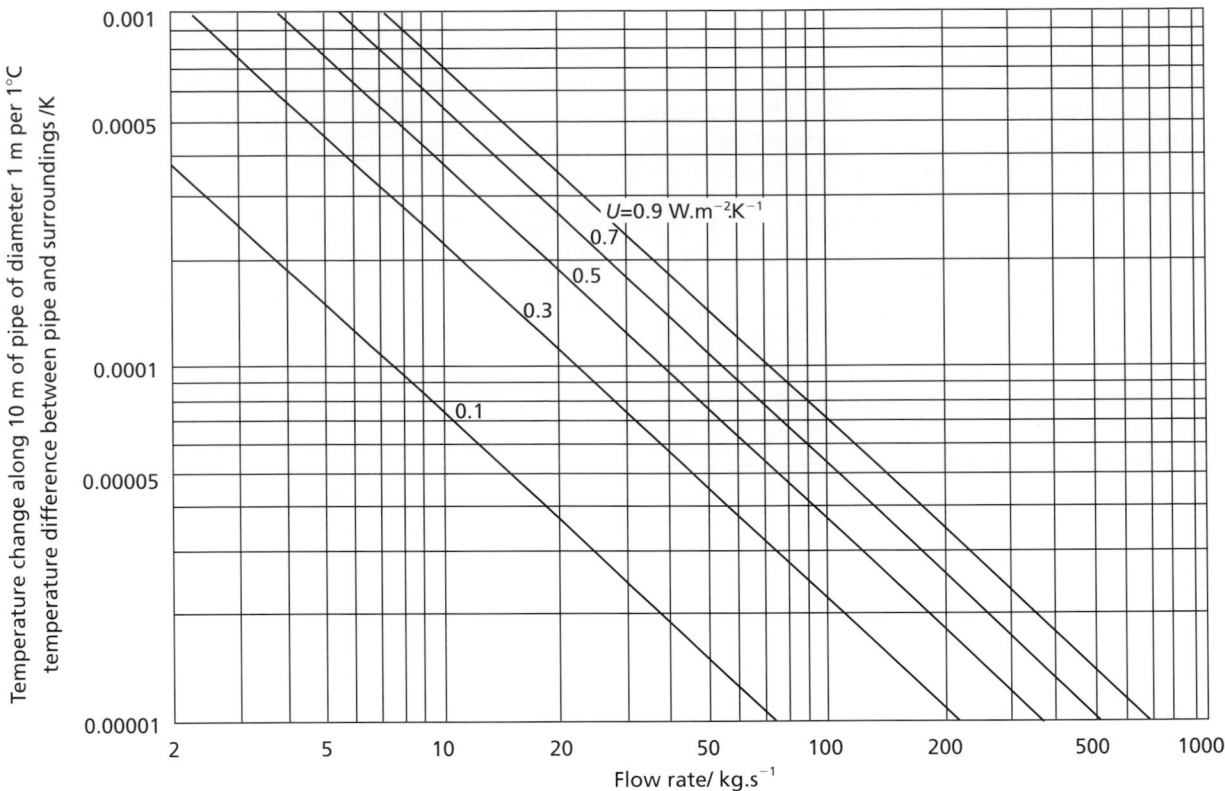

Figure 3.9 Temperature change along insulated pipes in air

The thermal resistance of the air gap is given by:

$$R_a = \frac{d_{op}}{h_{so}d_{on}} + \frac{d_{op}}{h_{si}d_{ic}} \qquad (3.122)$$

The thermal resistance of the earth cover is given by:

$$R_e = \frac{d_{op}}{2k_e} \ln\left(\left(\frac{2m}{d_{ic}}\right)\left\{1+\left[1-\left(\frac{d_{ic}}{2m}\right)^2\right]^{0.5}\right\}\right) \qquad (3.123)$$

If the burial depth m is greater than $2\,d_{on}$, then equation 3.123 reduces to approximately:

$$R_e = \frac{d_{op}}{2k_e} \ln\left(\frac{4m}{d_{ic}}\right) \qquad (3.124)$$

If there is no air gap, d_{ic} equals d_{on}.

The outside surface temperature of the insulation or pressure tight casing is often an important factor and may be calculated from:

$$t_c = t_s - R_n(t_s - t_e)U \qquad (3.125)$$

The overall thermal transmittance required to prevent the fluid temperature falling below a specified value over a given distance may be calculated from:

$$U = \frac{Mc_p \ln\left(\dfrac{t_{s_1} - t_e}{t_{s_2} - t_e}\right)}{\pi d_{op} L} \qquad (3.126)$$

In equations (3.119) to (3.126), it is assumed that the surface temperature of the pipe is equal to the fluid tempera-

ture and that the thermal resistances of the pressure tight casing and the surface of the ground are negligible.

Twin pipe arrangements

Underground flow and return mains are often run together in a pressure tight casing or concrete duct as illustrated in Figure 3.10. The following method of heat loss calculation gives answers agreeing closely with field test results:

(a)   Use equation (3.119) to calculate the heat losses from the flow pipe assuming it to be alone in the centre of a large casing.

(b)   Repeat this procedure for the return pipe.

(c)   Add (a) and (b) to give the total loss.

If the duct is of rectangular cross-section (dimension a_1 by a_2) then the equivalent diameter should be used, given by:

$$d_h = \frac{2a_1a_2}{a_1 + a_2} \qquad (3.127)$$

Thermal conductivity of insulation k_n

The values shown in Figure 3.8 are for dry insulants. If the insulation becomes wet, a considerable increase in the thermal conductivity may be expected. Thoroughly saturated insulation may approach the thermal conductivity of water (approximately 0.6 $W.m^{-1}.K^{-1}$). If evaporation occurs, the heat loss will be even greater.

Thermal resistance of air space R_a

Although an air space around an encased insulated pipe is essential for the detection of leaks and for the drainage and

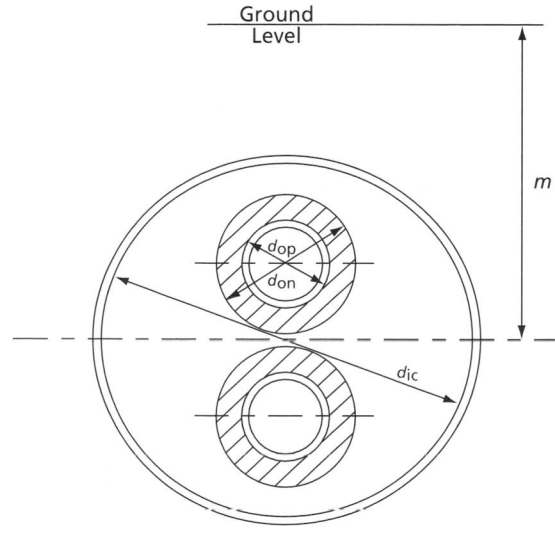

Figure 3.10 Twin pipe underground mains

drying of wet insulation, it has only a small insulation value compared with normal dry insulants. A typical value for resistance of the air space would be in the order of 0.06 $m^2.K^{-1}.W^{-1}$.

Thermal conductivity of earth cover k_e

The depth of the earth cover, the physical properties of the soil including the porosity and permeability and the external temperature of the casing are all factors affecting the thermal conductivity of the earth. Table 3.24 lists typical values of thermal conductivity for various soils[49].

Ground ambient temperature t_e

Table 3.25 lists mean values of ground ambient temperature for various parts of the UK, measured at a depth of one metre.

Table 3.24 Earth thermal conductivity k_e

Soil description	Summer		Winter	
	Moisture content %	Thermal conductivity / $W.m^{-1}.K^{-1}$	Moisture content %	Thermal conductivity / $W.m^{-1}.K^{-1}$
Pea ballast	2.3	0.7	11.8	1.8
Poorly graded sand (one predominant particle size)	2.4	0.5	7.7	1.6
Well graded sand	9.3	0.9	21.8	2.1
Sand/Clay (~20%) mixture	15.0	0.9	33.0	1.9
If under impervious cover, e.g. paving	—	—	—	1.3*
Predominantly clay	26.3	0.8	33.7	1.5
Chalk	18.2	0.9	30.4	1.2
Mean Values	Less than 12%	0.7	More than 25%	1.7

*Areas in the order of 10^3 to 10^4 m²

Table 3.25 Ground ambient temperature t_e at the depth of 1 m

Part of UK	Mean ground temp. (°C)	
	Summer	Winter
South of England and South Wales	18	8
Northern Ireland, North Wales and East Midlands	17	7
Central Midlands, North West England and Scotland	16	6
North East England and Scotland	15	5

1°C should be added to the above values in built-up areas

Tabulated data

Owing to the number of variables involved, tables of heat emissions are not presented. However, for convenience, values of $\ln(d_{on}/d_{op})$ and $\ln(4m/d_{ic})$ are listed in Tables 3.26 and 3.27 for different thicknesses of insulation and burial depths.

The following practical considerations should be taken into account.

(a) To use theoretical figures based on dry insulation could sometimes lead to technical or economic failures except in those systems which are pressure tight to at least 200 kPa and the insulation is capable of being dried out to its original thermal and physical condition.

Table 3.26 Solutions of $\ln(d_{on}/d_{op})$ for steel pipe

Nominal pipe size (mm)	Value of $\ln(d_{on}/d_{op})$ for stated thickness of insulation (mm)						
	12.5	19	25	38	50	75	100
15	0.77	1.0	1.2	1.5	1.7	2.1	2.3
20	0.66	0.88	1.1	1.3	1.6	1.9	2.1
25	0.55	0.75	0.91	1.2	1.4	1.7	1.9
32	0.46	0.64	0.78	1.0	1.2	1.5	1.7
40	0.42	0.58	0.71	0.94	1.1	1.4	1.6
50	0.35	0.49	0.60	0.82	0.98	1.2	1.5
65	0.28	0.41	0.51	0.69	0.84	1.1	1.3
80	0.25	0.36	0.45	0.62	0.75	0.99	1.2
100	0.20	0.29	0.36	0.51	0.63	0.84	1.0
125	0.16	0.24	0.31	0.43	0.54	0.73	0.89
150	0.14	0.21	0.26	0.38	0.47	0.65	0.79
200	0.11	0.16	0.21	0.30	0.38	0.52	0.65
250	0.088	0.13	0.17	0.25	0.31	0.44	0.55
300	0.074	0.11	0.14	0.21	0.27	0.38	0.48
350	0.068	0.10	0.13	0.19	0.25	0.35	0.45
400	0.060	0.089	0.12	0.17	0.22	0.31	0.40

Table 3.27 Solutions of $\ln(4m/d_{ic})$

Nominal pipe size (mm)	Value of $\ln(4m/d_{ic})$ for given values of burial depth (mm)					
	0.5	1.0	1.5	2.0	2.5	3.0
50	3.7	4.4	4.8	5.1	5.3	5.5
75	3.3	4.0	4.4	4.7	4.9	5.1
100	3.0	3.7	4.1	4.4	4.6	4.8
125	2.8	3.5	3.9	4.2	4.4	4.6
150	2.6	3.3	3.7	4.0	4.2	4.4
175	2.4	3.1	3.5	3.8	4.1	4.2
200	2.3	3.0	3.4	3.7	3.9	4.1
225	2.2	2.9	3.3	3.6	3.8	4.0
250	2.1	2.8	3.2	3.5	3.7	3.9
275	2.0	2.7	3.1	3.4	3.6	3.8
300	1.9	2.6	3.0	3.3	3.5	3.7

Table 3.28 U values for insulated air ducts

Thermal conductivity of insulation / W.m^{-1}.K^{-1}	U value / W.m^{-2}.K^{-1} for given thickness of insulation (mm)			
	25	50	75	100
0.025	0.89	0.47	0.32	0.24
0.03	1.04	0.56	0.38	0.29
0.035	1.19	0.64	0.44	0.34
0.04	1.33	0.73	0.50	0.38
0.045	1.47	0.81	0.56	0.43
0.05	1.60	0.89	0.61	0.47
0.055	1.72	0.97	0.67	0.51
0.06	1.84	1.04	0.73	0.56
0.07	2.07	1.19	0.83	0.64
0.08	2.28	1.33	0.94	0.73

(b) Damp insulation conditions are frequently observed and measurement of heat losses, even on installations which appear satisfactory, have shown that they can be several times greater than theoretical figures and of the same order as bare pipes in dry soils. Flooded ducts and wet soil will result in excessive heat losses.

(c) The thermal conductivity of the soil makes a relatively small contribution to the overall heat loss from a well-insulated pipe. Therefore, where the type and state of the earth cover are not known, the mean value may be used without undue error.

(d) There are now many proprietary underground piping systems available and reference should be made to the manufacturer's literature.

Cross conduction between pipes

Where flow and return pipes at different temperatures are close together, heat will flow between them. Whilst this is not a heat loss, it amounts to an extra pump load on the system involving extra running costs. For example, an increase of only 3 per cent in the mass flow rate to maintain flow temperature would result in 10 per cent extra pumping costs. Insulation systems which can positively ensure an adequate degree of insulation between flow and return will therefore show an additional economic benefit.

Air ducts

In an air duct system heat gains or losses of the ducted air can be significant, especially when passing through an untreated space in a supply system. This also has the effect of reducing the heating or cooling capacity of the air. The heat emission or absorption from air ducts is given by:

$$\Phi = UA\Delta t \tag{3.128}$$

or more conveniently per metre run of duct:

$$Q = UP\left[\frac{(t_u + t_d)}{2} - t_a\right] \tag{3.129}$$

The temperature change in a ducted air stream is given by:

$$t_u - t_d = \frac{t_u - t_a}{0.5 + f} \tag{3.130}$$

where:

$$f = \frac{c_p d_h \rho v \times 10^3}{4UL} \tag{3.131}$$

The hydraulic mean (equivalent) diameter is found from equation (3.127) or from the following equation:

$$d_h = 4A_c/P \tag{3.132}$$

For air at 20°C, $\rho = 1.2$ kg.m^{-3} and $c_p = 1.02$ kJ.kg^{-1}.K^{-1}, the temperature change per metre is:

$$\Delta t_m = \frac{U(t_u - t_a)}{306 \, d_h v} \tag{3.133}$$

This relationship is illustrated in Figure 3.11 for various thicknesses of insulation, where a thermal conductivity of 0.045 W.m^{-1}.K^{-1} has been used. Figure 3.12 relates the change in temperature to air volume flow for various air velocities and for 25 mm of insulation.

Overall thermal transmittance U

In this context, the overall thermal transmittance is given by:

$$U = \cfrac{1}{\cfrac{1}{h_{si}} + \cfrac{l_n}{k_n} + \cfrac{1}{h_{so}}} \tag{3.134}$$

Table 3.28 lists values of U for various thicknesses and thermal conductivities of insulation.

Inside surface heat transfer coefficient h_{si}

The internal surface heat transfer coefficient is a function of the Reynolds number as shown by equation (3.41). The

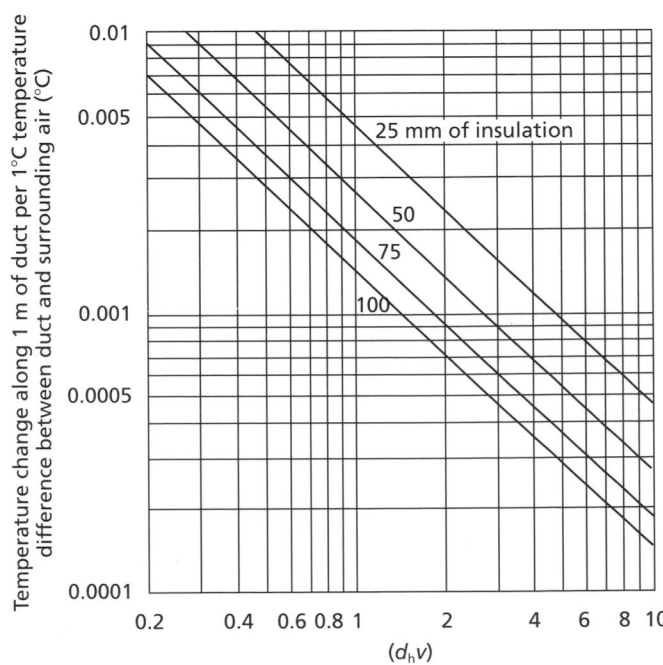

Figure 3.11 Temperature change along insulated ducts for various thicknesses of insulation

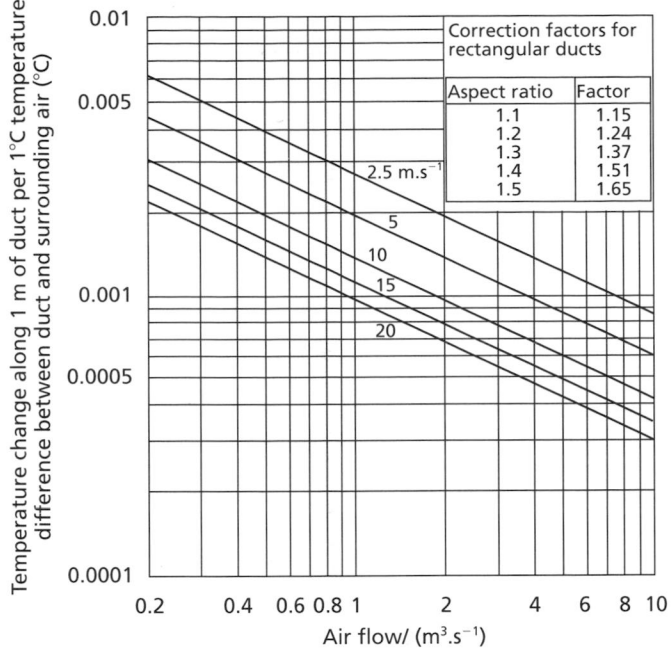

Figure 3.12 Temperature change along insulated ducts for various air flow velocities (for 25 mm of insulation)

value for h_{si} can be determined from equation (3.41), applied for the appropriate conditions.

Outside surface heat transfer coefficient h_{so}

A typical value for this coefficient is 10 W.m^{-2}.K^{-1} but this value may well be lower if the duct is in close proximity to its surroundings. Values for other conditions may be obtained from Table 3.22.

Practical considerations

As the temperature difference in equation (3.133) is expressed in terms of the initial temperature difference rather than the mean temperature difference, some error will be introduced if the value of the length of duct work chosen for calculation is too large. The smaller the value of $d_h \times v$, the larger the error. A maximum length of 10 m is recommended. It may be noted from Figure 3.11 that as the value of $d_h \times v$ falls below 1.5, the rate of temperature drop in ducts with 50 mm or less of insulation increases considerably. It is usually not practical to keep the value above this by changing d_h or v so extra insulation should be considered.

Heat exchangers

A heat exchanger transfers heat from one fluid to another by conduction, radiation or convection or by a combination of these. The two fluids can stay as liquids or gases or they can change from one state to the other as, for example, in evaporators and condensers. The commonest arrangement is for the two fluids to flow in separate channels with heat exchange between them and without change of state. This arrangement is discussed below.

The basic equation for heat transfer between the two fluids separated by a solid surface is:

$$\Phi = UA\Delta t_l \tag{3.135}$$

Overall heat transfer coefficient U

The overall heat transfer coefficient between two separated fluids can be calculated from:

$$U = \cfrac{1}{\cfrac{1}{h_{so}} + \cfrac{l_{zo}}{k_{zo}} + \cfrac{l_w}{k_w} + \cfrac{l_{zi}}{k_{zi}} + \cfrac{1}{h_{si}}} \tag{3.136}$$

The overall heat transfer coefficient for thin-walled tubes is:

$$U = \cfrac{1}{\cfrac{1}{h_{so}} + \cfrac{l_{zo}}{k_{zo}} + \cfrac{l_w}{k_w} + \left(\cfrac{l_{zi}}{k_{zi}} \cdot \cfrac{A_{so}}{A_{si}}\right) + \left(\cfrac{1}{h_{si}} \cdot \cfrac{A_{so}}{A_{si}}\right)} \tag{3.137}$$

In the case of clean tube surfaces and neglecting the tube wall resistance, equation (3.137) simplifies to:

$$U_c = \cfrac{1}{\cfrac{1}{h_{so}} + \cfrac{1}{h_{si}} \cdot \cfrac{A_{so}}{A_{si}}} \tag{3.138}$$

or more conveniently:

$$U_c = \cfrac{h_{so}h_{si}\cfrac{A_{si}}{A_{so}}}{h_{so} + h_{si}\cfrac{A_{si}}{A_{so}}} \tag{3.139}$$

Values for U_c can be obtained from Figure 3.13.

Values of h_{si} for water flow through tubes and h_{so} for forced water flow over tubes are shown in Tables 3.29 and 3.30. Values for l_{zo}/k_{zo} and l_{zi}/k_{zi} (fouling resistances) are usually determined by experience although the values given in Table 3.31 provide a guide[50]. The fouling resistance can be a significant proportion of the total resistance and hence should be taken into account in heat exchanger design.

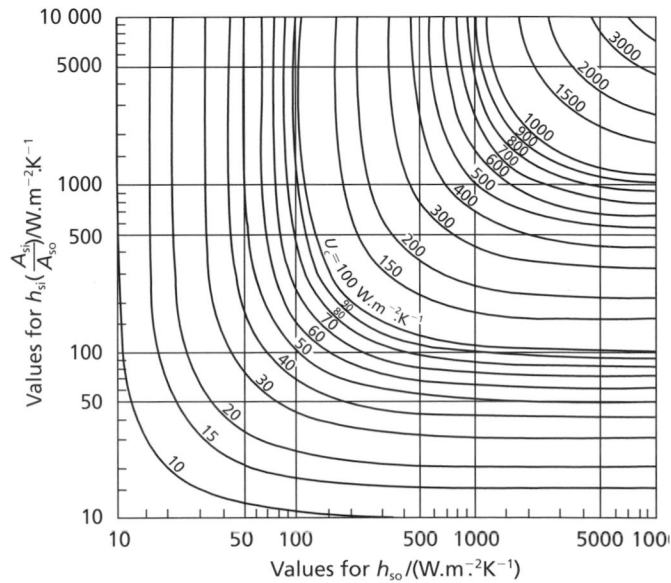

Figure 3.13 Values of U_c

Table 3.29 Convective film coefficient (h_{si}) for turbulent water flow through straight plain tubes

Tube inside diameter (mm)	Inside film coefficient / W.m^{-2}.K^{-1} for a water temperature of 75°C and the mean water velocities (m.s^{-1}) given								
	0.2	0.4	0.6	0.8	1.0	1.5	2	3	4
20	1760	3060	4240	5340	6380	8820	11 100	15 400	19 300
25	1680	2930	4060	5110	6100	8440	10 600	14 700	18 500
32	1600	2790	3860	4860	5810	8040	10 100	14 000	17 600
40	1530	2670	3690	4650	5560	7680	9620	13 400	16 800
50	1470	2550	3540	4450	5320	7350	9260	12 800	16 100
80	1330	2320	3210	4050	4830	6690	8420	11 600	14 700
100	1280	2220	3080	3870	4630	6400	8050	11 100	14 000

Correction factors for other water temperatures

Temperature (°C)			10	25	50	100	150	200
multiplying factor			0.65	0.65	0.84	1.11	1.34	1.37

Table 3.30 Convective film coefficient (h_{so}) for turbulent water flow over straight plain tubes

Tube outside diameter (mm)	Outside film coefficient / W.m^{-2}.K^{-1} for a water temperature of 75°C and the mean water velocity / m.s^{-1} given					
	0.1	0.2	0.3	0.4	0.5	0.6
20	2120	3100	3880	4540	5130	5680
25	1920	2810	3510	4110	4650	5140
32	1720	2520	3150	3680	4170	—
40	1550	2270	2840	3320	—	—
50	1400	2050	2570	—	—	—

Correction factors for other water temperatures

Temperature (°C)	10	25	50	100	150	200
multiplying factor	0.70	0.79	0.85	1.10	1.21	1.25

For heat transfer calculations in shell and tube heat exchangers, it is necessary to determine values for the surface heat transfer coefficients for particular flow configurations.

(a)   For forced convection flow inside tubes, use can be made of equation (3.41) or (3.43).

(b)   For free or forced correction flow outside single tubes, refer to Tables 3.2 and 3.4, respectively.

(c)   For forced convection flow outside tube bundles, details can be found in reference 4.

Logarithmic mean temperature difference

The logarithmic mean temperature difference can be calculated from:

$$\Delta t_l = \frac{\Delta t_{tg} - \Delta t_{ts}}{\ln\left(\dfrac{\Delta t_{tg}}{\Delta t_{ts}}\right)} \tag{3.140}$$

Equation (3.140) is not solvable if Δt_{tg} and Δt_{ts} are equal. In this case, and when the ratio $\Delta t_{tg}/\Delta t_{ts}$ is close to unity, Δt_{tl} can be taken as the arithmetic mean value of Δt_{tg} and Δt_{ts}. Values of Δt_l can be obtained directly from Figure 3.14. Here, the term 'terminal' temperature difference refers to the temperature difference between hot and cold fluid streams at a given end of the heat exchanger. The use of the logarithmic mean temperature difference in equation (3.135) is strictly correct only for constant U, constant specific heat capacity and parallel or counterflow arrangements under steady state conditions.

For multipass shell and tube heat exchangers, a correction factor should be applied to the logarithmic mean temperature difference. Typical values of this factor are given in Figures 3.15 and 3.16 and more detailed information can be obtained from references 50, 51 and 52, or more recently 53.

Effectiveness — NTU method

The logarithmic mean temperature difference method[53] is convenient for calculating the heat transfer rates in heat exchangers when terminal temperatures are known. More commonly, however, the terminal temperatures are unknown, rendering impractical the adoption of the logarithmic mean temperature difference method. In this situation, the alternative 'effectiveness-NTU' method can be used[53]. This method is based upon three dimensionless parameters:

$$\eta = \frac{\text{actual rate of heat transfer, } q}{\text{maximum possible rate of heat transfer, } q_{max}} \tag{3.141}$$

$$\text{NTU} = \frac{AU}{C_{min}} \tag{3.142}$$

Table 3.31 Fouling resistances l_z/k_z

Type of water	Fouling resistances / m^{-2}.K^{-1}.W^{-1} for stated temperatures and water velocities			
	Heating medium ≤120°C Water ≤50°C		Heating medium 120–200°C Water >50°C	
	≤1 m.s^{-1}	>1 m.s^{-1}	≤1 m.s^{-1}	>1 m.s^{-1}
Sea water	0.000 1	0.000 1	0.000 2	0.000 2
Brackish water	0.000 4	0.000 2	0.000 5	0.000 4
Cooling tower treated make-up	0.000 2	0.000 2	0.000 4	0.000 4
Cooling tower untreated make-up	0.000 5	0.000 5	0.000 9	0.000 7
Well water	0.000 2	0.000 2	0.000 4	0.000 4
River water (clean)	0.000 4	0.000 2	0.000 5	0.000 4
(polluted)	0.001 4	0.001 1	0.001 8	0.001 4
Boiler blowdown	0.000 4	0.000 4	0.000 4	0.000 4
Treated boiler feedwater	0.000 2	0.000 1	0.000 2	0.000 2

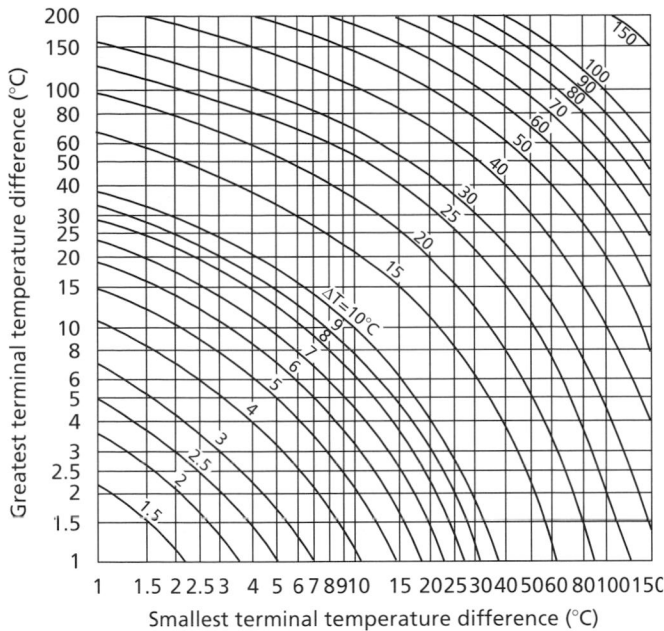

Figure 3.14 Values of logarithmic mean temperature difference

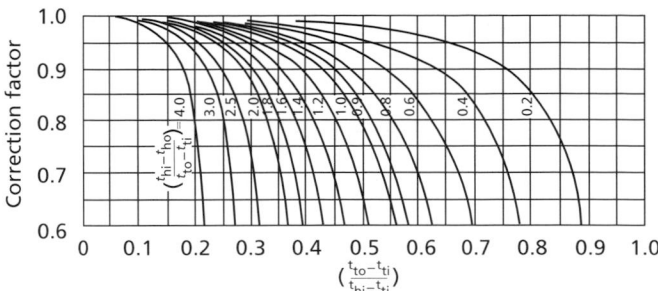

Figure 3.15 Correction factors for assessing mean temperature difference of multipass heat exchangers (one shell pass, two or more tube passes)

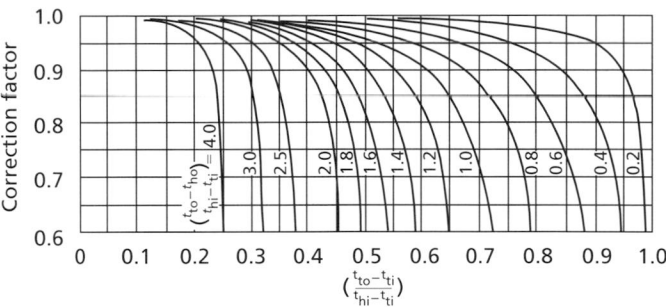

Figure 3.16 Correction factors for assessing mean temperature difference of multipass heat exhangers (two shell passes, four or more tube passes)

and:

$$Z = \frac{C_{min}}{C_{max}} \qquad (3.143)$$

where η is the heat exchanger effectiveness, NTU is the number of heat exchanger heat transfer units, Z is the heat capacity rate ratio, C_{min} and C_{max} are the smaller and greater, respectively, of the fluid heat capacity rates, i.e.

$$C_{(min, max)} = Mc_p \qquad (3.144)$$

The maximum possible heat transfer rate q_{max} is given by:

$$q_{max} = C_{min}(t_{hfi} - t_{cfi}) \qquad (3.145)$$

where t_{hfi} and t_{cfi} are the inlet temperatures of the hot and cold fluids, respectively. From a knowledge of these temperatures, together with values for C_{min} and η, the actual heat transfer rate of the heat exchanger can be determined from equation (3.141). Note that fluid inlet temperatures and mass flow rates are commonly known in a given application. The heat exchanger effectiveness η is a function of NTU, Z and heat exchanger configuration.

For the parallel flow configuration:

$$\eta = \frac{1 - \exp[-NTU(1 + Z)]}{1 + Z} \qquad (3.146)$$

For the counterflow configuration:

$$\eta = \frac{1 - \exp[-NTU(1 + Z)]}{1 - Z\exp[-NTU(1 - Z)]} \qquad (3.147)$$

Note that for the case where one fluid remains at a constant temperature throughout the heat exchanger (i.e. a change in phase occurs, as in condensers and evaporators in refrigeration), Z becomes zero. For any heat exchanger configuration where Z is zero, the heat exchanger effectiveness is:

$$\eta = 1 - \exp(-NTU) \qquad (3.148)$$

References 10 and 54 give relationships for η, NTU and Z for other configurations.

Example 3.4

Hot oil enters a counterflow heat exchanger at a mass flow rate of 2.0 kg.s⁻¹ with an inlet temperature of 150°C and an outlet temperature of 45°C. The hot oil is being cooled by water that enters at a temperature of 20°C and at a mass flow rate of 1.8 kg.s⁻¹. The diameter of the heat exchanger inner tube is 0.025 m and its length is 10.0 m. Determine the overall heat transfer coefficient, U, of this heat exchanger using (*a*) the LMTD method, and (*b*) the effectiveness-NTU method.

Data: specific heat capacity of oil is 2.1 kJ.kg⁻¹.K⁻¹, specific heat capacity of water is 4.2 kJ.kg⁻¹.K⁻¹.

Solution

(*a*)   Using the LMTD method.

The actual rate of heat transfer from the hot fluid (oil) is:

$$q = Mc_p(t_{hfi} - t_{hfo}) \qquad (3.149)$$
$$= 2.0 \times 2.1(150 - 45) = 441 \text{ kW}$$

The outlet temperature of the cold fluid (water) is:

$$t_{cfo} = t_{cfi} + (q/Mc_p) \qquad (3.150)$$
$$= 20 + (441/(1.8 \times 4.2)) = 78.3°C$$

Now that the outlet temperatures of both fluids are known, the LMTD method can be used to determine the overall heat transfer coefficient.

The greatest terminal temperature difference, Δt_{tg}, is:

$$\Delta t_{tg} = t_{hfi} - t_{cfo} \qquad (3.151)$$

$$= 150 - 78.3 = 71.7°C$$

The smallest terminal temperature difference, Δt_{ts}, is:

$$\Delta t_{ts} = t_{hfo} - t_{cfi} \qquad (3.152)$$

$$= 45 - 20 = 25°C$$

The logarithmic mean temperature difference, Δt_l, can then be determined either graphically from Figure 3.14 or analytically from equation (3.140) giving:

$$\Delta t_l = \frac{(71.7 - 25)}{\ln(71.7/25)} = 44.3°C$$

The overall heat transfer coefficient, U, is then given by:

$$U = q/A\Delta t_l \qquad (3.153)$$

$$= 441/(\pi \times 0.025 \times 10 \times 44.3)$$

$$= 12.7 \text{ kW.m}^{-2}.\text{K}^{-1}$$

(b)   Using the effectiveness — NTU method

The heat capacity rate of the hot fluid (oil) is:

$$Mc_p = 2.0 \times 2.1 = 4.2 \text{ kW.K}^{-1}.$$

Similarly, the heat capacity rate of the cold fluid (water) is:

$$1.8 \times 4.2 = 7.56 \text{ kW.K}^{-1}.$$

The heat capacity rate ratio, Z, from equation (3.143) is:

$$Z = 4.2/7.56 = 0.56$$

The maximum possible heat transfer rate, q_{max}, is given by equation (3.145):

$$q_{max} = 4.2(150 - 20) = 546 \text{ kW}$$

The actual rate of heat transfer from the hot fluid (oil) from equation (3.149) is:

$$Q = 2.0 \times 2.1(150 - 45) = 441 \text{ kW}$$

The heat exchanger effectiveness, η, is given by equation (3.141):

$$\eta = 441/546 = 0.81$$

For the counterflow configuration, equation (3.147) can be rearranged to give:

$$\text{NTU} = \frac{1}{(Z-1)} \ln\left(\frac{\eta - 1}{\eta Z - 1}\right) \qquad (3.154)$$

$$= \frac{1}{0.56 - 1} \ln\left(\frac{0.81 - 1}{(0.81 \times 0.56) - 1}\right) = 2.4$$

From equation (3.142), the overall heat transfer coefficient, U, can be calculated to give:

$$U = (2.4 \times 4.2)/(\pi \times 0.025 \times 10)$$

$$= 12.8 \text{ kW.m}^{-2}.\text{K}^{-1}$$

Evaporators and condensers

For heat exchangers where one or both of the fluids undergoes a change of state, the fundamental equations are complex. They depend on the mode of boiling and condensing and this, in turn, depends on the fluid conditions and mechanical design of the heat exchanger. Information on the design of heat exchangers for boiling and condensing is given in references 5 and 55.

Open water surfaces

Heat transfer from open water surfaces takes place by:

(a)   Evaporation, i.e. conversion of part of the water to vapour and the transfer of the vapour through diffusion and convection (ϕ_e).

(b)   Convection to, and from, the air in contact with the surface (ϕ_c).

(c)   Radiation to, and from, the surface (ϕ_r).

(d)   Conduction to, and from, the surrounds (ϕ_{cd}).

The actual part which each component contributes to the total heat transfer depends on prevailing conditions. In the case of indoor pools and tanks, this will normally depend on the environment provided by the heating and ventilating installation.

In the case of outdoor pools, reservoirs, cooling ponds, etc., it will depend on the prevailing weather conditions and hence the time of the year. During the hottest part of summer up to 90% or more of the total heat transfer from the water is by evaporation. In winter, at low air temperature, the surface evaporation is reduced to approximately 50% of the total, and convection increases proportionately.

Heat transfer by radiation can be an important factor in the performance of outdoor cooling ponds where solar radiation can seriously reduce the cooling effect. Conversely, solar radiation can contribute to the heating of outdoor swimming pools.

Conduction between the water and the surrounds when the containing walls are sunk in the ground is generally small, even when heating up or cooling down, because of the large masses involved. The mass of earth surrounding the pool eventually warms up to a temperature close to that of the water and acts as a stabiliser because of its large thermal capacity.

Where the containing walls are surrounded by air, conduction can be significant but is still small in relation to the total heat transfer.

The total heat loss rate from unit area of an open water surface, in W.m^{-2}, can be expressed as:

$$\sum \phi = \phi_e + \phi_c + \phi_r + \phi_{cd} \qquad (3.155)$$

where:

$$\phi_e = (91.5 + 77.6\,v_a)(p_{sw} - p_v) \qquad (3.156)$$

$$\phi_c = 3.18\,v_a^{0.8}(t_{sw} - t_a) \qquad (3.157)$$

$$\phi_r = 5.67 \times 10^{-8}\,\varepsilon_w(T_{sw}^4 - T_{rs}^4) - I \qquad (3.158)$$

$$\phi_{cd} = U_t(t_w - t_a) \qquad (3.159)$$

Values of ϕ_e are given in Figure 3.17 and values of convective heat transfer coefficient for estimation of ϕ_c are given in Figure 3.18. Where the tank walls and floor are surrounded by a large mass, ϕ_{cd} is generally negligible. The emissivity of water can be taken as 0.96 and radiation gains, for the UK, are given in Table 3.32[56].

The mass rate of evaporation from an open water surface is:

$$W = \frac{\phi_e}{H_l \times 10^3} \qquad (3.160)$$

The mean radiant absolute temperature of the sky can be approximated as:

$$T_{rs} = 253 + t_a \qquad (3.161)$$

(that is, 20 K below the air temperature).

Assumptions for water temperature values must be made in order to solve these equations. If there is an artificial heat source and this is thermostatically controlled to maintain the water at a constant temperature, no problem exists. In the case of a cooling pond the outlet temperature should be used.

If there is no artificial heat source, an equilibrium state is reached when the rate of heat loss by evaporation equals the rates of heat gain by convection plus radiation and conduction, i.e.:

$$\phi_e = \phi_c + \phi_r + \phi_{cd} \qquad (3.162)$$

If ϕ_r and ϕ_{cd} are approximately zero, then the water temperature will approach the wet-bulb temperature of the air.

The heat transfer by evaporation is based on empirical data. Various references give results covering a fairly wide range. Many of the experiments that have been carried out have been based on relatively small surface areas and errors are probably introduced when extrapolating to larger areas. The equation given appears to give reasonable results in practice.

Building components

The thermal performance of building components (e.g. walls, windows, doors, shutters, etc.) is evaluated for design purposes using standardised procedures. This is necessary because of the variability of conditions that can be encountered in practice. For example, for the purpose of Building Regulations, U values (thermal transmittances) are calculated for conditions of steady state heat transfer and on the basis of standardised heat transfer coefficients. The reader is therefore referred to references 57, 58, 59, 60, 61 and 62 which cover in more detail the measurement, simulation or calculation of parameters, such as U values, in accordance with accepted standards.

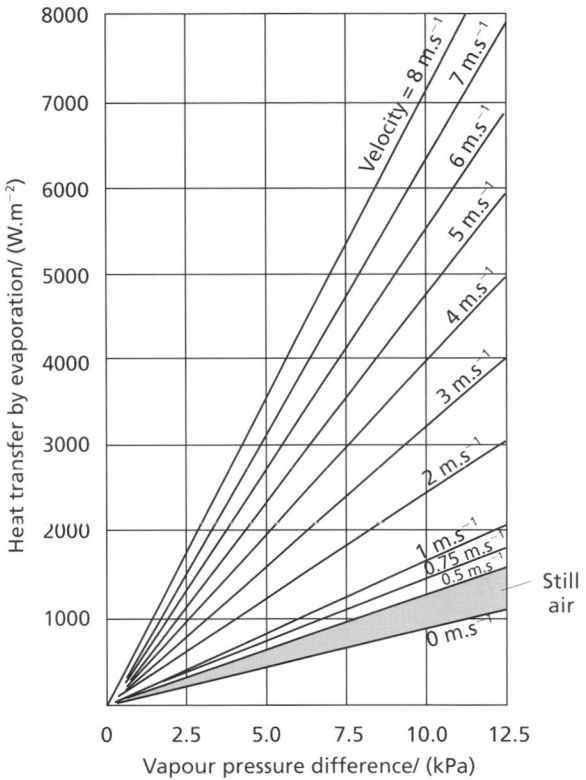

Figure 3.17 Values of evaporative heat transfer

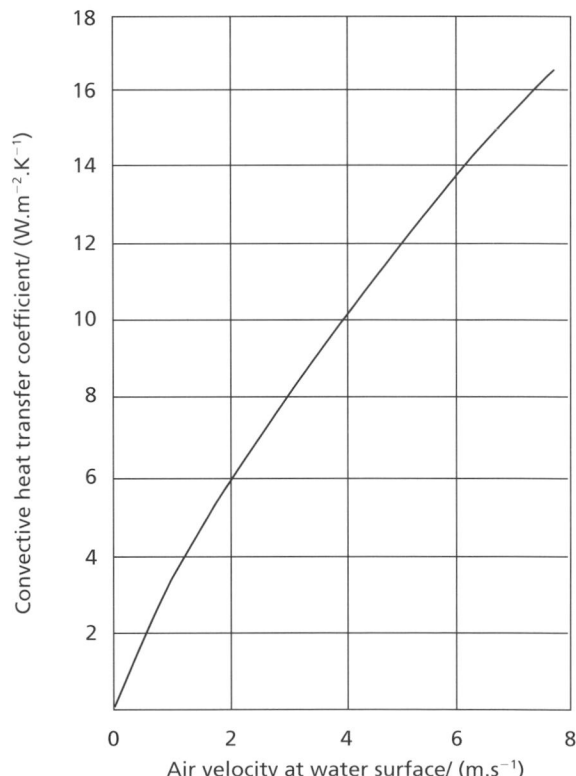

Figure 3.18 Values of convective heat transfer

Table 3.32 Radiation gains to outside pools and reservoirs in the UK

Period	Daily gain / MJ.m^{-2}	Average duration of radiation gain / h	Average intensity / W.m^{-2}	Average night loss rate / W.m^{-2}	Night loss / MJ.m^{-2}	Net radiation gain per 24 hours / MJ.m^{-2}
May–September	13.8	16	240	50	1.4	12.4
January–December	8.4	14	167	41	1.5	6.9

To consider one example, windows (glazing and framework) can have a significant influence upon the internal environment of a building. In this respect, the thermal, acoustic, solar and daylight performances of windows are important and need to be evaluated in an integrated manner. With respect to glazing only, such factors have been investigated for a number of configurations[63], and the reader is referred to the latter reference. For double glazed windows, factors such as temperature distribution of the inner glazing, condensation occurrence and the influence of framework are discussed in reference 64. For further information on overall window performance, refer to the standards cited above.

References

1   Eckert, E R G and Drake, R M, *Heat and Mass Transfer*, McGraw-Hill (1959)

2   Rogers, G F C and Mayhew, Y R, *Engineering Thermodynamics, Work and Heat Transfer*, 4th edition, Longman, Harlow, UK (1992)

3   Cengel, Y A, *Introduction to Thermodynamics and Heat Transfer*, McGraw-Hill, USA (1997)

4   Incropera, F P and De Witt, D P, *Fundamentals of Heat and Mass Transfer*, 3rd edition, Wiley (1990)

5   Jakob, M, *Heat Transfer*, Vol. 1, John Wiley and Sons, New York (1949)

6   Zhukauskas, A, 'Heat Transfer from Tubes in Cross Flow', in Hartnett, J P and Irvine, T F Jr, Eds, *Advances in Heat Transfer*, Vol. 8, Academic Press, New York (1972)

7   Gnielinski, V, 'New Equations for Heat and Mass Transfer in Turbulent Pipe and Channel Flow', *Int. Chemical Engineering*, Vol. 16, No. 2, 359 (1976)

8   Kreith, F and Bohm, M S, *Principles of Heat Transfer*, 5th edition, West (1993).

9   Holman, J P, *Heat Transfer*, 7th edition, McGraw-Hill (1992)

10   McAdams, W H, *Heat Transmission*, McGraw-Hill (1954)

11   Moon, P H, *Scientific Basis of Illuminating Engineering*, McGraw-Hill (1936)

12   Hamilton, D C and Morgan, W R, 'Radiant Interchange Configuration Factors', NACA Technical Note 2536, National Advisory Council on Aeronautics (USA) (1952)

13   Siegel, R and Howell, J R, *Thermal Radiation Heat Transfer*, 3rd edition, Hemisphere Publishing Corporation, USA (1992)

14   Kays, W M and Crawford, M E, *Convective Heat and Mass Transfer*, 3rd edition, McGraw-Hill, New York (1993)

15   Mills, A F, *Basic Heat and Mass Transfer*, 2nd edition, Prentice Hall Inc., New Jersey, USA (1999)

16   Threlkeld, J L, *Thermal Environmental Engineering*, 2nd edition, Prentice-Hall Inc., Englewood Cliffs, New Jersey, USA (1970)

17   Rowley, F B, Algren, A B and Blackshaw, J L, 'Effects of Air Velocities on Surface Coefficients', *ASHVE Trans.*, **36**, 426–446 (1930)

18   Jürges, W, 'Der Wärmeubergang an einem ebenen Wand (Heat Transfer at a Plane Wall)', *Beizh. Z Giesundh. Ing. Ser.*, **1** (19) (1924)

19   Loveday, D L and Taki, A H, 'Outside Surface Resistance: Proposed New Value for Building Design', *Proc. CIBSE A: Building Serv. Eng. Res. Technol.*, **19**(1), 23–29 (1998)

20   Awbi, H B and Hatton, A, 'Natural Convection from Heated Room Surfaces', *Energy and Buildings*, Vol. 30, No. 3 (1999)

21   Min, T C, Schutrum, L F, Parmelee, G V and Vouris, J D, 'Natural Convection and Radiation in a Panel Heated Room', *ASHVE Trans.*, **62**, 337 (1956)

22   Alamdari, F and Hammond, G P, 'Improved Correlations for Buoyancy — Driven Convection in Rooms', *Building Serv. Eng. Res. Technol.*, **4**, 106–112 (1983)

23   Khalifa, A J N and Marshall, R H, 'Validation of Heat Transfer Coefficients on Interior Building Surfaces Using a Real-Sized Indoor Test Cell', *Int. J. Heat and Mass Transfer*, **33**, 2219–2236 (1990)

24   *ASHRAE Handbook*, Fundamentals, American Society of Heating, Refrigerating and Air-conditioning Engineers, Atlanta, USA (1997)

25   Walton, G N, 'A New Algorithm for Radiant Interchange in Room Load Calculations', *ASHRAE Trans.*, (2), 190–208 (1980)

26   *ASHRAE Handbook*, HVAC Systems and Equipment, American Society of Heating, Refrigerating and Air-conditioning Engineers, Atlanta, USA (1996)

27   Fanger, P O, *Thermal Comfort Analysis and Applications in Environmental Engineering*, Danish Technical Press, Copenhagen (1970)

28   McIntyre, D A, *Indoor Climate*, Applied Science Publishers Ltd, London (1980)

29   Nishi, Y. and Gagge A P, 'Effective Temperature Scale Useful for Hyperbaric Environments', *Aviation, Space and Environmental Medicine*, **48**, 97–107 (1977)

30   Neilson, M and Pedersen, L, 'Studies on the Heat Loss by Radiation and Convection from the Clothed Human Body', *Gita Psychologica Scandinavia*, **27**, 272–294 (1952)

31   Rapp, G M, 'Convective Heat Transfer and Convective Coefficients of Nude Man, Cylinders and Spheres at Low Air Velocities', *ASHRAE Trans.*, **79**(1), 75–87 (1973)

32   Mitchell, D, 'Convective Heat Loss from Man and Other Animals', in Monteith, IL and Mount, LE, Eds, *Heat Loss from Animals and Man*, Butterworths, London (1974)

33   Winslow, C E A, Herrington, L P and Gagge, A P, 'Physiological Reactions of the Human Body to Varying Environmental Temperatures', *American Journal of Physiology*, **120**(1), 1–22 (1937)

34   Kerslake, D McK, *The Stress of Hot Environments*, Cambridge University Press (1972)

35   ISO7726, 'Thermal Environments—Instruments and Methods of Measuring Basic Physical Quantities', Geneva, 1994.

36   Parsons, K C, *Human Thermal Environments*, Taylor and Francis Ltd, London (1993)

37   BS1387:1967, Steel tubes and tubulars suitable for screwing to BS 21 pipe threads

38   BS3600:1973, Dimensions and masses per unit length of welded and seamless steel pipes and tubes for pressure purposes.

39   BS3600:1997, Specification for dimensions and masses per unit length of welded and seamless steel pipes and tubes for pressure purposes

40   BS2871:1971, Copper and copper alloy tubes (3 parts)

41   Gardner, K A, 'Efficiency of Extended Surface', *Trans. ASME*, **67**(8), 621 (November 1945)

42   Schneider, R J, *Conduction Heat Transfer*, Addison-Wesley (1955)

43   Norris, R H and Spofford, W A, 'High Performance Fins for Heat Transfer', *Trans. ASME*, **64**, 489 (1942)

44   Joyce, T F, 'Optimisation and Design of Fin-Tube Heat Exchangers,' *JIHVE*, **35**, 8 (April 1967)

45   Peach, J, 'Radiators and Other Convectors', *JIHVE*, **39**, 239 (February 1972) and **40**, 85 (July 1972)

46   BS5422:1990, Method for specifying thermal insulating materials on pipes, ductwork and equipment (in the temperature range −40°C to +700°C)

47   BS3958: Thermal insulation materials (6 parts)

48   BS4508: Thermally insulated underground piping systems (3 parts)

49   Mochlinski, K and Gosland, L, 'Field Evidence on Soil Properties Affecting Cable Ratings', *ERA* 70–88, Electrical Research Association (1970)

50   Standards of the Tubular Exchanger Manufacturers Association, 7th edition, TEMA Inc., Tarrytown, New York, USA (1988)

51   Smith, D M, 'Mean Temperature Difference in Cross Flow', *Engineering*, **138**, 479 and 606 (1934)

52   Bowman, R A, Mueller, A C and Nagle, W M, 'Mean Temperature Difference in Design,' *Trans. ASME*, 283, May (1940)

53   Cengel, Y A, *Heat Transfer: A Practical Approach*, WCB McGraw-Hill, USA (1998)

54   Kays, W M and London, A L, *Compact Heat Exchangers*, 3rd edition, McGraw-Hill, New York (1984)

55   Kern, D Q, *Process Heat Transfer*, McGraw-Hill (1950)

56   Holt, J S C, 'Some Aspects of Swimming Pool Design', HVRA Technical Note no. 10 (1962)

57   BS874, Methods for determining thermal insulating properties with definitions of thermal insulating terms, Part 3: Tests for thermal transmittance and conductance

58   BS6993, Thermal and radiometric properties of glazing, Part 1: Method for calculation of the steady state U value (thermal transmittance) (shortly to be superseded by BSEN673), and Part 2: Method of direct measurement of the steady state U value (thermal transmittance)

59   PrEN12412: Windows, doors and shutters; determination of thermal transmittance by hot-box method, Part 1: Windows and doors

60   PrEN10077: Windows and door components; thermal transmittance, Part 1: Simplified calculation method, and Part 2: Numerical calculation method

61   BSENISO6946: Building components and building elements; thermal resistance and thermal transmittance calculation method

62   BSENISO10211: Thermal bridges in building construction; heat flows and surface temperatures, Part 1: General calculation methods

63   Muneer, T and Han, B, 'Multiple Glazed Windows: Design Charts', *Proc. CIBSE A: Building Serv. Eng. Res. Technol.*, **17**(4), 223–229 (1996)

64   Muneer, T, Abodahab, N and Gilchrist, A, 'Combined Conduction, Convection and Radiation Heat Transfer Model for Double-Glazed Windows', *Proc. CIBSE A: Building Serv. Eng. Res. Technol.*, **18**(4), 183–191 (1997)

4   Flow of fluids in pipes and ducts

4.1   Introduction

For the correct selection of pumps and fans, and the sizing of pipes/ducts on a life-cycle cost-effective basis, it is essential that calculated predictions of pressure loss should be reasonably accurate. The range of data provided has been extended to enable this. This Guide is intended for everyday use. Complex as some of these data might appear to be, it is nevertheless often the result of simplifications in an effort to produce guidance which is at the same time both easy to use and of acceptable accuracy.

The major change concerns the section on pressure loss factors which has been considerably extended. Data on laminar flow for bends and tees of pipework have been introduced. An effort has been made to include all the components listed in the latest HVCA specification. Even now, conflicting or missing data have made it difficult to select appropriate values with complete confidence as to their accuracy. In the case of missing information some small measure of advice is included rather than none.

Many symbols have been changed. This has not been done lightly. Wherever possible CIBSE adheres to the British Standards (frequently ISO standards).

A further change concerns the Moody chart and its associated equation. Two versions of the Moody chart have been in circulation over the years, one using a factor $4f$, and another f or λ, even within the UK. In the new presentation we have changed over to λ to be in harmony with British hydraulics engineers and international practice. The new chart has the appropriate version of the D'Arcy equation printed on it to avoid any chance of misuse. Nevertheless, with PCs now so widespread, numerical calculations will be preferred to the more difficult and inaccurate graphical method, so additional information has been included to render the iterative calculations easier and quicker.

Where it has been simple to achieve, equations have been included in addition to tables, to facilitate computational methods and make interpolation easier. However, much work remains to be done on this.

CETIAT[22] is the only source having tested seemingly identical components but obtained from different manufacturers. CETIAT investigated the effect of six different manufacturers, but only for the 90° elbow of 400 mm diameter. The results showed a scatter of approximately ±40 per cent. This was largely due to the results of one manufacturer whose bend gave results very much out of step with all the rest. Nevertheless it highlights the problem of predicting the pressure loss at the design stage.

Mathematical methods unlikely to be used, and properties of fluids have been moved to appendices.

4.2   Notation

c = velocity $(m.s^{-1})$

c_p = specific thermal capacity (at constant pressure) $(kJ.kg^{-1}.K^{-1})$

d = diameter (m, mm)

d_e = equivalent diameter (mm)

d_h = hydraulic diameter = $4 \times$ hydraulic mean radius (mm)

g = gravitational acceleration $(= 9.807 \; (m.s^{-2})$

h = breadth of rectangular duct (perpendicular to the turning plane for bends) (mm)

k = equivalent roughness (mm)

l = length (m)

l_e = equivalent length of pipe, length for a pressure drop equal to velocity pressure p_v (m)

l_{ef} = equivalent length of a pipe to give the same pressure drop as the fitment (m)

p = pressure (Pa)

Δp = pressure difference (Pa)

Δp_b = pressure difference, buoyancy (Pa)

Δp_f = drop in total pressure, caused by friction (Pa)

p_v = velocity pressure $(= \frac{1}{2}\rho c^2)$ (Pa)

q_v = volume flow $(\mathrm{m^3.s^{-1}}$ or $\mathrm{l.s^{-1}})$

q_m = mass flow $(\mathrm{kg.s^{-1}})$

q_c = combined flow at tees

q_b = branch flow at tees

q_s = minor flow in the straight of a tee

r = mean radius of a bend (mm)

r_i = inner radius of a bend (mm)

r_o = outer radius of a bend (mm)

v = specific volume $(\mathrm{m^3.kg^{-1}})$

w = width of rectangular duct (in the turning plane for bends) (mm)

x = distance between blades of louvres (mm)

z = height or head of liquid (m or mm)

A = cross-section area of duct $(\mathrm{m^2})$

A_c = clear area of mesh screen $(\mathrm{m^2})$

C, C_1, C_2, etc. = correction factors

K = capacity (sometimes called 'flow capacity') $(\mathrm{l.h^{-1}.bar^{-1/2}}$ or $\mathrm{m^3.s^{-1}.Pa^{-1/2}})$

P = perimeter (mm)

P_f = fan power (W)

R = Gas constant $(\mathrm{kJ.kg^{-1}\,K^{-1}})$

Re = Reynolds number

T = temperature (absolute) (K)

V = volume $(\mathrm{m^3})$

Z = pressure factor for compressible flow (steam and air) $(\mathrm{kPa^{1.929}})$

a = angle turned (degrees)

θ = included angle of contractions or expansions (degrees)

Δ = difference in two values

λ = friction coefficient

λ_c = friction coefficient for a circular duct

λ_r = friction coefficient for a rectangular duct

η = dynamic viscosity $(\mathrm{kg.m^{-1}.s^{-1}})$

v = kinematic viscosity $(\mathrm{m^2.s^{-1}})$

ρ = density $(\mathrm{kg.m^{-3}})$

ζ = pressure loss factor

4.3 Fluid flow in straight pipes and ducts

4.3.1 General

In sections 4.4 and 4.5, pre-calculated values of friction pressure drop are given for the special cases of water flow and air flow.

This section gives the basic principles on which those values are based. It is these same principles which must be used when predicting the pressure drop resulting from the use of other fluids, or even for air and water at temperatures other than those used for the pre-calculated values.

The D'Arcy equation for pressure loss due to friction may be given as

$$\Delta p = \lambda \frac{l}{d} \tfrac{1}{2}\rho c^2 \qquad (4.1)$$

The friction factor, λ, may be obtained mathematically or from the Moody chart, Figure 4.1, and depends upon the values of Reynolds number, Re and relative roughness, where

$$Re = \frac{\rho c d}{\eta} = \frac{cd}{v} \qquad (4.2)$$

Relative roughness = k/d

Values of roughness k are given in Table 4.1.

Values of ρ, η, and v for some fluids are given in Appendix 4.A1.

4.3.2 Laminar flow

Laminar flow occurs for values of Reynolds number less than 2000. This is most unlikely to occur for water flow or air flow, but is very likely for more viscous fluids such as oil. Rather than use Figure 4.1, the value of λ is more easily obtained from the Poisseuille equation:

For $Re < 2000$

$$\lambda = \frac{64}{Re} \qquad (4.3)$$

With increasing velocity, Re increases and λ is seen to decrease. Nevertheless when substituted into equation (4.1), it will be found that the pressure drop increases with increasing velocity. With laminar flow, the pressure drop is directly proportional to velocity. This type of flow sometimes occurs for air passing through HEPA filters where the air passageways are particularly small, and with liquids of high viscosity.

Surface roughness of the duct or pipe is found to have no effect.

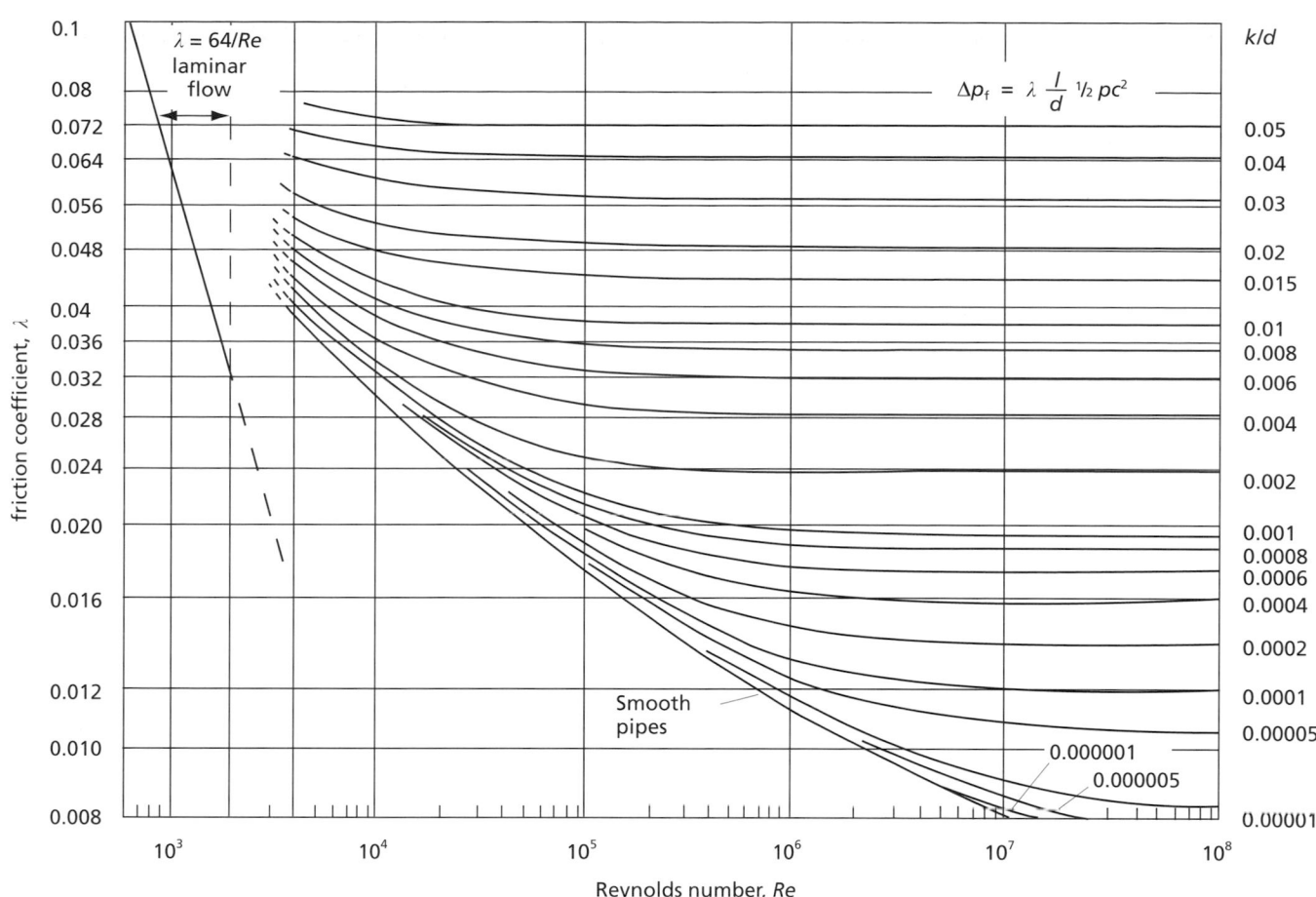

Figure 4.1 The Moody chart: the variation of friction coefficient λ, with Reynolds number and relative roughness

4.3.3 Turbulent flow

This occurs for values of Re greater than 3000. Since air flow in ducts is more likely to have a Reynolds number in the region of 100 000 (10^5), it is clear that air flow is almost invariably turbulent. Water flow is also likely to be turbulent.

It will be seen from Figure 4.1 that the friction coefficient depends upon values of both Reynolds number and relative roughness, k/d. The family of curves on the chart was generated from the following equation, developed by Colbrook–White, which may be used directly instead of using the chart:

$$\frac{1}{\sqrt{\lambda}} = -2 \log \left(\frac{2.51}{Re \sqrt{\lambda}} + \frac{k/d}{3.7} \right) \qquad (4.4)$$

(Note that the square of the above equation might appear more elegant, but the essential negative sign would thereby be lost. It is $\sqrt{\lambda}$ which is needed for the iteration.)

The Moody chart was constructed using this equation.

Several texts give abbreviated forms of equation (4.4) for particularly smooth pipes and for high values of Re. We consider it safer not to risk making false assumptions with the consequential risk of using an inappropriate equation. It is better to persevere with equation (4.4).

Although the solution is iterative, with practice it will be found to be almost as quick as using the Moody chart, Figure 4.1, and to be far more accurate. The worked solution below illustrates this. Computation is much reduced if the first estimate is somewhere in the right region. For such a first estimate we have the equation of Altshul to help us:

$$\lambda = 0.11 \left(\frac{k}{d} + \frac{68}{Re} \right)^{0.25} \qquad (4.5)$$

The Colbrook–White equation, (4.4), gives values of λ which are some 2 to 4 per cent greater than others and so can be considered to include a small margin of safety[2].

Example 4.1

Calculation of λ by iteration for

$$k/d = 0.0006 \quad \text{and} \quad Re = 10^5$$

A first estimate is obtained using Altshul's equation, (4.5)

$$\lambda = 0.11 \left(0.0006 + \frac{68}{10^5} \right)^{0.25}$$

First estimate: $\lambda = 0.02081$

$$\frac{1}{\sqrt{\lambda}} = 6.932$$

Table 4.1 Values of equivalent roughness k for various pipes and duct materials

Type of material	Condition	Roughness, k (mm)	Reference
Seamless: copper, brass, lead	Commercially smoothed	0.0015–0.0100	2
Cast iron	New	0.25–1.00	2
	Corroded	1.00–1.25	2
	With deposits	2.0–4.0	2
	Heavily corroded	up to 3.0	2
Steel, seamless	New	0.02–0.10	2
	Old but cleaned	0.04	2
	Moderately corroded	0.4	2
	Water pipelines, used	1.2–1.5	2
	Encrusted	0.5–2.0	11
	Poor condition	>5.0	2
Welded steel tubes	New	0.04–1.0	2
	With deposits	1.5	2
	Appreciable deposits	2.0–4.0	2
	Poor condition	>5.0	2
Galvanised steel tube	Bright, new	0.07–0.10	2
	Ordinary	0.10–0.15	2
	Longitudinal seams	0.05–0.10	4
	Spiral seams	0.06–0.12	4
Galvanised steel sheet	New	0.15	2
Coated steel	Glass enamel	0.001–0.01	2
	Asphalt	0.12–0.30	2
Glass		0.0015–0.010 (mm)	2
Brick	Fair-faced brickwork	1.3	
Plaster	New	0.05–0.15	2
Concrete	Tubes, new	0.25–0.34	2
	Carefully smoothed	0.50	2
	Brushed, air placed	2.30	2
	Non-smoothed, air placed	3.0–6.0	2
Polymers	PVC	0.01–0.05	4
	Polybutylene	0.0015–0.010	6
	Polyethylene (PE-X)	0.0015–0.010	6
Aluminium		0.05	4
Flexible duct	Fully extended	1.0–4.6	4
Fibrous glass duct	Spray coated	4.5	4
Rock tunnels	Blast-hewed, little jointing	100–140	2
	Roughly cut, highly uneven surface	500–1500	2

This can be used as a starting value in the right-hand side of equation (4.4):

$$\frac{1}{\sqrt{\lambda}} = -2 \log \left(\frac{2.51}{Re \sqrt{\lambda}} + \frac{k/d}{3.7} \right)$$

$$\frac{1}{\sqrt{\lambda}} = -2 \log \left(\frac{2.51 \times 6.932}{10^5} + \frac{0.0006}{3.7} \right)$$

Thus the second estimate: $1/\sqrt{\lambda} = 6.947$

Inserting this new value into the right-hand side of equation (4.4) gives:

$$\frac{1}{\sqrt{\lambda}} = -2 \log \left(\frac{2.51 \times 6.947}{10^5} + \frac{0.0006}{3.7} \right)$$

Thus the third estimate: $1/\sqrt{\lambda} = 6.946$

$$\lambda = 0.2073$$

It should be noted that even the first estimate was more accurate than could be read from the chart.

4.3.4 Unpredictable flow

In the region $2000 < Re < 3000$ the flow may be laminar or turbulent depending upon upstream conditions. The nature of the flow may even be unstable and oscillate between laminar and turbulent. Applying caution in pressure drop estimates, it would appear prudent to base calculations on turbulent flow in this region.

4.3.5 Non-circular ducts

In section 4.3.1 the basic equation (4.1), needs to be rewritten in terms of hydraulic mean diameter instead of diameter, where hydraulic mean diameter d_h is given by:

$$d_h = \frac{4A}{P} \tag{4.6}$$

where A is the cross-sectional area and P is the perimeter of the duct. (For a circular duct the hydraulic diameter d_h is equal to the real diameter d.)

$$\Delta p = \frac{\lambda l \rho c^2}{2 d_h} \tag{4.7}$$

Table 4.2 Internal diameters of metal pipes (manufacturing tolerances give variations of only ∓0.25–0.37%)

Nominal pipe size/mm	Mean internal diameter/mm					Nominal pipe size/mm	Mean internal diameter (mm)		
	Mild steel				Cast iron		Copper		
	BS1387			BS3600*	BS1211		BS2871: Part 1		
	Medium black	Heavy black	Heavy galvanised		Class C		Table X	Table Y	Table Z
10	12.4	11.3	10.8	—	—	6	4.80	4.40	5.00
15	16.1	14.9	14.4	—	—	8	6.80	6.40	7.00
20	21.6	20.4	19.9	—	—	10	8.80	8.40	9.00
25	27.3	25.7	25.2	—	—	12	10.80	10.40	11.00
32	36.0	34.4	33.9	—	—	15	13.60	13.00	14.00
40	41.9	40.3	39.8	—	—	22	20.22	19.62	20.82
50	53.0	51.3	50.8	—	—	28	26.22	25.62	26.82
65	68.7	67.0	66.5	—	—	35	32.63	32.03	33.63
80	80.7	79.1	78.6	—	80.8	42	39.63	39.03	40.43
90	93.15	91.55	91.05	—	—	54	51.63	50.03	52.23
100	105.1	103.5	102.8	—	116.3	67	64.27	63.07	64.67
125	129.95	128.85	128.35	—	130.8	75	73.22	72.22	73.82
150	155.4	154.3	153.8	—	159.2	110	105.12	103.12	105.72
175	—	—	—	183.0	184.1	133	130.38	—	130.38
200	—	—	—	207.3	210.2	160	155.38	—	156.38
225	—	—	—	232.7	236.1				
250	—	—	—	260.4	262.1				
300	—	—	—	311.3	328.6				
350	—	—	—	341.4	—				
400	—	—	—	392.2	—				
450	—	—	—	441.0	—				
500	—	—	—	492.0	—				
550	—	—	—	541.4	—				
600	—	—	—	592.4	—				

*BS3600 quotes such a wide range of possible sizes for large steel pipes that the values given should only be regarded as typical.

Table 4.3 Internal diameters of thermoplastic pipes (note that the nominal diameters correspond with the external diameters for BS sizes but this appears not to be the case for DIN sizes)

UPVC			PB and PE-X		PB and PE-X		PVC-C	
(BS3505) Nominal dia. (mm)	Class C Int dia. (mm)	Class E Int dia. (mm)	BS7291 / 2871 Nominal dia. (mm)	Int dia. (mm)	BS7291 / 5556 Nominal dia. (mm)	Int dia. (mm)	BS7291 Nominal dia. (mm)	Int dia. (mm)
10		13.4	10	6.7	10	6.8		
15		17.2	12	8.7	12	8.8	12	8.57
20		21.8	15	11.3	16	12.3	16	12.07
25		28.2	18	14.2	20	16.0	20	15.87
32		35.9	22	17.7	25	20.2	25	19.97
40		40.9	28	22.5	32	26	32	25.67
50	54.4	51.4	35	28.3			40	32.30
65	68.2	64.2					50	40.40
80	80.7	75.9					63	51.00
100	103.9	97.7						
							DIN16968	
125	127.6	120.0					65	61.4
150	153.2	144.0					80	73.6
175	176.3	165.9					100	90.0
200	201.5	190.9						
225	224.8	212.8						
250	251.2	238.0						
300	298.0	282.2						

Manufacturing tolerances are ±4.5% on the small sizes and ±1.5% on the larger.

4.3.6 Components and fittings: pressure loss factor, ζ

To obtain the extra pressure loss due to the installation of any fitting, data are generally presented in terms of a pressure loss factor ζ. The data obtained experimentally are complex but a simplified collection of the data is available in section 4.8. Whether for liquids in pipes, or gases in ducts, the same fundamental equation applies:

$$\Delta p = \zeta \tfrac{1}{2}\rho c^2 \qquad (4.8)$$

In particular it should be noted that where velocity changes occur due to either changes in section or flow splitting in tees, there may be instances where the static pressure increases despite a loss in total pressure due to friction. Δp is always the 'drop in *total pressure*'.

$$\Delta p = (p_1 + \tfrac{1}{2}\rho c_1^2) - (p_2 + \tfrac{1}{2}\rho c_2^2) \qquad (4.9)$$

4.3.7 Components and fittings: capacity K

Most valve manufacturers and damper manufacturers quote the performance of their components in terms of capacity K defined in the following relationship:

$$q_v = K\sqrt{\Delta p} \qquad (4.10)$$

This implies that K has units, usually of $m^3.h^{-1}.bar^{-1/2}$ for liquids, or $m^3.s^{-1}.Pa^{-1/2}$ for gases. Some manufacturers may quote values of K with different units so care is needed.

There is a relation between K and ζ, but it is not really necessary to convert one to the other. Pressure drops are more simply calculated separately for those components for which K is given.

K is also useful when dealing with the authority of a valve, and in the prediction of flows in complex circuits. Further information can be obtained from Appendix 4.A5.

4.3.8 Head and head loss

An alternative method of presenting pressure and pressure loss of liquids (but not of gases) is in terms of an imaginary column of the liquid with standard atmospheric pressure acting on the free surface. Technically this height or 'head' would also depend on the temperature of the liquid.

The equivalent head and head loss are then given by:

$$z = \frac{p}{\rho g}$$

$$\Delta z = \frac{\Delta p}{\rho g} \qquad (4.11)$$

It should be noted that several pump manufacturers quote 'pump head' or 'delivery pressure' when meaning the increased head or pressure of the outlet compared with the inlet.

4.3.9 Buoyancy

Natural circulation may occur whenever there are density differences in a circuit and a vertical height. Fan or pump pressures will usually render the buoyancy effect negligible. However, in the absence of a pump or fan, natural circulation will take place. This is sometimes called a 'thermosyphon'.

The pumping pressure difference due to buoyancy is given by the following formula where the densities ρ_c and ρ_h of the cold and hot parts of the fluid are the average of the downward and upward flowing parts of the circuit respectively. Note that it is only the vertical height, z, which is significant, not the length of pipe:

$$\Delta p = gz(\rho_c - \rho_h) \qquad (4.12)$$

4.3.10 Pressure measurements

With air flow, where pressure measurements are small, it has been common practice to use manometers and to quote the pressure in a height of the manometer fluid. Care

should be taken in the case of liquid flow where the density of the manometer fluid, ρ_m, may be of the same order of magnitude as that of the flowing fluid, ρ. The equation to be used is (4.13).

$$\Delta p = gz(\rho_m - \rho) \qquad (4.13)$$

4.4 Water flow in pipes

4.4.1 Pipe sizing: desirable velocities

The are no rules concerning pipe sizing. The most cost effective will be the design based on life-cycle costing including the pumping costs. The smaller the pipework, the greater the pumping power and energy consumption. Increasing the pipe diameter by one size can have a large effect in decreasing pumping power: smaller friction pressure drops of the basic circuit will require smaller pressure drops through control valves, for the same value of valve authority. The optimum sizing from the point of view of life-cycle costing must consider the length of the system, the capital cost, the mean pressure drop, the running time at full and partial flow, the efficiency of the pump–motor combination, and anticipated electrical tariffs (on-peak, off-peak operation).

To give a starting point in selecting pipe sizes, rule of thumb water velocities are reproduced from BSRIA[12] in Table 4.4. An alternative starting point might be to consider a typical pressure drop per unit length of 360 $Pa.m^{-1[12]}$ or 250 $Pa.m^{-1[4]}$ but this is arbitrary.

4.4.2 Noise

With small pipes, excessive velocities can lead to noise generation where, with hot water, cavitation may occur at elbows, valves, pumps and especially orifice plates. Ball and Webster[19] and Rogers[18] give some information on this. In this respect larger pipes should be able to tolerate higher velocities without a noise problem. Noise problems are all the more likely to occur if entrained air is not separated and vented. Arrangements should be made so that this is achieved easily. An upstand and air vent at the top of each vertical run of pipe is recommended; during pump-off periods, entrained air will separate out into the higher position. This will simultaneously reduce corrosion by eliminating oxygen as soon as possible.

4.4.3 Allowances for ageing

Corrosion and scaling of the internal diameter of pipework will occur with age depending on the chemical composition of the water. This will increase the surface

Table 4.4 Typical water velocities for pipework[12]

Situation	Velocity/m.s^{-1}	Total pressure drop/kPa
Small bore	< 1.0	
$d = 15 - 50$ mm	0.75–1.15	
$d > 50$ mm	1.25–3.0	
In heating and cooling coils	0.5–1.5	28

roughness of the pipe and decrease the internal diameter, both of which will increase the friction pressure drop. No firm recommendation can be made on the allowance to be made. A large allowance is more justifiable with small diameter pipes. Open systems will suffer more than closed systems. ASHRAE[4] reports the Carrier recommendation that for open systems, the friction factor should be 1.75 times higher than for closed systems and reports also the work of the Plastic Pipe Institute (1971) that there is little corrosion problem with plastic pipe.

4.4.4 Water hammer

Large pressures can arise when the fluid flow is stopped abruptly by the sudden closure of a valve. This pressure wave then reverberates within the pipework. The magnitude of the pressure wave is in proportion to the momentum of the flowing fluid and thus to its velocity.

4.4.5 Expansion

Between a heating system being cold (usually under the 'fill' situation), and warm under the design running condition, the water contained in the system will expand. The expansion, as a percentage, has been calculated with reference to a cold situation of 4°C using:

$$\frac{\Delta V}{V_4} = \left(\frac{\rho_4}{\rho} - 1 \right) \qquad (4.14)$$

The volumetric expansion of the pipework may be deducted from the volumetric expansion of the water, if desired.

4.4.6 Pre-calculated pressure drops for flow in straight pipes

In this section many pre-calculated values of pressure drop are given for water. In the previous edition of the Guide they were given at water temperatures of 10°C and 75°C. In the interests of boiler efficiency which, especially with condensing boilers, improves with lower water temperatures, designers should not choose 75°C merely for the convenience of using the pre-calculated tables of this Guide.

4.4.7 Components and fittings

Whether for liquids in pipes, or gases in ducts, the same fundamental equation applies:

$$\Delta p_f = \zeta \tfrac{1}{2} \rho c^2 \qquad (4.15)$$

Data for ζ are given in section 4.8.

Values of $\tfrac{1}{2} \rho c^2$ are given in Table 4.5, for water at 10°C. Values of mean water velocity c, should not be taken from tables of pre-calculated pressure drop (Tables 4.9 to 4.33) due to the unreliability of non-linear interpolation. The values of velocity pressure in Table 4.5 may be corrected for different temperatures by dividing by the density of water at 10°C (999.7 kg.m^{-3}) and multiplying by the density at the temperature being considered.

Since the additional pressure drop caused by a fitting is largely due to the internal friction of the fluid suffering an abrupt change of direction, rusting and scaling will not have a large effect on pressure drop. No allowance for ageing is therefore needed, for tees, though it is advisable for elbows, Koch[26].

An alternative method of calculating the pressure drop for components, unique to the CIBSE Guide, is to determine:

ζ from section 4.8.

$\Delta p/l$ for a straight pipe of the same dimension from the pre-calculated tables

l_e the 'equivalent length for $\zeta = 1$', from the tables

from which:

$$\Delta p = \zeta l_e \frac{\Delta p}{l} \qquad (4.16)$$

This 'equivalent length' is the length of pipe which will produce a drop in pressure equal to one velocity pressure ($\rho c^2/2$).

It will be noted that some of the tables, particularly for small pipes, have an asterisk and a dagger in the mass flow rate column. The asterisk signifies the condition when $Re = 2000$, and for flows lower than this the flow is laminar. In the unpredictable zone, $2000 < Re < 3000$, the pressure drop has been calculated on the basis of turbulent flow as this gives the higher, safer, figure.

4.4.8 Non-pre-calculated values

When conditions arise outside the range of pre-calculated values of Tables 4.9–4.33, or when other tube materials or non-BS pipe sizes are being used, it will be necessary to calculate the pressure loss using the method outlined in section 4.3. Useful property values for water are given in Table 4.6 and in Appendix 4.A1.

Table 4.5 Velocity pressures ($\tfrac{1}{2} \rho c^2$) for water at 10°C

c/m.s^{-1}	p_v/Pa	c/m.s^{-1}	p_v/Pa	c/m.s^{-1}	p_v/Pa	c/m.s^{-1}	p_v/Pa	c/m.s^{-1}	p_v/Pa
0.01	0.049 99	0.10	4.999	0.55	151.209	1.1	604.837	2	1999.5
0.02	0.199 95	0.15	11.247	0.6	179.951	1.2	719.806	2.5	3124.2
0.03	0.449 88	0.20	19.995	0.65	211.193	1.3	844.772	3	4498.8
0.04	0.799 78	0.25	31.242	0.7	244.934	1.4	979.735	3.5	6123.3
0.05	1.249 66	0.30	44.988	0.75	281.174	1.5	1124.69	4	7997.8
0.06	1.799 51	0.35	61.233	0.8	319.914	1.6	1279.65	4.5	10 122.3
0.07	2.449 34	0.40	79.978	0.85	361.152	1.7	1444.61	5	12 496.6
0.08	3.199 14	0.45	101.223	0.9	404.891	1.8	1619.56	5.5	15 120.9
0.09	4.048 91	0.50	124.966	0.95	451.128	1.9	1804.51	6	17 995.1
				1	499.865	2	1999.46	6.5	21 119.3

Table 4.6 Properties of water: density, dynamic and kinematic viscosity

T /°C	ρ /kg.m^{-3}	η /10^{-6} kg.m^{-1}.s^{-1}	v /10^{-6} m^2.s^{-1}	T /°C	ρ /kg.m^{-3}	η /10^{-6} kg.m^{-1}.s^{-1}	v /10^{-6} m^2.s^{-1}
0.001	999.8	1752	1.7524				
4	1000.0	1551	1.5510	100	958.4	279	0.2911
10	999.7	1300	1.3004	110	950.6	252	0.2651
20	999.8	1002	1.0022	120	943.4	230	0.2438
30	995.6	797	0.8005	130	934.6	211	0.2258
40	992.2	651	0.6561	140	925.9	195	0.2106
50	988.0	544	0.5506	150	916.6	181	0.1975
60	983.2	463	0.4709	160	907.4	169	0.1862
70	977.8	400	0.4091	170	897.7	158	0.1760
80	971.8	351	0.3612	180	886.5	149	0.1681
90	965.3	311	0.3222	190	875.6	141	0.1610
100	958.4	279	0.2911	200	864.3	134	0.1550

Table 4.7 The thermosyphon driving pressure for a gravity hot water system Δp_b using equation (4.12)

Flow temperature/°C	Circulating pressure/(Pa) per metre height for the following temperature differences (flow-return)/°C										
	2	4	6	8	10	12	14	16	18	20	22
40	7.37	14.4	21.2	27.6	33.7	39.5	44.9	49.9	54.6	58.8	62.7
45	8.07	15.9	23.4	30.6	37.5	44.1	50.4	56.3	61.9	67.1	72.0
50	8.73	17.2	25.4	33.3	41.0	48.4	55.4	62.2	68.6	74.7	80.5
55	9.36	18.5	27.3	35.9	44.3	52.3	60.2	67.7	74.9	81.8	88.4
60	9.95	19.7	29.1	38.4	47.4	56.1	64.6	72.8	80.7	88.4	95.7
65	10.5	20.8	30.9	40.7	50.3	59.7	68.8	77.6	86.2	94.6	103
70	11.1	21.9	32.6	43.0	53.1	63.1	72.8	82.3	91.5	101	109
75	11.6	23.0	34.2	45.1	55.9	66.4	76.7	86.8	96.6	106	116
80	12.1	24.0	35.8	47.3	58.5	69.6	80.5	91.1	102	112	122
85	12.6	25.1	37.3	49.3	61.1	72.7	84.1	95.3	106	117	128
90	13.1	26.0	38.7	51.3	63.6	75.7	87.6	99.3	111	122	133
95	13.5	26.9	40.1	53.1	65.9	78.5	91.0	103	115	127	139

Table 4.8 Expansion of water at different temperatures, relative to the volume at 4°C, in percentages

T/°C	40	50	60	70	80	90	100	110	120
%	0.786	1.21	1.71	2.27	2.90	3.63	4.34	5.20	6.00

T/°C	130	140	150	160	170	180	190	200	
%	7.00	8.00	9.10	10.2	11.4	12.8	14.2	15.7	

Table 4.9 Flow of water at 75°C in copper pipes

q_m = mass flow rate   kg.s^{-1}
c = velocity   m.s^{-1}
$\Delta p/l$ = pressure drop per unit length   Pa.m^{-1}
l_e = equivalent length of a component
       for $\zeta = 1$ m

† $(Re) = 3000$

MINIBORE COPPER

WATER AT 75°C

COPPER, TABLE X

$\Delta p/l$	6 mm q_m	l_e	8 mm q_m	l_e	10 mm q_m	l_e	c
50			0.004	0.1	0.009	0.2	
55			0.004	0.1	0.009	0.2	
60			0.005	0.1	0.010	0.2	
65			0.005	0.1	0.010	0.2	
70			0.005	0.1	0.011	0.2	
75			0.005	0.1	0.011	0.2	
80			0.006	0.2	0.011	0.2	
85			0.006	0.2	0.012	0.2	0.2
90			0.006†	0.2	0.012	0.2	
95			0.006	0.2	0.013	0.2	
100			0.006	0.2	0.013	0.2	
110			0.007	0.2	0.014	0.2	
120			0.007	0.2	0.014	0.2	
130			0.007	0.2	0.015	0.2	
140			0.008	0.2	0.016	0.2	
150			0.008	0.2	0.016	0.2	
160			0.008	0.2	0.017	0.3	0.3
170			0.009	0.2	0.018	0.3	
180			0.009	0.2	0.018	0.3	
190	0.004	0.1	0.009	0.2	0.019	0.3	
200	0.004	0.1	0.010	0.2	0.019	0.3	
225	0.004	0.1	0.010	0.2	0.021	0.3	
250	0.004†	0.1	0.011	0.2	0.022	0.3	
275	0.004	0.1	0.012	0.2	0.023	0.3	0.4
300	0.005	0.1	0.012	0.2	0.025	0.3	
325	0.005	0.1	0.013	0.2	0.026	0.3	
350	0.005	0.1	0.013	0.2	0.027	0.3	
375	0.005	0.1	0.014	0.2	0.028	0.3	
400	0.005	0.1	0.014	0.2	0.029	0.3	
425	0.006	0.1	0.015	0.2	0.030	0.3	
450	0.006	0.1	0.015	0.2	0.031	0.3	
475	0.006	0.1	0.016	0.2	0.032	0.3	
500	0.006	0.1	0.016	0.2	0.033	0.3	
550	0.007	0.1	0.017	0.2	0.035	0.3	0.6
600	0.007	0.1	0.018	0.2	0.036	0.3	
650	0.007	0.1	0.019	0.2	0.038	0.3	
700	0.008	0.1	0.020	0.2	0.040	0.3	
750	0.008	0.1	0.021	0.2	0.041	0.3	
800	0.008	0.1	0.021	0.2	0.043	0.3	
850	0.009	0.1	0.022	0.2	0.044	0.3	
900	0.009	0.1	0.023	0.2	0.046	0.3	
950	0.009	0.1	0.024	0.2	0.047	0.3	
1000	0.009	0.1	0.024	0.2	0.049	0.3	
1100	0.010	0.1	0.026	0.2	0.051	0.3	
1200	0.010	0.1	0.027	0.2	0.054	0.3	
1300	0.011	0.1	0.028	0.2	0.057	0.3	
1400	0.011	0.1	0.029	0.2	0.059	0.3	1.0
1500	0.012	0.1	0.031	0.2	0.061	0.3	
1600	0.012	0.1	0.032	0.2	0.064	0.4	
1700	0.013	0.1	0.033	0.2	0.066	0.4	
1800	0.013	0.1	0.034	0.2	0.068	0.4	
1900	0.014	0.2	0.035	0.2	0.070	0.4	
2000	0.014	0.2	0.036	0.3	0.072	0.4	
2250	0.015	0.2	0.038	0.3	0.077	0.4	
2500	0.016	0.2	0.041	0.3	0.082	0.4	
2750	0.017	0.2	0.043	0.3	0.086	0.4	1.5
3000	0.018	0.2	0.045	0.3	0.091	0.4	
3250	0.018	0.2	0.047	0.3	0.095	0.4	
3500	0.019	0.2	0.049	0.3	0.099	0.4	
3750	0.020	0.2	0.051	0.3	0.103	0.4	
4000	0.021	0.2	0.053	0.3	0.106	0.4	
4250	0.021	0.2	0.055	0.3	0.110	0.4	
4500	0.022	0.2	0.057	0.3	0.114	0.4	
4750	0.023	0.2	0.059	0.3	0.117	0.4	2.0
5000	0.023	0.2	0.060	0.3	0.121	0.4	

COPPER, TABLE Y

$\Delta p/l$	6 mm q_m	l_e	8 mm q_m	l_e	10 mm q_m	l_e	c
50			0.004	0.1	0.008	0.2	
55			0.004	0.1	0.008	0.2	
60			0.004	0.1	0.008	0.2	
65			0.004	0.1	0.009	0.2	
70			0.004	0.1	0.009	0.2	
75			0.005	0.1	0.010	0.2	
80			0.005	0.1	0.010	0.2	
85			0.005	0.1	0.010	0.2	
90			0.005	0.1	0.011	0.2	0.2
95			0.005	0.1	0.011	0.2	
100			0.005†	0.1	0.011	0.2	
110			0.006	0.1	0.012	0.2	
120			0.006	0.1	0.013	0.2	
130			0.006	0.2	0.013	0.2	
140			0.007	0.2	0.014	0.2	
150			0.007	0.2	0.014	0.2	
160			0.007	0.2	0.015	0.2	
170			0.007	0.2	0.016	0.2	0.3
180			0.008	0.2	0.016	0.2	
190			0.008	0.2	0.017	0.2	
200			0.008	0.2	0.017	0.2	
225			0.009	0.2	0.018	0.2	
250			0.009	0.2	0.019	0.3	
275			0.010	0.2	0.021	0.3	0.4
300	0.004	0.1	0.010	0.2	0.022	0.3	
325	0.004†	0.1	0.011	0.2	0.023	0.3	
350	0.004	0.1	0.011	0.2	0.024	0.3	
375	0.004	0.1	0.012	0.2	0.025	0.3	
400	0.004	0.1	0.012	0.2	0.025	0.3	
425	0.004	0.1	0.013	0.2	0.026	0.3	
450	0.005	0.1	0.013	0.2	0.027	0.3	
475	0.005	0.1	0.013	0.2	0.028	0.3	
500	0.005	0.1	0.014	0.2	0.029	0.3	
550	0.005	0.1	0.015	0.2	0.031	0.3	
600	0.005	0.1	0.015	0.2	0.032	0.3	0.6
650	0.006	0.1	0.016	0.2	0.034	0.3	
700	0.006	0.1	0.017	0.2	0.035	0.3	
750	0.006	0.1	0.017	0.2	0.036	0.3	
800	0.006	0.1	0.018	0.2	0.038	0.3	
850	0.007	0.1	0.019	0.2	0.039	0.3	
900	0.007	0.1	0.019	0.2	0.040	0.3	
950	0.007	0.1	0.020	0.2	0.042	0.3	
1000	0.007	0.1	0.021	0.2	0.043	0.3	
1100	0.008	0.1	0.022	0.2	0.045	0.3	
1200	0.008	0.1	0.023	0.2	0.048	0.3	
1300	0.009	0.1	0.024	0.2	0.050	0.3	
1400	0.009	0.1	0.025	0.2	0.052	0.3	
1500	0.009	0.1	0.026	0.2	0.054	0.3	1.0
1600	0.010	0.1	0.027	0.2	0.056	0.3	
1700	0.010	0.1	0.028	0.2	0.058	0.3	
1800	0.010	0.1	0.029	0.2	0.060	0.3	
1900	0.011	0.1	0.030	0.2	0.062	0.3	
2000	0.011	0.1	0.030	0.2	0.064	0.3	
2250	0.012	0.1	0.033	0.2	0.068	0.3	
2500	0.012	0.1	0.035	0.2	0.072	0.3	
2750	0.013	0.1	0.037	0.2	0.076	0.4	
3000	0.014	0.1	0.038	0.2	0.080	0.4	1.5
3250	0.015	0.1	0.040	0.2	0.084	0.4	
3500	0.015	0.1	0.042	0.2	0.087	0.4	
3750	0.016	0.1	0.043	0.2	0.091	0.4	
4000	0.016	0.1	0.045	0.3	0.094	0.4	
4250	0.017	0.1	0.047	0.3	0.097	0.4	
4500	0.017	0.2	0.048	0.3	0.100	0.4	
4750	0.018	0.2	0.050	0.3	0.103	0.4	
5000	0.019	0.2	0.051	0.3	0.106	0.4	

Table 4.10 Flow of water at 75°C in heavy grade steel pipes

q_m	= mass flow rate	kg.s^{-1}
c	= velocity	m.s^{-1}
$\Delta p/l$	= pressure drop per unit length	Pa.m^{-1}
l_e	= equivalent length of a component for $\zeta = 1$	m

* $(Re) = 2000$
† $(Re) = 3000$

> **HEAVY GRADE STEEL**
> **WATER AT 75°C**

$\Delta p/l$	c	10 mm q_m	l_e	15 mm q_m	l_e	20 mm q_m	l_e	25 mm q_m	l_e	32 mm q_m	l_e	40 mm q_m	l_e	50 mm q_m	l_e	c	$\Delta p/l$
0.1								0.003	0.1	0.009	0.5	0.017*	0.9	0.031	1.2		0.1
0.2								0.006	0.3	0.018*	0.9	0.024	0.9	0.044†	1.2		0.2
0.3						0.003	0.2	0.008	0.4	0.020	0.8	0.029	0.9	0.055	1.2		0.3
0.4						0.004	0.2	0.011	0.6	0.023	0.8	0.034†	0.9	0.065	1.3		0.4
0.5						0.005	0.3	0.014*	0.7	0.025	0.8	0.038	0.9	0.074	1.3		0.5
0.6						0.007	0.3	0.013	0.6	0.028	0.8	0.042	1.0	0.082	1.4		0.6
0.7						0.008	0.4	0.014	0.6	0.030†	0.8	0.046	1.0	0.090	1.4		0.7
0.8						0.009	0.5	0.015	0.6	0.032	0.8	0.050	1.0	0.097	1.4	0.05	0.8
0.9						0.010	0.5	0.016	0.6	0.035	0.8	0.054	1.0	0.104	1.5		0.9
1.0				0.003	0.2	0.011*	0.6	0.017	0.6	0.037	0.8	0.057	1.0	0.110	1.5		1.0
1.5				0.005	0.2	0.012	0.4	0.021†	0.6	0.047	0.9	0.072	1.1	0.139	1.6		1.5
2.0				0.006	0.3	0.014	0.4	0.025	0.6	0.055	0.9	0.085	1.1	0.164	1.6		2.0
2.5				0.008	0.4	0.015	0.4	0.028	0.6	0.062	0.9	0.096	1.2	0.186	1.7		2.5
3.0				0.009*	0.5	0.017	0.4	0.031	0.6	0.069	1.0	0.107	1.2	0.206	1.7		3.0
3.5				0.008	0.3	0.018†	0.4	0.034	0.6	0.076	1.0	0.116	1.2	0.224	1.7		3.5
4.0		0.004	0.2	0.009	0.3	0.020	0.4	0.037	0.7	0.082	1.0	0.126	1.3	0.242	1.8		4.0
4.5	0.05	0.005	0.2	0.009	0.3	0.021	0.5	0.039	0.7	0.087	1.0	0.134	1.3	0.258	1.8		4.5
5.0		0.005	0.3	0.010	0.3	0.022	0.5	0.042	0.7	0.093	1.0	0.142	1.3	0.274	1.8		5.0
5.5		0.006	0.3	0.010	0.3	0.023	0.5	0.044	0.7	0.098	1.0	0.150	1.3	0.289	1.8	0.15	5.5
6.0		0.006	0.3	0.011	0.3	0.025	0.5	0.046	0.7	0.103	1.1	0.158	1.3	0.303	1.9		6.0
6.5		0.007*	0.4	0.011	0.3	0.026	0.5	0.048	0.7	0.107	1.1	0.165	1.3	0.317	1.9		6.5
7.0		0.006	0.2	0.012	0.3	0.027	0.6	0.050	0.7	0.112	1.1	0.172	1.3	0.330	1.9		7.0
7.5		0.006	0.2	0.012	0.3	0.028	0.5	0.052	0.7	0.116	1.1	0.179	1.4	0.343	1.9		7.5
8.0		0.006	0.2	0.012	0.3	0.029	0.5	0.054	0.7	0.120	1.1	0.185	1.4	0.355	1.9		8.0
8.5		0.006	0.2	0.013	0.3	0.030	0.5	0.056	0.7	0.125	1.1	0.191	1.4	0.368	1.9		8.5
9.0		0.007	0.2	0.013	0.3	0.031	0.5	0.058	0.7	0.129	1.1	0.198	1.4	0.379	1.9		9.0
9.5		0.007	0.2	0.014	0.3	0.032	0.5	0.060	0.7	0.133	1.1	0.204	1.4	0.391	1.9		9.5
10.0		0.007	0.2	0.014	0.3	0.033	0.5	0.062	0.7	0.136	1.1	0.210	1.4	0.402	2.0		10.0
12.5		0.008	0.2	0.016	0.3	0.037	0.5	0.070	0.8	0.154	1.1	0.237	1.4	0.454	2.0		12.5
15.0		0.008	0.2	0.018	0.4	0.042	0.6	0.077	0.8	0.171	1.2	0.262	1.5	0.502	2.0		15.0
17.5		0.009	0.2	0.019	0.4	0.045	0.6	0.084	0.8	0.186	1.2	0.285	1.5	0.546	2.0		17.5
20.0		0.010	0.2	0.021	0.4	0.049	0.6	0.091	0.8	0.200	1.2	0.307	1.5	0.587	2.1	0.30	20.0
22.5		0.010†	0.2	0.022	0.4	0.052	0.6	0.097	0.8	0.214	1.2	0.327	1.5	0.626	2.1		22.5
25.0		0.011	0.3	0.023	0.4	0.055	0.6	0.103	0.8	0.226	1.2	0.347	1.5	0.663	2.1		25.0
27.5		0.012	0.3	0.025	0.4	0.058	0.6	0.108	0.8	0.238	1.2	0.365	1.5	0.698	2.1		27.5
30.0		0.012	0.3	0.026	0.4	0.061	0.6	0.114	0.8	0.250	1.2	0.383	1.6	0.731	2.2		30.0
32.5		0.013	0.3	0.027	0.4	0.064	0.6	0.119	0.8	0.261	1.3	0.400	1.6	0.763	2.2		32.5
35.0		0.013	0.3	0.028	0.4	0.067	0.6	0.124	0.8	0.272	1.3	0.416	1.6	0.794	2.2		35.0
37.5		0.014	0.3	0.029	0.4	0.069	0.6	0.129	0.9	0.282	1.3	0.432	1.6	0.824	2.2		37.5
40.0		0.014	0.3	0.031	0.4	0.072	0.6	0.133	0.9	0.292	1.3	0.447	1.6	0.853	2.2		40.0
42.5	0.15	0.015	0.3	0.032	0.4	0.074	0.6	0.138	0.9	0.302	1.3	0.462	1.6	0.882	2.2		42.5
45.0		0.015	0.3	0.033	0.4	0.077	0.6	0.142	0.9	0.312	1.3	0.477	1.6	0.909	2.2		45.0
47.5		0.016	0.3	0.034	0.4	0.079	0.6	0.146	0.9	0.321	1.3	0.491	1.6	0.936	2.2		47.5
50.0		0.016	0.3	0.035	0.4	0.081	0.6	0.150	0.9	0.330	1.3	0.504	1.6	0.962	2.2		50.0
52.5		0.017	0.3	0.036	0.4	0.083	0.6	0.155	0.9	0.339	1.3	0.518	1.6	0.987	2.2	0.50	52.5
55.0		0.017	0.3	0.036	0.4	0.085	0.6	0.159	0.9	0.347	1.3	0.531	1.6	1.01	2.2		55.0
57.5		0.018	0.3	0.037	0.4	0.088	0.6	0.162	0.9	0.356	1.3	0.544	1.6	1.04	2.3		57.5
60.0		0.018	0.3	0.038	0.4	0.090	0.6	0.166	0.9	0.364	1.3	0.556	1.6	1.06	2.3		60.0
62.5		0.018	0.3	0.039	0.4	0.092	0.7	0.170	0.9	0.372	1.3	0.569	1.6	1.08	2.3		62.5
65.0		0.019	0.3	0.040	0.4	0.094	0.7	0.174	0.9	0.380	1.3	0.581	1.6	1.11	2.3		65.0
67.5		0.019	0.3	0.041	0.4	0.096	0.7	0.177	0.9	0.388	1.3	0.592	1.7	1.13	2.3		67.5
70.0		0.020	0.3	0.042	0.4	0.098	0.7	0.181	0.9	0.395	1.3	0.604	1.7	1.15	2.3		70.0
72.5		0.020	0.3	0.042	0.4	0.099	0.7	0.184	0.9	0.403	1.3	0.616	1.7	1.17	2.3		72.5
75.0		0.020	0.3	0.043	0.4	0.101	0.7	0.188	0.9	0.410	1.3	0.627	1.7	1.19	2.3		75.0
77.5		0.021	0.3	0.044	0.4	0.103	0.7	0.191	0.9	0.418	1.3	0.638	1.7	1.21	2.3		77.5
80.0		0.021	0.3	0.045	0.4	0.105	0.7	0.194	0.9	0.425	1.4	0.649	1.7	1.24	2.3		80.0
82.5		0.021	0.3	0.046	0.4	0.107	0.7	0.197	0.9	0.432	1.4	0.659	1.7	1.26	2.3		82.5
85.0		0.022	0.3	0.046	0.4	0.108	0.7	0.201	0.9	0.439	1.4	0.670	1.7	1.28	2.3		85.0
87.5		0.022	0.3	0.047	0.4	0.110	0.7	0.204	0.9	0.446	1.4	0.680	1.7	1.30	2.3		87.5
90.0		0.023	0.3	0.048	0.4	0.112	0.7	0.207	0.9	0.452	1.4	0.691	1.7	1.31	2.3		90.0

Table 4.10 Flow of water at 75°C in heavy grade steel pipes — *continued*

q_m	= mass flow rate	kg.s^{-1}
c	= velocity	m.s^{-1}
$\Delta p/l$	= pressure drop per unit length	Pa.m^{-1}
l_e	= equivalent length of a component for $\zeta = 1$	m

HEAVY GRADE STEEL

WATER AT 75°C

$\Delta p/l$	c	10 mm q_m	l_e	15 mm q_m	l_e	20 mm q_m	l_e	25 mm q_m	l_e	32 mm q_m	l_e	40 mm q_m	l_e	50 mm q_m	l_e	c	$\Delta p/l$
92.5		0.023	0.3	0.049	0.4	0.113	0.7	0.210	0.9	0.459	1.4	0.701	1.7	1.33	2.3		92.5
95.0		0.023	0.3	0.049	0.4	0.115	0.7	0.213	0.9	0.466	1.4	0.711	1.7	1.35	2.3		95.0
97.5		0.024	0.3	0.050	0.4	0.117	0.7	0.216	0.9	0.472	1.4	0.721	1.7	1.37	2.3		97.5
100		0.024	0.3	0.051	0.4	0.118	0.7	0.219	0.9	0.479	1.4	0.731	1.7	1.39	2.3		100
120		0.026	0.3	0.056	0.4	0.131	0.7	0.242	0.9	0.527	1.4	0.805	1.7	1.53	2.4		120
140	0.3	0.029	0.3	0.061	0.5	0.142	0.7	0.262	0.9	0.572	1.4	0.873	1.7	1.66	2.4		140
160		0.031	0.3	0.065	0.5	0.152	0.7	0.282	1.0	0.614	1.4	0.937	1.7	1.78	2.4		160
180		0.033	0.3	0.070	0.5	0.162	0.7	0.300	1.0	0.654	1.4	0.997	1.8	1.89	2.4		180
200		0.035	0.3	0.074	0.5	0.172	0.7	0.317	1.0	0.691	1.4	1.05	1.8	2.00	2.4	1.0	200
220		0.037	0.3	0.078	0.5	0.181	0.7	0.334	1.0	0.727	1.4	1.11	1.8	2.10	2.4		220
240		0.039	0.3	0.081	0.5	0.189	0.7	0.349	1.0	0.761	1.4	1.16	1.8	2.20	2.4		240
260		0.040	0.3	0.085	0.5	0.198	0.7	0.364	1.0	0.793	1.5	1.21	1.8	2.29	2.4		260
280		0.042	0.3	0.088	0.5	0.206	0.7	0.379	1.0	0.825	1.5	1.26	1.8	2.38	2.4		280
300		0.044	0.3	0.092	0.5	0.213	0.7	0.393	1.0	0.855	1.5	1.30	1.8	2.47	2.5		300
320		0.045	0.3	0.095	0.5	0.221	0.7	0.407	1.0	0.884	1.5	1.35	1.8	2.55	2.5		320
340		0.047	0.3	0.098	0.5	0.228	0.7	0.420	1.0	0.913	1.5	1.39	1.8	2.64	2.5		340
360		0.048	0.3	0.101	0.5	0.235	0.7	0.433	1.0	0.941	1.5	1.43	1.8	2.71	2.5		360
380	0.5	0.049	0.3	0.104	0.5	0.242	0.7	0.445	1.0	0.970	1.5	1.47	1.8	2.79	2.5		380
400		0.051	0.3	0.107	0.5	0.248	0.7	0.457	1.0	0.994	1.5	1.51	1.8	2.87	2.5		400
420		0.052	0.3	0.110	0.5	0.255	0.7	0.469	1.0	1.02	1.5	1.55	1.8	2.94	2.5		420
440		0.054	0.3	0.113	0.5	0.261	0.7	0.481	1.0	1.04	1.5	1.59	1.8	3.01	2.5	1.5	440
460		0.055	0.3	0.115	0.5	0.267	0.7	0.492	1.0	1.07	1.5	1.63	1.8	3.08	2.5		460
480		0.056	0.3	0.118	0.5	0.273	0.8	0.503	1.0	1.09	1.5	1.66	1.8	3.15	2.5		480
500		0.057	0.3	0.120	0.5	0.279	0.8	0.514	1.0	1.12	1.5	1.69	1.8	3.22	2.5		500
520		0.059	0.3	0.123	0.5	0.285	0.8	0.524	1.0	1.14	1.5	1.73	1.8	3.28	2.5		520
540		0.060	0.3	0.125	0.5	0.291	0.8	0.535	1.0	1.16	1.5	1.77	1.8	3.35	2.5		540
560		0.061	0.3	0.128	0.5	0.296	0.8	0.545	1.0	1.17	1.5	1.80	1.8	3.41	2.5		560
580		0.062	0.3	0.130	0.5	0.302	0.8	0.555	1.0	1.21	1.5	1.83	1.8	3.47	2.5		580
600		0.063	0.3	0.133	0.5	0.307	0.8	0.565	1.0	1.23	1.5	1.87	1.8	3.53	2.5		600
620		0.064	0.3	0.135	0.5	0.312	0.8	0.575	1.0	1.25	1.5	1.90	1.8	3.59	2.5		620
640		0.065	0.3	0.137	0.5	0.318	0.8	0.584	1.0	1.27	1.5	1.93	1.8	3.65	2.5		640
660		0.066	0.3	0.139	0.5	0.323	0.8	0.594	1.0	1.29	1.5	1.96	1.8	3.71	2.5		660
680		0.067	0.3	0.142	0.5	0.328	0.8	0.603	1.0	1.31	1.5	1.99	1.9	3.77	2.5		680
700		0.069	0.3	0.144	0.5	0.333	0.8	0.612	1.0	1.33	1.5	2.02	1.9	3.83	2.5		700
720		0.070	0.3	0.146	0.5	0.338	0.8	0.621	1.0	1.35	1.5	2.05	1.9	3.88	2.5		720
740		0.071	0.3	0.148	0.5	0.343	0.8	0.630	1.0	1.37	1.5	2.08	1.9	3.94	2.5		740
760		0.072	0.3	0.150	0.5	0.347	0.8	0.639	1.0	1.39	1.5	2.10	1.9	3.99	2.5	2.0	760
780		0.073	0.3	0.152	0.5	0.352	0.8	0.648	1.0	1.41	1.5	2.14	1.9	4.04	2.5		780
800		0.074	0.3	0.154	0.5	0.357	0.8	0.656	1.0	1.42	1.5	2.17	1.9	4.10	2.5		800
820		0.075	0.4	0.156	0.5	0.362	0.8	0.665	1.0	1.44	1.5	2.19	1.9	4.15	2.5		820
840		0.075	0.4	0.158	0.5	0.366	0.8	0.673	1.0	1.46	1.5	2.22	1.9	4.20	2.5		840
860		0.076	0.4	0.160	0.5	0.371	0.8	0.681	1.0	1.48	1.5	2.25	1.9	4.25	2.5		860
880		0.077	0.4	0.162	0.5	0.375	0.8	0.689	1.0	1.50	1.5	2.27	1.9	4.30	2.5		880
900		0.078	0.4	0.164	0.5	0.379	0.8	0.698	1.0	1.51	1.5	2.30	1.9	4.35	2.5		900
920		0.079	0.4	0.166	0.5	0.384	0.8	0.706	1.0	1.53	1.5	2.33	1.9	4.40	2.5		920
940		0.080	0.4	0.168	0.5	0.388	0.8	0.713	1.0	1.55	1.5	2.35	1.9	4.45	2.5		940
960		0.081	0.4	0.170	0.5	0.392	0.8	0.721	1.0	1.56	1.5	2.38	1.9	4.50	2.5		960
980		0.082	0.4	0.172	0.5	0.397	0.8	0.729	1.0	1.58	1.5	2.40	1.9	4.55	2.5		980
1000		0.083	0.4	0.173	0.5	0.401	0.8	0.737	1.0	1.60	1.5	2.43	1.9	4.59	2.5		1000
1100		0.087	0.4	0.182	0.5	0.421	0.8	0.774	1.1	1.68	1.5	2.55	1.9	4.82	2.6		1100
1200		0.091	0.4	0.191	0.5	0.441	0.8	0.809	1.1	1.75	1.5	2.67	1.9	5.04	2.6		1200
1300	1.0	0.095	0.4	0.199	0.5	0.459	0.8	0.844	1.1	1.83	1.5	2.78	1.9	5.25	2.6		1300
1400		0.099	0.4	0.207	0.5	0.477	0.8	0.876	1.1	1.90	1.5	2.89	1.9	5.46	2.6		1400
1500		0.102	0.4	0.214	0.5	0.495	0.8	0.908	1.1	1.98	1.5	2.99	1.9	5.65	2.6		1500
1600		0.106	0.4	0.222	0.5	0.511	0.8	0.939	1.1	2.03	1.5	3.09	1.9	5.84	2.6		1600
1700		0.109	0.4	0.229	0.5	0.528	0.8	0.968	1.1	2.10	1.5	3.19	1.9	6.02	2.6	3.0	1700
1800		0.113	0.4	0.236	0.5	0.543	0.8	0.997	1.1	2.16	1.6	3.28	1.9				1800
1900		0.116	0.4	0.242	0.5	0.559	0.8	1.03	1.1	2.22	1.6	3.37	1.9				1900
2000		0.119	0.4	0.249	0.5	0.574	0.8	1.05	1.1	2.28	1.6	3.46	1.9				2000

Table 4.10 Flow of water at 75°C in heavy grade steel pipes — *continued*

q_m = mass flow rate   kg.s^{-1}
c = velocity   m.s^{-1}
$\Delta p/l$ = pressure drop per unit length   Pa.m^{-1}
l_e = equivalent length of a component
     for $\zeta = 1$ m

> **HEAVY GRADE STEEL**
> **WATER AT 75°C**

$\Delta p/l$	c	65 mm q_m	65 mm l_e	80 mm q_m	80 mm l_e	90 mm q_m	90 mm l_e	100 mm q_m	100 mm l_e	125 mm q_m	125 mm l_e	150 mm q_m	150 mm l_e	c	$\Delta p/l$
0.1		0.061	1.5	0.096	2.0	0.144	2.5	0.200	2.9	0.362	4.1	0.600	5.3		0.1
0.2		0.091	1.7	0.144	2.2	0.215	2.7	0.298	3.3	0.544	4.5	0.889	5.8	0.05	0.2
0.3		0.115	1.8	0.181	2.3	0.271	2.9	0.375	3.4	0.685	4.7	1.12	6.1		0.3
0.4		0.136	1.9	0.214	2.4	0.319	3.0	0.442	3.6	0.805	4.9	1.31	6.4		0.4
0.5		0.154	2.0	0.243	2.5	0.362	3.1	0.501	3.7	0.913	5.0	1.49	6.5		0.5
0.6	0.05	0.171	2.0	0.269	2.6	0.401	3.2	0.556	3.8	1.01	5.1	1.65	6.6		0.6
0.7		0.187	2.1	0.294	2.6	0.438	3.2	0.606	3.8	1.10	5.2	1.79	6.7		0.7
0.8		0.202	2.1	0.317	2.7	0.472	3.3	0.653	3.9	1.19	5.3	1.93	6.8		0.8
0.9		0.216	2.1	0.339	2.7	0.504	3.3	0.698	4.0	1.27	5.4	2.06	6.9		0.9
1.0		0.229	2.2	0.359	2.8	0.535	3.4	0.740	4.0	1.34	5.5	2.18	7.0		1.0
1.5		0.288	2.3	0.451	2.9	0.671	3.6	0.928	4.2	1.68	5.7	2.73	7.3	0.15	1.5
2.0		0.338	2.4	0.530	3.0	0.787	3.7	1.09	4.3	1.97	5.9	3.20	7.5		2.0
2.5		0.383	2.4	0.600	3.1	0.891	3.8	1.23	4.4	2.23	6.0	3.61	7.6		2.5
3.0		0.424	2.5	0.664	3.1	0.985	3.8	1.36	4.5	2.46	6.1	3.99	7.7		3.0
3.5		0.462	2.5	0.723	3.2	1.07	3.9	1.48	4.6	2.68	6.2	4.34	7.9		3.5
4.0	0.15	0.498	2.6	0.778	3.2	1.15	3.9	1.59	4.6	2.88	6.3	4.66	8.0		4.0
4.5		0.531	2.6	0.830	3.3	1.23	4.0	1.70	4.7	3.07	6.3	4.97	8.0		4.5
5.0		0.563	2.6	0.880	3.3	1.30	4.0	1.80	4.7	3.25	6.4	5.26	8.1	0.30	5.0
5.5		0.594	2.7	0.927	3.3	1.37	4.1	1.90	4.8	3.42	6.4	5.54	8.2		5.5
6.0		0.623	2.7	0.973	3.4	1.44	4.1	1.99	4.8	3.59	6.5	5.81	8.2		6.0
6.5		0.651	2.7	1.02	3.4	1.51	4.1	2.08	4.9	3.75	6.5	6.06	8.3		6.5
7.0		0.678	2.7	1.06	3.4	1.57	4.2	2.16	4.9	3.90	6.6	6.31	8.3		7.0
7.5		0.704	2.7	1.10	3.4	1.63	4.2	2.24	4.9	4.05	6.6	6.55	8.4		7.5
8.0		0.729	2.7	1.14	3.5	1.69	4.2	2.32	4.9	4.19	6.6	6.78	8.4		8.0
8.5		0.754	2.8	1.18	3.5	1.74	4.2	2.40	5.0	4.33	6.7	7.00	8.4		8.5
9.0		0.778	2.8	1.21	3.5	1.80	4.2	2.48	5.0	4.46	6.7	7.22	8.5		9.0
9.5		0.801	2.8	1.25	3.5	1.85	4.3	2.55	5.0	4.60	6.7	7.43	8.5		9.5
10.0		0.824	2.8	1.29	3.5	1.90	4.3	2.62	5.0	4.72	6.7	7.63	8.5		10.0
12.5		0.930	2.9	1.45	3.6	2.14	4.4	2.96	5.1	5.32	6.8	8.60	8.7	0.50	12.5
15.0	0.30	1.03	2.9	1.60	3.6	2.37	4.4	3.26	5.2	5.87	6.9	9.47	8.8		15.0
17.5		1.12	3.0	1.74	3.7	2.57	4.5	3.54	5.2	6.37	7.0	10.3	8.8		17.5
20.0		1.20	3.0	1.87	3.7	2.76	4.5	3.80	5.3	6.84	7.1	11.0	8.9		20.0
22.5		1.28	3.0	1.99	3.8	2.94	4.6	4.05	5.3	7.28	7.1	11.7	9.0		22.5
25.0		1.35	3.0	2.11	3.8	3.11	4.6	4.28	5.4	7.69	7.1	12.4	9.0		25.0
27.5		1.42	3.1	2.22	3.8	3.27	4.6	4.50	5.4	8.09	7.2	13.0	9.1		27.5
30.0		1.49	3.1	2.32	3.8	3.43	4.6	4.71	5.4	8.47	7.2	13.6	9.1		30.0
32.5		1.56	3.1	2.42	3.8	3.58	4.7	4.92	5.4	8.84	7.3	14.2	9.1		32.5
35.0		1.62	3.1	2.52	3.9	3.72	4.7	5.12	5.5	9.19	7.3	14.8	9.2		35.0
37.5	0.50	1.68	3.1	2.61	3.9	3.86	4.7	5.31	5.5	9.53	7.3	15.3	9.2		37.5
40.0		1.74	3.1	2.70	3.9	3.99	4.7	5.49	5.5	9.86	7.3	15.9	9.2		40.0
42.5		1.80	3.1	2.79	3.9	4.12	4.7	5.67	5.5	10.2	7.4	16.4	9.3		42.5
45.0		1.85	3.2	2.88	3.9	4.25	4.7	5.84	5.5	10.5	7.4	16.9	9.3		45.0
47.5		1.91	3.2	2.96	3.9	4.37	4.8	6.01	5.6	10.8	7.4	17.4	9.3		47.5
50.0		1.96	3.2	3.04	3.9	4.49	4.8	6.17	5.6	11.1	7.4	17.8	9.3	1.0	50.0
52.5		2.01	3.2	3.12	4.0	4.61	4.8	6.33	5.6	11.4	7.4	18.3	9.3		52.5
55.0		2.06	3.2	3.20	4.0	4.72	4.8	6.49	5.6	11.6	7.4	18.8	9.4		55.0
57.5		2.11	3.2	3.28	4.0	4.83	4.8	6.64	5.6	11.9	7.5	19.2	9.4		57.5
60.0		2.16	3.2	3.35	4.0	4.94	4.8	6.79	5.6	12.2	7.5	19.6	9.4		60.0
62.5		2.20	3.2	3.42	4.0	5.05	4.8	6.94	5.6	12.5	7.5	20.0	9.4		62.5
65.0		2.25	3.2	3.50	4.0	5.16	4.8	7.08	5.7	12.7	7.5	20.5	9.4		65.0
67.5		2.30	3.2	3.57	4.0	5.26	4.9	7.22	5.7	13.0	7.5	20.9	9.4		67.5
70.0		2.34	3.2	3.63	4.0	5.36	4.9	7.36	5.7	13.2	7.5	21.3	9.4		70.0
72.5		2.38	3.2	3.70	4.0	5.46	4.9	7.50	5.7	13.5	7.5	21.7	9.5		72.5
75.0		2.43	3.3	3.77	4.0	5.56	4.9	7.63	5.7	13.7	7.5	22.0	9.5		75.0
77.5		2.47	3.3	3.83	4.0	5.65	4.9	7.77	5.7	13.9	7.5	22.4	9.5		77.5
80.0		2.51	3.3	3.90	4.0	5.75	4.9	7.90	5.7	14.2	7.6	22.8	9.5		80.0
82.5		2.55	3.3	3.96	4.1	5.84	4.9	8.02	5.7	14.4	7.6	23.2	9.5		82.5
85.0		2.59	3.3	4.02	4.1	5.93	4.9	8.15	5.7	14.6	7.6	23.5	9.5		85.0
87.5		2.63	3.3	4.09	4.1	6.02	4.9	8.27	5.7	14.8	7.6	23.9	9.5		87.5
90.0		2.67	3.3	4.15	4.1	6.11	4.9	8.40	5.7	15.0	7.6	24.2	9.5		90.0

Table 4.10 Flow of water at 75°C in heavy grade steel pipes — *continued*

q_m = mass flow rate   kg.s^{-1}
c = velocity   m.s^{-1}
$\Delta p/l$ = pressure drop per unit length   Pa.m^{-1}
l_e = equivalent length of a component
     for $\zeta = 1$ m

HEAVY GRADE STEEL
WATER AT 75°C

$\Delta p/l$	c	65 mm		80 mm		90 mm		100 mm		125 mm		150 mm		c	$\Delta p/l$
		q_m	l_e	q_m	l_e	q_m	l_e	q_m	l_e	q_m	l_e	q_m	l_e		
92.5		2.71	3.3	4.21	4.1	6.20	4.9	8.52	5.7	15.3	7.6	24.6	9.5		92.5
95.0		2.75	3.3	4.27	4.1	6.29	4.9	8.64	5.7	15.5	7.6	24.9	9.6		95.0
97.5		2.79	3.3	4.32	4.1	6.37	4.9	8.75	5.8	15.7	7.6	25.2	9.6		97.5
100.0		2.82	3.3	4.38	4.1	6.46	4.9	8.87	5.8	15.9	7.6	25.6	9.6	1.5	100.0
120.0		3.11	3.3	4.82	4.1	7.10	5.0	9.75	5.8	17.5	7.7	28.1	9.6		120.0
140.0	1.0	3.37	3.4	5.22	4.2	7.69	5.0	10.6	5.8	18.9	7.7	30.4	9.7		140.0
160.0		3.61	3.4	5.60	4.2	8.25	5.0	11.3	5.9	20.3	7.7	32.6	9.7		160.0
180.0		3.84	3.4	5.95	4.2	8.76	5.0	12.0	5.9	21.6	7.8	34.6	9.7	2.0	180.0
200.0		4.05	3.4	6.29	4.2	9.25	5.0	12.7	5.9	22.7	7.8	36.5	9.8		200.0
220.0		4.26	3.4	6.60	4.2	9.72	5.1	13.3	5.9	23.9	7.8	38.4	9.8		220.0
240.0		4.46	3.4	6.91	4.2	10.2	5.1	14.0	5.9	25.0	7.8	40.1	9.8		240.0
260.0		4.65	3.4	7.20	4.2	10.6	5.1	14.5	6.0	26.0	7.9	41.8	9.8		260.0
280.0		4.83	3.4	7.48	4.3	11.0	5.1	15.1	6.0	27.0	7.9	43.4	9.9		280.0
300.0	1.5	5.00	3.5	7.75	4.3	11.4	5.1	15.6	6.0	28.0	7.9	45.0	9.9		300.0
320.0		5.17	3.5	8.01	4.3	11.8	5.1	16.2	6.0	29.0	7.9	46.5	9.9		320.0
340.0		5.34	3.5	8.27	4.3	12.2	5.2	16.7	6.0	29.8	7.9	47.9	9.9		340.0
360.0		5.50	3.5	8.51	4.3	12.5	5.2	17.2	6.0	30.7	7.9	49.4	9.9		360.0
380.0		5.65	3.5	8.75	4.3	12.8	5.2	17.7	6.0	31.6	7.9	50.7	9.9		380.0
400.0		5.80	3.5	8.99	4.3	13.2	5.2	18.1	6.0	32.4	7.9	52.1	9.9		400.0
420.0		5.95	3.5	9.22	4.3	13.6	5.2	18.6	6.0	33.2	7.9	53.4	9.9		420.0
440.0		6.09	3.5	9.44	4.3	13.9	5.2	19.0	6.0	34.0	7.9	54.7	9.9	3.0	440.0
460.0		6.24	3.5	9.66	4.3	14.2	5.2	19.5	6.0	34.8	8.0	55.9	9.9		460.0
480.0		6.37	3.5	9.87	4.3	14.5	5.2	19.9	6.0	35.6	8.0	57.2	10		480.0
500.0		6.51	3.5	10.1	4.3	14.8	5.2	20.3	6.0	36.3	8.0	58.4	10		500.0
520.0		6.64	3.5	10.3	4.3	15.1	5.2	20.7	6.1	37.1	8.0	59.5	10		520.0
540.0	2.0	6.77	3.5	10.5	4.3	15.4	5.2	21.1	6.1	37.8	8.0	60.7	10		540.0
560.0		6.90	3.5	10.7	4.3	15.7	5.2	21.5	6.1	38.5	8.0	61.8	10		560.0
580.0		7.02	3.5	10.9	4.3	16.0	5.2	21.9	6.1	39.2	8.0	62.9	10		580.0
600.0		7.15	3.5	11.1	4.3	16.3	5.2	22.3	6.1	39.9	8.0	64.0	10		600.0
620.0		7.27	3.5	11.3	4.4	16.6	5.2	22.7	6.1	40.5	8.0	65.1	10		620.0
640.0		7.39	3.5	11.4	4.4	16.8	5.2	23.1	6.1	41.2	8.0	66.2	10		640.0
660.0		7.50	3.5	11.6	4.4	17.1	5.2	23.4	6.1	41.9	8.0	67.2	10		660.0
680.0		7.62	3.5	11.8	4.4	17.3	5.2	23.8	6.1	42.5	8.0	68.2	10		680.0
700.0		7.73	3.5	12.0	4.4	17.6	5.2	24.1	6.1	43.1	8.0	69.2	10		700.0
720.0		7.85	3.5	12.2	4.4	17.8	5.2	24.5	6.1	43.7	8.0	70.2	10		720.0
740.0		7.96	3.5	12.3	4.4	18.1	5.2	24.8	6.1	44.4	8.0	71.2	10		740.0
760.0		8.07	3.5	12.4	4.4	18.4	5.3	25.1	6.1	45.0	8.0	72.2	10	4.0	760.0
780.0		8.17	3.5	12.6	4.4	18.6	5.3	25.5	6.1	45.6	8.0				780.0
800.0		8.28	3.6	12.8	4.4	18.8	5.3	25.8	6.1	46.2	8.0				800.0
820.0		8.39	3.6	12.9	4.4	19.1	5.3	26.2	6.1	46.7	8.0				820.0
840.0		8.49	3.6	13.1	4.4	19.3	5.3	26.5	6.1	47.3	8.0				840.0
860.0		8.59	3.6	13.3	4.4	19.6	5.3	26.8	6.1	47.9	8.0				860.0
880.0		8.69	3.6	13.5	4.4	19.8	5.3	27.1	6.1	48.4	8.0				880.0
900.0		8.80	3.6	13.6	4.4	20.0	5.3	27.4	6.1	49.0	8.0				900.0
920.0		8.89	3.6	13.8	4.4	20.2	5.3	27.7	6.1	49.6	8.1				920.0
940.0		8.99	3.6	13.9	4.4	20.5	5.3	28.0	6.1	50.1	8.1				940.0
960.0		9.09	3.6	14.1	4.4	20.7	5.3	28.3	6.1	50.6	8.1				960.0
980.0		9.19	3.6	14.2	4.4	20.9	5.3	28.6	6.1						980.0
1000.0		9.28	3.6	14.4	4.4	21.1	5.3	28.9	6.1						1000.0
1100.0		9.74	3.6	15.1	4.4	22.2	5.3	30.4	6.1						1100.0
1200.0	3.0	10.2	3.6	15.8	4.4	23.2	5.3	31.7	6.1						1200.0
1300.0		10.6	3.6	16.4	4.4	24.1	5.3								1300.0
1400.0		11.0	3.6	17.0	4.4	25.0	5.3								1400.0
1500.0		11.4	3.6	17.6	4.4										1500.0
1600.0		11.8	3.6	18.2	4.4										1600.0
1700.0		12..2	3.6	18.8	4.4										1700.0
1800.0		12.5	3.6												1800.0
1900.0		12.9	3.6												1900.0
2000.0		13.2	3.6												2000.0

Table 4.11 Flow of water at 75°C in medium grade steel pipes

q_m = mass flow rate $\quad$ kg.s^{-1}
c = velocity $\quad$ m.s^{-1}
$\Delta p/l$ = pressure drop per unit length $\quad$ Pa.m^{-1}
l_e = equivalent length of a component
$\quad$ for $\zeta = 1$ $\quad$ m

* $(Re) = 2000$
† $(Re) = 3000$

> MEDIUM GRADE STEEL
> WATER AT 75°C

$\Delta p/l$	c	10 mm q_m	l_e	15 mm q_m	l_e	20 mm q_m	l_e	25 mm q_m	l_e	32 mm q_m	l_e	40 mm q_m	l_e	50 mm q_m	l_e	c	$\Delta p/l$
0.1								0.004	0.2	0.011	0.6	0.020*	1.0	0.034†	1.2		0.1
0.2						0.003	0.1	0.007	0.4	0.021*	1.0	0.026	0.9	0.048	1.2		0.2
0.3						0.004	0.2	0.011	0.6	0.022	0.8	0.032	0.9	0.060	1.3		0.3
0.4						0.006	0.3	0.014*	0.7	0.025	0.8	0.037†	0.9	0.071	1.3		0.4
0.5						0.007	0.4	0.014	0.6	0.028	0.8	0.042	1.0	0.081	1.4		0.5
0.6						0.008	0.4	0.016	0.6	0.031†	0.8	0.047	1.0	0.090	1.4		0.6
0.7				0.003	0.2	0.010	0.5	0.017	0.6	0.034	0.8	0.052	1.0	0.098	1.5		0.7
0.8				0.003	0.2	0.011	0.6	0.018	0.6	0.037	0.8	0.056	1.1	0.106	1.5	0.05	0.8
0.9				0.004	0.2	0.012*	0.7	0.019	0.6	0.039	0.9	0.060	1.1	0.114	1.5		0.9
1.0				0.004	0.2	0.011	0.5	0.020†	0.6	0.042	0.9	0.063	1.1	0.121	1.5		1.0
1.5				0.006	0.3	0.014	0.5	0.025	0.6	0.053	0.9	0.080	1.1	0.152	1.6		1.5
2.0				0.009*	0.4	0.016	0.5	0.029	0.6	0.062	1.0	0.094	1.2	0.179	1.7		2.0
2.5		0.004	0.2	0.008	0.3	0.018	0.5	0.033	0.7	0.071	1.0	0.107	1.2	0.203	1.7		2.5
3.0		0.004	0.2	0.009	0.3	0.019†	0.5	0.037	0.7	0.078	1.0	0.119	1.3	0.225	1.8		3.0
3.5		0.005	0.3	0.010	0.3	0.021	0.5	0.040	0.7	0.086	1.0	0.130	1.3	0.245	1.8		3.5
4.0	0.05	0.006	0.3	0.011	0.3	0.023	0.5	0.043	0.7	0.092	1.1	0.140	1.3	0.264	1.8		4.0
4.5		0.007	0.4	0.011	0.3	0.024	0.5	0.046	0.7	0.099	1.1	0.149	1.3	0.282	1.9		4.5
5.0		0.007*	0.4	0.012	0.3	0.026	0.5	0.049	0.7	0.105	1.1	0.158	1.4	0.299	1.9		5.0
5.5		0.006	0.3	0.012	0.3	0.027	0.5	0.052	0.7	0.110	1.1	0.167	1.4	0.315	1.9	0.15	5.5
6.0		0.007	0.3	0.013	0.3	0.029	0.5	0.055	0.8	0.116	1.1	0.175	1.4	0.331	1.9		6.0
6.5		0.007	0.3	0.014	0.3	0.030	0.5	0.057	0.8	0.121	1.1	0.183	1.4	0.346	2.0		6.5
7.0		0.007	0.3	0.014†	0.4	0.032	0.5	0.060	0.8	0.127	1.1	0.191	1.4	0.361	2.0		7.0
7.5		0.008	0.3	0.015	0.4	0.033	0.6	0.062	0.8	0.131	1.2	0.198	1.4	0.375	2.0		7.5
8.0		0.008	0.3	0.015	0.4	0.034	0.6	0.064	0.8	0.136	1.2	0.206	1.4	0.388	2.0		8.0
8.5		0.008	0.3	0.016	0.4	0.035	0.6	0.066	0.8	0.141	1.2	0.213	1.4	0.401	2.0		8.5
9.0		0.008	0.3	0.016	0.4	0.036	0.6	0.069	0.8	0.146	1.2	0.220	1.5	0.414	2.0		9.0
9.5		0.008	0.3	0.017	0.4	0.037	0.6	0.071	0.8	0.150	1.2	0.226	1.5	0.427	2.0		9.5
10.0		0.009	0.3	0.017	0.4	0.039	0.6	0.073	0.8	0.154	1.2	0.233	1.5	0.439	2.0		10.0
12.5		0.010	0.3	0.020	0.4	0.044	0.6	0.082	0.8	0.175	1.2	0.263	1.5	0.496	2.1		12.5
15.0		0.011†	0.3	0.022	0.4	0.049	0.6	0.091	0.8	0.193	1.2	0.291	1.5	0.548	2.1		15.0
17.5		0.012	0.3	0.024	0.4	0.053	0.6	0.099	0.9	0.210	1.3	0.317	1.6	0.596	2.2		17.5
20.0		0.012	0.3	0.026	0.4	0.057	0.6	0.107	0.9	0.226	1.3	0.341	1.6	0.641	2.2	0.30	20.0
22.5		0.013	0.3	0.027	0.4	0.061	0.6	0.114	0.9	0.242	1.3	0.363	1.6	0.683	2.2		22.5
25.0		0.014	0.3	0.029	0.4	0.065	0.6	0.121	0.9	0.256	1.3	0.385	1.6	0.723	2.2		25.0
27.5		0.015	0.3	0.031	0.4	0.068	0.6	0.128	0.9	0.270	1.3	0.405	1.6	0.761	2.2		27.5
30.0		0.016	0.3	0.032	0.4	0.071	0.7	0.134	0.9	0.283	1.3	0.425	1.6	0.798	2.2		30.0
32.5		0.016	0.3	0.034	0.4	0.075	0.7	0.140	0.9	0.295	1.3	0.444	1.6	0.833	2.3		32.5
35.0		0.017	0.3	0.035	0.4	0.078	0.7	0.146	0.9	0.307	1.3	0.462	1.7	0.867	2.3		35.0
37.5	0.15	0.018	0.3	0.036	0.4	0.081	0.7	0.151	0.9	0.319	1.4	0.479	1.7	0.899	2.3		37.5
40.0		0.018	0.3	0.038	0.4	0.084	0.7	0.157	0.9	0.330	1.4	0.496	1.7	0.931	2.3		40.0
42.5		0.019	0.3	0.039	0.4	0.087	0.7	0.162	0.9	0.341	1.4	0.513	1.7	0.962	2.3		42.5
45.0		0.020	0.3	0.040	0.5	0.089	0.7	0.167	0.9	0.352	1.4	0.529	1.7	0.992	2.3		45.0
47.5		0.020	0.3	0.041	0.5	0.092	0.7	0.172	0.9	0.363	1.4	0.545	1.7	1.02	2.3		47.5
50.0		0.021	0.3	0.043	0.5	0.095	0.7	0.177	1.0	0.373	1.4	0.560	1.7	1.05	2.3	0.50	50.0
52.5		0.022	0.3	0.044	0.5	0.097	0.7	0.182	1.0	0.383	1.4	0.575	1.7	1.08	2.3		52.5
55.0		0.022	0.3	0.045	0.5	0.100	0.7	0.187	1.0	0.392	1.4	0.589	1.7	1.10	2.3		55.0
57.5		0.023	0.3	0.046	0.5	0.102	0.7	0.191	1.0	0.402	1.4	0.603	1.7	1.13	2.4		57.5
60.0		0.023	0.3	0.047	0.5	0.105	0.7	0.196	1.0	0.411	1.4	0.617	1.7	1.16	2.4		60.0
62.5		0.024	0.3	0.048	0.5	0.107	0.7	0.200	1.0	0.420	1.4	0.631	1.7	1.18	2.4		62.5
65.0		0.024	0.3	0.049	0.5	0.109	0.7	0.204	1.0	0.429	1.4	0.644	1.7	1.21	2.4		65.0
67.5		0.025	0.3	0.050	0.5	0.112	0.7	0.208	1.0	0.438	1.4	0.657	1.7	1.23	2.4		67.5
70.0		0.025	0.3	0.051	0.5	0.114	0.7	0.213	1.0	0.447	1.4	0.670	1.7	1.26	2.4		70.0
72.5		0.026	0.3	0.052	0.5	0.116	0.7	0.217	1.0	0.455	1.4	0.683	1.7	1.28	2.4		72.5
75.0		0.026	0.3	0.053	0.5	0.118	0.7	0.221	1.0	0.464	1.4	0.696	1.8	1.30	2.4		75.0
77.5		0.027	0.3	0.054	0.5	0.120	0.7	0.225	1.0	0.472	1.4	0.708	1.8	1.32	2.4		77.5
80.0		0.027	0.3	0.055	0.5	0.122	0.7	0.228	1.0	0.480	1.4	0.720	1.8	1.35	2.4		80.0
82.5		0.028	0.3	0.056	0.5	0.124	0.7	0.232	1.0	0.488	1.4	0.732	1.8	1.37	2.4		82.5
85.0		0.028	0.3	0.057	0.5	0.126	0.7	0.236	1.0	0.496	1.4	0.743	1.8	1.39	2.4		85.0
87.5		0.029	0.3	0.058	0.5	0.128	0.7	0.240	1.0	0.503	1.4	0.755	1.8	1.41	2.4		87.5
90.0		0.029	0.3	0.059	0.5	0.130	0.7	0.243	1.0	0.511	1.4	0.766	1.8	1.43	2.4		90.0

Table 4.11 Flow of water at 75°C in medium grade steel pipes — *continued*

q_m	= mass flow rate	kg.s^{-1}
c	= velocity	m.s^{-1}
$\Delta p/l$	= pressure drop per unit length	Pa.m^{-1}
l_e	= equivalent length of a component for $\zeta = 1$	m

MEDIUM GRADE STEEL

WATER AT 75°C

$\Delta p/l$	c	10 mm		15 mm		20 mm		25 mm		32 mm		40 mm		50 mm		c	$\Delta p/l$
		q_m	l_e	q_m	l_e	q_m	l_e	q_m	l_e	q_m	l_e	q_m	l_e	q_m	l_e		
92.5		0.029	0.3	0.060	0.5	0.132	0.7	0.247	1.0	0.518	1.5	0.778	1.8	1.45	2.4		92.5
95.0		0.030	0.3	0.061	0.5	0.134	0.7	0.251	1.0	0.526	1.5	0.789	1.8	1.48	2.4		95.0
97.5		0.030	0.3	0.062	0.5	0.136	0.7	0.254	1.0	0.533	1.5	0.800	1.8	1.50	2.4		97.5
100.0		0.031	0.3	0.062	0.5	0.138	0.7	0.258	1.0	0.540	1.5	0.810	1.8	1.52	2.4		100.0
120.0	0.30	0.034	0.3	0.069	0.5	0.152	0.7	0.284	1.0	0.595	1.5	0.893	1.8	1.67	2.4		120.0
140.0		0.037	0.3	0.075	0.5	0.165	0.8	0.308	1.0	0.646	1.5	0.968	1.8	1.81	2.5		140.0
160.0		0.040	0.4	0.081	0.5	0.178	0.8	0.331	1.0	0.693	1.5	1.04	1.8	1.94	2.5		160.0
180.0		0.042	0.4	0.086	0.5	0.189	0.8	0.353	1.0	0.738	1.5	1.11	1.8	2.06	2.5	1.0	180.0
200.0		0.045	0.4	0.091	0.5	0.200	0.8	0.373	1.1	0.780	1.5	1.17	1.9	2.18	2.5		200.0
220.0		0.047	0.4	0.096	0.5	0.211	0.8	0.392	1.1	0.820	1.5	1.28	1.9	2.29	2.5		220.0
240.0		0.050	0.4	0.100	0.5	0.221	0.8	0.411	1.1	0.858	1.5	1.29	1.9	2.40	2.5		240.0
260.0		0.052	0.4	0.105	0.5	0.230	0.8	0.428	1.1	0.895	1.5	1.34	1.9	2.50	2.5		260.0
280.0		0.054	0.4	0.109	0.5	0.239	0.8	0.445	1.1	0.931	1.5	1.39	1.9	2.60	2.6		280.0
300.0		0.056	0.4	0.113	0.5	0.248	0.8	0.462	1.1	0.965	1.5	1.44	1.9	2.69	2.6		300.0
320.0	0.50	0.058	0.4	0.117	0.5	0.257	0.8	0.478	1.1	0.998	1.6	1.49	1.9	2.78	2.6		320.0
340.0		0.060	0.4	0.121	0.5	0.265	0.8	0.493	1.1	1.03	1.6	1.54	1.9	2.87	2.6		340.0
360.0		0.062	0.4	0.125	0.5	0.273	0.8	0.508	1.1	1.06	1.6	1.59	1.9	2.96	2.6		360.0
380.0		0.064	0.4	0.128	0.5	0.281	0.8	0.523	1.1	1.09	1.6	1.63	1.9	3.04	2.6		380.0
400.0		0.065	0.4	0.132	0.5	0.289	0.8	0.537	1.1	1.12	1.6	1.68	1.9	3.12	2.6		400.0
420.0		0.067	0.4	0.135	0.5	0.297	0.8	0.551	1.1	1.15	1.6	1.72	1.9	3.20	2.6	1.5	420.0
440.0		0.069	0.4	0.139	0.5	0.304	0.8	0.564	1.1	1.18	1.6	1.76	1.9	3.28	2.6		440.0
460.0		0.070	0.4	0.142	0.5	0.311	0.8	0.578	1.1	1.21	1.6	1.80	1.9	3.36	2.6		460.0
480.0		0.072	0.4	0.145	0.5	0.318	0.8	0.591	1.1	1.23	1.6	1.84	1.9	3.43	2.6		480.0
500.0		0.074	0.4	0.148	0.5	0.325	0.8	0.603	1.1	1.25	1.6	1.88	1.9	3.51	2.6		500.0
520.0		0.075	0.4	0.151	0.5	0.332	0.8	0.616	1.1	1.29	1.6	1.92	1.9	3.58	2.6		520.0
540.0		0.077	0.4	0.154	0.6	0.338	0.8	0.628	1.1	1.31	1.6	1.96	1.9	3.65	2.6		540.0
560.0		0.078	0.4	0.157	0.6	0.345	0.8	0.640	1.1	1.34	1.6	2.00	1.9	3.72	2.6		560.0
580.0		0.080	0.4	0.160	0.6	0.351	0.8	0.652	1.1	1.36	1.6	2.03	1.9	3.78	2.6		580.0
600.0		0.081	0.4	0.163	0.6	0.355	0.8	0.664	1.1	1.38	1.6	2.07	1.9	3.85	2.6		600.0
620.0		0.082	0.4	0.166	0.6	0.364	0.8	0.675	1.1	1.41	1.6	2.10	1.9	3.92	2.6		620.0
640.0		0.084	0.4	0.169	0.6	0.370	0.8	0.686	1.1	1.43	1.6	2.14	1.9	3.98	2.6		640.0
660.0		0.085	0.4	0.172	0.6	0.376	0.8	0.697	1.1	1.45	1.6	2.17	1.9	4.04	2.6		660.0
680.0		0.087	0.4	0.174	0.6	0.382	0.8	0.708	1.1	1.48	1.6	2.21	1.9	4.11	2.6		680.0
700.0		0.088	0.4	0.177	0.6	0.388	0.8	0.719	1.1	1.50	1.6	2.24	1.9	4.17	2.6		700.0
720.0		0.089	0.4	0.180	0.6	0.393	0.8	0.730	1.1	1.52	1.6	2.27	1.9	4.23	2.6		720.0
740.0		0.091	0.4	0.182	0.6	0.399	0.8	0.740	1.1	1.54	1.6	2.31	2.0	4.29	2.6	2.0	740.0
760.0		0.092	0.4	0.185	0.6	0.405	0.8	0.750	1.1	1.56	1.6	2.34	2.0	4.35	2.6		760.0
780.0		0.093	0.4	0.187	0.6	0.410	0.8	0.761	1.1	1.59	1.6	2.37	2.0	4.41	2.6		780.0
800.0		0.094	0.4	0.190	0.6	0.416	0.8	0.771	1.1	1.61	1.6	2.40	2.0	4.46	2.6		800.0
820.0		0.096	0.4	0.192	0.6	0.421	0.8	0.780	1.1	1.63	1.6	2.43	2.0	4.52	2.6		820.0
840.0		0.097	0.4	0.195	0.6	0.426	0.8	0.790	1.1	1.65	1.6	2.46	2.0	4.58	2.6		840.0
860.0		0.098	0.4	0.197	0.6	0.431	0.8	0.800	1.1	1.67	1.6	2.49	2.0	4.63	2.6		860.0
880.0		0.099	0.4	0.200	0.6	0.437	0.8	0.810	1.1	1.69	1.6	2.52	2.0	4.69	2.6		880.0
900.0		0.100	0.4	0.202	0.6	0.442	0.8	0.819	1.1	1.71	1.6	2.55	2.0	4.74	2.6		900.0
920.0		0.102	0.4	0.204	0.6	0.447	0.8	0.828	1.1	1.73	1.6	2.58	2.0	4.80	2.6		920.0
940.0		0.103	0.4	0.207	0.6	0.452	0.8	0.838	1.1	1.75	1.6	2.61	2.0	4.85	2.6		940.0
960.0		0.104	0.4	0.209	0.6	0.457	0.8	0.847	1.1	1.76	1.6	2.64	2.0	4.90	2.6		960.0
980.0		0.105	0.4	0.211	0.6	0.462	0.8	0.856	1.1	1.78	1.6	2.66	2.0	4.95	2.6		980.0
1000.0		0.106	0.4	0.213	0.6	0.467	0.8	0.865	1.1	1.80	1.6	2.69	2.0	5.00	2.6		1000.0
1100.0		0.112	0.4	0.224	0.6	0.490	0.8	0.909	1.1	1.89	1.6	2.83	2.0	5.26	2.7		1100.0
1200.0	1.0	0.117	0.4	0.235	0.6	0.513	0.8	0.950	1.1	1.98	1.6	2.96	2.0	5.49	2.7		1200.0
1300.0		0.122	0.4	0.245	0.6	0.535	0.8	0.990	1.1	2.06	1.6	3.08	2.0	5.72	2.7		1300.0
1400.0		0.127	0.4	0.254	0.6	0.555	0.8	1.03	1.1	2.14	1.6	3.20	2.0	5.94	2.7		1400.0
1500.0		0.131	0.4	0.263	0.6	0.576	0.8	1.07	1.1	2.22	1.6	3.31	2.0	6.16	2.7		1500.0
1600.0		0.136	0.4	0.272	0.6	0.595	0.9	1.10	1.1	2.29	1.6	3.42	2.0	6.36	2.7	3.0	1600.0
1700.0		0.140	0.4	0.281	0.6	0.614	0.9	1.14	1.2	2.37	1.6	3.53	2.0	6.56	2.7		1700.0
1800.0		0.144	0.4	0.290	0.6	0.632	0.9	1.17	1.2	2.44	1.6	3.64	2.0	6.76	2.7		1800.0
1900.0		0.148	0.4	0.298	0.6	0.650	0.9	1.20	1.2	2.50	1.6	3.74	2.0	6.94	2.7		1900.0

Table 4.11 Flow of water at 75°C in medium grade steel pipes — *continued*

q_m = mass flow rate   kg.s⁻¹ — q_m = mass flow rate, kg.s^{-1}
c = velocity, m.s^{-1}
$\Delta p/l$ = pressure drop per unit length, Pa.m^{-1}
l_e = equivalent length of a component for $\zeta = 1$, m

> **MEDIUM GRADE STEEL**
> **WATER AT 75°C**

$\Delta p/l$	c	65 mm q_m	l_e	80 mm q_m	l_e	90 mm q_m	l_e	100 mm q_m	l_e	125 mm q_m	l_e	150 mm q_m	l_e	c	$\Delta p/l$
0.1		0.065	1.6	0.102	2.0	0.151	2.5	0.210	3.0	0.375	4.1	0.612	5.3		0.1
0.2		0.097	1.8	0.152	2.3	0.225	2.8	0.312	3.3	0.557	4.5	0.906	5.8	0.05	0.2
0.3		0.123	1.9	0.191	2.4	0.284	3.0	0.394	3.5	0.701	4.8	1.14	6.1		0.3
0.4		0.145	2.0	0.226	2.5	0.334	3.1	0.463	3.7	0.824	5.0	1.34	6.4		0.4
0.5	0.05	0.165	2.0	0.256	2.6	0.379	3.2	0.525	3.8	0.934	5.0	1.52	6.5		0.5
0.6		0.183	2.1	0.284	2.6	0.420	3.2	0.582	3.9	1.03	5.2	1.68	6.7		0.6
0.7		0.200	2.1	0.310	2.7	0.459	3.3	0.635	3.9	1.13	5.3	1.83	6.8		0.7
0.8		0.216	2.2	0.335	2.8	0.494	3.4	0.684	4.0	1.21	5.4	1.97	6.9		0.8
0.9		0.231	2.2	0.358	2.8	0.528	3.4	0.731	4.1	1.30	5.4	2.10	7.0		0.9
1.0		0.245	2.2	0.379	2.8	0.560	3.5	0.775	4.1	1.38	5.5	2.23	7.0	0.15	1.0
1.5		0.308	2.4	0.476	3.1	0.703	3.6	0.972	4.3	1.72	5.8	2.78	7.3		1.5
2.0		0.362	2.4	0.559	3.1	0.825	3.8	1.14	4.4	2.02	5.9	3.26	7.6		2.0
2.5		0.410	2.5	0.633	3.2	0.933	3.8	1.29	4.5	2.28	6.1	3.68	7.7		2.5
3.0		0.454	2.6	0.701	3.2	1.03	3.9	1.42	4.6	2.52	6.2	4.07	7.8		3.0
3.5		0.495	2.6	0.763	3.3	1.12	4.0	1.55	4.7	2.74	6.2	4.42	7.9		3.5
4.0	0.15	0.533	2.6	0.821	3.3	1.21	4.0	1.67	4.8	2.95	6.3	4.75	8.0		4.0
4.5		0.568	2.7	0.876	3.4	1.29	4.1	1.78	4.8	3.12	6.4	5.07	8.1		4.5
5.0		0.602	2.7	0.929	3.4	1.37	4.1	1.88	4.8	3.32	6.4	5.36	8.2	0.30	5.0
5.5		0.635	2.7	0.978	3.4	1.44	4.2	1.98	4.9	3.50	6.5	5.64	8.2		5.5
6.0		0.666	2.8	1.03	3.4	1.51	4.2	2.08	4.9	3.67	6.5	5.92	8.3		6.0
6.5		0.696	2.8	1.07	3.5	1.58	4.2	2.17	5.0	3.83	6.6	6.18	8.3		6.5
7.0		0.725	2.8	1.12	3.5	1.64	4.2	2.26	5.0	3.99	6.6	6.43	8.4		7.0
7.5		0.753	2.8	1.16	3.5	1.70	4.3	2.35	5.0	4.14	6.7	6.67	8.4		7.5
8.0		0.780	2.8	1.20	3.5	1.76	4.3	2.43	5.0	4.29	6.7	6.90	8.5		8.0
8.5		0.806	2.9	1.24	3.6	1.82	4.3	2.51	5.1	4.43	6.7	7.13	8.5		8.5
9.0		0.832	2.9	1.28	3.6	1.88	4.3	2.59	5.1	4.57	6.8	7.35	8.5		9.0
9.5		0.857	2.9	1.32	3.6	1.94	4.4	2.67	5.1	4.70	6.8	7.57	8.6		9.5
10.0		0.881	2.9	1.36	3.6	1.99	4.4	2.74	5.1	4.83	6.8	7.78	8.6		10.0
12.5	0.30	0.995	3.0	1.53	3.7	2.25	4.5	3.09	5.2	5.44	6.9	8.76	8.7	0.50	12.5
15.0		1.19	3.0	1.69	3.7	2.48	4.5	3.41	5.3	6.00	7.0	9.65	8.8		15.0
17.5		1.19	3.1	1.83	3.8	2.69	4.6	3.70	5.4	6.51	7.1	10.5	8.9		17.5
20.0		1.28	3.1	1.97	3.8	2.89	4.6	3.98	5.4	6.99	7.1	11.2	9.0		20.0
22.5		1.37	3.1	2.10	3.8	3.08	4.6	4.24	5.4	7.44	7.2	12.0	9.0		22.5
25.0		1.45	3.1	2.22	3.9	3.26	4.7	4.48	5.5	7.87	7.2	12.6	9.1		25.0
27.5		1.52	3.2	2.34	3.9	3.43	4.7	4.71	5.5	8.28	7.3	13.3	9.1		27.5
30.0		1.60	3.2	2.45	3.9	3.59	4.7	4.94	5.5	8.66	7.3	13.9	9.2		30.0
32.5		1.66	3.2	2.55	3.9	3.74	4.8	5.15	5.6	9.04	7.3	14.5	9.2		32.5
35.0		1.73	3.2	2.66	4.0	3.89	4.8	5.35	5.6	9.40	7.4	15.1	9.2		35.0
37.5	0.50	1.80	3.2	2.76	4.0	4.04	4.8	5.55	5.6	9.74	7.4	15.6	9.3		37.5
40.0		1.86	3.2	2.85	4.0	4.18	4.8	5.74	5.6	10.1	7.4	16.2	9.3		40.0
42.5		1.92	3.2	2.94	4.0	4.31	4.8	5.93	5.6	10.4	7.4	16.7	9.3		42.5
45.0		1.98	3.3	3.04	4.0	4.45	4.8	6.11	5.7	10.7	7.4	17.2	9.4		45.0
47.5		2.04	3.3	3.12	4.0	4.58	4.9	6.29	5.7	11.0	7.5	17.7	9.4		47.5
50.0		2.09	3.3	3.21	4.0	4.70	4.9	6.46	5.7	11.3	7.5	18.2	9.4	1.00	50.0
52.5		2.15	3.3	3.29	4.1	4.82	4.9	6.63	5.7	11.6	7.5	18.7	9.4		52.5
55.0		2.20	3.3	3.38	4.1	4.94	4.9	6.79	5.7	11.9	7.5	19.1	9.4		55.0
57.5		2.25	3.3	3.46	4.1	5.06	4.9	6.95	5.7	12.2	7.5	19.6	9.5		57.5
60.0		2.30	3.3	3.53	4.1	5.17	4.9	7.11	5.8	12.4	7.5	20.0	9.5		60.0
62.5		2.36	3.3	3.61	4.1	5.29	4.9	7.26	5.8	12.7	7.6	20.4	9.5		62.5
65.0		2.40	3.3	3.69	4.1	5.40	5.0	7.41	5.8	13.0	7.6	20.8	9.5		65.0
67.5		2.45	3.3	3.76	4.1	5.50	5.0	7.56	5.8	13.3	7.6	21.3	9.5		67.5
70.0		2.50	3.3	3.83	4.1	5.61	5.0	7.71	5.8	13.6	7.6	21.7	9.5		70.0
72.5		2.55	3.4	3.90	4.1	5.71	5.0	7.85	5.8	13.8	7.6	22.1	9.5		72.5
75.0		2.59	3.4	3.97	4.1	5.82	5.0	7.99	5.8	14.0	7.6	22.4	9.6		75.0
77.5		2.64	3.4	4.04	4.1	5.92	5.0	8.13	5.8	14.2	7.6	22.8	9.6		77.5
80.0		2.68	3.4	4.11	4.1	6.02	5.0	8.26	5.8	14.5	7.6	23.2	9.6		80.0
82.5		2.73	3.4	4.18	4.1	6.11	5.0	8.40	5.8	14.7	7.6	23.6	9.6		82.5
85.0		2.77	3.4	4.24	4.1	6.21	5.0	8.53	5.8	14.9	7.6	24.0	9.6		85.0
87.5		2.81	3.4	4.31	4.2	6.30	5.0	8.66	5.8	15.2	7.7	24.3	9.6		87.5
90.0		2.85	3.4	4.37	4.2	6.40	5.0	8.79	5.9	15.4	7.7	24.7	9.6		90.0

Table 4.11 Flow of water at 75°C in medium grade steel pipes — *continued*

q_m = mass flow rate   kg.s^{-1}
c = velocity   m.s^{-1}
$\Delta p/l$ = pressure drop per unit length   Pa.m^{-1}
l_e = equivalent length of a component
     for $\zeta = 1$ m

MEDIUM GRADE STEEL
WATER AT 75°C

$\Delta p/l$	c	65 mm		80 mm		90 mm		100 mm		125 mm		150 mm		c	$\Delta p/l$
		q_m	l_e	q_m	l_e	q_m	l_e	q_m	l_e	q_m	l_e	q_m	l_e		
92.5		2.90	3.4	4.44	4.2	6.49	5.0	8.91	5.9	15.6	7.7	25.0	9.6		92.5
95.0		2.94	3.4	4.50	4.2	6.58	5.0	9.04	5.9	15.8	7.7	25.4	9.6		95.0
97.5		2.98	3.4	4.56	4.2	6.67	5.0	9.16	5.9	16.0	7.7	25.7	9.6		97.5
100.0		3.02	3.4	4.62	4.2	6.76	5.0	9.28	5.9	16.3	7.7	26.1	9.6	1.5	100.0
120.0		3.32	3.4	5.08	4.2	7.43	5.1	10.2	5.9	17.9	7.8	28.6	9.7		120.0
140.0	1.0	3.60	3.5	5.51	4.2	8.05	5.1	11.1	6.0	19.3	7.8	31.0	9.8		140.0
160.0		3.86	3.5	5.90	4.3	8.63	5.1	11.8	6.0	20.7	7.8	33.2	9.8		160.0
180.0		4.10	3.5	6.27	4.3	9.17	5.2	12.6	6.0	22.0	7.8	35.3	9.8	2.0	180.0
200.0		4.33	3.5	6.63	4.3	9.69	5.2	13.3	6.0	23.2	7.9	37.2	9.8		200.0
220.0		4.55	3.5	6.96	4.3	10.2	5.2	14.0	6.0	24.4	7.9	39.1	9.9		220.0
240.0		4.76	3.5	7.28	4.3	10.6	5.2	14.6	6.1	25.5	7.9	40.9	9.9		240.0
260.0		4.96	3.6	7.59	4.4	11.1	5.2	15.2	6.1	26.6	7.9	42.6	9.9		260.0
280.0		5.16	3.6	7.88	4.4	11.5	5.2	15.8	6.1	27.6	8.0	44.2	9.9		280.0
300.0	1.5	5.34	3.6	8.17	4.4	11.9	5.2	16.4	6.1	28.6	8.0	45.8	10.0		300.0
320.0		5.52	3.6	8.45	4.4	12.3	5.2	16.9	6.1	29.6	8.0	47.4	10.0		320.0
340.0		5.70	3.6	8.71	4.4	12.7	5.3	17.5	6.1	30.5	8.0	48.8	10.0		340.0
360.0		5.87	3.6	8.97	4.4	13.1	5.3	18.0	6.1	31.4	8.0	50.3	10.0		360.0
380.0		6.06	3.6	9.23	4.4	13.5	5.3	18.5	6.1	32.3	8.0	51.7	10.0		380.0
400.0		6.20	3.6	9.47	4.4	13.8	5.3	19.0	6.1	33.2	8.0	53.1	10.0		400.0
420.0		6.36	3.6	9.71	4.4	14.2	5.3	19.5	6.2	34.0	8.0	54.4	10.0	3.0	420.0
440.0		6.51	3.6	9.95	4.4	14.5	5.3	19.9	6.2	34.8	8.0	55.7	10.0		440.0
460.0		6.66	3.6	10.2	4.4	14.9	5.3	20.4	6.2	35.6	8.0	57.0	10.0		460.0
480.0		6.81	3.6	10.4	4.4	15.2	5.3	20.8	6.2	36.3	8.0	58.2	10.0		480.0
500.0		6.95	3.6	10.6	4.4	15.5	5.3	21.3	6.2	37.1	8.0	59.4	10.0		500.0
520.0	2.0	7.09	3.6	10.8	4.4	15.8	5.3	21.7	6.2	37.9	8.0	60.6	10.1		520.0
540.0		7.23	3.6	11.0	4.4	16.1	5.3	22.1	6.2	38.6	8.1	61.8	10.1		540.0
560.0		7.37	3.6	11.3	4.4	16.4	5.3	22.5	6.2	39.4	8.1	63.0	10.1		560.0
580.0		7.50	3.6	11.5	4.4	16.7	5.3	22.9	6.2	40.1	8.1	64.1	10.1		580.0
600.0		7.63	3.6	11.7	4.4	17.0	5.3	23.3	6.2	40.8	8.1	65.2	10.1		600.0
620.0		7.76	3.6	11.9	4.5	17.3	5.3	23.7	6.2	41.5	8.1	66.3	10.1		620.0
640.0		7.89	3.6	12.1	4.5	17.6	5.3	24.1	6.2	42.1	8.1	67.4	10.1		640.0
660.0		8.02	3.6	12.2	4.5	17.9	5.4	24.5	6.2	42.8	8.1	68.5	10.1		660.0
680.0		8.14	3.6	12.4	4.5	18.2	5.4	24.9	6.2	43.5	8.1	69.5	10.1		680.0
700.0		8.26	3.6	12.6	4.5	18.4	5.4	25.3	6.2	44.1	8.1	70.5	10.1		700.0
720.0		8.38	3.6	12.8	4.5	18.7	5.4	25.6	6.2	44.7	8.1	71.6	10.1		720.0
740.0		8.50	3.6	13.0	4.5	19.0	5.4	26.0	6.2	45.4	8.1	72.6	10.1		740.0
760.0		8.61	3.7	13.2	4.5	19.2	5.4	26.3	6.2	46.0	8.1	73.5	10.1	4.0	760.0
780.0		8.73	3.7	13.3	4.5	19.5	5.4	26.7	6.2	46.6	8.1				780.0
800.0		8.84	3.7	13.6	4.5	19.7	5.4	27.0	6.2	47.2	8.1				800.0
820.0		8.96	3.7	13.7	4.5	20.0	5.4	27.4	6.2	47.8	8.1				820.0
840.0		9.07	3.7	13.8	4.5	20.2	5.4	27.7	6.2	48.4	8.1				840.0
860.0		9.18	3.7	14.0	4.5	20.4	5.4	28.0	6.2	49.0	8.1				860.0
880.0		9.29	3.7	14.1	4.5	20.7	5.4	28.4	6.2	49.5	8.1				880.0
900.0		9.39	3.7	14.3	4.5	20.9	5.4	28.7	6.2	50.1	8.1				900.0
920.0		9.50	3.7	14.5	4.5	21.2	5.4	29.0	6.2	50.7	8.1				920.0
940.0		9.60	3.7	14.7	4.5	21.4	5.4	29.3	6.2	51.2	8.1				940.0
960.0		9.71	3.7	14.8	4.5	21.6	5.4	29.7	6.2						960.0
980.0		9.81	3.7	15.0	4.5	21.9	5.4	30.0	6.3						980.0
1000.0		9.91	3.7	15.1	4.5	22.1	5.4	30.3	6.3						1000.0
1100.0	3.0	10.4	3.7	15.9	4.5	23.2	5.4	31.8	6.3						1100.0
1200.0		10.9	3.7	16.6	4.5	24.2	5.4	33.2	6.3						1200.0
1300.0		11.3	3.7	17.3	4.5	25.2	5.4								1300.0
1400.0		11.8	3.7	18.0	4.5	26.2	5.4								1400.0
1500.0		12.2	3.7	18.6	4.5										1500.0
1600.0		12.6	3.7	19.2	4.5										1600.0
1700.0		13.0	3.7	19.8	4.5										1700.0
1800.0		13.4	3.7												1800.0
1900.0		13.7	3.7												1900.0

Table 4.12 Flow of water at 75°C in large steel pipes

q_m = mass flow rate   kg.s^{-1}
c = velocity   m.s^{-1}
$\Delta p/l$ = pressure drop per unit length   Pa.m^{-1}
l_e = equivalent length of a component
       for $\zeta = 1$ m

LARGE SIZE STEEL

WATER AT 75°C

$\Delta p/l$	c	175 mm		200 mm		225 mm		250 mm		300 mm		c	$\Delta p/l$
		q_m	l_e	q_m	l_e	q_m	l_e	q_m	l_e	q_m	l_e		
0.1		0.95	6.8	1.33	8.1	1.82	9.5	2.47	11	3.99	14	0.05	0.1
0.2	0.05	1.41	7.4	1.97	8.8	2.69	10	3.64	12	5.87	15		0.2
0.3		1.77	7.8	2.47	9.2	3.37	11	4.55	13	7.34	16	0.1	0.3
0.4		2.07	8.0	2.90	9.5	3.95	11	5.34	13	8.61	16		0.4
0.5		2.35	8.2	3.28	9.7	4.47	11	6.04	13	9.73	17		0.5
0.6	0.1	2.60	8.4	3.63	9.9	4.94	12	6.68	14	10.8	17		0.6
0.7		2.83	8.5	3.95	10	5.38	12	7.27	14	11.7	17	0.15	0.7
0.8		3.04	8.6	4.25	10	5.79	12	7.82	14	12.6	18		0.8
0.9		3.25	8.7	4.54	10	6.18	12	8.34	14	13.4	18		0.9
1.0	0.15	3.44	8.8	4.81	11	6.54	12	8.83	14	14.2	18		1.0
1.5		4.30	9.2	6.00	11	8.16	13	11.0	15	17.7	19		1.5
2.0		5.03	9.4	7.02	11	9.54	13	12.9	15	20.7	19		2.0
2.5		5.68	9.6	7.92	11	10.8	13	14.5	15	23.3	19	0.3	2.5
3.0		6.27	9.8	8.74	12	11.9	13	16.0	16	25.7	20		3.0
3.5		6.81	9.9	9.50	12	12.9	14	17.4	16	27.9	20		3.5
4.0		7.32	10	10.2	12	13.9	14	18.7	16	29.9	20		4.0
4.5	0.3	7.80	10	10.9	12	14.8	14	19.9	16	31.9	20		4.5
5.0		8.26	10	11.5	12	15.6	14	21.0	16	33.7	20		5.0
5.5		8.69	10	12.1	12	16.4	14	22.1	16	35.5	20		5.5
6.0		9.11	10	12.7	12	17.2	14	23.2	16	37.2	20	0.5	6.0
6.5		9.51	10	13.2	12	18.0	14	24.2	16	38.8	21		6.5
7.0		9.89	10	13.8	12	18.7	14	25.2	16	40.3	21		7.0
7.5		10.3	10	14.3	12	19.4	14	26.1	16	41.8	21		7.5
8.0		10.6	11	14.8	12	20.1	14	27.0	17	43.3	21		8.0
8.5		11.0	11	15.3	12	20.7	14	27.9	17	44.7	21		8.5
9.0		11.3	11	15.7	12	21.4	14	28.8	17	46.0	21		9.0
9.5		11.6	11	16.2	12	22.0	14	29.6	17	47.4	21		9.5
10.0	0.5	12.0	11	16.7	13	22.6	15	30.4	17	48.7	21		10.0
12.5		13.5	11	18.7	13	25.4	15	34.2	17	54.7	21		12.5
15.0		14.8	11	20.6	13	28.0	15	37.6	17	60.2	21		15.0
17.5		16.1	11	22.4	13	30.3	15	40.8	17	65.2	22		17.5
20.0		17.3	11	24.0	13	32.5	15	43.8	17	69.9	22		20.0
22.5		18.4	11	25.5	13	34.6	15	46.5	17	74.4	22	1.0	22.5
25.0		19.4	11	27.0	13	36.6	15	49.2	18	78.6	22		25.0
27.5		20.4	11	28.4	13	38.4	15	51.7	18	82.5	22		27.5
30.0		21.4	11	29.7	13	40.2	15	54.1	18	86.3	22		30.0
32.5		22.3	11	31.0	13	41.9	15	56.4	18	90.0	22		32.5
35.0		23.2	11	32.2	13	43.6	15	58.6	18	93.5	22		35.0
37.5		24.0	11	33.4	13	45.2	15	60.7	18	96.9	22		37.5
40.0		24.8	11	34.5	13	46.7	16	62.8	18	100	22		40.0
42.5	1.0	25.6	12	35.6	13	48.2	16	64.8	18	103	22		42.5
45.0		26.4	12	36.7	14	49.7	16	66.7	18	107	22		45.0
47.5		27.2	12	37.7	14	51.1	16	68.6	18	110	22	1.5	47.5
50.0		27.9	12	38.7	14	52.5	16	70.5	18	112	22		50.0
52.5		28.6	12	39.7	14	53.8	16	72.3	18	115	22		52.5
55.0		29.3	12	40.7	14	55.1	16	74.0	18	118	23		55.0
57.5		30.0	12	41.6	14	56.4	16	75.7	18	121	23		57.5
60.0		30.7	12	42.6	14	57.6	16	77.4	18	124	23		60.0
62.5		31.3	12	43.5	14	58.9	16	79.1	18	126	23		62.5
65.0		32.0	12	44.4	14	60.1	16	80.7	18	129	23		65.0
67.5		32.6	12	45.2	14	61.3	16	82.3	18	131	23		67.5
70.0		33.2	12	46.1	14	62.4	16	83.8	18	134	23		70.0
72.5		33.8	12	46.9	14	63.6	16	85.3	18	136	23		72.5
75.0		34.4	12	47.8	14	64.7	16	86.8	18	139	23		75.0
77.5		35.0	12	48.6	14	65.8	16	88.3	18	141	23		77.5
80.0		35.6	12	49.4	14	66.9	16	89.8	18	143	23		80.0
82.5		36.2	12	50.2	14	67.9	16	91.2	18	145	23		82.5
85.0		36.7	12	51.0	14	69.0	16	92.6	18	148	23		85.0
87.5		37.3	12	51.7	14	70.0	16	94.0	18	150	23		87.5
90.0		37.8	12	52.5	14	71.0	16	95.4	18	152	23		90.0

Table 4.12 Flow of water at 75°C in large steel pipes — *continued*

q_m	=	mass flow rate	kg.s⁻¹	
c	=	velocity	m.s⁻¹	
$\Delta p/l$	=	pressure drop per unit length	Pa.m⁻¹	
l_e	=	equivalent length of a component for $\zeta = 1$	m	

LARGE SIZE STEEL

WATER AT 75°C

$\Delta p/l$	c	175 mm q_m	l_e	200 mm q_m	l_e	225 mm q_m	l_e	250 mm q_m	l_e	300 mm q_m	l_e	c	$\Delta p/l$
92.5	1.5	38.4	12	53.2	14	72.1	16	96.7	18	154	23		92.5
95.0		38.9	12	54.0	14	73.0	16	98.0	18	156	23		95.0
97.5		39.4	12	54.7	14	74.0	16	99.4	18	158	23		97.5
100.0		39.9	12	55.4	14	75.0	16	101	18	161	23		100.0
120.0		43.9	12	60.8	14	82.4	16	111	18	176	23		120.0
140.0		47.5	12	65.9	14	89.1	16	120	19	191	23		140.0
160.0		50.8	12	70.5	14	95.4	16	128	19	204	23		160.0
180.0		54.0	12	74.9	14	101	16	136	19	217	23	3.0	180.0
200.0		57.0	12	79.1	14	107	16	144	19	229	23		200.0
220.0		59.8	12	83.0	14	112	16	151	19	240	23		220.0
240.0		62.6	12	86.8	14	117	16	157	19	251	23		240.0
260.0		65.2	12	90.4	14	122	16	164	19	261	23		260.0
280.0		67.7	12	93.9	14	127	16	170	19	271	23		280.0
300.0		70.1	12	97.2	14	132	16	176	19	281	23		300.0
320.0		72.5	12	100	14	136	16	182	19	290	23		320.0
340.0		74.7	12	104	14	140	16	188	19	300	23	4.0	340.0
360.0	3.0	76.9	12	107	14	144	16	194	19	308	23		360.0
380.0		79.1	12	110	14	148	16	199	19	317	23		380.0
400.0		81.2	12	113	14	152	16	204	19	325	23		400.0
420.0		83.2	12	115	14	156	16	209	19	333	23		420.0
440.0		85.2	12	118	14	160	16	214	19	341	23		440.0
460.0		87.2	12	121	14	163	17	219	19	349	24		460.0
480.0		89.1	12	123	14	167	17	224	19	357	24		480.0
500.0		90.9	12	126	14	170	17	229	19	364	24		500.0
520.0		92.8	12	129	14	174	17	233	19	371	24	5.0	520.0
540.0		94.6	12	131	14	177	17	238	19	379	24		540.0
560.0		96.3	12	134	14	181	17	242	19	386	24		560.0
580.0		98.1	12	136	14	184	17	246	19	393	24		580.0
600.0		99.8	12	138	14	187	17	251	19	399	24		600.0
620.0		101	12	141	14	190	17	255	19	406	24		620.0
640.0	4.0	103	12	143	14	193	17	259	19	413	24		640.0
660.0		105	12	145	14	196	17	263	19	419	24		660.0
680.0		106	12	147	14	199	17	267	19	425	24		680.0
700.0		108	12	150	14	202	17	271	19	432	24		700.0
720.0		109	12	152	14	205	17	275	19	438	24		720.0
740.0		111	12	154	14	208	17	279	19	444	24	6.0	740.0
760.0		112	12	156	14	211	17	283	19				760.0
780.0		114	12	158	14	214	17	286	19				780.0
800.0		115	12	160	14	216	17	290	19				800.0
820.0		117	12	162	14	219	17	294	19				820.0
840.0		118	12	164	14	222	17	297	19				840.0
860.0		120	12	166	14	224	17	301	19				860.0
880.0		121	12	168	14	227	17	304	19				880.0
900.0		123	12	170	14	230	17	308	19				900.0
920.0		124	12	172	14	232	17	311	19				920.0
940.0		125	12	174	15	235	17						940.0
960.0		127	12	175	15	237	17						960.0
980.0	5.0	128	12	177	15	240	17						980.0
1000.0		129	12	179	15	242	17						1000.0
1100.0		136	12	188	15								1100.0
1200.0		142	12	196	15								1200.0
1300.0		148	12										1300.0
1400.0	6.0	153	12										1400.0
1500.0													1500.0
1600.0													1600.0
1700.0													1700.0
1800.0													1800.0
1900.0													1900.0
2000.0													2000.0

Table 4.12 Flow of water at 75°C in large steel pipes — *continued*

q_m = mass flow rate   kg.s^{-1}
c = velocity   m.s^{-1}
$\Delta p/l$ = pressure drop per unit length   Pa.m^{-1}
l_e = equivalent length of a component
     for $\zeta = 1$ m

LARGE SIZE STEEL
WATER AT 75°C

$\Delta p/l$	c	350 mm		400 mm		450 mm		500 mm		550 mm		600 mm		c	$\Delta p/l$
		q_m	l_e	q_m	l_e	q_m	l_e	q_m	l_e	q_m	l_e	q_m	l_e		
0.1		5.11	16	7.42	19	10.2	23	13.6	26	17.6	30	22.3	34	0.1	0.1
0.2		7.51	17	10.9	21	14.9	24	19.9	28	25.7	32	32.7	36		0.2
0.3	0.1	9.40	18	13.6	22	18.6	25	24.9	29	32.1	33	40.7	37	0.15	0.3
0.4		11.0	19	15.9	22	21.8	26	29.1	30	37.5	34	47.6	38		0.4
0.5		12.4	19	18.0	23	24.6	27	32.9	31	42.4	35	53.8	39		0.5
0.6	0.15	13.8	19	19.9	23	27.1	27	36.3	31	46.7	35	59.3	40		0.6
0.7		15.0	20	21.6	24	29.5	27	39.4	32	50.8	36	64.5	40		0.7
0.8		16.1	20	23.2	24	31.7	28	42.4	32	54.6	36	69.3	41		0.8
0.9		17.1	20	24.8	24	33.8	28	45.1	32	58.2	36	73.8	41		0.9
1.0		18.1	20	26.2	24	35.8	28	47.8	33	61.5	37	78.1	41	0.3	1.0
1.5		22.6	21	32.6	25	44.5	29	59.4	33	76.5	38	97.0	42		1.5
2.0	0.3	26.4	21	38.1	26	51.9	30	69.3	34	89.1	39	113	43		2.0
2.5		29.7	22	42.9	26	58.5	30	78.0	35	100	39	127	44		2.5
3.0		32.8	22	47.3	26	64.4	31	85.9	35	111	40	140	44	0.5	3.0
3.5		35.6	22	51.3	27	69.9	31	93.3	35	120	40	152	45		3.5
4.0		38.2	22	55.1	27	75.0	31	100	36	129	40	163	45		4.0
4.5		40.7	23	58.6	27	79.9	31	106	36	137	40	173	45		4.5
5.0		43.0	23	62.0	27	84.4	31	113	36	146	41	183	45		5.0
5.5	0.5	45.2	23	65.2	27	88.8	32	118	36	152	41	193	46		5.5
6.0		47.4	23	68.3	27	92.9	32	124	36	159	41	202	46		6.0
6.5		49.4	23	71.2	28	96.9	32	129	37	166	41	210	46		6.5
7.0		51.4	23	74.1	28	101	32	134	37	173	41	219	46		7.0
7.5		53.3	23	76.8	28	105	32	139	37	179	41	227	46		7.5
8.0		55.2	23	79.4	28	108	32	144	37	185	42	234	46		8.0
8.5		56.9	23	82.0	28	112	32	149	37	191	42	242	47		8.5
9.0		58.7	24	84.5	28	115	32	153	37	197	42	249	47		9.0
9.5		60.4	24	86.9	28	118	32	158	37	203	42	256	47		9.5
10.0		62.0	24	89.3	28	122	33	162	37	208	42	263	47	1.0	10.0
12.5		69.7	24	100	28	137	33	182	38	234	42	295	47		12.5
15.0		76.7	24	110	29	150	33	200	38	257	43	325	48		15.0
17.5		83.1	24	120	29	163	33	217	38	278	43	352	48		17.5
20.0	1.0	89.1	24	128	29	174	33	232	38	298	43	377	48		20.0
22.5		94.7	25	136	29	185	34	247	38	317	43	401	48	1.5	22.5
25.0		100	25	144	29	196	34	260	39	334	43	423	48		25.0
27.5		105	25	151	29	206	34	274	39	351	44	444	48		27.5
30.0		110	25	158	29	215	34	286	39	367	44	465	49		30.0
32.5		115	25	165	29	224	34	298	39	383	44	484	49		32.5
35.0		119	25	171	30	233	34	310	39	398	44	503	49		35.0
37.5		123	25	177	30	241	34	321	39	412	44	521	49		37.5
40.0		128	25	183	30	249	34	332	39	426	44	539	49		40.0
42.5	1.5	132	25	189	30	257	34	342	39	439	44	556	49		42.5
45.0		136	25	195	30	265	34	352	39	452	44	572	49		45.0
47.5		139	25	200	30	272	34	362	39	465	44	588	49		47.5
50.0		143	25	206	30	280	34	372	39	478	44	604	49		50.0
52.5		147	25	211	30	287	34	381	39	490	44	619	49		52.5
55.0		150	25	216	30	294	35	391	39	501	44	634	49		55.0
57.5		154	25	221	30	300	35	400	39	513	44	649	49		57.5
60.0		157	25	226	30	307	35	408	40	524	44	663	49		60.0
62.5		161	25	231	30	313	35	417	40	535	44	677	50		62.5
65.0		164	25	235	30	320	35	426	40	546	44	690	50		65.0
67.5		167	25	240	30	326	35	434	40	557	44	704	50		67.5
70.0		170	25	245	30	332	35	442	40	567	45	717	50		70.0
72.5		173	25	249	30	338	35	450	40	577	45	730	50		72.5
75.0		176	25	253	30	344	35	458	40	587	45	743	50		75.0
77.5		179	25	258	30	350	35	466	40	597	45	755	50		77.5
80.0		182	26	262	30	356	35	473	40	607	45	768	50		80.0
82.5		185	26	266	30	361	35	481	40	617	45	780	50		82.5
85.0		188	26	270	30	367	35	488	40	626	45	792	50		85.0
87.5		191	26	274	30	372	35	495	40	636	45	803	50	3.0	87.5
90.0		194	26	278	30	378	35	502	40	645	45	815	50		90.0

Table 4.12 Flow of water at 75°C in large steel pipes — *continued*

q_m = mass flow rate — kg.s^{-1}
c = velocity — m.s^{-1}
$\Delta p/l$ = pressure drop per unit length — Pa.m^{-1}
l_e = equivalent length of a component for $\zeta = 1$ — m

LARGE SIZE STEEL
WATER AT 75°C

$\Delta p/l$	c	350 mm		400 mm		450 mm		500 mm		550 mm		600 mm		c	$\Delta p/l$
		q_m	l_e	q_m	l_e	q_m	l_e	q_m	l_e	q_m	l_e	q_m	l_e		
92.5		196	26	282	30	383	35	509	40	654	45	826	50		92.5
95.0		199	26	286	30	388	35	516	40	663	45	838	50		95.0
97.5		202	26	290	30	393	35	523	40	672	45	849	50		97.5
100.0		204	26	294	30	399	35	530	40	680	45	860	50		100.0
120.0		224	26	322	30	437	35	582	40	746	45	943	50		120.0
140.0		243	26	348	31	473	35	629	40	807	45	1020	50		140.0
160.0		260	26	373	31	506	35	673	40	864	45	1090	50	4.0	160.0
180.0	3.0	276	26	396	31	538	35	715	40	917	45	1160	50		180.0
200.0		291	26	418	31	567	35	754	40	967	45	1220	51		200.0
220.0		305	26	439	31	595	35	791	40	1020	45	1280	51		220.0
240.0		319	26	458	31	622	36	827	41	1060	45	1340	51	5.0	240.0
260.0		333	26	477	31	648	36	861	41	1100	45	1400	51		260.0
280.0		345	26	496	31	673	36	894	41	1150	46	1450	51		280.0
300.0	4.0	358	26	513	31	696	36	926	41	1190	46	1500	51		300.0
320.0		369	26	530	31	720	36	957	41	1230	46	1550	51		320.0
340.0		381	26	547	31	742	36	986	41	1270	46	1600	51	6.0	340.0
360.0		392	26	563	31	764	36	1020	41	1300	46				360.0
380.0		403	26	578	31	785	36	1040	41	1340	46				380.0
400.0		414	26	594	31	806	36	1070	41						400.0
420.0		424	26	609	31	826	36	1100	41						420.0
440.0		434	26	623	31	845	36	1120	41						440.0
460.0	5.0	444	26	637	31	865	36								460.0
480.0		454	26	651	31	883	36								480.0
500.0		463	26	665	31										500.0
520.0		472	26	678	31										520.0
540.0		482	26	691	31										540.0
560.0		490	26	704	31										560.0
580.0		499	26												580.0
600.0		508	26												600.0
620.0		516	26												620.0
640.0		525	26												640.0
660.0	6.0	533	26												660.0
680.0															680.0
700.0															700.0
720.0															720.0
740.0															740.0
760.0															760.0
780.0															780.0
800.0															800.0
820.0															820.0
840.0															840.0
860.0															860.0
880.0															880.0
900.0															900.0
920.0															920.0
940.0															940.0
960.0															960.0
980.0															980.0
1000.0															1000.0
1100.0															1100.0
1200.0															1200.0
1300.0															1300.0
1400.0															1400.0
1500.0															1500.0
1600.0															1600.0
1700.0															1700.0
1800.0															1800.0
1900.0															1900.0
2000.0															2000.0

Table 4.13 Flow of water at 75°C in copper pipes

q_m = mass flow rate $\qquad\qquad$ kg.s^{-1}
c = velocity $\qquad\qquad\qquad$ m.s^{-1}
$\Delta p/l$ = pressure drop per unit length $\quad$ Pa.m^{-1}
l_e = equivalent length of a component
$\qquad$ for $\zeta = 1$ $\qquad\qquad\qquad$ m

* $(Re) = 2000$
† $(Re) = 3000$

COPPER, TABLE X
WATER AT 75°C

$\Delta p/l$	c	12 mm		15 mm		22 mm		28 mm		35 mm		42 mm		c	$\Delta p/l$
		q_m	l_e	q_m	l_e	q_m	l_e	q_m	l_e	q_m	l_e	q_m	l_e		
0.1								0.003	0.2	0.007	0.4	0.016*	0.8		0.1
0.2								0.006	0.3	0.014*	0.8	0.023	0.9		0.2
0.3						0.003	0.2	0.009	0.5	0.017	0.7	0.028	0.9		0.3
0.4						0.004	0.2	0.012	0.6	0.020	0.7	0.033†	0.9		0.4
0.5						0.005	0.3	0.015*	0.8	0.023	0.7	0.037	0.9		0.5
0.6						0.006	0.3	0.014	0.6	0.025	0.7	0.041	1.0		0.6
0.7						0.007	0.4	0.015	0.6	0.027	0.7	0.045	1.0		0.7
0.8						0.008	0.4	0.017	0.6	0.029†	0.7	0.049	1.0		0.8
0.9						0.010	0.5	0.018	0.6	0.031	0.8	0.052	1.0		0.9
1.0						0.011*	0.6	0.018	0.6	0.032	0.8	0.056	1.0	0.05	1.0
1.5				0.003	0.2	0.012	0.5	0.022†	0.6	0.041	0.8	0.070	1.1		1.5
2.0				0.004	0.2	0.014	0.5	0.027	0.6	0.049	0.9	0.083	1.2		2.0
2.5				0.005	0.3	0.015	0.5	0.030	0.7	0.056	0.9	0.095	1.2		2.5
3.0				0.006	0.3	0.017	0.5	0.034	0.7	0.062	0.9	0.105	1.2		3.0
3.5				0.008*	0.4	0.018†	0.5	0.037	0.7	0.068	1.0	0.115	1.3	0.10	3.5
4.0				0.007	0.3	0.020	0.5	0.040	0.7	0.073	1.0	0.124	1.3		4.0
4.5		0.004	0.2	0.008	0.3	0.021	0.5	0.043	0.7	0.078	1.0	0.133	1.3		4.5
5.0	0.05	0.004	0.2	0.008	0.3	0.022	0.5	0.046	0.7	0.083	1.0	0.141	1.3		5.0
5.5		0.005	0.2	0.008	0.3	0.024	0.5	0.048	0.7	0.088	1.0	0.149	1.4		5.5
6.0		0.005	0.3	0.009	0.3	0.025	0.5	0.051	0.8	0.092	1.0	0.157	1.4		6.0
6.5		0.006	0.3	0.009	0.3	0.026	0.5	0.053	0.8	0.097	1.1	0.164	1.4		6.5
7.0		0.006	0.3	0.009	0.3	0.027	0.5	0.055	0.8	0.101	1.1	0.171	1.4		7.0
7.5		0.006*	0.3	0.010	0.3	0.028	0.5	0.058	0.8	0.105	1.1	0.178	1.4	0.15	7.5
8.0		0.006	0.2	0.010	0.3	0.029	0.5	0.060	0.8	0.109	1.1	0.185	1.4		8.0
8.5		0.006	0.2	0.010	0.3	0.030	0.5	0.062	0.8	0.113	1.1	0.191	1.4		8.5
9.0		0.006	0.2	0.011†	0.3	0.031	0.5	0.064	0.8	0.116	1.1	0.198	1.5		9.0
9.5		0.006	0.2	0.011	0.3	0.032	0.6	0.066	0.8	0.120	1.1	0.204	1.5		9.5
10.0		0.006	0.2	0.011	0.3	0.033	0.6	0.068	0.8	0.124	1.1	0.210	1.5		10.0
12.5		0.007	0.2	0.013	0.3	0.038	0.6	0.077	0.8	0.141	1.2	0.238	1.5		12.5
15.0		0.008	0.2	0.014	0.3	0.042	0.6	0.086	0.9	0.156	1.2	0.264	1.6		15.0
17.5		0.008	0.2	0.016	0.3	0.046	0.6	0.094	0.9	0.170	1.2	0.288	1.6		17.5
20.0	0.10	0.009	0.2	0.017	0.3	0.050	0.6	0.101	0.9	0.184	1.2	0.311	1.6		20.0
22.5		0.010†	0.2	0.018	0.4	0.053	0.6	0.108	0.9	0.197	1.3	0.333	1.7		22.5
25.0		0.010	0.3	0.019	0.4	0.057	0.6	0.115	0.9	0.209	1.3	0.353	1.7	0.30	25.0
27.5		0.011	0.3	0.020	0.4	0.060	0.7	0.122	0.9	0.220	1.3	0.372	1.7		27.5
30.0		0.011	0.3	0.021	0.4	0.063	0.7	0.128	1.0	0.231	1.3	0.391	1.7		30.0
32.5		0.012	0.3	0.022	0.4	0.066	0.7	0.134	1.0	0.242	1.3	0.409	1.7		32.5
35.0		0.012	0.3	0.023	0.4	0.069	0.7	0.140	1.0	0.253	1.3	0.427	1.8		35.0
37.5	0.15	0.013	0.3	0.024	0.4	0.072	0.7	0.145	1.0	0.263	1.3	0.444	1.8		37.5
40.0		0.013	0.3	0.025	0.4	0.074	0.7	0.151	1.0	0.272	1.4	0.460	1.8		40.0
42.5		0.014	0.3	0.026	0.4	0.077	0.7	0.156	1.0	0.282	1.4	0.476	1.8		42.5
45.0		0.014	0.3	0.027	0.4	0.080	0.7	0.161	1.0	0.291	1.4	0.492	1.8		45.0
47.5		0.015	0.3	0.028	0.4	0.082	0.7	0.166	1.0	0.300	1.4	0.507	1.8		47.5
50.0		0.015	0.3	0.029	0.4	0.084	0.7	0.171	1.0	0.309	1.4	0.522	1.8		50.0
52.5		0.016	0.3	0.029	0.4	0.087	0.7	0.176	1.0	0.318	1.4	0.536	1.8		52.5
55.0		0.016	0.3	0.030	0.4	0.089	0.7	0.181	1.0	0.326	1.4	0.550	1.9		55.0
57.5		0.017	0.3	0.031	0.4	0.092	0.7	0.185	1.0	0.334	1.4	0.564	1.9		57.5
60.0		0.017	0.3	0.032	0.4	0.094	0.7	0.190	1.1	0.343	1.4	0.578	1.9		60.0
62.5		0.017	0.3	0.033	0.4	0.096	0.7	0.194	1.1	0.351	1.4	0.591	1.9		62.5
65.0		0.018	0.3	0.033	0.4	0.098	0.7	0.198	1.1	0.358	1.4	0.605	1.9	0.50	65.0
67.5		0.018	0.3	0.034	0.4	0.100	0.7	0.203	1.1	0.366	1.5	0.617	1.9		67.5
70.0		0.019	0.3	0.035	0.4	0.102	0.7	0.207	1.1	0.374	1.5	0.630	1.9		70.0
72.5		0.019	0.3	0.036	0.4	0.104	0.7	0.211	1.1	0.381	1.5	0.643	1.9		72.5
75.0		0.019	0.3	0.036	0.4	0.106	0.8	0.215	1.1	0.388	1.5	0.655	1.9		75.0
77.5		0.020	0.3	0.037	0.4	0.108	0.8	0.219	1.1	0.396	1.5	0.667	1.9		77.5
80.0		0.020	0.3	0.038	0.4	0.110	0.8	0.223	1.1	0.403	1.5	0.679	1.9		80.0
82.5		0.020	0.3	0.038	0.4	0.112	0.8	0.227	1.1	0.410	1.5	0.691	2.0		82.5
85.0		0.021	0.3	0.039	0.4	0.114	0.8	0.231	1.1	0.417	1.5	0.703	2.0		85.0
87.5		0.021	0.3	0.040	0.4	0.116	0.8	0.235	1.1	0.424	1.5	0.714	2.0		87.5
90.0		0.021	0.3	0.040	0.4	0.118	0.8	0.239	1.1	0.430	1.5	0.725	2.0		90.0

Table 4.13 Flow of water at 75°C in copper pipes — *continued*

q_m = mass flow rate   kg.s^{-1}
c = velocity   m.s^{-1}
$\Delta p/l$ = pressure drop per unit length   Pa.m^{-1}
l_e = equivalent length of a component
     for $\zeta = 1$ m

COPPER, TABLE X

WATER AT 75°C

$\Delta p/l$	c	12 mm		15 mm		22 mm		28 mm		35 mm		42 mm		c	$\Delta p/l$
		q_m	l_e	q_m	l_e	q_m	l_e	q_m	l_e	q_m	l_e	q_m	l_e		
92.5		0.022	0.3	0.041	0.4	0.120	0.8	0.242	1.1	0.437	1.5	0.737	2.0		92.5
95.0		0.222	0.3	0.042	0.4	0.122	0.8	0.246	1.1	0.444	1.5	0.748	2.0		95.0
97.5		0.022	0.3	0.042	0.4	0.124	0.8	0.250	1.1	0.450	1.5	0.759	2.0		97.5
100.0		0.023	0.3	0.043	0.4	0.125	0.8	0.253	1.1	0.457	1.5	0.769	2.0		100.0
120.0	0.3	0.025	0.3	0.047	0.5	0.139	0.8	0.281	1.2	0.506	1.6	0.852	2.0		120.0
140.0		0.028	0.3	0.052	0.5	0.152	0.8	0.306	1.2	0.551	1.6	0.928	2.1		140.0
160.0		0.030	0.3	0.056	0.5	0.164	0.8	0.330	1.2	0.594	1.6	1.00	2.1		160.0
180.0		0.032	0.3	0.060	0.5	0.175	0.8	0.352	1.2	0.635	1.6	1.07	2.1		180.0
200.0		0.034	0.4	0.064	0.5	0.186	0.9	0.374	1.2	0.673	1.7	1.13	2.2		200.0
220.0		0.036	0.4	0.067	0.5	0.196	0.9	0.394	1.2	0.710	1.7	1.19	2.2	1.0	220.0
240.0		0.038	0.4	0.071	0.5	0.206	0.9	0.414	1.3	0.745	1.7	1.25	2.2		240.0
260.0		0.039	0.4	0.074	0.5	0.215	0.9	0.433	1.3	0.779	1.7	1.31	2.2		260.0
280.0		0.041	0.4	0.077	0.5	0.224	0.9	0.451	1.3	0.812	1.7	1.37	2.2		280.0
300.0		0.043	0.4	0.080	0.5	0.233	0.9	0.469	1.3	0.844	1.7	1.42	2.3		300.0
320.0	0.5	0.044	0.4	0.083	0.5	0.242	0.9	0.486	1.3	0.874	1.8	1.47	2.3		320.0
340.0		0.046	0.4	0.086	0.5	0.250	0.9	0.503	1.3	0.904	1.8	1.52	2.3		340.0
360.0		0.048	0.4	0.089	0.5	0.258	0.9	0.519	1.3	0.934	1.8	1.57	2.3		360.0
380.0		0.049	0.4	0.092	0.5	0.266	0.9	0.535	1.3	0.962	1.8	1.62	2.3		380.0
400.0		0.050	0.4	0.094	0.5	0.274	0.9	0.551	1.3	0.990	1.8	1.66	2.3		400.0
420.0		0.052	0.4	0.097	0.5	0.282	0.9	0.566	1.3	1.02	1.8	1.71	2.3		420.0
440.0		0.053	0.4	0.099	0.5	0.289	0.9	0.581	1.4	1.04	1.8	1.75	2.4		440.0
460.0		0.055	0.4	0.102	0.5	0.297	1.0	0.595	1.4	1.07	1.8	1.80	2.4	1.5	460.0
480.0		0.056	0.4	0.104	0.6	0.304	1.0	0.610	1.4	1.10	1.8	1.84	2.4		480.0
500.0		0.057	0.4	0.107	0.6	0.311	1.0	0.624	1.4	1.12	1.8	1.88	2.4		500.0
520.0		0.059	0.4	0.109	0.6	0.318	1.0	0.637	1.4	1.15	1.8	1.92	2.4		520.0
540.0		0.060	0.4	0.112	0.6	0.324	1.0	0.651	1.4	1.17	1.9	1.96	2.4		540.0
560.0		0.061	0.4	0.114	0.6	0.331	1.0	0.664	1.4	1.19	1.9	2.00	2.4		560.0
580.0		0.062	0.4	0.116	0.6	0.338	1.0	0.677	1.4	1.22	1.9	2.04	2.4		580.0
600.0		0.064	0.4	0.119	0.6	0.344	1.0	0.690	1.4	1.24	1.9	2.08	2.4		600.0
620.0		0.065	0.4	0.121	0.6	0.350	1.0	0.703	1.4	1.26	1.9	2.12	2.4		620.0
640.0		0.066	0.4	0.123	0.6	0.357	1.0	0.715	1.4	1.28	1.9	2.16	2.4		640.0
660.0		0.067	0.4	0.125	0.6	0.363	1.0	0.728	1.4	1.31	1.9	2.19	2.5		660.0
680.0		0.068	0.4	0.127	0.6	0.369	1.0	0.740	1.4	1.33	1.9	2.23	2.5		680.0
700.0		0.069	0.4	0.129	0.6	0.375	1.0	0.752	1.4	1.35	1.9	2.26	2.5		700.0
720.0		0.070	0.4	0.131	0.6	0.381	1.0	0.764	1.4	1.37	1.9	2.30	2.5		720.0
740.0		0.072	0.4	0.133	0.6	0.387	1.0	0.775	1.4	1.39	1.9	2.33	2.5		740.0
760.0		0.073	0.4	0.135	0.6	0.392	1.0	0.787	1.4	1.41	1.9	2.37	2.5		760.0
780.0		0.074	0.4	0.137	0.6	0.398	1.0	0.798	1.4	1.43	1.9	2.40	2.5	2.0	780.0
800.0		0.075	0.4	0.139	0.6	0.404	1.0	0.810	1.4	1.45	1.9	2.44	2.5		800.0
820.0		0.076	0.4	0.141	0.6	0.409	1.0	0.821	1.4	1.47	1.9	2.48	2.5		820.0
840.0		0.077	0.4	0.143	0.6	0.415	1.0	0.832	1.5	1.49	1.9	2.50	2.5		840.0
860.0		0.078	0.4	0.145	0.6	0.420	1.0	0.843	1.5	1.51	1.9	2.54	2.5		860.0
880.0		0.079	0.4	0.147	0.6	0.426	1.0	0.853	1.5	1.53	2.0	2.57	2.5		880.0
900.0		0.080	0.4	0.149	0.6	0.431	1.0	0.864	1.5	1.55	2.0	2.60	2.5		900.0
920.0		0.081	0.4	0.151	0.6	0.437	1.0	0.875	1.5	1.57	2.0	2.63	2.5		920.0
940.0		0.082	0.4	0.153	0.6	0.442	1.0	0.885	1.5	1.59	2.0	2.66	2.5		940.0
960.0		0.083	0.4	0.154	0.6	0.447	1.0	0.896	1.5	1.61	2.0	2.69	2.5		960.0
980.0		0.084	0.4	0.156	0.6	0.452	1.0	0.906	1.5	1.62	2.0	2.72	2.6		980.0
1000.0	1.0	0.085	0.4	0.158	0.6	0.457	1.0	0.916	1.5	1.64	2.0	2.75	2.6		1000.0
1100.0		0.090	0.4	0.167	0.6	0.482	1.1	0.965	1.5	1.73	2.0	2.90	2.6		1100.0
1200.0		0.094	0.5	0.175	0.6	0.506	1.1	1.01	1.5	1.82	2.0	3.04	2.6		1200.0
1300.0		0.098	0.5	0.183	0.6	0.529	1.1	1.06	1.5	1.90	2.0	3.18	2.6		1300.0
1400.0		0.103	0.5	0.191	0.6	0.551	1.1	1.10	1.5	1.98	2.0	3.31	2.6		1400.0
1500.0		0.107	0.5	0.198	0.6	0.573	1.1	1.15	1.5	2.05	2.1	3.44	2.7		1500.0
1600.0		0.111	0.5	0.205	0.6	0.593	1.1	1.19	1.6	2.13	2.1	3.56	2.7	3.0	1600.0
1700.0		0.114	0.5	0.213	0.6	0.614	1.1	1.23	1.6	2.20	2.1	3.68	2.7		1700.0
1800.0		0.118	0.5	0.219	0.6	0.633	1.1	1.27	1.6	2.27	2.1	3.80	2.7		1800.0
1900.0		0.122	0.5	0.226	0.7	0.653	1.1	1.31	1.6	2.34	2.1	3.91	2.7		1900.0
2000.0		0.125	0.5	0.233	0.7	0.671	1.1	1.34	1.6	2.40	2.1	4.02	2.7		2000.0

Table 4.13 Flow of water at 75°C in copper pipes — *continued*

q_m = mass flow rate　　　　　　　kg.s^{-1}
c = velocity　　　　　　　　　　m.s^{-1}
$\Delta p/l$ = pressure drop per unit length　Pa.m^{-1}
l_e = equivalent length of a component
　　　for $\zeta = 1$　　　　　　　　m

COPPER, TABLE X

WATER AT 75°C

$\Delta p/l$	c	54 mm		67 mm		76 mm		108 mm		133 mm		159 mm		c	$\Delta p/l$
		q_m	l_e	q_m	l_e	q_m	l_e	q_m	l_e	q_m	l_e	q_m	l_e		
0.1		0.030	1.0	0.055	1.5	0.078	1.8	0.212	3.1	0.382	4.2	0.616	5.4	0.05	0.1
0.2		0.045	1.2	0.082	1.6	0.118	2.0	0.317	3.4	0.569	4.7	0.916	6.0		0.2
0.3		0.057	1.3	0.104	1.8	0.149	2.1	0.400	3.6	0.717	4.9	1.15	6.3		0.3
0.4		0.067	1.3	0.123	1.8	0.176	2.2	0.471	3.8	0.845	5.1	1.36	6.6		0.4
0.5		0.077	1.4	0.140	1.9	0.200	2.3	0.535	3.9	0.959	5.3	1.54	6.8		0.5
0.6		0.085	1.4	0.156	2.0	0.222	2.4	0.594	4.0	1.06	5.4	1.71	6.9	0.10	0.6
0.7		0.093	1.5	0.170	2.0	0.243	2.4	0.649	4.1	1.16	5.5	1.86	7.1		0.7
0.8	0.05	0.101	1.5	0.184	2.1	0.262	2.5	0.700	4.2	1.25	5.6	2.10	7.2		0.8
0.9		0.108	1.5	0.197	2.1	0.281	2.5	0.748	4.2	1.34	5.7	2.15	7.3	0.15	0.9
1.0		0.115	1.5	0.209	2.1	0.298	2.6	0.794	4.3	1.42	5.8	2.28	7.4		1.0
1.5		0.145	1.6	0.264	2.3	0.376	2.7	1.00	4.5	1.79	6.1	2.86	7.8		1.5
2.0		0.171	1.7	0.311	2.4	0.443	2.8	1.18	4.7	2.10	6.3	3.36	8.1		2.0
2.5	0.10	0.195	1.8	0.353	2.4	0.503	2.9	1.33	4.8	2.38	6.5	3.81	8.3		2.5
3.0		0.216	1.8	0.392	2.5	0.557	3.0	1.48	5.0	2.64	6.7	4.22	8.4		3.0
3.5		0.236	1.9	0.428	2.5	0.608	3.1	1.61	5.1	2.87	6.8	4.59	8.6		3.5
4.0		0.255	1.9	0.461	2.6	0.656	3.1	1.74	5.1	3.10	6.9	4.95	8.7		4.0
4.5		0.273	1.9	0.493	2.6	0.701	3.2	1.86	5.2	3.31	7.0	5.28	8.9	0.30	4.5
5.0		0.290	2.0	0.524	2.7	0.744	3.2	1.97	5.3	3.51	7.1	5.60	9.0		5.0
5.5	0.15	0.306	2.0	0.553	2.7	0.786	3.2	2.08	5.3	3.70	7.2	5.91	9.1		5.5
6.0		0.321	2.0	0.581	2.7	0.825	3.3	2.18	5.4	3.88	7.2	6.20	9.1		6.0
6.5		0.336	2.0	0.607	2.8	0.863	3.3	2.28	5.4	4.06	7.3	6.49	9.2		6.5
7.0		0.351	2.1	0.633	2.8	0.900	3.3	2.38	5.5	4.23	7.4	6.76	9.3		7.0
7.5		0.365	2.1	0.659	2.8	0.936	3.4	2.47	5.5	4.39	7.4	7.02	9.4		7.5
8.0		0.378	2.1	0.683	2.8	0.970	3.4	2.56	5.6	4.56	7.5	7.27	9.4		8.0
8.5		0.391	2.1	0.707	2.9	1.00	3.4	2.65	5.6	4.71	7.5	7.52	9.5		8.5
9.0		0.404	2.1	0.730	2.9	1.04	3.5	2.74	5.7	4.86	7.6	7.77	9.6		9.0
9.5		0.417	2.1	0.753	2.9	1.07	3.5	2.82	5.7	5.01	7.6	8.00	9.6		9.5
10.0		0.429	2.2	0.775	2.9	1.10	3.5	2.90	5.7	5.16	7.7	8.23	9.7	0.50	10.0
12.5		0.487	2.2	0.878	3.0	1.25	3.6	3.29	5.9	5.84	7.8	9.31	9.9		12.5
15.0		0.540	2.3	0.973	3.1	1.38	3.7	3.64	6.0	6.46	8.0	10.3	10.1		15.0
17.5	0.30	0.589	2.3	1.06	3.1	1.51	3.7	3.96	6.1	7.03	8.1	11.2	10.3		17.5
20.0		0.635	2.4	1.14	3.2	1.62	3.8	4.27	6.2	7.57	8.2	12.1	10.4		20.0
22.5		0.678	2.4	1.22	3.2	1.73	3.9	4.55	6.3	8.08	8.4	12.9	10.5		22.5
25.0		0.720	2.4	1.30	3.3	1.84	3.9	4.83	6.3	8.57	8.4	13.7	10.6		25.0
27.5		0.759	2.5	1.37	3.3	1.94	3.9	5.09	6.4	9.03	8.5	14.4	10.7		27.5
30.0		0.797	2.5	1.44	3.3	2.03	4.0	5.34	6.5	9.47	8.6	15.1	10.8		30.0
32.5		0.834	2.5	1.50	3.4	2.13	4.0	5.58	6.5	9.90	8.7	15.8	10.9		32.5
35.0		0.869	2.5	1.56	3.4	2.22	4.1	5.82	6.6	10.3	8.7	16.4	11.1		35.0
37.5		0.904	2.5	1.63	3.4	2.30	4.1	6.04	6.6	10.7	8.8	17.1	11.1		37.5
40.0		0.937	2.6	1.69	3.5	2.39	4.1	6.26	6.7	11.1	8.9	17.7	11.1		40.0
42.5		0.969	2.6	1.74	3.5	2.47	4.1	6.48	6.7	11.5	8.9	18.3	11.2	1.0	42.5
45.0	0.50	1.00	2.6	1.80	3.5	2.55	4.2	6.68	6.8	11.8	9.0	18.9	11.3		45.0
47.5		1.03	2.6	1.85	3.5	2.63	4.2	6.89	6.8	12.2	9.0	19.4	11.3		47.5
50.0		1.06	2.6	1.91	3.5	2.70	4.2	7.08	6.8	12.5	9.1	20.0	11.4		50.0
52.5		1.09	2.7	1.96	3.6	2.78	4.2	7.28	6.9	12.9	9.1	20.5	11.4		52.5
55.0		1.12	2.7	2.01	3.6	2.85	4.3	7.47	6.9	13.2	9.1	21.1	11.5		55.0
57.5		1.15	2.7	2.06	3.6	2.92	4.3	7.65	6.9	13.5	9.2	21.6	11.5		57.5
60.0		1.18	2.7	2.11	3.6	2.99	4.3	7.83	7.0	13.9	9.2	22.1	11.6		60.0
62.5		1.20	2.7	2.16	3.6	3.06	4.3	8.01	7.0	14.2	9.3	22.6	11.6		62.5
65.0		1.23	2.7	2.21	3.7	3.13	4.3	8.19	7.0	14.5	9.3	23.1	11.7		65.0
67.5		1.26	2.7	2.25	3.7	3.19	4.4	8.36	7.0	14.8	9.3	23.6	11.7		67.5
70.0		1.28	2.7	2.30	3.7	3.26	4.4	8.53	7.1	15.1	9.4	24.0	11.8		70.0
72.5		1.31	2.8	2.35	3.7	3.32	4.4	8.69	7.1	15.4	9.4	24.5	11.8		72.5
75.0		1.33	2.8	2.39	3.7	3.38	4.4	8.86	7.1	15.7	9.4	25.0	11.8		75.0
77.5		1.36	2.8	2.43	3.7	3.45	4.4	9.02	7.1	16.0	9.5	25.4	11.9		77.5
80.0		1.38	2.8	2.48	3.7	3.51	4.4	9.18	7.2	16.2	9.5	25.8	11.9		80.0
82.5		1.40	2.8	2.52	3.7	3.57	4.5	9.33	7.2	16.5	9.5	26.3	11.9		82.5
85.0		1.43	2.8	2.56	3.8	3.63	4.5	9.49	7.2	16.8	9.5	26.7	12.0		85.0
87.5		1.45	2.8	2.60	3.8	3.69	4.5	9.64	7.2	17.1	9.6	27.1	12.0		87.5
90.0		1.48	2.8	2.64	3.8	3.74	4.5	9.79	7.2	17.3	9.6	27.6	12.0		90.0

Table 4.13 Flow of water at 75°C in copper pipes — *continued*

q_m = mass flow rate   kg.s^{-1}
c = velocity   m.s^{-1}
$\Delta p/l$ = pressure drop per unit length   Pa.m^{-1}
l_e = equivalent length of a component
      for $\zeta = 1$ m

COPPER, TABLE X
WATER AT 75°C

$\Delta p/l$	c	54 mm		67 mm		76 mm		108 mm		133 mm		159 mm		c	$\Delta p/l$
		q_m	l_e	q_m	l_e	q_m	l_e	q_m	l_e	q_m	l_e	q_m	l_e		
92.5		1.50	2.8	2.68	3.8	3.80	4.5	9.94	7.3	17.6	9.6	27.9	12.1		92.5
95.0		1.52	2.8	2.72	3.8	3.86	4.5	10.1	7.3	17.8	9.6	28.4	12.1		95.0
97.5		1.54	2.8	2.76	3.8	3.91	4.5	10.2	7.3	18.1	9.7	28.8	12.1		97.5
100.0		1.56	2.9	2.80	3.8	3.97	4.6	10.4	7.3	18.4	9.7	29.2	12.2		100.0
120.0		1.73	2.9	3.10	3.9	4.39	4.6	11.5	7.5	20.3	9.9	32.2	12.4		120.0
140.0		1.88	3.0	3.38	4.0	4.78	4.7	12.5	7.6	22.1	10.0	35.1	12.5	2.0	140.0
160.0	1.0	2.03	3.0	3.63	4.0	5.14	4.8	13.4	7.7	23.7	10.1	37.7	12.7		160.0
180.0		2.16	3.0	3.88	4.1	5.49	4.8	14.3	7.7	25.3	10.2	40.2	12.8		180.0
200.0		2.29	3.1	4.11	4.1	5.81	4.9	15.2	7.8	26.8	10.3	42.6	12.9		200.0
220.0		2.42	3.1	4.33	4.2	6.12	4.9	16.0	7.9	28.2	10.4	44.8	13.0		220.0
240.0		2.54	3.1	4.54	4.2	6.42	5.0	16.7	8.0	29.6	10.5	47.0	13.1		240.0
260.0		2.65	3.2	4.75	4.2	6.72	5.0	17.5	8.0	30.9	10.6	49.1	13.2		260.0
280.0		2.76	3.2	4.95	4.3	6.99	5.0	18.2	8.1	32.2	10.6	51.1	13.3		280.0
300.0		2.87	3.2	5.14	4.3	7.26	5.1	18.9	8.1	33.4	10.7	53.0	13.4		300.0
320.0		2.97	3.2	5.32	4.3	7.52	5.1	19.6	8.2	34.6	10.8	55.0	13.5	3.0	320.0
340.0		3.08	3.3	5.50	4.3	7.78	5.1	20.2	8.2	35.7	10.8	56.8	13.5		340.0
360.0		3.17	3.3	5.68	4.4	8.02	5.2	20.9	8.3	36.9	10.9	58.5	13.6		360.0
380.0		3.27	3.3	5.85	4.4	8.26	5.2	21.5	8.3	38.0	10.9	60.3	13.6		380.0
400.0		3.36	3.3	6.02	4.4	8.50	5.2	22.1	8.3	39.0	11.0	62.0	13.7		400.0
420.0		3.45	3.3	6.18	4.4	8.73	5.2	22.7	8.4	40.1	11.0	63.6	13.8		420.0
440.0		3.54	3.3	6.34	4.4	8.95	5.3	23.3	8.4	41.1	11.1	65.2	13.8		440.0
460.0		3.63	3.4	6.49	4.5	9.17	5.3	23.9	8.4	42.1	11.1	66.8	13.9		460.0
480.0		3.72	3.4	6.65	4.5	9.39	5.3	24.4	8.5	43.1	11.1	68.4	13.9		480.0
500.0		3.80	3.4	6.80	4.5	9.60	5.3	25.0	8.5	44.1	11.2	69.9	13.9		500.0
520.0		3.88	3.4	6.94	4.5	9.81	5.3	25.5	8.5	45.0	11.2	71.4	14.0		520.0
540.0		3.97	3.4	7.09	4.5	10.0	5.4	26.0	8.5	46.0	11.2	72.9	14.0	4.0	540.0
560.0	2.0	4.05	3.4	7.23	4.5	10.2	5.4	26.6	8.6	46.8	11.3	74.3	14.1		560.0
580.0		4.12	3.4	7.37	4.6	10.4	5.4	27.1	8.6	47.7	11.3	75.7	14.1		580.0
600.0		4.20	3.4	7.51	4.6	10.6	5.4	27.6	8.6	48.6	11.3	77.1	14.1		600.0
620.0		4.28	3.5	7.64	4.6	10.8	5.4	28.1	8.6	49.5	11.4	78.5	14.2		620.0
640.0		4.35	3.5	7.78	4.6	11.0	5.5	28.5	8.7	50.3	11.4	79.8	14.2		640.0
660.0		4.43	3.5	7.91	4.6	11.2	5.5	29.0	8.7	51.2	11.4	81.2	14.2		660.0
680.0		4.50	3.5	8.04	4.6	11.4	5.5	29.5	8.7	52.0	11.4	82.5	14.3		680.0
700.0		4.57	3.5	8.17	4.6	11.6	5.5	30.0	8.7	52.8	11.5	83.8	14.3		700.0
720.0		4.64	3.5	8.29	4.7	11.7	5.5	30.4	8.8	53.6	11.5	85.1	14.3		720.0
740.0		4.71	3.5	8.42	4.7	11.9	5.5	30.9	8.8	54.4	11.5	86.3	14.4		740.0
760.0		4.78	3.5	8.54	4.7	12.1	5.5	31.3	8.8	55.2	11.5	87.6	14.4		760.0
780.0		4.85	3.5	8.66	4.7	12.2	5.5	31.8	8.8	55.9	11.6	88.8	14.4		780.0
800.0		4.92	3.5	8.78	4.7	12.4	5.6	32.2	8.8	56.8	11.6	90.0	14.5		800.0
820.0		4.98	3.5	8.90	4.7	12.6	5.6	32.6	8.8	57.5	11.6	91.2	14.5		820.0
840.0		5.05	3.6	9.02	4.7	12.7	5.6	33.1	8.9	58.3	11.6	92.4	14.5	5.0	840.0
860.0		5.12	3.6	9.14	4.7	12.9	5.6	33.5	8.9	59.0	11.7	93.9	14.5		860.0
880.0		5.18	3.6	9.25	4.7	13.1	5.6	33.9	8.9	59.7	11.7	94.8	14.6		880.0
900.0		5.24	3.6	9.37	4.7	13.2	5.6	34.3	8.9	60.5	11.7	95.9	14.6		900.0
920.0		5.31	3.6	9.48	4.8	13.4	5.6	34.7	8.9	61.2	11.7	97.0	14.6		920.0
940.0		5.37	3.6	9.59	4.8	13.5	5.6	35.1	8.9	61.9	11.7	98.2	14.6		940.0
960.0		5.43	3.6	9.70	4.8	13.7	5.6	35.5	9.0	62.6	11.8	99.3	14.6		960.0
980.0		5.49	3.6	9.81	4.8	13.8	5.7	35.9	9.0	63.3	11.8	100	14.7		980.0
1000.0		5.56	3.6	9.92	4.8	14.0	5.7	36.3	9.0	64.0	11.8	102	14.7		1000.0
1100.0	3.0	5.85	3.6	10.4	4.8	14.7	5.7	38.2	9.1	67.4	11.8	107	14.8	6.0	1100.0
1200.0		6.14	3.7	11.0	4.9	15.5	5.8	40.1	9.1	70.6	12.0				1200.0
1300.0		6.41	3.7	11.4	4.9	16.1	5.8	41.8	9.2	73.7	12.0				1300.0
1400.0		6.67	3.7	11.9	4.9	16.8	5.8	43.6	9.2	76.7	12.1				1400.0
1500.0		6.93	3.7	12.4	5.0	17.4	5.9	45.2	9.3	79.6	12.1				1500.0
1600.0		7.18	3.8	12.8	5.0	18.1	5.9	46.8	9.3						1600.0
1700.0		7.42	3.8	13.2	5.0	18.7	5.9	48.3	9.4						1700.0
1800.0		7.65	3.8	13.6	5.0	19.2	5.9	49.8	9.4						1800.0
1900.0		7.88	3.8	14.0	5.1	19.8	6.0								1900.0
2000.0		8.10	3.8	14.4	5.1	20.4	6.0								2000.0

Table 4.14 Flow of water at 75°C in copper pipes

q_m = mass flow rate   kg.s^{-1}
c = velocity   m.s^{-1}
$\Delta p/l$ = pressure drop per unit length   Pa.m^{-1}
l_e = equivalent length of a component
       for $\zeta = 1$ m

* $(Re) = 2000$
† $(Re) = 3000$

COPPER, TABLE Z
WATER AT 75°C

$\Delta p/l$	c	12 mm q_m	l_e	15 mm q_m	l_e	22 mm q_m	l_e	28 mm q_m	l_e	35 mm q_m	l_e	42 mm q_m	l_e	c	$\Delta p/l$
0.1								0.003	0.2	0.008	0.4	0.017*	0.9		0.1
0.2								0.007	0.3	0.016*	0.9	0.024	0.9		0.2
0.3						0.004	0.2	0.010	0.5	0.019	0.8	0.030	0.9		0.3
0.4						0.005	0.3	0.013*	0.7	0.022	0.8	0.035†	0.9		0.4
0.5						0.006	0.3	0.014	0.6	0.024	0.8	0.039	1.0		0.5
0.6						0.007	0.4	0.015	0.6	0.027	0.8	0.044	1.0		0.6
0.7						0.008	0.4	0.016	0.6	0.029†	0.8	0.048	1.0		0.7
0.8						0.010	0.5	0.017	0.6	0.031	0.8	0.052	1.0		0.8
0.9						0.011	0.6	0.019	0.6	0.033	0.8	0.055	1.1		0.9
1.0						0.012*	0.6	0.020†	0.6	0.035	0.8	0.059	1.1	0.05	1.0
1.5				0.004	0.2	0.013	0.5	0.024	0.6	0.045	0.9	0.074	1.1		1.5
2.0				0.005	0.3	0.015	0.5	0.028	0.6	0.053	0.9	0.088	1.2		2.0
2.5				0.006	0.3	0.016	0.5	0.032	0.7	0.060	0.9	0.100	1.2		2.5
3.0				0.007*	0.4	0.018†	0.5	0.036	0.7	0.067	1.0	0.111	1.3		3.0
3.5				0.007	0.3	0.020	0.5	0.039	0.7	0.073	1.0	0.121	1.3	0.10	3.5
4.0		0.004	0.2	0.008	0.3	0.021	0.5	0.043	0.7	0.079	1.0	0.131	1.3		4.0
4.5		0.004	0.2	0.008	0.3	0.023	0.5	0.046	0.7	0.085	1.0	0.140	1.4		4.5
5.0	0.05	0.005	0.2	0.009	0.3	0.024	0.5	0.048	0.8	0.090	1.1	0.149	1.4		5.0
5.5		0.005	0.3	0.009	0.3	0.026	0.5	0.051	0.8	0.095	1.1	0.157	1.4		5.5
6.0		0.006	0.3	0.009	0.3	0.027	0.5	0.054	0.8	0.100	1.1	0.165	1.4		6.0
6.5		0.006	0.3	0.010	0.3	0.028	0.5	0.056	0.8	0.105	1.1	0.173	1.4		6.5
7.0		0.006*	0.3	0.010	0.3	0.029	0.5	0.059	0.8	0.109	1.1	0.181	1.5	0.15	7.0
7.5		0.006	0.3	0.011	0.3	0.031	0.6	0.061	0.8	0.114	1.1	0.188	1.5		7.5
8.0		0.006	0.3	0.011	0.3	0.032	0.6	0.064	0.8	0.118	1.1	0.195	1.5		8.0
8.5		0.006	0.3	0.011	0.3	0.033	0.6	0.066	0.8	0.122	1.1	0.202	1.5		8.5
9.0		0.006	0.3	0.012	0.3	0.034	0.6	0.068	0.8	0.126	1.2	0.209	1.5		9.0
9.5		0.006	0.3	0.012	0.3	0.035	0.6	0.070	0.8	0.130	1.2	0.215	1.5		9.5
10.0		0.007	0.3	0.012†	0.3	0.036	0.6	0.072	0.8	0.134	1.2	0.221	1.5		10.0
12.5		0.007	0.3	0.014	0.3	0.041	0.6	0.082	0.9	0.153	1.2	0.251	1.6		12.5
15.0		0.008	0.3	0.015	0.3	0.046	0.6	0.091	0.9	0.169	1.2	0.279	1.6		15.0
17.5		0.009	0.3	0.017	0.4	0.050	0.6	0.100	0.9	0.185	1.3	0.304	1.6		17.5
20.0	0.10	0.009†	0.3	0.018	0.4	0.054	0.6	0.108	0.9	0.200	1.3	0.328	1.7		20.0
22.5		0.010	0.3	0.020	0.4	0.058	0.7	0.115	1.0	0.213	1.3	0.351	1.7		22.5
25.0		0.011	0.3	0.021	0.4	0.061	0.7	0.123	1.0	0.226	1.3	0.373	1.7	0.30	25.0
27.5		0.011	0.3	0.022	0.4	0.065	0.7	0.129	1.0	0.239	1.4	0.393	1.7		27.5
30.0		0.012	0.3	0.023	0.4	0.068	0.7	0.136	1.0	0.251	1.4	0.413	1.8		30.0
32.5		0.012	0.3	0.024	0.4	0.071	0.7	0.142	1.0	0.263	1.4	0.432	1.8		32.5
35.0	0.15	0.013	0.3	0.025	0.4	0.075	0.7	0.148	1.0	0.274	1.4	0.450	1.8		35.0
37.5		0.014	0.3	0.026	0.4	0.078	0.7	0.154	1.0	0.285	1.4	0.468	1.8		37.5
40.0		0.014	0.3	0.027	0.4	0.081	0.7	0.160	1.0	0.296	1.4	0.486	1.8		40.0
42.5		0.015	0.3	0.028	0.4	0.083	0.7	0.166	1.0	0.306	1.4	0.502	1.8		42.5
45.0		0.015	0.3	0.029	0.4	0.086	0.7	0.171	1.0	0.316	1.4	0.519	1.9		45.0
47.5		0.016	0.3	0.030	0.4	0.089	0.7	0.177	1.1	0.326	1.5	0.535	1.9		47.5
50.0		0.016	0.3	0.031	0.4	0.091	0.7	0.182	1.1	0.335	1.5	0.551	1.9		50.0
52.5		0.016	0.3	0.032	0.4	0.094	0.7	0.187	1.1	0.345	1.5	0.566	1.9		52.5
55.0		0.017	0.3	0.033	0.4	0.097	0.8	0.192	1.1	0.354	1.5	0.581	1.9		55.0
57.5		0.017	0.3	0.034	0.4	0.099	0.8	0.197	1.1	0.363	1.5	0.595	1.9		57.5
60.0		0.018	0.3	0.034	0.4	0.102	0.8	0.202	1.1	0.372	1.5	0.610	1.9		60.0
62.5		0.018	0.3	0.035	0.4	0.104	0.8	0.206	1.1	0.380	1.5	0.624	1.9	0.50	62.5
65.0		0.019	0.3	0.036	0.4	0.106	0.8	0.211	1.1	0.389	1.5	0.638	1.9		65.0
67.5		0.019	0.3	0.037	0.4	0.109	0.8	0.216	1.1	0.397	1.5	0.652	2.0		67.5
70.0		0.019	0.3	0.038	0.4	0.111	0.8	0.220	1.1	0.405	1.5	0.665	2.0		70.0
72.5		0.020	0.3	0.038	0.4	0.113	0.8	0.224	1.1	0.413	1.5	0.678	2.0		72.5
75.0		0.020	0.3	0.039	0.4	0.115	0.8	0.229	1.1	0.421	1.5	0.691	2.0		75.0
77.5		0.021	0.3	0.040	0.4	0.117	0.8	0.233	1.1	0.429	1.5	0.704	2.0		77.5
80.0		0.021	0.3	0.041	0.4	0.120	0.8	0.237	1.1	0.437	1.6	0.717	2.0		80.0
82.5		0.021	0.3	0.041	0.5	0.122	0.8	0.241	1.1	0.445	1.6	0.729	2.0		82.5
85.0		0.022	0.3	0.042	0.5	0.124	0.8	0.246	1.1	0.452	1.6	0.741	2.0		85.0
87.5		0.022	0.3	0.043	0.5	0.126	0.8	0.250	1.1	0.459	1.6	0.753	2.0		87.5
90.0		0.023	0.3	0.044	0.5	0.128	0.8	0.254	1.1	0.467	1.6	0.765	2.0		90.0

Table 4.14 Flow of water at 75°C in copper pipes — *continued*

q_m = mass flow rate   kg.s^{-1}
c = velocity   m.s^{-1}
$\Delta p/l$ = pressure drop per unit length   Pa.m^{-1}
l_e = equivalent length of a component
     for $\zeta = 1$ m

COPPER, TABLE Z

WATER AT 75°C

$\Delta p/l$	c	12 mm		15 mm		22 mm		28 mm		35 mm		42 mm		c	$\Delta p/l$
		q_m	l_e	q_m	l_e	q_m	l_e	q_m	l_e	q_m	l_e	q_m	l_e		
92.5		0.023	0.3	0.044	0.5	0.130	0.8	0.258	1.2	0.474	1.6	0.777	2.0		92.5
95.0		0.023	0.3	0.045	0.5	0.132	0.8	0.261	1.2	0.481	1.6	0.789	2.0		95.0
97.5		0.024	0.3	0.046	0.5	0.134	0.8	0.265	1.2	0.488	1.6	0.800	2.0		97.5
100.0		0.024	0.3	0.046	0.5	0.136	0.8	0.269	1.2	0.495	1.6	0.812	2.1		100.0
120.0	0.30	0.027	0.3	0.051	0.5	0.151	0.8	0.298	1.2	0.549	1.6	0.899	2.1		120.0
140.0		0.029	0.3	0.056	0.5	0.164	0.9	0.325	1.2	0.598	1.7	0.979	2.1		140.0
160.0		0.031	0.4	0.061	0.5	0.177	0.9	0.351	1.2	0.644	1.7	1.06	2.2		160.0
180.0		0.034	0.4	0.065	0.5	0.189	0.9	0.375	1.3	0.688	1.7	1.13	2.2		180.0
200.0		0.036	0.4	0.069	0.5	0.201	0.9	0.397	1.3	0.730	1.7	1.19	2.2	1.0	200.0
220.0		0.038	0.4	0.073	0.5	0.212	0.9	0.419	1.3	0.770	1.7	1.26	2.2		220.0
240.0		0.040	0.4	0.076	0.5	0.223	0.9	0.440	1.3	0.808	1.8	1.32	2.3		240.0
260.0		0.041	0.4	0.080	0.5	0.233	0.9	0.460	1.3	0.845	1.8	1.38	2.3		260.0
280.0		0.043	0.4	0.083	0.5	0.243	0.9	0.480	1.3	0.880	1.8	1.44	2.3		280.0
300.0	0.50	0.045	0.4	0.087	0.5	0.252	0.9	0.498	1.3	0.915	1.8	1.50	2.3		300.0
320.0		0.047	0.4	0.090	0.5	0.262	0.9	0.517	1.3	0.948	1.8	1.55	2.3		320.0
340.0		0.048	0.4	0.093	0.6	0.271	1.0	0.535	1.4	0.980	1.8	1.60	2.4		340.0
360.0		0.050	0.4	0.096	0.6	0.280	1.0	0.552	1.4	1.01	1.8	1.66	2.4		360.0
380.0		0.052	0.4	0.099	0.6	0.288	1.0	0.569	1.4	1.04	1.9	1.71	2.4		380.0
400.0		0.053	0.4	0.102	0.6	0.297	1.0	0.585	1.4	1.07	1.9	1.75	2.4		400.0
420.0		0.055	0.4	0.105	0.6	0.305	1.0	0.601	1.4	1.10	1.9	1.80	2.4		420.0
440.0		0.056	0.4	0.108	0.6	0.313	1.0	0.617	1.4	1.13	1.9	1.85	2.4	1.5	440.0
460.0		0.057	0.4	0.110	0.6	0.321	1.0	0.633	1.4	1.16	1.9	1.89	2.4		460.0
480.0		0.059	0.4	0.113	0.6	0.328	1.0	0.648	1.4	1.19	1.9	1.94	2.4		480.0
500.0		0.060	0.4	0.116	0.6	0.336	1.0	0.663	1.4	1.21	1.9	1.98	2.4		500.0
520.0		0.062	0.4	0.118	0.6	0.344	1.0	0.677	1.4	1.24	1.9	2.03	2.5		520.0
540.0		0.063	0.4	0.121	0.6	0.351	1.0	0.692	1.4	1.27	1.9	2.07	2.5		540.0
560.0		0.064	0.4	0.123	0.6	0.358	1.0	0.706	1.4	1.29	1.9	2.11	2.5		560.0
580.0		0.066	0.4	0.126	0.6	0.365	1.0	0.720	1.4	1.32	1.9	2.15	2.5		580.0
600.0		0.067	0.4	0.128	0.6	0.372	1.0	0.733	1.4	1.34	2.0	2.19	2.5		600.0
620.0		0.068	0.4	0.131	0.6	0.379	1.0	0.747	1.4	1.37	2.0	2.23	2.5		620.0
640.0		0.069	0.4	0.133	0.6	0.386	1.0	0.760	1.5	1.39	2.0	2.27	2.5		640.0
660.0		0.070	0.4	0.135	0.6	0.392	1.0	0.773	1.5	1.42	2.0	2.31	2.5		660.0
680.0		0.072	0.4	0.138	0.6	0.399	1.0	0.786	1.5	1.44	2.0	2.35	2.5		680.0
700.0		0.073	0.4	0.140	0.6	0.405	1.0	0.799	1.5	1.46	2.0	2.39	2.5		700.0
720.0		0.074	0.4	0.142	0.6	0.412	1.0	0.811	1.5	1.49	2.0	2.43	2.5		720.0
740.0		0.075	0.4	0.144	0.6	0.418	1.0	0.824	1.5	1.51	2.0	2.46	2.5		740.0
760.0		0.076	0.4	0.146	0.6	0.424	1.1	0.836	1.5	1.53	2.0	2.50	2.6	2.0	760.0
780.0		0.077	0.4	0.149	0.6	0.431	1.1	0.848	1.5	1.55	2.0	2.53	2.6		780.0
800.0		0.079	0.4	0.151	0.6	0.437	1.1	0.860	1.5	1.57	2.0	2.57	2.6		800.0
820.0		0.080	0.4	0.153	0.6	0.443	1.1	0.872	1.5	1.60	2.0	2.60	2.6		820.0
840.0		0.081	0.4	0.155	0.6	0.449	1.1	0.884	1.5	1.62	2.0	2.64	2.6		840.0
860.0		0.082	0.4	0.157	0.6	0.455	1.1	0.895	1.5	1.64	2.0	2.67	2.6		860.0
880.0		0.083	0.4	0.159	0.6	0.461	1.1	0.907	1.5	1.66	2.0	2.71	2.6		880.0
900.0		0.084	0.4	0.161	0.6	0.466	1.1	0.918	1.5	1.68	2.0	2.74	2.6		900.0
920.0		0.085	0.4	0.163	0.6	0.472	1.1	0.929	1.5	1.70	2.0	2.77	2.6		920.0
940.0		0.086	0.4	0.165	0.6	0.478	1.1	0.940	1.5	1.72	2.0	2.81	2.6		940.0
960.0		0.087	0.4	0.167	0.6	0.483	1.1	0.951	1.5	1.74	2.1	2.84	2.6		960.0
980.0		0.088	0.5	0.169	0.6	0.489	1.1	0.962	1.5	1.76	2.1	2.87	2.6		980.0
1000.0	1.0	0.089	0.5	0.171	0.6	0.494	1.1	0.973	1.5	1.78	2.1	2.90	2.6		1000.0
1100.0		0.094	0.5	0.180	0.6	0.521	1.1	1.03	1.5	1.88	2.1	3.06	2.6		1100.0
1200.0		0.099	0.5	0.189	0.6	0.547	1.1	1.08	1.6	1.97	2.1	3.21	2.7		1200.0
1300.0		0.103	0.5	0.198	0.7	0.572	1.1	1.13	1.6	2.06	2.1	3.35	2.7		1300.0
1400.0		0.108	0.5	0.206	0.7	0.596	1.1	1.17	1.6	2.14	2.1	3.49	2.7		1400.0
1500.0		0.112	0.5	0.214	0.7	0.619	1.1	1.22	1.6	2.22	2.1	3.63	2.7	3.0	1500.0
1600.0		0.116	0.5	0.222	0.7	0.642	1.1	1.26	1.6	2.30	2.2	3.76	2.7		1600.0
1700.0		0.120	0.5	0.230	0.7	0.663	1.1	1.30	1.6	2.38	2.2	3.88	2.8		1700.0
1800.0		0.124	0.5	0.237	0.7	0.685	1.2	1.35	1.6	2.46	2.2	4.00	2.8		1800.0
1900.0		0.128	0.5	0.244	0.7	0.705	1.2	1.39	1.6	2.53	2.2	4.12	2.8		1900.0
2000.0		0.132	0.5	0.251	0.7	0.726	1.2	1.43	1.6	2.60	2.2	4.24	2.8		2000.0

Table 4.14 Flow of water at 75°C in copper pipes — *continued*

q_m　=　mass flow rate　　　　　　　　　　　kg.s^{-1}
c　=　velocity　　　　　　　　　　　　　　　m.s^{-1}
$\Delta p/l$　=　pressure drop per unit length　　　　Pa.m^{-1}
l_e　=　equivalent length of a component
　　　　for $\zeta = 1$　　　　　　　　　　　　m

COPPER, TABLE Z
WATER AT 75°C

$\Delta p/l$	c	54 mm		67 mm		76 mm		108 mm		133 mm		159 mm		c	$\Delta p/l$
		q_m	l_e	q_m	l_e	q_m	l_e	q_m	l_e	q_m	l_e	q_m	l_e		
0.1		0.031	1.1	0.056	1.5	0.080	1.8	0.215	3.1	0.382	4.2	0.627	5.5		0.1
0.2		0.046	1.2	0.084	1.7	0.120	2.0	0.322	3.4	0.569	4.7	0.932	6.0	0.05	0.2
0.3		0.059	1.3	0.106	1.8	0.152	2.2	0.406	3.7	0.717	4.9	1.17	6.4		0.3
0.4		0.070	1.4	0.125	1.9	0.180	2.3	0.479	3.8	0.845	5.1	1.38	6.6		0.4
0.5		0.079	1.4	0.142	1.9	0.205	2.3	0.544	3.9	0.959	5.3	1.57	6.8		0.5
0.6		0.088	1.4	0.158	2.0	0.227	2.4	0.603	4.0	1.06	5.4	1.74	7.0	0.10	0.6
0.7		0.096	1.5	0.173	2.0	0.248	2.5	0.659	4.1	1.16	5.5	1.90	7.1		0.7
0.8	0.05	0.104	1.5	0.187	2.1	0.268	2.5	0.711	4.2	1.25	5.6	2.04	7.3		0.8
0.9		0.112	1.5	0.200	2.1	0.287	2.6	0.760	4.3	1.34	5.7	2.18	7.4		0.9
1.0		0.119	1.6	0.213	2.1	0.305	2.6	0.807	4.3	1.42	5.8	2.32	7.5	0.15	1.0
1.5		0.150	1.7	0.268	2.3	0.384	2.8	1.02	4.6	1.79	6.1	2.91	7.8		1.5
2.0		0.177	1.7	0.316	2.4	0.453	2.9	1.19	4.7	2.10	6.3	3.42	8.1		2.0
2.5	0.10	0.201	1.8	0.359	2.5	0.514	3.0	1.35	4.9	2.38	6.5	3.87	8.3		2.5
3.0		0.223	1.9	0.398	2.5	0.570	3.0	1.50	5.0	2.64	6.7	4.29	8.5		3.0
3.5		0.244	1.9	0.435	2.6	0.622	3.1	1.64	5.1	2.87	6.8	4.67	8.7		3.5
4.0		0.263	1.9	0.469	2.6	0.671	3.1	1.76	5.2	3.10	6.9	5.03	8.8		4.0
4.5		0.281	2.0	0.502	2.7	0.717	3.2	1.89	5.3	3.31	7.0	5.38	8.9	0.30	4.5
5.0	0.15	0.299	2.0	0.532	2.7	0.761	3.2	2.00	5.3	3.51	7.1	5.70	9.0		5.0
5.5		0.315	2.0	0.562	2.7	0.803	3.3	2.11	5.4	3.70	7.2	6.01	9.1		5.5
6.0		0.331	2.0	0.590	2.8	0.843	3.3	2.21	5.4	3.88	7.2	6.31	9.2		6.0
6.5		0.347	2.1	0.618	2.8	0.882	3.4	2.32	5.5	4.06	7.3	6.59	9.3		6.5
7.0		0.362	2.1	0.644	2.8	0.920	3.4	2.41	5.5	4.23	7.4	6.87	9.4		7.0
7.5		0.376	2.1	0.670	2.8	0.956	3.4	2.51	5.6	4.39	7.4	7.14	9.4		7.5
8.0		0.390	2.1	0.695	2.9	0.992	3.4	2.60	5.6	4.56	7.5	7.40	9.5		8.0
8.5		0.404	2.1	0.719	2.9	1.03	3.5	2.69	5.7	4.71	7.5	7.65	9.6		8.5
9.0		0.417	2.2	0.742	2.9	1.06	3.5	2.78	5.7	4.86	7.6	7.90	9.6		9.0
9.5		0.430	2.2	0.765	2.9	1.09	3.5	2.86	5.7	5.01	7.6	8.14	9.7		9.5
10.0		0.443	2.2	0.788	2.9	1.12	3.5	2.95	5.8	5.16	7.7	8.37	9.7	0.50	10.0
12.5		0.502	2.3	0.893	3.0	1.27	3.6	3.34	5.9	5.84	7.8	9.47	10.0		12.5
15.0		0.557	2.3	0.989	3.1	1.41	3.7	3.69	6.0	6.46	8.0	10.5	10.2		15.0
17.5	0.30	0.607	2.4	1.08	3.2	1.54	3.8	4.02	6.2	7.03	8.1	11.4	10.3		17.5
20.0		0.655	2.4	1.16	3.2	1.66	3.8	4.33	6.2	7.57	8.2	12.3	10.5		20.0
22.5		0.700	2.4	1.24	3.3	1.77	3.9	4.62	6.3	8.08	8.4	13.1	10.6		22.5
25.0		0.742	2.5	1.32	3.3	1.88	3.9	4.90	6.4	8.57	8.4	13.9	10.7		25.0
27.5		0.783	2.5	1.39	3.3	1.98	4.0	5.17	6.5	9.03	8.5	14.6	10.8		27.5
30.0		0.822	2.5	1.46	3.4	2.08	4.0	5.42	6.5	9.47	8.6	15.4	10.9		30.0
32.5		0.860	2.5	1.53	3.4	2.17	4.1	5.67	6.6	9.90	8.7	16.0	11.0		32.5
35.0		0.897	2.6	1.59	3.4	2.27	4.1	5.90	6.6	10.3	8.7	16.7	11.1		35.0
37.5		0.932	2.6	1.65	3.5	2.35	4.1	6.14	6.7	10.7	8.8	17.4	11.2		37.5
40.0		0.966	2.6	1.71	3.5	2.44	4.2	6.36	6.7	11.1	8.9	18.0	11.2		40.0
42.5		1.00	2.6	1.77	3.5	2.52	4.2	6.57	6.8	11.5	8.9	18.6	11.3	1.0	42.5
45.0	0.50	1.03	2.6	1.83	3.5	2.60	4.2	6.79	6.8	11.8	9.0	19.2	11.4		45.0
47.5		1.06	2.7	1.89	3.6	2.68	4.2	6.99	6.8	12.2	9.0	19.8	11.4		47.5
50.0		1.10	2.7	1.94	3.6	2.76	4.3	7.19	6.9	12.5	9.1	20.3	11.5		50.0
52.5		1.13	2.7	1.99	3.6	2.84	4.3	7.39	6.9	12.9	9.1	20.9	11.5		52.5
55.0		1.16	2.7	2.05	3.6	2.91	4.3	7.58	6.9	13.2	9.1	21.4	11.6		55.0
57.5		1.18	2.7	2.10	3.6	2.99	4.3	7.77	7.0	13.5	9.2	21.9	11.6		57.5
60.0		1.21	2.7	2.15	3.6	3.06	4.4	7.95	7.0	13.9	9.2	22.5	11.7		60.0
62.5		1.24	2.7	2.20	3.7	3.13	4.4	8.13	7.0	14.2	9.3	23.0	11.7		62.5
65.0		1.27	2.8	2.24	3.7	3.19	4.4	8.31	7.1	14.5	9.3	23.5	11.8		65.0
67.5		1.29	2.8	2.29	3.7	3.26	4.4	8.48	7.1	14.8	9.3	24.0	11.8		67.5
70.0		1.32	2.8	2.34	3.7	3.32	4.4	8.66	7.1	15.1	9.4	24.4	11.9		70.0
72.5		1.35	2.8	2.38	3.7	3.39	4.4	8.82	7.1	15.4	9.4	24.9	11.9		72.5
75.0		1.37	2.8	2.43	3.7	3.46	4.5	8.99	7.2	15.7	9.4	25.4	11.9		75.0
77.5		1.40	2.8	2.47	3.8	3.52	4.5	9.15	7.2	16.0	9.5	25.8	12.0		77.5
80.0		1.42	2.8	2.52	3.8	3.58	4.5	9.32	7.2	16.2	9.5	26.3	12.0		80.0
82.5		1.45	2.8	2.56	3.8	3.65	4.5	9.47	7.2	16.5	9.5	26.7	12.0		82.5
85.0		1.47	2.8	2.60	3.8	3.71	4.5	9.63	7.3	16.8	9.5	27.2	12.1		85.0
87.5		1.50	2.9	2.65	3.8	3.77	4.5	9.79	7.3	17.1	9.6	27.6	12.1		87.5
90.0		1.52	2.9	2.69	3.8	3.83	4.6	9.94	7.3	17.3	9.6	28.0	12.1	1.5	90.0

Table 4.14 Flow of water at 75°C in copper pipes — *continued*

q_m = mass flow rate   kg.s^{-1}
c = velocity   m.s^{-1}
$\Delta p/l$ = pressure drop per unit length   Pa.m^{-1}
l_e = equivalent length of a component
     for $\zeta = 1$ m

COPPER, TABLE Z
WATER AT 75°C

$\Delta p/l$	c	54 mm q_m	l_e	67 mm q_m	l_e	76 mm q_m	l_e	108 mm q_m	l_e	133 mm q_m	l_e	159 mm q_m	l_e	c	$\Delta p/l$
92.5		1.54	2.9	2.73	3.8	3.88	4.6	10.1	7.3	17.6	9.6	28.5	12.2		92.5
95.0		1.57	2.9	2.77	3.8	3.94	4.6	10.2	7.3	17.8	9.6	28.9	12.2		95.0
97.5		1.59	2.9	2.81	3.8	4.00	4.6	10.4	7.4	18.1	9.7	29.3	12.2		97.5
100.0		1.61	2.9	2.85	3.9	4.05	4.6	10.5	7.4	18.4	9.7	29.7	12.3		100.0
120.0		1.78	3.0	3.15	3.9	4.48	4.7	11.6	7.5	20.3	9.9	32.8	12.5		120.0
140.0	1.0	1.94	3.0	3.43	4.0	4.88	4.8	12.7	7.6	22.1	10.0	35.7	12.6	2.0	140.0
160.0		2.09	3.1	3.70	4.1	5.25	4.8	13.6	7.7	23.7	10.1	38.4	12.8		160.0
180.0		2.23	3.1	3.94	4.1	5.61	4.9	14.5	7.8	25.3	10.2	40.9	12.9		180.0
200.0		2.37	3.1	4.18	4.1	5.94	4.9	15.4	7.9	26.8	10.3	43.3	13.0		200.0
220.0		2.49	3.2	4.40	4.2	6.26	5.0	16.2	8.0	28.2	10.4	45.6	13.1		220.0
240.0		2.62	3.2	4.62	4.2	6.57	5.0	17.0	8.0	29.6	10.5	47.8	13.2		240.0
260.0		2.74	3.2	4.83	4.3	6.86	5.1	17.8	8.1	30.9	10.6	49.9	13.3		260.0
280.0		2.85	3.2	5.03	4.3	7.14	5.1	18.5	8.1	32.2	10.6	52.0	13.4		280.0
300.0		2.96	3.3	5.22	4.3	7.42	5.1	19.2	8.2	33.4	10.7	53.9	13.5		300.0
320.0	1.5	3.07	3.3	5.41	4.3	7.69	5.2	19.9	8.2	34.6	10.8	55.9	13.6	3.0	320.0
340.0		3.17	3.3	5.59	4.4	7.95	5.2	20.6	8.3	35.7	10.8	57.7	13.6		340.0
360.0		3.27	3.3	5.77	4.4	8.20	5.2	21.2	8.3	36.9	10.9	59.5	13.7		360.0
380.0		3.37	3.3	5.95	4.4	8.44	5.3	21.8	8.4	38.0	10.9	61.3	13.8		380.0
400.0		3.47	3.4	6.12	4.4	8.68	5.3	22.5	8.4	39.0	11.0	63.0	13.8		400.0
420.0		3.56	3.4	6.28	4.5	8.92	5.3	23.1	8.4	40.1	11.0	64.7	13.9		420.0
440.0		3.65	3.4	6.44	4.5	9.15	5.3	23.6	8.5	41.1	11.1	66.4	13.9		440.0
460.0		3.74	3.4	6.60	4.5	9.37	5.3	24.2	8.5	42.1	11.1	68.0	14.0		460.0
480.0		3.83	3.4	6.76	4.5	9.59	5.4	24.7	8.5	43.1	11.1	69.5	14.0		480.0
500.0		3.92	3.4	6.91	4.5	9.81	5.4	25.3	8.6	44.0	11.2	71.1	14.1		500.0
520.0		4.00	3.4	7.06	4.6	10.0	5.4	25.9	8.6	45.0	11.2	72.6	14.1		520.0
540.0		4.09	3.5	7.21	4.6	10.2	5.4	26.4	8.6	45.9	11.2	74.1	14.1	4.0	540.0
560.0	2.0	4.17	3.5	7.35	4.6	10.4	5.4	27.0	8.6	46.8	11.3	75.6	14.2		560.0
580.0		4.25	3.5	7.49	4.6	10.6	5.5	27.5	8.7	47.7	11.3	77.0	14.2		580.0
600.0		4.33	3.5	7.63	4.6	10.8	5.5	28.0	8.7	48.6	11.3	78.4	14.3		600.0
620.0		4.41	3.5	7.77	4.6	11.0	5.5	28.5	8.7	49.5	11.4	79.8	14.3		620.0
640.0		4.49	3.5	7.91	4.6	11.2	5.5	29.0	8.7	50.3	11.4	81.2	14.3		640.0
660.0		4.56	3.5	8.04	4.7	11.4	5.5	29.5	8.8	51.2	11.4	82.6	14.4		660.0
680.0		4.64	3.5	8.17	4.7	11.6	5.5	29.9	8.8	52.0	11.4	83.9	14.4		680.0
700.0		4.71	3.5	8.30	4.7	11.8	5.6	30.4	8.8	52.8	11.5	85.2	14.4		700.0
720.0		4.79	3.6	8.43	4.7	12.0	5.6	30.9	8.8	53.6	11.5	86.5	14.5		720.0
740.0		4.86	3.6	8.56	4.7	12.1	5.6	31.3	8.8	54.4	11.5	87.8	14.5		740.0
760.0		4.93	3.6	8.68	4.7	12.3	5.6	31.8	8.9	55.2	11.5	89.1	14.5		760.0
780.0		5.00	3.6	8.81	4.7	12.5	5.6	32.2	8.9	56.0	11.6	90.3	14.5		780.0
800.0		5.07	3.6	8.93	4.7	12.7	5.6	32.7	8.9	56.8	11.6	91.6	14.6		800.0
820.0		5.14	3.6	9.05	4.7	12.8	5.6	33.1	8.9	57.5	11.6	92.8	14.6	5.0	820.0
840.0		5.21	3.6	9.17	4.8	13.0	5.6	33.6	8.9	58.3	11.6	94.0	14.6		840.0
860.0		5.28	3.6	9.29	4.8	13.2	5.7	34.0	8.9	59.0	11.7	95.2	14.6		860.0
880.0		5.34	3.6	9.41	4.8	13.3	5.7	34.4	9.0	59.8	11.7	96.4	14.7		880.0
900.0		5.41	3.6	9.52	4.8	13.5	5.7	34.8	9.0	60.5	11.7	97.5	14.7		900.0
920.0		5.47	3.6	9.64	4.8	13.7	5.7	35.3	9.0	61.2	11.7	98.7	14.7		920.0
940.0		5.54	3.6	9.75	4.8	13.8	5.7	35.7	9.0	61.9	11.7	99.8	14.7		940.0
960.0		5.60	3.7	9.86	4.8	14.0	5.7	36.1	9.0	62.6	11.8	101	14.8		960.0
980.0		5.67	3.7	9.97	4.8	14.1	5.7	36.5	9.0	63.3	11.8	102	14.8		980.0
1000.0		5.73	3.7	10.1	4.8	14.3	5.7	36.9	9.0	64.0	11.8	103	14.8		1000.0
1100.0	3.0	6.03	3.7	10.6	4.9	15.1	5.8	38.8	9.1	67.4	11.9	108	14.9	6.0	1100.0
1200.0		6.33	3.7	11.1	4.9	15.8	5.8	40.7	9.2	70.6	12.0				1200.0
1300.0		6.61	3.8	11.6	4.9	16.5	5.9	42.5	9.2	73.7	12.0				1300.0
1400.0		6.88	3.8	12.1	5.0	17.1	5.9	44.2	9.3	76.7	12.1				1400.0
1500.0		7.14	3.8	12.6	5.0	17.8	5.9	45.9	9.3						1500.0
1600.0		7.40	3.8	13.0	5.0	18.4	6.0	47.5	9.4						1600.0
1700.0		7.65	3.8	13.4	5.1	19.1	6.0	49.1	9.4						1700.0
1800.0		7.89	3.9	13.9	5.1	19.7	6.0	50.6	9.5						1800.0
1900.0		8.12	3.9	14.3	5.1	20.2	6.0								1900.0
2000.0		8.35	3.9	14.7	5.1	20.8	6.1								2000.0

Table 4.15 Flow of water at 75°C in galvanised pipes

q_m = mass flow rate kg.s^{-1}
c = velocity m.s^{-1}
$\Delta p/l$ = pressure drop per unit length Pa.m^{-1}
l_e = equivalent length of a component
for $\zeta = 1$ m

GALVANISED STEEL

WATER AT 75°C

$\Delta p/l$	c	10 mm q_m	l_e	15 mm q_m	l_e	20 mm q_m	l_e	25 mm q_m	l_e	32 mm q_m	l_e	40 mm q_m	l_e	c	$\Delta p/l$
50		0.013	0.21	0.029	0.3	0.069	0.5	0.129	0.7	0.287	1.0	0.441	1.3		50
55		0.014	0.21	0.030	0.3	0.072	0.5	0.136	0.7	0.301	1.0	0.463	1.3		55
60		0.014	0.21	0.031	0.3	0.076	0.5	0.142	0.7	0.315	1.0	0.485	1.3		60
70		0.016	0.21	0.034	0.3	0.082	0.5	0.154	0.7	0.342	1.1	0.525	1.3		70
80		0.017	0.21	0.037	0.3	0.088	0.5	0.165	0.7	0.366	1.1	0.563	1.3		80
90		0.018	0.22	0.039	0.3	0.094	0.5	0.176	0.7	0.389	1.1	0.598	1.3		90
100		0.019	0.22	0.041	0.3	0.099	0.5	0.186	0.7	0.411	1.1	0.631	1.3		100
110		0.020	0.22	0.043	0.3	0.104	0.5	0.195	0.7	0.432	1.1	0.663	1.3		110
120		0.021	0.22	0.045	0.3	0.109	0.5	0.204	0.7	0.452	1.1	0.693	1.3		120
130		0.022	0.22	0.047	0.3	0.113	0.5	0.213	0.7	0.471	1.1	0.723	1.3		130
140		0.023	0.22	0.049	0.3	0.118	0.5	0.221	0.7	0.489	1.1	0.751	1.3		140
150		0.023	0.22	0.051	0.3	0.122	0.5	0.229	0.7	0.507	1.1	0.778	1.3		150
175		0.025	0.23	0.055	0.3	0.132	0.5	0.248	0.7	0.549	1.1	0.842	1.4		175
200		0.027	0.23	0.059	0.3	0.142	0.5	0.266	0.7	0.588	1.1	0.901	1.4		200
225		0.029	0.23	0.063	0.3	0.151	0.5	0.282	0.7	0.624	1.1	0.957	1.4		225
250		0.031	0.23	0.067	0.3	0.159	0.5	0.298	0.7	0.659	1.1	1.01	1.4		250
275		0.032	0.23	0.070	0.4	0.167	0.5	0.313	0.7	0.692	1.1	1.06	1.4		275
300		0.034	0.23	0.073	0.4	0.175	0.5	0.327	0.7	0.723	1.1	1.11	1.4		300
350		0.037	0.23	0.079	0.4	0.189	0.5	0.354	0.7	0.782	1.1	1.20	1.4	1.0	350
400		0.039	0.23	0.085	0.4	0.203	0.5	0.379	0.8	0.837	1.1	1.28	1.4		400
450		0.042	0.23	0.090	0.4	0.215	0.6	0.403	0.8	0.889	1.1	1.36	1.4		450
500		0.044	0.24	0.095	0.4	0.227	0.6	0.425	0.8	0.938	1.1	1.44	1.4		500
550		0.046	0.24	0.100	0.4	0.239	0.6	0.446	0.8	0.985	1.1	1.51	1.4		550
600		0.048	0.24	0.105	0.4	0.250	0.6	0.467	0.8	1.03	1.1	1.58	1.4		600
700		0.052	0.24	0.113	0.4	0.270	0.6	0.505	0.8	1.11	1.1	1.70	1.4		700
800		0.056	0.24	0.121	0.4	0.289	0.6	0.540	0.8	1.19	1.1	1.82	1.4		800
900		0.059	0.24	0.129	0.4	0.307	0.6	0.573	0.8	1.26	1.1	1.94	1.4		900
1000		0.063	0.24	0.136	0.4	0.324	0.6	0.605	0.8	1.33	1.1	2.04	1.4		1000
1250		0.070	0.24	0.153	0.4	0.363	0.6	0.677	0.8	1.49	1.1	2.29	1.4	2.0	1250
1500		0.077	0.24	0.167	0.4	0.398	0.6	0.743	0.8	1.64	1.1	2.51	1.4		1500
1750		0.084	0.24	0.181	0.4	0.430	0.6	0.803	0.8	1.77	1.1	2.71	1.4		1750
2000	1.0	0.089	0.24	0.194	0.4	0.460	0.6	0.859	0.8	1.89	1.1	2.90	1.4		2000
2250		0.095	0.24	0.206	0.4	0.489	0.6	0.912	0.8	2.01	1.1	3.08	1.4		2250
2500		0.100	0.25	0.217	0.4	0.515	0.6	0.962	0.8	2.12	1.1	3.24	1.4		2500
2750		0.105	0.25	0.228	0.4	0.541	0.6	1.01	0.8	2.22	1.1	3.40	1.4		2750
3000		0.110	0.25	0.238	0.4	0.565	0.6	1.06	0.8	2.32	1.1	3.56	1.4	3.0	3000
3500		0.119	0.25	0.257	0.4	0.611	0.6	1.14	0.8	2.51	1.1	3.84	1.4		3500
4000		0.127	0.25	0.275	0.4	0.653	0.6	1.22	0.8	2.68	1.1	4.11	1.4		4000
4500		0.135	0.25	0.292	0.4	0.693	0.6	1.29	0.8	2.85	1.1	4.36	1.4		4500
5000		0.142	0.25	0.308	0.4	0.731	0.6	1.36	0.8	3.00	1.1	4.60	1.4		5000
5500		0.149	0.25	0.323	0.4	0.767	0.6	1.43	0.8	3.15	1.1	4.82	1.4	4.0	5500
6000		0.156	0.25	0.338	0.4	0.802	0.6	1.50	0.8	3.29	1.1	5.04	1.4		6000
7000	2.0	0.169	0.25	0.365	0.4	0.866	0.6	1.62	0.8	3.56	1.1	5.44	1.4		7000
8000		0.181	0.25	0.391	0.4	0.927	0.6	1.73	0.8	3.80	1.1	5.82	1.4	5.0	8000
9000		0.192	0.25	0.414	0.4	0.983	0.6	1.83	0.8	4.04	1.2	6.18	1.4		9000
10 000		0.202	0.25	0.437	0.4	1.04	0.6	1.93	0.8	4.26	1.2	6.51	1.4		10 000
11 000		0.212	0.25	0.458	0.4	1.09	0.6	2.03	0.8	4.46	1.2	6.83	1.4		11 000
12 000		0.222	0.25	0.479	0.4	1.14	0.6	2.12	0.8	4.66	1.2	7.14	1.4		12 000

Table 4.15 Flow of water at 75°C in galvanised pipes — *continued*

q_m	=	mass flow rate	kg.s^{-1}
c	=	velocity	m.s^{-1}
$\Delta p/l$	=	pressure drop per unit length	Pa.m^{-1}
l_e	=	equivalent length of a component for $\zeta = 1$	m

GALVANISED STEEL

WATER AT 75°C

$\Delta p/l$	c	50 mm		65 mm		80 mm		100 mm		125 mm		150 mm		c	$\Delta p/l$
		q_m	l_e	q_m	l_e	q_m	l_e	q_m	l_e	q_m	l_e	q_m	l_e		
50		0.84	1.8	1.73	2.6	2.70	3.2	5.50	4.5	9.90	6.0	16.0	7.6		50
55		0.88	1.8	1.82	2.6	2.84	3.2	5.77	4.5	10.4	6.0	16.8	7.6		55
60		0.93	1.8	1.90	2.6	2.97	3.2	6.04	4.5	10.9	6.0	17.5	7.6	1.0	60
70		1.01	1.8	2.06	2.6	3.21	3.2	6.53	4.6	11.7	6.0	19.0	7.6		70
80		1.08	1.8	2.21	2.6	3.44	3.2	6.99	4.6	12.6	6.0	20.3	7.6		80
90		1.15	1.8	2.34	2.6	3.65	3.2	7.43	4.6	13.4	6.1	21.5	7.6		90
100		1.21	1.8	2.47	2.6	3.85	3.2	7.84	4.6	14.1	6.1	22.7	7.7		100
110		1.27	1.8	2.60	2.6	4.05	3.3	8.22	4.6	14.8	6.1	23.8	7.7		110
120		1.33	1.8	2.72	2.6	4.23	3.3	8.60	4.6	15.5	6.1	24.9	7.7		120
130		1.38	1.8	2.83	2.6	4.40	3.3	8.95	4.6	16.1	6.1	25.9	7.7		130
140		1.44	1.8	2.94	2.6	4.57	3.3	9.30	4.6	16.7	6.1	26.9	7.7		140
150		1.49	1.8	3.04	2.6	4.74	3.3	9.63	4.6	17.3	6.1	27.9	7.7		150
175		1.61	1.9	3.29	2.6	5.12	3.3	10.4	4.6	18.7	6.1	30.2	7.7		175
200		1.72	1.9	3.52	2.6	5.48	3.3	11.1	4.6	20.0	6.1	32.3	7.7		200
225		1.83	1.9	3.74	2.6	5.82	3.3	11.8	4.6	21.2	6.1	34.2	7.7		225
250	1.0	1.93	1.9	3.95	2.7	6.14	3.3	12.5	4.6	22.4	6.2	36.1	7.7	2.0	250
275		2.03	1.9	4.14	2.7	6.45	3.3	13.1	4.6	23.5	6.2	37.9	7.7		275
300		2.12	1.9	4.33	2.7	6.74	3.3	13.7	4.6	24.6	6.2	39.6	7.7		300
350		2.29	1.9	4.68	2.7	7.28	3.3	14.8	4.7	26.6	6.2	42.8	7.8		350
400		2.46	1.9	5.01	2.7	7.79	3.3	15.8	4.7	28.4	6.2	45.8	7.8		400
450		2.61	1.9	5.32	2.7	8.27	3.3	16.8	4.7	30.2	6.2	48.6	7.8		450
500		2.75	1.9	5.61	2.7	8.72	3.3	17.7	4.7	31.8	6.2	51.2	7.8		500
550		2.88	1.9	5.88	2.7	9.15	3.3	18.6	4.7	33.4	6.2	53.7	7.8	3.0	550
600		3.02	1.9	6.15	2.7	9.56	3.3	19.4	4.7	34.8	6.2	56.1	7.8		600
700		3.26	1.9	6.65	2.7	10.3	3.3	21.0	4.7	37.7	6.2	60.7	7.8		700
800		3.49	1.9	7.11	2.7	11.1	3.3	22.4	4.7	40.3	6.2	64.9	7.8		800
900		3.70	1.9	7.55	2.7	11.7	3.3	23.8	4.7	42.7	6.2	68.8	7.8	4.0	900
1000	2.0	3.90	1.9	7.96	2.7	12.4	3.3	25.1	4.7	45.1	6.2	72.6	7.8		1000
1250		4.37	1.9	8.90	2.7	13.8	3.3	28.1	4.7	50.4	6.2	81.2	7.8		1250
1500		4.79	1.9	9.76	2.7	15.2	3.4	30.8	4.7	55.2	6.2	89.0	7.8	5.0	1500
1750		5.18	1.9	10.6	2.7	16.4	3.4	33.3	4.7	59.7	6.2	96.1	7.8		1750
2000		5.54	1.9	11.3	2.7	17.5	3.4	35.6	4.7	63.8	6.2	103	7.8	6.0	2000
2250	3.0	5.87	1.9	12.0	2.7	18.6	3.4	37.7	4.7	67.7	6.2				2250
2500		6.19	1.9	12.6	2.7	19.6	3.4	39.8	4.7	71.4	6.2				2500
2750		6.50	1.9	13.2	2.7	20.6	3.4	41.7	4.7	74.9	6.2				2750
3000		6.79	1.9	13.8	2.7	21.5	3.4	43.6	4.7						3000
3500		7.34	1.9	14.9	2.7	23.2	3.4	47.1	4.7						3500
4000	4.0	7.84	1.9	16.0	2.7	24.8	3.4								4000
4500		8.32	1.9	17.0	2.7	26.4	3.4								4500
5000		8.78	1.9	17.9	2.7	27.8	3.4								5000
5500		9.20	1.9	18.8	2.7										5500
6000	5.0	9.62	1.9	19.6	2.7										6000
7000		10.4	1.9												7000
8000		11.1	1.9												8000
9000	6.0	11.8	1.9												9000
10 000															10 000
11 000															11 000
12 000															12 000

Table 4.16 Flow of water at 10°C in heavy grade steel pipes

q_m	= mass flow rate	kg.s^{-1}
c	= velocity	m.s^{-1}
$\Delta p/l$	= pressure drop per unit length	Pa.m^{-1}
l_e	= equivalent length of a component for $\zeta = 1$	m

HEAVY GRADE STEEL

WATER AT 10°C

* $(Re) = 2000$
† $(Re) = 3000$

$\Delta p/l$	c	10 mm		15 mm		20 mm		25 mm		32 mm		40 mm		c	$\Delta p/l$
		q_m	l_e	q_m	l_e	q_m	l_e	q_m	l_e	q_m	l_e	q_m	l_e		
0.1								0.001	0.0	0.003	0.1	0.005	0.1		0.1
0.2								0.002	0.0	0.005	0.1	0.010	0.2		0.2
0.3								0.002	0.1	0.008	0.1	0.015	0.2		0.3
0.4						0.001	0.0	0.003	0.1	0.010	0.2	0.020	0.3		0.4
0.5						0.002	0.0	0.004	0.1	0.013	0.2	0.025	0.4		0.5
0.6						0.002	0.0	0.005	0.1	0.016	0.3	0.029	0.5		0.6
0.7						0.002	0.1	0.006	0.1	0.018	0.3	0.034	0.5		0.7
0.8						0.003	0.1	0.007	0.1	0.021	0.3	0.039	0.6		0.8
0.9						0.003	0.1	0.007	0.1	0.023	0.4	0.044	0.7		0.9
1.0						0.003	0.1	0.008	0.1	0.026	0.4	0.049	0.8	0.05	1.0
1.5				0.001	0.0	0.005	0.1	0.012	0.2	0.039	0.6	0.074	1.1		1.5
2.0				0.002	0.0	0.007	0.1	0.016	0.3	0.052	0.8	0.086★	1.2		2.0
2.5				0.002	0.1	0.008	0.1	0.020	0.3	0.065	1.0	0.092	1.1		2.5
3.0				0.003	0.1	0.010	0.2	0.025	0.4	0.072	1.0	0.098	1.0		3.0
3.5				0.003	0.1	0.011	0.2	0.029	0.5	0.075★	1.0	0.103	1.0		3.5
4.0				0.004	0.1	0.013	0.2	0.033	0.5	0.078	0.9	0.109	0.9		4.0
4.5				0.004	0.1	0.015	0.2	0.037	0.6	0.081	0.9	0.115	0.9		4.5
5.0		0.001	0.0	0.005	0.1	0.016	0.3	0.041	0.6	0.084	0.8	0.121†	0.9		5.0
5.5		0.002	0.0	0.005	0.1	0.018	0.3	0.045	0.7	0.087	0.8	0.127	0.9		5.5
6.0		0.002	0.0	0.006	0.1	0.020	0.3	0.049	0.8	0.090	0.8	0.134	0.9		6.0
6.5		0.002	0.0	0.006	0.1	0.021	0.3	0.053	0.8	0.093	0.8	0.140	0.9		6.5
7.0		0.002	0.0	0.006	0.1	0.023	0.4	0.054	0.8	0.096	0.8	0.146	1.0		7.0
7.5		0.002	0.1	0.007	0.1	0.024	0.4	0.055★	0.8	0.099	0.8	0.152	1.0		7.5
8.0		0.002	0.1	0.007	0.1	0.026	0.4	0.056	0.7	0.103	0.8	0.158	1.0		8.0
8.5		0.003	0.1	0.008	0.1	0.028	0.4	0.057	0.7	0.106†	0.8	0.164	1.0		8.5
9.0		0.003	0.1	0.008	0.1	0.029	0.5	0.058	0.7	0.109	0.8	0.169	1.0		9.0
9.5		0.003	0.1	0.009	0.1	0.031	0.5	0.059	0.7	0.113	0.8	0.174	1.0		9.5
10.0		0.003	0.1	0.009	0.2	0.033	0.5	0.059	0.7	0.116	0.8	0.180	1.0	0.15	10.0
12.5		0.004	0.1	0.012	0.2	0.041	0.6	0.064	0.6	0.132	0.8	0.204	1.0		12.5
15.0	0.05	0.004	0.1	0.014	0.2	0.043	0.6	0.069	0.6	0.147	0.8	0.227	1.1		15.0
17.5		0.005	0.1	0.016	0.3	0.045★	0.6	0.074	0.6	0.160	0.9	0.248	1.1		17.5
20.0		0.006	0.1	0.019	0.3	0.047	0.5	0.078†	0.6	0.173	0.9	0.268	1.1		20.0
22.5		0.007	0.1	0.021	0.3	0.049	0.5	0.084	0.6	0.185	0.9	0.286	1.1		22.5
25.0		0.007	0.1	0.023	0.4	0.051	0.5	0.089	0.6	0.197	0.9	0.304	1.2		25.0
27.5		0.008	0.1	0.025	0.4	0.053	0.5	0.094	0.6	0.208	0.9	0.321	1.2		27.5
30.0		0.009	0.2	0.028	0.4	0.055	0.5	0.099	0.6	0.219	0.9	0.337	1.2		30.0
32.5		0.010	0.2	0.030	0.5	0.056	0.5	0.104	0.6	0.229	0.9	0.353	1.2		32.5
35.0		0.010	0.2	0.031★	0.5	0.058	0.5	0.108	0.6	0.239	1.0	0.368	1.2	0.30	35.0
37.5		0.011	0.2	0.032	0.4	0.060	0.5	0.112	0.6	0.248	1.0	0.383	1.2		37.5
40.0		0.012	0.2	0.032	0.4	0.062†	0.5	0.117	0.6	0.257	1.0	0.397	1.2		40.0
42.5		0.013	0.2	0.033	0.4	0.064	0.5	0.121	0.6	0.266	1.0	0.411	1.2		42.5
45.0		0.013	0.2	0.033	0.4	0.066	0.5	0.125	0.6	0.275	1.0	0.424	1.2		45.0
47.5		0.014	0.2	0.034	0.4	0.068	0.5	0.129	0.7	0.284	1.0	0.437	1.3		47.5
50.0	0.15	0.015	0.2	0.034	0.4	0.070	0.5	0.133	0.7	0.292	1.0	0.450	1.3		50.0
52.5		0.016	0.3	0.035	0.4	0.072	0.5	0.136	0.7	0.300	1.0	0.463	1.3		52.5
55.0		0.016	0.3	0.035	0.4	0.074	0.5	0.140	0.7	0.308	1.0	0.475	1.3		55.0
57.5		0.017	0.3	0.036	0.4	0.076	0.5	0.144	0.7	0.316	1.0	0.487	1.3		57.5
60.0		0.018	0.3	0.036	0.4	0.078	0.5	0.147	0.7	0.324	1.0	0.499	1.3		60.0
62.5		0.018	0.3	0.037	0.4	0.080	0.5	0.151	0.7	0.331	1.0	0.510	1.3		62.5
65.0		0.019	0.3	0.037	0.4	0.082	0.5	0.154	0.7	0.339	1.0	0.521	1.3		65.0
67.5		0.020	0.3	0.038	0.4	0.084	0.5	0.157	0.7	0.346	1.0	0.533	1.3		67.5
70.0		0.021	0.3	0.038	0.4	0.085	0.5	0.161	0.7	0.353	1.0	0.544	1.3		70.0
72.5		0.021	0.3	0.039	0.3	0.087	0.5	0.164	0.7	0.360	1.1	0.554	1.3		72.5
75.0		0.022	0.4	0.039	0.3	0.089	0.5	0.167	0.7	0.367	1.1	0.565	1.3		75.0
77.5		0.023	0.4	0.040	0.3	0.090	0.5	0.170	0.7	0.374	1.1	0.575	1.3		77.5
80.0		0.023	0.4	0.040	0.3	0.092	0.5	0.173	0.7	0.381	1.1	0.586	1.3		80.0
82.5		0.023★	0.4	0.041	0.3	0.094	0.5	0.176	0.7	0.387	1.1	0.596	1.3		82.5
85.0		0.023	0.3	0.041	0.3	0.095	0.5	0.179	0.7	0.394	1.1	0.606	1.3		85.0
87.5		0.024	0.3	0.042	0.3	0.097	0.5	0.182	0.7	0.400	1.1	0.616	1.3		87.5
90.0		0.024	0.3	0.042	0.3	0.098	0.5	0.185	0.7	0.407	1.1	0.625	1.4		90.0

Table 4.16 Flow of water at 10°C in heavy grade steel pipes — *continued*

q_m = mass flow rate kg.s^{-1}
c = velocity m.s^{-1}
$\Delta p/l$ = pressure drop per unit length Pa.m^{-1}
l_e = equivalent length of a component
    for $\zeta = 1$ m

* $(Re) = 2000$
† $(Re) = 3000$

HEAVY GRADE STEEL
WATER AT 10°C

$\Delta p/l$	c	10 mm		15 mm		20 mm		25 mm		32 mm		40 mm		c	$\Delta p/l$
		q_m	l_e	q_m	l_e	q_m	l_e	q_m	l_e	q_m	l_e	q_m	l_e		
92.5		0.024	0.3	0.043	0.3	0.100	0.5	0.188	0.7	0.413	1.1	0.635	1.4	0.5	92.5
95.0		0.024	0.3	0.043	0.3	0.102	0.5	0.191	0.7	0.419	1.1	0.645	1.4		95.0
97.5		0.024	0.3	0.044	0.3	0.103	0.5	0.194	0.7	0.425	1.1	0.654	1.4		97.5
100.0		0.024	0.3	0.044†	0.3	0.105	0.5	0.197	0.7	0.431	1.1	0.663	1.4		100.0
120.0		0.026	0.3	0.049	0.3	0.116	0.5	0.218	0.7	0.478	1.1	0.734	1.4		120.0
140.0		0.027	0.3	0.053	0.3	0.127	0.5	0.238	0.8	0.520	1.1	0.799	1.4		140.0
160.0		0.028	0.3	0.058	0.3	0.137	0.5	0.256	0.8	0.560	1.2	0.860	1.4		160.0
180.0	0.3	0.030	0.3	0.062	0.4	0.146	0.6	0.274	0.8	0.598	1.2	0.918	1.5		180.0
200.0		0.031	0.3	0.066	0.4	0.155	0.6	0.290	0.8	0.634	1.2	0.972	1.5		200.0
220.0		0.032	0.2	0.069	0.4	0.163	0.6	0.306	0.8	0.668	1.2	1.02	1.5		220.0
240.0		0.033	0.2	0.073	0.4	0.171	0.6	0.321	0.8	0.701	1.2	1.07	1.5		240.0
260.0		0.035†	0.2	0.076	0.4	0.179	0.6	0.336	0.8	0.732	1.2	1.12	1.5		260.0
280.0		0.036	0.2	0.079	0.4	0.187	0.6	0.350	0.8	0.763	1.2	1.17	1.5		280.0
300.0		0.038	0.2	0.082	0.4	0.194	0.6	0.363	0.8	0.792	1.2	1.21	1.5		300.0
320.0		0.039	0.3	0.085	0.4	0.201	0.6	0.377	0.8	0.820	1.2	1.26	1.5	1.0	320.0
340.0		0.040	0.3	0.088	0.4	0.208	0.6	0.389	0.8	0.848	1.2	1.30	1.5		340.0
360.0		0.042	0.3	0.091	0.4	0.215	0.6	0.402	0.8	0.875	1.2	1.34	1.5		360.0
380.0		0.043	0.3	0.094	0.4	0.221	0.6	0.414	0.8	0.901	1.3	1.33	1.6		380.0
400.0		0.044	0.3	0.097	0.4	0.228	0.6	0.426	0.8	0.926	1.3	1.42	1.6		400.0
420.0		0.046	0.3	0.100	0.4	0.234	0.6	0.437	0.8	0.951	1.3	1.46	1.6		420.0
440.0		0.047	0.3	0.102	0.4	0.240	0.6	0.449	0.9	0.976	1.3	1.49	1.6		440.0
460.0		0.048	0.3	0.105	0.4	0.246	0.6	0.460	0.9	1.00	1.3	1.53	1.6		460.0
480.0		0.049	0.3	0.107	0.4	0.252	0.6	0.470	0.9	1.02	1.3	1.57	1.6		480.0
500.0	0.5	0.050	0.3	0.110	0.4	0.258	0.6	0.481	0.9	1.05	1.3	1.60	1.6		500.0
520.0		0.051	0.3	0.112	0.4	0.263	0.6	0.491	0.9	1.07	1.3	1.64	1.6		520.0
540.0		0.053	0.3	0.115	0.4	0.269	0.6	0.502	0.9	1.09	1.3	1.67	1.6		540.0
560.0		0.054	0.3	0.117	0.4	0.274	0.6	0.512	0.9	1.11	1.3	1.70	1.6		560.0
580.0		0.055	0.3	0.119	0.4	0.280	0.6	0.522	0.9	1.13	1.3	1.73	1.6		580.0
600.0		0.056	0.3	0.121	0.4	0.285	0.6	0.531	0.9	1.15	1.3	1.77	1.6		600.0
620.0		0.057	0.3	0.124	0.4	0.290	0.6	0.541	0.9	1.17	1.3	1.80	1.6		620.0
640.0		0.058	0.3	0.126	0.4	0.295	0.6	0.550	0.9	1.20	1.3	1.83	1.6		640.0
660.0		0.059	0.3	0.128	0.4	0.300	0.6	0.560	0.9	1.22	1.3	1.86	1.6		660.0
680.0		0.060	0.3	0.130	0.4	0.305	0.6	0.569	0.9	1.24	1.3	1.89	1.6	1.5	680.0
700.0		0.061	0.3	0.132	0.4	0.310	0.6	0.578	0.9	1.25	1.3	1.92	1.6		700.0
720.0		0.062	0.3	0.134	0.4	0.315	0.6	0.587	0.9	1.27	1.3	1.95	1.6		720.0
740.0		0.063	0.3	0.136	0.4	0.320	0.6	0.596	0.9	1.29	1.3	1.98	1.6		740.0
760.0		0.064	0.3	0.138	0.4	0.324	0.6	0.604	0.9	1.31	1.3	2.00	1.6		760.0
780.0		0.065	0.3	0.140	0.4	0.329	0.7	0.613	0.9	1.33	1.3	2.03	1.6		780.0
800.0		0.065	0.3	0.142	0.4	0.333	0.7	0.621	0.9	1.35	1.3	2.06	1.6		800.0
820.0		0.066	0.3	0.144	0.4	0.338	0.7	0.630	0.9	1.37	1.3	2.09	1.7		820.0
840.0		0.067	0.3	0.146	0.4	0.342	0.7	0.638	0.9	1.38	1.3	2.11	1.7		840.0
860.0		0.068	0.3	0.148	0.4	0.347	0.7	0.646	0.9	1.40	1.3	2.14	1.7		860.0
880.0		0.069	0.3	0.150	0.4	0.351	0.7	0.654	0.9	1.41	1.3	2.17	1.7		880.0
900.0		0.070	0.3	0.152	0.4	0.356	0.7	0.662	0.9	1.44	1.3	2.19	1.7		900.0
920.0		0.071	0.3	0.154	0.4	0.360	0.7	0.670	0.9	1.45	1.3	2.22	1.7		920.0
940.0		0.072	0.3	0.156	0.4	0.364	0.7	0.678	0.9	1.47	1.3	2.25	1.7		940.0
960.0		0.072	0.3	0.157	0.4	0.368	0.7	0.686	0.9	1.49	1.3	2.27	1.7		960.0
980.0		0.073	0.3	0.159	0.4	0.372	0.7	0.693	0.9	1.50	1.4	2.30	1.7		980.0
1000.0		0.074	0.3	0.161	0.4	0.377	0.7	0.701	0.9	1.52	1.4	2.32	1.7		1000.0
1100.0		0.078	0.3	0.170	0.4	0.397	0.7	0.738	0.9	1.60	1.4	2.44	1.7	2.0	1100.0
1200.0		0.082	0.3	0.178	0.4	0.416	0.7	0.773	0.9	1.67	1.4	2.56	1.7		1200.0
1300.0		0.086	0.3	0.186	0.4	0.434	0.7	0.807	0.9	1.75	1.4	2.67	1.7		1300.0
1400.0		0.089	0.3	0.194	0.4	0.452	0.7	0.840	0.9	1.82	1.4	2.77	1.7		1400.0
1500.0		0.093	0.3	0.201	0.4	0.469	0.7	0.871	0.9	1.89	1.4	2.88	1.7		1500.0
1600.0		0.096	0.3	0.208	0.4	0.485	0.7	0.902	0.9	1.95	1.4	2.98	1.7		1600.0
1700.0	1.0	0.099	0.3	0.215	0.5	0.502	0.7	0.932	1.0	2.01	1.4	3.07	1.7		1700.0
1800.0		0.102	0.3	0.222	0.5	0.517	0.7	0.960	1.0	2.08	1.4	3.17	1.7		1800.0
1900.0		0.106	0.3	0.229	0.5	0.532	0.7	0.989	1.0	2.14	1.4	3.26	1.7		1900.0

Table 4.16 Flow of water at 10°C in heavy grade steel pipes — *continued*

q_m = mass flow rate $\quad\quad$ kg.s^{-1}
c = velocity $\quad\quad\quad\quad\quad$ m.s^{-1}
$\Delta p/l$ = pressure drop per unit length $\quad$ Pa.m^{-1}
l_e = equivalent length of a component
$\quad\quad$ for $\zeta = 1$ $\quad\quad\quad\quad\quad\quad$ m

* $(Re) = 2000$
† $(Re) = 3000$

HEAVY GRADE STEEL
WATER AT 10°C

$\Delta p/l$	c	50 mm		65 mm		80 mm		100 mm		125 mm		150 mm		c	$\Delta p/l$
		q_m	l_e	q_m	l_e	q_m	l_e	q_m	l_e	q_m	l_e	q_m	l_e		
0.1		0.013	0.2	0.038	0.6	0.073	1.1	0.213	2.2	0.319	3.1	0.500	3.6		0.1
0.2		0.026	0.4	0.076	1.2	0.146	2.2	0.264	2.5	0.445	3.0	0.751	4.0	0.05	0.2
0.3		0.039	0.6	0.114	1.7	0.176*	2.2	0.315†	2.4	0.565	3.2	0.950	4.3		0.3
0.4		0.052	0.8	0.141*	2.0	0.193	2.0	0.373	2.5	0.668	3.4	1.12	4.5		0.4
0.5		0.065	1.0	0.150	1.8	0.211	1.9	0.424	2.6	0.760	3.5	1.28	4.7		0.5
0.6		0.078	1.2	0.159	1.7	0.228	1.8	0.473	2.6	0.844	3.6	1.42	4.8		0.6
0.7	0.05	0.091	1.4	0.168	1.6	0.246†	1.8	0.516	2.7	0.923	3.7	1.55	4.9		0.7
0.8		0.104*	1.6	0.177	1.6	0.266	1.8	0.557	2.8	0.997	3.8	1.67	5.0		0.8
0.9		0.108	1.5	0.186	1.6	0.285	1.9	0.597	2.8	1.07	3.9	1.79	5.1		0.9
1.0		0.111	1.5	0.195†	1.5	0.303	1.9	0.634	2.9	1.13	3.9	1.90	5.2		1.0
1.5		0.127	1.3	0.244	1.6	0.384	2.1	0.801	3.1	1.43	4.2	2.39	5.5		1.5
2.0		0.142	1.2	0.288	1.7	0.454	2.1	0.945	3.2	1.68	4.3	2.81	5.7	0.15	2.0
2.5		0.157†	1.2	0.328	1.7	0.516	2.2	1.07	3.3	1.91	4.5	3.19	5.8		2.5
3.0		0.175	1.2	0.365	1.8	0.573	2.3	1.19	3.4	2.12	4.6	3.54	6.0		3.0
3.5		0.191	1.2	0.399	1.8	0.626	2.3	1.30	3.4	2.31	4.7	3.86	6.1		3.5
4.0		0.207	1.3	0.431	1.9	0.676	2.4	1.40	3.5	2.49	4.7	4.16	6.2		4.0
4.5		0.221	1.3	0.461	1.9	0.723	2.4	1.50	3.6	2.66	4.8	4.44	6.3		4.5
5.0		0.235	1.3	0.490	1.9	0.768	2.5	1.59	3.6	2.83	4.9	4.71	6.3		5.0
5.5		0.249	1.3	0.517	2.0	0.811	2.5	1.68	3.7	2.98	4.9	4.97	6.4		5.5
6.0		0.261	1.3	0.543	2.0	0.852	2.5	1.77	3.7	3.13	5.0	5.21	6.5		6.0
6.5		0.274	1.4	0.569	2.0	0.891	2.5	1.85	3.7	3.28	5.0	5.45	6.5	0.3	6.5
7.0		0.286	1.4	0.593	2.0	0.930	2.6	1.93	3.8	3.41	5.1	5.68	6.6		7.0
7.5		0.297	1.4	0.617	2.0	0.967	2.6	2.00	3.8	3.55	5.1	5.90	6.6		7.5
8.0	0.15	0.309	1.4	0.640	2.1	1.00	2.6	2.08	3.8	3.68	5.2	6.12	6.7		8.0
8.5		0.320	1.4	0.663	2.1	1.04	2.6	2.15	3.9	3.80	5.2	6.33	6.7		8.5
9.0		0.330	1.4	0.685	2.1	1.07	2.7	2.22	3.9	3.93	5.2	6.53	6.8		9.0
9.5		0.341	1.4	0.706	2.1	1.11	2.7	2.29	3.9	4.05	5.3	6.73	6.8		9.5
10.0		0.351	1.4	0.727	2.1	1.14	2.7	2.35	3.9	4.17	5.3	6.93	6.9		10.0
12.5		0.399	1.5	0.825	2.2	1.29	2.8	2.66	4.0	4.72	5.4	7.84	7.0		12.5
15.0		0.442	1.5	0.915	2.2	1.43	2.8	2.95	4.1	5.22	5.5	8.66	7.2	0.5	15.0
17.5		0.483	1.6	0.998	2.3	1.56	2.9	3.21	4.2	5.68	5.6	9.43	7.3		17.5
20.0		0.521	1.6	1.08	2.3	1.68	2.9	3.46	4.3	6.11	5.7	10.1	7.4		20.0
22.5		0.557	1.6	1.15	2.4	1.79	3.0	3.69	4.3	6.52	5.8	10.8	7.4		22.5
25.0	0.3	0.591	1.6	1.22	2.4	1.90	3.0	3.92	4.4	6.91	5.8	11.5	7.5		25.0
27.5		0.623	1.7	1.29	2.4	2.00	3.0	4.13	4.4	7.28	5.9	12.1	7.6		27.5
30.0		0.655	1.7	1.35	2.4	2.11	3.1	4.33	4.4	7.64	5.9	12.7	7.7		30.0
32.5		0.685	1.7	1.41	2.5	2.20	3.1	4.52	4.5	7.98	6.0	13.2	7.7		32.5
35.0		0.714	1.7	1.47	2.5	2.29	3.1	4.71	4.5	8.31	6.0	13.8	7.8		35.0
37.5		0.742	1.7	1.53	2.5	2.38	3.2	4.89	4.5	8.63	6.1	14.3	7.8		37.5
40.0		0.770	1.7	1.58	2.5	2.47	3.2	5.07	4.6	8.94	6.1	14.8	7.8		40.0
42.5		0.796	1.7	1.64	2.5	2.55	3.2	5.24	4.6	9.24	6.1	15.3	7.9		42.5
45.0		0.822	1.8	1.69	2.6	2.63	3.2	5.41	4.6	9.53	6.2	15.8	7.9		45.0
47.5		0.847	1.8	1.74	2.6	2.71	3.2	5.57	4.7	9.82	6.2	16.3	8.0		47.5
50.0		0.872	1.8	1.79	2.6	2.79	3.2	5.73	4.7	10.1	6.2	16.7	8.0		50.0
52.5		0.896	1.8	1.84	2.6	2.87	3.3	5.88	4.7	10.4	6.2	17.2	8.0		52.5
55.0		0.919	1.8	1.89	2.6	2.94	3.3	6.04	4.7	10.6	6.3	17.6	8.1		55.0
57.5		0.942	1.8	1.94	2.6	3.02	3.3	6.18	4.7	10.9	6.3	18.0	8.1		57.5
60.0		0.965	1.8	1.98	2.6	3.09	3.3	6.33	4.8	11.1	6.3	18.4	8.1	1.0	60.0
62.5		0.987	1.8	2.03	2.6	3.16	3.3	6.47	4.8	11.4	6.3	18.9	8.1		62.5
65.0		1.01	1.8	2.07	2.7	3.23	3.3	6.61	4.8	11.6	6.4	19.3	8.2		65.0
67.5	0.5	1.03	1.8	2.12	2.7	3.29	3.3	6.75	4.8	11.9	6.4	19.7	8.2		67.5
70.0		1.05	1.9	2.16	2.7	3.36	3.4	6.88	4.8	12.1	6.4	20.0	8.2		70.0
72.5		1.07	1.9	2.20	2.7	3.42	3.4	7.02	4.8	12.3	6.4	20.4	8.2		72.5
75.0		1.09	1.9	2.24	2.7	3.49	3.4	7.15	4.8	12.6	6.4	20.8	8.3		75.0
77.5		1.11	1.9	2.28	2.7	3.55	3.4	7.27	4.9	12.8	6.4	21.2	8.3		77.5
80.0		1.13	1.9	2.32	2.7	3.61	3.4	7.40	4.9	13.0	6.5	21.5	8.3		80.0
82.5		1.15	1.9	2.36	2.7	3.67	3.4	7.53	4.9	13.2	6.5	21.9	8.3		82.5
85.0		1.17	1.9	2.40	2.7	3.73	3.4	7.65	4.9	13.5	6.5	22.3	8.3		85.0
87.5		1.19	1.9	2.44	2.7	3.79	3.4	7.77	4.9	13.7	6.5	22.6	8.3		87.5
90.0		1.21	1.9	2.48	2.7	3.85	3.4	7.89	4.9	13.9	6.5	22.9	8.4		90.0

Table 4.16 Flow of water at 10°C in heavy grade steel pipes — *continued*

q_m = mass flow rate — kg.s^{-1}
c = velocity — m.s^{-1}
$\Delta p/l$ = pressure drop per unit length — Pa.m^{-1}
l_e = equivalent length of a component for $\zeta = 1$ — m

HEAVY GRADE STEEL
WATER AT 10°C

$\Delta p/l$	c	50 mm q_m	50 mm l_e	65 mm q_m	65 mm l_e	80 mm q_m	80 mm l_e	100 mm q_m	100 mm l_e	125 mm q_m	125 mm l_e	150 mm q_m	150 mm l_e	c	$\Delta p/l$
92.5		1.23	1.9	2.52	2.8	3.91	3.4	8.01	4.9	14.1	6.5	23.2	8.4		92.5
95.0		1.25	1.9	2.55	2.8	3.97	3.5	8.12	4.9	14.3	6.6	23.6	8.4		95.0
97.5		1.26	1.9	2.59	2.8	4.03	3.5	8.24	5.0	14.5	6.6	23.9	8.4		97.5
100.0		1.28	1.9	2.62	2.8	4.08	3.5	8.35	5.0	14.7	6.6	24.3	8.4		100.0
120.0		1.42	2.0	2.90	2.8	4.51	3.5	9.22	5.0	16.2	6.7	26.8	8.5	1.5	120.0
140.0		1.54	2.0	3.15	2.9	4.90	3.6	10.0	5.1	17.6	6.7	29.1	8.6		140.0
160.0		1.65	2.0	3.39	2.9	5.27	3.6	10.8	5.1	18.9	6.8	31.2	8.7		160.0
180.0		1.77	2.0	3.61	2.9	5.61	3.6	11.5	5.2	20.1	6.9	33.2	8.8		180.0
200.0		1.87	2.1	3.83	2.9	5.94	3.7	12.1	5.2	21.3	6.9	35.1	8.8		200.0
220.0	1.0	1.97	2.1	4.03	3.0	6.25	3.7	12.8	5.3	22.4	6.9	36.9	8.9	2.0	220.0
240.0		2.07	2.1	4.22	3.0	6.55	3.7	13.4	5.3	23.4	7.0	38.7	8.9		240.0
260.0		2.16	2.1	4.41	3.0	6.83	3.7	13.9	5.3	24.5	7.0	40.3	8.9		260.0
280.0		2.25	2.1	4.59	3.0	7.11	3.8	14.5	5.3	25.4	7.0	41.9	9.0		280.0
300.0		2.33	2.1	4.76	3.0	7.38	3.8	15.0	5.4	26.4	7.1	43.5	9.0		300.0
320.0		2.42	2.1	4.93	3.1	7.64	3.8	15.6	5.4	27.3	7.1	45.0	9.0		320.0
340.0		2.50	2.1	5.09	3.1	7.89	3.8	16.1	5.4	28.2	7.1	46.5	9.1		340.0
360.0		2.57	2.2	5.25	3.1	8.13	3.8	16.6	5.4	29.0	7.1	47.9	9.1		360.0
380.0		2.65	2.2	5.40	3.1	8.37	3.8	17.1	5.4	29.9	7.2	49.2	9.1		380.0
400.0		2.72	2.2	5.55	3.1	8.60	3.9	17.5	5.5	30.7	7.2	50.6	9.2		400.0
420.0		2.80	2.2	5.70	3.1	8.83	3.9	18.0	5.5	31.5	7.2	51.9	9.2		420.0
440.0		2.87	2.2	5.84	3.1	9.05	3.9	18.4	5.5	32.3	7.2	53.2	9.2		440.0
460.0		2.94	2.2	5.98	3.1	9.27	3.9	18.9	5.5	33.0	7.2	54.4	9.2		460.0
480.0		3.00	2.2	6.12	3.1	9.47	3.9	19.3	5.5	33.8	7.3	55.7	9.2	3.0	480.0
500.0	1.5	3.07	2.2	6.25	3.1	9.69	3.9	19.7	5.5	34.5	7.3	56.9	9.2		500.0
520.0		3.14	2.2	6.38	3.2	9.89	3.9	20.1	5.5	35.2	7.3	58.0	9.3		520.0
540.0		3.20	2.2	6.51	3.2	10.1	3.9	20.5	5.6	35.9	7.3	59.2	9.3		540.0
560.0		3.26	2.2	6.64	3.2	10.3	3.9	20.9	5.6	36.6	7.3	60.3	9.3		560.0
580.0		3.32	2.2	6.77	3.2	10.5	3.9	21.3	5.6	37.3	7.3	61.4	9.3		580.0
600.0		3.39	2.2	6.89	3.2	10.7	3.9	21.7	5.6	38.0	7.3	62.5	9.3		600.0
620.0		3.45	2.2	7.01	3.2	10.9	4.0	22.1	5.6	38.6	7.3	63.6	9.3		620.0
640.0		3.50	2.2	7.13	3.2	11.0	4.0	22.4	5.6	39.3	7.4	64.7	9.3		640.0
660.0		3.56	2.3	7.25	3.2	11.2	4.0	22.8	5.6	39.9	7.4	65.7	9.4		660.0
680.0		3.62	2.3	7.36	3.2	11.4	4.0	23.2	5.6	40.5	7.4	66.7	9.4		680.0
700.0		3.68	2.3	7.47	3.2	11.6	4.0	23.5	5.6	41.2	7.4	67.8	9.4		700.0
720.0		3.73	2.3	7.59	3.2	11.7	4.0	23.9	5.6	41.8	7.4	68.8	9.4		720.0
740.0		3.79	2.3	7.70	3.2	11.9	4.0	24.2	5.6	42.4	7.4	69.7	9.4		740.0
760.0		3.84	2.3	7.81	3.2	12.1	4.0	24.6	5.7	43.0	7.4	70.7	9.4		760.0
780.0		3.89	2.3	7.91	3.2	12.2	4.0	24.9	5.7	43.6	7.4	71.7	9.4		780.0
800.0		3.95	2.3	8.02	3.2	12.4	4.0	25.2	5.7	44.1	7.4	72.6	9.4		800.0
820.0		4.00	2.3	8.13	3.2	12.6	4.0	25.6	5.7	44.7	7.4	73.6	9.4		820.0
840.0		4.05	2.3	8.23	3.2	12.7	4.0	25.9	5.7	45.3	7.4	74.5	9.4		840.0
860.0	2.0	4.10	2.3	8.33	3.2	12.9	4.0	26.2	5.7	45.8	7.4	75.4	9.5		860.0
880.0		4.15	2.3	8.44	3.3	13.1	4.0	26.5	5.7	46.4	7.5	76.3	9.5		880.0
900.0		4.20	2.3	8.54	3.3	13.2	4.0	26.8	5.7	46.9	7.5	77.2	9.5		900.0
920.0		4.25	2.3	8.63	3.3	13.4	4.0	27.1	5.7	47.5	7.5	78.1	9.5		920.0
940.0		4.30	2.3	8.73	3.3	13.5	4.0	27.4	5.7	48.0	7.5	79.0	9.5		940.0
960.0		4.35	2.3	8.83	3.3	13.7	4.0	27.7	5.7	48.5	7.5	79.8	9.5		960.0
980.0		4.39	2.3	8.93	3.3	13.8	4.1	28.0	5.7	49.0	7.5	80.7	9.5		980.0
1000.0		4.44	2.3	9.02	3.3	14.0	4.1	28.3	5.7	49.6	7.5	81.5	9.5		1000.0
1100.0		4.67	2.3	9.49	3.3	14.7	4.1	29.8	5.7	52.1	7.5	85.7	9.5		1100.0
1200.0		4.89	2.3	9.93	3.3	15.4	4.1	31.2	5.8	54.5	7.5	89.6	9.6		1200.0
1300.0		5.10	2.3	10.4	3.3	16.0	4.1	32.5	5.8	56.8	7.6	93.4	9.6		1300.0
1400.0		5.30	2.4	10.8	3.3	16.6	4.1	33.8	5.8	59.0	7.6	97.0	9.6		1400.0
1500.0		5.50	2.4	11.2	3.3	17.3	4.1	35.0	5.8	61.1	7.6	101	9.6		1500.0
1600.0		5.69	2.4	11.5	3.4	17.8	4.1	36.2	5.8	63.2	7.6	104	9.7		1600.0
1700.0		5.87	2.4	11.9	3.4	18.4	4.2	37.3	5.8	65.2	7.6	107	9.7		1700.0
1800.0	3.0	6.05	2.4	12.3	3.4	19.0	4.2	38.5	5.9	67.2	7.6	110	9.7		1800.0
1900.0		6.23	2.4	12.6	3.4	19.5	4.2	39.6	5.9	69.1	7.7	114	9.7		1900.0

Table 4.17 Flow of water at 10°C in large steel pipes

q_m = mass flow rate   kg.s^{-1}
c = velocity   m.s^{-1}
$\Delta p/l$ = pressure drop per unit length   Pa.m^{-1}
l_e = equivalent length of a component
     for $\zeta = 1$ m

> **LARGE SIZE STEEL**
> **WATER AT 10°C**

$\Delta p/l$	c	175 mm		200 mm		225 mm		250 mm		300 mm		c	$\Delta p/l$
		q_m	l_e	q_m	l_e	q_m	l_e	q_m	l_e	q_m	l_e		
0.1		0.80	4.7	1.13	5.7	1.55	6.7	2.11	7.9	3.44	10	0.05	0.1
0.2	0.05	1.20	5.3	1.69	6.3	2.31	7.5	3.14	8.8	5.11	11		0.2
0.3		1.52	5.6	2.13	6.7	2.92	7.9	3.96	9.3	6.43	12		0.3
0.4		1.79	5.8	2.51	7.0	3.44	8.2	4.67	9.7	7.57	12	0.1	0.4
0.5		2.03	6.0	2.85	7.2	3.91	8.5	5.30	9.9	8.58	13		0.5
0.6		2.25	6.2	3.17	7.4	4.33	8.7	5.87	10	9.51	13		0.6
0.7		2.46	6.3	3.46	7.5	4.73	8.9	6.41	10	10.4	13		0.7
0.8	0.1	2.66	6.4	3.73	7.7	5.10	9.0	6.91	11	11.2	14	0.15	0.8
0.9		2.84	6.5	3.98	7.8	5.45	9.2	7.38	11	11.9	14		0.9
1.0		3.02	6.6	4.23	7.9	5.78	9.3	7.83	11	12.7	14		1.0
1.5	0.15	3.79	7.0	5.32	8.3	7.26	9.8	9.83	11	15.9	15		1.5
2.0		4.46	7.3	6.25	8.6	8.53	10	11.5	12	18.6	15		2.0
2.5		5.06	7.5	7.08	8.9	9.66	10	13.1	12	21.1	15		2.5
3.0		5.60	7.6	7.84	9.0	10.7	11	14.5	12	23.3	16	0.3	3.0
3.5		6.11	7.8	8.54	9.2	11.7	11	15.8	13	25.4	16		3.5
4.0		6.58	7.9	9.20	9.3	12.6	11	17.0	13	27.4	16		4.0
4.5		7.02	8.0	9.83	9.5	13.4	11	18.1	13	29.2	16		4.5
5.0		7.45	8.1	10.4	9.6	14.2	11	19.2	13	30.9	17		5.0
5.5	0.3	7.85	8.2	11.0	9.7	15.0	11	20.2	13	32.6	17		5.5
6.0		8.24	8.2	11.5	9.8	15.7	11	21.2	13	34.2	17		6.0
6.5		8.62	8.3	12.1	9.9	16.4	12	22.2	13	35.7	17		6.5
7.0		8.98	8.4	12.6	9.9	17.1	12	23.1	14	37.2	17		7.0
7.5		9.33	8.4	13.0	10	17.8	12	24.0	14	38.6	17	0.5	7.5
8.0		9.67	8.5	13.5	10	18.4	12	24.9	14	40.0	17		8.0
8.5		10.0	8.6	14.0	10	19.0	12	25.7	14	41.3	17		8.5
9.0		10.3	8.6	14.4	10	19.6	12	26.5	14	42.7	18		9.0
9.5		10.6	8.7	14.9	10	20.2	12	27.3	14	43.9	18		9.5
10.0		10.9	8.7	15.3	10	20.8	12	28.1	14	45.2	18		10.0
12.5		12.4	8.9	17.3	11	23.5	12	31.7	14	51.0	18		12.5
15.0	0.5	13.7	9.1	19.1	11	26.0	13	35.0	15	56.3	18		15.0
17.5		14.9	9.2	20.8	11	28.2	13	38.1	15	61.2	19		17.5
20.0		16.0	9.3	22.3	11	30.4	13	40.9	15	65.8	19		20.0
22.5		17.1	9.4	23.8	11	32.4	13	43.6	15	70.1	19		22.5
25.0		18.1	9.5	25.2	11	34.3	13	46.2	15	74.2	19		25.0
27.5		19.0	9.6	26.5	11	36.1	13	48.6	15	78.1	19	1.0	27.5
30.0		20.0	9.6	27.8	11	37.8	13	51.0	15	81.8	19		30.0
32.5		20.8	9.7	29.1	11	39.5	13	53.2	15	85.4	19		32.5
35.0		21.7	9.8	30.2	12	41.1	13	55.4	16	88.8	20		35.0
37.5		22.5	9.8	31.4	12	42.7	14	57.5	16	92.2	20		37.5
40.0		23.3	9.9	32.5	12	44.2	14	59.5	16	95.4	20		40.0
42.5		24.1	9.9	33.6	12	45.6	14	61.5	16	98.5	20		42.5
45.0		24.8	10	34.6	12	47.0	14	63.4	16	102	20		45.0
47.5		25.6	10	35.6	12	48.4	14	65.2	16	105	20		47.5
50.0	1.0	26.3	10	36.6	12	49.8	14	67.0	16	107	20		50.0
52.5		27.0	10	37.6	12	51.1	14	68.8	16	110	20		52.5
55.0		27.7	10	38.6	12	52.4	14	70.5	16	113	20	1.5	55.0
57.5		28.4	10	39.5	12	53.6	14	72.2	16	116	20		57.5
60.0		29.0	10	40.4	12	54.6	14	73.9	16	118	20		60.0
62.5		29.7	10	41.3	12	56.1	14	75.5	16	121	20		62.5
65.0		30.3	10	42.2	12	57.3	14	77.1	16	124	20		65.0
67.5		30.9	10	43.0	12	58.4	14	78.7	16	126	20		67.5
70.0		31.5	10	43.9	12	59.6	14	80.2	16	128	20		70.0
72.5		32.1	10	44.7	12	60.7	14	81.7	16	131	21		72.5
75.0		32.7	10	45.5	12	61.8	14	83.2	16	133	21		75.0
77.5		33.3	10	46.3	12	62.9	14	84.7	16	136	21		77.5
80.0		33.8	10	47.1	12	64.0	14	86.1	16	138	21		80.0
82.5		34.4	10	47.9	12	65.0	14	87.5	16	140	21		82.5
85.0		35.0	11	48.7	12	66.0	14	88.9	16	142	21		85.0
87.5		35.5	11	49.4	12	67.1	14	90.3	17	145	21		87.5
90.0		36.0	11	50.2	12	68.1	14	91.7	17	147	21		90.0

Table 4.17 Flow of water at 10°C in large steel pipes — *continued*

q_m	=	mass flow rate	kg.s^{-1}
c	=	velocity	m.s^{-1}
$\Delta p/l$	=	pressure drop per unit length	Pa.m^{-1}
l_e	=	equivalent length of a component for $\zeta = 1$	m

LARGE SIZE STEEL

WATER AT 10°C

$\Delta p/l$	c	175 mm		200 mm		225 mm		250 mm		300 mm		c	$\Delta p/l$
		q_m	l_e	q_m	l_e	q_m	l_e	q_m	l_e	q_m	l_e		
92.5		36.6	11	50.9	12	69.1	14	93.0	17	149	21		92.5
95.0		37.1	11	51.6	12	70.1	14	94.3	17	151	21		95.0
97.5		37.6	11	52.4	12	71.1	14	95.6	17	153	21		97.5
100.0	1.5	38.1	11	53.1	12	72.0	14	96.9	17	155	21		100.0
120.0		42.0	11	58.5	13	79.3	15	107	17	171	21		120.0
140.0		45.6	11	63.4	13	86.1	15	116	17	185	21		140.0
160.0		48.9	11	68.1	13	92.3	15	124	17	199	21		160.0
180.0		52.1	11	72.4	13	98.2	15	132	17	211	22		180.0
200.0		55.1	11	76.6	13	104	15	140	17	223	22	3.0	200.0
220.0		57.9	11	80.5	13	109	15	147	17	235	22		220.0
240.0		60.6	11	84.3	13	114	15	154	17	246	22		240.0
260.0		63.2	11	87.9	13	119	15	160	17	256	22		260.0
280.0		65.7	11	91.4	13	124	15	166	18	266	22		280.0
300.0		68.2	11	94.7	13	128	15	173	18	276	22		300.0
320.0		70.5	11	98.0	13	133	15	178	18	285	22		320.0
340.0		72.8	11	101	13	137	15	184	18	294	22		340.0
360.0		75.0	11	104	13	141	15	190	18	303	22	4.0	360.0
380.0		77.1	11	107	13	145	15	195	18	312	22		380.0
400.0	3.0	79.2	11	110	13	149	15	200	18	320	22		400.0
420.0		81.3	11	113	13	153	15	206	18	328	22		420.0
440.0		83.3	12	116	13	157	16	211	18	336	22		440.0
460.0		85.2	12	118	13	160	16	215	18	344	22		460.0
480.0		87.2	12	121	14	164	16	220	18	352	22		480.0
500.0		89.0	12	124	14	168	16	225	18	359	22		500.0
520.0		90.9	12	126	14	171	16	230	18	367	22		520.0
540.0		92.7	12	129	14	174	16	234	18	374	22		540.0
560.0		94.4	12	131	14	178	16	239	18	381	22	5.0	560.0
580.0		96.2	12	134	14	181	16	243	18	388	22		580.0
600.0		97.9	12	136	14	184	16	247	18	395	22		600.0
620.0		99.6	12	138	14	187	16	251	18	401	23		620.0
640.0		101	12	141	14	190	16	256	18	408	23		640.0
660.0		103	12	143	14	193	16	260	18	415	23		660.0
680.0		104	12	145	14	196	16	264	18	421	23		680.0
700.0	4.0	106	12	147	14	199	16	268	18	427	23		700.0
720.0		108	12	149	14	202	16	272	18	434	23		720.0
740.0		109	12	152	14	205	16	276	18	440	23		740.0
760.0		111	12	154	14	208	16	279	18	446	23		760.0
780.0		112	12	156	14	211	16	283	18	452	23		780.0
800.0		114	12	158	14	214	16	287	18	458	23	6.0	800.0
820.0		115	12	160	14	216	16	290	18				820.0
840.0		117	12	162	14	219	16	294	18				840.0
860.0		118	12	164	14	222	16	298	18				860.0
880.0		119	12	166	14	224	16	301	18				880.0
900.0		121	12	168	14	227	16	305	18				900.0
920.0		122	12	170	14	230	16	308	18				920.0
940.0		124	12	171	14	232	16	312	18				940.0
960.0		125	12	173	14	235	16	315	18				960.0
980.0		126	12	175	14	237	16	318	18				980.0
1000.0		128	12	177	14	240	16						1000.0
1100.0	5.0	134	12	186	14	252	16						1100.0
1200.0		140	12	194	14								1200.0
1300.0		146	12	203	14								1300.0
1400.0		157	12										1400.0
1500.0	6.0	157	12										1500.0
1600.0													1600.0
1700.0													1700.0
1800.0													1800.0
1900.0													1900.0
2000.0													2000.0

Table 4.17 Flow of water at 10°C in large steel pipes — *continued*

q_m = mass flow rate   kg.s^{-1}
c = velocity   m.s^{-1}
$\Delta p/l$ = pressure drop per unit length   Pa.m^{-1}
l_e = equivalent length of a component
    for $\zeta = 1$ m

LARGE SIZE STEEL

WATER AT 10°C

$\Delta p/l$	c	350 mm q_m	l_e	400 mm q_m	l_e	450 mm q_m	l_e	500 mm q_m	l_e	550 mm q_m	l_e	600 mm q_m	l_e	c	$\Delta p/l$
0.1	0.05	4.42	12	6.44	14	8.82	17	11.9	20	15.4	23	19.7	26		0.1
0.2		6.56	13	9.55	16	13.1	19	17.6	22	22.8	25	29.0	28	0.1	0.2
0.3		8.25	14	12.0	17	16.5	19	22.1	23	28.6	26	36.4	29		0.3
0.4	0.1	9.71	14	14.1	17	19.4	20	26.0	23	33.6	27	42.7	30	0.15	0.4
0.5		11.0	15	16.0	18	21.9	21	29.4	24	38.0	27	48.4	31		0.5
0.6		12.2	15	17.7	18	24.3	21	32.6	25	42.1	28	53.5	32		0.6
0.7	0.15	13.3	15	19.3	18	26.5	22	35.5	25	45.8	28	58.3	32		0.7
0.8		14.3	15	20.8	19	28.5	22	38.2	25	49.3	29	62.7	32		0.8
0.9		15.3	16	22.2	19	30.4	22	40.8	26	52.7	29	67.0	33		0.9
1.0		16.2	16	23.6	19	32.3	22	43.2	26	55.8	29	71.0	33		1.0
1.5		20.4	17	29.5	20	40.4	23	54.1	27	69.8	31	88.7	35	0.3	1.5
2.0		23.9	17	34.6	21	47.3	24	63.3	28	81.7	32	104	36		2.0
2.5	0.3	27.0	18	39.1	21	53.5	25	71.6	28	92.4	32	117	36		2.5
3.0		29.9	18	43.3	21	59.1	25	79.1	29	102	33	130	37		3.0
3.5		32.5	18	47.1	22	64.3	25	86.1	29	111	33	141	37	0.5	3.5
4.0		35.0	18	50.7	22	69.2	26	92.6	30	119	34	152	38		4.0
4.5		37.3	19	54.0	22	73.8	26	98.7	30	127	34	162	38		4.5
5.0		39.6	19	57.2	23	78.2	26	105	30	135	34	171	39		5.0
5.5		41.7	19	60.3	23	82.3	27	110	31	142	35	180	39		5.5
6.0		43.7	19	63.2	23	86.3	27	115	31	149	35	189	39		6.0
6.5	0.5	45.7	19	66.0	23	90.2	27	121	31	155	35	197	39		6.5
7.0		47.5	19	68.7	23	93.9	27	125	31	162	35	205	40		7.0
7.5		49.4	19	71.4	23	97.4	27	130	31	168	36	213	39		7.5
8.0		51.1	20	73.9	24	101	27	135	32	174	36	220	40		8.0
8.5		52.8	20	76.4	24	104	28	139	32	180	36	228	40		8.5
9.0		54.5	20	78.8	24	108	28	144	32	185	36	235	40		9.0
9.5		56.1	20	81.2	24	111	28	148	32	191	36	242	41		9.5
10.0		57.7	20	83.4	24	114	28	152	32	196	36	249	41	1.0	10.0
12.5		65.2	20	94.1	24	128	28	172	33	221	37	280	41		12.5
15.0		71.9	21	104	25	142	29	189	33	244	37	309	42		15.0
17.5		78.2	21	113	25	154	29	205	33	265	38	335	42		17.5
20.0		84.0	21	121	25	165	29	221	34	284	38	360	43		20.0
22.5	1.0	89.5	21	129	26	176	30	235	34	302	38	383	43		22.5
25.0		94.7	22	137	26	186	30	249	34	320	39	405	43	1.5	25.0
27.5		99.7	22	144	26	196	30	262	35	337	39	427	44		27.5
30.0		104	22	151	26	205	30	274	35	352	39	447	44		30.0
32.5		109	22	157	26	214	30	286	35	368	39	466	44		32.5
35.0		113	22	164	26	223	31	297	35	382	39	485	44		35.0
37.5		118	22	170	26	231	31	308	35	397	40	503	44		37.5
40.0		122	22	176	27	239	31	319	35	410	40	520	45		40.0
42.5		126	22	181	27	247	31	329	35	424	40	537	45		42.5
45.0		130	22	187	27	255	31	340	36	437	40	553	45		45.0
47.5		133	22	192	27	262	31	349	36	449	40	569	45		47.5
50.0	1.5	137	23	198	27	269	31	359	36	462	40	585	45		50.0
52.5		141	23	203	27	276	31	368	36	474	40	600	45		52.5
55.0		144	23	208	27	283	31	378	36	485	41	615	45		55.0
57.5		148	23	213	27	290	31	386	36	497	41	629	45		57.5
60.0		151	23	218	27	296	31	395	36	508	41	644	46		60.0
62.5		154	23	222	27	303	32	404	36	519	41	658	46		62.5
65.0		158	23	227	27	309	32	412	36	530	41	671	46		65.0
67.5		161	23	232	27	315	32	420	36	541	41	685	46		67.5
70.0		164	23	236	27	321	32	429	36	551	41	698	46		70.0
72.5		167	23	241	27	327	32	437	36	561	41	711	46		72.5
75.0		170	23	245	27	333	32	444	36	571	41	723	46		75.0
77.5		173	23	249	28	339	32	452	37	581	41	736	46		77.5
80.0		176	23	253	28	345	32	460	37	591	41	748	46		80.0
82.5		179	23	258	28	350	32	467	37	600	41	760	46		82.5
85.0		182	23	262	28	356	32	475	37	610	41	772	46		85.0
87.5		184	23	266	28	361	32	482	37	619	41	784	46		87.5
90.0		187	23	270	28	367	32	489	37	628	41	796	46		90.0

Table 4.17 Flow of water at 10°C in large steel pipes — *continued*

q_m = mass flow rate kg.s⁻¹
c = velocity m.s⁻¹
$\Delta p/l$ = pressure drop per unit length Pa.m⁻¹
l_e = equivalent length of a component
  for $\zeta = 1$ m

LARGE SIZE STEEL
WATER AT 10°C

$\Delta p/l$	c	350 mm q_m	l_e	400 mm q_m	l_e	450 mm q_m	l_e	500 mm q_m	l_e	550 mm q_m	l_e	600 mm q_m	l_e	c	$\Delta p/l$
92.5		190	23	273	28	372	32	496	37	638	42	807	46		92.5
95.0		193	23	277	28	377	32	503	37	646	42	818	46		95.0
97.5		195	23	281	28	383	32	510	37	655	42	830	47	3.0	97.5
100.0		198	23	285	28	388	32	517	37	664	42	841	47		100.0
120.0		218	24	313	28	427	33	568	37	730	42	924	47		120.0
140.0		236	24	340	28	462	33	616	38	791	42	1000	47		140.0
160.0		253	24	364	29	496	33	660	38	848	43	1070	47	4.0	160.0
180.0	3.0	269	24	387	29	527	33	702	38	901	43	1140	48		180.0
200.0		284	24	409	29	557	33	741	38	952	43	1200	48		200.0
220.0		299	24	430	29	585	33	779	38	1000	43	1270	48		220.0
240.0		313	24	450	29	612	34	815	38	1050	43	1320	48		240.0
260.0		326	24	469	29	638	34	849	38	1090	43	1380	48	5.0	260.0
280.0		339	25	487	29	663	34	882	39	1130	43	1430	48		280.0
300.0		351	25	505	29	687	34	914	39	1170	43	1480	48		300.0
320.0	4.0	363	25	522	29	710	34	945	39	1210	44	1540	48		320.0
340.0		375	25	539	29	732	34	975	39	1250	44	1580	49		340.0
360.0		386	25	555	29	754	34	1000	39	1290	44	1630	49	6.0	360.0
380.0		397	25	571	29	776	34	1030	39	1330	44				380.0
400.0		408	25	586	29	797	34	1060	39	1360	44				400.0
420.0		418	25	601	30	817	34	1090	39	1400	44				420.0
440.0		428	25	616	30	837	34	1116	39						440.0
460.0		438	25	630	30	856	34	1140	39						460.0
480.0		448	25	644	30	875	34								480.0
500.0	5.0	457	25	658	30	893	34								500.0
520.0		467	25	671	30	912	34								520.0
540.0		476	25	684	30										540.0
560.0		485	25	697	30										560.0
580.0		494	25	710	30										580.0
600.0		503	25	722	30										600.0
620.0		511	25	735	30										620.0
640.0		519	25												640.0
660.0		578	25												660.0
680.0		536	25												680.0
700.0		544	25												700.0
720.0	6.0	552	25												720.0
740.0															740.0
760.0															760.0
780.0															780.0
800.0															800.0
820.0															820.0
840.0															840.0
860.0															860.0
880.0															880.0
900.0															900.0
920.0															920.0
940.0															940.0
960.0															960.0
980.0															980.0
1000.0															1000.0
1100.0															1100.0
1200.0															1200.0
1300.0															1300.0
1400.0															1400.0
1500.0															1500.0
1600.0															1600.0
1700.0															1700.0
1800.0															1800.0
1900.0															1900.0
2000.0															2000.0

Table 4.18 Flow of water at 10°C in copper pipes

q_m = mass flow rate   kg.s^{-1}
c = velocity   m.s^{-1}
$\Delta p/l$ = pressure drop per unit length   Pa.m^{-1}
l_e = equivalent length of a component
     for $\zeta = 1$ m

* $(Re) = 2000$
† $(Re) = 3000$

COPPER, TABLE X
WATER AT 10°C

$\Delta p/l$	c	12 mm q_m	l_e	15 mm q_m	l_e	22 mm q_m	l_e	28 mm q_m	l_e	35 mm q_m	l_e	42 mm q_m	l_e	c	$\Delta p/l$
50		0.013	0.2	0.026	0.3	0.070	0.5	0.144	0.7	0.263	1.0	0.447	1.3		50
55		0.014	0.2	0.027	0.3	0.074	0.5	0.152	0.7	0.278	1.0	0.472	1.3		55
60		0.015	0.2	0.028	0.3	0.078	0.5	0.160	0.7	0.292	1.0	0.496	1.3		60
65		0.017	0.3	0.029	0.3	0.082	0.5	0.168	0.7	0.306	1.0	0.520	1.4		65
70	0.2	0.018	0.3	0.030	0.3	0.086	0.5	0.175	0.8	0.319	1.0	0.542	1.4		70
75		0.019	0.3	0.031	0.3	0.089	0.5	0.182	0.8	0.332	1.1	0.564	1.4		75
80		0.020	0.3	0.032	0.3	0.093	0.5	0.189	0.8	0.345	1.1	0.585	1.4		80
85		0.022*	0.3	0.033	0.3	0.096	0.5	0.196	0.8	0.357	1.1	0.606	1.4	0.5	85
90		0.019	0.2	0.034	0.3	0.099	0.5	0.203	0.8	0.369	1.1	0.626	1.4		90
95		0.020	0.2	0.035	0.3	0.103	0.5	0.209	0.8	0.381	1.1	0.646	1.4		95
100		0.020	0.2	0.036	0.3	0.106	0.5	0.216	0.8	0.392	1.1	0.665	1.5		100
110		0.021	0.2	0.038	0.3	0.112	0.6	0.228	0.8	0.414	1.1	0.702	1.5		110
120		0.022	0.2	0.040	0.3	0.117	0.6	0.239	0.8	0.435	1.1	0.738	1.5		120
130		0.023	0.2	0.041†	0.3	0.123	0.6	0.251	0.8	0.456	1.1	0.772	1.5		130
140		0.024	0.2	0.043	0.3	0.128	0.6	0.262	0.8	0.475	1.2	0.805	1.5		140
150		0.025	0.2	0.045	0.3	0.134	0.6	0.272	0.8	0.494	1.2	0.838	1.5		150
160		0.026	0.2	0.047	0.3	0.139	0.6	0.283	0.9	0.513	1.2	0.869	1.6		160
170		0.027	0.2	0.048	0.3	0.144	0.6	0.293	0.9	0.531	1.2	0.899	1.6		170
180	0.3	0.027	0.2	0.050	0.3	0.149	0.6	0.302	0.9	0.549	1.2	0.929	1.6		180
190		0.028	0.2	0.052	0.3	0.153	0.6	0.312	0.9	0.566	1.2	0.958	1.6		190
200		0.029	0.2	0.053	0.3	0.158	0.6	0.321	0.9	0.583	1.2	0.986	1.6		200
225		0.031	0.2	0.057	0.3	0.169	0.6	0.344	0.9	0.623	1.2	1.05	1.6		225
250		0.032†	0.2	0.061	0.3	0.180	0.6	0.365	0.9	0.662	1.3	1.12	1.6		250
275		0.034	0.2	0.064	0.4	0.190	0.6	0.385	0.9	0.698	1.3	1.18	1.7	1.0	275
300		0.036	0.3	0.067	0.4	0.200	0.6	0.405	0.9	0.734	1.3	1.24	1.7		300
325		0.037	0.3	0.071	0.4	0.209	0.7	0.424	1.0	0.768	1.3	1.30	1.7		325
350		0.039	0.3	0.074	0.4	0.218	0.7	0.442	1.0	0.801	1.3	1.35	1.7		350
375		0.041	0.3	0.077	0.4	0.227	0.7	0.460	1.0	0.833	1.3	1.41	1.7		375
400		0.042	0.3	0.080	0.4	0.236	0.7	0.477	1.0	0.864	1.3	1.46	1.8		400
425		0.044	0.3	0.083	0.4	0.244	0.7	0.494	1.0	0.894	1.3	1.51	1.8		425
450	0.5	0.045	0.3	0.085	0.4	0.252	0.7	0.511	1.0	0.924	1.4	1.56	1.8		450
475		0.047	0.3	0.088	0.4	0.260	0.7	0.527	1.0	0.952	1.4	1.61	1.8		475
500		0.048	0.3	0.091	0.4	0.268	0.7	0.542	1.0	0.980	1.4	1.66	1.8		500
550		0.051	0.3	0.096	0.4	0.283	0.7	0.572	1.0	1.04	1.4	1.75	1.8		550
600		0.054	0.3	0.101	0.4	0.297	0.7	0.601	1.0	1.09	1.4	1.83	1.8	1.5	600
650		0.056	0.3	0.106	0.4	0.311	0.7	0.629	1.0	1.14	1.4	1.92	1.9		650
700		0.059	0.3	0.110	0.4	0.324	0.7	0.656	1.1	1.19	1.4	2.00	1.9		700
750		0.061	0.3	0.115	0.4	0.337	0.7	0.682	1.1	1.23	1.4	2.08	1.9		750
800		0.063	0.3	0.119	0.4	0.350	0.7	0.708	1.1	1.28	1.5	2.16	1.9		800
850		0.066	0.3	0.123	0.4	0.362	0.8	0.732	1.1	1.32	1.5	2.23	1.9		850
900		0.068	0.3	0.127	0.4	0.374	0.8	0.757	1.1	1.37	1.5	2.30	1.9		900
950		0.070	0.3	0.131	0.4	0.386	0.8	0.780	1.1	1.41	1.5	2.37	1.9		950
1000		0.072	0.3	0.135	0.4	0.398	0.8	0.803	1.1	1.45	1.5	2.44	2.0	2.0	1000
1100		0.076	0.3	0.143	0.4	0.420	0.8	0.847	1.1	1.53	1.5	2.58	2.0		1100
1200		0.080	0.3	0.150	0.4	0.441	0.8	0.890	1.1	1.61	1.5	2.71	2.0		1200
1300		0.084	0.3	0.157	0.5	0.461	0.8	0.931	1.1	1.68	1.6	2.83	2.0		1300
1400		0.088	0.3	0.164	0.5	0.481	0.8	0.971	1.2	1.75	1.6	2.95	2.0		1400
1500	1.0	0.091	0.3	0.171	0.5	0.500	0.8	1.00	1.2	1.82	1.6	3.06	2.1		1500
1600		0.095	0.3	0.177	0.5	0.519	0.8	1.05	1.2	1.89	1.6	3.18	2.1		1600
1700		0.098	0.3	0.184	0.5	0.537	0.8	1.08	1.2	1.95	1.6	3.29	2.1		1700
1800		0.101	0.3	0.190	0.5	0.555	0.8	1.12	1.2	2.02	1.6	3.39	2.1		1800
1900		0.104	0.3	0.196	0.5	0.572	0.8	1.15	1.2	2.08	1.6	3.50	2.1		1900
2000		0.108	0.3	0.201	0.5	0.589	0.8	1.19	1.2	2.14	1.6	3.60	2.1	3.0	2000
2250		0.115	0.4	0.215	0.5	0.629	0.9	1.27	1.2	2.28	1.7	3.84	2.2		2250
2500		0.122	0.4	0.229	0.5	0.668	0.9	1.35	1.2	2.42	1.7	4.07	2.2		2500
2750		0.129	0.4	0.242	0.5	0.705	0.9	1.42	1.3	2.55	1.7	4.29	2.2		2750
3000	1.5	0.136	0.4	0.254	0.5	0.740	0.9	1.49	1.3	2.68	1.7	4.51	2.2		3000
3250		0.142	0.4	0.266	0.5	0.774	0.9	1.56	1.3	2.80	1.7	4.71	2.2		3250
3500		0.148	0.4	0.277	0.5	0.807	0.9	1.62	1.3	2.92	1.7	4.91	2.3	4.0	3500
3750		0.154	0.4	0.288	0.5	0.839	0.9	1.69	1.3	3.03	1.8	5.10	2.3		3750

Table 4.18 Flow of water at 10°C in copper pipes — *continued*

q_m = mass flow rate   kg.s^{-1}
c = velocity   m.s^{-1}
$\Delta p/l$ = pressure drop per unit length   Pa.m^{-1}
l_e = equivalent length of a component
     for $\zeta = 1$ m

> **COPPER, TABLE X**
>
> **WATER AT 10°C**

$\Delta p/l$	c	54 mm q_m	l_e	67 mm q_m	l_e	76 mm q_m	l_e	108 mm q_m	l_e	133 mm q_m	l_e	159 mm q_m	l_e	c	$\Delta p/l$
50		0.917	1.9	1.660	2.6	2.361	3.1	6.248	5.2	11.128	6.9	17.8	8.8		50
55		0.969	1.9	1.752	2.7	2.492	3.2	6.591	5.2	11.737	7.0	18.8	8.9	1.0	55
60	0.5	1.02	2.0	1.841	2.7	2.617	3.2	6.921	5.3	12.321	7.1	19.7	9.0		60
65		1.07	2.0	1.926	2.7	2.738	3.3	7.238	5.4	12.883	7.2	20.6	9.1		65
70		1.11	2.0	2.009	2.7	2.855	3.3	7.545	5.4	13.426	7.2	21.4	9.1		70
75		1.16	2.0	2.089	2.8	2.968	3.3	7.842	5.4	13.953	7.3	22.3	9.2		75
80		1.20	2.0	2.166	2.8	3.078	3.3	8.130	5.5	14.463	7.3	23.1	9.3		80
85		1.24	2.1	2.242	2.8	3.185	3.4	8.410	5.5	14.960	7.4	23.9	9.3		85
90		1.28	2.1	2.315	2.8	3.289	3.4	8.683	5.6	15.443	7.4	24.7	9.4		90
95		1.32	2.1	2.387	2.8	3.391	3.4	8.949	5.6	15.915	7.5	25.4	9.5		95
100		1.36	2.1	2.457	2.9	3.490	3.4	9.209	5.6	16.376	7.5	26.1	9.5		100
110		1.44	2.1	2.593	2.9	3.682	3.5	9.713	5.7	17.267	7.6	27.6	9.6	1.5	110
120		1.51	2.2	2.723	2.9	3.867	3.5	10.196	5.8	18.122	7.7	28.9	9.7		120
130		1.58	2.2	2.849	3.0	4.045	3.5	10.661	5.8	18.945	7.7	30.2	9.8		130
140		1.65	2.2	2.970	3.0	4.216	3.6	11.110	5.9	19.741	7.8	31.5	9.9		140
150		1.71	2.2	3.088	3.0	4.383	3.6	11.545	5.9	20.511	7.9	32.7	9.9		150
160		1.78	2.2	3.202	3.0	4.544	3.6	11.968	5.9	21.258	7.9	33.9	10.0		160
170		1.84	2.3	3.313	3.1	4.701	3.7	12.378	6.0	21.984	8.0	35.1	10.1		170
180		1.90	2.3	3.421	3.1	4.854	3.7	12.778	6.0	22.692	8.0	36.2	10.1		180
190	1.0	1.96	2.3	3.526	3.1	5.003	3.7	13.168	6.1	23.381	8.1	37.3	10.2	2.0	190
200		2.01	2.3	3.629	3.1	5.149	3.7	13.549	6.1	24.055	8.1	38.4	10.2		200
225		2.15	2.3	3.877	3.2	5.500	3.8	14.464	6.2	25.675	8.2	40.9	10.4		225
250		2.28	2.4	4.113	3.2	5.833	3.8	15.336	6.2	27.215	8.3	43.4	10.5		250
275		2.41	2.4	4.338	3.3	6.152	3.9	16.168	6.3	28.686	8.4	45.7	10.6		275
300		2.53	2.4	4.554	3.3	6.458	3.9	16.967	6.4	30.098	8.5	48.0	10.7		300
325		2.65	2.5	4.763	3.3	6.753	4.0	17.736	6.4	31.458	8.5	50.1	10.8		325
350		2.76	2.5	4.964	3.3	7.038	4.0	18.479	6.5	32.770	8.6	52.2	10.8		350
375		2.87	2.5	5.159	3.4	7.313	4.0	19.198	6.5	34.041	8.7	54.2	10.9		375
400		2.97	2.5	5.349	3.4	7.581	4.1	19.896	6.6	35.274	8.7	56.2	11.0	3.0	400
425	1.5	3.08	2.5	5.533	3.4	7.841	4.1	20.574	6.6	36.472	8.8	58.1	11.0		425
450		3.18	2.6	5.712	3.4	8.095	4.1	21.234	6.7	37.638	8.8	60.0	11.1		450
475		3.27	2.6	5.887	3.5	8.342	4.1	21.878	6.7	38.775	8.9	61.7	11.2		475
500		3.37	2.6	6.057	3.5	8.583	4.2	22.507	6.7	39.885	8.9	63.5	11.2		500
550		3.55	2.6	6.388	3.5	9.050	4.2	23.723	6.8	42.032	9.0	66.9	11.3		550
600		3.73	2.6	6.705	3.6	9.498	4.2	24.889	6.9	44.091	9.1	70.2	11.4		600
650		3.90	2.7	7.010	3.6	9.929	4.3	26.012	6.9	46.074	9.2	73.4	11.5	4.0	650
700	2.0	4.07	2.7	7.305	3.6	10.345	4.3	27.097	7.0	47.987	9.2	76.4	11.6		700
750		4.23	2.7	7.590	3.6	10.749	4.3	28.146	7.0	49.480	9.3	79.3	11.7		750
800		4.38	2.7	7.867	3.7	11.140	4.4	29.164	7.1	51.636	9.4	82.2	11.7		800
850		4.53	2.8	8.136	3.7	11.520	4.4	30.154	7.1	53.382	9.4	85.0	11.8		850
900		4.68	2.8	8.398	3.7	11.891	4.4	31.117	7.1	55.081	9.5	87.6	11.9		900
950		4.82	2.8	8.654	3.7	12.252	4.5	32.056	7.2	56.737	9.5	90.3	11.9		950
1000		4.96	2.8	8.904	3.8	12.604	4.5	32.972	7.2	58.354	9.6	92.8	12.0		1000
1100		5.23	2.8	9.387	3.8	13.286	4.5	34.745	7.3	61.480	9.6	97.8	12.1		1100
1200		5.49	2.9	9.850	3.8	13.941	4.6	36.445	7.3	64.478	9.7	103	12.2		1200
1300		5.74	2.9	10.296	3.9	14.570	4.6	38.082	7.4	67.363	9.8	107	12.3		1300
1400		5.98	2.9	10.727	3.9	15.179	4.6	39.661	7.5	70.148	9.9	112	12.4	6.0	1400
1500	3.0	6.22	2.9	11.144	3.9	15.767	4.7	41.190	7.5	72.842	9.9				1500
1600		6.44	3.0	11.548	4.0	16.338	4.7	42.673	7.6	75.455	10.0				1600
1700		6.66	3.0	11.942	4.0	16.893	4.7	44.113	7.6	77.994	10.0				1700
1800		6.88	3.0	12.324	4.0	17.434	4.8	45.516	7.6						1800
1900		7.09	3.0	12.698	4.0	17.960	4.8	46.882	7.7						1900
2000		7.29	3.0	13.062	4.1	18.474	4.8	48.217	7.7						2000
2250		7.78	3.1	13.938	4.1	19.711	4.9	51.423	7.8						2250
2500	4.0	8.25	3.1	14.771	4.1	20.886	4.9								2500
2750		8.69	3.1	15.566	4.2	22.008	5.0								2750
3000		9.12	3.2	16.329	4.2	23.084	5.0								3000
3250		9.53	3.2	17.064	4.3	24.120	5.0								3250
3500		9.30	3.2	17.773	4.3	25.119	5.1								3500
3750		10.3	3.2	18.459	4.3										3750

Table 4.19 Flow of water at 10°C in copper pipes

q_m = mass flow rate — kg.s^{-1}
c = velocity — m.s^{-1}
$\Delta p/l$ = pressure drop per unit length — Pa.m^{-1}
l_e = equivalent length of a component
for $\zeta = 1$ — m

* (Re) = 2000
† (Re) = 3000

COPPER, TABLE Y
WATER AT 10°C

$\Delta p/l$	c	12 mm		15 mm		22 mm		28 mm		35 mm		42 mm		c	$\Delta p/l$
		q_m	l_e	q_m	l_e	q_m	l_e	q_m	l_e	q_m	l_e	q_m	l_e		
50		0.011	0.2	0.023	0.3	0.065	0.5	0.135	0.7	0.250	1.0	0.429	1.3		50
55		0.012	0.2	0.024	0.3	0.068	0.5	0.143	0.7	0.264	1.0	0.453	1.3		55
60	0.15	0.013	0.2	0.025	0.3	0.072	0.5	0.150	0.7	0.278	1.0	0.476	1.3		60
65		0.014	0.2	0.026	0.3	0.076	0.5	0.158	0.7	0.291	1.0	0.498	1.3		65
70		0.015	0.2	0.027	0.3	0.079	0.5	0.164	0.7	0.303	1.0	0.520	1.3		70
75		0.016	0.3	0.028	0.3	0.082	0.5	0.171	0.7	0.316	1.0	0.541	1.4		75
80		0.018	0.3	0.029	0.3	0.085	0.5	0.178	0.7	0.328	1.0	0.561	1.4		80
85		0.019	0.3	0.030	0.3	0.088	0.5	0.184	0.8	0.339	1.0	0.581	1.4	0.50	85
90		0.020	0.3	0.031	0.3	0.091	0.5	0.190	0.8	0.351	1.1	0.601	1.4		90
95		0.021*	0.3	0.032	0.3	0.094	0.5	0.196	0.8	0.362	1.1	0.619	1.4		95
100		0.019	0.2	0.032	0.3	0.097	0.5	0.202	0.8	0.373	1.1	0.638	1.4		100
110		0.019	0.2	0.034	0.3	0.103	0.5	0.214	0.8	0.394	1.1	0.673	1.4		110
120		0.020	0.2	0.036	0.3	0.108	0.5	0.225	0.8	0.414	1.1	0.708	1.5		120
130		0.021	0.2	0.037	0.3	0.113	0.5	0.235	0.8	0.433	1.1	0.741	1.5		130
140		0.022	0.2	0.038	0.3	0.118	0.5	0.246	0.8	0.452	1.1	0.773	1.5		140
150		0.023	0.2	0.040†	0.3	0.123	0.6	0.256	0.8	0.470	1.1	0.804	1.5		150
160		0.023	0.2	0.041	0.3	0.128	0.6	0.265	0.8	0.488	1.1	0.834	1.5		160
170		0.024	0.2	0.043	0.3	0.132	0.6	0.275	0.8	0.505	1.2	0.863	1.5		170
180	0.30	0.025	0.2	0.044	0.3	0.137	0.6	0.284	0.8	0.522	1.2	0.891	1.5		180
190		0.026	0.2	0.046	0.3	0.141	0.6	0.293	0.9	0.538	1.2	0.919	1.6		190
200		0.026	0.2	0.047	0.3	0.145	0.6	0.302	0.9	0.554	1.2	0.946	1.6		200
225		0.028	0.2	0.050	0.3	0.156	0.6	0.323	0.9	0.593	1.2	1.01	1.6		225
250		0.029	0.2	0.054	0.3	0.165	0.6	0.343	0.9	0.629	1.2	1.07	1.6		250
275		0.031	0.2	0.057	0.3	0.175	0.6	0.362	0.9	0.664	1.2	1.13	1.6		275
300		0.032†	0.2	0.060	0.3	0.184	0.6	0.380	0.9	0.698	1.3	1.19	1.7	1.0	300
325		0.034	0.2	0.062	0.3	0.193	0.6	0.398	0.9	0.730	1.3	1.25	1.7		325
350		0.035	0.2	0.065	0.3	0.201	0.6	0.415	0.9	0.762	1.3	1.30	1.7		350
375		0.037	0.2	0.068	0.3	0.209	0.6	0.432	0.9	0.792	1.3	1.35	1.7		375
400		0.038	0.3	0.070	0.4	0.217	0.6	0.448	0.9	0.882	1.3	1.40	1.7		400
425		0.039	0.3	0.073	0.4	0.225	0.7	0.464	1.0	0.850	1.3	1.45	1.7		425
450		0.041	0.3	0.075	0.4	0.232	0.7	0.479	1.0	0.878	1.3	1.50	1.7		450
475	0.50	0.042	0.3	0.078	0.4	0.239	0.7	0.494	1.0	0.906	1.3	1.54	1.8		475
500		0.043	0.3	0.080	0.4	0.247	0.7	0.509	1.0	0.932	1.3	1.59	1.8		500
550		0.046	0.3	0.085	0.4	0.260	0.7	0.537	1.0	0.984	1.4	1.68	1.8		550
600		0.048	0.3	0.089	0.4	0.274	0.7	0.565	1.0	1.03	1.4	1.76	1.8	1.5	600
650		0.051	0.3	0.093	0.4	0.287	0.7	0.591	1.0	1.08	1.4	1.84	1.8		650
700		0.053	0.3	0.097	0.4	0.299	0.7	0.616	1.0	1.13	1.4	1.92	1.8		700
750		0.055	0.3	0.101	0.4	0.311	0.7	0.641	1.0	1.17	1.4	2.00	1.9		750
800		0.057	0.3	0.105	0.4	0.323	0.7	0.665	1.0	1.22	1.4	2.07	1.9		800
850		0.059	0.3	0.109	0.4	0.334	0.7	0.688	1.0	1.26	1.4	2.14	1.9		850
900		0.061	0.3	0.113	0.4	0.345	0.7	0.711	1.1	1.30	1.4	2.21	1.9		900
950		0.063	0.3	0.116	0.4	0.356	0.7	0.733	1.1	1.34	1.5	2.28	1.9		950
1000		0.065	0.3	0.120	0.4	0.366	0.7	0.754	1.1	1.38	1.5	2.34	1.9	2.0	1000
1100		0.069	0.3	0.126	0.4	0.387	0.7	0.796	1.1	1.45	1.5	2.47	1.9		1100
1200		0.072	0.3	0.133	0.4	0.406	0.8	0.836	1.1	1.53	1.5	2.60	2.0		1200
1300		0.076	0.3	0.139	0.4	0.425	0.8	0.875	1.1	1.60	1.5	2.72	2.0		1300
1400		0.079	0.3	0.145	0.4	0.444	0.8	0.912	1.1	1.67	1.5	2.83	2.0		1400
1500	1.0	0.082	0.3	0.151	0.4	0.461	0.8	0.948	1.1	1.73	1.5	2.94	2.0		1500
1600		0.085	0.3	0.157	0.4	0.478	0.8	0.983	1.1	1.79	1.5	3.05	2.0		1600
1700		0.088	0.3	0.162	0.4	0.495	0.8	1.02	1.1	1.86	1.6	3.15	2.0		1700
1800		0.091	0.3	0.168	0.4	0.511	0.8	1.05	1.2	1.92	1.6	3.26	2.1		1800
1900		0.094	0.3	0.173	0.4	0.527	0.8	1.08	1.2	1.98	1.6	3.36	2.1		1900
2000		0.097	0.3	0.178	0.5	0.543	0.8	1.11	1.2	2.03	1.6	3.45	2.1	3.0	2000
2250		0.104	0.3	0.191	0.5	0.580	0.8	1.19	1.2	2.17	1.6	3.69	2.1		2250
2500		0.110	0.3	0.202	0.5	0.616	0.8	1.26	1.2	2.30	1.6	3.91	2.1		2500
2750		0.116	0.3	0.214	0.5	0.650	0.8	1.33	1.2	2.43	1.7	4.12	2.2		2750
3000	1.5	0.122	0.3	0.225	0.5	0.682	0.9	1.40	1.2	2.55	1.7	4.33	2.2		3000
3250		0.128	0.4	0.235	0.5	0.714	0.9	1.46	1.2	2.67	1.7	4.52	2.2		3250
3500		0.134	0.4	0.245	0.5	0.744	0.9	1.53	1.3	2.78	1.7	4.71	2.2	4.0	3500
3750		0.139	0.4	0.255	0.5	0.774	0.9	1.59	1.3	2.89	1.7	4.89	2.2		3750

Table 4.19 Flow of water at 10°C in copper pipes — *continued*

q_m = mass flow rate   kg.s^{-1}
c = velocity   m.s^{-1}
$\Delta p/l$ = pressure drop per unit length   Pa.m^{-1}
l_e = equivalent length of a component
     for $\zeta = 1$ m

```
COPPER, TABLE Y
WATER AT 10°C
```

$\Delta p/l$	c	54 mm q_m	l_e	67 mm q_m	l_e	76 mm q_m	l_e	108 mm q_m	l_e	c	$\Delta p/l$
50		0.842	1.8	1.55	2.5	2.28	3.1	5.93	5.0		50
55		0.889	1.9	1.64	2.6	2.40	3.1	6.26	5.1		55
60		0.935	1.9	1.72	2.6	2.52	3.2	6.57	5.2		60
65	0.50	0.978	1.9	1.80	2.6	2.64	3.2	6.87	5.2		65
70		1.02	1.9	1.88	2.6	2.75	3.2	7.17	5.3		70
75		1.06	1.9	1.95	2.7	2.86	3.3	7.45	5.3		75
80		1.10	2.0	2.02	2.7	2.97	3.3	7.72	5.3		80
85		1.14	2.0	2.09	2.7	3.07	3.3	7.99	5.4		85
90		1.18	2.0	2.16	2.7	3.17	3.3	8.25	5.4	1.0	90
95		1.21	2.0	2.23	2.8	3.27	3.3	8.50	5.5		95
100		1.25	2.0	2.30	2.8	3.36	3.4	8.75	5.5		100
110		1.32	2.0	2.42	2.8	3.55	3.4	9.23	5.5		110
120		1.39	2.1	2.54	2.8	3.73	3.4	9.69	5.6		120
130		1.45	2.1	2.66	2.9	3.90	3.5	10.1	5.7		130
140		1.51	2.1	2.78	2.9	4.06	3.5	10.6	5.7		140
150		1.57	2.1	2.89	2.9	4.22	3.5	11.0	5.7		150
160		1.63	2.2	2.99	2.9	4.38	3.6	11.4	5.8		160
170		1.69	2.2	3.10	3.0	4.53	3.6	11.8	5.8		170
180		1.74	2.2	3.20	3.0	4.68	3.6	12.1	5.9		180
190		1.80	2.2	3.30	3.0	4.82	3.6	12.5	5.9	1.5	190
200	1.0	1.85	2.2	3.39	3.0	4.96	3.7	12.9	5.9		200
225		2.00	2.2	3.62	3.1	5.30	3.7	13.7	6.0		225
250		2.10	2.3	3.84	3.1	5.62	3.8	14.6	6.1		250
275		2.21	2.3	4.05	3.1	5.93	3.8	15.4	6.2		275
300		2.33	2.3	4.26	3.2	6.23	3.8	16.1	6.2	2.0	300
325		2.43	2.4	4.45	3.2	6.51	3.9	16.9	6.3		325
350		2.54	2.4	4.64	3.2	6.78	3.9	17.6	6.3		350
375		2.64	2.4	4.82	3.3	7.05	3.9	18.2	6.4		375
400		2.73	2.4	5.00	3.3	7.31	4.0	18.9	6.4		400
425		2.83	2.4	5.17	3.3	7.56	4.0	19.5	6.4		425
450	1.5	2.92	2.4	5.34	3.3	7.80	4.0	20.2	6.5		450
475		3.01	2.5	5.50	3.3	8.04	4.1	20.8	6.5		475
500		3.10	2.5	5.66	3.4	8.27	4.1	21.4	6.6		500
550		3.27	2.5	5.97	3.4	8.72	4.1	22.5	6.6		550
600		3.43	2.5	6.27	3.4	9.16	4.2	23.7	6.7		600
650		3.59	2.6	6.55	3.5	9.57	4.2	24.7	6.7	3.0	650
700		3.74	2.6	6.83	3.5	9.97	4.2	25.7	6.8		700
750	2.0	3.88	2.6	7.10	3.5	10.4	4.3	26.7	6.8		750
800		4.03	2.6	7.35	3.6	10.7	4.3	27.7	6.9		800
850		4.17	2.6	7.61	3.6	11.1	4.3	28.7	6.9		850
900		4.30	2.7	7.85	3.6	11.5	4.4	29.6	7.0		900
950		4.43	2.7	8.09	3.6	11.8	4.4	30.5	7.0		950
1000		4.56	2.7	8.32	3.6	12.2	4.4	31.3	7.0		1000
1100		4.81	2.7	8.78	3.7	12.8	4.4	33.0	7.1	4.0	1100
1200		5.05	2.7	9.21	3.7	13.4	4.5	34.6	7.2		1200
1300		5.28	2.8	9.63	3.7	14.0	4.5	36.2	7.2		1300
1400		5.50	2.8	10.0	3.8	14.6	4.6	37.7	7.3		1400
1500	3.0	5.71	2.8	10.4	3.8	15.2	4.6	39.1	7.3		1500
1600		5.92	2.8	10.8	3.8	15.8	4.6	40.6	7.4	5.0	1600
1700		6.12	2.9	11.2	3.9	16.3	4.6	41.9	7.4		1700
1800		6.32	2.9	11.5	3.9	16.8	4.7	43.3	7.5		1800
1900		6.51	2.9	11.9	3.9	17.3	4.7	44.6	7.5		1900
2000		6.70	2.9	12.2	3.9	17.8	4.7	45.8	7.5		2000
2250		7.15	2.9	13.0	4.0	19.0	4.8	48.9	7.6	6.0	2250
2500	4.0	7.58	3.0	13.8	4.0	20.1	4.8				2500
2750		7.99	3.0	14.6	4.0	21.2	4.9				2750
3000		8.39	3.0	15.3	4.1	22.3	4.9				3000
3250		8.77	3.1	16.0	4.1	23.3	5.0				3250
3500		9.13	3.1	16.6	4.1	24.2	5.0				3500
3750		9.49	3.1	17.3	4.2	25.2	5.0				3750

Table 4.20 Flow of water at 10°C in galvanised pipes

q_m = mass flow rate $\quad\quad\quad$ kg.s^{-1}
c = velocity $\quad\quad\quad\quad\quad\quad$ m.s^{-1}
$\Delta p/l$ = pressure drop per unit length $\quad$ Pa.m^{-1}
l_e = equivalent length of a component
$\quad\quad$ for $\zeta = 1$ $\quad\quad\quad\quad\quad\quad$ m

GALVANISED STEEL
WATER AT 10°C

$\Delta p/l$	c	10 mm		15 mm		20 mm		25 mm		32 mm		40 mm		c	$\Delta p/l$
		q_m	l_e	q_m	l_e	q_m	l_e	q_m	l_e	q_m	l_e	q_m	l_e		
50		0.011	0.15	0.025	0.24	0.062	0.40	0.118	0.57	0.266	0.88	0.412	1.1		50
55		0.012	0.16	0.027	0.25	0.066	0.40	0.125	0.57	0.281	0.88	0.434	1.1		55
60		0.013	0.16	0.028	0.25	0.069	0.41	0.131	0.58	0.294	0.89	0.455	1.1		60
70		0.014	0.16	0.031	0.25	0.075	0.42	0.143	0.59	0.320	0.90	0.495	1.1		70
80		0.015	0.17	0.033	0.26	0.081	0.42	0.154	0.60	0.344	0.92	0.532	1.2		80
90		0.016	0.17	0.035	0.26	0.086	0.43	0.164	0.60	0.367	0.92	0.567	1.2		90
100		0.017	0.17	0.038	0.27	0.092	0.43	0.173	0.61	0.389	0.93	0.600	1.2		100
110		0.018	0.17	0.040	0.27	0.096	0.44	0.183	0.62	0.409	0.94	0.631	1.2		110
120		0.019	0.17	0.042	0.27	0.101	0.44	0.192	0.62	0.429	0.95	0.662	1.2		120
130		0.020	0.18	0.044	0.27	0.106	0.44	0.200	0.62	0.448	0.95	0.691	1.2		130
140		0.020	0.18	0.045	0.28	0.110	0.45	0.208	0.63	0.466	0.96	0.719	1.2		140
150		0.021	0.18	0.047	0.28	0.114	0.45	0.216	0.63	0.483	0.96	0.746	1.2		150
175		0.023	0.18	0.051	0.28	0.124	0.46	0.235	0.64	0.525	0.97	0.809	1.2		175
200		0.025	0.19	0.055	0.29	0.134	0.46	0.253	0.65	0.564	0.98	0.869	1.2		200
225		0.027	0.19	0.059	0.29	0.143	0.47	0.269	0.65	0.600	0.99	0.925	1.2		225
250		0.028	0.19	0.062	0.29	0.151	0.47	0.285	0.66	0.635	1.0	0.978	1.2		250
275		0.030	0.19	0.066	0.30	0.159	0.47	0.300	0.66	0.668	1.0	1.03	1.2		275
300		0.031	0.19	0.069	0.30	0.167	0.48	0.314	0.67	0.699	1.0	1.08	1.2		300
350		0.034	0.20	0.075	0.30	0.181	0.48	0.341	0.67	0.759	1.0	1.17	1.3		350
400		0.037	0.20	0.081	0.31	0.194	0.49	0.366	0.68	0.814	1.0	1.25	1.3	1.0	400
450		0.039	0.20	0.086	0.31	0.207	0.49	0.390	0.68	0.866	1.0	1.33	1.3		450
500		0.041	0.20	0.091	0.31	0.219	0.50	0.412	0.69	0.915	1.0	1.41	1.3		500
550		0.044	0.21	0.096	0.31	0.230	0.50	0.433	0.69	0.962	1.0	1.48	1.3		550
600		0.046	0.21	0.100	0.32	0.241	0.50	0.453	0.69	1.01	1.0	1.55	1.3		600
700		0.050	0.21	0.109	0.32	0.262	0.51	0.492	0.70	1.09	1.0	1.68	1.3		700
800		0.053	0.21	0.117	0.32	0.281	0.51	0.527	0.70	1.17	1.1	1.80	1.3		800
900		0.057	0.21	0.124	0.32	0.299	0.51	0.561	0.71	1.24	1.1	1.91	1.3		900
1000		0.060	0.22	0.130	0.33	0.316	0.51	0.592	0.71	1.31	1.1	2.02	1.3		1000
1250		0.068	0.22	0.148	0.33	0.355	0.52	0.665	0.72	1.47	1.1	2.26	1.3		1250
1500		0.075	0.22	0.163	0.33	0.390	0.52	0.732	0.72	1.62	1.1	2.49	1.3	2.0	1500
1750		0.081	0.22	0.177	0.34	0.423	0.53	0.792	0.73	1.75	1.1	2.69	1.3		1750
2000		0.087	0.23	0.190	0.34	0.453	0.53	0.849	0.73	1.88	1.1	2.88	1.4		2000
2250	1.0	0.092	0.23	0.202	0.34	0.482	0.53	0.902	0.73	2.00	1.1	3.06	1.4		2250
2500		0.098	0.23	0.213	0.34	0.509	0.53	0.952	0.74	2.11	1.1	3.23	1.4		2500
2750		0.103	0.23	0.224	0.34	0.534	0.54	1.00	0.74	2.21	1.1	3.39	1.4		2750
3000		0.108	0.23	0.234	0.34	0.559	0.54	1.05	0.74	2.31	1.1	3.54	1.4		3000
3500		0.117	0.23	0.254	0.35	0.605	0.54	1.13	0.74	2.50	1.1	3.84	1.4	3.0	3500
4000		0.125	0.23	0.272	0.35	0.648	0.54	1.21	0.74	2.68	1.1	4.10	1.4		4000
4500		0.133	0.23	0.289	0.35	0.688	0.54	1.29	0.75	2.84	1.1	4.36	1.4		4500
5000		0.140	0.23	0.305	0.35	0.727	0.55	1.36	0.75	3.00	1.1	4.60	1.4		5000
5500		0.147	0.24	0.320	0.35	0.763	0.55	1.43	0.75	3.15	1.1	4.83	1.4		5500
6000		0.154	0.24	0.335	0.35	0.798	0.55	1.49	0.75	3.29	1.1	5.04	1.4	4.0	6000
7000		0.167	0.24	0.362	0.35	0.863	0.55	1.61	0.75	3.56	1.1	5.46	1.4		7000
8000	2.0	0.179	0.24	0.388	0.36	0.924	0.55	1.73	0.76	3.81	1.1	5.84	1.4		8000
9000		0.190	0.24	0.412	0.36	0.981	0.55	1.83	0.76	4.04	1.1	6.20	1.4	5.0	9000
10 000		0.201	0.24	0.435	0.36	1.04	0.55	1.94	0.76	4.27	1.1	6.54	1.4		10 000
11 000		0.211	0.24	0.457	0.36	1.04	0.56	2.03	0.76	4.48	1.1	6.86	1.4		11 000
12 000		0.220	0.24	0.478	0.36	1.14	0.56	2.12	0.76	4.68	1.1	7.17	1.4		12 000

Table 4.20 Flow of water at 10°C in galvanised pipes — *continued*

q_m = mass flow rate   kg.s^{-1}
c = velocity   m.s^{-1}
$\Delta p/l$ = pressure drop per unit length   Pa.m^{-1}
l_e = equivalent length of a component for $\zeta = 1$ m

GALVANISED STEEL
WATER AT 10°C

$\Delta p/l$	c	50 mm		65 mm		80 mm		100 mm		125 mm		150 mm		c	$\Delta p/l$
		q_m	l_e	q_m	l_e	q_m	l_e	q_m	l_e	q_m	l_e	q_m	l_e		
50		0.790	1.6	1.65	2.3	2.58	2.8	5.29	4.1	9.58	5.5	15.5	7.0		50
55		0.832	1.6	1.74	2.3	2.72	2.9	5.57	4.1	10.1	5.5	16.3	7.0		55
60		0.872	1.6	1.82	2.3	2.85	2.9	5.83	4.1	10.5	5.5	17.1	7.0		60
70		0.948	1.6	1.97	2.3	3.09	2.9	6.32	4.2	11.4	5.6	18.5	7.1	1.0	70
80		1.02	1.6	2.12	2.3	3.32	2.9	6.79	4.2	12.3	5.6	19.8	7.1		80
90		1.08	1.6	2.26	2.4	3.53	3.0	7.22	4.2	13.0	5.6	21.1	7.2		90
100		1.15	1.6	2.39	2.4	3.73	3.0	7.63	4.2	13.8	5.7	22.3	7.2		100
110		1.21	1.6	2.51	2.4	3.92	3.0	8.02	4.2	14.5	5.7	23.4	7.2		110
120		1.26	1.6	2.63	2.4	4.11	3.0	8.40	4.3	15.2	5.7	24.5	7.2		120
130		1.32	1.7	2.74	2.4	4.28	3.0	8.75	4.3	15.8	5.7	25.5	7.2		130
140		1.37	1.7	2.85	2.4	4.45	3.0	9.10	4.3	16.4	5.8	26.5	7.3		140
150		1.42	1.7	2.96	2.4	4.62	3.0	9.43	4.3	17.0	5.8	27.5	7.3		150
175		1.54	1.7	3.20	2.4	5.00	3.0	10.2	4.3	18.4	5.8	29.8	7.3		175
200		1.66	1.7	3.44	2.5	5.37	3.1	11.0	4.4	19.7	5.8	31.9	7.4		200
225		1.76	1.7	3.66	2.5	5.79	3.1	11.6	4.4	21.0	5.8	33.9	7.4		225
250		1.86	1.7	3.86	2.5	6.03	3.1	12.3	4.4	22.2	5.9	35.8	7.4		250
275	1.0	1.96	1.7	4.06	2.5	6.34	3.1	12.9	4.4	23.3	5.9	37.6	7.4	2.0	275
300		2.05	1.7	4.25	2.5	6.63	3.1	13.5	4.4	24.3	5.9	39.3	7.4		300
350		2.22	1.8	4.60	2.5	7.18	3.1	14.6	4.4	26.3	5.9	42.5	7.5		350
400		2.38	1.8	4.93	2.5	7.69	3.2	15.7	4.5	28.2	5.9	45.6	7.5		400
450		2.53	1.8	5.24	2.5	8.17	3.2	16.6	4.5	30.0	6.0	48.4	7.5		450
500		2.68	1.8	5.54	2.6	8.63	3.2	17.6	4.5	31.6	6.0	51.1	7.5		500
550		2.81	1.8	5.82	2.6	9.06	3.2	18.5	4.5	33.2	6.0	53.6	7.6		550
600		2.94	1.8	6.08	2.6	9.48	3.2	19.3	4.5	34.7	6.0	56.0	7.6	3.0	600
700		3.19	1.8	6.58	2.6	10.3	3.2	20.9	4.5	37.6	6.0	60.6	7.6		700
800		3.41	1.8	7.05	2.6	11.0	3.2	22.4	4.5	40.2	6.0	64.9	7.6		800
900		3.63	1.8	7.49	2.6	11.7	3.2	23.7	4.6	42.7	6.0	68.9	7.6		900
1000	2.0	3.83	1.8	7.91	2.6	12.3	3.2	25.1	4.6	45.0	6.1	72.7	7.6	4.0	1000
1250		4.30	1.8	8.86	2.6	13.8	3.2	28.1	4.6	50.5	6.1	81.4	7.6		1250
1500		4.72	1.8	9.73	2.6	15.1	3.3	30.8	4.6	55.4	6.1	89.2	7.7	5.0	1500
1750		5.10	1.8	10.5	2.6	16.3	3.3	33.3	4.6	59.9	6.1	96.5	7.7		1750
2000		5.46	1.8	11.3	2.6	17.5	3.3	35.6	4.6	64.0	6.1	103	7.7		2000
2250		5.80	1.8	12.0	2.6	18.6	3.3	37.8	4.6	68.0	6.1	110	7.7	6.0	2250
2500	3.0	6.12	1.9	12.6	2.6	19.6	3.3	39.9	4.6	71.7	6.1				2500
2750		6.43	1.9	13.2	2.6	20.6	3.3	41.9	4.6	75.2	6.2				2750
3000		6.72	1.9	13.8	2.7	21.5	3.3	43.8	4.6	78.6	6.2				3000
3500		7.27	1.9	15.0	2.7	23.3	3.3	47.3	4.6						3500
4000	4.0	7.78	1.9	16.0	2.7	24.9	3.3	50.6	4.7						4000
4500		8.26	1.9	17.0	2.7	26.4	3.3								4500
5000		8.71	1.9	17.9	2.7	27.9	3.3								5000
5500		9.14	1.9	18.8	2.7	29.2	3.3								5500
6000	5.0	9.55	1.9	19.7	2.7										6000
7000		10.33	1.9	21.2	2.7										7000
8000		11.05	1.9												8000
9000	6.0	11.73	1.9												9000
10 000															10 000
11 000															11 000
12 000															12 000

Table 4.21 Flow of water at 10°C in UPVC, Class C pipes

q_m	= mass flow rate	kg.s^{-1}
c	= velocity	m.s^{-1}
$\Delta p/l$	= pressure drop per unit length	Pa.m^{-1}
l_e	= equivalent length of a component	
	for $\zeta = 1$	m

UPVC, CLASS C	
WATER AT 10°C	

* $(Re) = 2000$
† $(Re) = 3000$

$\Delta p/l$	c	50 mm		65 mm		80 mm		100 mm		125 mm		c	$\Delta p/l$
		q_m	l_e	q_m	l_e	q_m	l_e	q_m	l_e	q_m	l_e		
0.1		0.02	0.4	0.04	0.6	0.08	1.2	0.20	2.4	0.31	2.9		0.1
0.2		0.03	0.6	0.08	1.2	0.16★	2.4	0.26	2.4	0.45	3.0		0.2
0.3		0.05	0.8	0.12★	1.8	0.18	1.9	0.32†	2.4	0.57	3.3		0.3
0.4		0.07	1.1	0.13	1.6	0.20	1.9	0.38	2.5	0.67	3.4	0.05	0.4
5.0		0.08	1.2	0.15	1.6	0.22	1.9	0.43	2.6	0.76	3.6		0.5
0.6		0.10	1.5	0.16	1.6	0.24†	1.9	0.48	2.7	0.85	3.7		0.6
0.7		0.11★	1.5	0.18	1.6	0.26	1.9	0.53	2.7	0.93	3.7		0.7
0.8		0.11	1.3	0.18	1.6	0.28	1.9	0.57	2.8	1.00	3.8		0.8
0.9	0.05	0.12	1.3	0.19	1.6	0.30	2.0	0.61	2.9	1.07	3.9		0.9
1.0		0.12	1.3	0.20†	1.6	0.32	2.0	0.65	2.9	1.14	4.0	0.1	1.0
1.5		0.14	1.3	0.26	1.7	0.41	2.1	0.82	3.1	1.44	4.2		1.5
2.0		0.16†	1.3	0.31	1.8	0.48	2.2	0.96	3.2	1.70	4.4		2.0
2.5		0.19	1.3	0.35	1.8	0.55	2.3	1.10	3.3	1.93	4.5	0.15	2.5
3.0		0.21	1.3	0.39	1.9	0.61	2.4	1.22	3.4	2.14	4.7		3.0
3.5	0.1	0.23	1.4	0.42	1.9	0.67	2.4	1.33	3.5	2.33	4.8		3.5
4.0		0.25	1.4	0.46	2.0	0.72	2.5	1.44	3.6	2.52	4.8		4.0
4.5		0.27	1.5	0.49	2.0	0.77	2.6	1.54	3.6	2.70	4.9		4.5
5.0		0.28	1.5	0.52	2.1	0.83	2.7	1.64	3.8	2.87	5.1		5.0
5.5		0.30	1.5	0.55	2.1	0.87	2.7	1.74	3.9	3.03	5.1		5.5
6.0		0.31	1.6	0.58	2.1	0.92	2.7	1.82	3.9	3.18	5.2		6.0
6.5		0.33	1.6	0.61	2.2	0.96	2.8	1.91	3.9	3.33	5.3		6.5
7.0		0.34	1.6	0.63	2.2	1.00	2.8	1.99	4.0	3.47	5.3		7.0
7.5	0.15	0.35	1.6	0.66	2.2	1.04	2.8	2.07	4.0	3.61	5.4		7.5
8.0		0.37	1.6	0.68	2.2	1.08	2.8	2.15	4.1	3.74	5.4		8.0
8.5		0.38	1.6	0.71	2.3	1.12	2.9	2.22	4.1	3.88	5.5	0.3	8.5
9.0		0.39	1.7	0.73	2.3	1.16	2.9	2.30	4.1	4.00	5.5		9.0
9.5		0.41	1.7	0.75	2.3	1.19	2.9	2.37	4.2	4.13	5.5		9.5
10.0		0.42	1.7	0.78	2.3	1.23	2.9	2.44	4.2	4.25	5.6		10.0
12.5		0.48	1.7	0.88	2.4	1.40	3.0	2.77	4.3	4.82	5.7		12.5
15.0		0.53	1.8	0.98	2.5	1.55	3.1	3.07	4.4	5.34	5.9		15.0
17.5		0.58	1.8	1.07	2.5	1.69	3.2	3.35	4.5	5.82	6.0		17.5
20.0		0.62	1.9	1.15	2.6	1.83	3.2	3.61	4.6	6.28	6.1	0.5	20.0
22.5		0.67	1.9	1.23	2.6	1.95	3.3	3.86	4.7	6.71	6.2		22.5
25.0	0.3	0.71	1.9	1.31	2.6	2.07	3.3	4.10	4.7	7.12	6.2		25.0
27.5		0.75	1.9	1.38	2.7	2.19	3.4	4.32	4.8	7.51	6.3		27.5
30.0		0.79	2.0	1.45	2.7	2.30	3.4	4.54	4.8	7.88	6.4		30.0
32.5		0.82	2.0	1.52	2.7	2.40	3.4	4.75	4.9	8.24	6.4		32.5
35.0		0.86	2.0	1.59	2.8	2.51	3.5	4.95	4.9	8.59	6.5		35.0
37.5		0.89	2.0	1.65	2.8	2.61	3.5	5.15	5.0	8.93	6.6		37.5
40.0		0.93	2.1	1.71	2.8	2.70	3.5	5.34	5.0	9.26	6.6		40.0
42.5		0.96	2.1	1.77	2.8	2.80	3.6	5.52	5.0	9.58	6.7		42.5
45.0		0.99	2.1	1.83	2.8	2.89	3.6	5.70	5.1	9.89	6.7		45.0
47.5		1.02	2.1	1.89	2.9	2.98	3.6	5.88	5.1	10.2	6.7		47.5
50.0		1.05	2.1	1.94	2.9	3.06	3.6	6.05	5.1	10.5	6.8		50.0
52.5		1.08	2.1	2.00	2.9	3.15	3.7	6.21	5.2	10.8	6.8		52.5
55.0		1.11	2.1	2.05	2.9	3.23	3.7	6.38	5.2	11.1	6.9		55.0
57.5		1.14	2.2	2.10	2.9	3.32	3.7	6.54	5.2	11.3	6.9		57.5
60.0	0.5	1.17	2.2	2.15	3.0	3.40	3.7	6.70	5.2	11.6	6.9		60.0
62.5		1.20	2.2	2.20	3.0	3.48	3.7	6.85	5.3	11.9	7.0		62.5
65.0		1.22	2.2	2.25	3.0	3.55	3.8	7.00	5.3	12.1	7.0		65.0
67.5		1.25	2.2	2.30	3.0	3.63	3.8	7.15	5.3	12.4	7.0		67.5
70.0		1.28	2.2	2.35	3.0	3.70	3.8	7.30	5.3	12.7	7.0	1.0	70.0
72.5		1.30	2.2	2.40	3.0	3.78	3.8	7.44	5.4	12.9	7.1		72.5
75.0		1.33	2.2	2.44	3.0	3.85	3.8	7.59	5.4	13.1	7.1		75.0
77.5		1.35	2.2	2.49	3.1	3.92	3.8	7.73	5.4	13.4	7.1		77.5
80.0		1.38	2.2	2.53	3.1	3.99	3.9	7.87	5.4	13.6	7.2		80.0
82.5		1.40	2.3	2.58	3.1	4.06	3.9	8.00	5.4	13.9	7.2		82.5
85.0		1.42	2.3	2.62	3.1	4.13	3.9	8.14	5.5	14.1	7.2		85.0
87.5		1.45	2.3	2.66	3.1	4.20	3.9	8.27	5.5	14.3	7.2		87.5
90.0		1.47	2.3	2.71	3.1	4.27	3.9	8.40	5.5	14.6	7.2		90.0

Table 4.21 Flow of water at 10°C in UPVC, Class C pipes — *continued*

q_m	= mass flow rate	kg.s^{-1}
c	= velocity	m.s^{-1}
$\Delta p/l$	= pressure drop per unit length	Pa.m^{-1}
l_e	= equivalent length of a component for $\zeta = 1$	m

		UPVC, CLASS C
		WATER AT 10°C

$\Delta p/l$	c	50 mm		65 mm		80 mm		100 mm		125 mm		c	$\Delta p/l$
		q_m	l_e	q_m	l_e	q_m	l_e	q_m	l_e	q_m	l_e		
92.5		1.49	2.3	2.75	3.1	4.33	3.9	8.53	5.5	14.8	7.3		92.5
95.0		1.52	2.3	2.79	3.1	4.40	3.9	8.66	5.5	15.0	7.3		95.0
97.5		1.54	2.3	2.83	3.1	4.46	4.0	8.78	5.6	15.2	7.3		97.5
100.0		1.56	2.3	2.87	3.2	4.52	4.0	8.91	5.6	15.4	7.3		100.0
120.0		1.73	2.4	3.18	3.2	5.01	4.1	9.86	5.7	17.1	7.5		120.0
140.0		1.89	2.4	3.47	3.3	5.46	4.1	10.7	5.8	18.6	7.6	1.5	140.0
160.0		2.04	2.5	3.74	3.3	5.89	4.2	11.6	5.9	20.0	7.7		160.0
180.0		2.17	2.5	4.00	3.4	6.29	4.2	12.4	5.9	21.4	7.8		180.0
200.0	1.0	2.31	2.5	4.24	3.4	6.67	4.3	13.1	6.0	22.7	7.9		200.0
220.0		2.43	2.6	4.47	3.5	7.03	4.3	13.8	6.1	23.9	8.0		220.0
240.0		2.56	2.6	4.69	3.5	7.38	4.4	14.5	6.1	25.1	8.0		240.0
260.0		2.67	2.6	4.91	3.5	7.71	4.4	15.2	6.2	26.2	8.1		260.0
280.0		2.79	2.6	5.11	3.6	8.04	4.5	15.8	6.2	27.3	8.2		280.0
300.0		2.90	2.6	5.31	3.6	8.35	4.5	16.4	6.3	28.3	8.2		300.0
320.0		3.00	2.7	5.51	3.6	8.66	4.5	17.0	6.3	29.4	8.3		320.0
340.0		3.11	2.7	5.70	3.6	8.95	4.6	17.6	6.4	30.4	8.3		340.0
360.0		3.21	2.7	5.88	3.7	9.24	4.6	18.1	6.4	31.3	8.4		360.0
380.0		3.31	2.7	6.06	3.7	9.52	4.6	18.7	6.4	32.3	8.4		380.0
400.0		3.40	2.7	6.24	3.7	9.80	4.6	19.2	6.5	33.2	8.5		400.0
420.0	1.5	3.50	2.8	6.41	3.7	10.1	4.7	19.7	6.5	34.1	8.5		420.0
440.0		3.59	2.8	6.58	3.7	10.3	4.7	20.3	6.5	35.0	8.5		440.0
460.0		3.68	2.8	6.74	3.8	10.6	4.7	20.8	6.6	35.8	8.6		460.0
480.0		3.77	2.8	6.90	3.8	10.8	4.7	21.3	6.6	36.7	8.6		480.0
500.0		3.85	2.8	7.06	3.8	11.1	4.7	21.7	6.6	37.5	8.7		500.0
520.0		3.94	2.8	7.22	3.8	11.3	4.8	22.2	6.6	38.3	8.7	3.0	520.0
540.0		4.02	2.8	7.37	3.8	11.6	4.8	22.7	6.7	39.1	8.7		540.0
560.0		4.11	2.8	7.52	3.8	11.8	4.8	23.1	6.7	39.9	8.8		560.0
580.0		4.19	2.9	7.67	3.9	12.0	4.8	23.6	6.7	40.7	8.8		580.0
600.0		4.27	2.9	7.81	3.9	12.3	4.8	24.0	6.7	41.5	8.8		600.0
620.0		4.34	2.9	7.95	3.9	12.5	4.9	24.5	6.8	42.2	8.8		620.0
640.0		4.42	2.9	8.10	3.9	12.7	4.9	24.9	6.8	42.9	8.9		640.0
660.0		4.50	2.9	8.23	3.9	12.9	4.9	25.3	6.8	43.7	8.9		660.0
680.0		4.57	2.9	8.37	3.9	13.1	4.9	25.7	6.8	44.4	8.9		680.0
700.0		4.65	2.9	8.51	3.9	13.4	4.9	26.2	6.8	45.1	8.9		700.0
720.0		4.72	2.9	8.64	3.9	13.6	4.9	26.6	6.9	45.8	9.0		720.0
740.0		4.79	2.9	8.77	4.0	13.8	4.9	27.0	6.9	46.5	9.0		740.0
760.0		4.87	2.9	8.90	4.0	14.0	5.0	27.4	6.9	47.2	9.0		760.0
780.0		4.94	3.0	9.03	4.0	14.2	5.0	27.8	6.9	47.9	9.0		780.0
800.0		5.01	3.0	9.16	4.0	14.4	5.0	28.1	6.9	48.5	9.0		800.0
820.0		5.07	3.0	9.29	4.0	14.6	5.0	28.5	6.9	49.2	9.1		820.0
840.0		5.14	3.0	9.41	4.0	14.8	5.0	28.9	7.0	49.8	9.1		840.0
860.0		5.21	3.0	9.53	4.0	15.0	5.0	29.3	7.0	50.5	9.1		860.0
880.0		5.28	3.0	9.65	4.0	15.1	5.0	29.6	7.0	51.1	9.1	4.0	880.0
900.0		5.34	3.0	9.78	4.0	15.3	5.0	30.0	7.0	51.7	9.1		900.0
920.0		5.41	3.0	9.89	4.0	15.5	5.1	30.4	7.0	52.4	9.2		920.0
940.0		5.47	3.0	10.0	4.1	15.7	5.1	30.7	7.0	53.0	9.2		940.0
960.0		5.54	3.0	10.1	4.1	15.9	5.1	31.1	7.1	53.6	9.2		960.0
980.0		5.60	3.0	10.3	4.1	16.1	5.1	31.4	7.1	54.2	9.2		980.0
1000.0		5.67	3.0	10.4	4.1	16.3	5.1	31.8	7.1	54.8	9.2		1000.0
1100.0		5.97	3.1	10.9	4.1	17.1	5.1	33.5	7.1	57.7	9.3		1100.0
1200.0		6.27	3.1	11.5	4.2	18.0	5.2	35.1	7.2	60.5	9.4		1200.0
1300.0		6.55	3.1	12.0	4.2	18.8	5.2	36.7	7.3	63.2	9.5	5.0	1300.0
1400.0	3.0	6.82	3.1	12.5	4.2	19.6	5.3	38.2	7.3	65.8	9.5		1400.0
1500.0		7.09	3.2	13.0	4.3	20.3	5.3	39.7	7.4	68.4	9.6		1500.0
1600.0		7.35	3.2	13.4	4.3	21.0	5.3	41.1	7.4	70.8	9.6		1600.0
1700.0		7.60	3.2	13.9	4.3	21.7	5.4	42.5	7.4	73.2	9.7		1700.0
1800.0		7.84	3.2	14.3	4.3	22.4	5.4	43.8	7.5	75.5	9.7		1800.0
1900.0		8.08	3.2	14.8	4.4	23.1	5.4	45.1	7.5	77.7	9.8		1900.0
2000.0		8.31	3.3	15.2	4.4	23.8	5.4	46.4	7.5	79.9	9.8		2000.0

Table 4.21 Flow of water at 10°C in UPVC, Class C pipes — *continued*

q_m = mass flow rate   kg.s⁻¹
c = velocity   m.s⁻¹
$\Delta p/l$ = pressure drop per unit length   Pa.m⁻¹
l_e = equivalent length of a component for $\zeta = 1$ m

UPVC, CLASS C

WATER AT 10°C

$\Delta p/l$	c	150 mm q_m	l_e	175 mm q_m	l_e	200 mm q_m	l_e	225 mm q_m	l_e	250 mm q_m	l_e	300 mm q_m	l_e	c	$\Delta p/l$
0.1		0.49	3.6	0.73	4.4	1.05	5.4	1.42	6.4	1.93	7.6	3.06	9.6		0.1
0.2		0.74	4.0	1.09	5.0	1.57	6.1	2.12	7.1	2.87	8.4	4.56	11		0.2
0.3		0.94	4.3	1.38	5.3	1.99	6.5	2.68	7.6	3.63	8.9	5.75	11		0.3
0.4		1.11	4.5	1.63	5.6	2.35	6.8	3.16	7.9	4.27	9.3	6.77	12	0.1	0.4
0.5		1.26	4.7	1.85	5.8	2.67	7.0	3.59	8.2	4.85	9.6	7.69	12		0.5
0.6		1.40	4.8	2.06	5.9	2.96	7.2	3.98	8.4	5.38	9.8	8.53	12		0.6
0.7		1.53	4.9	2.25	6.0	3.23	7.3	4.35	8.6	5.88	10	9.31	13		0.7
0.8		1.65	5.0	2.42	6.2	3.49	7.5	4.69	8.7	6.34	10	10.0	13		0.8
0.9		1.77	5.1	2.59	6.3	3.73	7.6	5.01	8.9	6.78	10	10.7	13	0.15	0.9
1.0	0.1	1.88	5.2	2.75	6.4	3.96	7.7	5.32	9.0	7.19	11	11.4	13		1.0
1.5		2.37	5.5	3.47	6.7	4.99	8.1	6.70	9.5	9.05	11	14.3	14		1.5
2.0	0.15	2.79	5.7	4.09	7.0	5.87	8.5	7.88	9.9	10.6	12	16.8	15		2.0
2.5		3.17	5.9	4.64	7.2	6.66	8.7	8.94	10	12.1	12	19.1	15		2.5
3.0		3.52	6.1	5.14	7.4	7.38	8.9	9.91	10	13.4	12	21.1	15	0.3	3.0
3.5		3.84	6.2	5.61	7.5	8.05	9.1	10.8	11	14.6	12	23.0	16		3.5
4.0		4.14	6.3	6.05	7.7	8.68	9.3	11.6	11	15.7	13	24.8	16		4.0
4.5		4.42	6.5	6.47	7.8	9.28	9.4	12.4	11	16.8	13	26.5	16		4.5
5.0		4.70	6.6	6.87	8.0	9.84	9.6	13.2	11	17.8	13	28.1	16		5.0
5.5		4.96	6.6	7.24	8.1	10.4	9.7	13.9	11	18.8	13	29.6	17		5.5
6.0		5.21	6.7	7.61	8.1	10.9	9.8	14.6	11	19.7	13	31.1	17		6.0
6.5	0.3	5.45	6.8	7.96	8.2	11.4	9.9	15.3	12	20.6	13	32.5	17		6.5
7.0		5.69	6.8	8.30	8.3	11.9	10	15.9	12	21.5	13	33.9	17	0.5	7.0
7.5		5.91	6.9	8.63	8.4	12.4	10	16.6	12	22.3	14	35.2	17		7.5
8.0		6.13	7.0	8.95	8.4	12.8	10	17.1	12	23.1	14	36.5	17		8.0
8.5		6.34	7.0	9.26	8.5	13.3	10	17.8	12	23.9	14	37.8	17		8.5
9.0		6.55	7.1	9.56	8.6	13.7	10	18.3	12	24.7	14	39.0	17		9.0
9.5		6.75	7.1	9.85	8.6	14.1	10	18.9	12	25.5	14	40.2	18		9.5
10.0		6.95	7.2	10.1	8.7	14.5	10	19.5	12	26.2	14	41.3	18		10.0
12.5		7.88	7.4	11.5	8.9	16.4	11	22.0	12	29.7	14	46.8	18		12.5
15.0		8.73	7.5	12.7	9.1	18.2	11	24.4	13	32.8	15	51.8	18		15.0
17.5	0.5	9.52	7.7	13.9	9.3	19.8	11	26.6	13	35.7	15	56.4	19		17.5
20.0		10.3	7.8	15.0	9.4	21.4	11	28.6	13	38.5	15	60.7	19		20.0
22.5		11.0	7.9	16.0	9.6	22.8	11	30.6	13	41.1	15	64.8	19		22.5
25.0		11.6	8.0	16.9	9.7	24.2	12	32.4	13	43.6	16	68.7	19	1.0	25.0
27.5		12.3	8.1	17.9	9.8	25.5	12	34.2	14	45.9	16	72.4	20		27.5
30.0		12.9	8.2	18.7	9.9	26.8	12	35.9	14	48.2	16	75.9	20		30.0
32.5		13.5	8.3	19.6	10	28.0	12	37.5	14	50.4	16	79.3	20		32.5
35.0		14.0	8.3	20.4	10	29.2	12	39.0	14	52.5	16	82.7	20		35.0
37.5		14.6	8.4	21.2	10	30.3	12	40.6	14	54.5	16	85.9	20		37.5
40.0		15.1	8.4	22.0	10	31.4	12	42.0	14	56.5	16	89.0	20		40.0
42.5		15.6	8.5	22.7	10	32.5	12	43.5	14	58.4	16	92.0	21		42.5
45.0		16.1	8.6	23.5	10	33.5	12	44.9	14	60.3	17	94.9	21		45.0
47.5		16.6	8.6	24.2	10	34.5	12	46.2	14	62.1	17	97.8	21		47.5
50.0		17.1	8.7	24.9	10	35.5	13	47.6	14	63.9	17	101	21		50.0
52.5		17.6	8.7	25.6	11	36.5	13	48.9	15	65.6	17	103	21		52.5
55.0		18.0	8.8	26.2	11	37.5	13	50.1	15	67.3	17	106	21	1.5	55.0
57.5	1.0	18.5	8.8	26.9	11	38.4	13	51.4	15	69.0	17	109	21		57.5
60.0		18.9	8.8	27.5	11	39.3	13	52.6	15	70.7	17	111	21		60.0
62.5		19.4	8.9	28.2	11	40.2	13	53.8	15	72.3	17	114	21		62.5
65.0		19.8	8.9	28.9	11	41.1	13	55.0	15	73.8	17	116	21		65.0
67.5		20.2	9.0	29.4	11	42.0	13	56.1	15	75.4	17	119	22		67.5
70.0		20.6	9.0	30.0	11	42.8	13	57.3	15	76.9	17	121	22		70.0
72.5		21.0	9.0	30.6	11	43.6	13	58.4	15	78.4	17	123	22		72.5
75.0		21.4	9.1	31.2	11	44.5	13	59.5	15	79.9	17	126	22		75.0
77.5		21.8	9.1	31.7	11	45.3	13	60.6	15	81.3	17	128	22		77.5
80.0		22.2	9.1	32.3	11	46.1	13	61.6	15	82.8	17	130	22		80.0
82.5		22.6	9.2	32.8	11	46.9	13	62.7	15	84.2	18	132	22		82.5
85.0		23.0	9.2	33.4	11	47.6	13	63.7	15	85.6	18	135	22		85.0
87.5		23.3	9.2	33.9	11	48.4	13	64.7	15	86.9	18	137	22		87.5
90.0		23.7	9.2	34.5	11	49.2	13	65.8	15	88.3	18	139	22		90.0

Table 4.21 Flow of water at 10°C in UPVC, Class C pipes — *continued*

q_m = mass flow rate   kg.s^{-1}
c = velocity   m.s^{-1}
$\Delta p/l$ = pressure drop per unit length   Pa.m^{-1}
l_e = equivalent length of a component
     for $\zeta = 1$ m

		UPVC, CLASS C
		WATER AT 10°C

$\Delta p/l$	c	150 mm		175 mm		200 mm		225 mm		250 mm		300 mm		c	$\Delta p/l$
		q_m	l_e	q_m	l_e	q_m	l_e	q_m	l_e	q_m	l_e	q_m	l_e		
92.5		24.1	9.3	35.0	11	49.9	13	66.8	15	89.6	18	141	22		92.5
95.0		24.4	9.3	35.5	11	50.7	13	67.7	15	91.0	18	143	22		95.0
97.5		24.8	9.3	36.0	11	51.4	13	68.7	15	92.3	18	145	22		97.5
100.0		25.1	9.3	36.5	11	52.1	13	69.7	16	93.6	18	147	22		100.0
120.0	1.5	27.8	9.5	40.4	12	57.6	14	77.0	16	103	18	162	23		120.0
140.0		30.3	9.7	44.0	12	62.7	14	83.8	16	113	19	177	23		140.0
160.0		32.6	9.8	47.3	12	67.5	14	90.2	16	121	19	190	23		160.0
180.0		34.8	9.9	50.5	12	72.0	14	96.2	16	129	19	203	24		180.0
200.0		36.8	10	53.5	12	76.3	14	102	17	137	19	215	24	3.0	200.0
220.0		38.8	10	56.4	12	80.4	15	107	17	144	19	226	24		220.0
240.0		40.7	10	59.1	12	84.3	15	113	17	151	19	237	24		240.0
260.0		42.6	10	61.8	12	88.0	15	118	17	158	20	248	24		260.0
280.0		44.3	10	64.4	12	91.7	15	122	17	164	20	258	25		280.0
300.0		46.0	11	66.8	13	95.2	15	127	17	171	20	268	25		300.0
320.0		47.7	11	69.2	13	98.6	15	132	17	177	20	277	25	4.0	320.0
340.0		49.3	11	71.6	13	102	15	136	17	183	20	287	25		340.0
360.0		50.9	11	73.8	13	105	15	140	17	188	20	296	25		360.0
380.0		52.4	11	76.1	13	108	15	145	18	194	20	305	25		380.0
400.0		53.9	11	78.2	13	111	15	149	18	199	20	313	25		400.0
420.0	3.0	55.4	11	80.3	13	114	15	153	18	205	20	321	25		420.0
440.0		56.8	11	82.4	13	117	15	157	18	210	21	330	25		440.0
460.0		58.2	11	84.4	13	120	16	161	18	215	21	338	26		460.0
480.0		59.6	11	86.4	13	123	16	164	18	220	21	346	26	5.0	480.0
500.0		60.9	11	88.4	13	126	16	168	18	225	21	353	26		500.0
520.0		62.2	11	90.3	13	129	16	172	18	230	21	361	26		520.0
540.0		63.5	11	92.2	13	131	16	175	18	235	21	368	26		540.0
560.0		64.8	11	94.0	13	134	16	179	18	239	21	376	26		560.0
580.0		66.1	11	95.8	13	136	16	182	18	244	21	383	26		580.0
600.0		67.3	11	97.6	13	139	16	185	18	249	21	390	26		600.0
620.0		68.5	11	99.4	13	141	16	189	18	253	21	397	26		620.0
640.0		69.7	11	101	13	144	16	192	18	257	21	404	26		640.0
660.0		70.9	11	103	13	146	16	195	18	262	21	411	26		660.0
680.0		72.1	11	104	14	149	16	199	18	266	21	417	26	6.0	680.0
700.0	4.0	73.2	11	106	14	151	16	202	18	270	21				700.0
720.0		74.3	11	108	14	153	16	205	19	274	21				720.0
740.0		75.5	11	109	14	156	16	208	19	278	21				740.0
760.0		76.6	11	111	14	158	16	211	19	283	21				760.0
780.0		77.7	11	113	14	160	16	214	19	287	21				780.0
800.0		78.7	11	114	14	162	16	217	19	290	22				800.0
820.0		79.8	11	116	14	165	16	220	19	294	22				820.0
840.0		80.9	12	117	14	167	16	223	19	248	22				840.0
860.0		81.9	12	119	14	169	16	225	19						860.0
880.0		82.9	12	120	14	171	16	228	19						880.0
900.0		84.0	12	122	14	173	16	231	19						900.0
920.0		85.0	12	123	14	175	16	234	19						920.0
940.0		86.0	12	125	14	177	16	237	19						940.0
960.0		87.0	12	126	14	179	17	239	19						960.0
980.0		87.9	12	127	14	181	17								980.0
1000.0		88.9	12	129	14	183	17								1000.0
1100.0	5.0	93.6	12	136	14	193	17								1100.0
1200.0		98.2	12	142	14										1200.0
1300.0		103	12	149	14										1300.0
1400.0		107	12												1400.0
1500.0	6.0	111	12												1500.0
1600.0															1600.0
1700.0															1700.0
1800.0															1800.0
1900.0															1900.0
2000.0															2000.0

Table 4.22 Flow of water at 10°C in UPVC, Class E pipes

q_m = mass flow rate — kg.s^{-1}
c = velocity — m.s^{-1}
$\Delta p/l$ = pressure drop per unit length — Pa.m^{-1}
l_e = equivalent length of a component
for $\zeta = 1$ — m

* $(Re) = 2000$
† $(Re) = 3000$

UPVC, CLASS E
WATER AT 10°C

$\Delta p/l$	c	10 mm q_m	l_e	15 mm q_m	l_e	20 mm q_m	l_e	25 mm q_m	l_e	32 mm q_m	l_e	40 mm q_m	l_e	c	$\Delta p/l$
0.1												0.005	0.1		0.1
0.2										0.006	0.1	0.011	0.2		0.2
0.3								0.004	0.1	0.009	0.1	0.016	0.2		0.3
0.4								0.005	0.1	0.012	0.2	0.021	0.3		0.4
0.5								0.006	0.1	0.016	0.2	0.026	0.4		0.5
0.6								0.007	0.1	0.019	0.3	0.032	0.5		0.6
0.7								0.008	0.1	0.022	0.3	0.037	0.6		0.7
0.8								0.010	0.1	0.025	0.4	0.042	0.6		0.8
0.9						0.004	0.1	0.011	0.2	0.028	0.4	0.047	0.7		0.9
1.0						0.004	0.1	0.012	0.2	0.031	0.5	0.053	0.8	0.05	1.0
1.5						0.006	0.1	0.018	0.3	0.047	0.7	0.079*	1.2		1.5
2.0						0.008	0.1	0.024	0.4	0.062*	1.0	0.083	0.9		2.0
2.5				0.004	0.1	0.011	0.2	0.030	0.5	0.069	0.8	0.091	0.9		2.5
3.0				0.005	0.1	0.013	0.2	0.036	0.5	0.074	0.8	0.098	0.9		3.0
3.5				0.006	0.1	0.015	0.2	0.042	0.6	0.078	0.8	0.105	0.9		3.5
4.0				0.007	0.1	0.017	0.3	0.048	0.7	0.082	0.8	0.113	0.9		4.0
4.5				0.007	0.1	0.019	0.3	0.053*	0.8	0.087	0.8	0.121	0.9		4.5
5.0				0.008	0.1	0.021	0.3	0.054	0.7	0.091	0.8	0.128†	0.9	1.0	5.0
5.5				0.009	0.1	0.023	0.4	0.056	0.7	0.095	0.8	0.135	1.0		5.5
6.0		0.004	0.1	0.010	0.2	0.025	0.4	0.057	0.7	0.099	0.8	0.142	1.0		6.0
6.5		0.004	0.1	0.011	0.2	0.028	0.4	0.059	0.7	0.104	0.8	0.149	1.0		6.5
7.0		0.004	0.1	0.012	0.2	0.030	0.5	0.060	0.7	0.109	0.8	0.155	1.0		7.0
7.5		0.005	0.1	0.012	0.2	0.032	0.5	0.062	0.7	0.113†	0.8	0.162	1.0		7.5
8.0		0.005	0.1	0.013	0.2	0.034	0.5	0.063	0.7	0.117	0.8	0.168	1.0		8.0
8.5		0.005	0.1	0.014	0.2	0.036	0.5	0.065	0.7	0.121	0.8	0.174	1.0		8.5
9.0		0.005	0.1	0.015	0.2	0.038	0.6	0.067	0.7	0.126	0.9	0.180	1.0		9.0
9.5		0.006	0.1	0.016	0.2	0.040	0.6	0.068	0.7	0.130	0.9	0.186	1.1		9.5
10.0		0.006	0.1	0.016	0.3	0.042*	0.6	0.070	0.7	0.134	0.9	0.192	1.1	0.15	10.0
12.5	0.05	0.008	0.1	0.021	0.3	0.043	0.4	0.078	0.7	0.152	0.9	0.218	1.1		12.5
15.0		0.009	0.1	0.025	0.4	0.046	0.4	0.087†	0.7	0.170	0.9	0.243	1.1		15.0
17.5		0.011	0.2	0.029	0.4	0.048	0.4	0.095	0.7	0.185	1.0	0.265	1.2		17.5
20.0		0.012	0.2	0.033*	0.5	0.051	0.4	0.103	0.7	0.201	1.0	0.287	1.2		20.0
22.5	0.1	0.014	0.2	0.033	0.4	0.054	0.4	0.110	0.7	0.215	1.0	0.307	1.2		22.5
25.0		0.015	0.2	0.034	0.4	0.057	0.4	0.117	0.7	0.228	1.0	0.326	1.2		25.0
27.5		0.017	0.3	0.035	0.4	0.061	0.4	0.124	0.7	0.241	1.0	0.345	1.3		27.5
30.0		0.018	0.3	0.036	0.4	0.064	0.4	0.131	0.7	0.254	1.0	0.363	1.3		30.0
32.5		0.020	0.3	0.037	0.4	0.067†	0.5	0.137	0.7	0.266	1.1	0.380	1.3		32.5
35.0	0.15	0.021	0.3	0.039	0.4	0.070	0.5	0.143	0.7	0.277	1.1	0.396	1.3	0.3	35.0
37.5		0.023	0.3	0.040	0.4	0.073	0.5	0.149	0.8	0.289	1.1	0.412	1.3		37.5
40.0		0.024	0.4	0.041	0.4	0.076	0.5	0.155	0.8	0.300	1.1	0.428	1.3		40.0
42.5		0.025	0.4	0.042	0.4	0.079	0.5	0.160	0.8	0.310	1.1	0.443	1.3		42.5
45.0		0.026*	0.4	0.043	0.4	0.081	0.5	0.166	0.8	0.321	1.1	0.458	1.3		45.0
47.5		0.026	0.3	0.044	0.4	0.084	0.5	0.171	0.8	0.331	1.1	0.472	1.4		47.5
50.0		0.026	0.3	0.045	0.4	0.087	0.5	0.176	0.8	0.341	1.1	0.486	1.4		50.0
52.5		0.026	0.3	0.047	0.4	0.089	0.5	0.181	0.8	0.351	1.1	0.500	1.4		52.5
55.0		0.026	0.3	0.048	0.4	0.092	0.5	0.186	0.8	0.360	1.2	0.514	1.4		55.0
57.5		0.027	0.3	0.050	0.4	0.094	0.6	0.191	0.8	0.369	1.2	0.527	1.4		57.5
60.0		0.027	0.3	0.051	0.4	0.096	0.6	0.196	0.8	0.379	1.2	0.540	1.4		60.0
62.5		0.027	0.3	0.052	0.4	0.099	0.6	0.200	0.8	0.388	1.2	0.553	1.4		62.5
65.0		0.028	0.3	0.053†	0.4	0.101	0.6	0.205	0.8	0.396	1.2	0.565	1.4		65.0
67.5		0.029	0.3	0.054	0.4	0.103	0.6	0.209	0.8	0.405	1.2	0.578	1.4		67.5
70.0		0.029	0.3	0.055	0.4	0.106	0.6	0.214	0.8	0.414	1.2	0.590	1.4		70.0
72.5		0.030	0.3	0.056	0.4	0.108	0.6	0.218	0.8	0.422	1.2	0.602	1.4		72.5
75.0		0.030	0.3	0.057	0.4	0.110	0.6	0.223	0.8	0.430	1.2	0.614	1.5		75.0
77.5		0.030	0.3	0.058	0.4	0.112	0.6	0.227	0.9	0.439	1.2	0.625	1.5		77.5
80.0		0.031	0.3	0.059	0.4	0.114	0.6	0.231	0.9	0.447	1.2	0.637	1.5		80.0
82.5		0.031	0.3	0.060	0.4	0.116	0.6	0.235	0.9	0.455	1.2	0.648	1.5		82.5
85.0		0.032	0.3	0.061	0.4	0.118	0.6	0.239	0.9	0.462	1.2	0.659	1.5	0.5	85.0
87.5		0.032	0.3	0.063	0.4	0.120	0.6	0.243	0.9	0.470	1.2	0.670	1.5		87.5
90.0		0.032	0.3	0.064	0.4	0.122	0.6	0.247	0.9	0.478	1.2	0.681	1.5		90.0

Table 4.22 Flow of water at 10°C in UPVC, Class E pipes — *continued*

q_m = mass flow rate $\quad$ kg.s^{-1}
c = velocity $\quad$ m.s^{-1}
$\Delta p/l$ = pressure drop per unit length $\quad$ Pa.m^{-1}
l_e = equivalent length of a component
$\quad$ for $\zeta = 1$ $\quad$ m

* $(Re) = 2000$
† $(Re) = 3000$

		UPVC, CLASS E
		WATER AT 10°C

$\Delta p/l$	c	10 mm		15 mm		20 mm		25 mm		32 mm		40 mm		c	$\Delta p/l$
		q_m	l_e	q_m	l_e	q_m	l_e	q_m	l_e	q_m	l_e	q_m	l_e		
92.5		0.033	0.3	0.065	0.4	0.124	0.6	0.251	0.9	0.485	1.2	0.692	1.5		92.5
95.0		0.033	0.3	0.066	0.4	0.126	0.6	0.255	0.9	0.493	1.2	0.702	1.5		95.0
97.5		0.034	0.3	0.067	0.4	0.128	0.6	0.259	0.9	0.500	1.3	0.713	1.5		97.5
100.0		0.035	0.3	0.068	0.4	0.130	0.6	0.263	0.9	0.507	1.3	0.723	1.5		100.0
120.0		0.038	0.3	0.075	0.4	0.144	0.6	0.292	0.9	0.563	1.3	0.802	1.6		120.0
140.0	0.3	0.041†	0.3	0.082	0.4	0.158	0.6	0.319	0.9	0.615	1.3	0.876	1.6		140.0
160.0		0.045	0.3	0.089	0.5	0.171	0.7	0.344	0.9	0.663	1.3	0.944	1.6		160.0
180.0		0.048	0.3	0.095	0.5	0.183	0.7	0.368	1.0	0.709	1.4	1.01	1.6		180.0
200.0		0.051	0.3	0.101	0.5	0.194	0.7	0.391	1.0	0.753	1.4	1.07	1.7		200.0
220.0		0.054	0.3	0.107	0.5	0.205	0.7	0.413	1.0	0.795	1.4	1.13	1.7		220.0
240.0		0.057	0.3	0.113	0.5	0.215	0.7	0.434	1.0	0.835	1.4	1.19	1.7		240.0
260.0		0.059	0.3	0.118	0.5	0.226	0.7	0.454	1.0	0.874	1.4	1.24	1.7		260.0
280.0		0.062	0.3	0.123	0.5	0.235	0.7	0.474	1.0	0.912	1.4	1.30	1.7	1.0	280.0
300.0		0.065	0.4	0.128	0.5	0.245	0.7	0.493	1.0	0.948	1.5	1.35	1.8		300.0
320.0		0.067	0.4	0.133	0.5	0.254	0.7	0.512	1.0	0.983	1.5	1.40	1.8		320.0
340.0	0.5	0.069	0.4	0.138	0.5	0.263	0.7	0.529	1.1	1.02	1.5	1.45	1.8		340.0
360.0		0.072	0.4	0.142	0.5	0.272	0.7	0.547	1.1	1.05	1.5	1.49	1.8		360.0
380.0		0.074	0.4	0.147	0.5	0.280	0.7	0.564	1.1	1.08	1.5	1.54	1.8		380.0
400.0		0.076	0.4	0.151	0.5	0.289	0.7	0.581	1.1	1.12	1.5	1.58	1.8		400.0
420.0		0.079	0.4	0.156	0.5	0.297	0.8	0.597	1.1	1.15	1.5	1.63	1.8		420.0
440.0		0.081	0.4	0.160	0.5	0.305	0.8	0.613	1.1	1.18	1.5	1.67	1.8		440.0
460.0		0.083	0.4	0.164	0.5	0.313	0.8	0.629	1.1	1.21	1.5	1.71	1.9		460.0
480.0		0.085	0.4	0.168	0.5	0.320	0.8	0.644	1.1	1.24	1.6	1.76	1.9		480.0
500.0		0.087	0.4	0.172	0.5	0.328	0.8	0.659	1.1	1.26	1.6	1.80	1.9		500.0
520.0		0.089	0.4	0.176	0.6	0.335	0.8	0.674	1.1	1.29	1.6	1.84	1.9		520.0
540.0		0.091	0.4	0.180	0.6	0.343	0.8	0.688	1.1	1.32	1.6	1.88	1.9		540.0
560.0		0.093	0.4	0.184	0.6	0.350	0.8	0.703	1.1	1.35	1.6	1.91	1.9		560.0
580.0		0.095	0.4	0.187	0.6	0.357	0.8	0.717	1.1	1.37	1.6	1.95	1.9		580.0
600.0		0.097	0.4	0.191	0.6	0.364	0.8	0.731	1.1	1.40	1.6	1.99	1.9	1.5	600.0
620.0		0.098	0.4	0.195	0.6	0.371	0.8	0.744	1.1	1.43	1.6	2.03	1.9		620.0
640.0		0.100	0.4	0.198	0.6	0.378	0.8	0.758	1.1	1.45	1.6	2.06	1.9		640.0
660.0		0.102	0.4	0.202	0.6	0.384	0.8	0.771	1.2	1.48	1.6	2.10	1.9		660.0
680.0		0.104	0.4	0.205	0.6	0.391	0.8	0.784	1.2	1.50	1.6	2.13	1.9		680.0
700.0		0.106	0.4	0.209	0.6	0.397	0.8	0.797	1.2	1.53	1.6	2.17	2.0		700.0
720.0		0.107	0.4	0.212	0.6	0.404	0.8	0.810	1.2	1.55	1.6	2.20	2.0		720.0
740.0		0.109	0.4	0.215	0.6	0.410	0.8	0.822	1.2	1.58	1.6	2.24	2.0		740.0
760.0		0.111	0.4	0.219	0.6	0.416	0.8	0.835	1.2	1.60	1.6	2.27	2.0		760.0
780.0		0.112	0.4	0.222	0.6	0.422	0.8	0.847	1.2	1.62	1.7	2.30	2.0		780.0
800.0		0.114	0.4	0.225	0.6	0.429	0.8	0.859	1.2	1.65	1.7	2.34	2.0		800.0
820.0		0.116	0.4	0.228	0.6	0.435	0.8	0.872	1.2	1.67	1.7	2.37	2.0		820.0
840.0		0.117	0.4	0.232	0.6	0.441	0.8	0.883	1.2	1.69	1.7	2.40	2.0		840.0
860.0		0.119	0.4	0.235	0.6	0.447	0.8	0.895	1.2	1.71	1.7	2.43	2.0		860.0
880.0		0.120	0.4	0.238	0.6	0.452	0.8	0.907	1.2	1.74	1.7	2.46	2.0		880.0
900.0		0.122	0.4	0.241	0.6	0.458	0.8	0.918	1.2	1.76	1.7	2.50	2.0		900.0
920.0		0.124	0.4	0.244	0.6	0.464	0.8	0.930	1.2	1.78	1.7	2.53	2.0		920.0
940.0		0.125	0.4	0.247	0.6	0.470	0.8	0.941	1.2	1.80	1.7	2.56	2.0		940.0
960.0		0.127	0.4	0.250	0.6	0.475	0.8	0.952	1.2	1.82	1.7	2.59	2.0		960.0
980.0		0.128	0.4	0.253	0.6	0.481	0.8	0.964	1.2	1.84	1.7	2.62	2.0		980.0
1000.0		0.130	0.4	0.256	0.6	0.486	0.8	0.975	1.2	1.87	1.7	2.65	2.0		1000.0
1100.0		0.137	0.4	0.270	0.6	0.513	0.9	1.03	1.2	1.97	1.7	2.79	2.1		1100.0
1200.0	1.0	0.144	0.4	0.284	0.6	0.539	0.9	1.08	1.2	2.07	1.7	2.93	2.1		1200.0
1300.0		0.151	0.4	0.297	0.6	0.564	0.9	1.13	1.3	2.16	1.8	3.06	2.1		1300.0
1400.0		0.157	0.4	0.310	0.6	0.588	0.9	1.18	1.3	2.25	1.8	3.19	2.1		1400.0
1500.0		0.164	0.4	0.322	0.6	0.612	0.9	1.22	1.3	2.34	1.8	3.32	2.1		1500.0
1600.0		0.170	0.5	0.334	0.6	0.634	0.9	1.27	1.3	2.42	1.8	3.44	2.1		1600.0
1700.0		0.176	0.5	0.346	0.7	0.656	0.9	1.31	1.3	2.51	1.8	3.56	2.2		1700.0
1800.0		0.181	0.5	0.357	0.7	0.678	0.9	1.36	1.3	2.59	1.8	3.67	2.2		1800.0
1900.0		0.187	0.5	0.368	0.7	0.699	0.9	1.40	1.3	2.67	1.8	3.78	2.2		1900.0

Table 4.22 Flow of water at 10°C in UPVC, Class E pipes — *continued*

q_m = mass flow rate $\qquad$ kg.s⁻¹ → $\mathrm{kg.s^{-1}}$
c = velocity $\qquad$ $\mathrm{m.s^{-1}}$
$\Delta p/l$ = pressure drop per unit length $\qquad$ $\mathrm{Pa.m^{-1}}$
l_e = equivalent length of a component for $\zeta = 1$ $\qquad$ m

* $(Re) = 2000$
† $(Re) = 3000$

UPVC, CLASS E
WATER AT 10°C

$\Delta p/l$	c	50 mm q_m	l_e	65 mm q_m	l_e	80 mm q_m	l_e	100 mm q_m	l_e	125 mm q_m	l_e	150 mm q_m	l_e	c	$\Delta p/l$
0.1		0.01	0.2	0.03	0.5	0.06	0.9	0.16	2.3	0.27	2.8	0.42	3.3		0.1
0.2		0.03	0.4	0.06	1.0	0.12*	1.8	0.23	2.3	0.38	2.8	0.62	3.7		0.2
0.3		0.04	0.6	0.10	1.5	0.15	1.8	0.28†	2.3	0.48	3.0	0.79	3.9	0.05	0.3
0.4		0.05	0.8	0.13*	2.0	0.17	1.8	0.32	2.3	0.57	3.1	0.93	4.1		0.4
0.5		0.07	1.0	0.13	1.5	0.19	1.8	0.37	2.4	0.64	3.2	1.06	4.3		0.5
0.6		0.08	1.2	0.14	1.5	0.21	1.8	0.41	2.5	0.72	3.3	1.18	4.4		0.6
0.7		0.09	1.4	0.15	1.5	0.22†	1.8	0.45	2.5	0.78	3.4	1.29	4.5		0.7
0.8	0.05	0.10*	1.6	0.16	1.5	0.24	1.8	0.48	2.6	0.85	3.5	1.39	4.6		0.8
0.9		0.10	1.2	0.17	1.5	0.26	1.8	0.52	2.6	0.91	3.6	1.49	4.7		0.9
1.0		0.11	1.2	0.18†	1.5	0.28	1.9	0.55	2.7	0.96	3.6	1.58	4.7	0.1	1.0
1.5		0.12	1.2	0.22	1.5	0.35	2.0	0.70	2.9	1.22	3.9	2.00	5.0		1.5
2.0		0.14	1.2	0.26	1.6	0.41	2.1	0.82	3.0	1.44	4.0	2.35	5.2	0.15	2.0
2.5		0.16†	1.2	0.29	1.7	0.47	2.2	0.93	3.1	1.63	4.2	2.67	5.4		2.5
3.0		0.18	1.2	0.33	1.7	0.52	2.2	1.04	3.2	1.81	4.3	2.96	5.5		3.0
3.5		0.19	1.3	0.36	1.8	0.57	2.3	1.13	3.3	1.98	4.4	3.24	5.6		3.5
4.0	0.1	0.21	1.3	0.39	1.8	0.62	2.3	1.22	3.3	2.13	4.5	3.49	5.7		4.0
4.5		0.23	1.3	0.41	1.8	0.66	2.4	1.31	3.4	2.28	4.6	3.73	5.8		4.5
5.0		0.24	1.4	0.44	1.9	0.70	2.4	1.39	3.5	2.43	4.7	3.98	6.0		5.0
5.5		0.25	1.4	0.47	1.9	0.74	2.5	1.47	3.5	2.56	4.7	4.20	6.1		5.5
6.0		0.27	1.4	0.49	2.0	0.78	2.5	1.54	3.6	2.69	4.8	4.41	6.2		6.0
6.5		0.28	1.4	0.51	2.0	0.81	2.5	1.62	3.6	2.82	4.8	4.61	6.2		6.5
7.0		0.29	1.5	0.54	2.0	0.85	2.6	1.69	3.7	2.94	4.9	4.81	6.3	0.3	7.0
7.5	0.15	0.30	1.5	0.56	2.0	0.88	2.6	1.75	3.7	3.06	4.9	5.00	6.3		7.5
8.0		0.31	1.5	0.58	2.1	0.92	2.6	1.82	3.7	3.17	5.0	5.19	6.4		8.0
8.5		0.33	1.5	0.60	2.1	0.95	2.6	1.88	3.8	3.28	5.0	5.37	6.4		8.5
9.0		0.34	1.5	0.62	2.1	0.98	2.7	1.94	3.8	3.39	5.0	5.55	6.5		9.0
9.5		0.35	1.5	0.64	2.1	1.01	2.7	2.01	3.8	3.50	5.1	5.72	6.5		9.5
10.0		0.36	1.5	0.66	2.1	1.04	2.7	2.06	3.8	3.60	5.1	5.88	6.6		10.0
12.5		0.41	1.6	0.75	2.2	1.18	2.8	2.34	4.0	4.08	5.3	6.67	6.8		12.5
15.0		0.45	1.6	0.83	2.3	1.31	2.9	2.60	4.1	4.53	5.4	7.39	6.9		15.0
17.5		0.50	1.7	0.91	2.3	1.43	2.9	2.84	4.1	4.94	5.5	8.06	7.1	0.5	17.5
20.0		0.53	1.7	0.98	2.3	1.55	3.0	3.06	4.2	5.32	5.6	8.69	7.2		20.0
22.5		0.57	1.7	1.05	2.4	1.65	3.0	3.27	4.3	5.69	5.7	9.28	7.3		22.5
25.0	0.3	0.61	1.8	1.11	2.4	1.75	3.1	3.47	4.3	6.03	5.7	9.84	7.4		25.0
27.5		0.64	1.8	1.17	2.5	1.85	3.1	3.66	4.4	6.37	5.8	10.4	7.4		27.5
30.0		0.67	1.8	1.23	2.5	1.95	3.1	3.85	4.4	6.68	5.9	10.9	7.5		30.0
32.5		0.71	1.8	1.29	2.5	2.04	3.2	4.02	4.5	6.99	5.9	11.4	7.6		32.5
35.0		0.74	1.9	1.35	2.5	2.12	3.2	4.19	4.5	7.29	6.0	11.9	7.7		35.0
37.5		0.77	1.9	1.40	2.6	2.21	3.2	4.36	4.6	7.58	6.0	12.6	7.7		37.5
40.0		0.80	1.9	1.45	2.6	2.29	3.3	4.52	4.6	7.85	6.1	12.8	7.8		40.0
42.5		0.82	1.9	1.50	2.6	2.37	3.3	4.68	4.6	8.13	6.1	13.2	7.8		42.5
45.0		0.85	1.9	1.55	2.6	2.45	3.3	4.83	4.7	8.39	6.2	13.7	7.9		45.0
47.5		0.88	1.9	1.60	2.6	2.52	3.3	4.98	4.7	8.65	6.2	14.1	7.9		47.5
50.0		0.90	2.0	1.65	2.7	2.60	3.3	5.13	4.7	8.90	6.2	14.5	8.0		50.0
52.5		0.93	2.0	1.70	2.7	2.67	3.4	5.27	4.8	9.14	6.3	14.9	8.0		52.5
55.0		0.95	2.0	1.74	2.7	2.74	3.4	5.41	4.8	9.39	6.3	15.3	8.1		55.0
57.5		0.98	2.0	1.79	2.7	2.81	3.4	5.54	4.8	9.62	6.3	15.7	8.1		57.5
60.0		1.00	2.0	1.83	2.7	2.88	3.4	5.68	4.8	9.85	6.4	16.0	8.1	1.0	60.0
62.5	0.5	1.03	2.0	1.87	2.7	2.95	3.4	5.81	4.9	10.1	6.4	16.4	8.2		62.5
65.0		1.05	2.0	1.91	2.7	3.01	3.5	5.94	4.9	10.3	6.4	16.8	8.2		65.0
67.5		1.07	2.0	1.96	2.8	3.08	3.5	6.06	4.9	10.5	6.5	17.1	8.2		67.5
70.0		1.09	2.0	2.00	2.8	3.14	3.5	6.19	4.9	10.7	6.5	17.5	8.3		70.0
72.5		1.12	2.1	2.04	2.8	3.20	3.5	6.31	4.9	11.0	6.5	17.8	8.3		72.5
75.0		1.14	2.1	2.08	2.8	3.26	3.5	6.43	5.0	11.2	6.5	18.2	8.3		75.0
77.5		1.16	2.1	2.11	2.8	3.33	3.5	6.55	5.0	11.4	6.6	18.5	8.4		77.5
80.0		1.18	2.1	2.15	2.8	3.39	3.6	6.67	5.0	11.6	6.6	18.8	8.4		80.0
82.5		1.20	2.1	2.19	2.8	3.44	3.6	6.79	5.0	11.8	6.6	19.2	8.4		82.5
85.0		1.22	2.1	2.23	2.8	3.50	3.6	6.90	5.0	12.0	6.6	19.5	8.5		85.0
87.5		1.24	2.1	2.26	2.9	3.56	3.6	7.01	5.1	12.2	6.7	19.8	8.5		87.5
90.0		1.26	2.1	2.30	2.9	3.62	3.6	7.12	5.1	12.4	6.7	20.1	8.5		90.0

Table 4.22 Flow of water at 10°C in UPVC, Class E pipes — *continued*

q_m = mass flow rate kg.s^{-1}
c = velocity m.s^{-1}
$\Delta p/l$ = pressure drop per unit length Pa.m^{-1}
l_e = equivalent length of a component
　　for $\zeta = 1$ m

UPVC, CLASS E

WATER AT 10°C

$\Delta p/l$	c	50 mm		65 mm		80 mm		100 mm		125 mm		150 mm		c	$\Delta p/l$
		q_m	l_e	q_m	l_e	q_m	l_e	q_m	l_e	q_m	l_e	q_m	l_e		
92.5		1.28	2.1	2.34	2.9	3.67	3.6	7.23	5.1	12.5	6.7	20.4	8.5		92.5
95.0		1.30	2.1	2.37	2.9	3.73	3.6	7.34	5.1	12.7	6.7	20.7	8.6		95.0
97.5		1.32	2.1	2.41	2.9	3.78	3.6	7.45	5.1	12.9	6.7	21.0	8.6		97.5
100.0		1.34	2.1	2.44	2.9	3.84	3.6	7.56	5.1	13.1	6.8	21.3	8.6		100.0
120.0		1.48	2.2	2.70	3.0	4.25	3.7	8.36	5.2	14.5	6.9	23.6	8.8	1.5	120.0
140.0		1.62	2.2	2.95	3.0	4.63	3.8	9.11	5.3	15.8	7.0	25.7	8.9		140.0
160.0		1.75	2.3	3.18	3.1	4.99	3.9	9.82	5.4	17.0	7.1	27.6	9.0		160.0
180.0		1.87	2.3	3.40	3.1	5.33	3.9	10.5	5.5	18.1	7.2	29.5	9.2		180.0
200.0		1.98	2.3	3.60	3.2	5.66	4.0	11.1	5.5	19.2	7.3	31.2	9.3		200.0
220.0	1.0	2.09	2.4	3.80	3.2	5.96	4.0	11.7	5.6	20.3	7.4	32.9	9.3		220.0
240.0		2.19	2.4	3.99	3.2	6.26	4.0	12.3	5.7	21.3	7.4	34.6	9.4		240.0
260.0		2.30	2.4	4.17	3.3	6.55	4.1	12.9	5.7	22.2	7.5	36.1	9.5		260.0
280.0		2.39	2.4	4.35	3.3	6.82	4.1	13.4	5.7	23.2	7.5	37.6	9.6		280.0
300.0		2.49	2.5	4.52	3.3	7.09	4.1	13.9	5.8	24.1	7.6	39.1	9.6		300.0
320.0		2.58	2.5	4.68	3.3	7.35	4.2	14.4	5.8	24.9	7.6	40.5	9.7		320.0
340.0		2.67	2.5	4.85	3.4	7.60	4.2	14.9	5.9	25.8	7.7	41.8	9.8		340.0
360.0		2.75	2.5	5.00	3.4	7.85	4.2	15.4	5.9	26.6	7.7	43.2	9.8		360.0
380.0		2.84	2.5	5.16	3.4	8.08	4.3	15.9	5.9	27.4	7.8	44.5	9.9		380.0
400.0		2.92	2.5	5.31	3.4	8.32	4.3	16.3	6.0	28.2	7.8	45.8	9.9		400.0
420.0		3.00	2.5	5.45	3.4	8.55	4.3	16.8	6.0	29.0	7.9	47.0	10		420.0
440.0	1.5	3.08	2.6	5.59	3.5	8.77	4.3	17.2	6.0	29.7	7.9	48.2	10		440.0
460.0		3.16	2.6	5.73	3.5	8.99	4.3	17.6	6.1	30.4	7.9	49.4	10	3.0	460.0
480.0		3.24	2.6	5.87	3.5	9.20	4.4	18.0	6.1	31.2	8.0	50.6	10		480.0
500.0		3.31	2.6	6.01	3.5	9.41	4.4	18.5	6.1	31.9	8.0	51.7	10		500.0
520.0		3.38	2.6	6.14	3.5	9.62	4.4	18.9	6.1	32.6	8.0	52.8	10		520.0
540.0		3.46	2.6	6.27	3.5	9.82	4.4	19.3	6.2	33.2	8.1	53.9	10		540.0
560.0		3.53	2.6	6.40	3.5	10.0	4.4	19.6	6.2	33.9	8.1	55.0	10		560.0
580.0		3.60	2.6	6.52	3.6	10.2	4.4	20.0	6.2	34.6	8.1	56.1	10		580.0
600.0		3.66	2.7	6.65	3.6	10.4	4.5	20.4	6.2	35.2	8.1	57.1	10		600.0
620.0		3.73	2.7	6.77	3.6	10.6	4.5	20.8	6.2	35.9	8.2	58.2	10		620.0
640.0		3.80	2.7	6.89	3.6	10.8	4.5	21.1	6.3	36.5	8.2	59.2	10		640.0
660.0		3.86	2.7	7.01	3.6	11.0	4.5	21.5	6.3	37.1	8.2	60.2	10		660.0
680.0		3.93	2.7	7.12	3.6	11.2	4.5	21.9	6.3	37.7	8.2	61.2	10		680.0
700.0		3.99	2.7	7.24	3.6	11.3	4.5	22.2	6.3	38.3	8.3	62.1	10		700.0
720.0		4.06	2.7	7.35	3.6	11.5	4.5	22.6	6.3	38.9	8.3	63.1	10		720.0
740.0		4.12	2.7	7.47	3.7	11.7	4.6	22.9	6.4	39.5	8.3	64.1	11		740.0
760.0		4.18	2.7	7.58	3.7	11.9	4.6	23.2	6.4	40.1	8.3	65.0	11	4.0	760.0
780.0		4.24	2.7	7.69	3.7	12.0	4.6	23.6	6.4	40.7	8.3	65.9	11		780.0
800.0		4.30	2.7	7.79	3.7	12.2	4.6	23.9	6.4	41.2	8.4	66.8	11		800.0
820.0		4.36	2.8	7.90	3.7	12.4	4.6	24.2	6.4	41.8	8.4	67.7	11		820.0
840.0		4.42	2.8	8.01	3.7	12.5	4.6	24.5	6.4	42.4	8.4	68.6	11		840.0
860.0		4.48	2.8	8.11	3.7	12.7	4.6	24.9	6.4	42.9	8.4	69.5	11		860.0
880.0		4.53	2.8	8.22	3.7	12.9	4.6	25.2	6.5	43.4	8.4	70.4	11		880.0
900.0		4.59	2.8	8.32	3.7	13.0	4.7	25.5	6.5	44.0	8.5	71.3	11		900.0
920.0		4.65	2.8	8.42	3.7	13.2	4.7	25.8	6.5	44.5	8.5	72.1	11		920.0
940.0		4.70	2.8	8.52	3.7	13.3	4.7	26.1	6.5	45.0	8.5	73.0	11		940.0
960.0		4.76	2.8	8.62	3.8	13.5	4.7	26.4	6.5	45.6	8.5	73.8	11		960.0
980.0		4.81	2.8	8.72	3.8	13.7	4.7	26.7	6.5	46.1	8.5	74.7	11		980.0
1000.0		4.87	2.8	8.82	3.8	13.8	4.7	27.0	6.5	46.6	8.5	75.5	11		1000.0
1100.0		5.13	2.8	9.29	3.8	14.6	4.7	28.5	6.6	49.1	8.6	79.5	11	5.0	1100.0
1200.0		5.39	2.9	9.75	3.8	15.3	4.8	29.8	6.7	51.5	8.7	83.3	11		1200.0
1300.0		5.63	2.9	10.2	3.9	15.9	4.8	31.2	6.7	53.8	8.7	87.1	11		1300.0
1400.0		5.87	2.9	10.6	3.9	16.6	4.9	32.4	6.7	56.0	8.8	90.6	11		1400.0
1500.0		6.09	2.9	11.0	3.9	17.3	4.9	33.7	6.8	58.1	8.8	94.1	11		1500.0
1600.0	3.0	6.31	3.0	11.4	4.0	17.9	4.9	34.9	6.8	60.2	8.9	97.4	11	6.0	1600.0
1700.0		6.53	3.0	11.8	4.0	18.5	5.0	36.1	6.9	62.2	8.9				1700.0
1800.0		6.74	3.0	12.2	4.0	19.1	5.0	37.2	6.9	64.2	9.0				1800.0
1900.0		6.94	3.0	12.6	4.0	19.6	5.0	38.4	6.9	66.1	9.0				1900.0

Table 4.22 Flow of water at 10°C in UPVC, Class E pipes — *continued*

q_m = mass flow rate $\quad$ kg.s^{-1}
c = velocity $\quad$ m.s^{-1}
$\Delta p/l$ = pressure drop per unit length $\quad$ Pa.m^{-1}
l_e = equivalent length of a component
  for $\zeta = 1$ $\quad$ m

$\Delta p/l$	c	175 mm		200 mm		225 mm		250 mm		300 mm		c	$\Delta p/l$
		q_m	l_e	q_m	l_e	q_m	l_e	q_m	l_e	q_m	l_e		
0.1		0.62	4.1	0.91	5.0	1.23	5.9	1.66	7.0	2.64	8.9	0.05	0.1
0.2		0.93	4.6	1.36	5.6	1.83	6.6	2.48	7.8	3.93	9.9		0.2
0.3	0.05	1.17	4.9	1.72	6.0	2.31	7.0	3.13	8.3	4.96	10		0.3
0.4		1.38	5.1	2.03	6.3	2.72	7.3	3.69	8.6	5.84	11	0.1	0.4
0.5		1.57	5.3	2.30	6.5	3.10	7.6	4.19	8.9	6.64	11		0.5
0.6		1.75	5.5	2.56	6.7	3.44	7.8	4.65	9.1	7.36	12		0.6
0.7		1.91	5.6	2.79	6.8	3.75	7.9	5.08	9.3	8.03	12		0.7
0.8		2.06	5.7	3.02	6.9	4.05	8.1	5.48	9.5	8.66	12		0.8
0.9	0.1	2.20	5.8	3.23	7.1	4.33	8.2	5.86	9.6	9.26	12	0.15	0.9
1.0		2.34	5.9	3.43	7.2	4.60	8.3	6.22	9.8	9.83	12		1.0
1.5		2.95	6.2	4.31	7.6	5.78	8.8	7.82	10	12.4	13		1.5
2.0	0.15	3.48	6.5	5.08	7.9	6.80	9.2	9.20	11	14.5	14		2.0
2.5		3.95	6.7	5.76	8.1	7.72	9.4	10.4	11	16.5	14		2.5
3.0		4.37	6.8	6.39	8.3	8.55	9.6	11.6	11	18.2	14	0.3	3.0
3.5		4.77	7.0	6.97	8.5	9.33	9.8	12.6	11	19.9	14		3.5
4.0		5.15	7.1	7.51	8.6	10.1	10	13.6	12	21.4	15		4.0
4.5		5.50	7.2	8.02	8.7	10.7	10	14.5	12	22.9	15		4.5
5.0		5.83	7.3	8.51	8.9	11.4	10	15.4	12	24.3	15		5.0
5.5		6.15	7.4	8.97	9.0	12.0	10	16.2	12	25.6	15		5.5
6.0	0.3	6.46	7.5	9.42	9.1	12.6	11	17.0	12	26.9	15		6.0
6.5		6.76	7.6	9.86	9.2	13.2	11	17.8	12	28.1	16		6.5
7.0		7.05	7.6	10.3	9.3	13.8	11	18.6	13	29.3	16		7.0
7.5		7.33	7.7	10.7	9.3	14.3	11	19.3	13	30.5	16		7.5
8.0		7.60	7.8	11.1	9.4	14.8	11	20.0	13	31.6	16	0.5	8.0
8.5		7.86	7.8	11.5	9.5	15.3	11	20.7	13	32.6	16		8.5
9.0		8.12	7.9	11.8	9.5	15.8	11	21.4	13	33.7	16		9.0
9.5		8.37	7.9	12.2	9.6	16.3	11	22.0	13	34.7	16		9.5
10.0		8.61	8.0	12.6	9.7	16.8	11	22.7	13	35.7	16		10.0
12.5		9.76	8.2	14.2	9.9	19.0	11	25.7	13	40.5	17		12.5
15.0	0.5	10.8	8.4	15.8	10	21.1	12	28.4	14	44.8	17		15.0
17.5		11.8	8.5	17.2	10	23.0	12	31.0	14	48.8	17		17.5
20.0		12.7	8.7	18.5	11	24.8	12	33.3	14	52.5	18		20.0
22.5		13.6	8.8	19.8	11	26.4	12	35.6	14	56.0	18		22.5
25.0		14.4	8.9	20.9	11	28.0	12	37.7	14	59.4	18		25.0
27.5		15.2	9.0	22.1	11	29.5	13	39.8	15	62.6	18	1.0	27.5
30.0		15.9	9.1	23.2	11	31.0	13	41.7	15	65.7	18		30.0
32.5		16.7	9.2	24.2	11	32.4	13	43.6	15	68.7	19		32.5
35.0		17.4	9.3	25.3	11	33.7	13	45.5	15	71.5	19		35.0
37.5		18.0	9.3	26.2	11	35.1	13	47.2	15	74.3	19		37.5
40.0		18.7	9.4	27.2	11	36.3	13	48.9	15	77.0	19		40.0
42.5		19.3	9.5	28.1	11	37.6	13	50.6	15	79.6	19		42.5
45.0		20.0	9.5	29.0	11	38.8	13	52.2	15	82.1	19		45.0
47.5		20.6	9.6	29.9	12	40.0	13	53.8	15	84.6	19		47.5
50.0		21.2	9.6	30.8	12	41.1	13	55.4	16	87.0	19		50.0
52.5	1.0	21.7	9.7	31.6	12	42.2	13	56.9	16	89.4	20		52.5
55.0		22.3	9.7	32.4	12	43.3	14	58.3	16	91.7	20		55.0
57.5		22.9	9.8	33.3	12	44.4	14	59.8	16	94.0	20	1.5	57.5
60.0		23.4	9.8	34.0	12	45.5	14	61.2	16	96.2	20		60.0
62.5		24.0	9.9	34.8	12	46.5	14	62.6	16	98.4	20		62.5
65.0		24.5	9.9	35.6	12	47.5	14	64.0	16	101	20		65.0
67.5		25.0	10	36.3	12	48.5	14	65.3	16	103	20		67.5
70.0		25.5	10	37.1	12	49.5	14	66.6	16	105	20		70.0
72.5		26.0	10	37.8	12	50.5	14	67.9	16	107	20		72.5
75.0		26.5	10	38.5	12	51.4	14	69.2	16	109	20		75.0
77.5		27.0	10	39.2	12	52.3	14	70.5	16	111	20		77.5
80.0		27.4	10	39.9	12	53.3	14	71.7	16	113	20		80.0
82.5		27.9	10	40.6	12	54.2	14	72.9	16	115	20		82.5
85.0		28.4	10	41.3	12	55.1	14	74.1	16	117	20		85.0
87.5		28.9	10	41.9	12	56.0	14	75.3	16	118	21		87.5
90.0		29.3	10	42.6	12	56.8	14	76.5	16	120	21		90.0

Table 4.22 Flow of water at 10°C in UPVC, Class E pipes — *continued*

q_m = mass flow rate $\quad$ kg.s^{-1}
c = velocity $\quad$ m.s^{-1}
$\Delta p/l$ = pressure drop per unit length $\quad$ Pa.m^{-1}
l_e = equivalent length of a component
$\quad$ for $\zeta = 1$ $\quad$ m

UPVC, CLASS E
WATER AT 10°C

$\Delta p/l$	c	175 mm		200 mm		225 mm		250 mm		300 mm		c	$\Delta p/l$
		q_m	l_e	q_m	l_e	q_m	l_e	q_m	l_e	q_m	l_e		
92.5		29.8	10	43.2	12	57.7	14	77.7	17	122	21		92.5
95.0		30.2	10	43.9	12	58.6	14	78.8	17	124	21		95.0
97.5		30.6	10	44.5	12	59.4	14	80.0	17	126	21		97.5
100.0	1.5	31.1	10	45.1	13	60.2	14	81.1	17	127	21		100.0
120.0		34.4	11	49.9	13	66.6	15	89.6	17	141	21		120.0
140.0		37.4	11	54.3	13	72.5	15	97.5	17	153	22		140.0
160.0		40.3	11	58.5	13	78.0	15	105	17	165	22		160.0
180.0		43.0	11	62.4	13	83.2	15	112	18	176	22		180.0
200.0		45.5	11	66.1	13	88.1	15	119	18	186	22	3.0	200.0
220.0		48.0	11	69.6	14	92.9	16	125	18	196	22		220.0
240.0		50.3	11	73.0	14	97.4	16	131	18	205	23		240.0
260.0		52.6	11	76.3	14	102	16	137	18	215	23		260.0
280.0		54.8	12	79.5	14	106	16	142	18	223	23		280.0
300.0		56.9	12	82.5	14	110	16	148	19	232	23		300.0
320.0		58.9	12	85.5	14	114	16	153	19	240	23		320.0
340.0		60.9	12	88.4	14	118	16	158	19	248	23	4.0	340.0
360.0		62.9	12	91.2	14	122	16	163	19	256	23		360.0
380.0	3.0	64.7	12	93.9	14	125	16	168	19	264	24		380.0
400.0		66.6	12	96.6	14	129	16	173	19	271	24		400.0
420.0		68.4	12	99.2	14	132	17	178	19	278	24		420.0
440.0		70.2	12	102	14	136	17	182	19	286	24		440.0
460.0		71.9	12	104	14	139	17	187	19	293	24		460.0
480.0		73.6	12	107	15	142	17	191	19	299	24		480.0
500.0		75.2	12	109	15	145	17	195	19	306	24		500.0
520.0		76.9	12	111	15	148	17	200	19	313	24	5.0	520.0
540.0		78.5	12	114	15	152	17	204	19	319	24		540.0
560.0		80.0	12	116	15	155	17	208	20	325	24		560.0
580.0		81.6	12	118	15	158	17	212	20	332	24		580.0
600.0		83.1	12	120	15	160	17	216	20	338	24		600.0
620.0		84.6	12	123	15	163	17	219	20	344	24		620.0
640.0	4.0	86.1	12	125	15	166	17	223	20	350	25		640.0
660.0		87.5	12	127	15	169	17	227	20	356	25		660.0
680.0		89.0	13	129	15	172	17	231	20	361	25		680.0
700.0		90.4	13	131	15	174	17	234	20	367	25		700.0
720.0		91.8	13	133	15	177	17	238	20	373	25	6.0	720.0
740.0		93.2	13	135	15	180	17	242	20				740.0
760.0		94.5	13	137	15	182	17	245	20				760.0
780.0		95.9	13	139	15	185	17	249	20				780.0
800.0		97.2	13	141	15	188	17	252	20				800.0
820.0		98.5	13	143	15	190	17	255	20				820.0
840.0		99.8	13	145	15	193	18	259	20				840.0
860.0		101	13	146	15	195	18	262	20				860.0
880.0		102	13	148	15	198	18	265	20				880.0
900.0		104	13	150	15	200	18	269	20				900.0
920.0		105	13	152	15	202	18						920.0
940.0		106	13	154	15	205	18						940.0
960.0		107	13	155	15	207	18						960.0
980.0	5.0	109	13	157	15	209	18						980.0
1000.0		110	13	159	16	212	18						1000.0
1100.0		116	13	167	16								1100.0
1200.0		121	13	175	16								1200.0
1300.0		127	13										1300.0
1400.0	6.0	132	13										1400.0
1500.0													1500.0
1600.0													1600.0
1700.0													1700.0
1800.0													1800.0
1900.0													1900.0

Table 4.23 Flow of water at 10°C in cast iron

q_m = mass flow rate — kg.s^{-1}
c = velocity — m.s^{-1}
$\Delta p/l$ = pressure drop per unit length — Pa.m^{-1}
l_e = equivalent length of a component for $\zeta = 1$ — m

CAST IRON
WATER AT 10°C

$\Delta p/l$	c	80 mm		100 mm		150 mm		200 mm		250 mm		300 mm		c	$\Delta p/l$
		q_m	l_e	q_m	l_e	q_m	l_e	q_m	l_e	q_m	l_e	q_m	l_e		
50		2.72	2.8	7.18	4.6	16.6	6.9	34.5	10	61.9	13	112	18		50
55		2.85	2.8	7.55	4.6	17.4	6.9	36.3	10	65.0	13	118	18		55
60		2.99	2.8	7.90	4.6	18.2	7.0	38.0	10	68.0	13	123	18		60
70		3.24	2.9	8.56	4.6	19.7	7.0	41.1	10	73.6	13	134	18		70
80		3.48	2.9	9.18	4.7	21.1	7.0	44.0	10	78.8	13	143	18		80
90		3.70	2.9	9.76	4.7	22.4	7.1	46.8	10	83.8	13	152	18		90
100		3.91	2.9	10.3	4.7	23.7	7.1	49.4	10	88.4	13	160	18		100
110		4.11	2.9	10.8	4.7	24.9	7.1	51.9	10	92.8	13	168	18	2.0	110
120		4.30	2.9	11.3	4.7	26.0	7.1	54.3	10	97.0	13	176	18		120
130		4.49	2.9	11.8	4.8	27.1	7.2	56.5	10	101	14	183	18		130
140		4.66	3.0	12.3	4.8	28.2	7.2	58.7	10	105	14	190	18		140
150	1.0	4.83	3.0	12.7	4.8	29.2	7.2	60.8	10	109	14	197	18		150
175		5.24	3.0	13.8	4.8	31.6	7.2	65.8	10	118	14	213	18		175
200		5.61	3.0	14.8	4.8	33.9	7.2	70.5	10	126	14	228	18		200
225		5.96	3.0	15.7	4.8	36.0	7.2	74.8	10	134	14	242	18	3.0	225
250		6.30	3.0	16.5	4.8	38.0	7.3	79.0	10	141	14	256	18		250
275		6.62	3.0	17.4	4.9	39.9	7.3	82.9	10	148	14	268	18		275
300		6.92	3.0	18.2	4.9	41.7	7.3	86.7	10	155	14	280	18		300
350		7.49	3.0	19.7	4.9	45.1	7.3	93.7	10	167	14	303	18		350
400		8.02	3.1	21.0	4.9	48.2	7.3	100	10	179	14	324	18	4.0	400
450		8.52	3.1	22.4	4.9	51.2	7.4	106	10	190	14	344	18		450
500		8.99	3.1	23.6	4.9	54.0	7.4	112	10	201	14	363	18		500
550		9.44	3.1	24.8	4.9	56.7	7.4	118	10	211	14	381	18		550
600	2.0	9.87	3.1	25.9	5.0	59.3	7.4	123	10	220	14	398	18	5.0	600
700		10.7	3.1	28.0	5.0	64.1	7.4	933	11	238	14	430	18		700
800		11.4	3.1	30.0	5.0	68.6	7.4	143	11	254	14	460	18		800
900		12.1	3.1	31.8	5.0	72.8	7.4	141	11	270	14	488	18	6.0	900
1000		12.8	3.1	33.6	5.0	76.8	7.4	160	11	285	14				1000
1250	3.0	14.4	3.1	37.6	5.0	86.0	7.5	179	11	319	14				1250
1500		15.7	3.1	41.2	5.0	94.3	7.5	196	11						1500
1750		17.0	3.2	44.5	5.0	102	7.5								1750
2000		18.2	3.2	47.7	5.0	109	7.5								2000
2250		19.3	3.2	50.6	5.0	116	7.5								2250
2500	4.0	20.4	3.2	53.4	5.0										2500
2570		21.4	3.2	56.0	5.0										2570
3000		22.4	3.2	58.5	5.0										3000
3500	5.0	24.2	3.2	63.2	5.1										3500
4000		25.9	3.2												4000
4500		27.4	3.2												4500
5000		28.9	3.2												5000
5500	6.0	30.4	3.2												5500
6000															6000
7000															7000
8000															8000
9000															9000
10 000															10 000
11 000															11 000
12 000															12 000

Table 4.24 Flow of water at 10°C in polybutylene (PB) pipes[21]

q_m = mass flow rate	kg.s^{-1}	
c = velocity	m.s^{-1}	
$\Delta p/l$ = pressure drop per unit length	Pa.m^{-1}	
l_e = equivalent length of a component for $\zeta = 1$	m	

POLYBUTYLENE (PB)
WATER AT 10°C
BS 2871

$\Delta p/l$	c	10 mm		12 mm		15 mm		18 mm		22 mm		28 mm		35 mm		c	$\Delta p/l$
		q_m	l_e	q_m	l_e	q_m	l_e	q_m	l_e	q_m	l_e	q_m	l_e	q_m	l_e		
0.1										0.001	0.07	0.002	0.12	0.004	0.19		0.1
0.2								0.001	0.06	0.002	0.09	0.003	0.16	0.006	0.24		0.2
0.3								0.001	0.07	0.002	0.11	0.004	0.18	0.008	0.27		0.3
0.4						0.001	0.05	0.001	0.08	0.002	0.12	0.005	0.19	0.010	0.29		0.4
0.5						0.001	0.05	0.001	0.08	0.003	0.13	0.006	0.21	0.011	0.31		0.5
0.6						0.001	0.06	0.002	0.09	0.003	0.14	0.006	0.22	0.012	0.33		0.6
0.7						0.001	0.06	0.002	0.09	0.004	0.14	0.007	0.23	0.014	0.34		0.7
0.8						0.001	0.06	0.002	0.10	0.004	0.15	0.008	0.23	0.015	0.35		0.8
0.9						0.001	0.06	0.002	0.10	0.004	0.16	0.008	0.24	0.016	0.36		0.9
1.0				0.001	0.04	0.001	0.07	0.002	0.11	0.004	0.16	0.009	0.25	0.017	0.37		1.0
1.5				0.001	0.04	0.002	0.08	0.003	0.12	0.006	0.18	0.011	0.28	0.022	0.41		1.5
2.0				0.001	0.05	0.002	0.08	0.004	0.13	0.007	0.19	0.014	0.30	0.026	0.44		2.0
2.5				0.001	0.05	0.002	0.09	0.004	0.14	0.008	0.21	0.016	0.31	0.030	0.46		2.5
3.0				0.001	0.06	0.002	0.10	0.005	0.15	0.009	0.22	0.018	0.32	0.034	0.48	0.05	3.0
3.5		0.001	0.04	0.001	0.06	0.003	0.10	0.005	0.15	0.010	0.22	0.019	0.34	0.037	0.49		3.5
4.0		0.001	0.04	0.001	0.06	0.003	0.10	0.006	0.16	0.011	0.23	0.021	0.35	0.040	0.51		4.0
4.5		0.001	0.04	0.001	0.07	0.003	0.11	0.006	0.16	0.011	0.24	0.022	0.35	0.043	0.52		4.5
5.0		0.001	0.04	0.002	0.07	0.003	0.11	0.006	0.17	0.012	0.24	0.024	0.36	0.046	0.53		5.0
5.5		0.001	0.04	0.002	0.07	0.004	0.11	0.007	0.17	0.013	0.25	0.025	0.37	0.048	0.54		5.5
6.0		0.001	0.04	0.002	0.07	0.004	0.12	0.007	0.17	0.014	0.25	0.027	0.38	0.051	0.55		6.0
6.5		0.001	0.04	0.002	0.07	0.004	0.12	0.008	0.18	0.014	0.26	0.028	0.38	0.054	0.56		6.5
7.0		0.001	0.05	0.002	0.07	0.004	0.12	0.008	0.18	0.015	0.26	0.029	0.39	0.056	0.57		7.0
7.5		0.001	0.05	0.002	0.08	0.004	0.12	0.008	0.18	0.016	0.27	0.031	0.40	0.058	0.57		7.5
8.0		0.001	0.05	0.002	0.08	0.004	0.13	0.009	0.19	0.016	0.27	0.032	0.40	0.061	0.58		8.0
8.5		0.001	0.05	0.002	0.08	0.005	0.13	0.009	0.19	0.017	0.28	0.033	0.41	0.063	0.59		8.5
9.0		0.001	0.05	0.002	0.08	0.005	0.13	0.009	0.19	0.017	0.28	0.034	0.41	0.065	0.59		9.0
9.5		0.001	0.05	0.002	0.08	0.005	0.13	0.010	0.19	0.018	0.28	0.035	0.41	0.067	0.60		9.5
10.0		0.001	0.05	0.002	0.08	0.005	0.13	0.010	0.20	0.019	0.28	0.036	0.42	0.069	0.60		10.0
12.5		0.001	0.05	0.003	0.09	0.006	0.14	0.011	0.21	0.021	0.30	0.042	0.44	0.079	0.63		12.5
15.0		0.001	0.06	0.003	0.09	0.007	0.15	0.013	0.22	0.024	0.31	0.046	0.45	0.088	0.65		15.0
17.5	0.05	0.002	0.06	0.003	0.10	0.007	0.15	0.014	0.22	0.026	0.32	0.051	0.47	0.096	0.67	0.15	17.5
20.0		0.002	0.06	0.004	0.10	0.008	0.16	0.015	0.23	0.028	0.33	0.055	0.48	0.104	0.69		20.0
22.5		0.002	0.06	0.004	0.10	0.009	0.16	0.016	0.24	0.030	0.34	0.059	0.49	0.112	0.70		22.5
25.0		0.002	0.07	0.004	0.11	0.009	0.17	0.017	0.24	0.032	0.34	0.063	0.50	0.119	0.71		25.0
27.5		0.002	0.07	0.005	0.11	0.010	0.17	0.018	0.25	0.034	0.35	0.066	0.51	0.126	0.73		27.5
30.0		0.002	0.07	0.005	0.11	0.010	0.17	0.019	0.25	0.036	0.36	0.070	0.51	0.132	0.74		30.0
32.5		0.002	0.07	0.005	0.11	0.011	0.18	0.020	0.25	0.038	0.36	0.073	0.52	0.139	0.75		32.5
35.0		0.003	0.07	0.005	0.11	0.011	0.18	0.021	0.26	0.039	0.37	0.077	0.53	0.145	0.76		35.0
37.5		0.003	0.07	0.006	0.12	0.012	0.18	0.022	0.26	0.041	0.37	0.080	0.54	0.151	0.76		37.5
40.0		0.003	0.08	0.006	0.12	0.012	0.18	0.023	0.27	0.043	0.38	0.083	0.54	0.156	0.77		40.0
42.5		0.003	0.08	0.006	0.12	0.013	0.19	0.024	0.27	0.044	0.38	0.086	0.55	0.162	0.78		42.5
45.0		0.003	0.08	0.006	0.12	0.013	0.19	0.025	0.27	0.046	0.39	0.089	0.55	0.168	0.79		45.0
47.5		0.003	0.08	0.006	0.12	0.013	0.19	0.026	0.28	0.047	0.39	0.092	0.56	0.173	0.79		47.5
50.0		0.003	0.08	0.007	0.12	0.014	0.19	0.026	0.28	0.049	0.39	0.094	0.56	0.178	0.80		50.0
52.5		0.003	0.08	0.007	0.13	0.014	0.19	0.027	0.28	0.050	0.40	0.097	0.57	0.183	0.81		52.5
55.0		0.003	0.08	0.007	0.13	0.015	0.20	0.028	0.28	0.052	0.40	0.100	0.57	0.188	0.81		55.0
57.5		0.003	0.08	0.007	0.13	0.015	0.20	0.029	0.29	0.053	0.40	0.102	0.58	0.193	0.82	0.30	57.5
60.0		0.004	0.08	0.007	0.13	0.016	0.20	0.029	0.29	0.054	0.41	0.105	0.58	0.198	0.83		60.0
62.5		0.004	0.08	0.008	0.13	0.016	0.20	0.030	0.29	0.056	0.41	0.107	0.58	0.203	0.83		62.5
65.0		0.004	0.08	0.008	0.13	0.016	0.20	0.031	0.29	0.057	0.41	0.110	0.59	0.207	0.84		65.0
67.5		0.004	0.09	0.008	0.13	0.017	0.20	0.032	0.29	0.058	0.41	0.112	0.59	0.212	0.84		67.5
70.0		0.004	0.09	0.008	0.13	0.017	0.21	0.032	0.30	0.060	0.42	0.115	0.59	0.217	0.85		70.0
72.5		0.004	0.09	0.008	0.14	0.017	0.21	0.033	0.30	0.061	0.42	0.117	0.60	0.221	0.85		72.5
75.0		0.004	0.09	0.009	0.14	0.018	0.21	0.034	0.30	0.062	0.42	0.119	0.60	0.225	0.86		75.0
77.5		0.004	0.09	0.009	0.14	0.018	0.21	0.034	0.30	0.063	0.43	0.122	0.60	0.230	0.86		77.5
80.0		0.004	0.09	0.009	0.14	0.018	0.21	0.035	0.30	0.064	0.43	0.124	0.61	0.234	0.86		80.0
82.5		0.004	0.09	0.009	0.14	0.019	0.21	0.036	0.31	0.066	0.43	0.126	0.61	0.238	0.87		82.5
85.0		0.004	0.09	0.009	0.14	0.019	0.21	0.036	0.31	0.067	0.43	0.128	0.61	0.242	0.87		85.0
87.5		0.004	0.09	0.009	0.14	0.019	0.22	0.037	0.31	0.068	0.43	0.131	0.62	0.246	0.88		87.5
90.0		0.005	0.09	0.010	0.14	0.020	0.22	0.037	0.31	0.069	0.44	0.133	0.62	0.250	0.88		90.0

Table 4.24 Flow of water at 10°C in polybutylene (PB) pipes[21] — *continued*

q_m = mass flow rate — kg.s^{-1}
c = velocity — m.s^{-1}
$\Delta p/l$ = pressure drop per unit length — Pa.m^{-1}
l_e = equivalent length of a component for $\zeta = 1$ — m

POLYBUTYLENE (PB)
WATER AT 10°C
BS 2871

$\Delta p/l$	c	10 mm q_m	l_e	12 mm q_m	l_e	15 mm q_m	l_e	18 mm q_m	l_e	22 mm q_m	l_e	28 mm q_m	l_e	35 mm q_m	l_e	c	$\Delta p/l$
92.5		0.005	0.09	0.010	0.14	0.020	0.22	0.038	0.31	0.070	0.44	0.135	0.62	0.254	0.88		92.5
95.0		0.005	0.09	0.010	0.14	0.020	0.22	0.039	0.31	0.071	0.44	0.137	0.62	0.258	0.89		95.0
97.5		0.005	0.09	0.010	0.14	0.021	0.22	0.039	0.32	0.072	0.44	0.139	0.63	0.262	0.89		97.5
100.0		0.005	0.09	0.010	0.15	0.021	0.22	0.040	0.32	0.073	0.44	0.141	0.63	0.266	0.89		100.0
120.0	0.15	0.005	0.10	0.011	0.15	0.024	0.23	0.044	0.33	0.082	0.46	0.157	0.65	0.296	0.92		120.0
140.0		0.006	0.10	0.012	0.16	0.026	0.24	0.049	0.34	0.089	0.47	0.171	0.66	0.323	0.94	0.50	140.0
160.0		0.006	0.10	0.013	0.16	0.028	0.24	0.053	0.35	0.097	0.48	0.185	0.68	0.349	0.96		160.0
180.0		0.007	0.11	0.014	0.16	0.030	0.25	0.056	0.35	0.103	0.49	0.198	0.69	0.373	0.98		180.0
200.0		0.007	0.11	0.015	0.17	0.032	0.25	0.060	0.36	0.110	0.50	0.210	0.70	0.397	0.99		200.0
220.0		0.008	0.11	0.016	0.17	0.034	0.26	0.063	0.36	0.116	0.51	0.222	0.71	0.419	1.01		220.0
240.0		0.008	0.11	0.017	0.17	0.036	0.26	0.067	0.37	0.122	0.51	0.234	0.72	0.440	1.02		240.0
260.0		0.009	0.12	0.018	0.18	0.037	0.26	0.070	0.38	0.128	0.52	0.245	0.73	0.461	1.03		260.0
280.0		0.009	0.12	0.019	0.18	0.039	0.27	0.073	0.38	0.134	0.53	0.255	0.74	0.481	1.04		280.0
300.0		0.009	0.12	0.020	0.18	0.041	0.27	0.076	0.38	0.139	0.53	0.265	0.74	0.500	1.05		300.0
320.0		0.010	0.12	0.020	0.18	0.042	0.27	0.079	0.39	0.145	0.54	0.275	0.75	0.519	1.06		320.0
340.0		0.010	0.12	0.021	0.19	0.044	0.28	0.082	0.39	0.150	0.54	0.285	0.76	0.537	1.07		340.0
360.0	0.30	0.011	0.12	0.022	0.19	0.045	0.28	0.085	0.40	0.155	0.55	0.295	0.76	0.555	1.08		360.0
380.0		0.011	0.13	0.023	0.19	0.047	0.28	0.087	0.40	0.160	0.55	0.304	0.77	0.572	1.09		380.0
400.0		0.011	0.13	0.023	0.19	0.048	0.29	0.090	0.40	0.164	0.56	0.313	0.77	0.589	1.10		400.0
420.0		0.012	0.13	0.024	0.19	0.049	0.29	0.093	0.41	0.169	0.56	0.322	0.78	0.606	1.10		420.0
440.0		0.012	0.13	0.025	0.19	0.051	0.29	0.095	0.41	0.174	0.57	0.330	0.78	0.622	1.11		440.0
460.0		0.012	0.13	0.025	0.20	0.052	0.29	0.098	0.41	0.178	0.57	0.339	0.79	0.638	1.12	1.00	460.0
480.0		0.013	0.13	0.026	0.20	0.053	0.29	0.100	0.41	0.183	0.57	0.347	0.79	0.654	1.12		480.0
500.0		0.013	0.13	0.027	0.20	0.055	0.30	0.102	0.42	0.187	0.58	0.355	0.80	0.669	1.13		500.0
520.0		0.013	0.13	0.027	0.20	0.056	0.30	0.105	0.42	0.191	0.58	0.363	0.80	0.684	1.14		520.0
540.0		0.013	0.13	0.028	0.20	0.057	0.30	0.107	0.42	0.195	0.58	0.371	0.81	0.699	1.14		540.0
560.0		0.014	0.14	0.028	0.20	0.058	0.30	0.109	0.43	0.200	0.59	0.379	0.81	0.714	1.15		560.0
580.0		0.014	0.14	0.029	0.20	0.060	0.30	0.112	0.43	0.204	0.59	0.386	0.81	0.728	1.15		580.0
600.0		0.014	0.14	0.030	0.21	0.061	0.31	0.114	0.43	0.208	0.59	0.394	0.82	0.742	1.16		600.0
620.0		0.015	0.14	0.030	0.21	0.062	0.31	0.116	0.43	0.212	0.60	0.401	0.82	0.756	1.16		620.0
640.0		0.015	0.14	0.031	0.21	0.063	0.31	0.118	0.43	0.216	0.60	0.408	0.82	0.770	1.17		640.0
660.0		0.015	0.14	0.031	0.21	0.064	0.31	0.120	0.44	0.219	0.60	0.415	0.83	0.783	1.17		660.0
680.0		0.015	0.14	0.032	0.21	0.065	0.31	0.122	0.44	0.223	0.60	0.423	0.83	0.797	1.18		680.0
700.0		0.016	0.14	0.032	0.21	0.067	0.31	0.124	0.44	0.227	0.61	0.429	0.83	0.810	1.18		700.0
720.0		0.016	0.14	0.033	0.21	0.068	0.32	0.126	0.44	0.231	0.61	0.436	0.84	0.823	1.19		720.0
740.0		0.016	0.14	0.033	0.21	0.069	0.32	0.128	0.44	0.234	0.61	0.443	0.84	0.836	1.19		740.0
760.0		0.016	0.14	0.034	0.21	0.070	0.32	0.130	0.45	0.238	0.61	0.450	0.84	0.849	1.20		760.0
780.0		0.017	0.14	0.034	0.22	0.071	0.32	0.132	0.45	0.241	0.62	0.457	0.84	0.861	1.20		780.0
800.0		0.017	0.14	0.035	0.22	0.072	0.32	0.134	0.45	0.245	0.62	0.463	0.85	0.874	1.21		800.0
820.0		0.017	0.15	0.036	0.22	0.073	0.32	0.136	0.45	0.248	0.62	0.470	0.85	0.886	1.21		820.0
840.0	0.50	0.017	0.15	0.036	0.22	0.074	0.32	0.138	0.45	0.252	0.62	0.476	0.85	0.898	1.21		840.0
860.0		0.018	0.15	0.037	0.22	0.075	0.32	0.140	0.45	0.255	0.63	0.482	0.86	0.910	1.22		860.0
880.0		0.018	0.15	0.037	0.22	0.076	0.33	0.142	0.46	0.259	0.63	0.489	0.86	0.922	1.22		880.0
900.0		0.018	0.15	0.038	0.22	0.077	0.33	0.144	0.46	0.262	0.63	0.495	0.86	0.934	1.22		900.0
920.0		0.018	0.15	0.038	0.22	0.078	0.33	0.146	0.46	0.265	0.63	0.501	0.86	0.946	1.23	1.50	920.0
940.0		0.019	0.15	0.038	0.22	0.079	0.33	0.147	0.46	0.269	0.63	0.507	0.86	0.957	1.23		940.0
960.0		0.019	0.15	0.039	0.22	0.080	0.33	0.149	0.46	0.272	0.64	0.513	0.87	0.969	1.23		960.0
980.0		0.019	0.15	0.039	0.22	0.081	0.33	0.151	0.46	0.275	0.64	0.519	0.87	0.980	1.24		980.0
1000.0		0.019	0.15	0.040	0.23	0.082	0.33	0.153	0.47	0.278	0.64	0.525	0.87	0.991	1.24		1000.0
1100.0		0.020	0.15	0.042	0.23	0.086	0.34	0.161	0.47	0.294	0.65	0.554	0.88	1.046	1.26		1100.0
1200.0		0.022	0.16	0.044	0.23	0.091	0.34	0.170	0.48	0.309	0.66	0.582	0.89	1.099	1.27		1200.0
1300.0		0.023	0.16	0.047	0.24	0.095	0.35	0.178	0.48	0.323	0.66	0.608	0.90	1.150	1.28		1300.0
1400.0		0.024	0.16	0.049	0.24	0.099	0.35	0.185	0.49	0.337	0.67	0.634	0.91	1.199	1.30		1400.0
1500.0		0.025	0.16	0.051	0.24	0.103	0.35	0.193	0.49	0.351	0.68	0.659	0.92	1.246	1.31		1500.0
1600.0		0.026	0.16	0.052	0.24	0.107	0.36	0.200	0.50	0.364	0.68	0.683	0.92	1.292	1.32	2.00	1600.0
1700.0		0.026	0.17	0.054	0.25	0.111	0.36	0.207	0.50	0.377	0.69	0.707	0.93	1.337	1.33		1700.0
1800.0		0.027	0.17	0.056	0.25	0.115	0.36	0.214	0.51	0.389	0.70	0.730	0.94	1.381	1.34		1800.0
1900.0		0.028	0.17	0.058	0.25	0.119	0.37	0.221	0.51	0.401	0.70	0.752	0.94	1.424	1.35		1900.0
2000.0		0.029	0.17	0.060	0.25	0.122	0.37	0.227	0.52	0.413	0.71	0.774	0.95	1.466	1.36		2000.0

Table 4.25 Flow of water at 75°C in polybutylene (PB) pipes[21]

POLYBUTYLENE (PB)		
WATER AT 75°C		
BS 2871		

q_m = mass flow rate kg.s^{-1}
c = velocity m.s^{-1}
$\Delta p/l$ = pressure drop per unit length Pa.m^{-1}
l_e = equivalent length of a component
for $\zeta = 1$ m

$\Delta p/l$	c	10 mm		12 mm		15 mm		18 mm		22 mm		28 mm		35 mm		c	$\Delta p/l$
		q_m	l_e	q_m	l_e	q_m	l_e	q_m	l_e	q_m	l_e	q_m	l_e	q_m	l_e		
0.1								0.001	0.11	0.001	0.17	0.003	0.26	0.005	0.39		0.1
0.2						0.001	0.09	0.001	0.14	0.002	0.20	0.004	0.31	0.008	0.45		0.2
0.3						0.001	0.10	0.001	0.15	0.003	0.22	0.006	0.34	0.011	0.49		0.3
0.4						0.001	0.11	0.002	0.16	0.003	0.24	0.007	0.36	0.013	0.52		0.4
0.5						0.001	0.12	0.002	0.17	0.004	0.25	0.008	0.37	0.015	0.54		0.5
0.6				0.001	0.07	0.001	0.12	0.002	0.18	0.004	0.26	0.008	0.39	0.016	0.56		0.6
0.7				0.001	0.08	0.001	0.13	0.003	0.19	0.005	0.27	0.009	0.40	0.018	0.58		0.7
0.8				0.001	0.08	0.001	0.13	0.003	0.19	0.005	0.28	0.010	0.41	0.019	0.60		0.8
0.9				0.001	0.08	0.002	0.13	0.003	0.20	0.006	0.29	0.011	0.42	0.021	0.61		0.9
1.0				0.001	0.09	0.002	0.14	0.003	0.20	0.006	0.29	0.012	0.43	0.022	0.62		1.0
1.5				0.001	0.10	0.002	0.15	0.004	0.22	0.008	0.32	0.015	0.47	0.028	0.67		1.5
2.0		0.001	0.06	0.001	0.10	0.003	0.16	0.005	0.24	0.009	0.34	0.017	0.49	0.033	0.70	0.05	2.0
2.5		0.001	0.07	0.001	0.11	0.003	0.17	0.006	0.25	0.010	0.35	0.020	0.51	0.038	0.73		2.5
3.0		0.001	0.07	0.002	0.11	0.003	0.18	0.006	0.26	0.011	0.37	0.022	0.53	0.042	0.76		3.0
3.5		0.001	0.08	0.002	0.12	0.004	0.18	0.007	0.27	0.013	0.38	0.024	0.54	0.046	0.77		3.5
4.0		0.001	0.08	0.002	0.12	0.004	0.19	0.007	0.27	0.014	0.39	0.026	0.56	0.049	0.79		4.0
4.5		0.001	0.08	0.002	0.13	0.004	0.19	0.008	0.28	0.015	0.40	0.028	0.57	0.053	0.81		4.5
5.0		0.001	0.08	0.002	0.13	0.004	0.20	0.008	0.29	0.015	0.40	0.030	0.58	0.056	0.82		5.0
5.5		0.001	0.08	0.002	0.13	0.005	0.20	0.009	0.29	0.016	0.41	0.032	0.59	0.060	0.83		5.5
6.0		0.001	0.09	0.002	0.13	0.005	0.21	0.009	0.30	0.017	0.42	0.033	0.59	0.063	0.85		6.0
6.5		0.001	0.09	0.002	0.14	0.005	0.21	0.010	0.30	0.018	0.42	0.035	0.60	0.066	0.86		6.5
7.0		0.001	0.09	0.003	0.14	0.005	0.21	0.010	0.31	0.019	0.43	0.036	0.61	0.068	0.87		7.0
7.5		0.001	0.09	0.003	0.14	0.006	0.22	0.011	0.31	0.020	0.43	0.038	0.62	0.071	0.88		7.5
8.0		0.001	0.09	0.003	0.14	0.006	0.22	0.011	0.31	0.020	0.44	0.039	0.62	0.074	0.88		8.0
8.5		0.001	0.09	0.003	0.14	0.006	0.22	0.011	0.32	0.021	0.44	0.041	0.63	0.077	0.89		8.5
9.0		0.001	0.10	0.003	0.15	0.006	0.22	0.012	0.32	0.022	0.45	0.042	0.63	0.079	0.90		9.0
9.5		0.001	0.10	0.003	0.15	0.006	0.23	0.012	0.32	0.023	0.45	0.043	0.64	0.082	0.91		9.5
10.0		0.002	0.10	0.003	0.15	0.007	0.23	0.013	0.33	0.023	0.46	0.045	0.64	0.084	0.92		10.0
12.5	0.05	0.002	0.10	0.004	0.16	0.008	0.24	0.014	0.34	0.026	0.47	0.051	0.67	0.096	0.95	0.15	12.5
15.0		0.002	0.11	0.004	0.16	0.009	0.25	0.016	0.35	0.029	0.49	0.056	0.69	0.106	0.97		15.0
17.5		0.002	0.11	0.005	0.17	0.009	0.25	0.018	0.36	0.032	0.50	0.062	0.70	0.116	1.00		17.5
20.0		0.002	0.11	0.005	0.17	0.010	0.26	0.019	0.37	0.035	0.51	0.066	0.72	0.125	1.02		20.0
22.5		0.003	0.12	0.005	0.18	0.011	0.27	0.020	0.38	0.037	0.52	0.071	0.73	0.134	1.03		22.5
25.0		0.003	0.12	0.006	0.18	0.012	0.27	0.022	0.38	0.040	0.53	0.075	0.74	0.142	1.05		25.0
27.5		0.003	0.12	0.006	0.18	0.012	0.27	0.023	0.39	0.042	0.54	0.080	0.75	0.150	1.06		27.5
30.0		0.003	0.12	0.006	0.19	0.013	0.28	0.024	0.39	0.044	0.55	0.084	0.76	0.158	1.08		30.0
32.5		0.003	0.13	0.007	0.19	0.013	0.28	0.025	0.40	0.046	0.55	0.088	0.77	0.165	1.09		32.5
35.0		0.003	0.13	0.007	0.19	0.014	0.29	0.026	0.40	0.048	0.56	0.091	0.77	0.172	1.10		35.0
37.5		0.003	0.13	0.007	0.19	0.015	0.29	0.027	0.41	0.050	0.57	0.095	0.78	0.179	1.11		37.5
40.0		0.004	0.13	0.007	0.20	0.015	0.29	0.028	0.41	0.052	0.57	0.099	0.79	0.186	1.12	0.30	40.0
42.5		0.004	0.13	0.008	0.20	0.016	0.30	0.029	0.42	0.054	0.58	0.102	0.80	0.192	1.13		42.5
45.0		0.004	0.13	0.008	0.20	0.016	0.30	0.030	0.42	0.056	0.58	0.105	0.80	0.199	1.14		45.0
47.5		0.004	0.13	0.008	0.20	0.017	0.30	0.031	0.42	0.057	0.59	0.109	0.81	0.205	1.15		47.5
50.0		0.004	0.14	0.008	0.20	0.017	0.30	0.032	0.43	0.059	0.59	0.112	0.81	0.211	1.15		50.0
52.5		0.004	0.14	0.009	0.21	0.018	0.31	0.033	0.43	0.061	0.59	0.115	0.82	0.217	1.16		52.5
55.0		0.004	0.14	0.009	0.21	0.018	0.31	0.034	0.43	0.062	0.60	0.118	0.82	0.223	1.17		55.0
57.5		0.004	0.14	0.009	0.21	0.019	0.31	0.035	0.44	0.064	0.60	0.121	0.83	0.229	1.18		57.5
60.0		0.005	0.14	0.009	0.21	0.019	0.31	0.036	0.44	0.066	0.61	0.124	0.83	0.234	1.18		60.0
62.5		0.005	0.14	0.010	0.21	0.020	0.32	0.037	0.44	0.067	0.61	0.127	0.84	0.240	1.19		62.5
65.0		0.005	0.14	0.010	0.21	0.020	0.32	0.038	0.45	0.069	0.61	0.130	0.84	0.245	1.20		65.0
67.5		0.005	0.14	0.010	0.22	0.021	0.32	0.038	0.45	0.070	0.62	0.133	0.85	0.250	1.20		67.5
70.0		0.005	0.15	0.010	0.22	0.021	0.32	0.039	0.45	0.072	0.62	0.135	0.85	0.255	1.21		70.0
72.5		0.005	0.15	0.010	0.22	0.021	0.32	0.040	0.45	0.073	0.62	0.138	0.85	0.261	1.21		72.5
75.0	0.15	0.005	0.15	0.011	0.22	0.022	0.33	0.041	0.46	0.075	0.63	0.141	0.86	0.266	1.22		75.0
77.5		0.005	0.15	0.011	0.22	0.022	0.33	0.042	0.46	0.076	0.63	0.143	0.86	0.271	1.22		77.5
80.0		0.005	0.15	0.011	0.22	0.023	0.33	0.042	0.46	0.077	0.63	0.146	0.86	0.276	1.23		80.0
82.5		0.005	0.15	0.011	0.22	0.023	0.33	0.043	0.46	0.079	0.64	0.149	0.87	0.280	1.23		82.5
85.0		0.006	0.15	0.011	0.22	0.024	0.33	0.044	0.46	0.080	0.64	0.151	0.87	0.285	1.24		85.0
87.5		0.006	0.15	0.012	0.23	0.024	0.33	0.045	0.47	0.081	0.64	0.154	0.87	0.290	1.24		87.5
90.0		0.006	0.15	0.012	0.23	0.024	0.34	0.045	0.47	0.083	0.64	0.156	0.88	0.295	1.25		90.0

Table 4.25 Flow of water at 75°C in polybutylene (PB) pipes[21] — *continued*

q_m	= mass flow rate	kg.s^{-1}
c	= velocity	m.s^{-1}
$\Delta p/l$	= pressure drop per unit length	Pa.m^{-1}
l_e	= equivalent length of a component for $\zeta = 1$	m

POLYBUTYLENE (PB)
WATER AT 75°C
BS 2871

$\Delta p/l$	c	10 mm		12 mm		15 mm		18 mm		22 mm		28 mm		35 mm		c	$\Delta p/l$
		q_m	l_e	q_m	l_e	q_m	l_e	q_m	l_e	q_m	l_e	q_m	l_e	q_m	l_e		
92.5		0.006	0.15	0.012	0.23	0.025	0.34	0.046	0.47	0.084	0.65	0.158	0.88	0.299	1.25		92.5
95.0		0.006	0.15	0.012	0.23	0.025	0.34	0.047	0.47	0.085	0.65	0.161	0.88	0.304	1.26		95.0
97.5		0.006	0.15	0.012	0.23	0.025	0.34	0.048	0.47	0.087	0.65	0.163	0.88	0.308	1.26	0.50	97.5
100.0		0.006	0.16	0.013	0.23	0.026	0.34	0.048	0.48	0.088	0.65	0.165	0.89	0.313	1.27		100.0
120.0		0.007	0.16	0.014	0.24	0.029	0.35	0.054	0.49	0.098	0.67	0.183	0.91	0.346	1.30		120.0
140.0		0.007	0.16	0.015	0.24	0.031	0.36	0.059	0.50	0.106	0.69	0.200	0.92	0.378	1.32		140.0
160.0		0.008	0.17	0.017	0.25	0.034	0.37	0.063	0.51	0.115	0.70	0.215	0.94	0.407	1.34		160.0
180.0		0.009	0.17	0.018	0.25	0.036	0.37	0.068	0.52	0.123	0.71	0.230	0.95	0.435	1.36		180.0
200.0		0.009	0.18	0.019	0.26	0.039	0.38	0.072	0.53	0.130	0.72	0.244	0.96	0.462	1.38		200.0
220.0		0.010	0.18	0.020	0.26	0.041	0.38	0.076	0.53	0.138	0.73	0.257	0.97	0.487	1.40		220.0
240.0		0.010	0.18	0.021	0.27	0.043	0.39	0.080	0.54	0.145	0.74	0.270	0.98	0.512	1.41		240.0
260.0	0.30	0.011	0.18	0.022	0.27	0.045	0.39	0.083	0.55	0.151	0.75	0.282	0.99	0.535	1.43		260.0
280.0		0.011	0.19	0.023	0.27	0.047	0.40	0.087	0.55	0.158	0.75	0.294	1.00	0.558	1.44		280.0
300.0		0.012	0.19	0.024	0.28	0.049	0.40	0.090	0.56	0.164	0.76	0.305	1.01	0.580	1.45		300.0
320.0		0.012	0.19	0.025	0.28	0.051	0.41	0.094	0.56	0.170	0.77	0.316	1.01	0.601	1.46		320.0
340.0		0.013	0.19	0.026	0.28	0.052	0.41	0.097	0.57	0.176	0.77	0.327	1.02	0.622	1.47	1.00	340.0
360.0		0.013	0.19	0.027	0.28	0.054	0.41	0.100	0.57	0.182	0.78	0.337	1.03	0.642	1.48		360.0
380.0		0.013	0.19	0.027	0.29	0.056	0.42	0.103	0.58	0.188	0.78	0.348	1.03	0.662	1.49		380.0
400.0		0.014	0.20	0.028	0.29	0.057	0.42	0.107	0.58	0.193	0.79	0.358	1.04	0.681	1.50		400.0
420.0		0.014	0.20	0.029	0.29	0.059	0.42	0.110	0.58	0.199	0.79	0.367	1.04	0.700	1.51		420.0
440.0		0.015	0.20	0.030	0.29	0.061	0.43	0.112	0.59	0.204	0.80	0.377	1.05	0.719	1.52		440.0
460.0		0.015	0.20	0.031	0.29	0.062	0.43	0.115	0.59	0.209	0.80	0.386	1.05	0.737	1.53		460.0
480.0		0.015	0.20	0.031	0.30	0.064	0.43	0.118	0.59	0.214	0.81	0.395	1.06	0.755	1.54		480.0
500.0		0.016	0.20	0.032	0.30	0.065	0.43	0.121	0.60	0.219	0.81	0.404	1.06	0.772	1.54		500.0
520.0		0.016	0.20	0.033	0.30	0.067	0.44	0.124	0.60	0.224	0.82	0.413	1.06	0.789	1.55		520.0
540.0		0.016	0.21	0.033	0.30	0.068	0.44	0.126	0.60	0.229	0.82	0.422	1.07	0.806	1.56		540.0
560.0		0.017	0.21	0.034	0.30	0.070	0.44	0.129	0.61	0.234	0.82	0.430	1.07	0.822	1.56		560.0
580.0		0.017	0.21	0.035	0.30	0.071	0.44	0.132	0.61	0.238	0.83	0.439	1.08	0.839	1.57		580.0
600.0	0.50	0.017	0.21	0.036	0.31	0.072	0.44	0.134	0.61	0.243	0.83	0.447	1.08	0.855	1.58		600.0
620.0		0.018	0.21	0.036	0.31	0.074	0.45	0.137	0.62	0.247	0.83	0.455	1.08	0.870	1.58		620.0
640.0		0.018	0.21	0.037	0.31	0.075	0.45	0.139	0.62	0.252	0.84	0.463	1.09	0.886	1.59		640.0
660.0		0.018	0.21	0.038	0.31	0.076	0.45	0.142	0.62	0.256	0.84	0.471	1.09	0.901	1.59		660.0
680.0		0.019	0.21	0.038	0.31	0.078	0.45	0.144	0.62	0.260	0.84	0.478	1.09	0.917	1.60		680.0
700.0		0.019	0.21	0.039	0.31	0.079	0.45	0.146	0.62	0.265	0.85	0.486	1.09	0.931	1.61	1.50	700.0
720.0		0.019	0.22	0.039	0.31	0.080	0.46	0.149	0.63	0.269	0.85	0.494	1.10	0.946	1.61		720.0
740.0		0.020	0.22	0.040	0.32	0.081	0.46	0.151	0.63	0.273	0.85	0.501	1.10	0.961	1.62		740.0
760.0		0.020	0.22	0.041	0.32	0.083	0.46	0.153	0.63	0.277	0.86	0.508	1.10	0.975	1.62		760.0
780.0		0.020	0.22	0.041	0.32	0.084	0.46	0.156	0.63	0.281	0.86	0.516	1.11	0.989	1.63		780.0
800.0		0.021	0.22	0.042	0.32	0.085	0.46	0.158	0.64	0.285	0.86	0.523	1.11	1.003	1.63		800.0
820.0		0.021	0.22	0.043	0.32	0.086	0.46	0.160	0.64	0.289	0.86	0.530	1.11	1.017	1.63		820.0
840.0		0.021	0.22	0.043	0.32	0.088	0.47	0.162	0.64	0.293	0.87	0.537	1.11	1.031	1.64		840.0
860.0		0.021	0.22	0.044	0.32	0.089	0.47	0.164	0.64	0.297	0.87	0.544	1.11	1.045	1.64		860.0
880.0		0.022	0.22	0.044	0.32	0.090	0.47	0.167	0.64	0.301	0.87	0.551	1.12	1.058	1.65		880.0
900.0		0.022	0.22	0.045	0.32	0.091	0.47	0.169	0.65	0.305	0.87	0.557	1.12	1.071	1.65		900.0
920.0		0.022	0.22	0.045	0.33	0.092	0.47	0.171	0.65	0.309	0.88	0.564	1.12	1.085	1.66		920.0
940.0		0.023	0.22	0.046	0.33	0.093	0.47	0.173	0.65	0.312	0.88	0.571	1.12	1.098	1.66		940.0
960.0		0.023	0.22	0.047	0.33	0.094	0.47	0.175	0.65	0.316	0.88	0.577	1.13	1.111	1.66		960.0
980.0		0.023	0.23	0.047	0.33	0.096	0.47	0.177	0.65	0.320	0.88	0.584	1.13	1.123	1.67		980.0
1000.0		0.023	0.23	0.048	0.33	0.097	0.48	0.179	0.65	0.324	0.89	0.590	1.13	1.136	1.67		1000.0
1100.0		0.025	0.23	0.050	0.33	0.102	0.48	0.189	0.66	0.341	0.90	0.622	1.14	1.198	1.69		1100.0
1200.0		0.026	0.23	0.053	0.34	0.107	0.49	0.198	0.67	0.358	0.91	0.652	1.15	1.257	1.71	2.00	1200.0
1300.0		0.027	0.24	0.055	0.34	0.112	0.49	0.207	0.68	0.375	0.91	0.681	1.15	1.315	1.72		1300.0
1400.0		0.028	0.24	0.058	0.34	0.117	0.50	0.216	0.68	0.391	0.92	0.708	1.16	1.370	1.74		1400.0
1500.0		0.030	0.24	0.060	0.35	0.122	0.50	0.225	0.69	0.406	0.93	0.735	1.17	1.423	1.75		1500.0
1600.0		0.031	0.24	0.062	0.35	0.126	0.51	0.233	0.69	0.421	0.94	0.761	1.17	1.475	1.76		1600.0
1700.0		0.032	0.24	0.064	0.35	0.131	0.51	0.241	0.70	0.436	0.94	0.787	1.18	1.526	1.77		1700.0
1800.0		0.033	0.25	0.067	0.36	0.135	0.51	0.249	0.70	0.450	0.95	0.811	1.19	1.575	1.79		1800.0
1900.0		0.034	0.25	0.069	0.36	0.139	0.52	0.257	0.71	0.464	0.96	0.835	1.19	1.623	1.80		1900.0
2000.0		0.035	0.25	0.071	0.36	0.143	0.52	0.264	0.71	0.477	0.96	0.858	1.19	1.670	1.81		2000.0

Table 4.26 Flow of water at 10°C in polyethylene (PE-X) pipes[21]

q_m = mass flow rate   kg.s^{-1}
c = velocity   m.s^{-1}
$\Delta p/l$ = pressure drop per unit length   Pa.m^{-1}
l_e = equivalent length of a component
     for $\zeta = 1$ m

POLYETHYLENE (PE-X)
WATER AT 10°C
BS 2871

$\Delta p/l$	c	10 mm		12 mm		15 mm		18 mm		22 mm		28 mm		35 mm		c	$\Delta p/l$
		q_m	l_e	q_m	l_e	q_m	l_e	q_m	l_e	q_m	l_e	q_m	l_e	q_m	l_e		
0.1										0.001	0.07	0.002	0.12	0.004	0.19		0.1
0.2								0.001	0.06	0.002	0.09	0.003	0.16	0.006	0.24		0.2
0.3						0.001	0.04	0.001	0.07	0.002	0.11	0.004	0.18	0.008	0.27		0.3
0.4						0.001	0.05	0.001	0.08	0.002	0.12	0.005	0.19	0.010	0.29		0.4
0.5						0.001	0.06	0.001	0.08	0.003	0.13	0.006	0.21	0.011	0.31		0.5
0.6						0.001	0.06	0.002	0.09	0.003	0.14	0.006	0.22	0.012	0.33		0.6
0.7						0.001	0.06	0.002	0.10	0.004	0.14	0.007	0.23	0.014	0.34		0.7
0.8						0.001	0.07	0.002	0.10	0.004	0.15	0.008	0.23	0.015	0.35		0.8
0.9						0.001	0.07	0.002	0.10	0.004	0.16	0.008	0.24	0.016	0.36		0.9
1.0				0.001	0.04	0.001	0.07	0.002	0.11	0.004	0.16	0.009	0.25	0.017	0.37		1.0
1.5				0.001	0.04	0.002	0.08	0.003	0.12	0.006	0.18	0.011	0.28	0.022	0.41		1.5
2.0				0.001	0.05	0.002	0.09	0.004	0.13	0.007	0.19	0.014	0.30	0.026	0.44		2.0
2.5				0.001	0.05	0.002	0.10	0.004	0.14	0.008	0.21	0.016	0.31	0.030	0.46		2.5
3.0				0.001	0.06	0.003	0.10	0.005	0.15	0.009	0.22	0.018	0.32	0.034	0.48	0.05	3.0
3.5		0.001	0.04	0.001	0.06	0.003	0.11	0.005	0.15	0.010	0.22	0.019	0.34	0.037	0.49		3.5
4.0		0.001	0.04	0.001	0.06	0.003	0.11	0.006	0.16	0.011	0.23	0.021	0.35	0.040	0.51		4.0
4.5		0.001	0.04	0.001	0.07	0.003	0.11	0.006	0.16	0.011	0.24	0.022	0.35	0.043	0.52		4.5
5.0		0.001	0.04	0.002	0.07	0.004	0.12	0.007	0.17	0.012	0.24	0.024	0.36	0.046	0.53		5.0
5.5		0.001	0.04	0.002	0.07	0.004	0.12	0.007	0.17	0.013	0.25	0.025	0.37	0.048	0.54		5.5
6.0		0.001	0.04	0.002	0.07	0.004	0.12	0.007	0.18	0.014	0.25	0.027	0.38	0.051	0.55		6.0
6.5		0.001	0.04	0.002	0.07	0.004	0.13	0.008	0.18	0.014	0.26	0.028	0.38	0.054	0.56		6.5
7.0		0.001	0.05	0.002	0.07	0.005	0.13	0.008	0.18	0.015	0.26	0.029	0.39	0.056	0.57		7.0
7.5		0.001	0.05	0.002	0.08	0.005	0.13	0.008	0.19	0.016	0.27	0.031	0.40	0.058	0.57		7.5
8.0		0.001	0.05	0.002	0.08	0.005	0.13	0.009	0.19	0.016	0.27	0.032	0.40	0.061	0.58		8.0
8.5		0.001	0.05	0.002	0.08	0.005	0.14	0.009	0.19	0.017	0.28	0.033	0.41	0.063	0.59		8.5
9.0		0.001	0.05	0.002	0.08	0.005	0.14	0.010	0.19	0.017	0.28	0.034	0.41	0.065	0.59		9.0
9.5		0.001	0.05	0.002	0.08	0.006	0.14	0.010	0.20	0.018	0.28	0.035	0.41	0.067	0.60		9.5
10.0		0.001	0.05	0.002	0.08	0.006	0.14	0.010	0.20	0.019	0.28	0.036	0.42	0.069	0.60		10.0
12.5		0.001	0.05	0.003	0.09	0.007	0.15	0.012	0.21	0.021	0.30	0.042	0.44	0.079	0.63		12.5
15.0		0.001	0.06	0.003	0.09	0.007	0.16	0.013	0.22	0.024	0.31	0.046	0.45	0.088	0.65		15.0
17.5	0.05	0.002	0.06	0.003	0.10	0.008	0.16	0.014	0.23	0.026	0.32	0.051	0.47	0.096	0.67	0.15	17.5
20.0		0.002	0.06	0.004	0.10	0.009	0.17	0.016	0.23	0.028	0.33	0.055	0.48	0.104	0.69		20.0
22.5		0.002	0.06	0.004	0.10	0.009	0.17	0.017	0.24	0.030	0.34	0.059	0.49	0.112	0.70		22.5
25.0		0.002	0.07	0.004	0.11	0.010	0.18	0.018	0.24	0.032	0.34	0.063	0.50	0.119	0.71		25.0
27.5		0.002	0.07	0.005	0.11	0.011	0.18	0.019	0.25	0.034	0.35	0.066	0.51	0.126	0.73		27.5
30.0		0.002	0.07	0.005	0.11	0.011	0.18	0.020	0.25	0.036	0.36	0.070	0.51	0.132	0.74		30.0
32.5		0.002	0.07	0.005	0.11	0.012	0.19	0.021	0.26	0.038	0.36	0.073	0.52	0.139	0.75		32.5
35.0		0.003	0.07	0.005	0.11	0.012	0.19	0.022	0.26	0.039	0.37	0.077	0.53	0.145	0.76		35.0
37.5		0.003	0.07	0.006	0.12	0.013	0.19	0.023	0.27	0.041	0.37	0.080	0.54	0.151	0.76		37.5
40.0		0.003	0.08	0.006	0.12	0.013	0.19	0.024	0.27	0.043	0.38	0.083	0.54	0.156	0.77		40.0
42.5		0.003	0.08	0.006	0.12	0.014	0.20	0.024	0.27	0.044	0.38	0.086	0.55	0.162	0.78		42.5
45.0		0.003	0.08	0.006	0.12	0.014	0.20	0.025	0.28	0.046	0.39	0.089	0.55	0.168	0.79		45.0
47.5		0.003	0.08	0.006	0.12	0.015	0.20	0.026	0.28	0.047	0.39	0.092	0.56	0.173	0.79		47.5
50.0		0.003	0.08	0.007	0.12	0.015	0.20	0.027	0.28	0.049	0.39	0.094	0.56	0.178	0.80		50.0
52.5		0.003	0.08	0.007	0.13	0.016	0.21	0.028	0.28	0.050	0.40	0.097	0.57	0.183	0.81		52.5
55.0		0.003	0.08	0.007	0.13	0.016	0.21	0.029	0.29	0.052	0.40	0.100	0.57	0.188	0.81		55.0
57.5		0.003	0.08	0.007	0.13	0.017	0.21	0.029	0.29	0.053	0.40	0.102	0.58	0.193	0.82	0.30	57.5
60.0		0.004	0.08	0.007	0.13	0.017	0.21	0.030	0.29	0.054	0.41	0.105	0.58	0.198	0.83		60.0
62.5		0.004	0.08	0.008	0.13	0.018	0.21	0.031	0.29	0.056	0.41	0.107	0.58	0.203	0.83		62.5
65.0		0.004	0.08	0.008	0.13	0.018	0.21	0.031	0.30	0.057	0.41	0.110	0.59	0.207	0.84		65.0
67.5		0.004	0.09	0.008	0.13	0.018	0.22	0.032	0.30	0.058	0.41	0.112	0.59	0.212	0.84		67.5
70.0		0.004	0.09	0.008	0.13	0.019	0.22	0.033	0.30	0.060	0.42	0.115	0.59	0.217	0.85		70.0
72.5		0.004	0.09	0.008	0.14	0.019	0.22	0.034	0.30	0.061	0.42	0.117	0.60	0.221	0.85		72.5
75.0		0.004	0.09	0.009	0.14	0.020	0.22	0.034	0.30	0.062	0.42	0.119	0.60	0.225	0.86		75.0
77.5		0.004	0.09	0.009	0.14	0.020	0.22	0.035	0.31	0.063	0.43	0.122	0.60	0.230	0.86		77.5
80.0		0.004	0.09	0.009	0.14	0.020	0.22	0.036	0.31	0.064	0.43	0.124	0.61	0.234	0.86		80.0
82.5		0.004	0.09	0.009	0.14	0.021	0.23	0.036	0.31	0.066	0.43	0.126	0.61	0.238	0.87		82.5
85.0		0.004	0.09	0.009	0.14	0.021	0.23	0.037	0.31	0.067	0.43	0.128	0.61	0.242	0.87		85.0
87.5		0.004	0.09	0.009	0.14	0.021	0.23	0.038	0.31	0.068	0.43	0.131	0.62	0.246	0.88		87.5
90.0		0.005	0.09	0.010	0.14	0.022	0.23	0.038	0.31	0.069	0.44	0.133	0.62	0.250	0.88		90.0

Table 4.26 Flow of water at 10°C in Polyethylene (PE-X) pipes[21] — *continued*

q_m = mass flow rate   kg.s^{-1}
c = velocity   m.s^{-1}
$\Delta p/l$ = pressure drop per unit length   Pa.m^{-1}
l_e = equivalent length of a component
     for $\zeta = 1$ m

POLYETHYLENE (PE-X)

WATER AT 10°C

BS 2871

$\Delta p/l$	c	10 mm		12 mm		15 mm		18 mm		22 mm		28 mm		35 mm		c	$\Delta p/l$
		q_m	l_e	q_m	l_e	q_m	l_e	q_m	l_e	q_m	l_e	q_m	l_e	q_m	l_e		
92.5		0.005	0.09	0.010	0.14	0.022	0.23	0.039	0.32	0.070	0.44	0.135	0.62	0.254	0.88		92.5
95.0		0.005	0.09	0.010	0.14	0.023	0.23	0.039	0.32	0.071	0.44	0.137	0.62	0.258	0.89		95.0
97.5		0.005	0.09	0.010	0.14	0.023	0.23	0.040	0.32	0.072	0.44	0.139	0.63	0.262	0.89		97.5
100.0		0.005	0.09	0.010	0.15	0.023	0.23	0.041	0.32	0.073	0.44	0.141	0.63	0.266	0.89		100.0
120.0	0.15	0.005	0.10	0.011	0.15	0.026	0.24	0.045	0.33	0.082	0.46	0.157	0.65	0.296	0.92		120.0
140.0		0.006	0.10	0.012	0.16	0.028	0.25	0.050	0.34	0.089	0.47	0.171	0.66	0.323	0.94	0.50	140.0
160.0		0.006	0.10	0.013	0.16	0.031	0.26	0.054	0.35	0.097	0.48	0.185	0.68	0.349	0.96		160.0
180.0		0.007	0.11	0.014	0.16	0.033	0.26	0.058	0.36	0.103	0.49	0.198	0.69	0.373	0.98		180.0
200.0		0.007	0.11	0.015	0.17	0.035	0.27	0.061	0.36	0.110	0.50	0.210	0.70	0.397	0.99		200.0
220.0		0.008	0.11	0.016	0.17	0.037	0.27	0.065	0.37	0.116	0.51	0.222	0.71	0.419	1.01		220.0
240.0		0.008	0.11	0.017	0.17	0.039	0.28	0.068	0.37	0.122	0.51	0.234	0.72	0.440	1.02		240.0
260.0		0.009	0.12	0.018	0.18	0.041	0.28	0.071	0.38	0.128	0.52	0.245	0.73	0.461	1.03		260.0
280.0		0.009	0.12	0.019	0.18	0.043	0.28	0.074	0.38	0.134	0.53	0.255	0.74	0.481	1.04		280.0
300.0		0.009	0.12	0.020	0.18	0.045	0.29	0.078	0.39	0.139	0.53	0.265	0.74	0.500	1.05		300.0
320.0		0.010	0.12	0.020	0.18	0.046	0.29	0.081	0.39	0.145	0.54	0.275	0.75	0.519	1.06		320.0
340.0		0.010	0.12	0.021	0.19	0.048	0.29	0.083	0.40	0.150	0.54	0.285	0.76	0.537	1.07		340.0
360.0	0.30	0.011	0.12	0.022	0.19	0.050	0.30	0.086	0.40	0.155	0.55	0.295	0.76	0.555	1.08		360.0
380.0		0.011	0.13	0.023	0.19	0.051	0.30	0.089	0.40	0.160	0.55	0.304	0.77	0.572	1.09		380.0
400.0		0.011	0.13	0.023	0.19	0.053	0.30	0.092	0.41	0.164	0.56	0.313	0.77	0.589	1.10		400.0
420.0		0.012	0.13	0.024	0.19	0.054	0.30	0.094	0.41	0.169	0.56	0.322	0.78	0.606	1.10		420.0
440.0		0.012	0.13	0.025	0.19	0.056	0.31	0.097	0.41	0.174	0.57	0.330	0.78	0.622	1.11		440.0
460.0		0.012	0.13	0.025	0.20	0.057	0.31	0.099	0.42	0.178	0.57	0.339	0.79	0.638	1.12	1.00	460.0
480.0		0.013	0.13	0.026	0.20	0.059	0.31	0.102	0.42	0.183	0.57	0.347	0.79	0.654	1.12		480.0
500.0		0.013	0.13	0.027	0.20	0.060	0.31	0.104	0.42	0.187	0.58	0.355	0.80	0.669	1.13		500.0
520.0		0.013	0.13	0.027	0.20	0.062	0.31	0.107	0.42	0.191	0.58	0.363	0.80	0.684	1.14		520.0
540.0		0.013	0.13	0.028	0.20	0.063	0.32	0.109	0.43	0.195	0.58	0.371	0.81	0.699	1.14		540.0
560.0		0.014	0.14	0.028	0.20	0.064	0.32	0.111	0.43	0.200	0.59	0.379	0.81	0.714	1.15		560.0
580.0		0.014	0.14	0.029	0.20	0.066	0.32	0.114	0.43	0.204	0.59	0.386	0.81	0.728	1.15		580.0
600.0		0.014	0.14	0.030	0.21	0.067	0.32	0.116	0.43	0.208	0.59	0.394	0.82	0.742	1.16		600.0
620.0		0.015	0.14	0.030	0.21	0.068	0.32	0.118	0.44	0.212	0.60	0.401	0.82	0.756	1.16		620.0
640.0		0.015	0.14	0.031	0.21	0.069	0.33	0.120	0.44	0.216	0.60	0.408	0.82	0.770	1.17		640.0
660.0		0.015	0.14	0.031	0.21	0.071	0.33	0.123	0.44	0.219	0.60	0.415	0.83	0.783	1.17		660.0
680.0		0.015	0.14	0.032	0.21	0.072	0.33	0.125	0.44	0.223	0.60	0.423	0.83	0.797	1.18		680.0
700.0		0.016	0.14	0.032	0.21	0.073	0.33	0.127	0.45	0.227	0.61	0.429	0.83	0.810	1.18		700.0
720.0		0.016	0.14	0.033	0.21	0.074	0.33	0.129	0.45	0.231	0.61	0.436	0.84	0.823	1.19		720.0
740.0		0.016	0.14	0.033	0.21	0.076	0.33	0.131	0.45	0.234	0.61	0.443	0.84	0.836	1.19		740.0
760.0		0.016	0.14	0.034	0.21	0.077	0.34	0.133	0.45	0.238	0.61	0.450	0.84	0.849	1.20		760.0
780.0		0.017	0.14	0.034	0.22	0.078	0.34	0.135	0.45	0.241	0.62	0.457	0.84	0.861	1.20		780.0
800.0		0.017	0.14	0.035	0.22	0.079	0.34	0.137	0.45	0.245	0.62	0.463	0.85	0.874	1.21		800.0
820.0		0.017	0.15	0.036	0.22	0.080	0.34	0.139	0.46	0.248	0.62	0.470	0.85	0.886	1.21		820.0
840.0	0.50	0.017	0.15	0.036	0.22	0.081	0.34	0.141	0.46	0.252	0.62	0.476	0.85	0.898	1.21		840.0
860.0		0.018	0.15	0.037	0.22	0.082	0.34	0.143	0.46	0.255	0.63	0.482	0.86	0.910	1.22		860.0
880.0		0.018	0.15	0.037	0.22	0.084	0.34	0.145	0.46	0.259	0.63	0.489	0.86	0.922	1.22		880.0
900.0		0.018	0.15	0.038	0.22	0.085	0.34	0.147	0.46	0.262	0.63	0.495	0.86	0.934	1.22		900.0
920.0		0.018	0.15	0.038	0.22	0.086	0.35	0.148	0.46	0.265	0.63	0.501	0.86	0.946	1.23	1.50	920.0
940.0		0.019	0.15	0.038	0.22	0.087	0.35	0.150	0.47	0.269	0.63	0.507	0.86	0.957	1.23		940.0
960.0		0.019	0.15	0.039	0.22	0.088	0.35	0.152	0.47	0.272	0.64	0.513	0.87	0.969	1.23		960.0
980.0		0.019	0.15	0.039	0.22	0.089	0.35	0.154	0.47	0.275	0.64	0.519	0.87	0.980	1.24		980.0
1000.0		0.019	0.15	0.040	0.23	0.090	0.35	0.156	0.47	0.278	0.64	0.525	0.87	0.991	1.24		1000.0
1100.0		0.020	0.15	0.042	0.23	0.095	0.36	0.165	0.48	0.294	0.65	0.554	0.88	1.046	1.26		1100.0
1200.0		0.022	0.16	0.044	0.23	0.100	0.36	0.173	0.48	0.309	0.66	0.582	0.89	1.099	1.27		1200.0
1300.0		0.023	0.16	0.047	0.24	0.105	0.37	0.181	0.49	0.323	0.66	0.608	0.90	1.150	1.28		1300.0
1400.0		0.024	0.16	0.049	0.24	0.109	0.37	0.189	0.49	0.337	0.67	0.634	0.91	1.199	1.30		1400.0
1500.0		0.025	0.16	0.051	0.24	0.114	0.37	0.197	0.50	0.351	0.68	0.659	0.92	1.246	1.31		1500.0
1600.0		0.026	0.16	0.052	0.24	0.118	0.38	0.204	0.50	0.364	0.68	0.683	0.92	1.292	1.32	2.00	1600.0
1700.0		0.026	0.17	0.054	0.25	0.122	0.38	0.211	0.51	0.377	0.69	0.707	0.93	1.337	1.33		1700.0
1800.0		0.027	0.17	0.056	0.25	0.126	0.38	0.218	0.51	0.389	0.70	0.730	0.94	1.381	1.34		1800.0
1900.0		0.028	0.17	0.058	0.25	0.130	0.39	0.225	0.52	0.401	0.70	0.752	0.94	1.424	1.35		1900.0
2000.0		0.029	0.17	0.060	0.25	0.134	0.39	0.232	0.52	0.413	0.71	0.774	0.95	1.466	1.36		2000.0

Table 4.27 Flow of water at 75°C in polyethylene (PE-X)[21] pipes

q_m = mass flow rate	kg.s^{-1}
c = velocity	m.s^{-1}
$\Delta p/l$ = pressure drop per unit length	Pa.m^{-1}
l_e = equivalent length of a component for $\zeta = 1$	m

POLYETHYLENE (PE-X)
WATER AT 75°C
BS 2871

$\Delta p/l$	c	10 mm		12 mm		15 mm		18 mm		22 mm		28 mm		35 mm		c	$\Delta p/l$
		q_m	l_e	q_m	l_e	q_m	l_e	q_m	l_e	q_m	l_e	q_m	l_e	q_m	l_e		
0.1								0.001	0.11	0.001	0.17	0.003	0.26	0.005	0.39		0.1
0.2						0.001	0.09	0.001	0.14	0.002	0.20	0.004	0.31	0.008	0.45		0.2
0.3						0.001	0.11	0.002	0.15	0.003	0.22	0.006	0.34	0.011	0.49		0.3
0.4						0.001	0.12	0.002	0.17	0.003	0.24	0.007	0.36	0.013	0.52		0.4
0.5						0.001	0.12	0.002	0.18	0.004	0.25	0.008	0.37	0.015	0.54		0.5
0.6				0.001	0.07	0.001	0.13	0.002	0.18	0.004	0.26	0.008	0.39	0.016	0.56		0.6
0.7				0.001	0.08	0.001	0.13	0.003	0.19	0.005	0.27	0.009	0.40	0.018	0.58		0.7
0.8				0.001	0.08	0.002	0.14	0.003	0.20	0.005	0.28	0.010	0.41	0.019	0.60		0.8
0.9				0.001	0.08	0.002	0.14	0.003	0.20	0.006	0.29	0.011	0.42	0.021	0.61		0.9
1.0				0.001	0.09	0.002	0.15	0.003	0.21	0.006	0.29	0.012	0.43	0.022	0.62		1.0
1.5				0.001	0.10	0.002	0.16	0.004	0.23	0.008	0.32	0.015	0.47	0.028	0.67		1.5
2.0		0.001	0.06	0.001	0.10	0.003	0.17	0.005	0.24	0.009	0.34	0.017	0.49	0.033	0.70	0.05	2.0
2.5		0.001	0.07	0.001	0.11	0.003	0.18	0.006	0.25	0.010	0.35	0.020	0.51	0.038	0.73		2.5
3.0		0.001	0.07	0.002	0.11	0.004	0.19	0.006	0.26	0.011	0.37	0.022	0.53	0.042	0.76		3.0
3.5		0.001	0.08	0.002	0.12	0.004	0.19	0.007	0.27	0.013	0.38	0.024	0.54	0.046	0.77		3.5
4.0		0.001	0.08	0.002	0.12	0.004	0.20	0.007	0.28	0.014	0.39	0.026	0.56	0.049	0.79		4.0
4.5		0.001	0.08	0.002	0.13	0.005	0.21	0.008	0.28	0.015	0.40	0.028	0.57	0.053	0.81		4.5
5.0		0.001	0.08	0.002	0.13	0.005	0.21	0.009	0.29	0.015	0.40	0.030	0.58	0.056	0.82		5.0
5.5		0.001	0.08	0.002	0.13	0.005	0.21	0.009	0.29	0.016	0.41	0.032	0.59	0.060	0.83		5.5
6.0		0.001	0.09	0.002	0.13	0.005	0.22	0.010	0.30	0.017	0.42	0.033	0.59	0.063	0.85		6.0
6.5		0.001	0.09	0.002	0.14	0.006	0.22	0.010	0.30	0.018	0.42	0.035	0.60	0.066	0.86		6.5
7.0		0.001	0.09	0.003	0.14	0.006	0.22	0.010	0.31	0.019	0.43	0.036	0.61	0.068	0.87		7.0
7.5		0.001	0.09	0.003	0.14	0.006	0.23	0.011	0.31	0.020	0.43	0.038	0.62	0.071	0.88		7.5
8.0		0.001	0.09	0.003	0.14	0.006	0.23	0.011	0.32	0.020	0.44	0.039	0.62	0.074	0.88		8.0
8.5		0.001	0.09	0.003	0.14	0.007	0.23	0.012	0.32	0.021	0.44	0.041	0.63	0.077	0.89		8.5
9.0		0.001	0.10	0.003	0.15	0.007	0.24	0.012	0.32	0.022	0.45	0.042	0.63	0.079	0.90		9.0
9.5		0.001	0.10	0.003	0.15	0.007	0.24	0.012	0.33	0.023	0.45	0.043	0.64	0.082	0.91		9.5
10.0		0.002	0.10	0.003	0.15	0.007	0.24	0.013	0.33	0.023	0.46	0.045	0.64	0.084	0.92		10.0
12.5	0.05	0.002	0.10	0.004	0.16	0.008	0.25	0.015	0.34	0.026	0.47	0.051	0.67	0.096	0.95	0.15	12.5
15.0		0.002	0.11	0.004	0.16	0.009	0.26	0.016	0.35	0.029	0.49	0.056	0.69	0.106	0.97		15.0
17.5		0.002	0.11	0.005	0.17	0.010	0.27	0.018	0.36	0.032	0.50	0.062	0.70	0.116	1.00		17.5
20.0		0.002	0.11	0.005	0.17	0.011	0.27	0.019	0.37	0.035	0.51	0.066	0.72	0.125	1.02		20.0
22.5		0.003	0.12	0.005	0.18	0.012	0.28	0.021	0.38	0.037	0.52	0.071	0.73	0.134	1.03		22.5
25.0		0.003	0.12	0.006	0.18	0.013	0.28	0.022	0.39	0.040	0.53	0.075	0.74	0.142	1.05		25.0
27.5		0.003	0.12	0.006	0.18	0.013	0.29	0.023	0.39	0.042	0.54	0.080	0.75	0.150	1.06		27.5
30.0		0.003	0.12	0.006	0.19	0.014	0.29	0.025	0.40	0.044	0.55	0.084	0.76	0.158	1.08		30.0
32.5		0.003	0.13	0.007	0.19	0.015	0.30	0.026	0.40	0.046	0.55	0.088	0.77	0.165	1.09		32.5
35.0		0.003	0.13	0.007	0.19	0.015	0.30	0.027	0.41	0.048	0.56	0.091	0.77	0.172	1.10		35.0
37.5		0.003	0.13	0.007	0.19	0.016	0.31	0.028	0.41	0.050	0.57	0.095	0.78	0.179	1.11		37.5
40.0		0.004	0.13	0.007	0.20	0.017	0.31	0.029	0.42	0.052	0.57	0.099	0.79	0.186	1.12	0.30	40.0
42.5		0.004	0.13	0.008	0.20	0.017	0.31	0.030	0.42	0.054	0.58	0.102	0.80	0.192	1.13		42.5
45.0		0.004	0.13	0.008	0.20	0.018	0.32	0.031	0.43	0.056	0.58	0.105	0.80	0.199	1.14		45.0
47.5		0.004	0.13	0.008	0.20	0.018	0.32	0.032	0.43	0.057	0.59	0.109	0.81	0.205	1.15		47.5
50.0		0.004	0.14	0.008	0.20	0.019	0.32	0.033	0.43	0.059	0.59	0.112	0.81	0.211	1.15		50.0
52.5		0.004	0.14	0.009	0.21	0.020	0.32	0.034	0.44	0.061	0.59	0.115	0.82	0.217	1.16		52.5
55.0		0.004	0.14	0.009	0.21	0.020	0.33	0.035	0.44	0.062	0.60	0.118	0.82	0.223	1.17		55.0
57.5		0.004	0.14	0.009	0.21	0.021	0.33	0.036	0.44	0.064	0.60	0.121	0.83	0.229	1.18		57.5
60.0		0.005	0.14	0.009	0.21	0.021	0.33	0.037	0.44	0.066	0.61	0.124	0.83	0.234	1.18		60.0
62.5		0.005	0.14	0.010	0.21	0.022	0.33	0.038	0.45	0.067	0.61	0.127	0.84	0.240	1.19		62.5
65.0		0.005	0.14	0.010	0.21	0.022	0.33	0.038	0.45	0.069	0.61	0.130	0.84	0.245	1.20		65.0
67.5		0.005	0.14	0.010	0.22	0.023	0.34	0.039	0.45	0.070	0.62	0.133	0.85	0.250	1.20		67.5
70.0		0.005	0.15	0.010	0.22	0.023	0.34	0.040	0.46	0.072	0.62	0.135	0.85	0.255	1.21		70.0
72.5		0.005	0.15	0.010	0.22	0.024	0.34	0.041	0.46	0.073	0.62	0.138	0.85	0.261	1.21		72.5
75.0	0.15	0.005	0.15	0.011	0.22	0.024	0.34	0.042	0.46	0.075	0.63	0.141	0.86	0.266	1.22		75.0
77.5		0.005	0.15	0.011	0.22	0.025	0.34	0.042	0.46	0.076	0.63	0.143	0.86	0.271	1.22		77.5
80.0		0.005	0.15	0.011	0.22	0.025	0.35	0.043	0.46	0.077	0.63	0.146	0.86	0.276	1.23		80.0
82.5		0.005	0.15	0.011	0.22	0.025	0.35	0.044	0.47	0.079	0.64	0.149	0.87	0.280	1.23		82.5
85.0		0.006	0.15	0.011	0.22	0.026	0.35	0.045	0.47	0.080	0.64	0.151	0.87	0.285	1.24		85.0
87.5		0.006	0.15	0.012	0.23	0.026	0.35	0.046	0.47	0.081	0.64	0.154	0.87	0.290	1.24		87.5
90.0		0.006	0.15	0.012	0.23	0.027	0.35	0.046	0.47	0.083	0.64	0.156	0.88	0.295	1.25		90.0

Table 4.27 Flow of water at 75°C in polyethylene (PE-X)[21] pipes — *continued*

q_m = mass flow rate   kg.s^{-1}
c = velocity   m.s^{-1}
$\Delta p/l$ = pressure drop per unit length   Pa.m^{-1}
l_e = equivalent length of a component
     for $\zeta = 1$ m

POLYETHYLENE (PE-X)
WATER AT 75°C
BS 2871

$\Delta p/l$	c	10 mm		12 mm		15 mm		18 mm		22 mm		28 mm		35 mm		c	$\Delta p/l$
		q_m	l_e	q_m	l_e	q_m	l_e	q_m	l_e	q_m	l_e	q_m	l_e	q_m	l_e		
92.5		0.006	0.15	0.012	0.23	0.027	0.35	0.047	0.48	0.084	0.65	0.158	0.88	0.299	1.25		92.5
95.0		0.006	0.15	0.012	0.23	0.028	0.36	0.048	0.48	0.085	0.65	0.161	0.88	0.304	1.26		95.0
97.5		0.006	0.15	0.012	0.23	0.028	0.36	0.048	0.48	0.087	0.65	0.163	0.88	0.308	1.26	0.50	97.5
100.0		0.006	0.16	0.013	0.23	0.028	0.36	0.049	0.48	0.088	0.65	0.165	0.89	0.313	1.27		100.0
120.0		0.007	0.16	0.014	0.24	0.032	0.37	0.055	0.49	0.098	0.67	0.183	0.91	0.346	1.30		120.0
140.0		0.007	0.16	0.015	0.24	0.035	0.38	0.060	0.51	0.106	0.69	0.200	0.92	0.378	1.32		140.0
160.0		0.008	0.17	0.017	0.25	0.037	0.39	0.064	0.52	0.115	0.70	0.215	0.94	0.407	1.34		160.0
180.0		0.009	0.17	0.018	0.25	0.040	0.39	0.069	0.52	0.123	0.71	0.230	0.95	0.435	1.36		180.0
200.0		0.009	0.18	0.019	0.26	0.042	0.40	0.073	0.53	0.130	0.72	0.244	0.96	0.462	1.38		200.0
220.0		0.010	0.18	0.020	0.26	0.045	0.40	0.077	0.54	0.138	0.73	0.257	0.97	0.487	1.40		220.0
240.0		0.010	0.18	0.021	0.27	0.047	0.41	0.081	0.55	0.145	0.74	0.270	0.98	0.512	1.41		240.0
260.0	0.30	0.011	0.18	0.022	0.27	0.049	0.41	0.085	0.55	0.151	0.75	0.282	0.99	0.535	1.43		260.0
280.0		0.011	0.19	0.023	0.27	0.051	0.42	0.089	0.56	0.158	0.75	0.294	1.00	0.558	1.44		280.0
300.0		0.012	0.19	0.024	0.28	0.054	0.42	0.092	0.56	0.164	0.76	0.305	1.01	0.580	1.45		300.0
320.0		0.012	0.19	0.025	0.28	0.056	0.43	0.096	0.57	0.170	0.77	0.316	1.01	0.601	1.46		320.0
340.0		0.013	0.19	0.026	0.28	0.057	0.43	0.099	0.57	0.176	0.77	0.327	1.02	0.622	1.47	1.00	340.0
360.0		0.013	0.19	0.027	0.28	0.059	0.43	0.102	0.58	0.182	0.78	0.337	1.03	0.642	1.48		360.0
380.0		0.013	0.19	0.027	0.29	0.061	0.44	0.105	0.58	0.188	0.78	0.348	1.03	0.662	1.49		380.0
400.0		0.014	0.20	0.028	0.29	0.063	0.44	0.109	0.59	0.193	0.79	0.358	1.04	0.681	1.50		400.0
420.0		0.014	0.20	0.029	0.29	0.065	0.44	0.112	0.59	0.199	0.79	0.367	1.04	0.700	1.51		420.0
440.0		0.015	0.20	0.030	0.29	0.067	0.45	0.115	0.59	0.204	0.80	0.377	1.05	0.719	1.52		440.0
460.0		0.015	0.20	0.031	0.29	0.068	0.45	0.118	0.60	0.209	0.80	0.386	1.05	0.737	1.53		460.0
480.0		0.015	0.20	0.031	0.30	0.070	0.45	0.120	0.60	0.214	0.81	0.395	1.06	0.755	1.54		480.0
500.0		0.016	0.20	0.032	0.30	0.072	0.45	0.123	0.60	0.219	0.81	0.404	1.06	0.772	1.54		500.0
520.0		0.016	0.20	0.033	0.30	0.073	0.46	0.126	0.61	0.224	0.82	0.413	1.06	0.789	1.55		520.0
540.0		0.016	0.21	0.033	0.30	0.075	0.46	0.129	0.61	0.229	0.82	0.422	1.07	0.806	1.56		540.0
560.0		0.017	0.21	0.034	0.30	0.076	0.46	0.131	0.61	0.234	0.82	0.430	1.07	0.822	1.56		560.0
580.0		0.017	0.21	0.035	0.30	0.078	0.46	0.134	0.62	0.238	0.83	0.439	1.08	0.839	1.57		580.0
600.0	0.50	0.017	0.21	0.036	0.31	0.079	0.47	0.137	0.62	0.243	0.83	0.447	1.08	0.855	1.58		600.0
620.0		0.018	0.21	0.036	0.31	0.081	0.47	0.139	0.62	0.247	0.83	0.455	1.08	0.870	1.58		620.0
640.0		0.018	0.21	0.037	0.31	0.082	0.47	0.142	0.62	0.252	0.84	0.463	1.09	0.886	1.59		640.0
660.0		0.018	0.21	0.038	0.31	0.084	0.47	0.144	0.63	0.256	0.84	0.471	1.09	0.901	1.59		660.0
680.0		0.019	0.21	0.038	0.31	0.085	0.47	0.147	0.63	0.260	0.84	0.478	1.09	0.917	1.60		680.0
700.0		0.019	0.21	0.039	0.31	0.087	0.48	0.149	0.63	0.265	0.85	0.486	1.09	0.931	1.61	1.50	700.0
720.0		0.019	0.22	0.039	0.31	0.088	0.48	0.151	0.63	0.269	0.85	0.494	1.10	0.946	1.61		720.0
740.0		0.020	0.22	0.040	0.32	0.090	0.48	0.154	0.64	0.273	0.85	0.501	1.10	0.961	1.62		740.0
760.0		0.020	0.22	0.041	0.32	0.091	0.48	0.156	0.64	0.277	0.86	0.508	1.10	0.975	1.62		760.0
780.0		0.020	0.22	0.041	0.32	0.092	0.48	0.159	0.64	0.281	0.86	0.516	1.11	0.989	1.63		780.0
800.0		0.021	0.22	0.042	0.32	0.094	0.49	0.161	0.64	0.285	0.86	0.523	1.11	1.003	1.63		800.0
820.0		0.021	0.22	0.043	0.32	0.095	0.49	0.163	0.64	0.289	0.86	0.530	1.11	1.017	1.63		820.0
840.0		0.021	0.22	0.043	0.32	0.096	0.49	0.165	0.65	0.293	0.87	0.537	1.11	1.031	1.64		840.0
860.0		0.021	0.22	0.044	0.32	0.097	0.49	0.167	0.65	0.297	0.87	0.544	1.11	1.045	1.64		860.0
880.0		0.022	0.22	0.044	0.32	0.099	0.49	0.170	0.65	0.301	0.87	0.551	1.12	1.058	1.65		880.0
900.0		0.022	0.22	0.045	0.32	0.100	0.49	0.172	0.65	0.305	0.87	0.564	1.12	1.071	1.65		900.0
920.0		0.022	0.22	0.045	0.33	0.101	0.49	0.174	0.65	0.309	0.88	0.564	1.12	1.085	1.66		920.0
940.0		0.023	0.22	0.046	0.33	0.103	0.50	0.176	0.66	0.312	0.88	0.571	1.12	1.098	1.66		940.0
960.0		0.023	0.22	0.047	0.33	0.104	0.50	0.178	0.66	0.316	0.88	0.577	1.13	1.111	1.66		960.0
980.0		0.023	0.23	0.047	0.33	0.105	0.50	0.180	0.66	0.320	0.88	0.584	1.13	1.123	1.67		980.0
1000.0		0.023	0.23	0.048	0.33	0.106	0.50	0.182	0.66	0.324	0.89	0.590	1.13	1.136	1.67		1000.0
1100.0		0.025	0.23	0.050	0.33	0.112	0.51	0.192	0.67	0.341	0.90	0.622	1.14	1.198	1.69		1100.0
1200.0		0.026	0.23	0.053	0.34	0.118	0.51	0.202	0.68	0.358	0.91	0.652	1.15	1.257	1.71	2.00	1200.0
1300.0		0.027	0.24	0.055	0.34	0.123	0.52	0.211	0.68	0.375	0.91	0.681	1.15	1.315	1.72		1300.0
1400.0		0.028	0.24	0.058	0.34	0.128	0.52	0.220	0.69	0.391	0.92	0.708	1.16	1.370	1.74		1400.0
1500.0		0.030	0.24	0.060	0.35	0.134	0.53	0.229	0.70	0.406	0.93	0.735	1.17	1.423	1.75		1500.0
1600.0		0.031	0.24	0.062	0.35	0.139	0.53	0.238	0.70	0.421	0.94	0.761	1.17	1.475	1.76		1600.0
1700.0		0.032	0.24	0.064	0.35	0.143	0.54	0.246	0.71	0.436	0.94	0.787	1.18	1.526	1.77		1700.0
1800.0		0.033	0.25	0.067	0.36	0.148	0.54	0.254	0.71	0.450	0.95	0.811	1.19	1.575	1.79		1800.0
1900.0		0.034	0.25	0.069	0.36	0.153	0.54	0.262	0.72	0.464	0.96	0.835	1.19	1.623	1.80		1900.0
2000.0		0.035	0.25	0.071	0.36	0.157	0.55	0.269	0.72	0.477	0.96	0.858	1.19	1.670	1.81		2000.0

Table 4.28 Flow of water at 10°C in polybutylene (PB) pipes[21]

q_m	= mass flow rate	kg.s^{-1}
c	= velocity	m.s^{-1}
$\Delta p/l$	= pressure drop per unit length	Pa.m^{-1}
l_e	= equivalent length of a component	
	for $\zeta = 1$	m

POLYBUTYLENE (PB)

WATER AT 10°C

BS 5556

$\Delta p/l$	c	10 mm		12 mm		16 mm		20 mm		25 mm		32 mm		c	$\Delta p/l$
		q_m	l_e	q_m	l_e	q_m	l_e	q_m	l_e	q_m	l_e	q_m	l_e		
0.1								0.001	0.06	0.001	0.10	0.003	0.16		0.1
0.2								0.001	0.08	0.002	0.12	0.005	0.21		0.2
0.3						0.001	0.05	0.001	0.09	0.003	0.14	0.006	0.23		0.3
0.4						0.001	0.06	0.002	0.10	0.004	0.16	0.008	0.25		0.4
0.5						0.001	0.06	0.002	0.11	0.004	0.17	0.009	0.27		0.5
0.6						0.001	0.07	0.002	0.11	0.005	0.18	0.010	0.28		0.6
0.7						0.001	0.07	0.003	0.12	0.005	0.19	0.011	0.29		0.7
0.8						0.001	0.07	0.003	0.12	0.006	0.19	0.012	0.30		0.8
0.9						0.001	0.08	0.003	0.13	0.006	0.20	0.013	0.31		0.9
1.0				0.001	0.04	0.001	0.08	0.003	0.13	0.006	0.20	0.013	0.32		1.0
1.5				0.001	0.05	0.002	0.09	0.004	0.15	0.008	0.23	0.017	0.35		1.5
2.0				0.001	0.05	0.002	0.10	0.005	0.16	0.010	0.25	0.021	0.38		2.0
2.5				0.001	0.06	0.003	0.11	0.006	0.17	0.012	0.26	0.024	0.40		2.5
3.0		0.001	0.03	0.001	0.06	0.003	0.11	0.007	0.18	0.013	0.27	0.026	0.41	0.05	3.0
3.5		0.001	0.04	0.001	0.06	0.003	0.12	0.007	0.19	0.014	0.28	0.029	0.43		3.5
4.0		0.001	0.04	0.001	0.06	0.004	0.12	0.008	0.19	0.015	0.29	0.031	0.44		4.0
4.5		0.001	0.04	0.001	0.07	0.004	0.13	0.009	0.20	0.017	0.30	0.034	0.45		4.5
5.0		0.001	0.04	0.002	0.07	0.004	0.13	0.009	0.21	0.018	0.30	0.036	0.46		5.0
5.5		0.001	0.04	0.002	0.07	0.005	0.13	0.010	0.21	0.019	0.31	0.038	0.47		5.5
6.0		0.001	0.04	0.002	0.07	0.005	0.14	0.010	0.21	0.020	0.32	0.040	0.48		6.0
6.5		0.001	0.05	0.002	0.07	0.005	0.14	0.011	0.22	0.021	0.32	0.042	0.48		6.5
7.0		0.001	0.05	0.002	0.08	0.005	0.14	0.011	0.22	0.022	0.33	0.044	0.49		7.0
7.5		0.001	0.05	0.002	0.08	0.006	0.14	0.012	0.23	0.023	0.33	0.046	0.50		7.5
8.0		0.001	0.05	0.002	0.08	0.006	0.15	0.012	0.23	0.024	0.34	0.048	0.50		8.0
8.5		0.001	0.05	0.002	0.08	0.006	0.15	0.013	0.23	0.024	0.34	0.049	0.51		8.5
9.0		0.001	0.05	0.002	0.08	0.006	0.15	0.013	0.24	0.025	0.35	0.051	0.52		9.0
9.5		0.001	0.05	0.002	0.08	0.006	0.15	0.014	0.24	0.026	0.35	0.053	0.52		9.5
10.0		0.001	0.05	0.003	0.08	0.007	0.15	0.014	0.24	0.027	0.35	0.055	0.53		10.0
12.5		0.001	0.06	0.003	0.09	0.008	0.16	0.016	0.25	0.031	0.37	0.062	0.55		12.5
15.0		0.002	0.06	0.003	0.09	0.008	0.17	0.018	0.26	0.034	0.38	0.069	0.57		15.0
17.5	0.05	0.002	0.06	0.004	0.10	0.009	0.18	0.020	0.27	0.038	0.40	0.076	0.58		17.5
20.0		0.002	0.06	0.004	0.10	0.010	0.18	0.021	0.28	0.041	0.41	0.082	0.60	0.15	20.0
22.5		0.002	0.07	0.004	0.11	0.011	0.19	0.023	0.29	0.044	0.42	0.088	0.61		22.5
25.0		0.002	0.07	0.004	0.11	0.012	0.19	0.024	0.29	0.047	0.42	0.094	0.62		25.0
27.5		0.002	0.07	0.005	0.11	0.012	0.19	0.026	0.30	0.049	0.43	0.099	0.63		27.5
30.0		0.002	0.07	0.005	0.11	0.013	0.20	0.027	0.30	0.052	0.44	0.104	0.64		30.0
32.5		0.003	0.07	0.005	0.11	0.014	0.20	0.028	0.31	0.055	0.45	0.109	0.65		32.5
35.0		0.003	0.07	0.006	0.12	0.014	0.20	0.030	0.31	0.057	0.45	0.114	0.66		35.0
37.5		0.003	0.08	0.006	0.12	0.015	0.21	0.031	0.32	0.059	0.46	0.119	0.67		37.5
40.0		0.003	0.08	0.006	0.12	0.015	0.21	0.032	0.32	0.062	0.46	0.123	0.67		40.0
42.5		0.003	0.08	0.006	0.12	0.016	0.21	0.033	0.33	0.064	0.47	0.128	0.68		42.5
45.0		0.003	0.08	0.006	0.12	0.017	0.22	0.035	0.33	0.066	0.47	0.132	0.69		45.0
47.5		0.003	0.08	0.007	0.13	0.017	0.22	0.036	0.33	0.068	0.48	0.136	0.69		47.5
50.0		0.003	0.08	0.007	0.13	0.018	0.22	0.037	0.34	0.070	0.48	0.140	0.70		50.0
52.5		0.003	0.08	0.007	0.13	0.018	0.22	0.038	0.34	0.072	0.49	0.144	0.71		52.5
55.0		0.003	0.08	0.007	0.13	0.019	0.23	0.039	0.34	0.074	0.49	0.148	0.71		55.0
57.5		0.004	0.08	0.007	0.13	0.019	0.23	0.040	0.34	0.076	0.49	0.152	0.72		57.5
60.0		0.004	0.09	0.008	0.13	0.020	0.23	0.041	0.35	0.078	0.50	0.156	0.72		60.0
62.5		0.004	0.09	0.008	0.13	0.020	0.23	0.042	0.35	0.080	0.50	0.160	0.73	0.30	62.5
65.0		0.004	0.09	0.008	0.13	0.021	0.23	0.043	0.35	0.082	0.50	0.164	0.73		65.0
67.5		0.004	0.09	0.008	0.14	0.021	0.23	0.044	0.35	0.084	0.51	0.167	0.73		67.5
70.0		0.004	0.09	0.008	0.14	0.022	0.24	0.045	0.36	0.086	0.51	0.171	0.74		70.0
72.5		0.004	0.09	0.009	0.14	0.022	0.24	0.046	0.36	0.088	0.51	0.174	0.74		72.5
75.0		0.004	0.09	0.009	0.14	0.023	0.24	0.047	0.36	0.089	0.52	0.178	0.75		75.0
77.5		0.004	0.09	0.009	0.14	0.023	0.24	0.048	0.36	0.091	0.52	0.181	0.75		77.5
80.0		0.004	0.09	0.009	0.14	0.023	0.24	0.049	0.37	0.093	0.52	0.184	0.75		80.0
82.5		0.004	0.09	0.009	0.14	0.024	0.24	0.050	0.37	0.094	0.53	0.188	0.76		82.5
85.0		0.005	0.09	0.009	0.14	0.024	0.25	0.050	0.37	0.096	0.53	0.191	0.76		85.0
87.5		0.005	0.09	0.010	0.14	0.025	0.25	0.051	0.37	0.098	0.53	0.194	0.76		87.5
90.0		0.005	0.09	0.010	0.14	0.025	0.25	0.052	0.37	0.099	0.53	0.197	0.77		90.0

Table 4.28 Flow of water at 10°C in polybutylene (PB) pipes[21] — *continued*

q_m = mass flow rate $\qquad$ kg.s^{-1}
c = velocity $\qquad$ m.s^{-1}
$\Delta p/l$ = pressure drop per unit length $\qquad$ Pa.m^{-1}
l_e = equivalent length of a component
$\qquad$ for $\zeta = 1$ $\qquad$ m

POLYBUTYLENE (PB)
WATER AT 10°C
BS 5556

$\Delta p/l$	c	10 mm		12 mm		16 mm		20 mm		25 mm		32 mm		c	$\Delta p/l$
		q_m	l_e	q_m	l_e	q_m	l_e	q_m	l_e	q_m	l_e	q_m	l_e		
92.5		0.005	0.09	0.010	0.15	0.026	0.25	0.053	0.38	0.101	0.54	0.201	0.77		92.5
95.0		0.005	0.10	0.010	0.15	0.026	0.25	0.054	0.38	0.103	0.54	0.204	0.77		95.0
97.5		0.005	0.10	0.010	0.15	0.026	0.25	0.055	0.38	0.104	0.54	0.207	0.78		97.5
100.0	0.15	0.005	0.10	0.010	0.15	0.027	0.25	0.056	0.38	0.106	0.54	0.210	0.78		100.0
120.0		0.006	0.10	0.012	0.15	0.030	0.26	0.062	0.39	0.117	0.56	0.233	0.80		120.0
140.0		0.006	0.10	0.013	0.16	0.033	0.27	0.068	0.40	0.128	0.57	0.255	0.82		140.0
160.0		0.007	0.11	0.014	0.16	0.035	0.28	0.073	0.41	0.139	0.59	0.275	0.84	0.50	160.0
180.0		0.007	0.11	0.015	0.17	0.038	0.28	0.078	0.42	0.149	0.60	0.294	0.85		180.0
200.0		0.008	0.11	0.016	0.17	0.040	0.29	0.083	0.43	0.158	0.61	0.312	0.86		200.0
220.0		0.008	0.11	0.017	0.17	0.043	0.29	0.088	0.44	0.167	0.62	0.330	0.88		220.0
240.0		0.009	0.12	0.018	0.18	0.045	0.30	0.093	0.44	0.176	0.63	0.346	0.89		240.0
260.0		0.009	0.12	0.19	0.18	0.047	0.30	0.097	0.45	0.184	0.63	0.363	0.90		260.0
280.0		0.009	0.12	0.019	0.18	0.049	0.31	0.101	0.45	0.192	0.64	0.378	0.91		280.0
300.0		0.010	0.12	0.020	0.18	0.051	0.31	0.106	0.46	0.200	0.65	0.393	0.91		300.0
320.0		0.010	0.12	0.021	0.19	0.0534	0.31	0.110	0.46	0.207	0.65	0.408	0.92		320.0
340.0		0.011	0.13	0.022	0.19	0.055	0.32	0.114	0.47	0.215	0.66	0.422	0.93		340.0
360.0	0.30	0.011	0.13	0.023	0.19	0.057	0.32	0.117	0.47	0.222	0.67	0.436	0.94		360.0
380.0		0.011	0.13	0.023	0.19	0.059	0.32	0.121	0.48	0.229	0.67	0.450	0.94		380.0
400.0		0.012	0.13	0.024	0.19	0.061	0.32	0.125	0.48	0.236	0.68	0.463	0.95		400.0
420.0		0.012	0.13	0.025	0.20	0.062	0.33	0.128	0.48	0.242	0.68	0.476	0.96		420.0
440.0		0.012	0.13	0.025	0.20	0.064	0.33	0.132	0.49	0.249	0.69	0.489	0.96		440.0
460.0		0.013	0.13	0.026	0.20	0.066	0.33	0.135	0.49	0.255	0.69	0.501	0.97		460.0
480.0		0.013	0.13	0.027	0.20	0.067	0.34	0.139	0.49	0.262	0.69	0.513	0.97		480.0
500.0		0.013	0.14	0.027	0.20	0.069	0.34	0.142	0.50	0.268	0.70	0.525	0.98		500.0
520.0		0.014	0.14	0.028	0.20	0.071	0.34	0.145	0.50	0.274	0.70	0.537	0.98	1.00	520.0
540.0		0.014	0.14	0.029	0.21	0.072	0.34	0.148	0.50	0.280	0.71	0.549	0.99		540.0
560.0		0.014	0.14	0.029	0.21	0.074	0.34	0.152	0.51	0.286	0.71	0.560	0.99		560.0
580.0		0.015	0.14	0.030	0.21	0.075	0.35	0.155	0.51	0.292	0.71	0.571	1.00		580.0
600.0		0.015	0.14	0.031	0.21	0.077	0.35	0.158	0.51	0.298	0.72	0.582	1.00		600.0
620.0		0.015	0.14	0.031	0.21	0.078	0.35	0.161	0.51	0.303	0.72	0.593	1.01		620.0
640.0		0.015	0.14	0.032	0.21	0.080	0.35	0.164	0.52	0.309	0.72	0.604	1.01		640.0
660.0		0.016	0.14	0.032	0.21	0.081	0.35	0.167	0.52	0.314	0.73	0.614	1.01		660.0
680.0		0.016	0.14	0.033	0.21	0.083	0.35	0.169	0.52	0.320	0.73	0.625	1.02		680.0
700.0		0.016	0.14	0.033	0.22	0.084	0.36	0.172	0.52	0.325	0.73	0.635	1.02		700.0
720.0		0.017	0.15	0.034	0.22	0.085	0.36	0.175	0.53	0.330	0.74	0.645	1.02		720.0
740.0		0.017	0.15	0.035	0.22	0.087	0.36	0.178	0.53	0.335	0.74	0.655	1.03		740.0
760.0		0.017	0.15	0.035	0.22	0.088	0.36	0.181	0.53	0.341	0.74	0.665	1.03		760.0
780.0		0.017	0.15	0.036	0.22	0.089	0.36	0.183	0.53	0.346	0.75	0.675	1.03		780.0
800.0		0.018	0.15	0.036	0.22	0.091	0.36	0.186	0.54	0.351	0.75	0.684	1.04		800.0
820.0		0.018	0.15	0.037	0.22	0.092	0.37	0.189	0.54	0.356	0.75	0.694	1.04		820.0
840.0	0.50	0.018	0.15	0.037	0.22	0.093	0.37	0.191	0.54	0.361	0.75	0.703	1.04		840.0
860.0		0.018	0.15	0.038	0.22	0.095	0.37	0.194	0.54	0.366	0.76	0.713	1.05		860.0
880.0		0.019	0.15	0.038	0.22	0.096	0.37	0.197	0.54	0.370	0.76	0.722	1.05		880.0
900.0		0.019	0.15	0.039	0.22	0.097	0.37	0.199	0.54	0.375	0.76	0.731	1.05		900.0
920.0		0.019	0.15	0.039	0.23	0.098	0.37	0.202	0.55	0.380	0.76	0.740	1.06		920.0
940.0		0.019	0.15	0.040	0.23	0.100	0.37	0.204	0.55	0.385	0.77	0.749	1.06		940.0
960.0		0.020	0.15	0.040	0.23	0.101	0.37	0.207	0.55	0.389	0.77	0.758	1.06		960.0
980.0		0.020	0.15	0.041	0.23	0.102	0.38	0.209	0.55	0.394	0.77	0.767	1.06		980.0
1000.0		0.020	0.15	0.041	0.23	0.103	0.38	0.212	0.55	0.398	0.77	0.776	1.07		1000.0
1100.0		0.021	0.16	0.044	0.23	0.109	0.38	0.223	0.56	0.421	0.78	0.818	1.08	1.50	1100.0
1200.0		0.022	0.16	0.046	0.24	0.115	0.39	0.235	0.57	0.442	0.79	0.859	1.09		1200.0
1300.0		0.024	0.16	0.048	0.24	0.120	0.39	0.246	0.57	0.463	0.80	0.898	1.10		1300.0
1400.0		0.025	0.16	0.050	0.24	0.125	0.40	0.257	0.58	0.483	0.81	0.936	1.11		1400.0
1500.0		0.026	0.17	0.052	0.25	0.130	0.40	0.267	0.59	0.502	0.82	0.973	1.12		1500.0
1600.0		0.027	0.17	0.054	0.25	0.135	0.41	0.277	0.59	0.521	0.82	1.008	1.13		1600.0
1700.0		0.028	0.17	0.056	kg.25	0.140	0.41	0.287	0.60	0.539	0.83	1.043	1.13		1700.0
1800.0		0.029	0.17	0.058	m.25	0.145	0.41	0.296	0.60	0.557	0.84	1.077	1.14	2.00	1800.0
1900.0		0.029	0.17	0.060	0.25	0.149	0.42	0.305	0.61	0.574	0.84	1.109	1.15		1900.0
2000.0		0.030	0.17	0.062	0.26	0.154	0.42	0.314	0.61	0.591	0.85	1.141	1.16		2000.0

Table 4.29 Flow of water at 75°C in polybutylene (PB) pipes[21]

POLYBUTYLENE (PB)		
WATER AT 75°C		
BS 5556		

q_m = mass flow rate — kg.s⁻¹ → q_m = mass flow rate, kg.s^{-1}

q_m = mass flow rate — kg.s^{-1}
c = velocity — m.s^{-1}
$\Delta p/l$ = pressure drop per unit length — Pa.m^{-1}
l_e = equivalent length of a component for $\zeta = 1$ — m

$\Delta p/l$	c	10 mm		12 mm		16 mm		20 mm		25 mm		32 mm		c	$\Delta p/l$
		q_m	l_e	q_m	l_e	q_m	l_e	q_m	l_e	q_m	l_e	q_m	l_e		
0.1								0.001	0.14	0.002	0.21	0.004	0.33		0.1
0.2						0.001	0.10	0.002	0.17	0.003	0.25	0.007	0.39		0.2
0.3						0.001	0.12	0.002	0.19	0.004	0.28	0.008	0.43		0.3
0.4						0.001	0.13	0.003	0.20	0.005	0.30	0.010	0.45		0.4
0.5				0.001	0.07	0.001	0.13	0.003	0.21	0.006	0.31	0.011	0.47		0.5
0.6				0.001	0.08	0.002	0.14	0.003	0.22	0.006	0.33	0.013	0.49		0.6
0.7				0.001	0.08	0.002	0.15	0.004	0.23	0.007	0.34	0.014	0.51		0.7
0.8				0.001	0.08	0.002	0.15	0.004	0.24	0.007	0.35	0.015	0.52		0.8
0.9				0.001	0.09	0.002	0.16	0.004	0.24	0.008	0.36	0.016	0.53		0.9
1.0				0.001	0.09	0.002	0.16	0.004	0.25	0.009	0.36	0.017	0.54		1.0
1.5				0.001	0.10	0.003	0.18	0.006	0.27	0.011	0.40	0.022	0.58		1.5
2.0		0.001	0.07	0.001	0.11	0.003	0.19	0.007	0.29	0.013	0.42	0.026	0.61	0.05	2.0
2.5		0.001	0.07	0.001	0.11	0.004	0.20	0.008	0.30	0.015	0.44	0.030	0.64		2.5
3.0		0.001	0.07	0.002	0.12	0.004	0.20	0.009	0.31	0.016	0.45	0.033	0.66		3.0
3.5		0.001	0.08	0.002	0.12	0.005	0.21	0.009	0.32	0.018	0.46	0.036	0.68		3.5
4.0		0.001	0.08	0.002	0.12	0.005	0.22	0.010	0.33	0.020	0.48	0.039	0.69		4.0
4.5		0.001	0.08	0.002	0.13	0.005	0.22	0.011	0.34	0.021	0.49	0.042	0.70		4.5
5.0		0.001	0.08	0.002	0.13	0.006	0.23	0.012	0.35	0.022	0.49	0.044	0.72		5.0
5.5		0.001	0.09	0.002	0.13	0.006	0.23	0.012	0.35	0.024	0.50	0.047	0.73		5.5
6.0		0.001	0.09	0.002	0.14	0.006	0.24	0.013	0.36	0.025	0.51	0.049	0.74		6.0
6.5		0.001	0.09	0.003	0.14	0.007	0.24	0.014	0.36	0.026	0.52	0.052	0.75		6.5
7.0		0.001	0.09	0.003	0.14	0.007	0.24	0.014	0.37	0.027	0.52	0.054	0.76		7.0
7.5		0.001	0.09	0.003	0.14	0.007	0.25	0.015	0.37	0.028	0.53	0.056	0.76		7.5
8.0		0.001	0.09	0.003	0.15	0.007	0.25	0.015	0.38	0.029	0.54	0.058	0.77		8.0
8.5		0.001	0.10	0.003	0.15	0.008	0.25	0.016	0.38	0.030	0.54	0.060	0.78		8.5
9.0		0.002	0.10	0.003	0.15	0.008	0.26	0.017	0.38	0.031	0.55	0.062	0.79		9.0
9.5		0.002	0.10	0.003	0.15	0.008	0.26	0.017	0.39	0.032	0.55	0.064	0.79		9.5
10.0	0.05	0.002	0.10	0.003	0.15	0.008	0.26	0.018	0.39	0.033	0.56	0.066	0.80		10.0
12.5		0.002	0.10	0.004	0.16	0.010	0.27	0.020	0.41	0.038	0.58	0.075	0.82	0.15	12.5
15.0		0.002	0.11	0.004	0.17	0.011	0.28	0.022	0.42	0.042	0.59	0.084	0.85		15.0
17.5		0.002	0.11	0.005	0.17	0.012	0.29	0.024	0.43	0.046	0.61	0.091	0.87		17.5
20.0		0.002	0.12	0.005	0.18	0.013	0.30	0.026	0.44	0.050	0.62	0.099	0.88		20.0
22.5		0.003	0.12	0.005	0.18	0.014	0.30	0.028	0.45	0.053	0.63	0.105	0.90		22.5
25.0		0.003	0.12	0.006	0.18	0.015	0.31	0.030	0.46	0.057	0.64	0.112	0.91		25.0
27.5		0.003	0.12	0.006	0.19	0.015	0.31	0.032	0.46	0.060	0.65	0.118	0.92		27.5
30.0		0.003	0.13	0.006	0.19	0.016	0.32	0.033	0.47	0.063	0.66	0.124	0.93		30.0
32.5		0.003	0.13	0.007	0.19	0.017	0.32	0.035	0.48	0.066	0.67	0.130	0.94		32.5
35.0		0.003	0.13	0.007	0.20	0.018	0.33	0.036	0.48	0.069	0.68	0.135	0.95		35.0
37.5		0.004	0.13	0.007	0.20	0.018	0.33	0.038	0.49	0.072	0.68	0.141	0.96		37.5
40.0		0.004	0.13	0.008	0.20	0.019	0.33	0.039	0.49	0.074	0.69	0.146	0.97		40.0
42.5		0.004	0.14	0.008	0.20	0.020	0.34	0.041	0.50	0.077	0.70	0.151	0.98		42.5
45.0		0.004	0.14	0.008	0.20	0.021	0.34	0.042	0.50	0.080	0.70	0.156	0.98	3.00	45.0
47.5		0.004	0.14	0.008	0.21	0.021	0.34	0.044	0.51	0.082	0.71	0.161	0.99		47.5
50.0		0.004	0.14	0.009	0.21	0.022	0.35	0.045	0.51	0.085	0.71	0.166	1.00		50.0
52.5		0.004	0.14	0.009	0.21	0.022	0.35	0.046	0.51	0.087	0.72	0.170	1.00		52.5
55.0		0.004	0.14	0.009	0.21	0.023	0.35	0.047	0.52	0.089	0.72	0.175	1.01		55.0
57.5		0.005	0.14	0.009	0.21	0.024	0.35	0.049	0.52	0.092	0.73	0.179	1.02		57.5
60.0		0.005	0.14	0.010	0.22	0.024	0.36	0.050	0.52	0.094	0.73	0.184	1.02		60.0
62.5		0.005	0.15	0.010	0.22	0.025	0.36	0.051	0.53	0.096	0.74	0.188	1.03		62.5
65.0		0.005	0.15	0.010	0.22	0.025	0.36	0.052	0.53	0.098	0.74	0.192	1.03		65.0
67.5		0.005	0.15	0.010	0.22	0.026	0.36	0.053	0.53	0.100	0.75	0.196	1.04		67.5
70.0	0.15	0.005	0.15	0.011	0.22	0.027	0.36	0.054	0.54	0.103	0.75	0.200	1.04		70.0
72.5		0.005	0.15	0.011	0.22	0.027	0.37	0.056	0.54	0.105	0.75	0.204	1.04		72.5
75.0		0.005	0.15	0.011	0.22	0.028	0.37	0.057	0.54	0.107	0.76	0.208	1.05		75.0
77.5		0.005	0.15	0.011	0.22	0.028	0.37	0.058	0.54	0.109	0.76	0.212	1.05		77.5
80.0		0.006	0.15	0.011	0.23	0.029	0.37	0.059	0.55	0.111	0.76	0.216	1.06		80.0
82.5		0.006	0.15	0.012	0.23	0.029	0.37	0.060	0.55	0.113	0.77	0.219	1.06		82.5
85.0		0.006	0.15	0.012	0.23	0.030	0.38	0.061	0.55	0.115	0.77	0.223	1.06		85.0
87.5		0.006	0.15	0.012	0.23	0.030	0.38	0.062	0.55	0.116	0.77	0.227	1.07		87.5
90.0		0.006	0.16	0.012	0.23	0.031	0.38	0.063	0.56	0.118	0.78	0.230	1.07		90.0

Table 4.29 Flow of water at 75°C in polybutylene (PB) pipes[21] — *continued*

q_m	= mass flow rate	kg.s^{-1}	**POLYBUTYLENE (PB)**
c	= velocity	m.s^{-1}	**WATER AT 75°C**
$\Delta p/l$	= pressure drop per unit length	Pa.m^{-1}	
l_e	= equivalent length of a component for $\zeta = 1$	m	**BS 5556**

$\Delta p/l$	c	10 mm q_m	l_e	12 mm q_m	l_e	16 mm q_m	l_e	20 mm q_m	l_e	25 mm q_m	l_e	32 mm q_m	l_e	c	$\Delta p/l$
92.5		0.006	0.16	0.012	0.23	0.031	0.38	0.064	0.56	0.120	0.78	0.234	1.08		92.5
95.0		0.006	0.16	0.013	0.23	0.032	0.38	0.065	0.56	0.122	0.78	0.237	1.08		95.0
97.5		0.006	0.16	0.013	0.23	0.032	0.38	0.066	0.56	0.124	0.79	0.241	1.08		97.5
100.0		0.006	0.16	0.013	0.23	0.033	0.39	0.067	0.57	0.126	0.79	0.244	1.09		100.0
120.0		0.007	0.16	0.014	0.24	0.036	0.40	0.074	0.58	0.139	0.81	0.271	1.11	0.50	120.0
140.0		0.008	0.17	0.016	0.25	0.040	0.41	0.081	0.59	0.152	0.83	0.295	1.13		140.0
160.0		0.008	0.17	0.017	0.25	0.043	0.41	0.087	0.60	0.164	0.84	0.318	1.15		160.0
180.0		0.009	0.18	0.018	0.26	0.046	0.42	0.093	0.62	0.176	0.85	0.339	1.16		180.0
200.0		0.010	0.18	0.019	0.26	0.049	0.43	0.099	0.62	0.186	0.87	0.359	1.17		200.0
220.0		0.010	0.18	0.021	0.27	0.051	0.43	0.105	0.63	0.197	0.88	0.379	1.19		220.0
240.0	0.30	0.011	0.18	0.022	0.27	0.054	0.44	0.110	0.64	0.207	0.89	0.397	1.20		240.0
260.0		0.011	0.19	0.023	0.27	0.056	0.45	0.115	0.65	0.216	0.90	0.415	1.21		260.0
280.0		0.012	0.19	0.024	0.28	0.059	0.45	0.120	0.65	0.225	0.91	0.433	1.22		280.0
300.0		0.012	0.19	0.025	0.28	0.061	0.45	0.125	0.66	0.234	0.91	0.450	1.22		300.0
320.0		0.013	0.19	0.026	0.28	0.064	0.46	0.130	0.67	0.243	0.92	0.466	1.23		320.0
340.0		0.013	0.20	0.026	0.29	0.066	0.46	0.134	0.67	0.251	0.93	0.482	1.24		340.0
360.0		0.014	0.20	0.027	0.29	0.068	0.47	0.139	0.68	0.260	0.94	0.497	1.25		360.0
380.0		0.014	0.20	0.028	0.29	0.070	0.47	0.143	0.68	0.268	0.94	0.512	1.25		380.0
400.0		0.014	0.20	0.029	0.29	0.072	0.47	0.147	0.69	0.276	0.95	0.527	1.26	1.00	400.0
420.0		0.015	0.20	0.030	0.29	0.074	0.48	0.151	0.69	0.283	0.95	0.541	1.27		420.0
440.0		0.015	0.20	0.031	0.30	0.076	0.48	0.155	0.69	0.291	0.96	0.555	1.27		440.0
460.0		0.016	0.21	0.032	0.30	0.078	0.48	0.159	0.70	0.298	0.96	0.568	1.28		460.0
480.0		0.016	0.21	0.032	0.30	0.080	0.49	0.163	0.70	0.305	0.97	0.582	1.28		480.0
500.0		0.016	0.21	0.033	0.30	0.082	0.49	0.167	0.71	0.313	0.97	0.595	1.29		500.0
520.0		0.017	0.21	0.034	0.30	0.084	0.49	0.171	0.71	0.319	0.98	0.608	1.29		520.0
540.0		0.017	0.21	0.035	0.31	0.086	0.49	0.174	0.71	0.326	0.98	0.621	1.30		540.0
560.0		0.017	0.21	0.035	0.31	0.087	0.50	0.178	0.72	0.333	0.99	0.633	1.30		560.0
580.0	0.50	0.018	0.21	0.036	0.31	0.089	0.50	0.181	0.72	0.340	0.99	0.645	1.31		580.0
600.0		0.018	0.21	0.037	0.31	0.091	0.50	0.185	0.72	0.346	1.00	0.657	1.31		600.0
620.0		0.019	0.21	0.037	0.31	0.093	0.50	0.188	0.73	0.353	1.00	0.669	1.31		620.0
640.0		0.019	0.22	0.038	0.31	0.094	0.51	0.192	0.73	0.359	1.01	0.681	1.32		640.0
660.0		0.019	0.22	0.039	0.32	0.096	0.51	0.195	0.73	0.365	1.01	0.692	1.32		660.0
680.0		0.020	0.22	0.039	0.32	0.098	0.51	0.199	0.73	0.371	1.01	0.704	1.32		680.0
700.0		0.020	0.22	0.040	0.32	0.099	0.51	0.202	0.74	0.378	1.02	0.715	1.33		700.0
720.0		0.020	0.22	0.041	0.32	0.101	0.51	0.205	0.74	0.384	1.02	0.726	1.33		720.0
740.0		0.021	0.22	0.041	0.32	0.102	0.52	0.208	0.74	0.390	1.02	0.737	1.33		740.0
760.0		0.021	0.22	0.042	0.32	0.104	0.52	0.211	0.75	0.395	1.03	0.748	1.34		760.0
780.0		0.021	0.22	0.043	0.32	0.106	0.52	0.214	0.75	0.401	1.03	0.758	1.34		780.0
800.0		0.021	0.22	0.043	0.32	0.107	0.52	0.218	0.75	0.407	1.03	0.769	1.34		800.0
820.0		0.022	0.22	0.044	0.33	0.109	0.52	0.221	0.75	0.413	1.04	0.779	1.35	1.50	820.0
840.0		0.022	0.23	0.044	0.33	0.110	0.52	0.224	0.75	0.418	1.04	0.789	1.35		840.0
860.0		0.022	0.23	0.045	0.33	0.112	0.53	0.227	0.76	0.424	1.04	0.79	1.35		860.0
880.0		0.023	0.23	0.046	0.33	0.113	0.53	0.230	0.76	0.429	1.04	0.810	1.35		880.0
900.0		0.023	0.23	0.046	0.33	0.114	0.53	0.232	0.76	0.435	1.05	0.819	1.36		900.0
920.0		0.023	0.23	0.047	0.33	0.116	0.53	0.235	0.76	0.440	1.05	0.829	1.36		920.0
940.0		0.024	0.23	0.047	0.33	0.117	0.53	0.238	0.77	0.445	1.05	0.839	1.36		940.0
960.0		0.024	0.23	0.048	0.33	0.119	0.53	0.241	0.77	0.451	1.06	0.849	1.36		960.0
980.0		0.024	0.23	0.049	0.33	0.120	0.53	0.244	0.77	0.456	1.06	0.858	1.37		980.0
1000.0		0.024	0.23	0.049	0.33	0.122	0.54	0.247	0.77	0.461	1.06	0.868	1.37		1000.0
1100.0		0.026	0.23	0.052	0.34	0.128	0.54	0.260	0.78	0.486	1.07	0.913	1.38		1100.0
1200.0		0.027	0.24	0.055	0.34	0.135	0.55	0.273	0.79	0.511	1.08	0.957	1.39		1200.0
1300.0		0.028	0.24	0.057	0.35	0.141	0.55	0.286	0.80	0.534	1.09	1.000	1.40		1300.0
1400.0		0.030	0.24	0.060	0.35	0.147	0.56	0.298	0.80	0.557	1.10	1.040	1.41	2.00	1400.0
1500.0		0.031	0.25	0.062	0.35	0.153	0.57	0.310	0.81	0.578	1.11	1.080	1.41		1500.0
1600.0		0.032	0.25	0.064	0.36	0.158	0.57	0.321	0.82	0.600	1.12	1.118	1.42		1600.0
1700.0		0.033	0.25	0.066	0.36	0.164	0.57	0.332	0.82	0.620	1.13	1.155	1.43		1700.0
1800.0		0.034	0.25	0.069	0.36	0.169	0.58	0.343	0.83	0.640	1.14	1.191	1.43		1800.0
1900.0		0.035	0.25	0.071	0.37	0.175	0.58	0.354	0.83	0.660	1.14	1.226	1.44		1900.0
2000.0		0.036	0.26	0.073	0.37	0.180	0.59	0.364	0.84	0.679	1.15	1.260	1.44		2000.0

Table 4.30 Flow of water at 10°C in polyethylene (PE-X) pipes[21]

q_m	= mass flow rate	$kg.s^{-1}$
c	= velocity	$m.s^{-1}$
$\Delta p/l$	= pressure drop per unit length	$Pa.m^{-1}$
l_e	= equivalent length of a component for $\zeta = 1$	m

POLYETHYLENE (PE-X)
WATER AT 10°C
BS 5556

$\Delta p/l$	c	10 mm		12 mm		16 mm		20 mm		25 mm		32 mm		c	$\Delta p/l$
		q_m	l_e	q_m	l_e	q_m	l_e	q_m	l_e	q_m	l_e	q_m	l_e		
0.1								0.001	0.06	0.001	0.10	0.003	0.16		0.1
0.2						0.001	0.05	0.001	0.08	0.002	0.12	0.005	0.21		0.2
0.3						0.001	0.05	0.001	0.09	0.003	0.14	0.006	0.23		0.3
0.4						0.001	0.06	0.002	0.10	0.004	0.16	0.008	0.25		0.4
0.5						0.001	0.07	0.002	0.11	0.004	0.17	0.009	0.27		0.5
0.6						0.001	0.07	0.002	0.11	0.005	0.18	0.010	0.28		0.6
0.7						0.001	0.08	0.003	0.12	0.005	0.19	0.011	0.29		0.7
0.8						0.001	0.08	0.003	0.12	0.006	0.19	0.012	0.30		0.8
0.9						0.002	0.08	0.003	0.13	0.006	0.20	0.013	0.31		0.9
1.0				0.001	0.04	0.002	0.09	0.003	0.13	0.006	0.20	0.013	0.32		1.0
1.5				0.001	0.05	0.002	0.10	0.004	0.15	0.008	0.23	0.017	0.35		1.5
2.0				0.001	0.05	0.003	0.11	0.005	0.16	0.010	0.25	0.021	0.38		2.0
2.5				0.001	0.06	0.003	0.11	0.006	0.17	0.012	0.26	0.024	0.40		2.5
3.0		0.001	0.03	0.001	0.06	0.003	0.12	0.007	0.18	0.013	0.27	0.026	0.41		3.0
3.5		0.001	0.04	0.001	0.06	0.004	0.13	0.007	0.19	0.014	0.28	0.029	0.43	0.05	3.5
4.0		0.001	0.04	0.001	0.06	0.004	0.13	0.008	0.19	0.015	0.29	0.031	0.44		4.0
4.5		0.001	0.04	0.001	0.07	0.004	0.13	0.009	0.20	0.017	0.30	0.034	0.45		4.5
5.0		0.001	0.04	0.002	0.07	0.005	0.14	0.009	0.21	0.018	0.30	0.036	0.46		5.0
5.5		0.001	0.04	0.002	0.07	0.005	0.14	0.010	0.21	0.019	0.31	0.038	0.47		5.5
6.0		0.001	0.04	0.002	0.07	0.005	0.15	0.010	0.21	0.020	0.32	0.040	0.48		6.0
6.5		0.001	0.05	0.002	0.07	0.006	0.15	0.011	0.22	0.021	0.32	0.042	0.48		6.5
7.0		0.001	0.05	0.002	0.08	0.006	0.15	0.011	0.22	0.022	0.33	0.044	0.49		7.0
7.5		0.001	0.05	0.002	0.08	0.006	0.15	0.012	0.23	0.023	0.33	0.046	0.50		7.5
8.0		0.001	0.05	0.002	0.08	0.006	0.16	0.012	0.23	0.024	0.34	0.048	0.50		8.0
8.5		0.001	0.05	0.002	0.08	0.007	0.16	0.013	0.23	0.024	0.34	0.049	0.51		8.5
9.0		0.001	0.05	0.002	0.08	0.007	0.16	0.013	0.24	0.025	0.35	0.051	0.52		9.0
9.5		0.001	0.05	0.002	0.08	0.007	0.16	0.014	0.24	0.026	0.35	0.053	0.52		9.5
10.0		0.001	0.05	0.003	0.08	0.007	0.16	0.014	0.24	0.027	0.35	0.055	0.53		10.0
12.5		0.001	0.06	0.003	0.09	0.008	0.17	0.016	0.25	0.031	0.37	0.062	0.55		12.5
15.0		0.002	0.06	0.003	0.09	0.009	0.18	0.018	0.26	0.034	0.38	0.069	0.57		15.0
17.5	0.05	0.002	0.06	0.004	0.10	0.010	0.19	0.020	0.27	0.038	0.40	0.076	0.58		17.5
20.0		0.002	0.06	0.004	0.10	0.011	0.19	0.021	0.28	0.041	0.41	0.082	0.60	0.15	20.0
22.5		0.002	0.07	0.004	0.11	0.012	0.20	0.023	0.29	0.044	0.42	0.088	0.61		22.5
25.0		0.002	0.07	0.004	0.11	0.013	0.20	0.024	0.29	0.047	0.42	0.094	0.62		25.0
27.5		0.002	0.07	0.005	0.11	0.014	0.21	0.026	0.30	0.049	0.43	0.099	0.63		27.5
30.0		0.002	0.07	0.005	0.11	0.015	0.21	0.027	0.30	0.052	0.44	0.104	0.64		30.0
32.5		0.003	0.07	0.005	0.11	0.015	0.22	0.028	0.31	0.055	0.45	0.109	0.65		32.5
35.0		0.003	0.07	0.006	0.12	0.016	0.22	0.030	0.31	0.057	0.45	0.114	0.66		35.0
37.5		0.003	0.08	0.006	0.12	0.017	0.22	0.031	0.32	0.059	0.46	0.119	0.67		37.5
40.0		0.003	0.08	0.006	0.12	0.017	0.22	0.032	0.32	0.062	0.46	0.123	0.67		40.0
42.5		0.003	0.08	0.006	0.12	0.018	0.23	0.033	0.33	0.064	0.47	0.128	0.68		42.5
45.0		0.003	0.08	0.006	0.12	0.019	0.23	0.035	0.33	0.066	0.47	0.132	0.69		45.0
47.5		0.003	0.08	0.007	0.13	0.019	0.23	0.036	0.33	0.068	0.48	0.136	0.69		47.5
50.0		0.003	0.08	0.007	0.13	0.020	0.24	0.037	0.34	0.070	0.48	0.140	0.70		50.0
52.5		0.003	0.08	0.007	0.13	0.020	0.24	0.038	0.34	0.072	0.49	0.144	0.71		52.5
55.0		0.003	0.08	0.007	0.13	0.021	0.24	0.039	0.34	0.074	0.49	0.148	0.71		55.0
57.5		0.004	0.08	0.007	0.13	0.021	0.24	0.040	0.34	0.076	0.49	0.152	0.72		57.5
60.0		0.004	0.09	0.008	0.13	0.022	0.24	0.041	0.35	0.078	0.50	0.156	0.72		60.0
62.5		0.004	0.09	0.008	0.13	0.023	0.25	0.042	0.35	0.080	0.50	0.160	0.73	0.30	62.5
65.0		0.004	0.09	0.008	0.13	0.023	0.25	0.043	0.35	0.082	0.50	0.164	0.73		65.0
67.5		0.004	0.09	0.008	0.14	0.024	0.25	0.044	0.35	0.084	0.51	0.167	0.73		67.5
70.0		0.004	0.09	0.008	0.14	0.024	0.25	0.045	0.36	0.086	0.51	0.171	0.74		70.0
72.5		0.004	0.09	0.009	0.14	0.025	0.25	0.046	0.36	0.088	0.51	0.174	0.74		72.5
75.0		0.004	0.09	0.009	0.14	0.025	0.26	0.047	0.36	0.089	0.52	0.178	0.75		75.0
77.5		0.004	0.09	0.009	0.14	0.026	0.26	0.048	0.36	0.091	0.52	0.181	0.75		77.5
80.0		0.004	0.09	0.009	0.14	0.026	0.26	0.049	0.37	0.093	0.52	0.184	0.75		80.0
82.5		0.004	0.09	0.009	0.14	0.027	0.26	0.050	0.37	0.094	0.53	0.188	0.76		82.5
85.0		0.005	0.09	0.009	0.14	0.027	0.26	0.050	0.37	0.096	0.53	0.191	0.76		85.0
87.5		0.005	0.09	0.010	0.14	0.028	0.26	0.051	0.37	0.098	0.53	0.194	0.76		87.5
90.0		0.005	0.09	0.010	0.14	0.028	0.26	0.052	0.37	0.099	0.53	0.197	0.77		90.0

Table 4.30 Flow of water at 10°C in polyethylene (PE-X) pipes[21] — *continued*

q_m	= mass flow rate	kg.s^{-1}
c	= velocity	m.s^{-1}
$\Delta p/l$	= pressure drop per unit length	Pa.m^{-1}
l_e	= equivalent length of a component	
	for $\zeta = 1$	m

POLYETHYLENE (PE-X)
WATER AT 10°C
BS 5556

$\Delta p/l$	c	10 mm q_m	l_e	12 mm q_m	l_e	16 mm q_m	l_e	20 mm q_m	l_e	25 mm q_m	l_e	32 mm q_m	l_e	c	$\Delta p/l$
92.5		0.005	0.09	0.010	0.15	0.029	0.27	0.053	0.38	0.101	0.54	0.201	0.77		92.5
95.0		0.005	0.10	0.010	0.15	0.029	0.27	0.054	0.38	0.103	0.54	0.204	0.77		95.0
97.5	0.15	0.005	0.10	0.010	0.15	0.029	0.27	0.055	0.38	0.104	0.54	0.207	0.78		97.5
100.0		0.005	0.10	0.010	0.15	0.030	0.27	0.056	0.38	0.106	0.54	0.210	0.78		100.0
120.0		0.006	0.10	0.012	0.15	0.033	0.28	0.062	0.39	0.117	0.56	0.233	0.80		120.0
140.0		0.006	0.10	0.013	0.16	0.036	0.29	0.068	0.40	0.128	0.57	0.255	0.82		140.0
160.0		0.007	0.11	0.014	0.16	0.039	0.29	0.073	0.41	0.139	0.59	0.275	0.84	0.50	160.0
180.0		0.007	0.11	0.015	0.17	0.042	0.30	0.078	0.42	0.149	0.60	0.294	0.85		180.0
200.0		0.008	0.11	0.016	0.17	0.045	0.31	0.083	0.43	0.158	0.61	0.312	0.86		200.0
220.0		0.008	0.11	0.017	0.17	0.048	0.31	0.088	0.44	0.167	0.62	0.330	0.88		220.0
240.0		0.009	0.12	0.018	0.18	0.050	0.32	0.093	0.44	0.176	0.63	0.346	0.89		240.0
260.0		0.009	0.12	0.019	0.18	0.053	0.32	0.097	0.45	0.184	0.63	0.363	0.90		260.0
280.0		0.009	0.12	0.019	0.18	0.055	0.32	0.101	0.45	0.192	0.64	0.378	0.91		280.0
300.0		0.010	0.12	0.020	0.18	0.057	0.33	0.106	0.46	0.200	0.65	0.393	0.91		300.0
320.0		0.010	0.12	0.021	0.19	0.059	0.33	0.110	0.46	0.207	0.65	0.408	0.92		320.0
340.0		0.011	0.13	0.022	0.19	0.061	0.34	0.114	0.47	0.215	0.66	0.422	0.93		340.0
360.0	0.30	0.011	0.13	0.023	0.19	0.064	0.34	0.117	0.47	0.222	0.67	0.436	0.94		360.0
380.0		0.011	0.13	0.023	0.19	0.066	0.34	0.121	0.48	0.229	0.67	0.450	0.94		380.0
400.0		0.012	0.13	0.024	0.19	0.068	0.35	0.125	0.48	0.236	0.68	0.463	0.95		400.0
420.0		0.012	0.13	0.025	0.20	0.070	0.35	0.128	0.48	0.242	0.68	0.476	0.96		420.0
440.0		0.012	0.13	0.025	0.20	0.071	0.35	0.132	0.49	0.249	0.69	0.489	0.96		440.0
460.0		0.013	0.13	0.026	0.20	0.073	0.35	0.135	0.49	0.255	0.69	0.501	0.97		460.0
480.0		0.013	0.13	0.027	0.20	0.075	0.36	0.139	0.49	0.262	0.69	0.513	0.97		480.0
500.0		0.013	0.14	0.027	0.20	0.077	0.36	0.142	0.50	0.268	0.70	0.525	0.98		500.0
520.0		0.014	0.14	0.028	0.20	0.079	0.36	0.145	0.50	0.274	0.70	0.537	0.98	1.00	520.0
540.0		0.014	0.14	0.029	0.21	0.081	0.36	0.148	0.50	0.280	0.71	0.549	0.99		540.0
560.0		0.014	0.14	0.029	0.21	0.082	0.36	0.152	0.51	0.286	0.71	0.560	0.99		560.0
580.0		0.015	0.14	0.030	0.21	0.084	0.37	0.155	0.51	0.292	0.71	0.571	1.00		580.0
600.0		0.015	0.14	0.031	0.21	0.086	0.37	0.158	0.51	0.298	0.72	0.582	1.00		600.0
620.0		0.015	0.14	0.031	0.21	0.087	0.37	0.161	0.51	0.303	0.72	0.593	1.01		620.0
640.0		0.015	0.14	0.032	0.21	0.089	0.37	0.164	0.52	0.309	0.72	0.604	1.01		640.0
660.0		0.016	0.14	0.032	0.21	0.090	0.37	0.167	0.52	0.314	0.73	0.614	1.01		660.0
680.0		0.016	0.14	0.033	0.21	0.092	0.38	0.169	0.52	0.320	0.73	0.625	1.02		680.0
700.0		0.016	0.14	0.033	0.22	0.094	0.38	0.172	0.52	0.325	0.73	0.635	1.02		700.0
720.0		0.017	0.15	0.034	0.22	0.095	0.38	0.175	0.53	0.330	0.74	0.645	1.02		720.0
740.0		0.017	0.15	0.035	0.22	0.097	0.38	0.178	0.53	0.335	0.74	0.655	1.03		740.0
760.0		0.017	0.15	0.035	0.22	0.098	0.38	0.181	0.53	0.341	0.74	0.665	1.03		760.0
780.0		0.017	0.15	0.036	0.22	0.100	0.38	0.183	0.53	0.346	0.75	0.675	1.03		780.0
800.0		0.018	0.15	0.036	0.22	0.101	0.39	0.186	0.54	0.351	0.75	0.684	1.04		800.0
820.0		0.018	0.15	0.037	0.22	0.103	0.39	0.189	0.54	0.356	0.75	0.694	1.04		820.0
840.0	0.50	0.018	0.15	0.037	0.22	0.104	0.39	0.191	0.54	0.361	0.75	0.703	1.04		840.0
860.0		0.018	0.15	0.038	0.22	0.105	0.39	0.194	0.54	0.366	0.76	0.713	1.05		860.0
880.0		0.019	0.15	0.038	0.22	0.107	0.39	0.197	0.54	0.370	0.76	0.722	1.05		880.0
900.0		0.019	0.15	0.039	0.22	0.108	0.39	0.199	0.54	0.375	0.76	0.731	1.05		900.0
920.0		0.019	0.15	0.039	0.23	0.110	0.39	0.202	0.55	0.380	0.76	0.740	1.06		920.0
940.0		0.019	0.15	0.040	0.23	0.111	0.40	0.204	0.55	0.385	0.77	0.749	1.06		940.0
960.0		0.020	0.15	0.040	0.23	0.112	0.40	0.207	0.55	0.389	0.77	0.758	1.06		960.0
980.0		0.020	0.15	0.041	0.23	0.114	0.40	0.209	0.55	0.394	0.77	0.767	1.06		980.0
1000.0		0.020	0.15	0.041	0.23	0.115	0.40	0.212	0.55	0.398	0.77	0.776	1.07		1000.0
1100.0		0.021	0.16	0.044	0.23	0.122	0.41	0.223	0.56	0.421	0.78	0.818	1.08	1.50	1100.0
1200.0		0.022	0.16	0.046	0.24	0.128	0.41	0.235	0.57	0.442	0.79	0.859	1.09		1200.0
1300.0		0.024	0.16	0.048	0.24	0.134	0.42	0.246	0.57	0.463	0.80	0.898	1.10		1300.0
1400.0		0.025	0.16	0.050	0.24	0.140	0.42	0.257	0.58	0.483	0.81	0.936	1.11		1400.0
1500.0		0.026	0.17	0.052	0.25	0.145	0.43	0.267	0.59	0.502	0.82	0.973	1.12		1500.0
1600.0		0.027	0.17	0.054	0.25	0.151	0.43	0.277	0.59	0.521	0.82	1.008	1.13		1600.0
1700.0		0.028	0.17	0.056	0.25	0.156	0.43	0.287	0.60	0.539	0.83	1.043	1.13		1700.0
1800.0		0.029	0.17	0.058	0.25	0.161	0.44	0.296	0.60	0.557	0.84	1.077	1.14	2.00	1800.0
1900.0		0.029	0.17	0.060	0.25	0.167	0.44	0.305	0.61	0.574	0.84	1.109	1.15		1900.0
2000.0		0.030	0.17	0.062	0.26	0.172	0.44	0.314	0.61	0.591	0.85	1.141	1.16		2000.0

Table 4.31 Flow of water at 75°C in polyethylene (PE-X) pipes[21]

q_m = mass flow rate — kg.s^{-1}
c = velocity — m.s^{-1}
$\Delta p/l$ = pressure drop per unit length — Pa.m^{-1}
l_e = equivalent length of a component
for $\zeta = 1$ — m

POLYETHYLENE (PE-X)
WATER AT 75°C
BS 5556

$\Delta p/l$	c	10 mm		12 mm		16 mm		20 mm		25 mm		32 mm		c	$\Delta p/l$
		q_m	l_e	q_m	l_e	q_m	l_e	q_m	l_e	q_m	l_e	q_m	l_e		
0.1						0.001	0.09	0.001	0.14	0.002	0.21	0.004	0.33		0.1
0.2						0.001	0.11	0.002	0.17	0.003	0.25	0.007	0.39		0.2
0.3						0.001	0.13	0.002	0.19	0.004	0.28	0.008	0.43		0.3
0.4						0.001	0.14	0.003	0.20	0.005	0.30	0.010	0.45		0.4
0.5				0.001	0.07	0.002	0.14	0.003	0.21	0.006	0.31	0.011	0.47		0.5
0.6				0.001	0.08	0.002	0.15	0.003	0.22	0.006	0.33	0.013	0.49		0.6
0.7				0.001	0.08	0.002	0.16	0.004	0.23	0.007	0.34	0.014	0.51		0.7
0.8				0.001	0.08	0.002	0.16	0.004	0.24	0.007	0.35	0.015	0.52		0.8
0.9				0.001	0.09	0.002	0.17	0.004	0.24	0.008	0.36	0.016	0.53		0.9
1.0				0.001	0.09	0.002	0.17	0.004	0.25	0.009	0.36	0.017	0.54		1.0
1.5				0.001	0.10	0.003	0.19	0.006	0.27	0.011	0.40	0.022	0.58		1.5
2.0		0.001	0.07	0.001	0.11	0.004	0.20	0.007	0.29	0.013	0.42	0.026	0.61	0.05	2.0
2.5		0.001	0.07	0.001	0.11	0.004	0.21	0.008	0.30	0.015	0.44	0.030	0.64		2.5
3.0		0.001	0.07	0.002	0.12	0.005	0.22	0.009	0.31	0.016	0.45	0.033	0.66		3.0
3.5		0.001	0.08	0.002	0.12	0.005	0.23	0.009	0.32	0.018	0.46	0.036	0.68		3.5
4.0		0.001	0.08	0.002	0.12	0.005	0.23	0.010	0.33	0.020	0.48	0.039	0.69		4.0
4.5		0.001	0.08	0.002	0.13	0.006	0.24	0.011	0.34	0.021	0.49	0.042	0.70		4.5
5.0		0.001	0.08	0.002	0.13	0.006	0.24	0.012	0.35	0.022	0.49	0.044	0.72		5.0
5.5		0.001	0.09	0.002	0.13	0.007	0.25	0.012	0.35	0.024	0.50	0.047	0.73		5.5
6.0		0.001	0.09	0.002	0.14	0.007	0.25	0.013	0.36	0.025	0.51	0.049	0.74		6.0
6.5		0.001	0.09	0.003	0.14	0.007	0.26	0.014	0.36	0.026	0.52	0.052	0.75		6.5
7.0		0.001	0.09	0.003	0.14	0.008	0.26	0.014	0.37	0.027	0.52	0.054	0.76		7.0
7.5		0.001	0.09	0.003	0.14	0.008	0.26	0.015	0.37	0.028	0.53	0.056	0.76		7.5
8.0		0.001	0.09	0.003	0.15	0.008	0.27	0.015	0.38	0.029	0.54	0.058	0.77		8.0
8.5		0.001	0.10	0.003	0.15	0.009	0.27	0.016	0.38	0.030	0.54	0.060	0.78		8.5
9.0		0.002	0.10	0.003	0.15	0.009	0.27	0.017	0.38	0.031	0.55	0.062	0.79		9.0
9.5		0.002	0.10	0.003	0.15	0.009	0.27	0.017	0.39	0.032	0.55	0.064	0.79		9.5
10.0	0.05	0.002	0.10	0.003	0.15	0.009	0.28	0.018	0.39	0.033	0.56	0.066	0.80		10.0
12.5		0.002	0.10	0.004	0.16	0.011	0.29	0.020	0.41	0.038	0.58	0.075	0.82	0.15	12.5
15.0		0.002	0.11	0.004	0.17	0.012	0.30	0.022	0.42	0.042	0.59	0.084	0.85		15.0
17.5		0.002	0.11	0.005	0.17	0.013	0.31	0.024	0.43	0.046	0.61	0.091	0.87		17.5
20.0		0.002	0.12	0.005	0.18	0.014	0.31	0.026	0.44	0.050	0.62	0.099	0.88		20.0
22.5		0.003	0.12	0.005	0.18	0.015	0.32	0.028	0.45	0.053	0.63	0.105	0.90		22.5
25.0		0.003	0.12	0.006	0.18	0.016	0.33	0.030	0.46	0.057	0.64	0.112	0.91		25.0
27.5		0.003	0.12	0.006	0.19	0.017	0.33	0.032	0.46	0.060	0.65	0.118	0.92		27.5
30.0		0.003	0.13	0.006	0.19	0.018	0.34	0.033	0.47	0.063	0.66	0.124	0.93		30.0
32.5		0.003	0.13	0.007	0.19	0.019	0.34	0.035	0.48	0.066	0.67	0.130	0.94		32.5
35.0		0.003	0.13	0.007	0.20	0.020	0.35	0.036	0.48	0.069	0.68	0.135	0.95		35.0
37.5		0.004	0.13	0.007	0.20	0.021	0.35	0.038	0.49	0.072	0.68	0.141	0.96		37.5
40.0		0.004	0.13	0.008	0.20	0.021	0.35	0.039	0.49	0.074	0.69	0.146	0.97		40.0
42.5		0.004	0.14	0.008	0.20	0.022	0.36	0.041	0.50	0.077	0.70	0.151	0.98		42.5
45.0		0.004	0.14	0.008	0.20	0.023	0.36	0.042	0.50	0.080	0.70	0.156	0.98	0.30	45.0
47.5		0.004	0.14	0.008	0.21	0.024	0.36	0.044	0.51	0.082	0.71	0.161	0.99		47.5
50.0		0.004	0.14	0.009	0.21	0.024	0.37	0.045	0.51	0.085	0.71	0.166	1.00		50.0
52.5		0.004	0.14	0.009	0.21	0.025	0.37	0.046	0.51	0.087	0.72	0.170	1.00		52.5
55.0		0.004	0.14	0.009	0.21	0.026	0.37	0.047	0.52	0.089	0.72	0.175	1.01		55.0
57.5		0.005	0.14	0.009	0.21	0.026	0.38	0.049	0.52	0.092	0.73	0.179	1.02		57.5
60.0		0.005	0.14	0.010	0.22	0.027	0.38	0.050	0.52	0.094	0.73	0.184	1.02		60.0
62.5		0.005	0.15	0.010	0.22	0.028	0.38	0.051	0.53	0.096	0.74	0.188	1.03		62.5
65.0		0.005	0.15	0.010	0.22	0.028	0.38	0.052	0.53	0.098	0.74	0.192	1.03		65.0
67.5		0.005	0.15	0.010	0.22	0.029	0.38	0.053	0.53	0.100	0.75	0.196	1.04		67.5
70.0	0.15	0.005	0.15	0.011	0.22	0.030	0.39	0.054	0.54	0.103	0.75	0.200	1.04		70.0
72.5		0.005	0.15	0.011	0.22	0.030	0.39	0.056	0.54	0.105	0.75	0.204	1.04		72.5
75.0		0.005	0.15	0.011	0.22	0.031	0.39	0.057	0.54	0.107	0.76	0.208	1.05		75.0
77.5		0.005	0.15	0.011	0.22	0.03	0.39	0.058	0.54	0.109	0.76	0.212	1.05		77.5
80.0		0.006	0.15	0.011	0.23	0.032	0.40	0.059	0.55	0.111	0.76	0.216	1.06		80.0
82.5		0.006	0.15	0.012	0.23	0.033	0.40	0.060	0.55	0.113	0.77	0.219	1.06		82.5
85.0		0.006	0.15	0.012	0.23	0.033	0.40	0.061	0.55	0.115	0.77	0.223	1.06		85.0
87.5		0.006	0.15	0.012	0.23	0.034	0.40	0.062	0.55	0.117	0.77	0.227	1.07		87.5
90.0		0.006	0.16	0.012	0.23	0.034	0.40	0.063	0.56	0.118	0.78	0.230	1.07		90.0

Table 4.31 Flow of water at 75°C in polyethylene (PE-X) pipes[21] — *continued*

q_m = mass flow rate　　　　　　　kg.s^{-1}
c = velocity　　　　　　　　　　m.s^{-1}
$\Delta p/l$ = pressure drop per unit length　Pa.m^{-1}
l_e = equivalent length of a component
　　　for $\zeta = 1$　　　　　　　m

POLYETHYLENE (PE-X)
WATER AT 75°C
BS 5556

$\Delta p/l$	c	10 mm		12 mm		16 mm		20 mm		25 mm		32 mm		c	$\Delta p/l$
		q_m	l_e	q_m	l_e	q_m	l_e	q_m	l_e	q_m	l_e	q_m	l_e		
92.5		0.006	0.16	0.012	0.23	0.035	0.40	0.064	0.56	0.120	0.78	0.234	1.08		92.5
95.0		0.006	0.16	0.013	0.23	0.035	0.41	0.065	0.56	0.122	0.78	0.237	1.08		95.0
97.5		0.006	0.16	0.013	0.23	0.036	0.41	0.066	0.56	0.124	0.79	0.241	1.08		97.5
100.0		0.006	0.16	0.013	0.23	0.036	0.41	0.067	0.57	0.126	0.79	0.244	1.09		100.0
120.0		0.007	0.16	0.014	0.24	0.040	0.42	0.074	0.58	0.139	0.81	0.271	1.11	0.50	120.0
140.0		0.008	0.17	0.016	0.25	0.044	0.43	0.081	0.59	0.152	0.83	0.295	1.13		140.0
160.0		0.008	0.17	0.017	0.25	0.048	0.44	0.087	0.60	0.164	0.84	0.318	1.15		160.0
180.0		0.009	0.18	0.018	0.26	0.051	0.45	0.093	0.62	0.176	0.85	0.339	1.16		180.0
200.0		0.010	0.18	0.019	0.26	0.054	0.45	0.099	0.62	0.186	0.87	0.359	1.17		200.0
220.0		0.010	0.18	0.021	0.27	0.057	0.46	0.105	0.63	0.197	0.88	0.379	1.19		220.0
240.0	0.30	0.011	0.18	0.022	0.27	0.060	0.47	0.110	0.64	0.207	0.89	0.397	1.20		240.0
260.0		0.011	0.19	0.023	0.27	0.063	0.47	0.115	0.65	0.216	0.90	0.415	1.21		260.0
280.0		0.012	0.19	0.024	0.28	0.066	0.48	0.120	0.65	0.225	0.91	0.433	1.22		280.0
300.0		0.012	0.19	0.025	0.28	0.068	0.48	0.125	0.66	0.234	0.91	0.450	1.22		300.0
320.0		0.013	0.19	0.026	0.28	0.071	0.49	0.130	0.67	0.243	0.92	0.466	1.23		320.0
340.0		0.013	0.20	0.026	0.29	0.073	0.49	0.134	0.67	0.252	0.93	0.482	1.24		340.0
360.0		0.014	0.20	0.027	0.29	0.076	0.49	0.139	0.68	0.260	0.94	0.497	1.25		360.0
380.0		0.014	0.20	0.028	0.29	0.078	0.50	0.143	0.68	0.268	0.94	0.512	1.25		380.0
400.0		0.014	0.20	0.029	0.29	0.080	0.50	0.147	0.69	0.276	0.95	0.527	1.26	1.00	400.0
420.0		0.015	0.20	0.030	0.29	0.083	0.50	0.151	0.69	0.283	0.95	0.541	1.27		420.0
440.0		0.015	0.20	0.031	0.30	0.085	0.51	0.155	0.69	0.291	0.96	0.555	1.27		440.0
460.0		0.016	0.21	0.032	0.30	0.087	0.51	0.159	0.70	0.298	0.96	0.568	1.28		460.0
480.0		0.016	0.21	0.032	0.30	0.089	0.51	0.163	0.70	0.305	0.97	0.582	1.28		480.0
500.0		0.016	0.21	0.033	0.30	0.091	0.52	0.167	0.71	0.313	0.97	0.595	1.29		500.0
520.0		0.017	0.21	0.034	0.30	0.093	0.52	0.171	0.71	0.319	0.98	0.608	1.29		520.0
540.0		0.017	0.21	0.035	0.31	0.095	0.52	0.174	0.71	0.326	0.98	0.621	1.30		540.0
560.0		0.017	0.21	0.035	0.31	0.097	0.52	0.178	0.72	0.333	0.99	0.633	1.30		560.0
580.0	0.50	0.018	0.21	0.036	0.31	0.099	0.53	0.181	0.72	0.340	0.99	0.645	1.31		580.0
600.0		0.018	0.21	0.037	0.31	0.101	0.53	0.185	0.72	0.346	1.00	0.657	1.31		600.0
620.0		0.019	0.21	0.037	0.31	0.103	0.53	0.188	0.73	0.353	1.00	0.669	1.31		620.0
640.0		0.019	0.22	0.038	0.31	0.105	0.53	0.192	0.73	0.359	1.01	0.681	1.32		640.0
660.0		0.019	0.22	0.039	0.32	0.107	0.54	0.195	0.73	0.265	1.01	0.692	1.32		660.0
680.0		0.020	0.22	0.039	0.32	0.109	0.54	0.199	0.73	0.371	1.01	0.704	1.32		680.0
700.0		0.020	0.22	0.040	0.32	0.111	0.54	0.202	0.74	0.378	1.02	0.715	1.33		700.0
720.0		0.020	0.22	0.041	0.32	0.112	0.54	0.205	0.74	0.384	1.02	0.726	1.33		720.0
740.0		0.021	0.22	0.041	0.32	0.114	0.54	0.208	0.74	0.390	1.02	0.737	1.33		740.0
760.0		0.021	0.22	0.042	0.32	0.116	0.55	0.211	0.75	0.395	1.03	0.748	1.34		760.0
780.0		0.021	0.22	0.043	0.32	0.118	0.55	0.214	0.75	0.401	1.03	0.758	1.34		780.0
800.0		0.021	0.22	0.043	0.32	0.119	0.55	0.218	0.75	0.407	1.03	0.769	1.34		800.0
820.0		0.022	0.22	0.044	0.33	0.121	0.55	0.221	0.75	0.413	1.04	0.779	1.35		820.0
840.0		0.022	0.23	0.044	0.33	0.123	0.55	0.224	0.75	0.418	1.04	0.789	1.35	1.50	840.0
860.0		0.022	0.23	0.045	0.33	0.124	0.56	0.227	0.76	0.424	1.04	0.799	1.35		860.0
880.0		0.023	0.23	0.046	0.33	0.126	0.56	0.230	0.76	0.429	1.04	0.810	1.35		880.0
900.0		0.023	0.23	0.046	0.33	0.127	0.56	0.232	0.76	0.435	1.05	0.819	1.36		900.0
920.0		0.023	0.23	0.047	0.33	0.129	0.56	0.235	0.76	0.440	1.05	0.829	1.36		920.0
940.0		0.024	0.23	0.047	0.33	0.131	0.56	0.238	0.77	0.445	1.05	0.839	1.36		940.0
960.0		0.024	0.23	0.048	0.33	0.132	0.56	0.241	0.77	0.451	1.06	0.849	1.36		960.0
980.0		0.024	0.23	0.049	0.33	0.134	0.57	0.244	0.77	0.456	1.06	0.858	1.37		980.0
1000.0		0.024	0.23	0.049	0.33	0.135	0.57	0.247	0.77	0.461	1.06	0.868	1.37		1000.0
1100.0		0.026	0.23	0.052	0.34	0.143	0.57	0.260	0.78	0.486	1.07	0.913	1.38		1100.0
1200.0		0.027	0.24	0.055	0.34	0.150	0.58	0.273	0.79	0.511	1.08	0.957	1.39		1200.0
1300.0		0.028	0.24	0.057	0.35	0.157	0.59	0.286	0.80	0.534	1.09	1.000	1.40		1300.0
1400.0		0.030	0.24	0.060	0.35	0.164	0.59	0.298	0.80	0.557	1.10	1.040	1.41	2.00	1400.0
1500.0		0.031	0.25	0.062	0.35	0.170	0.60	0.310	0.81	0.578	1.11	1.080	1.41		1500.0
1600.0		0.032	0.25	0.064	0.36	0.176	0.60	0.321	0.82	0.600	1.12	1.118	1.42		1600.0
1700.0		0.033	0.25	0.066	0.36	0.183	0.61	0.332	0.82	0.620	1.13	1.155	1.43		1700.0
1800.0		0.034	0.25	0.069	0.36	0.188	0.61	0.343	0.83	0.640	1.14	1.191	1.43		1800.0
1900.0		0.035	0.25	0.071	0.37	0.194	0.62	0.354	0.83	0.660	1.14	1.226	1.44		1900.0
2000.0		0.036	0.26	0.073	0.37	0.200	0.62	0.364	0.84	0.679	1.15	1.260	1.44		2000.0

Table 4.32 Flow of water at 10°C in polyvinyl chloride (PVC-C) pipes[21]

q_m	= mass flow rate	$kg.s^{-1}$
c	= velocity	$m.s^{-1}$
$\Delta p/l$	= pressure drop per unit length	$Pa.m^{-1}$
l_e	= equivalent length of a component	
	for $\zeta = 1$	m

POLYVINYL CHLORIDE (PVC-C)
WATER AT 10°C
BS 5556

$\Delta p/l$	c	12 mm q_m	12 mm l_e	16 mm q_m	16 mm l_e	20 mm q_m	20 mm l_e	25 mm q_m	25 mm l_e	32 mm q_m	32 mm l_e	40 mm q_m	40 mm l_e	50 mm q_m	50 mm l_e	63 mm q_m	63 mm l_e	c	$\Delta p/l$
0.1						0.001	0.06	0.001	0.09	0.003	0.16	0.006	0.25	0.011	0.39	0.022	0.59		0.1
0.2						0.001	0.07	0.002	0.12	0.005	0.20	0.009	0.31	0.017	0.46	0.034	0.70		0.2
0.3				0.001	0.05	0.001	0.09	0.003	0.14	0.006	0.23	0.012	0.34	0.023	0.51	0.044	0.76		0.3
0.4				0.001	0.05	0.002	0.10	0.003	0.15	0.007	0.25	0.014	0.37	0.027	0.55	0.052	0.81		0.4
0.5				0.001	0.06	0.002	0.10	0.004	0.16	0.008	0.26	0.016	0.39	0.031	0.58	0.060	0.85		0.5
0.6				0.001	0.06	0.002	0.11	0.005	0.17	0.009	0.28	0.018	0.41	0.035	0.60	0.067	0.89		0.6
0.7				0.001	0.07	0.003	0.12	0.005	0.18	0.010	0.29	0.020	0.42	0.038	0.62	0.073	0.92		0.7
0.8				0.001	0.07	0.003	0.12	0.005	0.19	0.011	0.30	0.022	0.44	0.041	0.64	0.079	0.94		0.8
0.9				0.001	0.07	0.003	0.13	0.006	0.20	0.012	0.31	0.023	0.45	0.044	0.66	0.085	0.96		0.9
1.0				0.001	0.08	0.003	0.13	0.006	0.20	0.013	0.31	0.025	0.46	0.047	0.68	0.091	0.98		1.0
1.5		0.001	0.04	0.002	0.09	0.004	0.15	0.008	0.22	0.017	0.35	0.032	0.50	0.060	0.73	0.115	1.06	0.05	1.5
2.0		0.001	0.05	0.002	0.10	0.005	0.16	0.010	0.24	0.020	0.37	0.038	0.54	0.072	0.78	0.137	1.12		2.0
2.5		0.001	0.05	0.003	0.10	0.006	0.17	0.011	0.25	0.023	0.39	0.043	0.56	0.082	0.81	0.156	1.17		2.5
3.0		0.001	0.06	0.003	0.11	0.006	0.18	0.013	0.27	0.026	0.41	0.048	0.58	0.091	0.84	0.174	1.21		3.0
3.5		0.001	0.06	0.003	0.11	0.007	0.19	0.014	0.28	0.028	0.42	0.053	0.60	0.100	0.87	0.191	1.24		3.5
4.0		0.001	0.06	0.004	0.12	0.008	0.19	0.015	0.28	0.030	0.43	0.057	0.62	0.108	0.89	0.206	1.27		4.0
4.5		0.001	0.06	0.004	0.12	0.008	0.20	0.016	0.29	0.033	0.44	0.062	0.63	0.116	0.91	0.221	1.30		4.5
5.0		0.001	0.07	0.004	0.12	0.009	0.20	0.017	0.30	0.035	0.45	0.066	0.64	0.123	0.93	0.235	1.32		5.0
5.5		0.002	0.07	0.004	0.13	0.009	0.21	0.018	0.31	0.037	0.46	0.069	0.65	0.130	0.94	0.248	1.34		5.5
6.0		0.002	0.07	0.005	0.13	0.010	0.21	0.019	0.31	0.039	0.47	0.073	0.66	0.137	0.96	0.261	1.36		6.0
6.5		0.002	0.07	0.005	0.13	0.010	0.22	0.020	0.32	0.041	0.48	0.077	0.67	0.144	0.97	0.274	1.38		6.5
7.0		0.002	0.07	0.005	0.14	0.011	0.22	0.021	0.32	0.043	0.48	0.080	0.68	0.150	0.98	0.286	1.40		7.0
7.5		0.002	0.07	0.005	0.14	0.011	0.22	0.022	0.33	0.044	0.49	0.083	0.69	0.157	0.99	0.298	1.41		7.5
8.0		0.002	0.08	0.005	0.14	0.012	0.23	0.023	0.33	0.046	0.50	0.087	0.70	0.163	1.01	0.309	1.43	0.15	8.0
8.5		0.002	0.08	0.006	0.14	0.012	0.23	0.024	0.34	0.048	0.50	0.090	0.71	0.169	1.02	0.320	1.44		8.5
9.0		0.002	0.08	0.006	0.15	0.013	0.23	0.025	0.34	0.050	0.51	0.093	0.71	0.174	1.03	0.331	1.46		9.0
9.5		0.002	0.08	0.006	0.15	0.013	0.24	0.025	0.34	0.051	0.51	0.096	0.72	0.180	1.04	0.341	1.47		9.5
10.0		0.002	0.08	0.006	0.15	0.014	0.24	0.026	0.35	0.053	0.52	0.099	0.73	0.185	1.05	0.352	1.48		10.0
12.5		0.003	0.09	0.007	0.16	0.016	0.25	0.030	0.36	0.060	0.54	0.113	0.75	0.211	1.09	0.400	1.54		12.5
15.0	0.05	0.003	0.09	0.008	0.16	0.017	0.26	0.033	0.38	0.067	0.56	0.125	0.78	0.235	1.12	0.445	1.58		15.0
17.5		0.003	0.09	0.009	0.17	0.019	0.27	0.037	0.39	0.074	0.58	0.137	0.80	0.257	1.15	0.486	1.62		17.5
20.0		0.004	0.10	0.010	0.18	0.021	0.28	0.040	0.40	0.080	0.59	0.148	0.81	0.278	1.17	0.525	1.65		20.0
22.5		0.004	0.10	0.010	0.18	0.022	0.28	0.043	0.41	0.085	0.60	0.158	0.83	0.297	1.20	0.562	1.68		22.5
25.0		0.004	0.10	0.011	0.18	0.024	0.29	0.045	0.42	0.091	0.62	0.168	0.84	0.316	1.22	0.597	1.71		25.0
27.5		0.004	0.11	0.012	0.19	0.025	0.30	0.048	0.42	0.096	0.63	0.178	0.86	0.334	1.23	0.631	1.73	0.30	27.5
30.0		0.005	0.11	0.012	0.19	0.027	0.30	0.050	0.43	0.101	0.64	0.187	0.87	0.351	1.25	0.663	1.76		30.0
32.5		0.005	0.11	0.013	0.20	0.028	0.31	0.053	0.44	0.106	0.64	0.196	0.88	0.368	1.27	0.694	1.78		32.5
35.0		0.005	0.11	0.014	0.20	0.029	0.31	0.055	0.44	0.111	0.65	0.204	0.89	0.384	1.28	0.724	1.80		35.0
37.5		0.005	0.11	0.014	0.20	0.030	0.31	0.058	0.45	0.115	0.66	0.213	0.90	0.400	1.29	0.754	1.81		37.5
40.0		0.006	0.12	0.015	0.20	0.032	0.32	0.060	0.45	0.120	0.67	0.221	0.91	0.415	1.31	0.782	1.83		40.0
42.5		0.006	0.12	0.015	0.21	0.033	0.32	0.062	0.46	0.124	0.67	0.228	0.91	0.429	1.32	0.810	1.85		42.5
45.0		0.006	0.12	0.016	0.21	0.034	0.33	0.064	0.46	0.128	0.68	0.236	0.92	0.444	1.33	0.837	1.86		45.0
47.5		0.006	0.12	0.016	0.21	0.035	0.33	0.066	0.47	0.132	0.69	0.243	0.93	0.458	1.34	0.863	1.88		47.5
50.0		0.006	0.12	0.017	0.21	0.036	0.33	0.068	0.47	0.136	0.69	0.251	0.94	0.472	1.35	0.888	1.89		50.0
52.5		0.007	0.12	0.017	0.22	0.037	0.33	0.070	0.48	0.140	0.70	0.258	0.94	0.485	1.36	0.914	1.90		52.5
55.0		0.007	0.12	0.018	0.22	0.038	0.34	0.072	0.48	0.144	0.70	0.265	0.95	0.498	1.37	0.938	1.92		55.0
57.5		0.007	0.13	0.018	0.22	0.039	0.34	0.074	0.49	0.148	0.71	0.271	0.95	0.511	1.38	0.962	1.93		57.5
60.0		0.007	0.13	0.019	0.22	0.040	0.34	0.076	0.49	0.152	0.71	0.278	0.96	0.524	1.39	0.986	1.94		60.0
62.5		0.007	0.13	0.019	0.22	0.041	0.35	0.078	0.49	0.155	0.72	0.285	0.96	0.536	1.40	1.009	1.95		62.5
65.0		0.007	0.13	0.020	0.23	0.042	0.35	0.080	0.50	0.159	0.72	0.291	0.97	0.548	1.41	1.032	1.96	0.50	65.0
67.5		0.008	0.13	0.020	0.23	0.043	0.35	0.081	0.50	0.162	0.73	0.297	0.97	0.560	1.41	1.054	1.97		67.5
70.0		0.008	0.13	0.021	0.23	0.044	0.35	0.083	0.50	0.166	0.73	0.304	0.98	0.572	1.42	1.076	1.98		70.0
72.5		0.008	0.13	0.021	0.23	0.045	0.36	0.085	0.51	0.169	0.74	0.310	0.98	0.584	1.43	1.098	1.99		72.5
75.0		0.008	0.13	0.021	0.23	0.046	0.36	0.087	0.51	0.173	0.74	0.316	0.99	0.595	1.44	1.120	2.00		75.0
77.5		0.008	0.13	0.022	0.23	0.047	0.36	0.088	0.51	0.176	0.74	0.322	0.99	0.606	1.44	1.141	2.01		77.5
80.0		0.009	0.14	0.022	0.24	0.048	0.36	0.090	0.51	0.179	0.75	0.327	1.00	0.617	1.45	1.162	2.02		80.0
82.5	0.15	0.009	0.14	0.023	0.24	0.049	0.36	0.092	0.52	0.183	0.75	0.333	1.00	0.628	1.46	1.182	2.03		82.5
85.0		0.009	0.14	0.023	0.24	0.049	0.37	0.093	0.52	0.186	0.76	0.339	1.01	0.639	1.46	1.202	2.04		85.0
87.5		0.009	0.14	0.023	0.24	0.050	0.37	0.095	0.52	0.189	0.76	0.344	1.01	0.650	1.47	1.222	2.04		87.5
90.0		0.009	0.14	0.024	0.24	0.051	0.37	0.096	0.52	0.192	0.76	0.350	1.01	0.660	1.47	1.242	2.05		90.0

Table 4.32 Flow of water at 10°C in polyvinyl chloride (PVC-C) pipes[21] — *continued*

q_m	= mass flow rate	kg.s^{-1}
c	= velocity	m.s^{-1}
$\Delta p/l$	= pressure drop per unit length	Pa.m^{-1}
l_e	= equivalent length of a component for $\zeta = 1$	m

POLYVINYL CHLORIDE (PVC-C)	
WATER AT 10°C	
BS 5556	

$\Delta p/l$	c	12 mm		16 mm		20 mm		25 mm		32 mm		40 mm		50 mm		63 mm		c	$\Delta p/l$
		q_m	l_e	q_m	l_e	q_m	l_e	q_m	l_e	q_m	l_e	q_m	l_e	q_m	l_e	q_m	l_e		
92.5		0.009	0.14	0.024	0.24	0.052	0.37	0.098	0.53	0.195	0.77	0.355	1.02	0.671	1.48	1.261	2.06		92.5
95.0		0.009	0.14	0.025	0.24	0.053	0.37	0.099	0.53	0.198	0.77	0.361	1.02	0.681	1.49	1.281	2.07		95.0
97.5		0.010	0.14	0.025	0.24	0.054	0.38	0.101	0.53	0.201	0.77	0.366	1.02	0.691	1.49	1.300	2.08		97.5
100.0		0.010	0.14	0.025	0.25	0.054	0.38	0.102	0.53	0.204	0.78	0.371	1.03	0.701	1.50	0.319	2.08		100.0
120.0		0.011	0.15	0.028	0.25	0.061	0.39	0.114	0.55	0.227	0.80	0.412	1.05	0.778	1.54	1.462	2.13		120.0
140.0		0.012	0.15	0.031	0.26	0.066	0.40	0.125	0.56	0.248	0.82	0.449	1.07	0.850	1.57	1.596	2.18		140.0
160.0		0.013	0.16	0.034	0.27	0.072	0.41	0.135	0.58	0.268	0.83	0.484	1.09	0.917	1.60	1.721	2.22		160.0
180.0		0.014	0.16	0.036	0.27	0.077	0.42	0.144	0.59	0.286	0.85	0.517	1.10	0.981	1.62	1.840	2.25		180.0
200.0		0.015	0.16	0.038	0.28	0.082	0.42	0.153	0.60	0.304	0.86	0.548	1.12	1.041	1.65	1.953	2.28		200.0
220.0		0.016	0.17	0.041	0.28	0.086	0.43	0.162	0.61	0.322	0.88	0.578	1.13	1.099	1.67	2.060	2.31	1.00	220.0
240.0		0.017	0.17	0.043	0.29	0.091	0.44	0.170	0.62	0.338	0.89	0.606	1.14	1.155	1.69	2.164	2.34		240.0
260.0		0.017	0.17	0.045	0.29	0.095	0.44	0.179	0.62	0.354	0.90	0.634	1.15	1.208	1.71	2.264	2.36		260.0
280.0	0.30	0.018	0.17	0.047	0.30	0.099	0.45	0.186	0.63	0.369	0.91	0.661	1.16	1.260	1.72	2.361	2.38		280.0
300.0		0.019	0.18	0.049	0.30	0.103	0.45	0.194	0.64	0.384	0.92	0.686	1.17	1.310	1.74	2.454	2.40		300.0
320.0		0.020	0.18	0.051	0.30	0.107	0.46	0.201	0.64	0.399	0.92	0.711	1.18	1.359	1.76	2.545	2.42		320.0
340.0		0.020	0.18	0.052	0.31	0.111	0.46	0.208	0.65	0.413	0.93	0.736	1.18	1.406	1.77	2.633	2.44		340.0
360.0		0.021	0.18	0.054	0.31	0.115	0.47	0.215	0.66	0.426	0.94	0.759	1.19	1.453	1.78	2.719	2.46		360.0
380.0		0.022	0.19	0.056	0.31	0.119	0.47	0.222	0.66	0.440	0.95	0.782	1.20	1.498	1.80	2.803	2.48		380.0
400.0		0.022	0.19	0.058	0.32	0.122	0.48	0.229	0.67	0.453	0.96	0.805	1.20	1.542	1.81	2.885	2.49		400.0
420.0		0.023	0.19	0.059	0.32	0.126	0.48	0.235	0.67	0.466	0.96	0.826	1.21	1.585	1.82	2.966	2.51		420.0
440.0		0.024	0.19	0.061	0.32	0.129	0.48	0.242	0.68	0.478	0.97	0.848	1.22	1.627	1.83	3.044	2.52		440.0
460.0		0.024	0.19	0.063	0.32	0.132	0.49	0.248	0.68	0.490	0.97	0.869	1.22	1.668	1.84	3.121	2.54	1.50	460.0
480.0		0.025	0.19	0.064	0.33	0.136	0.49	0.254	0.68	0.502	0.98	0.889	1.23	1.709	1.85	3.196	2.55		480.0
500.0		0.026	0.20	0.066	0.33	0.139	0.49	0.260	0.69	0.514	0.99	0.910	1.23	1.749	1.86	3.271	2.56		500.0
520.0		0.026	0.20	0.067	0.33	0.142	0.50	0.266	0.69	0.526	0.99	0.929	1.24	1.788	1.87	3.343	2.57		520.0
540.0		0.027	0.20	0.069	0.33	0.145	0.50	0.272	0.70	0.537	1.00	0.949	1.24	1.826	1.88	3.415	2.59		540.0
560.0		0.027	0.20	0.070	0.33	0.148	0.50	0.278	0.70	0.549	1.00	0.968	1.25	1.864	1.89	3.485	2.60		560.0
580.0		0.028	0.20	0.072	0.34	0.151	0.50	0.283	0.70	0.560	1.01	0.987	1.25	1.901	1.90	3.554	2.61		580.0
600.0		0.028	0.20	0.073	0.34	0.154	0.51	0.289	0.71	0.571	1.01	1.005	1.25	1.938	1.90	3.623	2.62		600.0
620.0	0.50	0.029	0.20	0.074	0.34	0.157	0.51	0.294	0.71	0.581	1.02	1.023	1.26	1.974	1.91	3.690	2.63		620.0
640.0		0.030	0.20	0.076	0.34	0.160	0.51	0.300	0.71	0.592	1.02	1.041	1.26	2.009	1.92	3.756	2.64		640.0
660.0		0.030	0.21	0.077	0.34	0.163	0.51	0.305	0.72	0.602	1.02	1.059	1.26	2.045	1.93	3.821	2.65		660.0
680.0		0.031	0.21	0.079	0.35	0.166	0.52	0.310	0.72	0.613	1.03	1.076	1.27	2.079	1.93	3.886	2.66		680.0
700.0		0.031	0.21	0.080	0.35	0.169	0.52	0.315	0.72	0.623	1.03	1.093	1.27	2.113	1.94	3.949	2.67		700.0
720.0		0.032	0.21	0.081	0.35	0.172	0.51	0.321	0.73	0.633	1.04	1.110	1.27	2.147	1.95	4.012	2.68		720.0
740.0		0.032	0.21	0.083	0.35	0.174	0.52	0.326	0.73	0.643	1.04	1.127	1.28	2.181	1.95	4.074	2.69		740.0
760.0		0.033	0.21	0.084	0.35	0.177	0.53	0.331	0.73	0.653	1.04	1.143	1.28	2.213	1.96	4.135	2.69	2.00	760.0
780.0		0.033	0.21	0.085	0.35	0.180	0.53	0.336	0.73	0.662	1.05	1.160	1.28	2.246	1.97	4.196	2.70		780.0
800.0		0.034	0.21	0.086	0.35	0.182	0.53	0.340	0.74	0.672	1.05	1.176	1.29	2.278	1.97	4.256	2.71		800.0
820.0		0.034	0.21	0.088	0.36	0.185	0.53	0.345	0.74	0.681	1.06	1.192	1.29	2.310	1.98	4.315	2.72		820.0
840.0		0.035	0.21	0.089	0.36	0.187	0.53	0.350	0.74	0.691	1.06	1.207	1.29	2.342	1.99	4.374	2.73		840.0
860.0		0.035	0.21	0.090	0.36	0.190	0.54	0.355	0.74	0.700	1.06	1.223	1.29	2.373	1.99	4.432	2.73		860.0
880.0		0.036	0.22	0.091	0.36	0.193	0.54	0.360	0.75	0.709	1.07	1.238	1.30	2.404	2.00	4.489	2.74		880.0
900.0		0.036	0.22	0.092	0.36	0.195	0.54	0.364	0.75	0.718	1.07	1.254	1.30	2.434	2.00	4.546	2.75		900.0
920.0		0.037	0.22	0.094	0.36	0.198	0.54	0.369	0.75	0.727	1.07	1.269	1.30	2.464	2.01	4.602	2.76		920.0
940.0		0.037	0.22	0.095	0.36	0.200	0.54	0.373	0.75	0.736	1.07	0.283	1.30	2.494	2.01	4.657	2.76		940.0
960.0		0.037	0.22	0.096	0.36	0.202	0.54	0.378	0.76	0.745	1.08	1.298	1.31	2.524	2.02	4.713	2.77		960.0
980.0		0.038	0.22	0.097	0.37	0.205	0.55	0.382	0.76	0.754	1.08	1.313	1.31	2.553	2.02	4.767	2.78		980.0
1000.0		0.038	0.22	0.098	0.37	0.207	0.55	0.387	0.76	0.763	1.08	1.327	1.31	2.582	2.03	4.821	2.78		1000.0
1100.0		0.041	0.22	0.104	0.37	0.219	0.55	0.408	0.77	0.805	1.10	1.398	1.32	2.742	2.05	5.085	2.82		1100.0
1200.0		0.043	0.23	0.109	0.38	0.230	0.56	0.429	0.78	0.846	1.11	1.465	1.33	2.860	2.07	5.338	2.84		1200.0
1300.0		0.045	0.23	0.114	0.38	0.241	0.57	0.449	0.79	0.885	1.12	1.529	1.34	2.992	2.09	5.582	2.87		1300.0
1400.0		0.047	0.23	0.119	0.39	0.251	0.57	0.468	0.80	0.923	1.13	1.592	1.35	3.118	2.11	5.817	2.89		1400.0
1500.0		0.049	0.24	0.124	0.39	0.261	0.58	0.487	0.80	0.959	1.14	1.652	1.35	3.241	2.13	6.045	2.92		1500.0
1600.0		0.051	0.24	0.129	0.39	0.271	0.59	0.505	0.81	0.995	1.15	1.710	1.36	3.360	2.15	6.267	2.94	3.00	1600.0
1700.0		0.052	0.24	0.133	0.40	0.281	0.59	0.523	0.82	1.030	1.16	1.767	1.37	3.476	2.16	6.482	2.96		1700.0
1800.0		0.054	0.24	0.138	0.40	0.290	0.60	0.540	0.82	1.063	1.17	1.822	1.37	3.589	2.18	6.691	2.98		1800.0
1900.0		0.056	0.25	0.142	0.41	0.299	0.60	0.557	0.83	1.096	1.18	1.875	1.38	3.699	2.19	6.896	3.00		1900.0
2000.0		0.058	0.25	0.147	0.41	0.308	0.60	0.574	0.84	1.129	1.19	1.928	1.38	3.807	2.20	7.095	3.01		2000.0

Table 4.33 Flow of water at 75°C in polyvinyl chloride (PVC-C) pipes[21]

q_m = mass flow rate $\qquad$ kg.s^{-1}
c = velocity $\qquad$ m.s^{-1}
$\Delta p/l$ = pressure drop per unit length $\qquad$ Pa.m^{-1}
l_e = equivalent length of a component
$\qquad$ for $\zeta = 1$ $\qquad$ m

		POLYVINYL CHLORIDE (PVC-C)
		WATER AT 75°C
		BS 5556

$\Delta p/l$	c	12 mm		16 mm		20 mm		25 mm		32 mm		40 mm		50 mm		63 mm		c	$\Delta p/l$
		q_m	l_e	q_m	l_e	q_m	l_e	q_m	le	q_m	l_e	q_m	l_e	q_m	l_e	q_m	l_e		
0.1						0.001	0.14	0.002	0.21	0.004	0.33	0.008	0.48	0.015	0.70	0.029	1.01		0.1
0.2				0.001	0.10	0.002	0.17	0.003	0.25	0.006	0.38	0.012	0.55	0.023	0.80	0.043	1.15		0.2
0.3				0.001	0.11	0.002	0.18	0.004	0.28	0.008	0.42	0.015	0.60	0.029	0.87	0.055	1.24		0.3
0.4				0.001	0.12	0.002	0.20	0.005	0.29	0.010	0.45	0.018	0.63	0.034	0.91	0.065	1.30		0.4
0.5				0.001	0.13	0.003	0.21	0.005	0.31	0.011	0.47	0.021	0.66	0.039	0.95	0.074	1.35		0.5
0.6		0.001	0.07	0.001	0.14	0.003	0.22	0.006	0.32	0.012	0.48	0.023	0.68	0.043	0.98	0.083	1.40		0.6
0.7		0.001	0.08	0.002	0.14	0.003	0.23	0.007	0.33	0.014	0.50	0.025	0.70	0.048	1.01	0.090	1.43		0.7
0.8		0.001	0.08	0.002	0.15	0.004	0.23	0.007	0.34	0.015	0.51	0.027	0.72	0.051	1.03	0.098	1.46		0.8
0.9		0.001	0.08	0.002	0.15	0.004	0.24	0.008	0.35	0.016	0.52	0.029	0.73	0.055	1.05	0.105	1.49	0.05	0.9
1.0		0.001	0.08	0.002	0.15	0.004	0.25	0.008	0.36	0.017	0.53	0.031	0.75	0.059	1.07	0.111	1.52		1.0
1.5		0.001	0.09	0.003	0.17	0.006	0.27	0.011	0.39	0.021	0.58	0.040	0.80	0.074	1.15	0.141	1.62		1.5
2.0		0.001	0.10	0.003	0.18	0.007	0.29	0.013	0.41	0.025	0.61	0.047	0.83	0.088	1.20	0.166	1.69		2.0
2.5		0.001	0.11	0.003	0.19	0.008	0.30	0.014	0.43	0.029	0.63	0.053	0.86	0.100	1.24	0.189	1.75		2.5
3.0		0.001	0.11	0.004	0.20	0.008	0.31	0.016	0.44	0.032	0.65	0.059	0.89	0.111	1.28	0.209	1.79		3.0
3.5		0.002	0.12	0.004	0.21	0.009	0.32	0.018	0.46	0.035	0.67	0.065	0.91	0.121	1.31	0.229	1.84		3.5
4.0		0.002	0.12	0.005	0.21	0.010	0.33	0.019	0.47	0.038	0.68	0.070	0.93	0.131	1.34	0.247	1.87		4.0
4.5		0.002	0.12	0.005	0.22	0.011	0.33	0.020	0.48	0.041	0.70	0.074	0.94	0.140	1.36	0.264	1.90		4.5
5.0		0.002	0.13	0.005	0.22	0.011	0.34	0.022	0.49	0.043	0.71	0.079	0.95	0.149	1.38	0.280	1.93		5.0
5.5		0.002	0.13	0.006	0.23	0.012	0.35	0.023	0.49	0.046	0.72	0.083	0.97	0.157	1.40	0.296	1.96		5.5
6.0		0.002	0.13	0.006	0.23	0.013	0.35	0.024	0.50	0.048	0.73	0.088	0.98	0.165	1.42	0.311	1.98	0.15	6.0
6.5		0.002	0.13	0.006	0.23	0.013	0.36	0.025	0.51	0.050	0.74	0.092	0.99	0.173	1.44	0.326	2.00		6.5
7.0		0.002	0.14	0.007	0.24	0.014	0.36	0.026	0.52	0.052	0.75	0.096	1.00	0.181	1.45	0.340	2.02		7.0
7.5		0.003	0.14	0.007	0.24	0.015	0.37	0.027	0.52	0.055	0.76	0.100	1.01	0.188	1.47	0.353	2.04		7.5
8.0		0.003	0.14	0.007	0.24	0.015	0.37	0.028	0.53	0.057	0.77	0.103	1.02	0.195	1.48	0.366	2.06		8.0
8.5		0.003	0.14	0.007	0.25	0.016	0.38	0.029	0.53	0.059	0.77	0.107	1.02	0.202	1.49	0.379	2.08		8.5
9.0	0.05	0.003	0.14	0.008	0.25	0.016	0.38	0.030	0.54	0.061	0.78	0.110	1.03	0.208	1.51	0.392	2.09		9.0
9.5		0.003	0.15	0.008	0.25	0.017	0.38	0.031	0.54	0.063	0.79	0.114	1.04	0.215	1.52	0.404	2.11		9.5
10.0		0.003	0.15	0.008	0.25	0.017	0.39	0.032	0.55	0.064	0.79	0.117	1.05	0.221	1.53	0.416	2.12		10.0
12.5		0.004	0.15	0.009	0.26	0.020	0.40	0.037	0.57	0.073	0.82	0.133	1.08	0.251	1.58	0.472	2.19		12.5
15.0		0.004	0.16	0.010	0.27	0.022	0.41	0.041	0.58	0.081	0.85	0.147	1.10	0.279	1.62	0.523	2.24		15.0
17.5		0.004	0.16	0.011	0.28	0.024	0.43	0.045	0.60	0.089	0.87	0.160	1.12	0.304	1.65	0.571	2.29		17.5
20.0		0.005	0.17	0.012	0.29	0.026	0.44	0.048	0.61	0.096	0.88	0.173	1.14	0.328	1.68	0.616	2.33	0.30	20.0
22.5		0.005	0.17	0.013	0.29	0.028	0.44	0.052	0.62	0.103	0.90	0.184	1.15	0.351	1.71	0.658	2.36		22.5
25.0		0.005	0.18	0.014	0.30	0.029	0.45	0.055	0.63	0.109	0.91	0.195	1.16	0.373	1.73	0.698	2.39		25.0
27.5		0.006	0.18	0.015	0.30	0.031	0.46	0.058	0.64	0.115	0.92	0.206	1.18	0.393	1.75	0.737	2.42		27.5
30.0		0.006	0.18	0.015	0.31	0.033	0.47	0.061	0.65	0.121	0.94	0.216	1.19	0.413	1.77	0.774	2.45		30.0
32.5		0.006	0.19	0.016	0.31	0.034	0.47	0.064	0.66	0.127	0.95	0.226	1.20	0.432	1.79	0.809	2.47		32.5
35.0		0.007	0.19	0.017	0.32	0.036	0.48	0.067	0.67	0.132	0.96	0.235	1.21	0.451	1.81	0.844	2.50		35.0
37.5		0.007	0.19	0.018	0.32	0.037	0.48	0.070	0.67	0.138	0.97	0.244	1.21	0.469	1.83	0.877	2.52		37.5
40.0		0.007	0.19	0.018	0.32	0.039	0.49	0.072	0.68	0.143	0.98	0.253	1.22	0.486	1.84	0.909	2.54		40.0
42.5		0.007	0.19	0.019	0.33	0.040	0.49	0.075	0.69	0.148	0.98	0.262	1.23	0.503	1.86	0.941	2.56		42.5
45.0		0.008	0.20	0.020	0.33	0.041	0.50	0.077	0.69	0.153	0.99	0.270	1.24	0.519	1.87	0.971	2.58		45.0
47.5		0.008	0.20	0.020	0.33	0.043	0.50	0.080	0.70	0.158	1.00	0.278	1.24	0.536	1.88	1.001	2.59	0.50	47.5
50.0		0.008	0.20	0.021	0.34	0.044	0.50	0.082	0.70	0.162	1.01	0.286	1.25	0.551	1.90	1.031	2.61		50.0
52.5		0.008	0.20	0.021	0.34	0.045	0.51	0.084	0.71	0.167	1.01	0.294	1.25	0.567	1.91	1.059	2.62		52.5
55.0	0.15	0.009	0.20	0.022	0.34	0.046	0.51	0.087	0.71	0.171	1.02	0.301	1.26	0.582	1.92	1.087	2.64		55.0
57.5		0.009	0.21	0.023	0.34	0.048	0.51	0.089	0.72	0.176	1.03	0.309	1.27	0.596	1.93	1.115	2.65		57.5
60.0		0.009	0.21	0.023	0.35	0.049	0.52	0.091	0.72	0.180	1.03	0.316	1.27	0.611	1.94	1.141	2.67		60.0
62.5		0.009	0.21	0.024	0.35	0.050	0.52	0.093	0.73	0.184	1.04	0.323	1.28	0.625	1.95	1.168	2.68		62.5
65.0		0.009	0.21	0.024	0.35	0.051	0.52	0.095	0.73	0.188	1.04	0.330	1.28	0.639	1.96	1.194	2.69		65.0
67.5		0.010	0.21	0.025	0.35	0.052	0.53	0.098	0.73	0.192	1.05	0.337	1.28	0.653	1.97	1.219	2.70		67.5
70.0		0.010	0.21	0.025	0.36	0.053	0.53	0.100	0.74	0.196	1.05	0.344	1.29	0.666	1.98	1.244	2.72		70.0
72.5		0.010	0.21	0.026	0.36	0.054	0.53	0.102	0.74	0.200	1.06	0.350	1.29	0.679	1.99	1.269	2.73		72.5
75.0		0.010	0.22	0.026	0.36	0.055	0.54	0.104	0.75	0.204	1.06	0.357	1.30	0.692	1.99	1.293	2.74		75.0
77.5		0.010	0.22	0.027	0.36	0.057	0.54	0.106	0.75	0.208	1.07	0.363	1.30	0.705	2.00	1.317	2.75		77.5
80.0		0.011	0.22	0.027	0.36	0.058	0.54	0.107	0.75	0.212	1.07	0.370	1.30	0.718	2.01	1.341	2.76		80.0
82.5		0.011	0.22	0.028	0.36	0.059	0.54	0.109	0.76	0.216	1.08	0.376	1.31	0.731	2.02	1.364	2.77		82.5
85.0		0.011	0.22	0.028	0.37	0.060	0.55	0.111	0.76	0.219	1.08	0.382	1.31	0.743	2.02	1.387	2.78		85.0
87.5		0.011	0.22	0.029	0.37	0.061	0.55	0.113	0.76	0.223	1.09	0.388	1.31	0.755	2.03	1.410	2.79		87.5
90.0		0.011	0.22	0.029	0.37	0.062	0.55	0.115	0.77	0.227	1.09	0.394	1.32	0.767	2.04	1.432	2.80		90.0

Table 4.33 Flow of water at 75°C in polyvinyl chloride (PVC-C) pipes[21] — *continued*

q_m = mass flow rate	kg.s^{-1}	
c = velocity	m.s^{-1}	
$\Delta p/l$ = pressure drop per unit length	Pa.m^{-1}	
l_e = equivalent length of a component		
for $\zeta = 1$	m	

POLYVINYL CHLORIDE (PVC-C)
WATER AT 75°C
BS 5556

$\Delta p/l$	c	12 mm		16 mm		20 mm		25 mm		32 mm		40 mm		50 mm		63 mm		c	$\Delta p/l$
		q_m	l_e	q_m	l_e	q_m	l_e	q_m	l_e	q_m	l_e	q_m	l_e	q_m	l_e	q_m	l_e		
92.5		0.012	0.22	0.030	0.37	0.063	0.55	0.117	0.77	0.230	1.09	0.400	1.32	0.779	2.05	1.454	2.81		92.5
95.0		0.012	0.22	0.030	0.37	0.064	0.56	0.119	0.77	0.234	1.10	0.406	1.32	0.791	2.05	1.476	2.82		95.0
97.5		0.012	0.23	0.031	0.37	0.064	0.56	0.120	0.77	0.237	1.10	0.411	1.32	0.802	2.06	1.497	2.82		97.5
100.0		0.012	0.23	0.031	0.38	0.065	0.56	0.122	0.78	0.241	1.11	0.417	1.33	0.814	2.07	1.519	2.83		100.0
120.0		0.014	0.23	0.034	0.39	0.073	0.57	0.135	0.80	0.267	1.13	0.460	1.35	0.901	2.11	1.681	2.89		120.0
140.0		0.015	0.24	0.038	0.40	0.079	0.59	0.148	0.81	0.291	1.16	0.500	1.36	0.982	2.15	1.832	2.94		140.0
160.0		0.016	0.24	0.041	0.40	0.086	0.60	0.159	0.83	0.314	1.18	0.537	1.38	1.059	2.18	1.973	2.99		160.0
180.0	0.30	0.017	0.25	0.044	0.41	0.092	0.61	0.170	0.84	0.335	1.19	0.572	1.39	1.131	2.22	2.107	3.03	1.00	180.0
200.0		0.018	0.25	0.046	0.42	0.097	0.62	0.181	0.85	0.356	1.21	0.605	1.40	1.199	2.24	2.234	3.07		200.0
220.0		0.019	0.26	0.049	0.42	0.103	0.63	0.191	0.86	0.375	1.22	0.637	1.41	1.265	2.27	2.356	3.10		220.0
240.0		0.020	0.26	0.051	0.43	0.108	0.63	0.201	0.87	0.394	1.24	0.667	1.41	1.328	2.29	2.472	3.13		240.0
260.0		0.021	0.26	0.054	0.43	0.113	0.64	0.210	0.88	0.412	1.25	0.696	1.42	1.388	2.31	2.585	3.16		260.0
280.0		0.022	0.27	0.056	0.44	0.118	0.65	0.219	0.89	0.430	1.26	0.724	1.43	1.447	2.33	2.693	3.18		280.0
300.0		0.023	0.27	0.058	0.44	0.122	0.65	0.228	0.90	0.447	1.27	0.751	1.43	1.503	2.35	2.798	3.21		300.0
320.0		0.024	0.27	0.061	0.45	0.127	0.66	0.236	0.91	0.463	1.28	0.777	1.44	1.558	2.37	2.900	3.23		320.0
340.0		0.025	0.28	0.063	0.45	0.131	0.66	0.244	0.91	0.479	1.29	0.802	1.44	1.612	2.38	2.999	3.25	1.50	340.0
360.0		0.026	0.28	0.065	0.45	0.136	0.67	0.252	0.92	0.495	1.30	0.827	1.45	1.664	2.40	3.096	3.27		360.0
380.0		0.026	0.28	0.067	0.46	0.140	0.67	0.260	0.93	0.510	1.31	0.851	1.45	1.715	2.41	3.190	3.29		380.0
400.0		0.027	0.28	0.069	0.46	0.144	0.68	0.268	0.93	0.525	1.32	0.874	1.46	1.764	2.43	3.282	3.31		400.0
420.0		0.028	0.28	0.071	0.46	0.148	0.68	0.275	0.94	0.540	1.33	0.897	1.46	1.813	2.44	3.372	3.32		420.0
440.0	0.50	0.029	0.29	0.073	0.47	0.152	0.69	0.282	0.94	0.554	1.33	0.919	1.47	1.860	2.45	3.460	3.34		440.0
460.0		0.029	0.29	0.074	0.47	0.156	0.69	0.290	0.95	0.568	1.34	0.941	1.47	1.907	2.47	3.546	3.36		460.0
480.0		0.030	0.29	0.076	0.47	0.160	0.70	0.297	0.96	0.582	1.35	0.962	1.47	1.952	2.48	3.630	3.37		480.0
500.0		0.031	0.29	0.078	0.48	0.164	0.70	0.303	0.96	0.595	1.35	0.983	1.48	1.997	2.49	3.713	3.39		500.0
520.0		0.032	0.29	0.080	0.48	0.167	0.70	0.310	0.96	0.608	1.36	1.003	1.48	2.041	2.50	3.795	3.40		520.0
540.0		0.032	0.30	0.082	0.48	0.171	0.71	0.317	0.97	0.621	1.37	1.023	1.48	2.084	2.51	3.875	3.41		540.0
560.0		0.033	0.30	0.083	0.48	0.174	0.71	0.323	0.97	0.634	1.37	1.043	1.48	2.127	2.52	3.953	3.43		560.0
580.0		0.034	0.30	0.085	0.49	0.178	0.71	0.330	0.98	0.647	1.38	1.063	1.49	2.169	2.53	4.031	3.44	2.00	580.0
600.0		0.034	0.30	0.087	0.49	0.181	0.72	0.336	0.98	0.659	1.38	1.082	1.49	2.210	2.54	4.107	3.45		600.0
620.0		0.035	0.30	0.088	0.49	0.185	0.72	0.342	0.99	0.671	1.39	1.100	1.49	2.250	2.55	4.182	3.46		620.0
640.0		0.036	0.30	0.090	0.49	0.188	0.72	0.349	0.99	0.683	1.39	1.119	1.49	2.290	2.56	4.256	3.48		640.0
660.0		0.036	0.30	0.091	0.49	0.191	0.72	0.355	0.99	0.695	1.40	1.137	1.50	2.330	2.56	4.329	3.49		660.0
680.0		0.037	0.31	0.093	0.50	0.195	0.73	0.361	1.00	0.707	1.40	1.155	1.50	2.368	2.57	4.401	3.50		680.0
700.0		0.037	0.31	0.095	0.50	0.198	0.73	0.367	1.00	0.718	1.41	1.173	1.50	2.407	2.58	4.472	3.51		700.0
720.0		0.038	0.31	0.096	0.50	0.201	0.73	0.372	1.00	0.730	1.41	1.190	1.50	2.445	2.59	4.542	3.52		720.0
740.0		0.039	0.31	0.098	0.50	0.204	0.74	0.378	1.01	0.741	1.42	1.207	1.50	2.482	2.60	4.611	3.53		740.0
760.0		0.039	0.31	0.099	0.50	0.207	0.74	0.384	1.01	0.752	1.42	1.224	1.50	2.519	2.60	4.679	3.54		760.0
780.0		0.040	0.31	0.101	0.51	0.210	0.74	0.390	1.01	0.763	1.43	1.241	1.51	2.556	2.61	4.747	3.55		780.0
800.0		0.040	0.31	0.102	0.51	0.213	0.74	0.395	1.02	0.774	1.43	1.257	1.51	2.592	2.62	4.814	3.56		800.0
820.0		0.041	0.31	0.103	0.51	0.216	0.74	0.401	1.02	0.785	1.43	1.274	1.51	2.627	2.63	4.880	3.57		820.0
840.0		0.042	0.31	0.105	0.51	0.219	0.75	0.406	1.02	0.795	1.44	1.290	1.51	2.663	2.63	4.945	3.58		840.0
860.0		0.042	0.32	0.106	0.51	0.222	0.75	0.411	1.03	0.806	1.44	1.306	1.51	2.697	2.64	5.010	3.58		860.0
880.0		0.043	0.32	0.108	0.51	0.225	0.75	0.417	1.03	0.816	1.45	1.321	1.51	2.732	2.65	5.074	3.59		880.0
900.0		0.043	0.32	0.109	0.52	0.228	0.75	0.422	1.03	0.826	1.45	1.337	1.52	2.766	2.65	5.137	3.60		900.0
920.0		0.044	0.32	0.110	0.52	0.231	0.76	0.427	1.03	0.837	1.45	1.352	1.52	2.800	2.66	5.199	3.61		920.0
940.0		0.044	0.32	0.112	0.52	0.233	0.76	0.433	1.04	0.847	1.46	1.367	1.52	2.834	2.66	5.261	3.62		940.0
960.0		0.045	0.32	0.113	0.52	0.236	0.76	0.438	1.04	0.857	1.46	1.383	1.52	2.867	2.67	5.323	3.62		960.0
980.0		0.045	0.32	0.114	0.52	0.239	0.76	0.443	1.04	0.867	1.46	1.397	1.52	2.900	2.68	5.384	3.63		980.0
1000.0		0.046	0.32	0.116	0.52	0.242	0.76	0.448	1.05	0.876	1.47	1.412	1.52	2.932	2.68	5.444	3.64		1000.0
1100.0		0.048	0.33	0.122	0.53	0.255	0.77	0.472	1.06	0.924	1.48	1.484	1.53	3.091	2.71	5.737	3.68		1100.0
1200.0		0.051	0.33	0.128	0.54	0.268	0.78	0.496	1.07	0.970	1.50	1.552	1.53	3.243	2.73	6.019	3.71	3.00	1200.0
1300.0		0.053	0.33	0.134	0.54	0.280	0.79	0.519	1.08	1.014	1.51	1.618	1.54	3.390	2.76	6.290	3.74		1300.0
1400.0		0.056	0.34	0.140	0.55	0.292	0.80	0.540	1.09	1.057	1.52	1.682	1.54	3.531	2.78	6.552	3.77		1400.0
1500.0	1.00	0.058	0.34	0.146	0.55	0.304	0.80	0.562	1.10	1.098	1.54	1.743	1.55	3.669	2.80	6.805	3.79		1500.0
1600.0		0.060	0.34	0.151	0.56	0.315	0.81	0.582	1.11	1.138	1.55	1.802	1.55	3.802	2.82	7.051	3.82		1600.0
1700.0		0.062	0.35	0.156	0.56	0.326	0.82	0.602	1.11	1.177	1.56	1.859	1.55	3.931	2.83	7.290	3.84		1700.0
1800.0		0.064	0.35	0.161	0.56	0.336	0.82	0.622	1.12	1.215	1.57	1.915	1.56	4.057	2.85	7.523	3.86		1800.0
1900.0		0.066	0.35	0.166	0.57	0.347	0.83	0.641	1.13	1.253	1.58	1.969	1.56	4.179	2.87	7.750	3.88		1900.0
2000.0		0.068	0.36	0.171	0.57	0.357	0.83	0.660	1.13	1.289	1.59	2.022	1.56	4.299	2.88	7.971	3.90		2000.0

4.5 Air flow in ducts

4.5.1 Duct sizing: desirable velocities

The are no rules concerning duct sizing. The most cost effective will be the design based on life-cycle costing including the fan running costs. The smaller the duct work, the greater the fan power and energy consumption. Increasing the duct size can have a large effect in decreasing fan power: smaller friction pressure drops of the basic circuit will require smaller pressure drops through control dampers, for the same value of control authority thus leading to a further saving. The optimum sizing from the point of view of life-cycle costing must consider the length of the system, the capital cost, the mean pressure drop, the running time at full and partial flow, the efficiency of the fan–motor combination and anticipated electrical tariffs (on-peak, off-peak operation).

To give a starting point in selecting duct sizes, rule of thumb air velocities are reproduced from BSRIA[12] in Table 4.34. An alternative starting point might be to consider a typical pressure drop per unit length of 1 Pa.m^{-1} for low velocity systems and 8 Pa.m^{-1} for high velocity systems (BSRIA[12]).

Table 4.34 Typical air velocities for duct work[12,4]

Situation	Velocity/ m.s^{-1}	Maximum pressure drop/length/(Pa.m^{-1})	Total pressure drop/kPa
Low velocity systems	3–6	1	0.900 supply 0.400 extract
High velocity systems	7.5–15	8	1.5–2.0 supply

4.5.2 Noise

The major source of duct-generated noise is caused by the vortices created in diffusers, grilles, fittings and the fan itself. Higher air velocities will create more noise. The ductwork and sharp elbows can have an attenuating effect especially of the higher frequencies. Frequently a noise problem may be due to noise from one zone being able to be propagated to another either via grilles and the ductwork, or by 'break-in' to the ductwork itself or 'break-out'. The subject is too complex to be dealt with in this section.

4.5.3 Pressure drop for circular ducts

The pressure drop per unit length can be calculated for any duct material, and for any air condition using the pressure loss factor λ, as explained in section 4.3.

Repeating the D'Arcy (equation (4.1)) for pressure loss due to friction:

$$\Delta p = \lambda \frac{l}{d} \tfrac{1}{2} \rho c^2$$

Useful property values for air are given below in Table 4.36, and in Appendix 4.A1.

Note that values of density, being the reciprocal of the specific volume, are best obtained from the psychrometric chart which covers any value of humidity.

Table 4.36 Some properties of air at a relative humidity of 50% and at a pressure of 1.013 25 bar

$T/$ °C	$\rho/$ kg.m^{-3}	$\eta/$ 10^{-6} kg.m^{-1}	$c_p/$ kJ.kg^{-1}.K^{-1}
0	1.29	17.15	1.006
5	1.27	17.39	1.009
10	1.24	17.63	1.011
15	1.22	17.88	1.014
20	1.20	18.12	1.018
25	1.18	18.36	1.022
30	1.16	18.55	1.030
35	1.14	18.78	1.039
40	1.11	19.01	1.050

The variation of density with pressure can be obtained using a value ρ_0 from Table 4.36 or the psychrometric chart, and the ideal gas equation, namely

$$\rho = \rho_0 \left(\frac{p}{1.01325} \right) \tag{4.17}$$

Values of viscosity and specific thermal capacity do not vary significantly with pressure.

4.5.4 Pre-calculated values of pressure drop for circular ducts (Figure 4.2)

For clean circular ducts made from galvanised sheet steel the pressure drops have been calculated and are presented in graphical form as Figure 4.2.

Table 4.35 Typical air velocities (face velocities) for air handling units and other components[11,4]

Situation	Velocity/m.s^{-1}		Pressure drop/Pa
Heating system	2.5–4	(through face area)	50–125
Cooling system	1.5–2.5	(through face area)	60–180
Inlet louvres	2.5	(through free area)	35 max[4]
Extract louvres	2.5	(through free area)	60 max[4]
Filters[4]:			
— flat panel	As duct		
— pleated	< 3.8		
— HEPA	1.3		
— moving curtain viscous	2.5		
— moving curtain dry	1.0		
— electronic, ionising	0.8–1.8		

The values for the appropriate variables have been taken as follows:

(a) Air at 20°C, 101.325 kPa, 43 per cent saturation.

(b) Air density $\rho = 1.200$ kg.m^{-3}, viscosity $\eta = 18.2 \times 10^{-6}$ kg.m^{-1}.s^{-1}.

(c) Roughness $k = 0.15$ mm (longitudinal seams).

For air at a temperature other than 20°C, the pressure loss read from the chart may be corrected by use of the following expression, where T must be in kelvins:

$$\Delta p = \Delta p_{20} \left(\frac{293}{T} \right)^{0.86} \tag{4.18}$$

4.5.5 Ducts of other materials

Tables 4.37, 4.38 and 4.39 may be used to obtain a correction factor C. The value of Δp_g read from the chart for galvanised ductwork (Figure 4.2), may then be multiplied by this factor to obtain the true Δp.

4.5.6 Non-circular ducts

In section 4.3 the basic equation (4.1) needs to be rewritten in terms of 'hydraulic diameter' instead of diameter, where hydraulic diameter d_h is given by:

$$d_h = \frac{4A}{P} \tag{4.19}$$

where A is the cross-sectional area and P is the perimeter of the duct. (For a circular duct the hydraulic diameter d_h is equal to the real diameter d.)

$$\Delta p = \frac{\lambda \rho c^2}{2d_h} \tag{4.20}$$

4.5.6.1 Rectangular ducts

The pressure drop chart (Figure 4.2) can also be used for rectangular ducts using the concept of an 'equivalent diameter'. There are two definitions of the equivalent diameter for a rectangular duct:

(a) the circular duct which would give the same pressure drop as the rectangular duct, for the same volume flow

(b) the circular duct which would give the same pressure drop as the rectangular duct, for the same velocity.

In most systems volume flow is the primary parameter, so use of (a) will lead to less confusion. The notional velocity in the notional circular equivalent duct will differ from the real air velocity, so it is better not to calculate it or use it. The pressure drop for fittings must be based on the actual mean velocity of air within the rectangular duct.

To determine the pressure drop along straight lengths of rectangular ductwork, equation 4.21 should be used for determining the 'equivalent diameter' of circular ductwork (as for (a) above):

Table 4.37 Values of correction factor C for ducts made of:
(a) neat cement or plaster ($k_s = 0.25$ mm)
(b) spiral wound galvanised ($k_s = 0.075$ mm)
(c) sheet aluminium ($k_s = 0.05$ mm)
(d) plastic ducts ($k_s = 0.005$ mm)

Pressure loss from chart/ (Pa.m^{-1})	Correction factor for ducts of all diameters			
	Plastered ducts	Spiral wound ducts	Aluminium ducts	Plastic ducts
0.1	1.03	0.97	0.96	0.90
0.2	1.05	0.96	0.95	0.90
0.5	1.07	0.95	0.93	0.88
1.0	1.08	0.94	0.91	0.85
2.0	1.08	0.93	0.90	0.84
5.0	1.09	0.92	0.88	0.80
10.0	1.09	0.91	0.86	0.77
20.0	1.10	0.90	0.85	0.75
50.0	1.10	0.88	0.83	0.71
100.0	1.11	0.86	0.79	0.68

Table 4.38 Values of correction factor C for ducts made of fair faced brickwork or concrete ($k_s = 1.3$ mm)

Pressure loss from chart/ (Pa.m^{-1})	Correction factors for ducts having the following diameters				
	100 mm	200 mm	500 mm	1000 mm	2000 mm
0.1	—	1.32	1.32	1.32	1.32
0.2	—	1.35	1.35	1.35	1.35
0.5	—	1.42	1.42	1.42	1.41
1.0	1.50	1.48	1.47	1.46	1.45
2.0	1.54	1.53	1.51	1.49	1.48
5.0	1.63	1.60	1.56	1.54	—
10.0	1.66	1.63	1.59	1.56	—
20.0	1.73	1.68	1.61	1.57	—
50.0	1.79	1.72	1.62	1.58	—
100.0	1.80	1.74	1.64	1.60	—

Table 4.39 Values of correction factor C for ducts made of rough brickwork ($k_s = 5$ mm)

Pressure loss from chart/ (Pa.m^{-1})	Correction factors for ducts having the following diameters				
	100 mm	200 mm	500 mm	1000 mm	2000 mm
0.1	—	1.88	1.88	1.87	1.85
0.2	—	2.02	1.99	1.96	1.90
0.5	—	2.18	2.11	2.05	1.97
1.0	2.46	2.34	2.20	2.12	2.04
2.0	2.62	2.45	2.28	2.17	2.12
5.0	2.76	2.58	2.36	2.23	—
10.0	2.91	2.67	2.42	2.25	—
20.0	3.01	2.73	2.46	2.28	—
50.0	3.10	2.83	2.51	2.30	—
100.0	3.18	2.92	2.53	2.31	—

$$d_e = \left[\frac{32\lambda_c \, w^3 \, h^3}{\pi^2 \, \lambda_r \, (w + h)} \right]^{0.2} \tag{4.21}$$

where λ_c refers to the pressure loss coefficient of the circular galvanised duct equivalent, and λ_r refers to the pressure loss coefficient of the rectangular duct of any material. The dimensions of the rectangular duct are given by w and h.

Table 4.40 gives pre-calculated values of equivalent diameter for galvanised rectangular ductwork.

Figure 4.2 may then be used, using the equivalent diameter and *the same volume flow*, to determine the pressure drop per unit length.

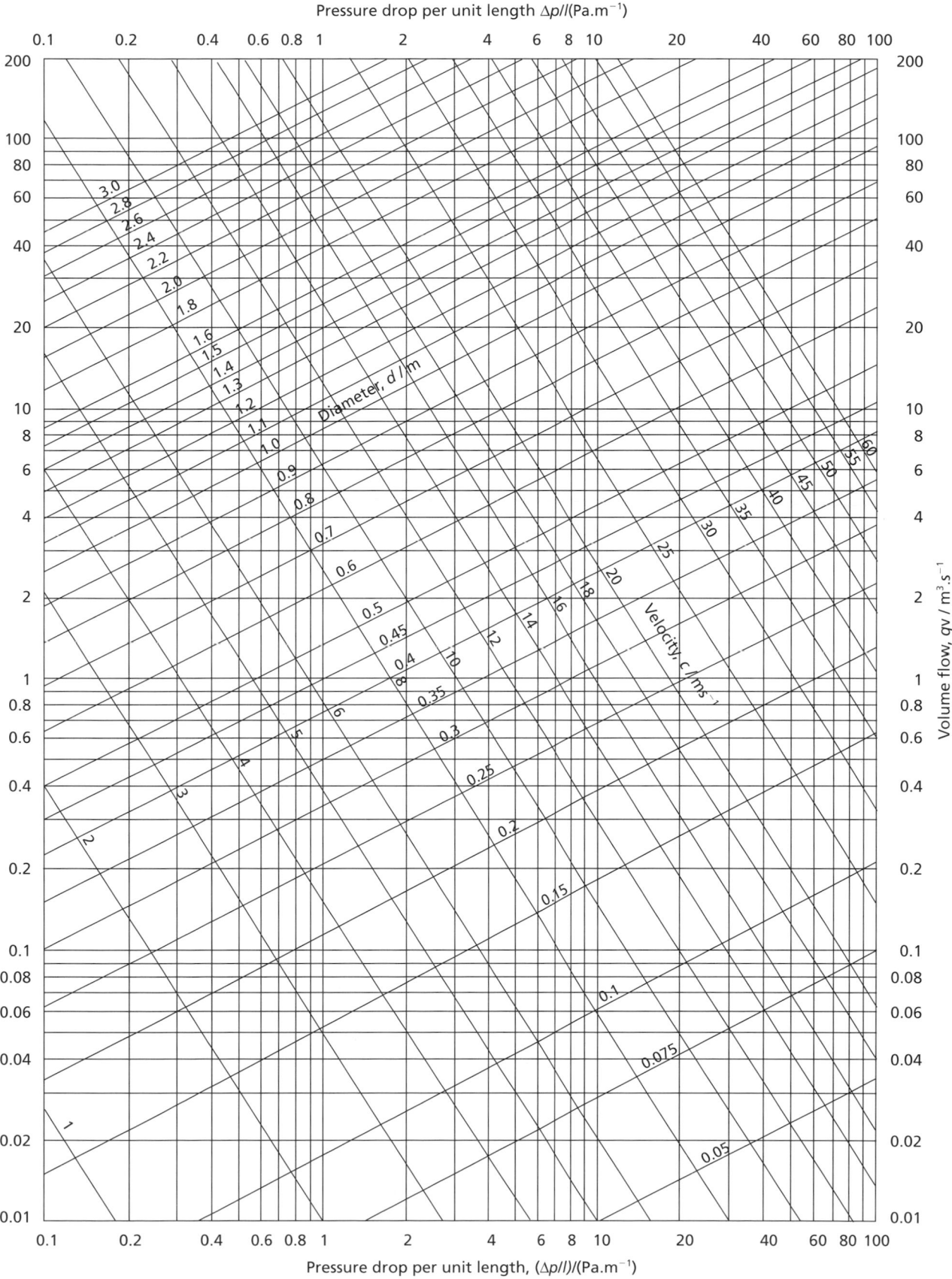

Figure 4.2 Pressure drop for air in galvanised circular ducts ($\rho = 1.2$ kg.m^{-3}; $T = 293$ K)

Table 4.40 Equivalent diameters for rectangular ductwork, to give the same pressure drop for the same volume flow and the same surface roughness. (For the same values of friction coefficient λ)

Dimen. of side, h	100	125	150	175	200	225	250	300	350	400	450	500	550	600	650	700	750	800	850	900	Dimen. of side, h
100	110	123	134	145	154	163	171	185	199	211	222	232	242	251	260	268	276	284	291	298	100
		138	151	162	173	183	192	209	225	239	251	263	275	285	295	305	314	323	331	339	125
			165	178	190	202	212	231	248	264	278	291	304	316	327	338	348	358	368	377	150
				193	206	218	230	251	269	287	303	317	331	344	357	369	380	391	401	411	175
					220	233	246	269	289	308	325	341	356	371	384	397	409	421	433	444	200
						248	261	286	308	328	346	364	380	395	410	424	437	450	462	474	225
							275	301	325	346	366	385	402	419	434	449	463	477	490	503	250
								330	357	381	403	424	443	462	479	496	512	527	542	556	300
									385	412	436	459	481	501	520	539	556	573	589	605	350
										441	467	492	515	537	558	578	597	616	633	650	400
125	347										496	522	547	571	594	615	636	655	674	693	450
150	385	394										551	577	603	627	650	672	693	713	733	500
175	421	430	440										606	633	658	682	706	728	749	770	550
200	454	464	474	484										661	688	713	738	761	784	806	600
225	485	496	507	517	527										716	743	769	793	817	840	650
250	515	527	538	549	560	570										771	798	824	849	873	700
300	570	583	596	608	620	632	643										826	853	879	904	750
350	620	635	649	662	676	689	701	714										881	908	934	800
400	667	683	698	713	727	741	755	768	794										936	963	850
450	710	727	744	760	776	791	806	820	848	874										991	900
500	751	770	787	804	821	837	853	869	898	927	954										
550	790	810	828	847	864	882	898	915	946	976	1005	1033									
600	827	848	867	887	905	924	941	959	992	1024	1054	1084	1112								
650	862	884	905	925	945	964	982	1001	1036	1069	1101	1132	1162	1191							
700	896	919	940	962	982	1002	1022	1041	1077	1113	1146	1179	1210	1240	1269						
750	928	952	975	997	1018	1039	1059	1079	1118	1154	1190	1223	1256	1287	1318	1347					
800	959	984	1008	1031	1053	1075	1096	1117	1157	1195	1231	1267	1301	1333	1365	1396	1426				
850	989	1015	1039	1063	1086	1109	1131	1152	1194	1234	1272	1308	1344	1378	1411	1443	1474	1504			
900	1018	1044	1070	1095	1119	1142	1165	1187	1230	1271	1311	1349	1385	1421	1455	1488	1520	1551	1582		
950	1046	1073	1100	1125	1150	1174	1198	1221	1265	1308	1349	1388	1426	1462	1498	1532	1565	1597	1629	1659	950
1000		1101	1128	1155	1180	1205	1230	1254	1299	1343	1385	1426	1465	1503	1539	1575	1609	1642	1675	1706	1000
1050			1156	1184	1210	1236	1261	1285	1332	1378	1421	1463	1503	1542	1580	1616	1652	1686	1719	1752	1050
1100				1211	1239	1265	1291	1316	1365	1411	1456	1499	1540	1580	1619	1657	1693	1729	1763	1797	1100
1150					1266	1294	1320	1346	1396	1444	1490	1534	1577	1618	1658	1696	1734	1770	1806	1840	1150
1200						1322	1349	1375	1427	1476	1523	1568	1612	1654	1695	1735	1773	1811	1847	1882	1200
1250							1377	1404	1456	1507	1555	1602	1646	1690	1732	1772	1812	1850	1888	1924	1250
1300								1432	1486	1537	1587	1634	1680	1725	1768	1809	1850	1889	1927	1965	1300
1400									1542	1596	1648	1697	1745	1792	1837	1881	1923	1964	2004	2043	1400
1500										1652	1706	1758	1808	1857	1904	1949	1993	2036	2078	2119	1500
1600											1762	1816	1868	1919	1968	2015	2061	2106	2149	2192	1600
1700												1872	1926	1979	2029	2078	2126	2173	2218	2262	1700
1800													1982	2036	2089	2140	2189	2237	2284	2330	1800
1900														2092	2147	2199	2250	2300	2348	2396	1900
2000															2203	2257	2310	2361	2411	2459	2000
2100																2313	2367	2420	2471	2521	2100
2200																	2423	2477	2530	2582	2200
2300																		2533	2587	2640	2300
2400																			2643	2697	2400
2500																				2753	2500
Dimen. of side, h	950	1000	1050	1100	1150	1200	1250	1300	1400	1500	1600	1700	1800	1900	2000	2100	2200	2300	2400	2500	Dimen. of side, h

$$d_e = \left[\frac{32 w^3 h^3}{\pi^2 (w + h)} \right]^{0.2}$$

4.5.6.2 Flat-oval spirally wound rectangular ducts

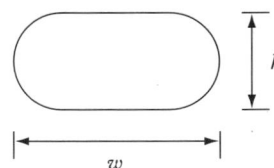

A mathematical relationship can be obtained, equation (4.22), for the equivalent diameter of circular duct which would give the same pressure drop as the flat-oval duct, for the same volume flow. However, it is to be expected that the spirally wound seam of the flat-oval duct will in reality give a slightly higher pressure drop than the longitudinally seamed circular equivalent.

$$d_e = \left\{ \frac{64\lambda_c \left[hw + h^2\left(\frac{\pi}{4} - 1\right)\right]^3}{\pi^2 \lambda_r \left[\pi h + 2(w - h)\right]} \right\}^{0.2} \quad (4.22)$$

4.5.7 Components and fittings

Whether for liquids in pipes, or gases in ducts, the same fundamental equation applies:

$$\Delta p_f = \zeta \tfrac{1}{2}\rho c^2 \quad (4.23)$$

Data for ζ are given in sections 4.10.2 (rectangular ducts) and 4.10.5 (circular ducts).

Table 4.41 Equivalent diameters for flat-oval ductwork, to give the same pressure drop for the same volume flow and the same surface roughness. (For the same values of friction coefficient λ.) The values given are for preferred sizes of ductwork, HVCA[5]

| | Height, h/mm | | | | | | | | | | | Perimeter, P/m |
	75	100	125	150	200	250	300	350	400	450	500	
w	320											0.718
d_e	160											
w	360	350	330	320								0.798
d_e	169	195	213	230								
w	400	390	370	360								0.878
d_e	177	205	225	244								
w	440	430	410	400								0.958
d_e	185	214	236	257								
w	480	470	450	440								1.037
d_e	192	223	247	269								
w	520	505	490	480								1.117
d_e	199	231	257	280								
w		545	530	520								1.197
d_e		239	266	291								
w				555	525							1.277
d_e				300	340							
w				635	605	580						1.436
d_e				319	364	400						
w				715	690	660	630					1.596
d_e				337	387	426	456					
w				800	770	740	710	685	655			1.756
d_e				354	408	450	485	513	534			
w				880	845	825	790	765	735	705	680	1.915
d_e				370	425	474	511	543	567	586	602	
w				960	930	900	875	845	815	785	755	2.075
d_e				384	444	494	537	571	599	621	638	
w				1040	1010	985	955	925	895	865	835	2.238
d_e				398	461	515	560	597	628	654	674	
w				1120	1090	1065	1035	1005	975	945	915	2.394
d_e				411	477	534	581	621	655	684	708	
w				1200	1170	1145	1115	1085	1055	1025	1000	2.553
d_e				423	492	552	602	645	682	713	741	
w					1335	1305	1275	1245	1215	1190	1160	2.873
d_e					576	585	640	688	730	768	800	
w						1465	1435	1405	1375	1350	1320	3.192
d_e						616	675	727	774	816	852	
w						1625	1595	1570	1540	1510	1480	3.511
d_e						644	708	765	815	860	901	
w						1785	1760	1730	1700	1670	1640	3.83
d_e						671	739	799	853	902	946	

Equivalent diameter, d_e (mm) for stated height, h (mm) and width, w (mm)

4.5.8 Velocity pressures of air

Table 4.42 Velocity pressures ($\frac{1}{2}\rho c^2$) in Pa for air having a density $\rho = 1.20$ kg.m^{-3}. This is the case at 20°C, percentage saturation 50% and a pressure of 101.325 bar. For other conditions see section 4.5.3 and Table 4.36

Velocity c/m.s^{-1}	Velocity pressure / Pa for stated velocity c/m.s^{-1}									
	0.0	0.1	0.2	0.3	0.4	0.5	0.6	0.7	0.8	0.9
0	0.00	0.01	0.02	0.05	0.10	0.15	0.22	0.29	0.38	0.49
1	0.60	0.73	0.86	1.01	1.18	1.35	1.54	1.73	1.94	2.17
2	2.40	2.65	2.90	3.17	3.46	3.75	4.06	4.37	4.70	5.05
3	5.40	5.77	6.14	6.53	6.94	7.35	7.78	8.21	8.66	9.13
4	9.60	10.09	10.58	11.09	11.62	12.15	12.70	13.25	13.82	14.41
5	15.00	15.61	16.22	16.85	17.50	18.15	18.82	19.49	20.18	20.89
6	21.60	22.33	23.06	23.81	24.58	25.35	26.14	26.93	27.74	28.57
7	29.40	30.25	31.10	31.97	32.86	33.75	34.66	35.57	36.50	37.45
8	38.40	39.37	40.34	41.33	42.34	43.35	44.38	45.41	46.46	47.53
9	48.60	49.69	50.78	51.89	53.02	54.15	55.30	56.45	57.62	58.81
10	60.00	61.21	62.42	63.65	64.90	66.15	67.42	68.69	69.98	71.29
11	72.60	73.93	75.26	76.61	77.98	79.35	80.74	82.13	83.54	84.97
12	86.40	87.85	89.30	90.77	92.26	93.75	95.26	96.77	98.30	99.85
13	101.40	102.97	104.54	106.13	107.74	109.35	110.98	112.61	114.26	115.93
14	117.60	119.29	120.98	122.69	124.42	126.15	127.90	129.65	131.42	133.21
15	135.00	136.81	138.62	140.45	142.30	144.15	146.02	147.89	149.78	151.69
16	153.60	155.53	157.46	159.41	161.38	163.35	165.34	167.33	169.34	171.37
17	173.40	175.45	177.50	179.57	181.66	183.75	185.86	187.97	109.10	192.25
18	194.40	196.57	198.74	200.93	203.14	205.35	207.58	209.81	212.06	214.33
19	216.60	218.89	221.18	223.49	225.82	228.15	230.50	232.85	235.22	237.61
20	240.00	242.41	244.82	247.25	249.70	252.15	254.62	257.09	259.58	262.09
21	264.60	367.13	269.66	272.21	274.78	277.35	279.94	282.53	285.14	287.77
22	290.40	293.05	295.70	298.37	301.06	303.75	306.46	309.17	311.90	413.65
23	317.40	320.17	322.94	325.73	328.54	331.35	334.18	337.01	339.86	342.73
24	345.60	348.49	351.38	354.29	357.22	360.15	363.10	366.05	369.02	372.01
25	375.00	378.01	381.02	384.05	387.10	390.15	393.22	396.29	399.38	402.49
26	405.60	408.73	411.86	415.01	418.18	421.35	424.54	427.37	430.94	434.17
27	437.40	440.65	443.90	447.17	450.46	453.75	457.06	460.37	463.70	467.05
28	470.40	473.77	477.14	480.53	483.94	487.35	490.78	494.21	497.66	501.13
29	504.60	508.09	511.58	515.09	518.62	522.15	525.70	529.25	532.82	536.41
30	540.00	543.61	547.22	550.85	554.50	558.15	561.82	565.49	569.18	572.89
31	576.60	580.33	584.06	587.81	591.58	595.35	599.14	602.93	606.74	610.57
32	614.40	618.25	622.10	625.97	629.86	633.75	637.66	641.57	645.50	649.45
33	653.40	657.37	661.34	665.33	669.34	673.35	677.38	681.41	685.46	689.53
34	693.60	697.69	701.78	705.89	710.02	714.15	718.30	722.45	726.62	730.81
35	735.00	739.21	743.42	747.65	751.90	756.15	760.42	764.69	768.98	773.29
36	777.60	781.93	786.26	790.61	794.98	799.35	803.74	808.13	812.54	816.97
37	821.40	825.85	830.30	834.77	839.26	843.75	848.26	852.77	857.30	861.85
38	866.40	870.97	875.54	880.13	884.74	889.35	893.98	898.61	903.26	907.93
39	912.60	917.29	921.98	926.69	931.42	936.15	940.90	945.65	950.42	955.21
40	960.00	964.81	969.62	974.45	979.30	984.15	989.02	993.78	998.78	1003.69

4.6 Steam in pipes

4.6.1 General

The pressure drop due to friction in steel pipes may be calculated from:

$$\frac{Z_1 - Z_2}{l} = \frac{30.32 q_m^{1.889}}{10^3 d^{5.027}} \tag{4.24}$$

and Z is given by:

$$Z = p^{1.929} \tag{4.25}$$

where the following units must be used: q_m in kg.s^{-1}, d in m, l in m, p in kPa (absolute) and Z in kPa$^{1.929}$.

These equations have been developed to cover initial steam pressures of between 100 kPa and 1000 kPa and for velocities ranging from 5 m.s^{-1} to 50 m.s^{-1}. They may be used with less accuracy for conditions outside these limits.

Pre-calculated values of pressure drop are given in Table 4.43.

Values of the pressure factor Z are given in Table 4.44.

The values used for density ρ and kinematic viscosity v were obtained from the following relations:

$$\rho = 7.83 \times 10^{-3}\, p^{0.94}$$
$$v = 9.79 \times 10^{-4}\, p^{-0.842} \tag{4.26}$$

where the following units must be used: p in kPa (absolute), ρ in kg.m^{-3}, and v in m^2.s^{-1}.

Table 4.43 Flow of saturated steam in heavy steel pipes

q_m = mass flow rate — kg.s^{-1}
p = pressure (absolute) — kPa
$\Delta Z/l$ = pressure drop per unit length — Pa$^{1.929}$.m^{-1}
l_e = equivalent length of a component
for $\zeta = 1$ — m

HEAVY GRADE STEEL
SATURATED STEAM

$\Delta Z/l$	p	10 mm q_m	l_e	15 mm q_m	l_e	20 mm q_m	l_e	25 mm q_m	l_e	32 mm q_m	l_e	40 mm q_m	l_e	p	$\Delta Z/l$
1		0.0001	0.2	0.0003	0.3	0.0007	0.5	0.001	0.7	0.003	1.0	0.004	1.3		1
2	10	0.0002	0.2	0.0004	0.4	0.001	0.5	0.002	0.7	0.004	1.1	0.006	1.3	30	2
4		0.0003	0.3	0.0006	0.4	0.001	0.6	0.003	0.8	0.006	1.1	0.009	1.4		4
6		0.0004	0.3	0.0008	0.4	0.001	0.6	0.003	0.8	0.007	1.1	0.011	1.4	50	6
8		0.0004	0.3	0.0009	0.4	0.002	0.6	0.004	0.8	0.008	1.2	0.013	1.4		8
10		0.0005	0.3	0.001	0.4	0.002	0.6	0.004	0.8	0.009	1.2	0.014	1.4		10
12		0.0005	0.3	0.001	0.4	0.002	0.6	0.005	0.8	0.010	1.2	0.016	1.5		12
14	30	0.0006	0.3	0.001	0.4	0.003	0.6	0.005	0.8	0.011	1.2	0.017	1.5		14
16		0.0006	0.3	0.001	0.4	0.003	0.6	0.005	0.8	0.012	1.2	0.018	1.5		16
18		0.0007	0.3	0.001	0.4	0.003	0.6	0.006	0.8	0.013	1.2	0.019	1.5		18
20		0.0007	0.3	0.001	0.4	0.003	0.6	0.006	0.8	0.013	1.2	0.020	1.5		20
25		0.0008	0.3	0.002	0.4	0.004	0.6	0.007	0.8	0.015	1.2	0.023	1.5	100	25
30		0.0009	0.3	0.002	0.4	0.004	0.6	0.008	0.8	0.017	1.2	0.025	1.5		30
35	50	0.0009	0.3	0.002	0.4	0.004	0.6	0.008	0.9	0.018	1.3	0.027	1.5		35
40		0.001	0.3	0.002	0.4	0.005	0.6	0.009	0.9	0.019	1.3	0.029	1.6		40
45		0.001	0.3	0.002	0.4	0.005	0.6	0.009	0.9	0.021	1.3	0.031	1.6		45
50		0.001	0.3	0.002	0.4	0.005	0.6	0.010	0.9	0.022	1.3	0.033	1.6		50
55		0.001	0.3	0.002	0.4	0.006	0.6	0.011	0.9	0.023	1.3	0.035	1.6		55
60		0.001	0.3	0.003	0.4	0.006	0.6	0.011	0.9	0.024	1.3	0.037	1.6		60
65		0.001	0.3	0.003	0.4	0.006	0.7	0.012	0.9	0.025	1.3	0.038	1.6		65
70		0.001	0.3	0.003	0.4	0.006	0.7	0.012	0.9	0.026	1.3	0.040	1.6		70
75		0.001	0.3	0.003	0.4	0.007	0.7	0.012	0.9	0.027	1.3	0.041	1.6		75
80		0.001	0.3	0.003	0.4	0.007	0.7	0.013	0.9	0.028	1.3	0.043	1.6		80
85		0.001	0.3	0.003	0.4	0.007	0.7	0.013	0.9	0.029	1.3	0.044	1.6		85
90		0.002	0.3	0.003	0.4	0.007	0.7	0.014	0.9	0.030	1.3	0.045	1.6		90
95	100	0.002	0.3	0.003	0.4	0.008	0.7	0.014	0.9	0.031	1.3	0.047	1.6		95
100		0.002	0.3	0.003	0.4	0.008	0.7	0.014	0.9	0.031	1.3	0.048	1.6		100
200		0.002	0.3	0.005	0.5	0.011	0.7	0.021	0.9	0.045	1.4	0.069	1.7	300	200
300		0.003	0.3	0.006	0.5	0.014	0.7	0.026	1.0	0.056	1.4	0.086	1.8		300
400		0.003	0.3	0.007	0.5	0.016	0.7	0.030	1.0	0.065	1.4	0.100	1.8	500	400
500		0.004	0.3	0.008	0.5	0.018	0.7	0.034	1.0	0.074	1.5	0.112	1.8		500
600		0.004	0.3	0.009	0.5	0.020	0.7	0.037	1.0	0.081	1.5	0.124	1.8		600
700		0.005	0.3	0.010	0.5	0.022	0.8	0.041	1.0	0.088	1.5	0.134	1.8		700
800	300	0.005	0.3	0.010	0.5	0.024	0.8	0.043	1.0	0.094	1.5	0.144	1.9		800
900		0.005	0.3	0.011	0.5	0.025	0.8	0.046	1.0	0.101	1.5	0.153	1.9		900
1000		0.005	0.4	0.011	0.5	0.026	0.8	0.049	1.0	0.106	1.5	0.162	1.9	1000	1000
2000	500	0.008	0.4	0.017	0.5	0.038	0.8	0.071	1.1	0.153	1.6	0.234	2.0		2000
3000		0.010	0.4	0.021	0.5	0.047	0.8	0.086	1.1	0.190	1.6	0.290	2.0		3000
4000		0.011	0.4	0.024	0.5	0.055	0.8	0.102	1.1	0.222	1.7	0.338	2.0		4000
5000		0.013	0.4	0.027	0.6	0.062	0.8	0.115	1.1	0.249	1.7	0.380	2.1		5000
6000		0.014	0.4	0.030	0.6	0.068	0.9	0.126	1.2	0.275	1.7	0.418	2.1		6000
7000	1000	0.015	0.4	0.032	0.6	0.074	0.9	0.137	1.2	0.298	1.7	0.454	2.1		7000
8000		0.017	0.4	0.034	0.6	0.080	0.9	0.147	1.2	0.320	1.7	0.487	2.1		8000
9000		0.018	0.4	0.037	0.6	0.085	0.9	0.157	1.2	0.340	1.7	0.519	2.1		9000
10 000		0.019	0.4	0.039	0.6	0.090	0.9	0.166	1.2	0.360	1.8	0.548	2.2	3000	10 000
20 000		0.027	0.4	0.056	0.6	0.129	0.9	0.239	1.2	0.519	1.8	0.791	2.2		20 000
30 000		0.033	0.4	0.069	0.6	0.160	0.9	0.296	1.3	0.644	1.9	0.981	2.3		30 000
40 000		0.039	0.4	0.081	0.6	0.187	1.0	0.345	1.3	0.750	1.9	1.14	2.3		40 000
50 000	3000	0.044	0.4	0.091	0.6	0.210	1.0	0.388	1.3	0.844	1.9	1.29	2.4		50 000
60 000		0.048	0.4	0.100	0.6	0.231	1.0	0.428	1.3	0.929	1.9	1.42	2.4		60 000

Table 4.43 Flow of saturated steam in heavy steel pipes — *continued*

q_m = mass flow rate   kg.s^{-1}
p = pressure (absolute)   kPa
$\Delta Z/l$ = pressure drop per unit length   Pa$^{1.929}$.m^{-1}
l_e = equivalent length of a component for $\zeta = 1$ m

$\Delta Z/l$	p	50 mm q_m	l_e	65 mm q_m	l_e	80 mm q_m	l_e	90 mm q_m	l_e	100 mm q_m	l_e	125 mm q_m	l_e	p	$\Delta Z/l$
1		0.008	1.7	0.016	2.5	0.025	3.1	0.037	3.7	0.051	4.4	0.092	5.8		1
2		0.011	1.8	0.023	2.6	0.036	3.2	0.054	3.9	0.074	4.5	0.133	6.1	50	2
4	50	0.017	1.9	0.034	2.7	0.052	3.3	0.077	4.0	0.107	4.7	0.192	6.3		4
6		0.021	1.9	0.042	2.7	0.065	3.4	0.096	4.1	0.132	4.8	0.238	6.5	100	6
8		0.024	2.0	0.049	2.8	0.076	3.5	0.112	4.2	0.154	4.9	0.277	6.6		8
10		0.027	2.0	0.055	2.8	0.085	3.5	0.126	4.3	0.173	5.0	0.312	6.7		10
12		0.030	2.0	0.060	2.8	0.094	3.5	0.138	4.3	0.191	5.0	0.344	6.8		12
14		0.032	2.0	0.065	2.9	0.102	3.6	0.150	4.3	0.207	5.1	0.373	6.8		14
16		0.035	2.0	0.070	2.9	0.109	3.6	0.161	4.4	0.222	5.1	0.400	6.9		16
18	100	0.037	2.0	0.075	2.9	0.116	3.6	0.172	4.4	0.237	5.2	0.426	6.9		18
20		0.039	2.1	0.079	2.9	0.123	3.7	0.181	4.4	0.250	5.2	0.450	7.0		20
25		0.044	2.1	0.089	3.0	0.138	3.7	0.204	4.5	0.281	5.3	0.507	7.1		25
30		0.048	2.1	0.098	3.0	0.152	3.7	0.225	4.5	0.310	5.3	0.558	7.1		30
35		0.052	2.1	0.106	3.0	0.165	3.8	0.244	4.6	0.336	5.4	0.606	7.2		35
40		0.056	2.1	0.114	3.1	0.177	3.8	0.262	4.6	0.361	5.4	0.650	7.3	300	40
45		0.060	2.2	0.121	3.1	0.189	3.8	0.279	4.6	0.384	5.5	0.692	7.3		45
50		0.063	2.2	0.128	3.1	0.200	3.9	0.295	4.7	0.406	5.5	0.732	7.4		50
55		0.066	2.2	0.135	3.1	0.210	3.9	0.310	4.7	0.427	5.5	0.769	7.4		55
60		0.069	2.2	0.141	3.1	0.220	3.9	0.324	4.7	0.447	5.5	0.806	7.4		60
65		0.072	2.2	0.147	3.1	0.229	3.9	0.339	4.8	0.467	5.6	0.841	7.5		65
70		0.075	2.2	0.153	3.2	0.239	3.9	0.352	4.8	0.485	5.6	0.874	7.5		70
75		0.078	2.2	0.159	3.2	0.247	3.9	0.365	4.8	0.504	5.6	0.907	7.5		75
80		0.081	2.2	0.165	3.2	0.256	4.0	0.378	4.8	0.521	5.6	0.938	7.6		80
85		0.084	2.2	0.170	3.2	0.264	4.0	0.390	4.8	0.538	5.7	0.969	7.6		85
90		0.086	2.3	0.175	3.2	0.273	4.0	0.402	4.8	0.555	5.7	0.999	7.6		90
95		0.089	2.3	0.180	3.2	0.280	4.0	0.414	4.9	0.571	5.7	1.03	7.6	500	95
100	300	0.091	2.3	0.185	3.2	0.288	4.0	0.425	4.9	0.586	5.7	1.06	7.7		100
200		0.131	2.4	0.267	3.4	0.416	4.2	0.614	5.1	0.846	6.0	1.52	8.0		200
300	500	0.163	2.4	0.331	3.4	0.516	4.3	0.761	5.2	1.05	6.1	1.89	8.2		300
400		0.190	2.5	0.386	3.5	0.600	4.4	0.886	5.3	1.22	6.2	2.20	8.3	1000	400
500		0.213	2.5	0.434	3.5	0.676	4.4	0.997	5.4	1.37	6.3	2.48	8.4		500
600		0.235	2.5	0.478	3.6	0.744	4.5	1.10	5.4	1.51	6.4	2.73	8.5		600
700		0.255	2.5	0.519	3.6	0.807	4.5	1.12	5.5	1.64	6.4	2.96	8.6		700
800		0.274	2.6	0.557	3.6	0.866	4.5	1.28	5.5	1.76	6.5	3.17	8.7		800
900		0.291	2.6	0.593	3.7	0.922	4.6	1.36	5.5	1.88	6.5	3.38	8.7		900
1000	1000	0.308	2.6	0.627	3.7	0.975	4.6	1.44	5.6	1.98	6.5	3.57	8.8		1000
2000		0.445	2.7	0.905	3.8	1.41	4.8	2.08	5.8	2.86	6.8	5.16	9.1		2000
3000		0.551	2.8	1.12	3.9	1.74	4.9	2.57	6.0	3.55	7.0	6.39	9.4	3000	3000
4000		0.642	2.8	1.31	4.0	2.03	5.0	3.00	6.1	4.13	7.1	7.44	9.5		4000
5000		0.722	2.9	1.47	4.1	2.29	5.1	3.34	6.1	4.65	7.2	8.38	9.6		5000
6000		0.795	2.9	1.62	4.1	2.52	5.1	3.71	6.2	5.12	7.3	9.22	9.7		6000
7000		0.863	2.9	1.76	4.1	2.73	5.2	4.03	6.3	5.56	7.3	10.0	9.8		7000
8000	3000	0.926	2.9	1.88	4.2	2.93	5.2	4.33	6.3	5.97	7.4	10.7	9.9		8000
9000		0.986	3.0	2.01	4.2	3.12	5.2	4.60	6.3	6.35	7.4	11.4	10		9000
10 000		1.04	3.0	2.12	4.2	3.23	5.3	4.87	6.4	6.71	7.5	12.1	10		10 000
20 000		1.50	3.1	3.06	4.4	4.76	5.5	7.03	6.7	9.69	7.8	17.4	11		20 000
30 000		1.86	3.2	3.79	4.5	5.90	5.6	8.71	6.8	12.0	8.0	21.6	11		30 000
40 000		2.17	3.2	4.42	4.6	6.87	5.7	10.1	6.9	14.0	8.1	25.2	11		40 000
50 000		2.44	3.3	4.97	4.6	7.74	5.8	11.4	6.9	15.7	8.2	28.3	11		50 000
60 000		2.69	3.3	5.48	4.7	8.52	5.8	12.6	7.1	17.3	8.3	31.2	11		60 000

Table 4.43 Flow of saturated steam in heavy steel pipes — *continued*

q_m	=	mass flow rate		$kg.s^{-1}$
p	=	pressure (absolute)		kPa
$\Delta Z/l$	=	pressure drop per unit length		$Pa^{1.929}.m^{-1}$
l_e	=	equivalent length of a component		
		for $\zeta = 1$		m

HEAVY GRADE STEEL

SATURATED STEAM

$\Delta Z/l$	p	150 mm q_m	l_e	175 mm q_m	l_e	200 mm q_m	l_e	225 mm q_m	l_e	250 mm q_m	l_e	300 mm q_m	l_e	p	$\Delta Z/l$
1		0.149	7.4	0.235	9.3	0.327	11	0.445	13	0.600	15	0.965	19		1
2		0.215	7.7	0.339	9.7	0.472	11	0.642	13	0.866	15	1.39	20	100	2
4	100	0.310	8.0	0.489	10	0.681	12	0.926	14	1.25	16	2.01	20		4
6		0.385	8.2	0.606	10	0.844	12	1.15	14	1.55	16	2.49	21		6
8		0.448	8.4	0.705	11	0.983	12	1.34	14	1.80	17	2.90	21		8
10		0.504	8.5	0.794	11	1.11	13	1.50	15	2.03	17	3.26	22		10
12		0.555	8.6	0.874	11	1.22	13	1.66	15	2.24	17	3.59	22		12
14		0.602	8.7	0.949	11	1.32	13	1.80	15	2.43	17	3.90	22	300	14
16		0.647	8.7	1.02	11	1.42	13	1.93	15	2.60	17	4.19	22		16
18		0.688	8.8	1.08	11	1.51	13	2.05	15	2.77	18	4.46	22		18
20		0.728	8.8	1.15	11	1.60	13	2.17	15	2.93	18	4.71	22		20
25		0.819	9.0	1.29	11	1.80	13	2.44	15	3.30	18	5.30	23		25
30	300	0.902	9.1	1.42	11	1.98	13	2.69	16	3.63	18	5.84	23		30
35		0.979	9.1	1.54	11	2.15	14	2.92	16	3.94	18	6.34	23	500	35
40		1.05	9.2	1.65	12	2.30	14	3.13	16	4.23	18	6.80	23		40
45		1.12	9.3	1.76	12	2.45	14	3.34	16	4.50	19	7.24	23		45
50		1.18	9.3	1.86	12	2.59	14	3.53	16	4.76	19	7.65	24		50
55		1.24	9.4	1.96	12	2.73	14	3.71	16	5.00	19	8.05	24		55
60		1.30	9.4	2.05	12	2.86	14	3.88	16	5.24	19	8.43	24		60
65		1.36	9.5	2.14	12	2.98	14	4.05	16	5.47	19	8.79	24		65
70		1.41	9.5	2.22	12	3.10	14	4.22	16	5.69	19	9.14	24		70
75		1.46	9.6	2.31	12	3.21	14	4.37	16	5.90	19	9.48	24		75
80	500	1.52	9.6	2.39	12	3.33	14	4.52	16	6.10	19	9.81	24		80
85		1.57	9.6	2.46	12	3.43	14	4.67	17	6.30	19	10.1	24		85
90		1.61	9.7	2.54	12	3.54	14	4.81	17	6.49	19	10.4	25		90
95		1.66	9.7	2.61	12	3.64	14	4.95	17	6.68	19	10.7	25		95
100		1.71	9.7	2.69	12	3.74	14	5.09	17	6.87	19	11.0	25	1000	100
200		2.46	10	3.88	13	5.40	15	7.35	17	9.91	20	15.9	26		200
300	1000	3.05	10	4.81	13	6.70	15	9.11	18	12.3	21	19.8	26		300
400		3.55	11	5.60	13	7.80	16	10.6	18	14.3	21	23.0	27		400
500		4.00	11	6.30	13	8.77	16	11.9	18	16.1	21	25.9	27		500
600		4.40	11	6.94	14	9.66	16	13.1	19	17.7	22	28.5	27		600
700		4.78	11	7.52	14	10.5	16	14.3	19	19.2	22	30.9	28		700
800		5.13	11	8.08	14	11.3	16	15.3	19	20.6	22	33.2	28		800
900		5.46	11	8.60	14	12.0	16	16.3	19	22.0	22	35.3	28	3000	900
1000		5.77	11	9.09	14	12.7	17	17.2	19	23.2	22	37.4	28		1000
2000	3000	8.33	12	13.1	15	18.3	17	24.9	20	33.5	23	53.9	29		2000
3000		10.3	12	16.3	15	22.7	18	30.8	21	41.6	24	66.8	30		3000
4000		12.0	12	18.9	15	26.4	18	35.9	21	48.4	24	77.8	31		4000
5000		13.5	12	21.3	15	29.7	18	40.4	21	54.5	25	87.6	31		5000
6000		14.9	12	23.5	15	32.7	18	44.5	21	60.0	25	96.5	31		6000
7000		16.2	13	25.5	16	35.5	19	48.3	22	65.1	25	105	32		7000
8000		17.4	13	27.3	16	38.1	19	51.8	22	69.9	25	112	32		8000
9000		18.5	13	29.1	16	40.5	19	55.1	22	74.4	25	120	32		9000
10 000		19.5	13	30.8	16	42.9	19	58.3	22	78.6	26	126	32		10 000
20 000		28.2	13	44.4	17	61.9	20	84.1	23	113	27	182	34		20 000
30 000		34.9	14	55.0	17	76.7	20	104	23	141	27	226	34		30 000
40 000		40.7	14	64.1	17	89.3	20	121	24	164	28	263	35		40 000
50 000		45.8	14	72.1	18	101	21	137	24	184	28	296	35		50 000
60 000		50.4	14	79.4	18	111	21	151	24	203	28	326	36		60 000

Table 4.44 Pressure factors Z for compressible flow, using $Z = p^{1.929}$

Pressure, p /kPa	Z /kPa$^{1.929}$	Pressure, p /kPa	Z /kPa$^{1.929}$	Pressure, p /kPa	Z /kPa$^{1.929}$	Pressure, p /kPa	Z /kPa$^{1.929}$	Pressure, p /kPa	Z /kPa$^{1.929}$
10	85	210	30 200	410	110 000	610	236 000	810	408 000
20	323	220	33 000	420	115 000	620	244 000	820	418 000
30	707	230	36 000	430	120 000	630	251 000	830	427 000
40	1 230	240	39 000	440	126 000	640	259 000	840	437 000
50	1 890	250	42 230	450	131 000	650	267 000	850	448 000
60	2 690	260	45 500	460	137 000	660	275 000	860	458 000
70	3 620	270	49 000	470	143 000	670	283 000	870	468 000
80	4 690	280	52 500	480	149 000	680	291 000	880	479 000
90	5 880	290	56 200	490	155 000	690	299 000	890	489 000
100	7 210	300	60 000	500	161 000	700	308 000	900	500 000
110	8 670	310	63 900	510	167 000	710	316 000	910	510 000
120	10 300	320	68 000	520	174 000	720	325 000	920	521 000
130	12 000	330	72 100	530	180 000	730	334 000	930	532 000
140	13 800	340	76 400	540	187 000	740	343 000	940	543 000
150	15 800	350	80 800	550	193 000	750	351 000	950	555 000
160	17 900	360	85 300	560	200 000	760	361 000	960	566 000
170	20 100	370	90 000	570	207 000	770	370 000	970	577 000
180	22 400	380	94 700	580	214 000	780	379 000	980	589 000
190	24 900	390	99 600	590	221 000	790	389 000	990	601 000
200	27 500	400	105 000	600	229 000	800	398 000	1000	612 000

4.6.2　Flow of condensate in pipes

The pressure drop of the condensate flow may be obtained in the same way as for hot water where:

(a)　thermostatic traps, having small pressure differentials between inlet and outlet, are employed and

(b)　air and other gases are prevented from entering the condensate drain by the use of automatic vents.

If air is required to traverse the condensate main and if flash steam is produced by the pressure drops in the trap and system then the resistance is greatly increased. The pressure drop in two-phase flow is always greater than for either phase individually, the calculation for which is beyond the scope of this Guide. In the absence of definite data it is recommended that where flash steam can occur the condensate mains should be sized for three times the normal hot water discharge.

4.7　Natural gas in pipes

4.7.1　General

It should be noted that gases are highly compressible and that the density therefore varies considerably with pressure and temperature. Although the viscosity varies little with pressure, that too varies with temperature. Thus pressure drops are therefore best obtained by direct calculation using the method explained in section 4.3. Although section 4.3 assumes incompressible flow (ρ = constant), the method may be used with reasonable accuracy so long as the drop in pressure along the pipe does not exceed 10 per cent of the initial (absolute) inlet pressure.

Nevertheless values have been calculated for a density ρ = 0.68 kg m^{-3} being reasonable for a pressure of 100 kPa and a temperature of 10°C. These are given in Table 4.45.

The table covers the regions of laminar and turbulent flows. In the intermediate zone, $2000 < Re < 3000$ the nature of the flow is unpredictable and the pressure drop similarly unpredictable. For $Re > 2000$ turbulent flow has been assumed in the calculation.

Table 4.45 Flow of natural gas in medium grade steel pipes

q_v	= volume flow rate	$m^3.s^{-1}$
c	= velocity	$m.s^{-1}$
$\Delta p/l$	= pressure drop per unit length	$Pa.m^{-1}$
l_e	= equivalent length of a component for $\zeta = 1$	m

★ $(Re) = 2000$ for $\rho = 0.68$ kg.m^{-3}
† $(Re) = 3000$

MEDIUM GRADE STEEL
NATURAL GAS

$\Delta p/l$	c	10 mm q_v	l_e	15 mm q_v	l_e	20 mm q_v	l_e	25 mm q_v	l_e	c	$\Delta p/l$
0.5		0.000028	0.1	0.000078	0.1	0.000250	0.4	0.000638	0.9		0.5
0.6		0.000033	0.1	0.000093	0.2	0.000300	0.4	0.000696★	0.9		0.6
0.7		0.000038	0.1	0.000108	0.2	0.000350	0.5	0.000725	0.9		0.7
0.8		0.000044	0.1	0.000124	0.2	0.000400	0.6	0.000755	0.9		0.8
0.9		0.000049	0.1	0.000139	0.2	0.000450	0.6	0.000784	0.9		0.9
1.0	0.5	0.000055	0.1	0.000155	0.2	0.000500★	0.7	0.000814	0.9		1.0
1.5		0.000082	0.2	0.000232	0.3	0.000583	0.6	0.000961†	1.0	1.5	1.5
2.0		0.000109	0.2	0.000309	0.4	0.000640	0.6	0.00112	1.0		2.0
2.5		0.000136	0.2	0.000386★	0.5	0.000697	0.5	0.00128	1.0		2.5
3.0		0.000163	0.3	0.000413	0.5	0.000754	0.5	0.00142	1.1		3.0
3.5	1.5	0.000190	0.3	0.000430	0.5	0.000813†	0.5	0.00155	1.1		3.5
4.0		0.000217	0.3	0.000447	0.5	0.000878	0.5	0.00167	1.1		4.0
4.5		0.000245	0.4	0.000465	0.4	0.000940	0.5	0.00179	1.1	3.0	4.5
5.0		0.000272	0.4	0.000482	0.4	0.000999	0.6	0.00190	1.1		5.0
5.5		0.000299	0.4	0.000499	0.4	0.00106	0.6	0.00201	1.1		5.5
6.0		0.000311★	0.4	0.000516	0.4	0.00111	0.6	0.00211	1.2		6.0
6.5		0.000317	0.4	0.000534	0.4	0.00116	0.6	0.00221	1.2		6.5
7.0		0.000323	0.4	0.000551	0.4	0.00121	0.6	0.00230	1.2		7.0
7.5		0.000329	0.4	0.000568	0.4	0.00126	0.6	0.00239	1.2		7.5
8.0		0.000335	0.4	0.000585	0.4	0.00131	0.6	0.00248	1.2		8.0
8.5		0.000341	0.4	0.000603†	0.4	0.00135	0.6	0.00257	1.2		8.5
9.0		0.000347	0.4	0.000624	0.4	0.00140	0.6	0.00265	1.2		9.0
9.5		0.000353	0.4	0.000643	0.4	0.00144	0.6	0.00274	1.2		9.5
10.0	3.0	0.000359	0.3	0.000663	0.4	0.00149	0.6	0.00282	1.2		10.0
12.5		0.000388	0.3	0.000753	0.4	0.00169	0.6	0.00319	1.2		12.5
15.0		0.000418	0.3	0.000836	0.4	0.00187	0.6	0.00354	1.3		15.0
17.5		0.000448†	0.3	0.000912	0.4	0.00204	0.7	0.00385	1.3		17.5
20.0		0.000480	0.3	0.000984	0.4	0.00220	0.7	0.00415	1.3		20.0
22.5		0.000514	0.3	0.00105	0.5	0.00235	0.7	0.00443	1.3		22.5
25.0		0.000545	0.3	0.00112	0.5	0.00249	0.7	0.00470	1.3		25.0

Table 4.45 Flow of natural gas in medium grade steel pipes — *continued*

$\Delta p/l$	c	32 mm q_v	l_e	40 mm q_v	l_e	50 mm q_v	l_e	65 mm q_v	l_e	c	$\Delta p/l$
0.5		0.00113	0.9	0.00164	1.0	0.00313	1.4	0.00637	2.1		0.5
0.6		0.00122	0.9	0.00182	1.0	0.00347	1.5	0.00707	2.1		0.6
0.7		0.00131†	0.9	0.00199	1.1	0.00380	1.5	0.00772	2.2		0.7
0.8		0.00141	0.9	0.00215	1.1	0.00410	1.5	0.00833	2.2		0.8
0.9	1.5	0.00151	0.9	0.00230	1.1	0.00439	1.5	0.00890	2.2		0.9
1.0		0.00161	0.9	0.00245	1.1	0.00466	1.6	0.00945	2.3		1.0
1.5		0.00204	1.0	0.00309	1.2	0.00587	1.7	0.0119	2.4	3.0	1.5
2.0		0.00240	1.0	0.00364	1.2	0.00691	1.7	0.0140	2.5		2.0
2.5		0.00273	1.0	0.00413	1.3	0.00784	1.8	0.0158	2.5		2.5
3.0	3.0	0.00303	1.1	0.00458	1.3	0.00869	1.8	0.0175	2.6		3.0
3.5		0.00331	1.1	0.00500	1.3	0.00947	1.8	0.0191	2.6	5.0	3.5
4.0		0.00357	1.1	0.00539	1.4	0.0102	1.9	0.0206	2.7		4.0
4.5		0.00381	1.1	0.00576	1.4	0.0109	1.9	0.0220	2.7		4.5
5.0		0.00405	1.1	0.00612	1.4	0.0116	1.9	0.0233	2.7		5.0
5.5		0.00427	1.1	0.00645	1.4	0.0122	1.9	0.0245	2.8		5.5
6.0		0.00449	1.2	0.00677	1.4	0.0128	2.0	0.0257	2.8		6.0
6.5		0.00469	1.2	0.00708	1.4	0.0134	2.0	0.0269	2.8		6.5
7.0		0.00489	1.2	0.00738	1.4	0.0139	2.0	0.0280	2.8		7.0
7.5	5.0	0.00508	1.2	0.00767	1.5	0.0145	2.0	0.0291	2.8		7.5
8.0		0.00527	1.2	0.00795	1.5	0.0150	2.0	0.0301	2.9		8.0
8.5		0.00545	1.2	0.00823	1.5	0.0155	2.0	0.0312	2.9		8.5
9.0		0.00563	1.2	0.00849	1.5	0.0160	2.1	0.0322	2.9		9.0
9.5		0.00580	1.2	0.00875	1.5	0.0165	2.1	0.0331	2.9		9.5
10.0		0.00597	1.2	0.00900	1.5	0.0170	2.1	0.0341	2.9		10.0
12.5		0.00676	1.2	0.0102	1.5	0.0192	2.1	0.0385	3.0	10.0	12.5
15.0		0.00748	1.3	0.0113	1.6	0.0212	2.1	0.0425	3.0		15.0
17.5		0.00814	1.3	0.0123	1.6	0.0231	2.2	0.0462	3.1		17.5
20.0		0.00877	1.3	0.0132	1.6	0.0248	2.2	0.0496	3.1		20.0
22.5		0.00935	1.3	0.0141	1.6	0.0264	2.2	0.0529	3.1		22.5
25.0		0.00991	1.3	0.0149	1.6	0.0280	2.2	0.0560	3.2		25.0

Table 4.45 Flow of natural gas in medium grade steel pipes — *continued*

q_v = volume flow rate $\quad\quad$ m^3.s^{-1}
c = velocity $\quad\quad\quad\quad\quad\quad$ m.s^{-1}
$\Delta p/l$ = pressure drop per unit length $\quad$ Pa.m^{-1}
l_e = equivalent length of a component
$\quad\quad$ for $\zeta = 1$ $\quad\quad\quad\quad\quad\quad$ m

$\star$ $(Re) = 2000$ $\quad\quad\quad$ for $\rho = 0.68$ kg.m^{-3}
$\dagger$ $(Re) = 3000$

$\Delta p/l$	c	80 mm		90 mm		100 mm		125 mm		150 mm		c	$\Delta p/l$
		q_v	l_e	q_v	l_e	q_v	l_e	q_v	l_e	q_v	l_e		
0.5		0.00988	2.6	0.0146	3.2	0.0203	3.8	0.0360	5.1	0.0583	6.5	3.0	0.5
0.6		0.0110	2.7	0.0162	3.2	0.0224	3.8	0.0400	5.2	0.0645	6.6		0.6
0.7		0.0120	2.7	0.0177	3.3	0.0245	3.9	0.0434	5.3	0.0703	6.7		0.7
0.8		0.0129	2.8	0.0190	3.4	0.0264	4.0	0.0468	5.3	0.0757	6.8		0.8
0.9		0.0138	2.8	0.0203	3.4	0.0282	4.0	0.0500	5.4	0.0808	6.9		0.9
1.0	3.0	0.0146	2.8	0.0216	3.5	0.0299	4.1	0.0530	5.5	0.0857	7.0		1.0
1.5		0.0184	3.0	0.0271	3.6	0.0375	4.3	0.0663	5.7	0.107	7.3	5.0	1.5
2.0		0.0216	3.1	0.0318	3.7	0.0440	4.4	0.0778	5.9	0.126	7.5		2.0
2.5	5.0	0.0245	3.2	0.0360	3.8	0.0498	4.5	0.0879	6.0	0.142	7.7		2.5
3.0		0.0271	3.2	0.0398	3.9	0.0550	4.6	0.0972	6.1	0.157	7.8		3.0
3.5		0.0295	3.3	0.0433	4.0	0.0599	4.7	0.106	6.2	0.170	7.9		3.5
4.0		0.0317	3.3	0.0466	4.0	0.0645	4.7	0.114	6.3	0.183	8.0	10.0	4.0
4.5		0.0339	3.4	0.0498	4.1	0.0688	4.8	0.121	6.4	0.195	8.1		4.5
5.0		0.0359	3.4	0.0572	4.1	0.0728	4.8	0.128	6.4	0.207	8.1		5.0
5.5		0.0378	3.4	0.0555	4.2	0.0767	4.9	0.135	6.5	0.218	8.2		5.5
6.0		0.0397	3.5	0.0583	4.2	0.0804	4.9	0.142	6.5	0.228	8.2		6.0
6.5		0.0414	3.5	0.0609	4.2	0.0840	5.0	0.148	6.6	0.238	8.3		6.5
7.0		0.0432	3.5	0.0634	4.2	0.0875	5.0	0.154	6.6	0.248	8.3		7.0
7.5		0.0448	3.5	0.0658	4.3	0.0908	5.0	0.160	6.6	0.257	8.4		7.5
8.0		0.0464	3.6	0.0681	4.3	0.0941	5.0	0.166	6.7	0.266	8.4		8.0
8.5		0.0480	3.6	0.0704	4.3	0.0972	5.1	0.171	6.7	0.275	8.5		8.5
9.0		0.0495	3.6	0.0726	4.3	0.100	5.1	0.176	6.7	0.284	8.5	15.0	9.0
9.5	10.0	0.0510	3.6	0.0748	4.4	0.103	5.1	0.182	6.8	0.292	8.5		9.5
10.0		0.0524	3.6	0.0769	4.4	0.106	5.1	0.187	6.8	0.300	8.6		10.0
12.5		0.0592	3.7	0.0868	4.5	0.120	5.2	0.210	6.9	0.338	8.7		12.5
15.0		0.0653	3.7	0.0957	4.5	0.132	5.3	0.232	7.0	0.373	8.8		15.0
17.5	15.0	0.0710	3.8	0.104	4.6	0.143	5.4	0.252	7.1	0.404	8.9		17.5
20.0		0.0763	3.8	0.112	4.6	0.154	5.4	0.270	7.1	0.434	9.0		20.0
22.5		0.0812	3.9	0.119	4.7	0.164	5.4	0.288	7.2	0.462	9.0		22.5
25.0		0.0860	3.9	0.126	4.7	0.173	5.5	0.304	7.2	0.488	9.1		25.0

Table 4.46 Flow of natural gas in copper pipes

q_v	= volume flow rate	$m^3.s^{-1}$
c	= velocity	$m.s^{-1}$
$\Delta p/l$	= pressure drop per unit length	$Pa.m^{-1}$
l_e	= equivalent length of a component	
	for $\zeta = 1$	m

$\star$ $(Re) = 2000$ for $\rho = 0.68$ kg.m^{-3}
$\dagger$ $(Re) = 3000$

COPPER, TABLE X
NATURAL GAS

$\Delta p/l$	c	6 mm q_v	l_e	8 mm q_v	l_e	10 mm q_v	l_e	12 mm q_v	l_e	15 mm q_v	l_e	c	$\Delta p/l$
0.5		0.000001	−0.0	0.000003	0.0	0.000007	0.0	0.000016	0.0	0.000040	0.1	0.3	0.5
0.6		0.000001	0.0	0.000003	0.0	0.000009	0.0	0.000019	0.0	0.000048	0.1		0.6
0.7		0.000001	0.0	0.000004	0.0	0.000010	0.0	0.000022	0.0	0.000055	0.1		0.7
0.8		0.000001	0.0	0.000004	0.0	0.000012	0.0	0.000025	0.0	0.000063	0.1		0.8
0.9		0.000002	0.0	0.000005	0.0	0.000013	0.0	0.000029	0.0	0.000071	0.1		0.9
1.0	0.1	0.000002	0.0	0.000005	0.0	0.000014	0.0	0.000032	0.0	0.000079	0.2		1.0
1.5		0.000002	0.0	0.000008	0.0	0.000021	0.0	0.000047	0.1	0.000118	0.2		1.5
2.0	0.15	0.000003	0.0	0.000010	0.0	0.000028	0.0	0.000063	0.1	0.000157	0.2	1.0	2.0
2.5		0.000004	0.0	0.000013	0.0	0.000035	0.0	0.000079	0.1	0.000197	0.3		2.5
3.0		0.000004	0.0	0.000015	0.0	0.000042	0.1	0.000094	0.1	0.000236	0.3		3.0
3.5		0.000005	0.0	0.000018	0.0	0.000049	0.1	0.000110	0.1	0.000275	0.4		3.5
4.0	0.3	0.000005	0.0	0.000020	0.0	0.000056	0.1	0.000125	0.2	0.000314	0.4		4.0
4.5		0.000006	0.0	0.000023	0.0	0.000062	0.1	0.000141	0.2	0.000341$\star$	0.5		4.5
5.0		0.000007	0.0	0.000025	0.0	0.000069	0.1	0.000157	0.2	0.000350	0.4		5.0
5.5		0.000007	0.0	0.000027	0.0	0.000076	0.1	0.000172	0.2	0.000360	0.4		5.5
6.0		0.000008	0.0	0.000030	0.0	0.000083	0.2	0.000188	0.2	0.000369	0.4		6.0
6.5		0.000008	0.0	0.000032	0.0	0.000090	0.2	0.000203	0.3	0.000379	0.4		6.5
7.0		0.000009	0.0	0.000035	0.0	0.000097	0.2	0.000219	0.3	0.000388	0.4		7.0
7.5		0.000010	0.0	0.000037	0.1	0.000104	0.2	0.000235	0.3	0.000398	0.4		7.5
8.0		0.000010	0.0	0.000040	0.1	0.000111	0.2	0.000250	0.4	0.000407	0.4		8.0
8.5		0.000011	0.0	0.000042	0.1	0.000117	0.2	0.000266$\star$	0.4	0.000417	0.4		8.5
9.0		0.000011	0.0	0.000045	0.1	0.000124	0.2	0.000271	0.3	0.000426	0.4		9.0
9.5		0.000012	0.0	0.000047	0.1	0.000131	0.2	0.000275	0.3	0.000436	0.4	3.0	9.5
10.0		0.000013	0.0	0.000050	0.1	0.000138	0.2	0.000278	0.3	0.000445	0.4		10.0
12.5		0.000016	0.0	0.000062	0.1	0.000172	0.3	0.000297	0.3	0.000493$\dagger$	0.4		12.5
15.0	1.0	0.000019	0.0	0.000074	0.1	0.000207$\star$	0.3	0.000316	0.3	0.000546	0.4		15.0
17.5		0.000022	0.0	0.000086	0.1	0.000224	0.3	0.000335	0.3	0.000597	0.4		17.5
20.0		0.000025	0.0	0.000099	0.1	0.000232	0.3	0.000354	0.3	0.000646	0.4		20.0
22.5		0.000028	0.0	0.000111	0.2	0.000240	0.3	0.000373	0.3	0.000692	0.4		22.5
25.0		0.000031	0.0	0.000123	0.2	0.000249	0.3	0.000392	0.3	0.000736	0.4		25.0

Table 4.46 Flow of natural gas in copper pipes — *continued*

$\Delta p/l$	c	22 mm q_v	l_e	28 mm q_v	l_e	35 mm q_v	l_e	42 mm q_v	l_e	54 mm q_v	l_e	c	$\Delta p/l$
0.5		0.000192	0.3	0.000543	0.7	0.000927	0.8	0.00143	1.0	0.00295	1.4		0.5
0.6		0.000231	0.3	0.000649$\star$	0.9	0.000990	0.8	0.00158	1.0	0.00328	1.4		0.6
0.7	1.0	0.000269	0.4	0.000675	0.8	0.00105	0.8	0.00173	1.0	0.00359	1.5		0.7
0.8		0.000307	0.4	0.000702	0.8	0.00112	0.8	0.00187	1.0	0.00388	1.5		0.8
0.9		0.000346	0.5	0.000728	0.7	0.00118$\dagger$	0.8	0.00201	1.0	0.00415	1.5		0.9
1.0		0.000384	0.5	0.000754	0.7	0.00125	0.8	0.00213	1.1	0.00442	1.6		1.0
1.5		0.000519$\star$	0.6	0.000886	0.7	0.00158	0.9	0.00270	1.1	0.00558	1.7		1.5
2.0		0.000565	0.5	0.00102$\dagger$	0.7	0.00187	0.9	0.00320	1.2	0.00659	1.7	3.0	2.0
2.5		0.000611	0.5	0.00117	0.7	0.00213	0.9	0.00364	1.2	0.00749	1.8		2.5
3.0		0.000658	0.5	0.00130	0.7	0.00237	1.0	0.00404	1.3	0.00832	1.8		3.0
3.5		0.000704	0.5	0.00142	0.7	0.00260	1.0	0.00442	1.3	0.00909	1.9		3.5
4.0		0.000751$\dagger$	0.5	0.00154	0.7	0.00281	1.0	0.00477	1.3	0.00981	1.9		4.0
4.5		0.000805	0.5	0.00165	0.8	0.00300	1.0	0.00510	1.3	0.0105	1.9	5.0	4.5
5.0		0.000856	0.5	0.00175	0.8	0.00319	1.0	0.00543	1.4	0.0111	2.0		5.0
5.5		0.000906	0.5	0.00185	0.8	0.00337	1.1	0.00573	1.4	0.0118	2.0		5.5
6.0	3.0	0.000953	0.5	0.00195	0.8	0.00355	1.1	0.00603	1.4	0.0124	2.0		6.0
6.5		0.000999	0.6	0.00204	0.8	0.00371	1.1	0.00631	1.4	0.0129	2.0		6.5
7.0		0.00104	0.6	0.00213	0.8	0.00388	1.1	0.00658	1.4	0.0135	2.1		7.0
7.5		0.00109	0.6	0.00222	0.8	0.00403	1.1	0.00685	1.4	0.0140	2.1		7.5
8.0		0.00113	0.6	0.00230	0.8	0.00419	1.1	0.00710	1.5	0.0146	2.1		8.0
8.5		0.00117	0.6	0.00239	0.8	0.00433	1.1	0.00736	1.5	0.0151	2.1		8.5
9.0		0.00121	0.6	0.00247	0.8	0.00448	1.1	0.00760	1.5	0.0156	2.1		9.0
9.5		0.00125	0.6	0.00254	0.8	0.00462	1.1	0.00784	1.5	0.0160	2.2		9.5
10.0		0.00129	0.6	0.00262	0.9	0.00476	1.2	0.00807	1.5	0.0165	2.2		10.0
12.5		0.00146	0.6	0.00298	0.9	0.00541	1.2	0.00917	1.6	0.0187	2.2		12.5
15.0	5.0	0.00163	0.6	0.00331	0.9	0.00600	1.2	0.0102	1.6	0.0208	2.3	10.0	15.0
17.5		0.00178	0.6	0.00361	0.9	0.00656	1.2	0.0111	1.6	0.0227	2.3		17.5
20.0		0.00192	0.7	0.00390	0.9	0.00707	1.3	0.0120	1.7	0.0244	2.4		20.0
22.5		0.00206	0.7	0.00418	1.0	0.00756	1.3	0.0128	1.7	0.0261	2.4		22.5
25.0		0.00219	0.7	0.00444	1.0	0.00803	1.3	0.0136	1.7	0.0277	2.4		25.0

4.8   Pressure loss factors

An extensive review of pressure loss factors has been undertaken. Many sources give conflicting information, and the most reliable sources quote research results of many years ago. The data presented here are those which are considered most reliable.

Recognising that engineers now use computer aids and programmable calculators, equations are included wherever these are readily available.

The pressure loss due to the insertion of a component such as an elbow is predominantly due to the vortices created downstream. Practical measurements close to the component would therefore be highly unrepeatable, and therefore unreliable. Experimental measurements of pressure are therefore made well upstream and downstream of the disturbance (i.e. 20 diameters downstream). The results are always quoted as the '*extra pressure drop due to the insertion of the component*'.

The pressure drop calculated for a component is always to be added to the pressure drop of the full length of the ducting (unless otherwise stated).

If the distance between one component and another (entries and exits included) is less than 20 diameters, no firm information is available. Thus the pressure drop could be more or less than the calculated figure depending on the type of component and the type of flow disturbances created, especially by the first.

With tees, there are three flows and three velocity pressures. Use of the velocity pressure of the combined flow is sometimes queried by engineers, and in that sense the use of branch velocity pressures for branch pressure loss might seem clearer. However, if branch velocity pressures were used, the value of the appropriate ζ would result in very large variations. When ζ is based upon the velocity pressure of the combined flow, the value of ζ along the straight is found to remain reasonably constant. This is clearly more convenient. In tabular form interpolations will be simpler and more accurate.

With elbows the predominant source of friction pressure drop is the flow separation and vortices downstream of the elbow or bend. This is very dependent upon the surface roughness and shape of the inner surface. Thus it is found that the values of ζ depend upon the diameter and on material. This applies to elbows of both pipework and ductwork CETIAT[22]. There is however little data on ductwork in this regard. On the other hand, the predominant source of friction pressure drop in tees is due to internal fluid friction resulting from the mixing or separation of different fluid streams. Here then the friction is not so dependent on surface effects, and the values of ζ do not depend on diameter, neither for pipework nor ductwork.

4.9 Pressure loss factors for pipework components

The guidance given on ageing allowances (section 4.4.3) should be applied to fittings as well as to straight pipes.

Care should be taken with any continental piping for which nominal diameters correspond closely to internal diameters.

4.9.1 Tees

Note that all values of ζ must be used with the velocity pressure of the combined flow. All the information below on tees is selected from Idelchik[2] and reproduced with the kind permission of his publishers.

Tees: laminar flow

In the laminar region, ζ is sensibly constant for flow to or from a branch, being rather higher than for turbulent flow.

Summarised from Idelchik[2] it could be said that for $Re < 2000$:

— laminar flow, converging flow branch $\zeta = 2.5$

— laminar flow, diverging flow branch $\zeta = 3.4$

Surprisingly, compared with turbulent flow, it is the pressure loss factor for the straight flow across the tee which is said to be most complex. Idelchik gives very complex relationships depending on the relative branch size, the relative flows and the Reynolds number. Carrying out a sample calculation for $Re = 100$, and with 50 per cent of the flow to or from a branch of the same diameter, revealed that, approximately:

— for converging flow, straight ζ is four times that for turbulent flow

— for diverging flow, straight ζ is four times that for turbulent flow.

Tees: turbulent flow

The value of ζ is seen to vary considerably with the ratio of the respective flows so no simplistic values can be given. The friction loss differs considerably between converging and diverging flows. The effect of branch diameter relative to the diameter of the part carrying the combined flow is appreciable and cannot be ignored.

For flows straight across the tee, the values of ζ do not vary significantly with the ratio of flows.

Some of the data are specifically for threaded pipe, some unspecified. Those which are unspecified must be assumed to be welded, clean bore.

A clear distinction is made between converging and diverging flows:

(a) Converging flows

The data are presented in terms of the fraction of flow arriving via the branch (q_b/q_c). For the flow arriving from the branch, the value of ζ varies considerably with both the ratio of the flows and on any change in pipe size between branch and combined straight.

The values of pressure loss factor for flows straight across the tee do not vary greatly with the ratio of flows (± 15 per cent).

When the flow from the branch is more than 20 per cent of the combined flow, threaded branches give losses 10–20 per cent more than for smooth connections. Conversely it would appear that when the branch flow is greater than 20 per cent of the total, threaded branches give losses 10–20 per cent less than for smooth connections.

(b) Diverging flows

For losses round to the branch, there is little difference between tees of smooth joints and those of screwed joints.

(c) The information as presented

Though strictly speaking the following data were obtained for malleable iron tees, in the absence of other reliable information there is little option but to accept them as valid also for other materials. Data for converging flow for tees have also been published by EDSU[23] which largely confirms that of Idelchik[2], though the values of ζ are smaller. The data of Idelchik are preferred and reproduced in Table 4.47, in the interests of consistency with the data for diverging flow (not tested by EDSU), and because the tests were conducted on real pipework, albeit only for malleable iron.

All values of ζ are quoted with respect to the velocity pressure of the combined flow.

Symbols:
ζ pressure loss factor
c velocity
d diameter
q volume or mass flow

Suffixes:
b   branch flow
c   combined flow
s   straight flow (single, not combined)

It will be observed that in the case of converging flows, it is possible under certain flow conditions for the flow from the branch to experience a negative pressure loss factor, i.e. to experience a pressure gain.

For those who are unfamiliar with the concept of a negative pressure loss factor, a small explanation is in order. With most of the flow going along the straight, this flow has the greater momentum. The flow arriving from the branch must be accelerated by frictional contact with the straight-flowing fluid, the effect more than counteracting the bend friction loss effect, and resulting in a pressure increase.

Table 4.47 Pressure loss coefficients for tees of various configurations[2] (reproduced from *Handbook of Hydraulic Resistance* by IE Idelchik (1994) by permission of CRC Press Inc., Boca Raton, Florida, USA ©CRC Press)

Converging flows: Values must be used with the velocity pressure of the combined flow.

(a) ζ_{b-c} From the branch

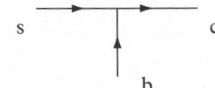

d_b/d_c	q_b/q_c									
	0.1	0.2	0.3	0.4	0.5	0.6	0.7	0.8	0.9	1.0
0.3	−0.50	2.97	9.9	19.7	32.4	48.8	66.5	86.9	110	136
0.44	−0.53	0.53	2.14	4.23	7.3	11.4	15.6	20.3	25.8	31.8
0.52	−0.69	0	1.11	2.18	3.76	5.90	8.38	11.3	14.6	18.4
0.59	−0.65	−0.09	0.59	1.31	2.24	3.52	5.20	7.28	9.23	12.2
0.66	−0.80	−0.27	0.26	0.84	1.59	2.66	4.00	5.73	7.40	6.60
0.74	−0.88	−0.48	0	0.53	1.15	1.89	2.92	4.00	5.36	6.00
1.00	−0.65	−0.40	−0.24	0.10	0.50	0.83	1.13	1.47	1.86	2.30

(b) ζ_{s-c} Along the straight

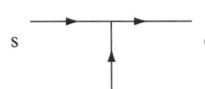

For all values of d_b/d_c,

d_b/d_c	q_b/q_c									
	0.1	0.2	0.3	0.4	0.5	0.6	0.7	0.8	0.9	1.0
All	0.70	0.64	0.60	0.65	0.75	0.85	0.92	0.96	0.99	1.00

(c) ζ_{1-c} From either side

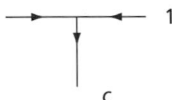

The pressure loss factor for one flow (1) depends on the proportion of that flow relative to the combined flow (q_1/q_c), and on the ratio of diameters, of branch 1 to that of the combined flow (d_1/d_c).

d_b/d_c	q_1/q_c										
	0	0.1	0.2	0.3	0.4	0.5	0.6	0.7	0.8	0.9	1.0
0.5	17.0	12.7	9.30	6.92	5.48	5.00	5.48	6.92	9.32	12.7	17
0.71	5.02	3.94	3.10	2.50	2.14	2.00	2.14	2.50	3.10	3.94	5.00
0.87	2.78	2.30	1.92	1.66	1.50	1.57	1.56	1.66	1.92	2.30	2.78
1.00	2.00	1.73	1.52	1.37	1.28	1.25	1.28	1.37	1.52	1.73	2.00

Diverging flows: Values must be used with the velocity pressure of the combined flow.

(d) ζ_{c-b} To the branch

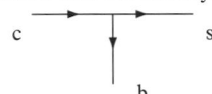

d_b/d_c	q_b/q_c									
	0.1	0.2	0.3	0.4	0.5	0.6	0.7	0.8	0.9	1.0
0.3	2.80	4.50	6.00	7.88	9.40	11.1	13.0	15.8	20.0	24.7
0.44	1.41	2.00	2.50	3.20	3.97	4.95	6.50	8.45	10.8	13.3
0.52	1.37	1.81	2.30	2.83	3.40	4.07	4.80	6.00	7.18	8.90
0.59	1.10	1.54	1.90	2.35	2.73	3.22	3.80	4.32	5.28	6.53
0.66	1.22	1.45	1.67	1.89	2.11	2.38	2.58	3.04	3.84	4.75
0.74	1.09	1.20	1.40	1.59	1.65	1.77	1.94	2.20	2.68	3.30
1.00	0.90	1.00	1.13	1.20	1.40	1.50	1.60	1.80	2.06	2.80

(e) ζ_{c-s} Along the straight

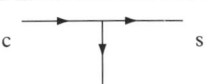

For all values of d_b/d_c,

d_b/d_c	q_b/q_c									
	0.1	0.2	0.3	0.4	0.5	0.6	0.7	0.8	0.9	1.0
All	0.70	0.64	0.60	0.57	0.55	0.51	0.49	0.55	0.62	0.70

(f) ζ_{c-1} To either side

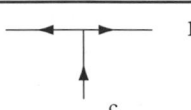

Note that in this case the data are more simply presented in terms of the ratio of velocity of the branch flow to that of the combined flow (c_1/c_c), thus eliminating variation with diameter. Note also that in this instance the data are available for screwthread and welded (smooth joint).

Joint type	c_1/c_c													
	0.1	0.2	0.3	0.4	0.5	0.6	0.7	0.8	0.9	1.0	1.2	1.4	1.6	2.0
Screwed	1.02	1.06	1.14	1.24	1.38	1.54	1.74	1.96	2.22	2.50	3.16	3.94	4.84	7.0
Welded	1.0	1.01	1.03	1.05	1.08	1.11	1.15	1.19	1.24	1.30	1.43	1.59	1.77	2.2

4.9.2 Elbows and bends

Laminar flow

As with the friction factor λ for straight pipes, the pressure loss factor ζ for elbows/bends is found to be much higher in laminar flow regime. This fact, generally overlooked, is important since pipes may often carry fluids of high viscosity which results in laminar flow.

Using Idelchik[2], to quote a simplistic value of ζ for laminar flow ($Re < 1000$), suggests that:

— $\zeta = 2.30$ could be taken for smooth elbows (laminar)

— $\zeta = 2.35$ could be taken for rough elbows (laminar)

It is interesting to note that in the laminar flow regime neither relative roughness nor pipe diameter have any great effect on the value of ζ.

Turbulent flow

Since the primary cause of friction pressure drop around elbows and tees is internal fluid friction resulting from the change of direction, it is sometimes considered that surface roughness will play little part. However, the work of Idelchik shows that surface roughness, particularly at the inner wall, does indeed have a very significant effect on pressure loss at elbows. Ageing allowances would therefore seem appropriate.

For the same reason it is found that the values of ζ for elbows depend upon diameter (unlike those for tees). As they also depend strongly on the internal shape of the elbow, they are likely to differ for elbows of different manufacturers and different materials. Available information is far from adequate so the values of Table 4.48 should be taken only as guidance.

Idelchik states that welded elbows give a greater pressure loss than smooth elbows since the welding seam upstream causes a discontinuity equivalent to increasing the effective roughness substantially. However, these differences are not specified. He shows that the effect of roughness is less for sharp elbows for which flow separation is the greatest factor.

In the ASHRAE Handbook, screwed elbows are quoted as giving greater pressure loss than flanged elbows. This would be due to the appreciable discontinuity at the upstream joint as with Idelchik's statement on welded elbows. In the absence of further information, the screwed elbow data could be extended to cover welded elbows.

The ASHRAE data are reproduced from the Hydraulic Institute publication of 1979[30]. Data are not given for different types of pipe material, but are given for:

— ζ for regular and long elbows, screwed pipe fittings

— ζ for regular and long elbows, flanged pipe fittings

Return bends

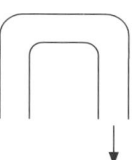

It is difficult to make accurate interpretations from the available conflicting data. For a close coupled return bend of 180°, Idelhik[2] quotes as a generality a total pressure drop equal to 1.6 × the pressure drop of the single 90° bend. Clearly the pressure drop will depend very much on the individual geometry of the bend. Table 4.49 of total ζ for the return 180° bend taken from Idelchik diagram 6–5, can therefore only be taken as rough guidance.

Table 4.49 Combined value of ζ for a return 180° bend (from Idelchik[2])

d_i/mm	13	25	37	50
ζ	1.23	0.70	0.65	0.58

Sudden contractions

For flow in the laminar regime, values of ζ are appreciably larger than for turbulent flow. The value varies considerably with Reynolds number in a very non-linear manner (e.g. for a value of $Re = 10$, $\zeta \approx 4.9$, referring to the velocity pressure at the smaller dimension.

Table 4.50 gives values for a sudden contraction, calculated from the equation given below. Standard pipe fittings would normally have a more rounded transition with slightly lower pressure drop.

Values of ζ are to be used with the velocity pressure at the smallest dimension, (0).

$$\Delta p = \zeta \tfrac{1}{2}\rho c_0^2$$

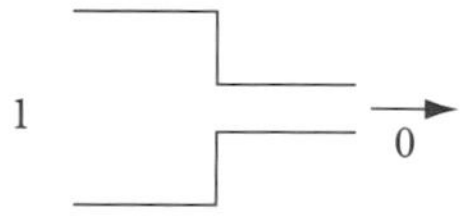

The following equation has been found by Idelchik[2] to fit the available data best for $Re > 10^4$:

$$\zeta = 0.5\left(1 - \frac{A_0}{A_1}\right)^{0.75}$$

Table 4.48 Values of pressure loss factor for elbows, (from ASHRAE [4] and Miller[29])

Type	10	15	20	25	32	40	50	75	100
(a) Sharp elbows									
Screwed fitting[4]	2.5	2.1	1.7	1.5	1.3	1.2	1.0	0.82	0.70
Rough, sharp inner edge[29]	1.56	1.45	1.35	1.3	1.24	1.18	1.15	1.10	1.10
Smooth radiused inner[29]	1.10	0.93	0.75	0.8	0.75	0.72	0.70	0.70	0.70
(b) Long elbows/bends									
Screwed fitting[4]			0.92	0.78	0.65	0.54	0.42	0.33	0.24
Smooth, $r/d > 1.5$[29]	0.57	0.53	0.49	0.46	0.43	0.42	0.4	0.4	0.4

Table heading spanning columns 10–100: *d*/mm

Table 4.50 Values of ζ for a sudden contraction (calculated from the equation given)

d_0/d_1	0	0.2	0.3	0.4	0.5	0.6	0.7	0.8	0.9	0.95
ζ		0.50	0.485	0.466	0.439	0.403	0.358	0.302	0.232	0.144 0.087

Sudden enlargements

The Borda–Carnot equation may be used for $Re > 10^4$. Values of ζ are to be used with the velocity pressure at the smallest dimension (0).

$$\zeta = \left(1 - \frac{A_0}{A_2}\right)^2$$

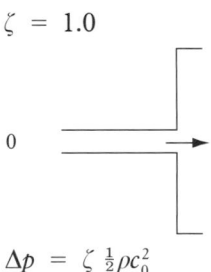

Table 4.51 Values of ζ for a sudden expansion (calculated from the equation given)

d_0/d_2	0	0.2	0.3	0.4	0.5	0.6	0.7	0.8	0.9	0.95
ζ	1	0.92	0.828	0.706	0.562	0.41	0.26	0.13	0.036	0.01

Sudden exit from a pipe into vessel

$$\zeta = 1.0$$

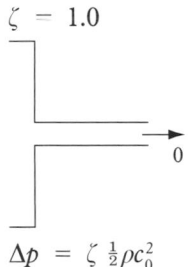

$$\Delta p = \zeta \tfrac{1}{2}\rho c_0^2$$

Sudden entry to a pipe from vessel

$$\zeta = 1.0$$

$$\Delta p = \zeta \tfrac{1}{2}\rho c_0^2$$

4.9.3 Valves

The value of pressure loss factor for a valve will depend upon the type of valve and the configuration used by the manufacturer. The following approximate data are given to help in initial design calculations. The actual values, supplied by the manufacturer, should be used as soon as they are available.

A single value can only be given for the valve in the fully open position. (The value of ζ will be infinite when the valve is closed.) For regulating valves therefore, the manufacturer will generally prefer to give a value of K, the valve capacity, for the valve fully open, and then a graph of the variation of the relative capacity with relative valve opening. (The value of K will be 0 when the valve is closed.) This is more useful than ζ when establishing the valve authority and overall control characteristic. The pressure drop due to the valve is easily calculated from equation (4.10) (section 4.37):

$$q_v = K\sqrt{\Delta p}$$

Further information can be obtained from Appendix 4.A5.

Globe valves

As with balancing valves, these are designed to give a better control characteristic, for use in either balancing or control. Values of ζ are for the valve fully open. These vary with the internal design of the valve so are included here only for guidance, so should be used with care.

Table 4.52 Approximate values of ζ (taken from Idelchik[2])

Standard globe valve; angular dividing walls							
dia/mm	20	40	80	100	150	200	300
ζ	8	4.9	4	4.1	4.4	4.7	5.4

Angle globe valve						
dia/mm	60	80	100	150	200	300
ζ	2.7	2.4	2.2	1.86	1.65	1.4

Gate valves

These should be installed for use in the fully open position. They are designed to give a clear bore when fully open. In operation, therefore, the pressure drop through them should be quite small.

Although designs may vary, the following should help give a rough estimate of pressure drop:

— spherical-seal gates, $\zeta = 0.03$

— plain-parallel gates, $\zeta = 0.3$

Non-return valve

A single value of ζ cannot be given.

For the gravity flap type, the greater the flow, the more the valve flap opens, but the higher becomes the value of ζ.

Spring-loaded non-return valves also behave very non-linearly, and so the manufacturer's characteristic must be used. The data in Table 4.54, taken from part of the performance characteristic of valves of one manufacturer, may be used for the purpose of first estimates.

Table 4.53 Approximate values of ζ for various flap positions of a non-return valve (from Idelchik[2])

a/degrees	20	30	40	50	60	70	75
ζ		1.7	3.2	6.6	14	30	62	90

Table 4.54 Approximate values of K for spring-loaded non-return valves

Nominal dia. / mm	25	38	50
K/m^3.h^{-1}.bar$^{-0.5}$	15	38	55

Pipe joints

(*a*) Welded metal tubes

The joints do not have a great effect and will generally be small in relation to the long tube lengths used. Therefore only brief guidance is given in Table 4.55, from Idelchik[2].

Table 4.55 Values of ζ for joints in welded metal pipe (from Idelchik[2])

dia./mm	200	300	400	500	600
ζ	0.026	0.0135	0.009	0.006	0.004

(b) Plastic joints

Tubes are likely to be shorter and smaller and the effects of joints can be significant. All of the values in Table 4.56 are taken from Idelchik[2] for values $1.8 \times 10^5 < Re < 5 \times 10^5$.

Table 4.56 Values of ζ for joints in plastic pipes (from Idelchik[2])

Joint	Diameter / mm						
	50	75	100	150	200	250	300
Welded	0.411	0.224	0.146	0.079	0.057	0.037	0.028
Flanged	0.131	0.13	0.114	0.096	0.079	0.062	0.045

Flexible steel-reinforced smooth rubber hoses

The pressure drop due to highly corrugated flexible hoses will of course depend on the length. The data are included here as the hose is frequently an extra component, but calculations of pressure drop will require the value of λ and use of the usual pipe equation (4.1).

As these hoses are usually used under pressure, the internal diameter will extend slightly with pressure.

The value of λ is found to increase considerably with pressure although the effect of Reynolds number is not great. Since lengths are not likely to be great, only a few guidance figures are given in Table 4.57, taken from Idelchik[2].

Table 4.57 Values of λ for flexible rubber hose (from Idelchik[2])

Pressure/kPa	25	50	100	150	200	250
$d = 65$ mm	0.03	0.03	0.04	0.07	0.09	0.11
$d = 100$ mm	0.02	0.02	0.03	0.03	0.04	0.06

4.9.4 Orifices

Orifice plates are generally used for flow measurement but may sometimes be installed to aid the balancing of flow. Although there is a gradual pressure recovery downstream of the orifice, they do incur a permanent pressure loss and permanent pumping costs. Thus if used for balancing purposes, consideration should be given to reducing the resistance of the parallel circuit instead.

Idelchik[2] gives data for various shapes of orifice and for the combination with a sudden contraction. In Table 4.58 the data for only one are given, namely for a thin sharp-edged orifice for which Re_o (within the orifice) $\geq 10^5$ and with the plate thickness l, $l/d_o \leq 0.0075$.

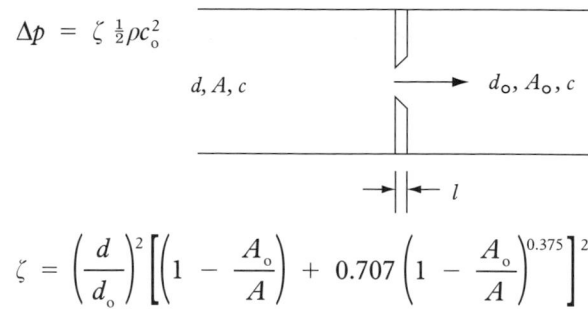

$$\Delta p = \zeta \tfrac{1}{2}\rho c_o^2$$

$$\zeta = \left(\frac{d}{d_o}\right)^2\left[\left(1 - \frac{A_o}{A}\right) + 0.707\left(1 - \frac{A_o}{A}\right)^{0.375}\right]^2$$

Table 4.58 Values of ζ calculated from the equation given for a sharp-edged orifice (from Idelchik[2], diagram 4–14)

A_o/A	0.02	0.05	0.1	0.3	0.4	0.5	0.6	0.7	0.8	0.9
ζ	7000	1050	245	18.2	8.25	4.00	2.00	0.97	0.42	0.13

Note that the values of ζ are to be used with the velocity pressure at the orifice.

Laminar flow through orifices

Idelchik[2] provides complex data on this. Some shows that laminar flow does not exist at:

$$Re > 10 \text{ for small orifices } (A_o/A = 0.05)$$

$$Re > 1000 \text{ for large orifices } (A_o/A = 0.64)$$

Some of these data are simplified and presented in Table 4.59.

Table 4.59 Values of ζ for laminar flow through an orifice with any edge (derived from Idelchik[2], diagram 4–19)

A_o/A	Re							
	30	40	60	100	200	400	1000	2000
0	3.00	2.46	2.31	2.20	2.20	2.33	2.28	2.31
0.2	2.54	2.14	1.89	1.80	1.76	1.79	1.73	1.71
0.3	2.20	1.80	1.58	1.53	1.47	1.44	1.43	1.42
0.4	1.85	1.51	1.35	1.20	1.13	1.09	1.10	1.11
0.5	1.49	1.15	0.99	0.83	0.76	0.75	0.78	0.80
0.6	1.14	0.86	0.63	0.56	0.52	0.53	0.56	0.58
0.7	0.78	0.58	0.46	0.36	0.35	0.36	0.40	0.41
0.8	0.52	0.37	0.27	0.21	0.21	0.22	0.23	0.24
0.9	0.26	0.19	0.14	0.10	0.09	0.09	0.10	0.11
0.95	0.06	0.06	0.05	0.04	0.04	0.05	0.05	0.05

4.10 Pressure loss factors for ductwork

4.10.1 General

An extensive review of pressure loss factors has been undertaken. Many sources give conflicting information. Note has been taken of recommended standard components given by HVCA[5] (1998), though in some instances no experimental data is given for these. Best estimates are presented and marked as such.

Technical names for components can vary, but for consistency HVCA labels are used.

The radius of curvature of bends is standardised as that of the mean air stream (HVCA being the exception).

The data presented here are those which are considered most reliable.

Much of the experimental data were obtained long ago and original source data are therefore difficult to obtain. Tolerance on the published values is therefore impossible to estimate. However, recent work by CETIAT[22] (1990) involved testing the same component, a simple elbow, made by six different manufacturers. A scatter of approximately ±40 per cent was obtained in the values of pressure loss factor, ζ. Though this was largely due to one rogue elbow it illustrates the problem of repeatability and accuracy.

Recognising that engineers now use computer aids and programmable calculators, equations are included wherever these are readily available.

No data are presented for laminar flow ($Re < 2 \times 10^3$). The data of SMACNA[3] only go down to $Re = 1 \times 10^4$. Laminar flow is in any case unlikely for ventilation air flow in ducts.

For tees ζ is to be used in conjunction with the velocity pressure of the combined flow, but the formula is included with each item as a reminder.

The first section deals with components for rectangular ductwork only.

4.10.2 Components for rectangular ductwork

When Reynolds number is required for rectangular ducts, the hydraulic diameter d_h is to be used, being four times the hydraulic radius:

$$d_h = \frac{2wh}{(w + h)} \qquad Re = \frac{\rho c d_h}{\eta}$$

where w and h are the width and height of the duct section.

4.10.3 Pressure loss factors for ductwork components and fittings

4.10.3.1 90° radius bends without vanes: rectangular (HVCA 86, 87)

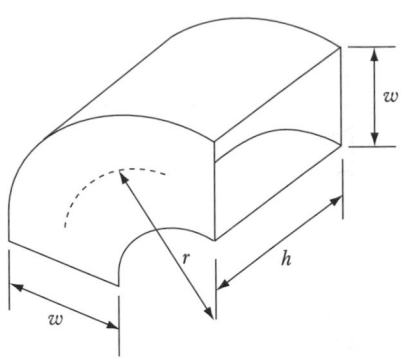

Table 4.60 Values of ζ for 90° bends (derived from data presented in diagram 6–1 of Idelchik[2])

r/w	h/w									
	0.25	0.5	0.75	1	1.15	2	3	4	6	8
0.5	1.28*	1.12*	1.04*	0.94*	0.84*	0.80*	0.80*	0.84*	0.85*	0.83*
0.75	0.64	0.56	0.52	0.47	0.42	0.40	0.40	0.42	0.45	0.46
1.0	0.35	0.29	0.26	0.24	0.21	0.20	0.20	0.21	0.22	0.23
1.5	0.33	0.27	0.24	0.21	0.19	0.18	0.17	0.18	0.19	0.20
2.0	0.35	0.26	0.23	0.21	0.18	0.17	0.17	0.17	0.18	0.18
2.5	0.35	0.26	0.23	0.20	0.18	0.17	0.16	0.16	0.17	0.17
4.0	0.44	0.31	0.26	0.23	0.20	0.18	0.17	0.17	0.18	0.18

* These values, being for the case of a sharp inner corner, are best estimates only. The values for ζ must be greater than those for $r/w = 0.75$. Equally, they must be less than for a mitred elbow.

The values have been calculated for $Re = 10^5$. For values of Re other than 10^5 the values in Table 4.60 should be multiplied by the correction factor C_1 given in Table 4.61.

Table 4.61 Correction factor C_1 for different values of Re (Koch[27] derived from Idelchik[2])

$Re \times 10^{-5}$	0.1	0.2	0.4	0.6	1	2
C_1 (for $r/w = 1$)	1.08	1.05	1.025	1.01	1	0.99

It should be noted that for the same duct area $w \times h$ and for the same radius of the inner part of the bend. 'Easy' bends ($h > w$) give values of ζ which are appreciably less than for 'difficult' bends ($h < w$).

For bends of angles a other than 90°, no simple table is possible, so the full calculation will need to be made using the following equation and data (Table 4.62) from Idelchik[2].

$$\zeta = C_2 \zeta_r C_3 + [3.29 \times 10^{-4} \, a \, r/d_h]$$

Table 4.62 Values of C_2, ζ_r and C_3

a	0°	20°	30°	45°	60°	75°	90°		
C_2	0	0.31	0.45	0.60	0.78	0.90	1.00		

r/w	0.5	0.75	1.0	1.5	2.0				
ζ_r	1.18*	0.45	0.21	0.17	0.15				

h/w	0.25	0.50	0.75	1.0	1.5	2	3	4	6	8
C_3	1.30	1.17	1.09	1.00	0.90	0.85	0.85	0.90	0.98	1.00

* This value seems high and leads to values of ζ higher than those for mitred elbows.

4.10.3.2 Short radius bends with vanes (splitters): rectangular (HVCA 88)

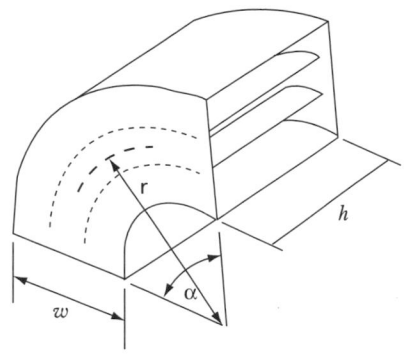

Table 4.63 Values of ζ for short radius 90° bends with vanes[3] (reproduced from *HVAC Systems Duct Design* (1990) by permission of the Sheet Metal and Air-Conditioning Contractors' Association (SMACNA), Chantilly, Virginia, USA)

r/w	h/w										
	0.25	0.5	1.0	1.5	2.0	3.0	4.0	5.0	6.0	7.0	8.0
(a) 1 turning vane											
0.55	0.52	0.40	0.43	0.49	0.55	0.66	0.75	0.84	0.93	1.0	1.1
0.60	0.36	0.27	0.25	0.28	0.30	0.35	0.39	0.42	0.46	0.49	0.52
0.65	0.28	0.21	0.18	0.19	0.20	0.22	0.25	0.26	0.28	0.30	0.32
0.70	0.22	0.16	0.14	0.14	0.15	0.16	0.17	0.18	0.19	0.20	0.21
0.75	0.18	0.13	0.11	0.11	0.11	0.12	0.13	0.14	0.14	0.15	0.15
0.80	0.15	0.11	0.09	0.09	0.09	0.09	0.10	0.10	0.11	0.11	0.12
0.90	0.11	0.08	0.07	0.06	0.06	0.06	0.06	0.07	0.07	0.07	0.07
1.00	0.09	0.06	0.05	0.05	0.04	0.04	0.04	0.05	0.05	0.05	0.05
(b) 2 turning vanes											
0.55	0.26	0.20	0.22	0.25	0.28	0.33	0.37	0.41	0.45	0.48	0.51
0.60	0.17	0.13	0.11	0.12	0.13	0.15	0.16	0.17	0.19	0.20	0.21
0.65	0.12	0.09	0.08	0.08	0.08	0.09	0.10	0.10	0.11	0.11	0.11
0.70	0.09	0.07	0.06	0.05	0.06	0.06	0.06	0.06	0.07	0.07	0.07
0.75	0.08	0.05	0.04	0.04	0.04	0.04	0.05	0.05	0.05	0.05	0.05
0.80	0.06	0.04	0.03	0.03	0.03	0.03	0.03	0.03	0.04	0.04	0.04
0.90	0.05	0.03	0.03	0.02	0.02	0.02	0.02	0.02	0.02	0.02	0.02
1.00	0.03	0.02	0.02	0.02	0.02	0.01	0.01	0.01	0.01	0.01	0.01
(c) 3 turning vanes											
0.55	0.11	0.10	0.12	0.13	0.14	0.16	0.18	0.19	0.21	0.22	0.23
0.60	0.07	0.05	0.06	0.06	0.06	0.07	0.07	0.08	0.08	0.08	0.09
0.65	0.05	0.04	0.04	0.04	0.04	0.04	0.04	0.04	0.04	0.05	0.05
0.70	0.03	0.03	0.03	0.03	0.03	0.03	0.03	0.03	0.03	0.03	0.03
0.75	0.03	0.02	0.02	0.02	0.02	0.02	0.02	0.02	0.02	0.02	0.02
0.80	0.03	0.02	0.02	0.02	0.02	0.01	0.01	0.01	0.01	0.01	0.01
0.90	0.02	0.01	0.01	0.01	0.01	0.01	0.01	0.01	0.01	0.01	0.01
1.00	0.01	0.01	0.01	0.01	0.01	0.01	0.01	0.01	0.01	0.01	0.01

Table 4.64 Correction factor C_1 for bends of other angles (Idelchik[2])

a	0°	20°	30°	45°	60°	75°	90°	110°	130°	150°	180°
C_1	0	0.31	0.45	0.60	0.78	0.90	1.00	1.13	1.20	1.28	1.40

Table 4.65 Advised positions of splitters, from HVCA DW/144

w/mm	No. of splitters	Splitter position		
		A	B	C
400–800	1	$w/3$		
801–1600	2	$w/4$	$w/4$	
1601–2000	3	$w/8$	$w/3$	$w/2$

4.10.3.3 90° radius bends of varying area: rectangular ('drop cheek bends') (HVCA 118)

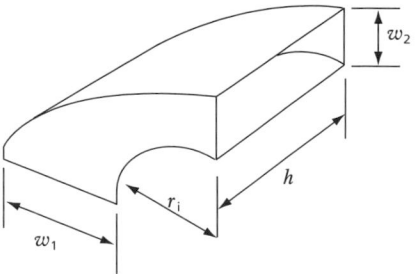

Table 4.66 gives values of the correction factor C_3 to be applied to the values of ζ given in Table 4.60 for bends of equal areas.

$$\zeta^\star = C_3 \times \zeta$$

The values of C_3 given in Table 4.66 are derived from information provided by Idelchik and are presented as a best estimate. It should be noted that in this instance the radius of curvature is that of the inner surface.

Note that the values of $\zeta^\star$ which result are to be applied to the velocity pressure at the position of the smallest area.

Table 4.66 Values of C_3 (derived from Idelchik[2])

r_i/w_{min}	A_2/A_1							
	0.2	0.5	1	1.5	2.0	3.0	4.0	5.0
0.6	0.34	0.68	1	1.12	1.19	1.26	1.30	1.32
0.65	0.29	0.60	1	1.18	1.31	1.42	1.48	1.52
0.7	0.26	0.50	1	1.26	1.44	1.62	1.72	1.80
0.8	0.30	0.50	1	1.43	1.74	2.06	2.26	2.40
0.9	0.30	0.50	1	1.50	1.87	2.30	2.50	2.70
1.5	0.36	0.54	1	1.60	2.00	2.46	2.77	3.00
	$\Delta p = C_3 \zeta \tfrac{1}{2}\rho c_2^2$				$\Delta p = C_3 \zeta \tfrac{1}{2}\rho c_1^2$			

4.10.3.4 90° mitred throat bend (up to 400 mm wide) (HVCA 85)

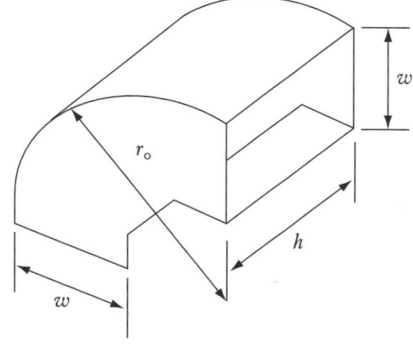

Table 4.67 Values for ζ for 90° mitred throat bend

h/w	0.25	0.5	0.75	1	1.5	2	3	4	5	6	8
ζ	0.31★	0.29★	0.27★	0.26★	0.24★	0.23★	0.22★	0.22★	0.22★	0.23★	0.23★

★ These values are estimates. No experimental values exist to justify them. They are based on lying between a mitred elbow with bevel inner, and a normal round bend. A confidence tolerance of ±30 per cent would seem appropriate.

The figures have been based on an assumption of $t/w = 0.5$, where t is the length of the bevel, and $r_o/w = 1.5$, where r_o is the radius of the outer surface.

4.10.3.5   Elbow, mitred, rectangular, any angle

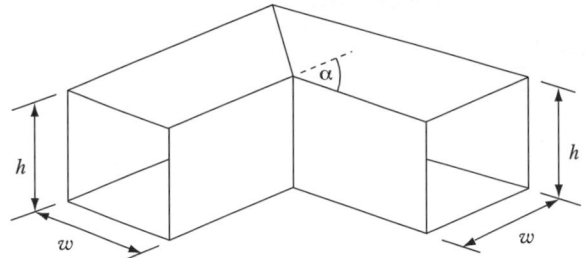

Table 4.68 has been derived from data of Idelchik[2].

Table 4.68 Values of ζ for mitred elbow (from Idelchik[2])

a	h / w										
	0.25	0.5	0.75	1	1.5	2	3	4	5	6	8
20°	0.14	0.13	0.13	0.12	0.12	0.11	0.1	0.1	0.09	0.09	0.09
30°	0.17	0.17	0.16	0.16	0.15	0.14	0.13	0.12	0.12	0.11	0.11
45°	0.35	0.34	0.33	0.32	0.3	0.29	0.26	0.25	0.24	0.23	0.22
60°	0.61	0.59	0.58	0.56	0.53	0.5	0.46	0.43	0.42	0.4	0.39
75°	0.89	0.86	0.84	0.81	0.77	0.73	0.67	0.63	0.6	0.58	0.56
90°	1.31	1.27	1.24	1.19	1.13	1.07	0.99	0.93	0.89	0.86	0.83

With the exception of small angles, similar values of ζ may be obtained using the following algorithm, being adapted from curve-fits by Idelchik[2].

$$\zeta = \left[0.97 - 0.13 \ln \frac{h}{w}\right]\left[0.89 + \frac{40}{a}\cos^2(a - 45)\right]$$
$$\times \left[0.95 \sin^2\left(\frac{a}{2}\right) + 2.05 \sin^4\left(\frac{a}{2}\right)\right]$$

4.10.3.6   Elbow, 90° rectangular, rounded inner corner

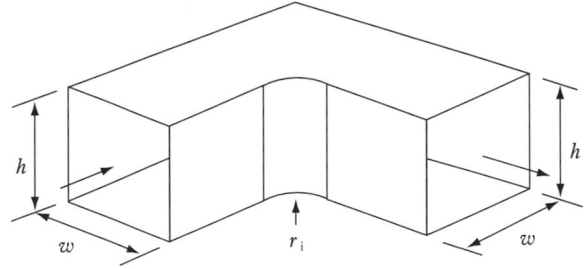

Table 4.69 has been derived from data of Idelchik[2].

Table 4.69 Values of ζ for 90° elbow with rounded inner corner (from Idelchik[2])

r_i/w	h/w						
	0.5	0.75	1	2	3	4	6
0.05	1.31	1.22	1.12	0.95	0.95	1.01	1.10
0.1	1.05	0.98	0.90	0.77	0.77	0.81	0.88
0.2	0.85	0.79	0.73	0.62	0.62	0.66	0.72
0.3	0.67	0.64	0.59	0.51	0.51	0.53	0.58
0.5	0.60	0.56	0.52	0.45	0.45	0.47	0.51
0.7	0.65	0.51	0.47	0.41	0.41	0.43	0.46

It is to be expected that the values of ζ for the rounded inner corner should be lower than for a 90° mitred corner, section 4.10.3.5 above. In most instances this is the case, but there are a few contradictions. Although both sets of data have been derived from Idelchik, his information is derived from many other sources so some discrepancies can be expected. Comparing the values for the sharpest inner corner above, $r_i/w = 0.05$, with those of a sharp corner, section 4.10.3.5, section 4.10.3.6 shows a different trend. There is therefore some doubt about the validity of the values in Table 4.69 but it is the only source of data available.

4.10.3.7   90° rectangular mitred elbows of unequal areas

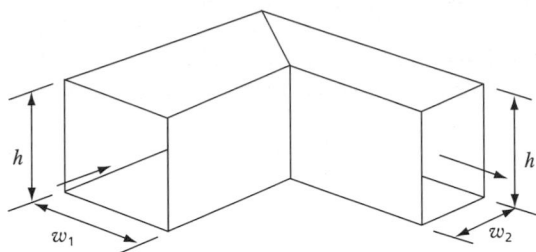

Table 4.70 for ζ is a modification of that from diagram 6–6 of Idelchik[2].

Table 4.70 Values of ζ for rectangular mitred elbow of unequal area (from Idelchik[2])

h/w_1	w_1/w_2						
	0.6	0.8	1	1.2	1.4	1.6	2
0.25	1.76	1.43	1.24	1.14	1.09	1.06	1.06
1	1.70	1.36	1.15	1.02	0.95	0.90	0.84
4	1.46	1.10	0.90	0.81	0.76	0.72	0.66
∞	1.50	1.04	0.79	0.69	0.63	0.6	0.55

Note that the value of ζ must be applied to the upstream velocity pressure, i.e.:

$$\Delta p = \zeta \frac{1}{2} \rho c_1^2$$

Note that in the case of equal areas, $w_1/w_2 = 1$, there are slight discrepancies compared with the 90° mitred elbow of section 4.10.3.5.

4.10.3.8   Rectangular mitred elbows with vanes

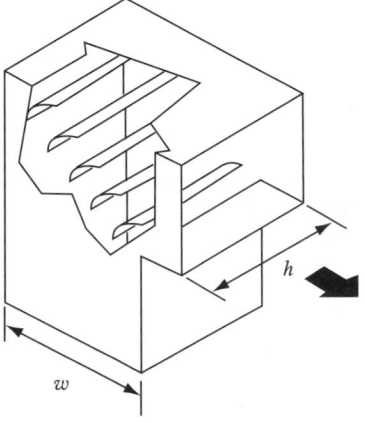

For blades of 50 mm radius (HVCA DW/144) and a spacing of 38 mm, SMACNA[3] gives the data shown in Table 4.71.

Table 4.71 Values of ζ for rectangular mitred elbow with vanes (SMACNA[2])

Type	Velocity/m.s^{-1}			
	5	7.5	10	12.5
Single skin vane	0.24	0.23	0.22	0.20
Double skin vane	0.43	0.42	0.41	0.40

Note that no data are available for vanes in elbows of unequal areas. The blades would need to have appropriate angles to ensure that downstream velocity distribution does not have eddies on either wall, since it is the avoidance of these eddies which reduces the pressure loss.

4.10.3.9 Bends in close proximity, rectangular ('gooseneck')

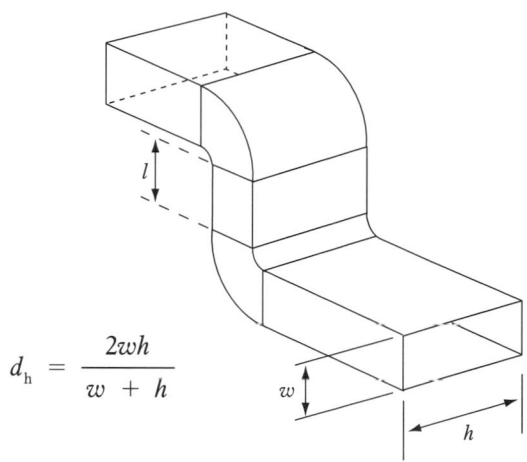

$$d_h = \frac{2wh}{w + h}$$

Derived from Idelchik[2], diagram 6–18.

The combined effect of the two bends is given below:

For:
$l / d_h < 1$: $\zeta = 1.7 \times \zeta$ of a single bend (section 4.10.3.1)

$1 < l / d_h < 3$: $\zeta = 1.6 \times \zeta$ of a single bend

$3 < l / d_h < 9$: $\zeta = 1.5 \times \zeta$ of a single bend

$9 < l / d_h < 11$: $\zeta = 1.6 \times \zeta$ of a single bend

$11 < l / d_h < 13$: $\zeta = 1.7 \times \zeta$ of a single bend

$13 < l / d_h < 17$: $\zeta = 1.8 \times \zeta$ of a single bend

$17 < l / d_h < 18$: $\zeta = 1.9 \times \zeta$ of a single bend

$18 < l / d_h$: $\zeta = 2.0 \times \zeta$ of a single bend

The effect of radius r is small, but the effect of separation l is not. Variation of h/w should also have an effect but to what extent is not clear.

Note that the above does not include for the pressure drop of the length of separation. The separation l, should be added to the length of straight ductwork of the same size.

4.10.3.10 Bends in close proximity, rectangular, through perpendicular plane

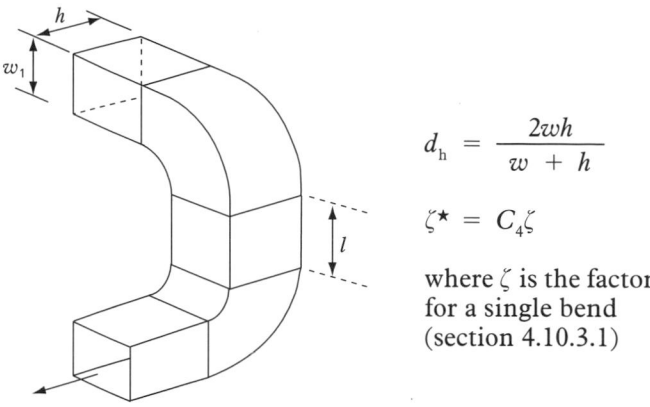

$$d_h = \frac{2wh}{w + h}$$

$$\zeta^\star = C_4\zeta$$

where ζ is the factor for a single bend (section 4.10.3.1)

Table 4.72 Values of correction factor C_4 (from Idelchik[2] (diagram 6–19))

l/d_h	1	2	3	4	6	8	10	12	14	>14
C_4	1.80	1.60	1.55	1.55	1.65	1.8	1.9	1.93	1.99	2

Since bends of this type involve one bend which is 'easy' and the other 'difficult', it is difficult to see which bend should be used for ζ. Use of ζ for a 'difficult' bend will give a conservative value.

Note that the above does not include the pressure drop of the length of separation. The separation l, should be added to the length of straight ductwork of the same size.

4.10.3.11 Elbows in close proximity, rectangular

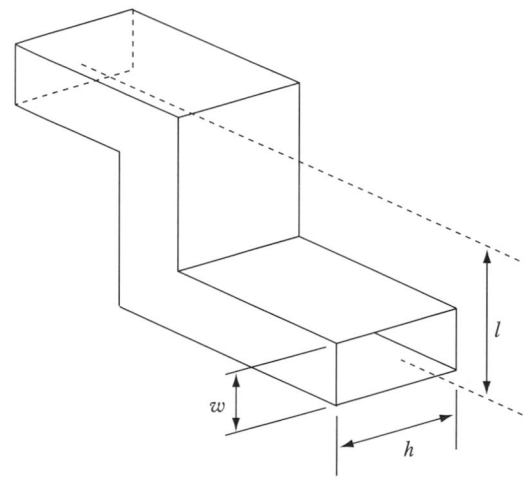

Table 4.73 Values of ζ for the whole combined double elbow, including the separating length (derived from Idelchik[2], diagram 6–13)

Offset step	h/w							
l/w	0.25	0.5	1	1.5	2	3	5	8
0.6	1	0.98	0.91	0.87	0.82	0.79	0.69	0.64
0.8	1.8	1.7	1.6	1.5	1.5	1.3	1.22	1.14
1	2.9	2.8	2.7	2.5	2.4	2.2	2	1.9
1.8	4.8	4.6	4.3	4	3.8	3.5	3.2	3
2	4.6	4.5	4.2	4	3.8	3.5	3.2	3
2.4	4.1	4	3.7	3.5	3.3	3.1	2.8	2.6
3	3.7	3.6	3.4	3.2	3	2.8	2.5	2.3
4	3.5	3.4	3.2	3	2.9	2.6	2.4	2.2
6	3.3	3.2	3	2.9	2.7	2.5	2.3	2.2
10	2.9	2.8	2.7	2.5	2.4	2.2	2	1.9

Notes

(1) The maximum friction pressure loss occurs when the offset is around $l/w = 1.8$.

(2) In this instance l is defined as the step and not the separation.

4.10.3.12 Elbows in close proximity, rectangular, through perpendicular plane

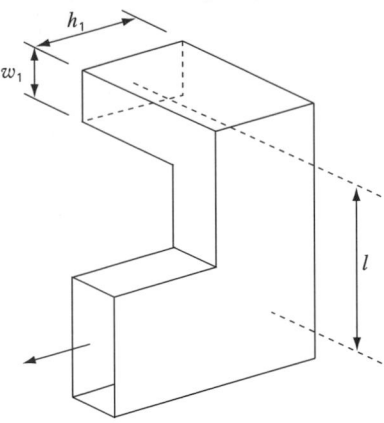

Table 4.74 Values of ζ for the whole combined double elbow, including the separating length (derived from Idelchik[2], diagram 6–13)

Step	h_1/w_1							
l/w_1	0.25	0.5	1	1.5	2	3	5	8
0	1.26	1.23	1.15	1.09	1.04	0.95	0.86	0.81
0.6	2.7	2.6	2.4	2.3	2.2	2	1.8	1.7
1	3.8	2.6	3.5	2.3	2.2	2	1.8	1.7
1.5	3.6	3.5	3.2	3	2.9	2.7	2.4	2.3
2	3.5	3.4	3.2	3	2.8	2.6	2.4	2.2
3	3.5	3.5	3.2	3.1	2.9	2.7	2.4	2.3
5	3.3	3.2	3	2.8	2.6	2.5	2.3	2.1
7	3.1	3	2.8	2.7	2.6	2.4	2.2	2
10	2.9	2.8	2.6	2.5	2.4	2.2	2	1.9

Notes

(1) In this instance l is defined as the step and not the separation.
(2) The outlet has the same dimensions h and w, but ζ is given in terms of w_1 and h_1.
(3) Although the data available are for any aspect ratio h/w, it is perhaps surprising that high values, implying that the second elbow is 'difficult', do not result in higher values of ζ. It is easier to have confidence in the values for a square or almost square duct.

4.10.3.13 Angled offsets (HVCA 96, 97, 98)

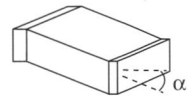 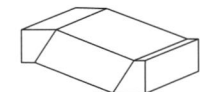 

Whether angled offset (96), mitred offset (97) or radiussed offset (98), no data are available.

A maximum angle of 30° is recommended. The values of ζ in Table 4.74 are derived from those for two mitred elbows with no reduction made for close proximity. It is expected that the radiussed offset would give lower pressure drop, but it is difficult to estimate the reduction.

Table 4.75 Values of ζ for angled offsets

a	h/w										
	0.25	0.5	0.75	1	1.5	2	3	4	5	6	8
20°	0.28	0.26	0.26	0.24	0.24	0.22	0.20	0.20	0.18	0.18	0.18
30°	0.34	0.34	0.32	0.32	0.30	0.28	0.26	0.24	0.24	0.22	0.22

4.10.3.14 Depression to avoid an obstruction

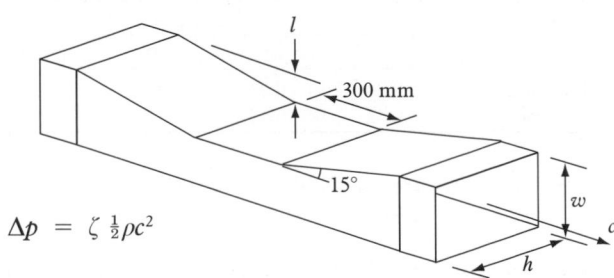

$$\Delta p = \zeta \tfrac{1}{2}\rho c^2$$

Table 4.76 Values of ζ for depression (SMACNA[3])

h/w	l/h			
	0.125	0.15	0.25	0.3
1	0.26	0.3	0.33	0.35
4	0.1	0.14	0.22	0.3

4.10.3.15 4 × 45° radius bends to avoid an obstruction

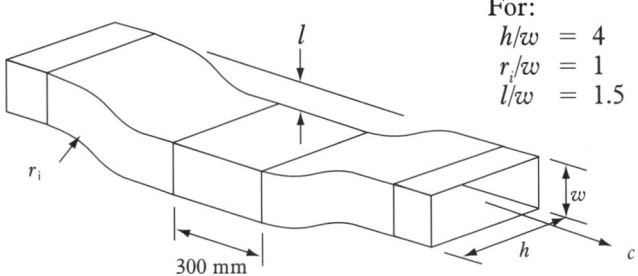

For:
$h/w = 4$
$r_i/w = 1$
$l/w = 1.5$

Table 4.77 Values of ζ for 4 × 45° radius bends (SMACNA[3])

c/m.s^{-1}	4	6	8	10	12
ζ	0.18	0.22	0.24	0.25	0.26

4.10.3.16 Opposed blade dampers

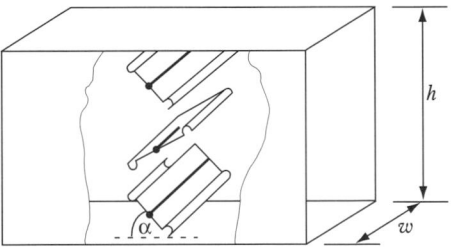

n = number of blades

Crimped leaf edges

$$x = \frac{nw}{2(h + w)}$$

Table 4.78 Values of ζ for opposed blade damper (from ASHRAE[4])

x	Angle of blades from axis of duct a								
	0°	10°	20°	30°	40°	50°	60°	70°	80°
0.3	0.52	0.79	1.91	3.77	8.55	19.5	70.1	295	807
0.4	0.52	0.85	2.07	4.61	10.4	26.7	92.9	346	926
0.5	0.52	0.93	2.25	5.44	12.3	34.0	119	393	1045
0.6	0.52	1.00	2.46	5.99	14.1	41.3	144	440	1163
0.8	0.52	1.08	2.66	6.96	18.2	56.5	194	520	1325
1	0.52	1.17	2.91	7.31	20.2	71.7	245	576	1521
1.5	0.52	1.38	3.16	9.51	27.6	104.4	361	717	1804

4.10.3.17 Symmetrical expansions

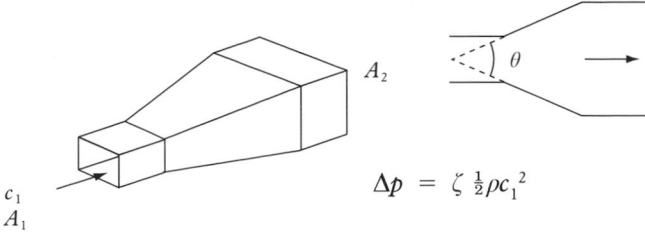

$$\Delta p = \zeta \tfrac{1}{2}\rho c_1^2$$

The values of ζ in Table 4.78 are taken from Idelchik (diagram 5–4). Values of ζ are shown to vary considerably with Reynolds number, Re, particularly for small angles of expansion.

Table 4.79 Values of ζ for symmetrical expansion (from Idelchik[2])

Re_1	A_2/A_1	Included angle θ								
		10°	15°	20°	30°	45°	60°	90°	120°	180°
0.5×10^5	10	0.47	0.585	0.670	0.760	0.850	0.900	0.960	0.920	0.880
	6	0.45	0.545	0.615	0.685	0.750	0.755	0.795	0.785	0.760
	4	0.40	0.468	0.510	0.565	0.610	0.635	0.655	0.650	0.640
	2	0.28	0.308	0.325	0.340	0.355	0.355	0.350	0.340	0.310
1×10^5	10	0.455	0.560	0.640	0.730	0.830	0.880	0.940	0.910	0.880
	6	0.405	0.515	0.580	0.650	0.720	0.775	0.780	0.775	0.760
	4	0.365	0.443	0.495	0.550	0.600	0.630	0.650	0.650	0.640
	2	0.235	0.268	0.290	0.310	0.330	0.340	0.340	0.320	0.310
$\geq 4 \times 10^5$	10	0.345	0.480	0.570	0.680	0.790	0.855	0.930	0.910	0.880
	6	0.320	0.443	0.520	0.615	0.695	0.740	0.770	0.775	0.760
	4	0.270	0.365	0.430	0.500	0.580	0.620	0.650	0.650	0.640
	2	0.160	0.203	0.235	0.265	0.300	0.320	0.335	0.320	0.310

For example:

for air at 21°C, A_1 being square, 1.0 m $\times$ 1.0 m, and $c_1 = 5.4$ m.s^{-1}, $Re_1 = 4 \times 10^5$

for air at 21°C, A_1 being square, 0.4 m $\times$ 0.4 m, and $c_1 = 3.4$ m.s^{-1}, $Re_1 = 1 \times 10^5$

4.10.3.18 Symmetrical contractions

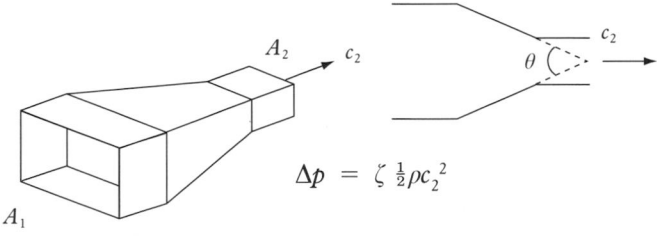

$$\Delta p = \zeta \tfrac{1}{2}\rho c_2^2$$

Table 4.80 Values of ζ for symmetrical contractions (from SMACNA[3])

A_2/A_1	Included angle θ						
	10°	15–40°	50–60°	90°	120°	150°	180°
0.50	0.05	0.05	0.06	0.12	0.18	0.24	0.26
0.25	0.05	0.04	0.07	0.17	0.27	0.35	0.41
0.17	0.05	0.04	0.07	0.18	0.28	0.36	0.42
0.10	0.05	0.05	0.08	0.19	0.29	0.37	0.43

Notes
(1) SMACNA gives the same values for symmetrical conical contractions.
(2) In the absence of further data it is reasonable to suppose similar values for contractions with 2 sides parallel.

4.10.3.19 Eccentric and offset taper contractions (HVCA 100 and 101)

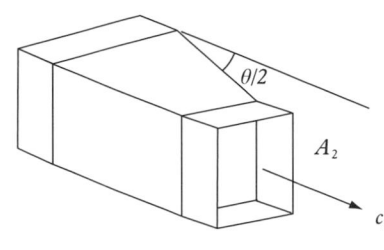

In the absence of specific data , it is suggested that Δp be estimated as the sum of Δp_1 for an angled offset based on inlet conditions, plus Δp_2 for a contraction. The angle θ to be used should be twice the angle of the side deviating most.

4.10.3.20 Y tees: rectangular (HVCA 95)

Note that data are only available for the case of:

$$A_{b1} = A_{b2} = \tfrac{1}{2}A_c$$

(a) Converging flows:

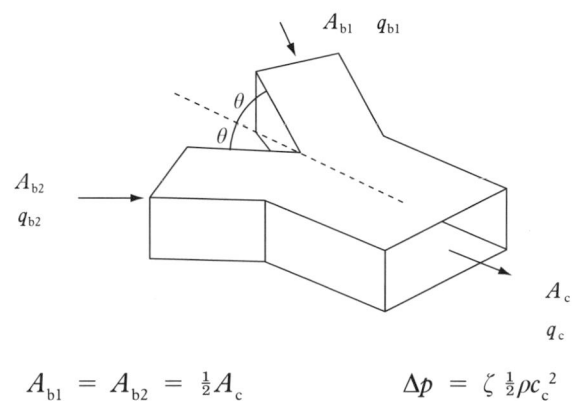

$$A_{b1} = A_{b2} = \tfrac{1}{2}A_c \qquad \Delta p = \zeta \tfrac{1}{2}\rho c_c^2$$

Table 4.81 Values of ζ_{b-c} for rectangular Y tees from (SMACNA[3])

θ	q_1/q_c or q_2/q_c										
	0	0.1	0.2	0.3	0.4	0.5	0.6	0.7	0.8	0.9	1
15°	−2.6	−1.9	−1.3	−0.77	−0.3	0.1	0.41	0.67	0.85	0.97	1
30°	−2.1	−1.5	−1	−0.53	−0.1	0.28	0.69	0.91	1.1	1.4	1.6
45°	−1.3	−0.93	−0.55	−0.16	0.2	0.56	0.92	1.26	1.6	2	2.3

(b) Diverging flows

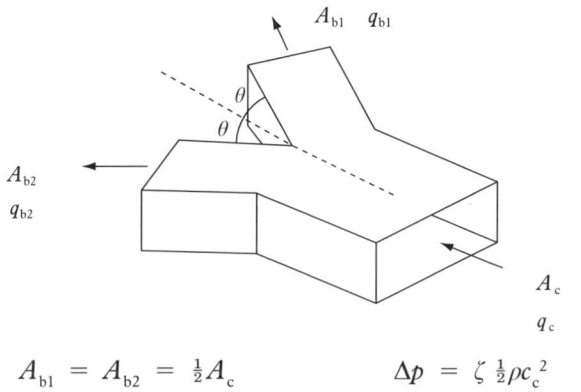

$$A_{b1} = A_{b2} = \tfrac{1}{2}A_c \qquad \Delta p = \zeta \tfrac{1}{2}\rho c_c^2$$

Table 4.82 Values of ζ_{c-bl} and ζ_{c-b2} for rectangular Y tees, diverging flows (from SMACNA[3])

θ	q_1/q_c or q_2/q_c							
	0.1	0.2	0.3	0.4	0.5	0.6	0.8	1
15°	0.81	0.65	0.51	0.38	0.28	0.2	0.11	0.06
30°	0.84	0.69	0.56	0.44	0.34	0.26	0.19	0.15
45°	0.87	0.74	0.63	0.54	0.45	0.38	0.28	0.24
60°	0.9	0.82	0.79	0.66	0.59	0.53	0.43	0.36
90°	1	1	1	1	1	1	1	1

θ	q_1/q_c or q_2/q_c				
	1.2	1.4	1.6	1.8	2
15°	0.14	0.3	0.51	0.76	1
30°	0.15	0.3	0.51	0.76	1
45°	0.23	0.3	0.51	0.76	1
60°	0.33	0.39	0.51	0.76	1
90°	1	1	1	1	1

4.10.3.21 90° branch tees: rectangular from rectangular (HVCA 104)

(a) Converging flows $(A_c = A_s)$

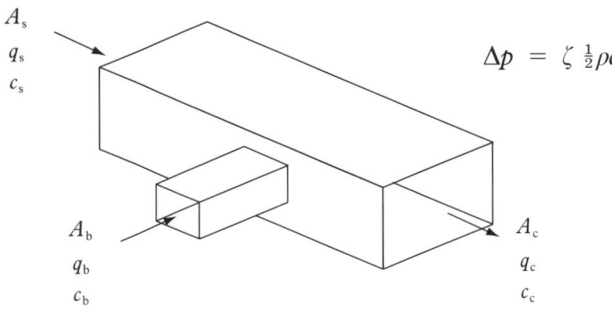

$$\Delta p = \zeta \tfrac{1}{2}\rho c_c^2$$

Table 4.83 Values of the straight factor ζ_{s-c} for all values of A_b/A_c (from Idelchik[2])

q_s/q_c	0	0.1	0.2	0.3	0.4	0.5	0.6	0.7	0.8	0.9	1
ζ_{s-c}	0.55	0.59	0.60	0.59	0.57	0.53	0.46	0.38	0.27	0.16	0

$$\zeta_{s-c} = 1.55\,\frac{q_b}{q_c} - \left(\frac{q_b}{q_c}\right)^2$$

$$= 1.55\left(1 - \frac{q_s}{q_c}\right) - \left(1 - \frac{q_s}{q_c}\right)^2$$

Table 4.84 Values for the branch factor, ζ_{b-c} (from SMACNA[3]; valid for $A_s = A_c = 2A_b$)

$c_c/$ m.s^{-1}	q_b/q_c									
	0.1	0.2	0.3	0.4	0.5	0.6	0.7	0.8	0.9	1
< 6	−0.75	−0.53	−0.03	0.33	0.80	1.40	2.15	2.93	4.18	4.78
> 6	−0.69	−0.21	0.23	0.67	1.17	1.66	2.67	3.36	3.93	5.13

Idelchik[2] gives slightly different values but his data extends to cover other branch areas. It is not clear whether the data covers rectangular or circular ducts so in the absence of other information it could be used for either. It is covered by the following set of equations:

for $A_b/A_c \leq 0.35$: $C = 1.0$

for $A_b/A_c > 0.35$ and $q_b/q_c \leq 0.4$: $C = 0.9(1.0 - q_b/q_c)$

for $A_b/A_c > 0.35$ and $q_b/q_c > 0.4$: $C = 0.55$

$$\zeta_{b-c} = C\left[1 + \left(\frac{q_b A_c}{q_c A_b}\right)^2 - 2\left(1 - \frac{q_b}{q_c}\right)^2\right]$$

(b) Diverging flows $(A_c = A_s)$

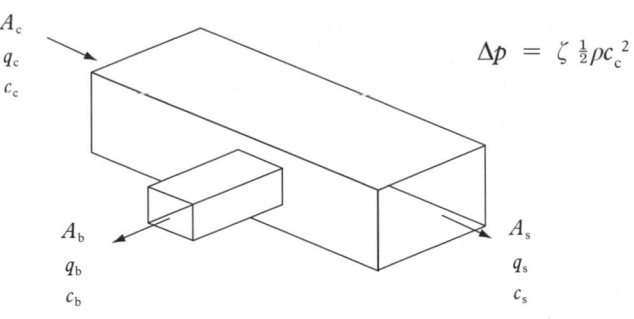

$$\Delta p = \zeta \tfrac{1}{2}\rho c_c^2$$

Table 4.85 Values of ζ_{c-s} (from Idelchik[2])

q_s/q_c	0	0.1	0.2	0.3	0.4	0.5	0.6	0.7	0.8	0.9	1
ζ_{c-s} for $A_b/A_c < 0.4$	0.40	0.324	0.256	0.196	0.144	0.1	0.064	0.036	0.016	0.004	0
ζ_{c-s} for $A_b/A_c > 0.4$	0.30	0.194	0.115	0.059	0.021	0	−0.064	−0.072	−0.048	−0.016	0

Table 4.86 Values for the branch factor ζ_{c-b} (from Idelchik[2])

c_b/c_c	0	0.1	0.2	0.4	0.6	0.8	1	1.2	1.4	1.6	2
ζ_{c-b} for $A_b/A_c < 2/3$	1	1.01	1.04	1.16	1.35	1.64	2.0	2.44	2.96	3.54	4.6
ζ_{c-b} for $A_b/A_c = 1$	1	1	1.01	1.05	1.11	1.19	1.3	1.43	1.59	1.77	2.2

Table 4.87 Values for the branch factor ζ_{c-b} (derived from SMACNA[3])

A_b/A_c	q_b/q_c								
	0.1	0.2	0.3	0.4	0.5	0.6	0.7	0.8	0.9
0.1	1.38	3							
0.2	1.08	1.4	2	3.6					
0.3	1.04	1.1	1.3	2.18	3.22				
0.4	1.03	1.02	1.2	1.48	1.90	2.46	3.0		
0.5	1.03	1.01	1.05	1.12	1.27	1.66	1.95	2.2	2.57
1.0*	0.93	0.87	0.82	0.81	0.83	0.88	0.94	1.02	*1.1*

* from Miller[29], square ducts
Note: italic denotes extrapolated value

4.10.3.22 90° shoe branch tees: rectangular from rectangular (HVCA 106)

(a) Converging flows $(A_c = A_s)$

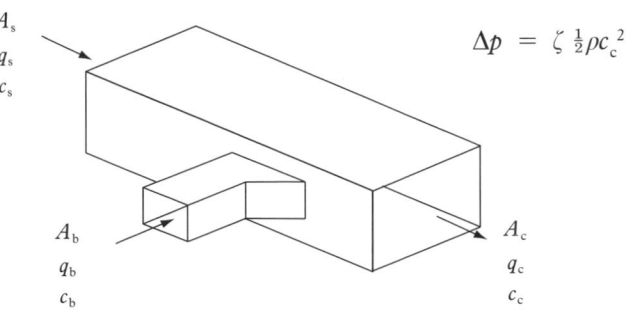

$$\Delta p = \zeta \tfrac{1}{2}\rho c_c^2$$

Table 4.88 Values of the straight factor ζ_{s-c} (from Miller[28])

q_s/q_c	0	0.1	0.2	0.3	0.4	0.5	0.6	0.7	0.8	0.9	1
ζ_{s-c}		0.09	0.16	0.22	0.27	0.29	0.3	0.27			

The data was for a square duct with $A_b/A_c = 1.0$. Miller shows that it makes no difference whether the depth of the shoe is $w/2$ or w.

Table 4.89 Values for the branch factor ζ_{b-c} (from SMACNA[3])

$c_c/$					q_b/q_c					
m.s^{-1}	0.1	0.2	0.3	0.4	0.5	0.6	0.7	0.8	0.9	1
< 6	−0.83	−0.68	−0.30	0.28	0.55	1.03	1.50	1.93	2.5	3.03
> 6	−0.72	−0.52	−0.23	0.34	0.76	1.14	1.83	2.01	2.9	3.63

(b) Diverging flows ($A_c = A_s$)

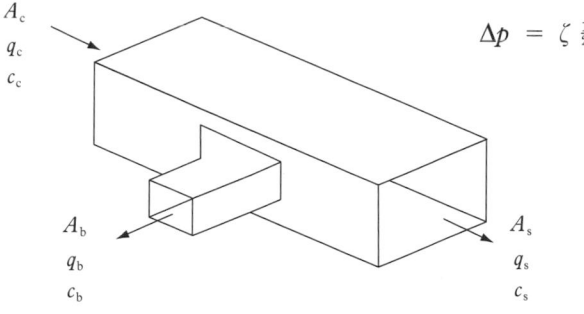

$$\Delta p = \zeta \tfrac{1}{2}\rho c_c^2$$

For the straight flow, SMACNA suggests no difference in values from those for a square branch.

Table 4.90 Table of values of ζ_{c-s} (from Idelchik[2])

q_s/q_c	0	0.1	0.2	0.3	0.4	0.5	0.6	0.7	0.8	0.9	1
ζ_{c-s} for $A_b/A_c < 0.4$	0.40	0.324	0.256	0.196	0.144	0.1	0.064	0.036	0.016	0.004	0
ζ_{c-s} for $A_b/A_c > 0.4$	0.30	0.194	0.115	0.059	0.021	0	−0.064	−0.072	−0.048	−0.016	0

No data are available for a shoe on a rectangular branch. However Miller[28] shows that a reduction of approximately 10% occurs for a leading bevel on a circular branch on a rectangular duct and that a 25% reduction occurs when a small radius ($r/d = 0.09$) is placed at the junction of a circular branch on a main circular duct. The figures in Table 4.90 have not taken account of any possible reduction.

Table 4.91 Values for the branch factor ζ_{c-b} (derived from SMACNA[3])

A_b/A_c					q_b/q_c					
	0.1	0.2	0.3	0.4	0.5	0.6	0.7	0.8	0.9
0.1	0.78	1.52							
0.2	0.79	0.95	1.40	1.9					
0.3	0.82	0.71	0.83	1.22	1.67				
0.4	0.86	0.78	0.69	0.79	1.15	1.40	1.55		
0.5	0.91	0.79	0.70	0.66	0.74	0.86	0.92	1.09	1.17
1.0*	1.0	0.78	0.66	0.59	0.54	0.49	0.46	0.42	0.40
1.0**	1.0	0.73	0.59	0.50	0.44	0.39	0.34	0.31	0.28

* from Miller[29], depth of shoe = $w/2$
** from Miller[29], depth of shoe = w
Note: italic denotes extrapolated values

4.10.3.23 Angled branch tees: rectangular from rectangular (HVCA 105)

Information from Idelchik[2].

(a) Converging flows ($A_c = A_s$)

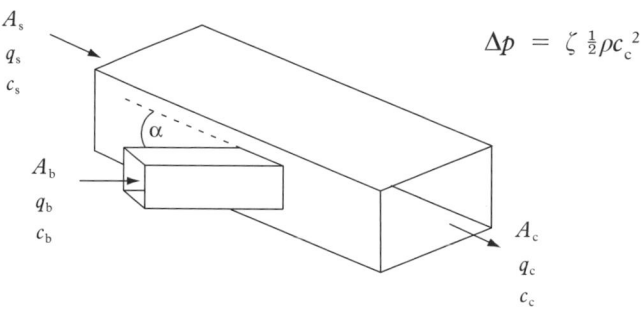

$$\Delta p = \zeta \tfrac{1}{2}\rho c_c^2$$

The values of ζ are found to vary with A_b/A_c, angle α, and q_b/q_c. A sample of values is given in the Tables 4.91 and 4.92, but useful equations are also given.

Table 4.92 Values for the straight factor ζ_{s-c} (Idelchik[2])

α	A_b/A_c					q_s/q_c				
		0.2	0.3	0.4	0.5	0.6	0.7	0.8	0.9	1
30°	0.1	−10.2	−7.6	−5.4	−3.6	−2.15	−1.06	−0.34	0.02	0
	0.2	−4.61	−3.35	−2.29	−1.43	−0.75	−0.27	0.01	0.1	0
	0.3	−2.74	−1.95	−1.25	−0.70	−0.3	−0.01	0.13	0.13	0
	0.4	−1.82	−1.2	−0.73	−0.35	−0.06	0.12	0.19	0.15	0
	0.6	−0.9	−0.5	−0.2	0	0.17	0.22	0.44	0.16	0
	0.8	−0.43	−0.15	0.06	0.21	0.29	0.30	0.27	0.17	0
	1.0	−0.15	0.06	0.21	0.32	0.36	0.35	0.29	0.17	0
45°	0.1	−8.10	−6.05	−4.3	−2.77	−1.65	−0.76	−0.02	0.05	0
	0.2	−3.56	−2.60	−1.70	−1.00	−0.50	−0.13	0.17	0.12	0
	0.3	−2.10	−1.40	−0.87	−0.49	−0.12	0.08	0.22	0.14	0
	0.4	−1.30	−0.85	−0.45	−0.13	0.08	0.20	0.27	0.16	0
	0.6	−0.55	−0.25	−0.04	0.16	0.26	0.28	0.27	0.17	0
	0.8	−0.17	0.08	0.20	0.30	0.36	0.32	0.29	0.17	0
	1.0	0.06	0.25	0.33	0.40	0.41	0.40	0.31	0.17	0
60°	0.1	−5.44	−4.00	−2.80	−1.75	−1.00	−0.40	0	0.09	0
	0.2	−2.24	−1.55	−0.95	−0.50	−0.16	0.06	0.16	0.14	0
	0.3	−1.17	−0.07	−0.35	−0.08	0.11	0.22	0.23	0.16	0
	0.4	−0.64	−0.30	−0.10	0.13	0.24	0.30	0.26	0.17	0
	0.6	−0.11	0.08	0.25	0.33	0.37	0.32	0.29	0.17	0
	0.8	0.16	0.28	0.40	0.44	0.44	0.41	0.31	0.18	0
	1.0	0.32	0.42	0.48	0.50	0.48	0.42	0.32	0.18	0

Values of straight factor, ζ_{s-c}, for the full range of straight flows can be obtained from the following equations provided by Idelchik[2].

For 30°,

$$\zeta_{s-c} = 1 - \left(\frac{q_s}{q_c}\right)^2 - 1.74\frac{A_c}{A_b}\left(1 - \frac{q_s}{q_c}\right)^2$$

For 45°,

$$\zeta_{s-c} = 1 - \left(\frac{q_s}{q_c}\right)^2 - 1.41\frac{A_c}{A_b}\left(1 - \frac{q_s}{q_c}\right)^2$$

For 60°,

$$\zeta_{s-c} = 1 - \left(\frac{q_s}{q_c}\right)^2 - \frac{A_c}{A_b}\left(1 - \frac{q_s}{q_c}\right)^2$$

Table 4.93 Values of the branch factor ζ_{b-c} (Idelchik[2])

a	A_b/A_c	q_b/q_c							
		0	0.1	0.2	0.3	0.4	0.5	0.6	0.7
30°	0.1	−1.00	0.21	3.02	7.45	13.5	21.2	30.4	41.3
	0.2	−1.00	−0.46	0.37	1.50	2.89	4.58	6.55	8.81
	0.3	−1.00	−0.57	−0.07	0.50	1.15	1.83	2.60	3.40
	0.4	−1.00	−0.60	−0.20	0.20	0.58	0.97	1.37	1.77
	0.6	−1.00	−0.62	−0.28	−0.01	0.26	0.47	0.64	0.76
	0.8	−1.00	−0.63	−0.30	−0.04	0.18	0.35	0.46	0.52
	1.0	−1.00	−0.63	−0.31	−0.05	0.16	0.32	0.41	0.46
45°	0.1	−1.00	0.24	3.15	8.00	14.00	21.9	31.6	42.9
	0.2	−1.00	−0.45	0.54	1.64	3.15	5.00	6.90	9.20
	0.3	−1.00	−0.56	−0.02	0.60	1.30	2.10	2.97	3.90
	0.4	−1.00	−0.59	−0.17	0.30	0.72	1.18	1.65	2.15
	0.6	−1.00	−0.61	−0.26	0.08	0.35	0.60	0.85	1.02
	0.8	−1.00	−0.62	−0.28	0	0.25	0.45	0.60	0.70
	1.0	−1.00	−0.62	−0.29	−0.03	0.21	0.40	0.53	0.60
60°	0.1	−1.00	0.26	3.35	8.20	14.7	23.0	33.1	44.9
	0.2	−1.00	−0.42	0.55	1.85	3.50	5.50	7.90	10.0
	0.3	−1.00	−0.54	0.03	0.75	1.55	2.40	3.50	4.60
	0.4	−1.00	−0.58	−0.13	0.40	0.92	1.44	2.05	2.70
	0.6	−1.00	−0.61	−0.23	0.10	0.45	0.78	1.08	1.40
	0.8	−1.00	−0.62	−0.26	0	0.35	0.58	0.80	0.98
	1.0	−1.00	−0.62	−0.26	−0.01	0.28	0.50	0.68	0.84

Idelchik does not make clear wheteher the above values are for tees of circular or rectangular ducts, but they are the only data available. It would seem reasonable to suppose that if the values are valid for square ducts with square branches, the values would be different for rectangular ducts and might depend on w/h.

Values of branch factor, ζ_{b-c}, for the full range of branch flows can be obtained from the following equations provided by Idelchik[2], where C is first obtained from the following:

for $A_b/A_c \leq 0.35$: $\qquad\qquad C = 1.0$

for $A_b/A_c > 0.35$ and $q_b/q_c \leq 0.4$: $C = 0.9\,(1 - q_b/q_c)$

for $A_b/A_c > 0.35$ and $q_b/q_c > 0.4$: $C = 0.55$

For $a = 30°$,

$$\zeta_{b-c} = C\left[1 + \left(\frac{q_b A_c}{q_c A_b}\right)^2 - 2\left(1 - \frac{q_b}{q_c}\right)^2 - 1.7\frac{A_c}{A_b}\left(\frac{q_b}{q_c}\right)^2\right]$$

For $a = 45°$,

$$\zeta_{b-c} = C\left[1 + \left(\frac{q_b A_c}{q_c A_b}\right)^2 - 2\left(1 - \frac{q_b}{q_c}\right)^2 - 1.41\frac{A_c}{A_b}\left(\frac{q_b}{q_c}\right)^2\right]$$

For $a = 60°$,

$$\zeta_{b-c} = C\left[1 + \left(\frac{q_b A_c}{q_c A_b}\right)^2 - 2\left(1 - \frac{q_b}{q_c}\right)^2 - 1.0\frac{A_c}{A_b}\left(\frac{q_b}{q_c}\right)^2\right]$$

(b) Diverging flows $(A_c = A_s)$

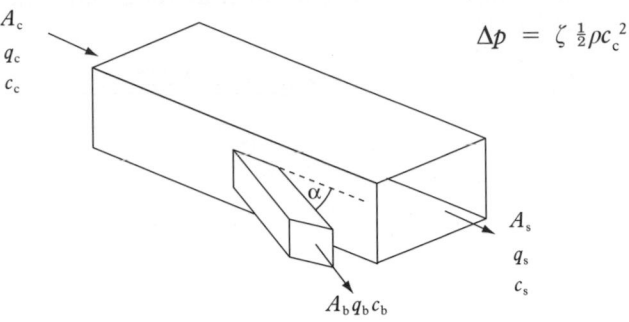

$$\Delta p = \zeta\,\tfrac{1}{2}\rho c_c^2$$

Table 4.94 Values of the straight factor ζ_{c-s}, for all values of A_b/A_c and for any angle $0 < a < 90°$ (from Idelchik[2])

q_s/q_c	0	0.1	0.2	0.3	0.4	0.5	0.6	0.7	0.8	0.9	1
ζ_{c-s} for $A_b/A_c < 0.4$	0.40	0.324	0.256	0.196	0.144	0.1	0.064	0.036	0.016	0.004	0
ζ_{c-s} for $A_b/A_c > 0.4$	0.30	0.194	0.115	0.059	0.021	0	−0.064	−0.072	−0.048	−0.016	0

Table 4.95 Values of the branch factor ζ_{c-b} (derived from Idelchik[2])

a	A_b/A_c	q_b/q_c							
		0	0.1	0.2	0.3	0.4	0.5	0.6	0.7
30°	0.1	1.10	0.28	1.46	6.60	12.3	20.0	29.3	—
	0.2	1.10	0.57	0.26	0.67	1.33	2.45	6.29	9.16
	0.3	1.10	0.55	0.27	0.24	0.51	0.61	1.29	2.10
	0.4	1.00	0.60	0.33	0.22	0.21	0.31	0.45	0.64
	0.6	1.00	0.70	0.48	0.30	0.22	0.18	0.16	0.21
	0.8	1.00	0.76	0.56	0.40	0.28	0.21	0.16	0.16
	1.0	1.00	0.79	0.62	0.48	0.35	0.26	0.19	0.17
45°	0.1	1.10	0.60	2.07	6.94	12.8	20.2	29.7	—
	0.2	1.10	0.57	0.56	1.02	1.88	3.21	6.63	9.61
	0.3	1.10	0.67	0.48	0.52	0.79	1.23	1.84	2.75
	0.4	1.00	0.67	0.45	0.42	0.44	0.56	0.68	0.97
	0.6	1.00	0.74	0.57	0.45	0.38	0.37	0.35	0.43
	0.8	1.00	0.79	0.63	0.51	0.38	0.36	0.31	0.32
	1.0	1.00	0.82	0.66	0.55	0.46	0.39	0.30	0.30
60°	0.1	1.10	1.03	2.88	7.21	13.05	20.4	29.8	—
	0.2	1.10	0.73	0.96	1.57	2.61	4.07	6.89	9.82
	0.3	1.10	0.74	0.67	0.89	1.25	1.83	2.55	3.57
	0.4	1.00	0.77	0.62	0.61	0.76	0.91	1.06	1.42
	0.6	1.00	0.81	0.69	0.58	0.53	0.58	0.60	0.72
	0.8	1.00	0.84	0.72	0.63	0.54	0.47	0.46	0.53
	1.0	1.00	0.86	0.74	0.66	0.58	0.49	0.39	0.44

4.10.3.24   90° Tees: rectangular from rectangular; bell-mouth branch (HVCA 107)

Converging and diverging flows $(A_c = A_s)$

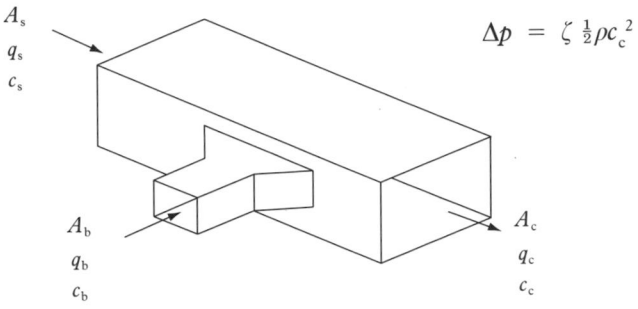

$$\Delta p = \zeta\,\tfrac{1}{2}\rho c_c^2$$

No data can be found on this component.

There is no reason to believe that the values of ζ will be any less than for a shoe branch.

4.10.3.25 90° Radiussed twin bend: rectangular (HVCA 91)

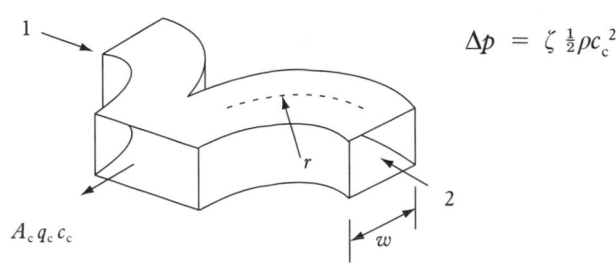

$$\Delta p = \zeta \tfrac{1}{2}\rho c_c^2$$

Information from SMACNA for $r/w = 1.5$ and $q_1 = q_2 = 0.5q_c$.

Table 4.96 Values of ζ for 90° radiussed twin bend (from SMACNA[3])

A_1/A_c or A_2/A_c		0.5	1.0
Converging	ζ_{1-c}	0.23	0.07
Diverging	ζ_{c-1}	0.30	0.25

4.10.3.26 90° swept branch

(a) Converging flows

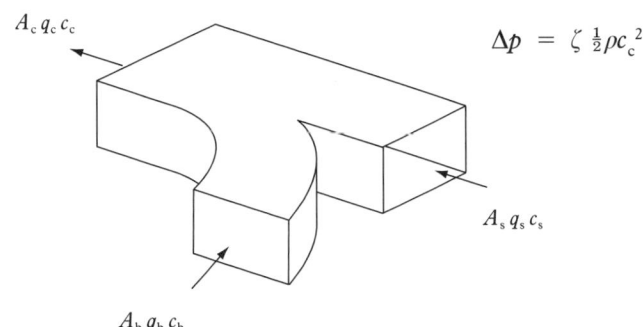

$$\Delta p = \zeta \tfrac{1}{2}\rho c_c^2$$

Table 4.97 Values for the straight factor ζ_{s-c} (from SMACNA[3])

A_s/A_c	A_b/A_c	q_s/q_c								
		0.1	0.2	0.3	0.4	0.5	0.6	0.7	0.8	0.9
0.75	0.25	−2.6	−2	−1.5	−0.92	−0.45	−0.1	0.2	0.3	0.3
1	0.5	−0.46	−0.37	−0.27	−0.18	−0.08	0	0.1	0.16	0.17
0.75	0.5	−0.58	−0.42	−0.23	−0.03	0.12	0.25	0.32	0.35	0.27
0.5	0.5	−1.3	−0.8	−0.4	0	0.35	0.65	0.9	1.1	1.2
1	1	−0.12	0	0.1	0.18	0.23	0.26	0.27	0.24	0.18
0.75	1	−0.22	−0.08	0.05	0.18	0.27	0.35	0.38	0.36	0.75
0.5	1	−0.1	0.08	0.25	0.4	0.55	0.68	0.8	0.87	0.8

Table 4.98 Values for the branch factor ζ_{b-c} (from SMACNA[3])

A_s/A_c	A_b/A_c	q_b/q_c								
		0.1	0.2	0.3	0.4	0.5	0.6	0.7	0.8	0.9
1	0.25	−0.5	0	0.5	1.2	2.2	3.7	5.8	8.4	11
0.75	0.25	−1.2	−0.4	0.4	1.6	3	4.8	6.8	8.9	11
1	0.5	−0.5	−0.2	0	0.25	0.45	0.7	1	1.5	2
0.75	0.5	−1	−0.6	−0.2	0.1	0.3	0.6	1	1.5	2
0.5	0.5	−2.2	−1.5	−0.95	−0.5	0	0.4	0.8	1.3	1.9
1	1	−0.6	−0.3	−0.1	−0.04	0.13	0.21	0.29	0.36	0.42
0.75	1	−1.2	−0.8	−0.4	−0.2	0	0.16	0.24	0.32	0.38
0.5	1	−2.1	−1.4	−0.9	−0.5	−0.2	0	0.2	0.25	0.3

(b) Diverging flows

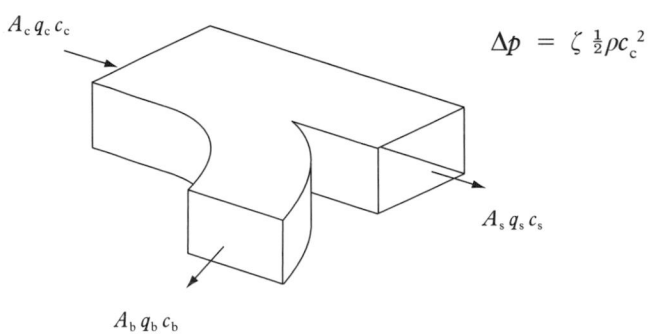

$$\Delta p = \zeta \tfrac{1}{2}\rho c_c^2$$

Table 4.99 Values for the straight factor ζ_{c-s} (from SMACNA[3])

A_s/A_c	A_b/A_c	q_s/q_c								
		0.1	0.2	0.3	0.4	0.5	0.6	0.7	0.8	0.9
1	0.25	0.46	0.38	0.29	0.21	0.13	0.05	−0.01	−0.03	−0.01
0.75	0.25	0.34	0.24	0.16	0.08	0.02	−0.01	−0.02	0	0.08
1	0.5	0.35	0.27	0.19	0.12	0.06	0	−0.05	−0.06	−0.03
0.75	0.5	0.37	0.23	0.12	0.04	−0.01	−0.03	−0.04	−0.02	0.04
0.5	0.5	0.3	0.18	0.09	0.04	0.05	0.13	0.28	0.48	0.72
1	1	0.38	0.3	0.22	0.13	0.06	−0.01	−0.04	−0.04	−0.02
0.75	1	0.3	0.2	0.1	0.03	−0.01	−0.03	0.01	0	0.01
0.5	1	0.2	0.1	0.06	0.05	0.08	0.13	0.23	0.38	0.62

Table 4.100 Values for the branch factor ζ_{c-b}

A_s/A_c	A_b/A_c	q_b/q_c								
		0.1	0.2	0.3	0.4	0.5	0.6	0.7	0.8	0.9
1	0.25	0.55	0.5	0.6	0.85	1.2	1.8	3.1	4.4	6
0.75	0.25	0.35	0.35	0.5	0.8	1.3	2	2.8	3.8	5
1	0.5	0.62	0.48	0.4	0.4	0.48	0.6	0.78	1.1	1.5
0.75	0.5	0.52	0.4	0.32	0.3	0.34	0.44	0.62	0.92	1.4
0.5	0.5	0.44	0.38	0.38	0.41	0.52	0.68	0.92	1.2	1.6
1	1	0.67	0.55	0.46	0.37	0.32	0.29	0.29	0.3	0.37
0.75	1	0.7	0.6	0.51	0.42	0.34	0.28	0.26	0.26	0.29
0.5	1	0.6	0.52	0.43	0.33	0.24	0.17	0.15	0.17	0.21

4.10.3.27 Exhaust vents; lateral openings with and without side louvres

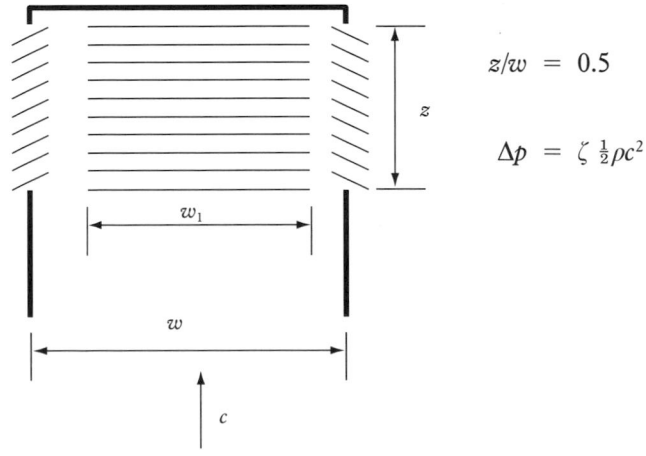

$$z/w = 0.5$$

$$\Delta p = \zeta \tfrac{1}{2}\rho c^2$$

The pressure drop through such exhaust vents will depend strongly on the geometry of the louvres so manufacturer's guidance should be sought. Table 4.101 taken from Idelchik[2] is intended for initial guidance only, when doing initial feasibility calculations.

Table 4.101 Values of ζ for exhaust vents (from Idelchik[2])

Number of openings	Layout	w_1/z	Values of ζ		
			Without louvres	30° louvres	45° louvres
One		1.5	15.5	22.0	
Two		1.5	5.0	7.2	
Three		1.5	3.5	5	
Four		1.5	2.2	2.6	3.5
		1.0	5.3	7.0	10.0
		0.5	15.6	19.6	29

4.10.3.28   Plain outlets (from Idelchik[2])

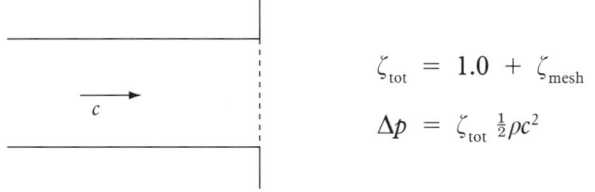

$$\zeta_{tot} = 1.0 + \zeta_{mesh}$$

$$\Delta p = \zeta_{tot}\,\tfrac{1}{2}\rho c^2$$

4.10.3.29   Mesh screens; grids of circular metal wire (from Idelchik[2])

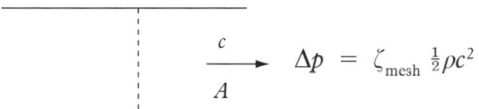

$$\Delta p = \zeta_{mesh}\,\tfrac{1}{2}\rho c^2$$

The value of ζ_{mesh} depends very much on the closeness of the mesh, or rather on the free area ratio, being A_c/A.

For turbulent flow ($Re < 1000$) through the mesh, defined in this instance by:

$$Re = \rho c_m d/\eta$$

where c_m is the mean velocity of air through the mesh and d is the diameter of the wire.

Table 4.102 Values of ζ for mesh screens (from Idelchik[2])

A_c/A	0.1	0.2	0.3	0.4	0.5	0.6	0.7	0.8
ζ_{mesh}	82	17	6.4	3.03	1.65	0.97	0.58	0.32

Idelchik gives the following formula as a reasonable curve-fit for low values of A_c/A:

$$\zeta_{mesh} = 1.3(1 - A_c/A) + (A/A_c - 1)^2$$

4.10.3.30   Plane duct entries (from Idelchik[2])

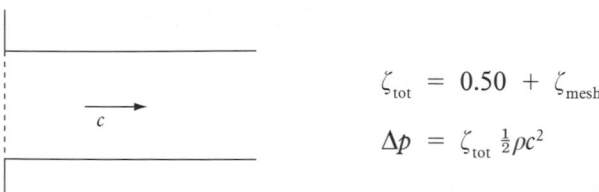

$$\zeta_{tot} = 0.50 + \zeta_{mesh}$$

$$\Delta p = \zeta_{tot}\,\tfrac{1}{2}\rho c^2$$

4.10.3.32   Inlet vents; lateral openings with and without side louvres

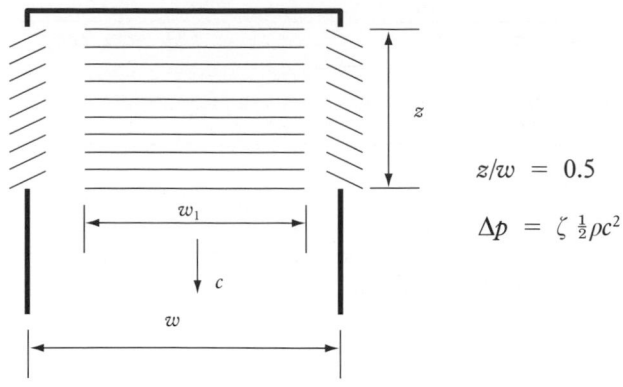

$$z/w = 0.5$$

$$\Delta p = \zeta\,\tfrac{1}{2}\rho c^2$$

The pressure drop through such exhaust vents will depend strongly on the geometry of the louvres so manufacturer's guidance should be sought. Table 4.103 taken from Idelchik[2] is intended for initial guidance only, when doing initial feasibility calculations.

Table 4.103 Values of ζ for inlet vents (from Idelchik[2])

Number of openings	Layout	w_1/z	Values of ζ		
			Without louvres	30° louvres	45° louvres
One		1.5	12.6	17.5	
Two		1.5	3.6	5.4	
Three		1.5	1.8	3.2	
Four		1.5	1.2	2.5	3.8
		1.0	2.0	3.6	6.0
		0.5	8.0	13.7	21.5

4.10.3.32   Louvred duct entries

The pressure drop through such exhaust vents will depend strongly on the geometry of the louvres so manufacturer's guidance should be sought. Table 4.104 derived from Idelchik[2] is intended for initial guidance only, when doing initial feasibility calculations.

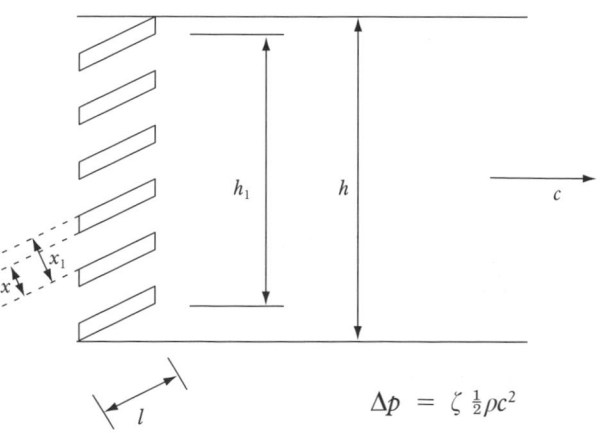

$$\Delta p = \zeta\,\tfrac{1}{2}\rho c^2$$

Table 4.104 Values of ζ for louvred duct entries (from Idelchik[2])

x_1/x		h_1/h					
		0.6	0.7	0.8	0.9	0.95	1.0
Case a	0.8	7.9	5.7	4.3	3.3	3.0	2.7
	0.9	6.7	4.8	3.7	2.9	2.6	2.3
Case b	0.8	4.8	3.4	2.6	2.0	1.8	1.6
	0.9	4.0	2.9	2.2	1.7	1.5	1.4

Table 4.104 gives values of ζ for the two idealised shapes of louvre blade, for values of $l/x_1 > 2.2$. Note that the angle of the blades would appear not to play a part, except insofar at it affects the value of h_1/h.

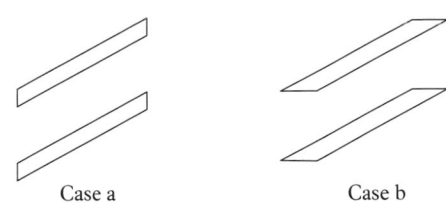

Case a   Case b

4.10.4 Rectangular to circular and circular to rectangular

4.10.4.1 Expansion, round to rectangular, and rectangular to round

Data from SMACNA.

To determine the appropriate included angle θ use:

(a) Circular to rectangular

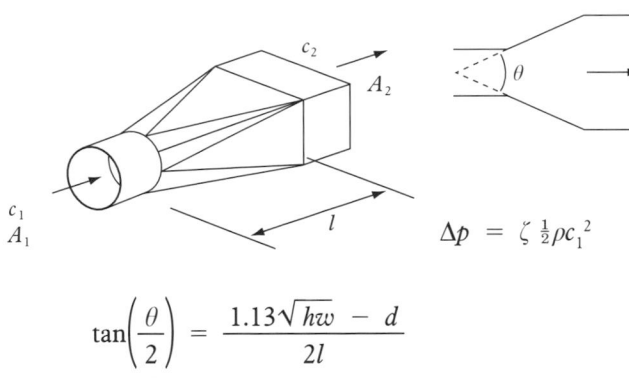

$$\Delta p = \zeta \tfrac{1}{2}\rho c_1^2$$

$$\tan\left(\frac{\theta}{2}\right) = \frac{1.13\sqrt{hw} - d}{2l}$$

(b) Rectangular to round

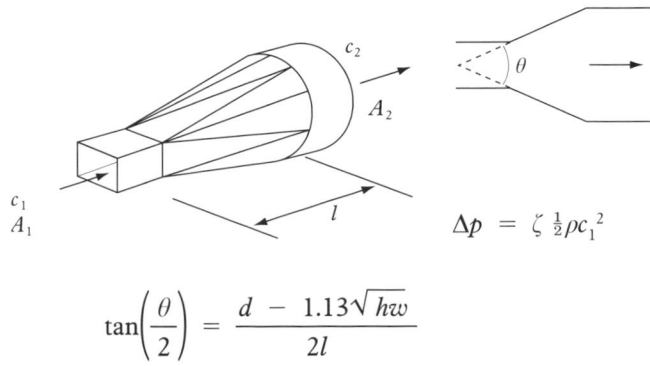

$$\Delta p = \zeta \tfrac{1}{2}\rho c_1^2$$

$$\tan\left(\frac{\theta}{2}\right) = \frac{d - 1.13\sqrt{hw}}{2l}$$

ζ can then be obtained from Table 4.79 for symmetrical rectangular expansions.

4.10.4.2 Contractions, round to rectangular, and rectangular to round

No data can be found.

It is suggested that the same equations as in section 4.10.4.1 above be used to determine the equivalent included angle θ, and that the data for symmetrical or eccentric contractions be used (section 4.10.3.18 or 4.10.3.19) as appropriate.

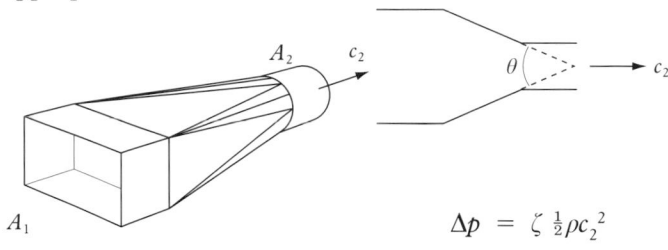

$$\Delta p = \zeta \tfrac{1}{2}\rho c_2^2$$

4.10.4.3 Rectangular contraction, slot to round

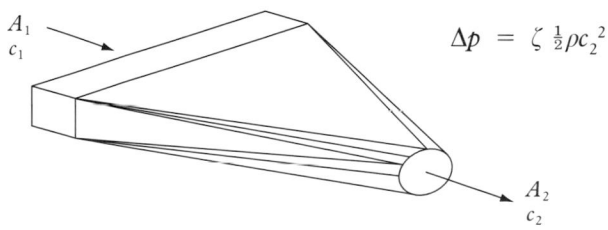

$$\Delta p = \zeta \tfrac{1}{2}\rho c_2^2$$

Table 4.105 Values of ζ for rectangular contraction, slot to round (from SMACNA[3])

$Re_2 \times 10^{-5}$	0.1	0.2	0.4	0.6	0.8	1	2	>4
ζ		0.27	0.25	0.2	0.17	0.14	0.11	0.04	0

4.10.4.4 90° branch tees, circular from rectangular (HVCA 135)

(a) Converging flows ($A_c = A_s$)

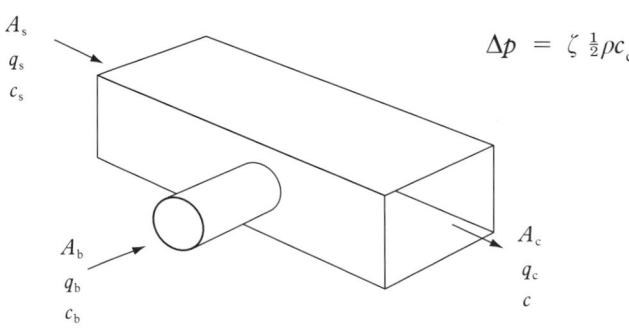

$$\Delta p = \zeta \tfrac{1}{2}\rho c_c^2$$

Table 4.106 Values of the straight factor ζ_{s-c} (from Miller[29])

A_b/A_c	q_s/q_c								
	0.1	0.2	0.3	0.4	0.5	0.6	0.7	0.8	0.9
0.79		0.64	0.61	0.58	0.54	0.49	0.43	0.33	
0.35			0.71	0.64	0.57	0.50	0.43	0.35	

Table 4.107 Values of the branch factor ζ_{b-c} (from Miller[28])

A_b/A_c	q_b/q_c								
	0.1	0.2	0.3	0.4	0.5	0.6	0.7	0.8	0.9
0.79		−0.14	0.15	0.39	0.58	0.78	0.96	*1.13*	*1.3*
0.35		0.08	0.50	0.95	1.48	2.17			

Note: italic denotes extrapolated values

(b) Diverging flows ($A_c = A_s$)

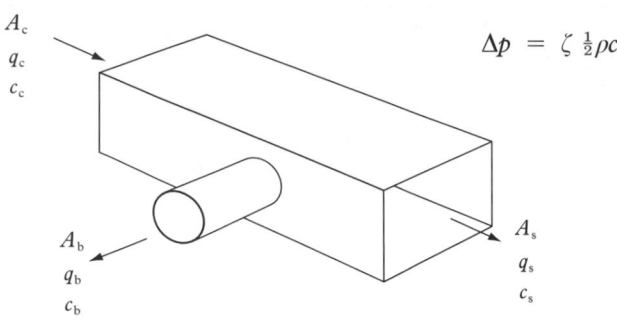

$$\Delta p = \zeta \tfrac{1}{2}\rho c_c^2$$

Table 4.108 Values of the straight factor ζ_{c-s} (information from Idelchik[2])

q_s/q_c	0	0.1	0.2	0.3	0.4	0.5	0.6	0.7	0.8	0.9	1
ζ_{c-s} for $A_b/A_c < 0.4$	0.40	0.324	0.256	0.196	0.144	0.1	0.064	0.036	0.016	0.004	0
ζ_{c-s} for $A_b/A_c > 0.4$	0.30	0.194	0.115	0.059	0.021	0	−0.064	−0.072	−0.048	−0.016	0

Table 4.109 Values for the branch factor ζ_{c-b} (information from Idelchik[2])

c_b/c_c	0	0.1	0.2	0.4	0.6	0.8	1	1.2	1.4	1.6	2
ζ_{c-b} for $A_b/A_c < 2/3$	1	1.01	1.04	1.16	1.35	1.64	2.0	2.44	2.96	3.54	4.6
ζ_{c-b} for $A_b/A_c = 1$	1	1	1.01	1.05	1.11	1.19	1.3	1.43	1.59	1.77	2.2

Idelchik[2] makes no distinction between circular and rectangular branches and ducts, so the above values are the same as for sections 4.10.3.22 and 4.10.5.11. However, SMACNA[3] does make a distinction, the information being given in Table 4.110.

Table 4.110 Values for the branch factor ζ_{c-b} (information derived from SMACNA[3])

A_b/A_c	q_b/q_c								
	0.1	0.2	0.3	0.4	0.5	0.6	0.7	0.8	0.9
0.1	1.30								
0.2	1.09	1.38	1.85						
0.3	1.00	1.20	1.20	1.45	1.9				
0.4	1.00	1.09	1.11	1.23	1.44	1.70	1.96		
0.5	1.00	1.07	1.08	1.13	1.26	1.48	1.71	1.88	2.07
0.79*		0.85	0.83	0.86	0.92	1.00	1.11	1.23	
0.35*		1.05	1.28	1.63	2.05	2.60			

*from Miller[28]

4.10.4.5   90° shoe branch tees, circular from rectangular (HVCA 136)

(a) Converging flows ($A_c = A_s$)

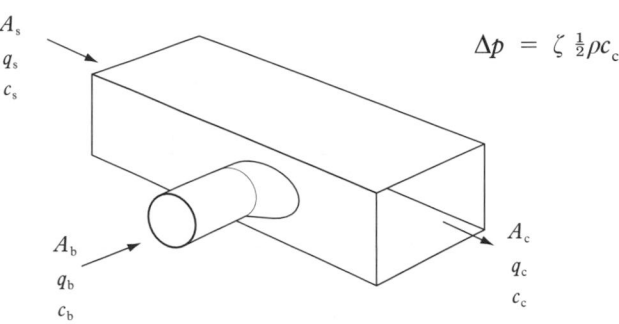

$$\Delta p = \zeta \tfrac{1}{2}\rho c_c^2$$

Table 4.111 Values for the straight factor ζ_{s-c} (from Miller[29])

A_b/A_c	Shoe depth	q_s/q_c								
		0.1	0.2	0.3	0.4	0.5	0.6	0.7	0.8	0.9
1.0	w/8			0.36	0.35	0.31	0.28	0.30	0.30	0.26
1.0	w/2			0.08	0.16	0.22	0.27	0.29	0.30	0.25
0.79	w/8			0.37	0.37	0.36	0.33	0.31	0.28	0.23
0.35	w/8				0.325	0.31	0.29	0.28	0.26	0.24

Table 4.112 Values for the branch factor ζ_{b-c} (from Miller[29])

A_b/A_c	Shoe depth	q_b/q_c								
		0.1	0.2	0.3	0.4	0.5	0.6	0.7	0.8	0.9
0.79	w/8			−0.02	0.18	0.36	0.52	0.69	*0.83*	
0.35	w/8		0.25	0.74	1.27	1.89				

Note: italic denotes extrapolated values

(b) Diverging flows ($A_c = A_s$)

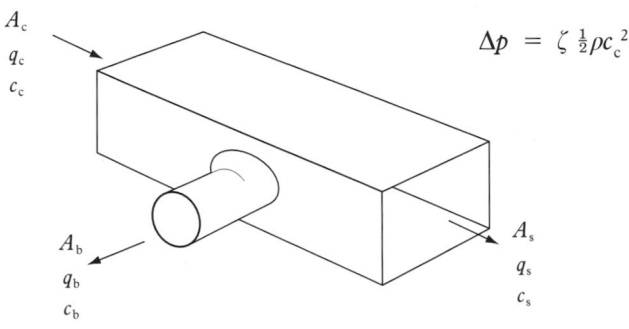

$$\Delta p = \zeta \tfrac{1}{2}\rho c_c^2$$

Table 4.113 Values for the straight factor ζ_{c-s} (information from Idelchik[2])

q_s/q_c	0	0.1	0.2	0.3	0.4	0.5	0.6	0.7	0.8	0.9	1
ζ_{c-s} for $A_b/A_c < 0.4$	0.40	0.324	0.256	0.196	0.144	0.1	0.064	0.036	0.016	0.004	0
ζ_{c-s} for $A_b/A_c > 0.4$	0.30	0.194	0.115	0.059	0.021	0	−0.064	−0.072	−0.048	−0.016	0

No data are available for a shoe on a circular branch. However Miller[28] shows that a reduction of approximately 10% occurs for a leading bevel on a circular branch on a rectangular duct and that a 25% reduction occurs when a small radius ($r/d = 0.09$) is placed at the juction of a circular branch on a circular main duct. The figures given in Table 4.113 have not taken into account any reduction.

Table 4.114 Values for the branch factor ζ_{c-b} (information derived from SMACNA[3])

A_b/A_c	q_b/q_c								
	0.1	0.2	0.3	0.4	0.5	0.6	0.7	0.8	0.9
0.1	0.74								
0.2	0.78	0.94	1.30						
0.3	0.80	0.78	0.77	0.81	0.99				
0.4	0.83	0.83	0.64	0.65	0.88	0.97	0.98		
0.5	0.88	0.72	0.72	0.67	0.73	0.77	0.80	0.94	0.95
0.79*		0.78	0.71	0.69	0.71	0.74	0.76	0.77	
0.35*		0.70	0.68	0.73	0.85	1.10			

*From Miller[28] for a shoe depth of w/8

4.10.5 Components for circular ductwork

4.10.5.1 90° Smooth radius round elbow (HVCA 127, 'pressed bend')

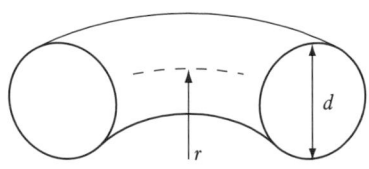

For bends of angles other than right angles, use the correction factor C_1 and

$$\zeta^\star = C_1 \zeta$$

Table 4.115 Variation of ζ with radius of bend (values derived from Idelchik[2], diagram 6–1)

r/d	0.5	0.75	1	1.5	2	4
ζ	1.2	0.42	0.24	0.22	0.21	0.24

Table 4.116 Correction factor C_1 for bends of other angles (Idelchik[2])

a	0	20°	30°	45°	60°	75°	90°	110°	130°	150°	180°
C_1	0	0.31	0.45	0.60	0.78	0.90	1.00	1.13	1.20	1.28	1.40

Table 4.117 Variation of ζ with Re for 90° bend (CETIAT[22]) ($r/d = 1$; $d = 250$ mm)

Re	4×10^4	8×10^4	1×10^5	2×10^5	4×10^5
ζ	0.385	0.29	0.26	0.235	0.225

Unfortunately CETIAT only tested for $d = 250$ mm and 400 mm.

As will be seen later with segmented elbows, those of larger diameters give slightly lower values of ζ. The implication is that a smaller diameter pressed bend may give higher values of ζ.

4.10.5.2 Segmented bend (HVCA 128)

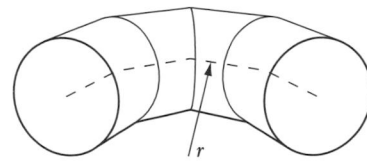

Medium radius, $r = d$ (HVCA standard radius).

Recent work by CETIAT[22] only covered the case of $r/d = 1$, but the results are considered to be the most reliable. The values for 90°, six sections are taken from AiCVF[17].

Table 4.115 Variation of ζ with radius of 90° bend

r/d	0.75	1	1.5	2	4
6 sections[17]		0.39	0.24	0.21	0.245
4 sections	0.5	0.37	0.27	0.24	

Table 4.119 Variation of ζ with diameter and Reynolds number for $r/d = 1$ (CETIAT[22])

Re	4×10^4	8×10^4	1×10^5	2×10^5	4×10^5	6.6×10^5
$d = 250$ mm 90°, 4 sections	0.38	0.31	0.285	0.26	0.24	
$d = 400$ mm 90°, 4 sections		0.31	0.28	0.24	0.23	0.213
$d = 250$ mm 60°	0.26	0.22	0.21	0.195	0.185	
$d = 400$ mm 60°		0.20	0.185	0.16	0.145	0.135
$d = 250$ mm 45°, 3 sections	0.185	0.14	0.135	0.125	0.12	
$d = 400$ mm 45°, 3 sections		0.13	0.11	0.08	0.07	0.05
$d = 250$ mm 30°	0.16	0.125	0.12	0.11	0.10	
$d = 400$ mm 30°		0.07	0.055	0.04	0.04	0.04

CETIAT also investigated the effect of six different manufacturers, but only for the 90° elbow of 400 mm. The results showed a scatter of approximately ±40°. This was largely due to the results of one manufacturer whose bend gave results very much out of step with all the rest. Although it is useful to note the large variation which can occur, in calculating the mean value for inclusion in the table above, the results of the rogue bend were ignored.

It is worth noting that the HVCA standard is:

90°, 4 sections; 60°, 3 sections; 45°, 3 sections; 30°, 2 sections.

4.10.5.3 Mitred elbow

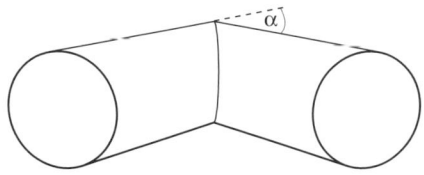

Table 4.120 Values of ζ for mitred elbow (data from AiCVF[17])

a	20°	30°	45°	60°	75°	90°
$Re = 10^5$	0.04	0.086	0.22	0.43	0.77	1.16
$Re = 10^6$	0.03	0.072	0.18	0.36	0.64	0.97

The figures for $Re = 10^6$ may be obtained using the following equation:

$$\zeta = 0.95 \sin^2\left(\frac{a}{2}\right) + 2.05 \sin^4\left(\frac{a}{2}\right)$$

For $Re = 10^5$, AiCVF suggests adding 20 per cent.

Table 4.121 Variation of ζ with Re for $a = 45°$ (CETIAT[22])

Re	0.4×10^5	0.8×10^5	1×10^5	2×10^5	4×10^5
ζ	0.375	0.26	0.25	0.23	0.225

Note: the values are valid for $d = 250$ mm

4.10.5.4　Segmented bends in close proximity, circular ('gooseneck')

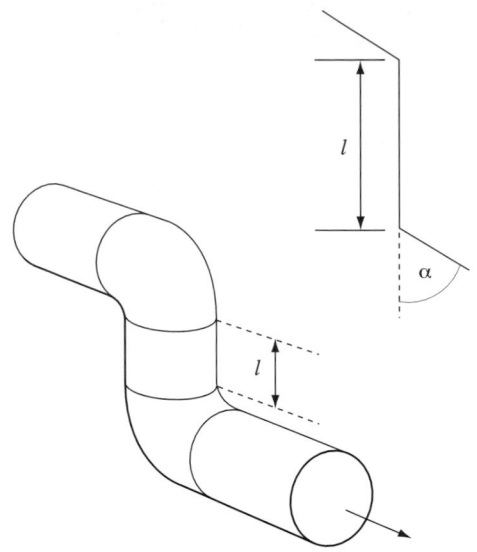

$$\zeta^\star = C_2 \zeta_1$$

where ζ_1 is the factor for a single bend (section 4.10.5.2)

Table 4.122 Values of ζ for circular gooseneck (from CETIAT[22])

a	Re	l/d		
		1	3	5
90° 4 elements	3.8×10^5	2.21	2.16	2.19
	2.5×10^5	2.13	2.08	2.11
	0.4×10^5	1.79	1.74	1.71
60° 3 elements	3.8×10^5	2.21	2.05	2.10
	2.5×10^5	2.21	2.04	2.07
	0.4×10^5	2.10	1.94	1.95
45° 3 elements	3.8×10^5	1.89	1.80	1.82
	2.5×10^5	1.90	1.82	1.35
	0.4×10^5	1.76	1.69	1.73

Note that Table 4.122 does not include the pressure drop of the length of separation. The separation l, should be added to the length of straight ductwork of the same size.

(CETIAT only tested the above configurations for bends of 250 mm diameter. Alternative sources of information do not specify the dimensions of the ductwork investigated.)

4.10.5.5　Bends in close proximity, circular, through perpendicular plane

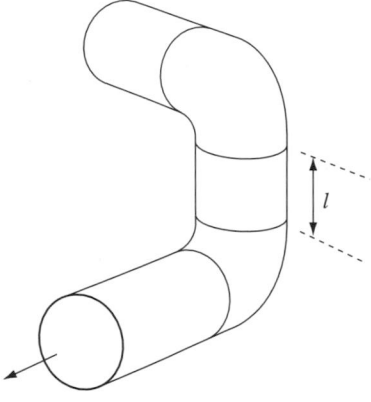

$$\zeta^\star = C_2 C_3 \zeta_1$$

where ζ_1 is the factor for a single bend, section 4.10.5.1, C_2 is as for section 4.10.10.4 and C_3 is given below.

No information is available. Idelchik (diagram 6–19) gives data for smooth elbows of this configuration and of the S bend[39]. The pressure loss factors for this configuration are consistently greater than for the gooseneck configuration.

In the absence of further information, values of the factor C_3 are given in Table 4.123 for double bends of 90°.

Table 4.123 Values of C_3 for double 90° bends

l/d	1	2	3	4	6	8	10	12	14	>14
C_3	1.06	1.00	1.00	1.03	1.1	1.2	1.19	1.14	1.11	1

4.10.5.6　Angled off-sets, circular (HVCA 134)

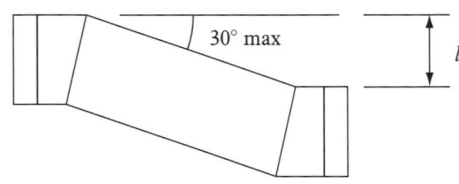

Table 4.124 Values of ζ for angled offsets (data from SMACNA[3])

l/d	0.5	1.0	1.5	2.0	2.5	3.0
ζ	0.15	0.15	0.16	0.16	0.16	0.16

4.10.5.7　Depression to avoid an obstruction

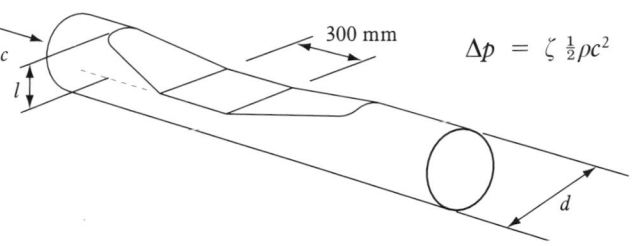

$$\Delta p = \zeta \tfrac{1}{2}\rho c^2$$

Data from SMACNA[3].

For $l/d = 0.33$, $\zeta = 0.24$

4.10.5.8　Symmetrical expansion (HVCA taper 132)

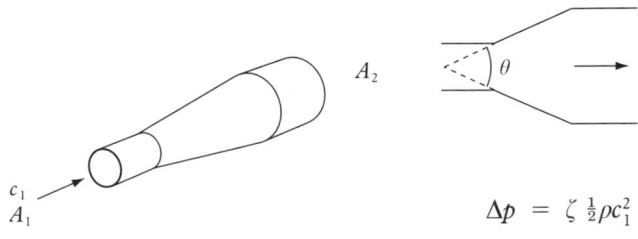

$$\Delta p = \zeta \tfrac{1}{2}\rho c_1^2$$

Values of ζ are very dependent upon flow separation which in turn is dependent upon the upstream flow conditions, by the velocity profile uniform, symmetrical, or asymmetrical. For small angles of divergence, the length of travel of an unseparated core of flow will depend very much upon the asymmetry. Thus for small angles of expansion, prediction of the pressure drop becomes more difficult. As a rule flow separation from the walls occurs for expansion angles greater than 40°.

The information presented below is derived from the data of Idelchik[2] (diagram 5–2). In the main this is 'worst case' data, fully developed flow being assumed as was the case for rectangular expansions. The values of ζ ought to be lower than those for rectangular expansions (section 4.10.3.17); however, several of the derived values are higher. No explanation has been found for this.

Values of ζ are shown to vary considerably with Reynolds number Re, particularly for small angles of expansion:

Table 4.125 Values of ζ for symmetrical expansion (derived from Idelchik[2])

Re_1	A_2/A_1	Included angle θ/deg.								
		10	15	20	30	45	60	90	120	180
0.5×10^5	10	0.420	0.543	0.646	0.807	0.958	0.960	0.915*	0.882	0.868
	6	0.376	0.505	0.561	0.665	0.806	0.919	0.779	0.767	0.758
	4	0.330	0.420	0.507	0.636	0.764	0.816	0.675	0.662	0.646
	2	0.221	0.255	0.306	0.353	0.367	0.359	0.347	0.339	0.328
1×10^5	10	0.340	0.450	0.564	0.789	1.02	0.960	0.861	0.860	0.856
	6	0.277	0.408	0.486	0.673	0.857	0.906	0.768	0.756	0.742
	4	0.252	0.336	0.445	0.661	0.802*	0.754	0.630	0.623	0.614
	2	0.234	0.255*	0.274	0.298	0.336	0.329	0.325	0.321	0.317
$\geq 6 \times 10^5$	10	0.275	0.439	0.608	0.692	0.732	0.897	0.883	0.899	0.883
	6	0.256	0.420*	0.523	0.606	0.732	0.814	0.756	0.749	0.744
	4	0.210*	0.359	0.439	0.508	0.617	0.661	0.623	0.617	0.607
	2	0.141	0.148*	0.150	0.194	0.352	0.335	0.299	0.295	0.295

* Denotes values massaged due to lack of confidence in the derived values:
   e.g. for air at 21°C, d_1 being 1.0 m, and $c_1 = 8.1$ m.s^{-1}, $Re_1 = 6 \times 10^5$
   for air at 21°C, d_1 being 0.4 m, and $c_1 = 3.4$ m.s^{-1}, $Re_1 = 1 \times 10^5$

4.10.5.9 Symmetrical contractions, and short tapers (HVCA 132 and 133)

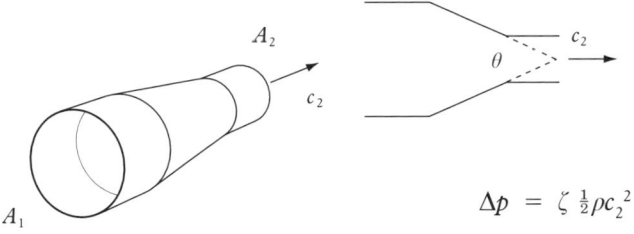

$$\Delta p = \zeta \tfrac{1}{2} \rho c_2^2$$

The values of ζ in Table 4.126 are taken from SMACNA[3].

Table 4.126 Values of ζ for symmetrical contraction (from SMACNA[3])

A_2/A_1	Included angle θ/deg.						
	10	15–40	50–60	90	120	150	180
0.50	0.05	0.05	0.06	0.12	0.18	0.24	0.26
0.25	0.05	0.04	0.07	0.17	0.27	0.35	0.41
0.17	0.05	0.04	0.07	0.18	0.28	0.36	0.42
0.10	0.05	0.05	0.08	0.19	0.29	0.37	0.43

Note: SMACNA gives the same values for a symmetrical rectangular contractions.

4.10.5.10 Offset taper contractions (HVCA 131)

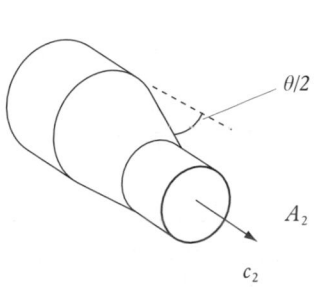

In the absence of specific data, it is suggested that Δp be estimated as the sum of Δp_1 for an angle offset based on inlet conditions, plus Δp_1 for a contraction. The angle θ to be used should be twice the angle of the part deviating most.

4.10.5.11 Circular tees: Y, breeches piece (HVCA 150)

SMACNA[3] provides data for the variation of ζ with the angle of the Y whereas ASHRAE[4] gives much more comprehensive data but only for an included angle of 60° for the Y.

Information from SMACNA[3]

Note that the following data are for the simple case of:

$$A_{b1} = A_{b2} = \tfrac{1}{2} A_c$$

(Values are the same as for rectangular Ys, section 4.10.3.21)

(a) Converging flows

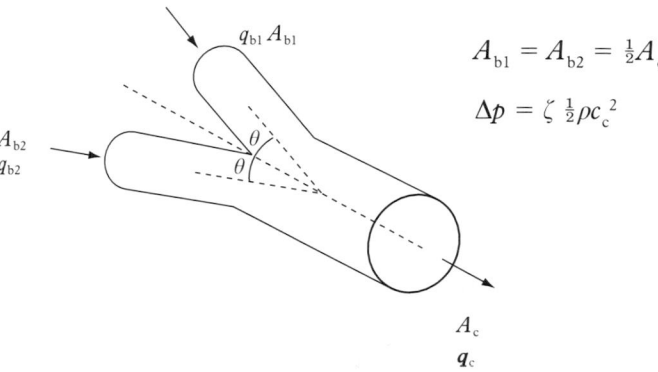

$$A_{b1} = A_{b2} = \tfrac{1}{2} A_c$$
$$\Delta p = \zeta \tfrac{1}{2} \rho c_c^2$$

Table 4.127 Values of ζ_{b-c} for converging circular tees (from SMACNA[3])

θ	q_{b1}/q_c or q_{b2}/q_c										
	0	0.1	0.2	0.3	0.4	0.5	0.6	0.7	0.8	0.9	1
15°	−2.6	−1.9	−1.3	−0.77	−0.3	0.1	0.41	0.67	0.85	0.97	1
30°	−2.1	−1.5	−1	−0.53	−0.1	0.28	0.69	0.91	1.1	1.4	1.6
45°	−1.3	−0.93	−0.55	−0.16	0.2	0.56	0.92	1.26	1.6	2	2.3

(b) Diverging flows

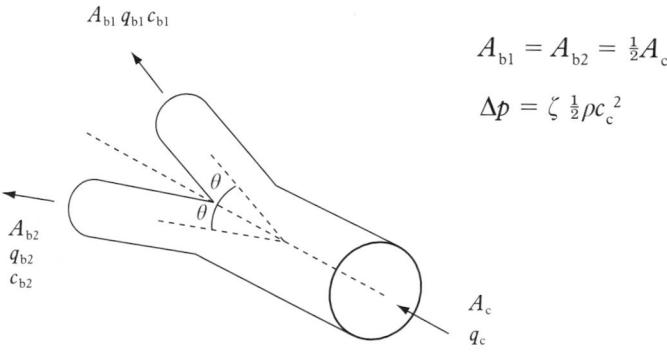

$$A_{b1} = A_{b2} = \tfrac{1}{2} A_c$$
$$\Delta p = \zeta \tfrac{1}{2} \rho c_c^2$$

Table 4.128 Values of ζ_{c-b1} and ζ_{c-b2} for diverging circular tees (from SMACNA[3])

θ	c_1/c_c or c_2/c_c							
	0.1	0.2	0.3	0.4	0.5	0.6	0.8	1
15°	0.81	0.65	0.51	0.38	0.28	0.2	0.11	0.06
30°	0.84	0.69	0.56	0.44	0.34	0.26	0.19	0.15
45°	0.87	0.74	0.63	0.54	0.45	0.38	0.29	0.24
60°	0.9	0.82	0.79	0.66	0.59	0.53	0.43	0.36
90°	1	1	1	1	1	1	1	1
θ	1.2	1.4	1.6	1.8	2			
15°	0.14	0.3	0.51	0.76	1			
30°	0.15	0.3	0.51	0.76	1			
45°	0.23	0.3	0.51	0.76	1			
60°	0.33	0.39	0.51	0.76	1			
90°	1	1	1	1	1			

Information provided by ASHRAE [4]

The following information from ASHRAE[4] is only available for converging flows and only for the case of a Y with total included angle of 60°.

Converging flows:

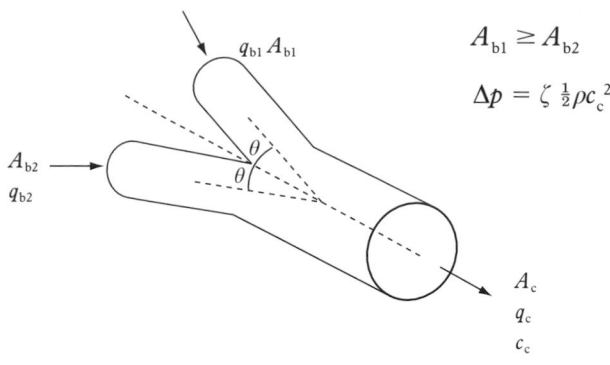

$$A_{b1} \geq A_{b2}$$

$$\Delta p = \zeta \tfrac{1}{2}\rho c_c^2$$

Table 4.129 Values of ζ_{b1-c} for converging circular tee ($A_{b1} \geq A_{b2}$) (from ASHRAE[4])

A_{b1}/A_c	A_{b2}/A_c	q_{b1}/q_c								
		0.1	0.2	0.3	0.4	0.5	0.6	0.7	0.8	0.9
0.2	0.20	−0.09	−0.38	−0.04	0.33	0.78	1.33	1.96	2.56	3.20
0.3	0.20	−0.31	−1.25	−0.73	−0.22	0.28	0.82	1.45	2.20	3.43
	0.30	−0.10	−0.39	−0.03	0.31	0.72	1.22	1.78	2.35	2.92
0.4	0.20	−0.35	−1.40	−0.96	−0.50	−0.06	0.30	0.61	0.91	12.76
	0.40	−0.16	−0.62	−0.48	0.15	0.12	0.50	0.88	1.26	1.72
0.5	0.20	−0.48	−1.90	−1.34	−0.84	−0.43	−0.09	0.16	0.33	0.45
	0.50	−0.10	−0.42	−0.19	0.00	0.16	0.31	0.46	0.60	0.66
0.6	0.20	−0.66	−2.62	−1.82	−1.16	−0.64	−0.24	0.05	0.23	0.31
	0.60	−0.13	−0.52	−0.39	0.30	−0.11	0.00	0.11	0.22	0.34
0.7	0.20	−0.78	−3.12	−2.19	−1.42	−0.81	−0.35	−0.03	0.17	0.24
	0.70	−0.24	−0.95	−0.89	0.55	−0.58	−0.41	−0.20	0.02	0.25
0.8	0.20	−0.87	−3.49	−2.47	−1.62	−0.94	−0.43	−0.08	0.12	0.20
	0.80	−0.32	−1.27	−1.26	0.70	−0.93	−0.72	−0.44	−0.14	0.19
0.9	0.20	−1.00	−3.98	−2.84	−1.88	−1.12	−0.56	−0.20	0.01	0.08
	0.90	−0.39	−1.56	−1.61	0.89	−1.44	−1.28	−1.02	−0.77	−0.48
1	0.20	−1.09	−1.39	−1.18	−0.81	−0.47	−0.21	−0.06	−0.01	−0.01
	1.00	−0.45	−1.80	−1.89	1.05	−1.85	−1.72	−1.49	−1.27	−0.85

Table 4.130 Values of ζ_{b2-c} for converging circular tee ($A_{b1} \geq A_{b2}$) (from ASHRAE[4])

A_{b1}/A_c	A_{b2}/A_c	q_{b2}/q_c								
		0.1	0.2	0.3	0.4	0.5	0.6	0.7	0.8	0.9
0.2	0.20	−0.38	−0.38	0.02	0.33	0.78	1.33	1.96	2.56	3.20
0.3	0.20	−0.41	−0.24	0.02	0.34	0.76	1.31	1.91	2.46	3.08
	0.30	−0.56	−0.39	−0.03	0.31	0.72	1.22	1.78	2.35	2.92
0.4	0.20	−0.03	−0.20	0.07	0.42	0.84	1.28	1.76	2.30	2.88
	0.40	−0.72	−0.62	−0.48	−0.23	0.12	0.50	0.88	1.26	1.72
0.5	0.20	−0.24	0.00	0.25	0.57	1.03	1.60	2.25	2.88	3.60
	0.50	−0.61	−0.42	−0.19	0.00	0.16	0.31	0.46	0.60	0.66
0.6	0.20	−0.18	0.01	0.35	0.78	1.33	1.94	2.65	3.39	4.21
	0.60	−0.68	−0.52	−0.39	−0.26	−0.11	0.00	0.11	0.22	0.34
0.7	0.20	−0.27	−0.08	0.25	0.69	1.23	1.84	2.55	3.30	4.13
	0.70	−1.04	−0.95	−0.89	−0.75	−0.58	−0.41	−0.20	0.02	0.25
0.8	0.20	−0.36	−0.17	0.15	0.58	1.13	1.75	2.45	3.20	4.01
	0.80	−1.32	−1.27	−1.26	−1.12	−0.93	−0.72	−0.44	−0.14	0.19
0.9	0.20	−0.25	−0.07	0.24	0.68	1.21	1.85	2.55	3.30	4.09
	0.90	−1.45	−1.56	−1.61	−1.55	−1.44	−1.28	−1.02	−0.77	−0.48
1	0.20	−0.14	0.04	0.34	0.77	1.31	1.94	2.65	3.39	4.21
	1.00	−1.55	−1.80	−1.89	−1.89	−1.85	−1.72	−1.49	−1.27	−0.85

4.10.5.12   90° branch tees, circular from circular (HVCA 139) and pressed equal tee (HVCA 130)

(*a*) Converging flows ($A_c = A_s$)

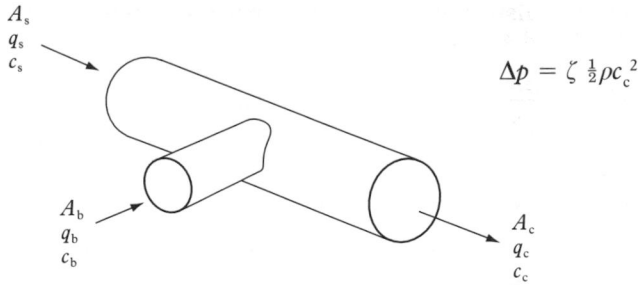

$$\Delta p = \zeta \tfrac{1}{2}\rho c_c^2$$

Table 4.131 Values for the straight factor ζ_{s-c}

q_s/q_c	0.2	0.3	0.4	0.5	0.6	0.7	0.8	0.9	1.0
ζ	0.79	0.71	0.63	0.55	0.47	0.38	0.28	0.16	0

The values are a mean of data from CETIAT[22], Miller[29] and Miller[28] for values of A_b/A_c between 0.1 and 1.0. There does not appear to be a consistent trend with A_b/A_c so the data above is valid for all values of A_b/A_c.

Table 4.132 Values for the branch factor ζ_{b-c} (from SMACNA[3])

A_b/A_c	q_b/q_c									
	0.1	0.2	0.3	0.4	0.5	0.6	0.7	0.8	0.9	1
0.1	0.40	3.8	9.2	16	26	37	43	65	82	101
0.2	−0.37	0.72	2.3	4.3	6.8	9.7	13	17	21	26
0.3	−0.41	0.17	1.0	2.1	3.2	4.7	6.3	7.9	9.7	12
0.4	−0.46	−0.10	0.25	0.66	1.1	1.6	2.1	2.7	3.4	4
0.5	−0.50	−0.20	0.14	0.42	0.80	1.15	1.5	1.9	2.3	2.85
0.6	−0.50	−0.25	0	0.26	0.66	0.92	1.2	1.5	1.8	2.1
0.8	−0.51	−0.25	0	0.20	0.49	0.69	0.88	1.1	1.2	1.4
1	−0.52	−0.25	−0.05	0.20	0.42	0.57	0.72	0.86	0.99	1.1

Information derived from SMACNA[3]. Where applicable the small amount of CETIAT[22] data largely confirm the SMACNA data but do show up some anomalous kinks in the SMACNA table. These have been ironed out. The disagreements are not sufficient to override the SMACNA data.

(*b*) Diverging flows ($A_c = A_s$)

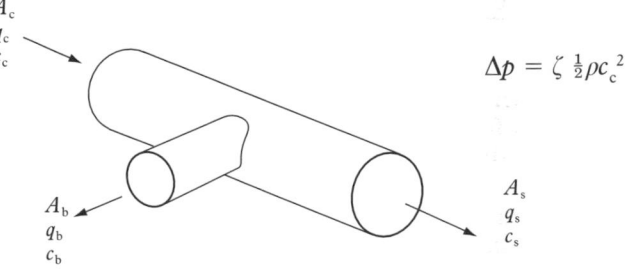

$$\Delta p = \zeta \tfrac{1}{2}\rho c_c^2$$

Table 4.133 Values of ζ_{c-s} (information from Idelchik[2])

q_s/q_c	0	0.1	0.2	0.3	0.4	0.5	0.6	0.7	0.8	0.9	1
ζ_{c-s} for $A_b/A_c < 0.4$	0.40	0.324	0.256	0.196	0.144	0.1	0.064	0.036	0.016	0.004	0
ζ_{c-s} for $A_b/A_c > 0.4$	0.30	0.194	0.115	0.059	0.021	0	−0.064	−0.072	−0.048	−0.016	0

Idelchik[2] makes no distinction as to whether the above data for ζ_{c-s} for the straight flow is for circular or rectangular ductwork. In the absence of further information, the same data should be used as for a rectangular branch (section 4.10.3.22), and for a circular branch off a rectangular duct (section 4.10.4.4). SMACNA gives slightly different data but also makes no distinction between sections 4.10.4.4 and 4.10.5.11 for the straight flow factor.

Table 4.134 Values for the branch factor ζ_{c-b} (information from Idelchik[2])

c_b/c_c	0	0.1	0.2	0.4	0.6	0.8	1	1.2	1.4	1.6	2
ζ_{c-b} for $A_b/A_c < 2/3$	1	1.01	1.04	1.16	1.35	1.64	2.0	2.44	2.96	3.54	4.6
ζ_{c-b} for $A_b/A_c = 1$	1	1	1.01	1.05	1.11	1.19	1.3	1.43	1.59	1.77	2.2

Table 4.135 Values for branch factor ζ_{c-b} (information from SMACNA[3])

A_b/A_c	q_b/q_c								
	0.1	0.2	0.3	0.4	0.5	0.6	0.7	0.8	0.9
0.1	2.1								
0.2	1.3	1.9	2.9						
0.3	1.1	1.4	1.8	2.3					
0.4	0.99	1.1	1.3	1.5	1.7	2.0	2.4		
0.5	0.97	1.0	1.1	1.2	1.4	1.5	1.8	2.1	2.5
0.8	0.95	0.92	0.92	0.93	0.94	0.95	1.1	1.2	1.4
1.0*	0.95	0.89	0.85	0.825	0.82	0.85	0.89	0.94	1.01

Note: Values in italic are extrapolated values, taking into consideration similiar results by others.
* Miller[28], who shows that the values of ζ are not significantly different

4.10.5.13 90° shoe branch tees, circular from circular (HVCA 139)

(a) Converging flows ($A_c = A_s$)

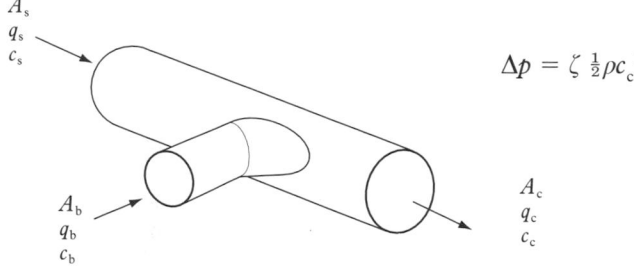

$$\Delta p = \zeta \tfrac{1}{2}\rho c_c^2$$

Table 4.136 Values of the straight factor ζ_{s-c}

q_s/q_c	0.2	0.3	0.4	0.5	0.6	0.7	0.8	0.9	1.0
ζ	0.47	0.43	0.38	0.33	0.28	0.23	0.17	0.10	0

No data are available for a shoe on a circular branch. However Miller[28] shows that a reduction of approximately 40% occurs for a trailing bevel on a circular branch on a rectangular duct and that a similar reduction occurs when a small radius ($r/d = 0.09$) is placed at the junction of a circular branch on a circular main duct. The figures in the above table have been obtained by reducing those of Table 4.131 by 40%.

Table 4.137 Values for the branch factor ζ_{b-c}

A_b/A_c	q_b/q_c									
	0.1	0.2	0.3	0.4	0.5	0.6	0.7	0.8	0.9	1
0.1	0.32	2.5	5.5	10	16	22	26	39	50	60
0.2	−0.52	0.5	1.6	2.6	4.1	5.8	7.8	10	13	16
0.3	−0.57	0.19	0.72	1.5	1.9	2.8	3.8	4.7	5.8	7.2
0.4	−0.58	−0.02	0.22	0.5	0.83	1.1	1.5	1.8	2.1	2.4
0.5	−0.55	−0.28	0.14	0.34	0.68	0.85	1.1	1.4	1.6	2.0
0.6	−0.64	−0.22	0	0.23	0.50	0.69	0.9	1.1	1.3	1.5
0.8	−0.65	0.22	0	0.16	0.37	0.50	0.63	0.82	0.90	1.04
1	−0.67	−0.22	−0.05	0.16	0.34	0.43	0.50	0.60	0.69	0.77

Note. No experimental data are available. Table 4.137 gives values which are only estimates. Values for the branch factor for a shoe branch ought to be lower than those for an abrupt branch, section 4.10.5.12. The values have been obtained by reducing the values of section 4.10.5.12 by a factor obtained from a comparison of the effects of a shoe branch on rectangular ducts (SMACNA[3]). However it should be noted that the SMACNA comparison was only for a branch area of half the main duct area.

(b) Diverging flows ($A_c = A_s$)

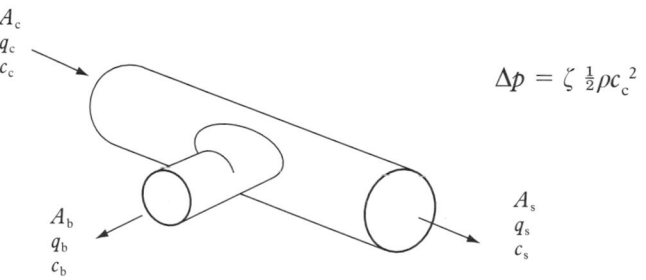

$$\Delta p = \zeta \tfrac{1}{2}\rho c_c^2$$

Table 4.138 Values of ζ_{c-s} for diverging 90° shoe branch tees (information from Idelchik[2])

q_s/q_c	0	0.1	0.2	0.3	0.4	0.5	0.6	0.7	0.8	0.9	1
ζ_{c-s} for $A_b/A_c<0.4$	0.40	0.324	0.256	0.196	0.144	0.1	0.064	0.036	0.016	0.004	0
ζ_{c-s} for $A_b/A_c>0.4$	0.30	0.194	0.115	0.059	0.021	0	−0.064	−0.072	−0.048	−0.016	0

No data are available for a shoe on a circular branch. However Miller[28] shows that a reduction of approximately 10% occurs for a trailing bevel on a circular branch on a rectangular duct and that a 25% reduction occurs when a small radius ($r/d = 0.09$) is placed at the junction of a circular branch on a circular main duct. The figures in the above table have not taken account of any reduction.

Table 4.139 Values for the branch factor ζ_{c-b} (information derived from SMACNA[3])

A_b/A_c	q_b/q_c								
	0.1	0.2	0.3	0.4	0.5	0.6	0.7	0.8	0.9
0.1	1.20								
0.2	0.94	1.29	2.0						
0.3	0.88	0.91	1.2	1.29					
0.4	0.82	0.84	0.75	0.80	1.0	1.1	1.2		
0.5	0.85	0.78	0.74	0.71	0.81	0.86	0.85	1.0	1.4
0.8	0.86	0.74	0.64	0.56	0.54	0.49	0.52	0.6	0.64
1.0*	0.84	0.77	0.72	0.68	0.66	0.68	0.71	0.75	0.81

* Miller[28], for a radius on the branch junction of $r/d = 0.09$ with $A_b/A_c = 1.0$

Note No experimental data are available. Table 4.139 gives values which are only estimates. Values for the branch factor for a shoe branch ought to be lower than those for an abrupt branch, section 4.10.5.12. The values have been obtained by reducing the values of section 4.10.5.12 by a factor obtained from a comparison of the effects of a shoe branch on rectangular ducts. However it should be noted that the SMACNA comparison was only for a branch area of half the main duct area.

4.10.5.14 Angle branch tees, circular from circular (HVCA 138)

(a) Converging flows ($A_c = A_s$)

Data from SMACNA[3] are only available for an angle of 45°.

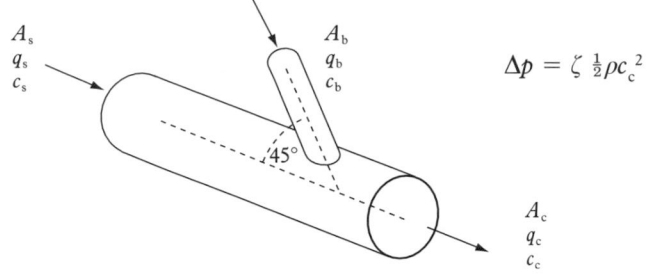

$$\Delta p = \zeta \tfrac{1}{2}\rho c_c^2$$

Table 4.140 Values for ζ_{s-c} for all values of A_b/A_c (information from SMACNA[3])

A_b/A_c	q_s/q_c									
	0.1	0.2	0.3	0.4	0.5	0.6	0.7	0.8	0.9	1
0.1	−8.6	−6.7	−5	−3.5	−2.3	−1.3	−0.63	−0.18	−0.3	−0.01
0.2	−4.1	−3.1	−2.2	−1.5	−0.95	−0.5	−0.18	0.01	0.07	0
0.3	−2.5	−1.9	−1.3	−0.88	−0.51	−0.22	−0.03	0.07	0.08	0
0.4	−1.7	−1.3	−0.88	−0.55	−0.28	−0.09	0.04	0.1	0.09	0.01
0.6	−0.97	−0.67	−0.42	−0.21	−0.06	0.05	0.12	0.13	0.1	0.02
0.8	−0.58	−0.36	−0.19	−0.05	0.06	0.12	0.16	0.15	0.11	0.04
1	−0.34	−0.18	−0.05	0.05	0.13	0.17	0.18	0.17	0.13	0.05

Some of the above data are confirmed by CETIAT[22].

Table 4.141 Values for ζ_{b-c} for all values of A_b/A_c (information derived from SMACNA[3])

A_b/A_c	q_s/q_c									
	0.1	0.2	0.3	0.4	0.5	0.6	0.7	0.8	0.9	1
0.1	0.22	3.1	8							
0.2	−0.37	0.31	1.5	3.2	5.3					
0.3		−0.12	0.38	1.11	2.1	3.2	4.6	6.2	8	
0.4		−0.21	0.08	0.44	1.02	1.6	2.4	3.2	4.3	5.4
0.6			−0.09	0.07	0.28	0.53	0.93	1.3	1.7	2.2
0.8				0.02	0.13	0.26	0.43	0.62	0.9	1.1
1				0.05	0.11	0.18	0.28	0.4	0.53	0.69

Some of the above data are confirmed by CETIAT[22].

CETIAT[22] carried out tests for three different branch area ratios in 1990. Two sets of the results agreed very accurately with those of SMACNA in Tables 4.140 and 4.141. The third set was so wildly different as to be presumed in error.

(b) Diverging flows ($A_c = A_s$)

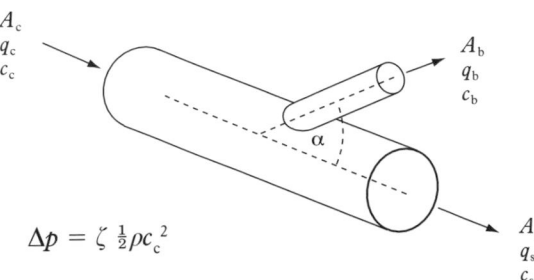

$$\Delta p = \zeta \tfrac{1}{2}\rho c_c^2$$

Table 4.142 Values of the straight factor ζ_{c-s} (information from Idelchik[2])

q_s/q_c	0	0.1	0.2	0.3	0.4	0.5	0.6	0.7	0.8	0.9	1
ζ_{c-s} for $A_b/A_c<0.4$	0.40	0.324	0.256	0.196	0.144	0.1	0.064	0.036	0.016	0.004	0
ζ_{c-s} for $A_b/A_c>0.4$	0.30	0.194	0.115	0.059	0.021	0	−0.064	−0.072	−0.048	−0.016	0

Idelchik[2] makes no distinction as to whether the above data for ζ_{c-s} for the straight flow are for circular or rectangular ductwork. In the absence of further information, the same data should be used as for a rectangular branch (section 4.10.3.23). SMACNA gives slightly different data but also makes no distinction between sections 4.10.3.23 and 4.10.5.12 for the straight flow factor.

Table 4.143 Values for the branch factor ζ_{c-b} (information derived from SMACNA[3])

a	A_b/A_c	q_b/q_c								
		0.1	0.2	0.3	0.4	0.5	0.6	0.7	0.8	0.9
30°	0.1	0.28	1.5							
	0.2	0.4	0.26	0.58	1.3	2.5				
	0.4	0.59	0.33	0.21	0.20	0.27	0.40	0.62	0.92	1.3
	0.6	0.69	0.46	0.31	0.21	0.17	0.16	0.2	0.28	0.39
	0.8	0.75	0.55	0.4	0.28	0.21	0.16	0.15	0.16	0.19
45°	0.1	0.6	2.1							
	0.2	0.56	0.56	1	1.8					
	0.4	0.66	0.47	0.4	0.43	0.54	0.69	0.95	1.3	1.7
	0.6	0.74	0.56	0.44	0.37	0.35	0.36	0.43	0.54	0.68
	0.8	0.78	0.62	0.49	0.4	0.34	0.31	0.32	0.35	0.4
60°	0.1	1	2.9							
	0.2	0.77	0.96	1.6	2.5					
	0.4	0.76	0.65	0.65	0.74	0.89	1.1	1.4	1.8	2.3
	0.6	0.81	0.68	0.6	0.58	0.58	0.61	0.72	0.87	1.1
	0.8	0.83	0.71	0.62	0.56	0.52	0.5	0.53	0.6	0.68

4.10.5.15 90° conical branch tees, circular from circular (HVCA 137 and 141)

(a) Converging flows ($A_c = A_s$)

No data are available.

(b) Diverging flows ($A_c = A_s$)

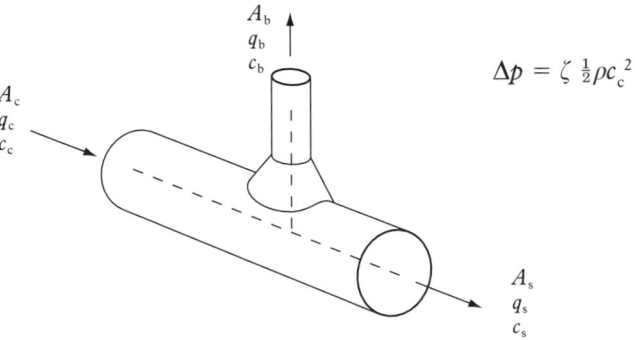

$$\Delta p = \zeta \tfrac{1}{2}\rho c_c^2$$

Table 4.144 Values for the straight factor ζ_{c-s} (information from Idelchik[2])

q_s/q_c	0	0.1	0.2	0.3	0.4	0.5	0.6	0.7	0.8	0.9	1
ζ_{c-s} for $A_b/A_c<0.4$	0.40	0.324	0.256	0.196	0.144	0.1	0.064	0.036	0.016	0.004	0
ζ_{c-s} for $A_b/A_c>0.4$	0.30	0.194	0.115	0.059	0.021	0	−0.064	−0.072	−0.048	−0.016	0

For the straight flow factor SMACNA[3] makes no distinction between the data for the branch being with or without a conical connection. For consistency the data from Idelchik[2] are reproduced in Table 4.144. A_b should be taken as the area of the duct, not of the cone.

Table 4.145 Values for the branch factor ζ_{c-b} (information from SMACNA[3])

c_b/c_c	0	0.2	0.4	0.6	0.8	1.0	1.2	1.4	1.6	1.8	2
ζ_{c-b}	1	0.85	0.74	0.62	0.52	0.42	0.36	0.32	0.32	0.37	0.52

4.10.5.16 45° conical branch tees, circular from circular (HVCA 137 and 141)

(*a*) Converging flows ($A_c = A_s$)

No data are available.

(*b*) Diverging flows ($A_c = A_s$)

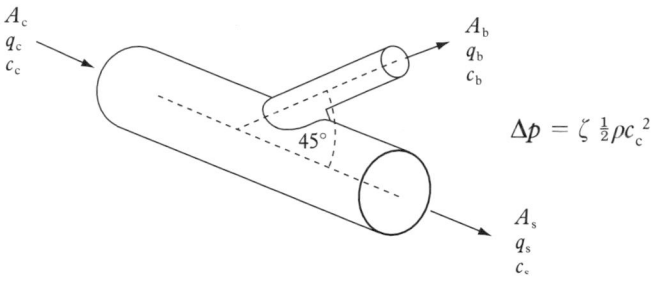

$$\Delta p = \zeta\,\tfrac{1}{2}\rho c_c^2$$

Table 4.146 Values of the straight factor ζ_{c-s} (information from Idelchik[2])

q_s/q_c	0	0.1	0.2	0.3	0.4	0.5	0.6	0.7	0.8	0.9	1
ζ_{c-s} for $A_b/A_c < 0.4$	0.40	0.324	0.256	0.196	0.144	0.1	0.064	0.036	0.016	0.004	0
ζ_{c-s} for $A_b/A_c > 0.4$	0.30	0.194	0.115	0.059	0.021	0	−0.064	−0.072	−0.048	−0.016	0

For the straight flow factor SMACNA[3] makes no distinction between the data for the branch being with or without a conical connection. For consistency the data from Idelchik[2] are reproduced in Table 4.146. A_b should be taken as the area of the duct, not of the cone.

Table 4.147 Values for the branch factor ζ_{c-b} (information from SMACNA[3])

c_b/c_c	0	0.2	0.4	0.6	0.8	1.0	1.2	1.4	1.6	1.8	2
ζ_{c-b}	1	0.84	0.61	0.41	0.27	0.17	0.12	0.12	0.14	0.18	0.27

4.10.5.17 Mitred branch tees, circular from circular (HVCA 143)

No reliable information is available.

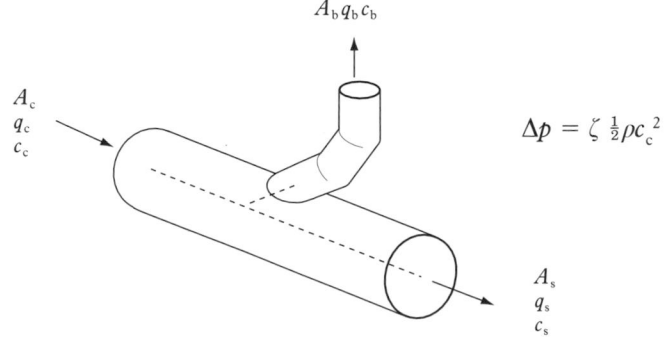

$$\Delta p = \zeta\,\tfrac{1}{2}\rho c_c^2$$

Until further information is available:

— The straight flow factors, ζ_{c-s} and ζ_{s-c}, should be taken to be the same as for an angle branch tee (section 4.10.5.9).

— The branch factors, ζ_{c-b} and ζ_{b-c}, should be taken to be the same as the sum of:

ζ for an angle branch tee (section 4.10.5.14) and

ζ for a 45° segmented bend (section 4.10.5.4).

4.10.5.18 Exhaust vents with hood

(*a*) Chinaman's hat

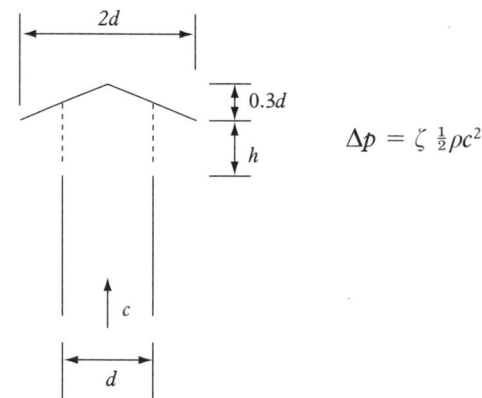

$$\Delta p = \zeta\,\tfrac{1}{2}\rho c^2$$

Table 4.148 Values of ζ for exhaust vent with hood (Chinaman's hat) (information from Idelchik[2])

h/d	0.1	0.2	0.25	0.3	0.35	0.4	0.5	0.6	0.8	1
ζ	4	2.3	1.90	1.60	1.40	1.30	1.15	1.10	1.00	1.00

(*b*) Plane baffle

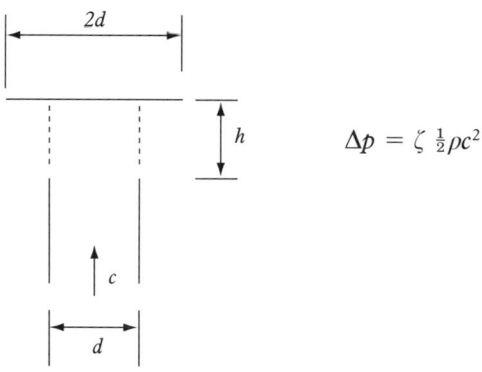

$$\Delta p = \zeta\,\tfrac{1}{2}\rho c^2$$

Table 4.149 Values of ζ for exhaust vent with hood (plane baffle) (information from Idelchik[2])

h/d	0.25	0.3	0.35	0.4	0.5	0.6	0.8	1
ζ	3.4	2.6	2.10	1.70	1.40	1.20	1.10	1.00

4.10.5.19 Plane extracts (from Idelchik[2])

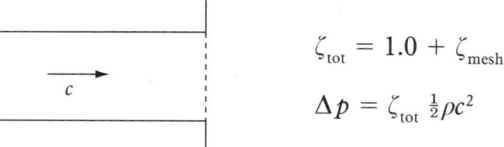

$$\zeta_{tot} = 1.0 + \zeta_{mesh}$$

$$\Delta p = \zeta_{tot}\,\tfrac{1}{2}\rho c^2$$

4.10.5.20 Mesh screens; grids of circular metal wire (from Idelchik[2])

$$\Delta p = \zeta_{mesh}\,\tfrac{1}{2}\rho c^2$$

The value of ζ_{mesh} depends very much on the closeness of the mesh, or rather on the free or clear area ratio, being A_c/A.

For turbulent flow ($Re > 1000$) through the mesh, defined in this instance by:

$$Re = \rho c_\text{m} d / \eta$$

where c_m is the mean velocity of air through the mesh and d is the diameter of the wire.

Table 4.150 Values of ζ for mesh screen (from Idelchik[2])

A_c/A	0.1	0.2	0.3	0.4	0.5	0.6	0.7	0.8
ζ_mesh	82	17	6.4	3.03	1.65	0.97	0.58	0.32

Idelchik gives the following formula as a reasonable curve-fit for low values of A_c/A:

$$\zeta_\text{mesh} = 1.3 \left(1 - \frac{A_c}{A} \right) + \left(\frac{A}{A_c} - 1 \right)^2$$

4.10.5.21   Plain duct entries (from Idelchik[2])

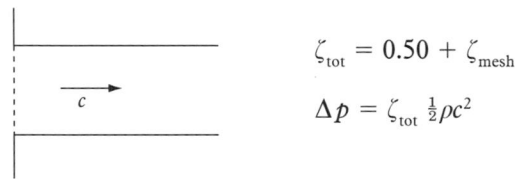

$$\zeta_\text{tot} = 0.50 + \zeta_\text{mesh}$$

$$\Delta p = \zeta_\text{tot} \tfrac{1}{2} \rho c^2$$

4.10.5.22   Inlet vents with hood

(*a*) Chinaman's hat

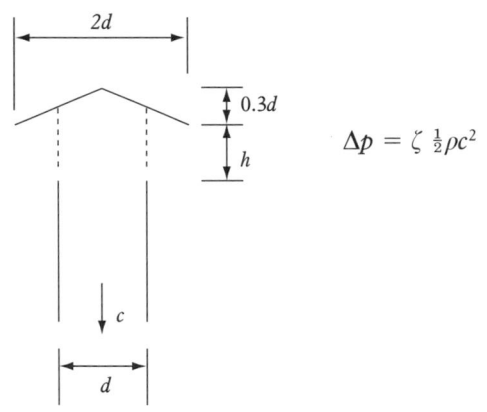

$$\Delta p = \zeta \tfrac{1}{2} \rho c^2$$

Table 4.151 Values of ζ for inlet vent with hood (Chinaman's hat) (information from Idelchik[2])

h/d	0.1	0.2	0.3	0.4	0.5	0.6	0.7	0.8	0.9	1.0
ζ	2.63	1.83	1.53	1.39	1.31	1.19	1.15	1.08	1.07	1.06

(*b*) Plane baffle

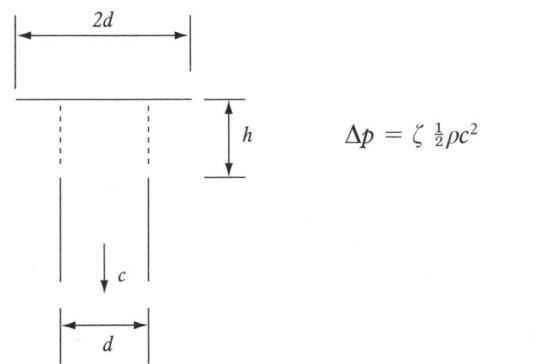

$$\Delta p = \zeta \tfrac{1}{2} \rho c^2$$

Table 4.152 Values for ζ for inlet vent with hood (plane baffle) (information from Idelchik[2])

h/d	0.2	0.3	0.4	0.5	0.6	0.7	0.8	0.9	1
ζ	4.4	2.15	1.78	1.57	1.35	1.23	1.13	1.10	1.06

4.10.5.23   Orifices

Some information is published in section 4.9.4. Since the source of that information, Idelchik[2], makes no distinction between orifices for pipework or ductwork, the same information can be used for ductwork.

References

1   1992, BS5775: 1993, Specification for quantities, units and symbols

2   Idelchik I E, *Handbook of Hydraulic Resistance*. 3rd edition. IBSN 0–8493–9908–4, CRC Press, Inc., Florida, USA (1994)

3   Smacna, *HVAC Systems Duct Design*, 3rd Edition 1990

4   ASHRAE, *Fundamentals Handbook* (SI), Chapter 32 (1997)

5   HVCA, *DW/144: Specification for sheet metal ductwork* (1998)

6   Schneider K J, *WIT 40. Bautabellen*, Vol.9 Auflage. Werner-Verlag Dusseldorf, Germany (1990)

7   Porcher G, *Cours de climatisation: bases du calcul des installations de climatisation*. Chaud-Froid-Plomberie, France (1990)

8   Dixon R, *A critical analysis and selection of pressure loss data for flow in pipe fittings*, Undergraduate thesis, Coventry University (1995)

9   Moody L F, Friction factors for pipe flow. *Trans. ASME*, **66**, 671 (1944)

10   HVRA, *Pressure loss data for ducting*, Information Circular 26 (1973)

11   BS1042: 1991, Measurement of fluid flow in closed circuits, Part 1, Pressure differential devices ISO5167–1: 1991

12   BSRIA, *Rules of thumb UK/France*, Technical note TN 18/1995

13   prEN1505, Ventilation for buildings — ductwork — rectangular sheet metal air duct fittings — dimensions, Draft document (1994)

14   prEN1506, Ventilation for buildings — ductwork — circular sheet metal air duct fittings — dimensions, Draft document (1994)

15   prEN1506, Ventilation for buildings — ductwork — rectangular sheet metal air ducts — strength and leakage, Draft document (1994)

16   *A review of the velocity pressure loss factors for HVAC duct fittings*, BSRIA (1998)

17   AiCVF, Guide 4: *Aéraulique: Principes de l'aérauliques appliqués au génie climatique*. Association d'Ingénieurs de Climatisation Ventilation, et Froid, ISBN 2 911008–29–4 (1991)

18   Rogers W L, 'Noise and vibration in water piping systems', *ASHRAE Journal*, **1**(3), 83–86 (1959)

19   Ball E F and Webster C, 'Some measurements of water-flow noise in copper and abs pipes', *Building Services Engineer*, **44**(5), 33–40 (May 1976)

20   Barnett J, Airflow in small ducts, Dissertation, Brunel University (1995)

21   Sprenger F, Friction pressure drops in ducts and pipes, Undergraduate thesis, Coventry University (1995)

22   CETIAT Catalogue des coefficients de perte d'énergie mécanique (Pressure loss coefficients) Becirspahic S et Dagonnot J, NTO Nr 90.307. Centre Technique des Industries Aérauliques et Thermiques, 69603 Villeurbanne, France (1990)

23   ESDU, Pressure losses in three-leg pipe junctions: combining flows, Engineering Sciences Data Unit; Item 73023 (1973)

24   ESDU, Pressure losses in curved ducts: interaction factors for two bends in series, Engineering Sciences Data Unit; Item 77009 (1977)

25   Eckert and Drake *Analysis of Heat and Mass Transfer.* McGraw-Hill, New York. (1972)

26   Koch P, 'A survey of available data for pressure loss coefficients, ζ for elbows and tees of pipework', *BSER&T*, **21**(3), 153–161 (2000)

27   Koch P, 'Comparisons and choice of pressure loss coefficients, ζ for ductwork components', submitted June 2000, to BSER&T

for publication,. Meanwhile copy available from the author at the Department of the Built Environment, Coventry University, CV1 5FB, UK

28   Miller D S *Internal flow: A guide to losses in pipe duct systems*, British Hydromechanics Research Association, Cranfield MK43 0AJ, UK (1971)

29   Miller D S *Internal flow systems* British Hydromechanics Research Association, Cranfield, MK43 0AJ, UK (1990) ISBN 0-947711-77-5

30   *Engineering Data Book.* Hydraulic Institute, Parsippany, NJ, USA (1979)

Appendix 4.A1: Properties of various fluids

Table 4.A1.1 Properties of water

T /°C	ρ /kg.m^{-3}	η /10^{-6} kg.m^{-1}.s^{-1}	ν /10^{-6} m^2.s^{-1}
0.01	999.8	1752	1.7524
4	1000.0	1551	1.5510
10	999.7	1300	1.3004
20	999.8	1002	1.0022
30	995.6	797	0.8005
40	992.2	651	0.6561
50	988.0	544	0.5506
60	983.2	463	0.4709
70	977.8	400	0.4091
80	971.8	351	0.3612
90	965.3	311	0.3222
100	958.4	279	0.2911
110	950.6	252	0.2651
120	943.4	230	0.2438
130	934.6	216	0.2258
140	925.9	195	0.2106
150	916.6	181	0.1975
160	907.4	169	0.1862
170	897.7	158	0.1760
180	886.5	149	0.1681
190	875.6	141	0.1610
200	864.3	134	0.1550

Table 4.A1.2 Some properties of air at a relative humidity of 50% and at a pressure of 1.012 bar

T /°C	ρ /kg.m^{-3}	η /10^{-6} kg.m^{-1}.s^{-1}	c_p /kJ.kg^{-1}.K^{-1}
0	1.29	17.15	1.006
5	1.27	17.39	1.009
10	1.24	17.63	1.011
15	1.22	17.88	1.014
20	1.20	18.12	1.018
25	1.18	18.36	1.022
30	1.16	18.55	1.030
35	1.14	18.78	1.039
40	1.11	19.01	1.050

Note that values of density, being the reciprocal of the specific volume, are best obtained from the psychrometric chart, see section 1, which covers any value of humidity.

The variation of density with pressure can be obtained using a value ρ_0 from Table 4.A1.2 or the psychrometric chart, and ideal gas equation, namely:

$$\rho = \rho_0 \left(\frac{p}{1.01325} \right) \qquad (4.A1.1)$$

Values of viscosity and specific thermal capacity do not vary significantly with pressure.

Tables 4.A1.4–4.A1.8 have all been adapted from reference 24.

Table 4.A1.3 Density of water–glycol mixture in kg.m^{-3} (S denotes, solid)

Temp. /°C	% glycol by mass									
	0	10	20	30	40	50	60	70	80	90
−40	S	S	S	S	S	S	1103	1121	1133	—
−30	S	S	S	S	S	1086	1100	1117	1128	1138
−20	S	S	S	S	1068	1082	1097	1112	1123	1132
−10	S	S	S	1047	1064	1079	1092	1107	1118	1126
0	1000	1016	1030	1045	1061	1075	1088	1101	1111	1120
10	1000	1012	1028	1042	1058	1070	1082	1094	1104	1112
20	998	1010	1026	1038	1052	1065	1077	1088	1098	1105
30	996	1008	1022	1033	1048	1059	1071	1080	1090	1098

Table 4.A1.4 Some properties of oxygen (gas at $p = 1$ atm)

T /K	ρ /kg m^{-3}	η /10^{-6} kg.m^{-1}.s^{-1}	c_p /kJ kg^{-1}.K^{-1}
150	2.619	11.49	0.9178
200	1.956	14.85	0.9131
250	1.562	17.87	0.9157
300	1.301	20.63	0.9203
350	1.113	23.16	0.9291
400	0.975	25.54	0.9420

Table 4.A1.5 Some properties of nitrogen (gas at $p = 1$ atm)

T /K	ρ /kg m^{-3}	η /10^{-6} kg.m^{-1}.s^{-1}	c_p /kJ kg^{-1}.K^{-1}
100	3.481	6.862	1.0722
200	1.711	12.95	1.0429
300	1.142	17.84	1.0408
400	0.8538	21.98	1.0459

Table 4.A1.6 Some properties of carbon dioxide (gas at $p = 1$ atm)

T /K	ρ /kg.m^{-3}	η /10^{-6} kg m^{-1}. s^{-1}	c_p /kJ kg^{-1}. K^{-1}
220	2.4733	11.10	0.783
250	2.1667	12.59	0.804
300	1.7973	14.96	0.871
350	1.5362	17.20	0.900
400	1.3424	19.32	0.942

Table 4.A1.7 Some properties of carbon monoxide (gas at $p = 1$ atm)

T /K	ρ /k gm^{-3}	η /10^{-6} kg.m^{-1}.s^{-1}	c_p /kJ.kg^{-1}.K^{-1}
220	1.5536	13.83	1.043
250	1.3668	15.40	1.042
300	1.1387	17.84	1.042
350	0.9742	20.09	1.043
400	0.8536	22.19	1.048

Table 4.A1.8 Some properties of ammonia (gas at $p = 1$ atm)

T /K	ρ /kg.m^{-3}	η /10^{-6} kg. m^{-1}. s^{-1}	c_p /kJ.kg^{-1}.K^{-1}
220	0.3828	7.255	2.198
273	0.7929	9.353	2.177
323	0.6487	11.03	2.177
373	0.5590	12.89	2.236
423	0.4934	14.67	2.315

Table 4.A1.9 Some properties of fuel gases

Gas	Density, ρ /kg.m^{-3}	Dynamic viscosity, η /10^{-6} kg.m^{-1}.s^{-1}
Natural	0.68	10.7
Butane	2.48	7.4
Propane	1.85	8.0

The viscosity varies considerably with temperature. Even within one grade of oil, properties can vary within a band. Thus property values are best obtained from the manufacturer. Some approximate values are given in Table 4.A1.10, taken from the graphs in section 5.

Table 4.A1.10 Viscosity of fuel oils

Class	Density, ρ /kg.m^{-3} at 15°C	Kinematic viscosity, v /10^{-6}.m^2.s^{-1}						
		−10°C	0°C	20°C	40°C	60°C	80°C	100°C
C2, kerosene	803	0.34	2.75	1.9	1.0–2.0			
D, gas oil	850	11	7.8	4.5	1.5–5.5			
E, light fuel oil	940		600	160	58	25	13.5	8.2
F, medium fuel oil	970			850	220	75	32	20.0 max
G, heavy fuel oil	980			3400	705	205	75	40.0 max

Appendix 4.A2: Mathematical basis for tables of pressure drop

4.A2.1 Mathematical basis on which tables for pressure drop were calculated for water

In section 4.3 the basic equations were given for calculating the pressure loss per unit length, $\Delta p/l$. For turbulent flow the evaluation of λ could only be obtained by iterative means, though an equation for speeding up the iteration was also given.

The tables of pre-calculated pressure drops for common British pipes required a different approach since the vertical axes of these tables are in round figures of $\Delta p/l$. Thus the calculations had to be done in the reverse direction — what mass flow would be required to give a particular value of $\Delta p/l$?

For laminar flow

Equations (4.1) and (4.3) of section 4.3 can be rearranged to give:

$$q_{\mathrm{m}} = \frac{\left(\dfrac{\Delta p}{l}\right) \pi d^2 \, \rho}{128 \eta} \tag{4.A2.1}$$

For turbulent flow

Equation (4.1) can be rearranged to give:

$$q_{\mathrm{m}} = \left[\frac{\rho \pi^2 d^5}{8} \, \frac{\Delta p}{l} \right]^{0.5} \frac{1}{\sqrt{\lambda}} \tag{4.A2.2}$$

Equation (4.1) can also be rearranged to give

$$\frac{1}{\sqrt{\lambda}} = \left[\frac{\rho c^2}{2d} \, \frac{l}{\Delta p} \right]^{0.5} \tag{4.A2.3}$$

Rewriting the Colbrook–White equation, (4.4):

$$\frac{1}{\sqrt{\lambda}} = -2 \log \left(\frac{2.51}{Re\sqrt{\lambda}} + \frac{\dfrac{k}{d}}{3.7} \right)$$

Into the right-hand side of this equation can be substituted equation 4.A2.3, giving

$$\frac{1}{\sqrt{\lambda}} = -2 \log \left(2.51 \eta \left\{ \frac{1}{2\rho d^3} \, \frac{1}{\Delta p} \right\}^{0.5} + \frac{\dfrac{k}{d}}{3.7} \right) \tag{4.A2.4}$$

Using equation (4.A2.4), the value of $1/\sqrt{\lambda}$ can be obtained directly for any required value of $\Delta p/l$. This can then be used in equation (4.A2.2) to obtain the value of mass flow q_{m}.

4.A2.2 Mathematical basis on which values of equivalent length were calculated for steam pipes

An alternative method of calculating the pressure drop for components, unique to the CIBSE Guide, is to use the 'equivalent length' method (see Appendix 4.A3), determining:

ζ from section 4.8;

$\Delta Z/l$ for a straight pipe of the same dimension from the pre-calculated tables;

l_{e} the 'equivalent length for $\zeta = 1$', from the tables

from which

$$\Delta p = \zeta \, l_{\mathrm{e}} \, \frac{\Delta p}{l} \tag{4.A2.5}$$

This 'equivalent length' is the length of pipe which will produce a drop in pressure equal to one velocity pressure ($\frac{1}{2} \rho c^2$). Values have been calculated using the following equation:

$$l_{\mathrm{e}} = 62.5 d^{1.027} \, q_{\mathrm{m}}^{0.111} \tag{4.A2.6}$$

Appendix 4.A3: Equivalent length

Although the 'equivalent length' concept is very popular with technicians, it should only be used with caution.

The pressure drop along a straight pipe is totally dependent on the value of λ, whereas according to the various formulae for ζ for elbows and tees by Idelchik, the value of λ has little effect on ζ. Thus if there is not a close link between ζ and λ, a change in the value of one does not cause a corresponding change in the other. It follows that the 'equivalent length' method for calculating the pressure drop due to fittings is inherently flawed.

Any 'equivalent length' must therefore vary with velocity and diameter. Fortunately the ASHRAE data for elbows does consider the variation with velocity. This is more than can be said of the 'equivalent length' data from other publications which oversimplify this method.

It is debatable whether this method is worth pursuing since a simple value of ζ can be used instead of a complex table of 'equivalent lengths'.

No 'equivalent lengths' of fittings are given in this Guide.

The equation (4.6) for pressure drop due to fittings, given in section 4.3:

$$\Delta p = \zeta \, \tfrac{1}{2} \rho c^2 \qquad\qquad (4.A3.1)$$

can be used for duct fittings or pipe fittings.

Traditionally, CIBSE members have sought to use a different method for pipe fittings and data for this are included in the Guide.

Instead of using the value of ζ in equation (4.A3.1) above, it can be used with a pre-calculated value of l_e, being the length of pipe over which the pressure drop Δp is equal to the velocity pressure p_v.

Then the 'equivalent pipe length of a fitting', l_{ef} is calculated from:

$$l_{ef} = l_e \zeta \qquad\qquad (4.A3.2)$$

Note. The following explains the manner in which the values of l_e have been calculated:

Equation (4.1) gives

$$\Delta p = \lambda \frac{l}{d} \, \tfrac{1}{2} \rho c^2 \qquad\qquad (4.A3.3)$$

For a pressure drop Δp equal to $\tfrac{1}{2}\rho c^2$ it follows that:

$$l_e = \frac{d}{\lambda} \qquad\qquad (4.A3.4)$$

Since λ has already been determined in calculating $\Delta p/l$ in equation (4.3), the substitution can be made that:

$$\frac{1}{\lambda} = \frac{1}{d} \, \frac{l}{\Delta p} \, \tfrac{1}{2}\rho c^2$$

And hence

$$l_e = \frac{8}{\pi^2} \, \frac{q_m^2}{\rho d^4} \left(\frac{l}{\Delta p} \right) \qquad\qquad (4.A3.5)$$

Appendix 4.A4: Compressible flow

In the case of compressible fluids flowing under conditions where the pressure drop is considerable in proportion to the initial pressure (greater than 10 per cent), the change in density between the initial and final condition should be taken into account.

For determining Reynolds number ($Re = \rho cd/\eta$) there is no problem. For a particular mass flow and constant pipe diameter, c is inversely proportional to density; thus density change has no effect. Viscosity η varies little with pressure (though varying considerably with temperature). Thus despite large changes in pressure, Re and λ will remain constant, and the calculation of total pressure drop need only take into account the variation of density and velocity between the inlet and outlet:

$$\int_1^2 \mathrm{d}p = \frac{\lambda}{2d} \int_1^2 \rho c^2 \, \mathrm{d}l \tag{4.A4.1}$$

Still considering isothermal flow, where despite pressure loss there is sufficient heat transfer to maintain the fluid at the temperature of the surroundings, the pressure loss may be represented by:

$$p_1^2 - p_2^2 = \frac{32 q_\mathrm{m}^2 RT}{\pi^2 d^4} \left[\frac{\lambda l}{2d} + \ln \frac{p_1}{p_2} \right] \tag{4.A4.2}$$

In other cases the theoretical equations become complex and use is made of approximate formula derived empirically.

In all of the following equations, x, y, K_1, K_2 and K_3 are experimentally determined.

The equations are generally of the form:

$$\frac{\mathrm{d}p}{\mathrm{d}l} = K_1 \frac{q_\mathrm{m}^x}{d^{(2x+y)} \rho^{(x-1)}} \tag{4.A4.3}$$

Variation of density is of the form:

$$\rho = K_2 p^n \tag{4.A4.4}$$

Substitution into the previous equation gives:

$$\frac{p_1^\mathrm{m} - p_2^\mathrm{m}}{l} = K_3 \frac{q_\mathrm{m}^x}{d^{(2x+y)}} \tag{4.A4.5}$$

Table 4.A4.1 gives some values of the constants for steam and compressed air.

Table 4.A4.1 Values of constants for steam and compressed air

Fluid	x	y	m	K_3
Steam	1.889	1.249	1.929	3.032×10^{-3}
Compressed air	1.889	1.249	1.929	4.268×10^{-3}

Appendix 4.A5 Capacity *K*, and complex networks

Whilst it is most common practice for components and fittings to have their pressure drop characteristic given in terms of the pressure loss factor ζ, most manufacturers of valves and dampers quote the performance in terms of the 'capacity, *K*', defined in the following relationship:

$$q_v = K\sqrt{\Delta p} \qquad (4.A5.1)$$

This implies that *K* has units, usually of $m^3/h \sqrt{bar}$ for liquids, or $m^3/s \sqrt{Pa}$ for gases. Some manufacturers may quote values of *K* with different units so care is needed.

There is a relation between *K* and ζ, but it is not really necessary to convert one to the other. Pressure drops are more simply calculated separately for those components for which *K* is given.

The use of *K* is also useful when dealing with the authority of a valve, and in the prediction of flows in complex circuits.

For every branch of a circuit, once the value of pressure drop is known for a particular volume flow, the value of *K* for that branch can be obtained from equation (4.A5.1).

(*a*) In parallel

For elements of the circuit in parallel, the overall capacity K_o is given by:

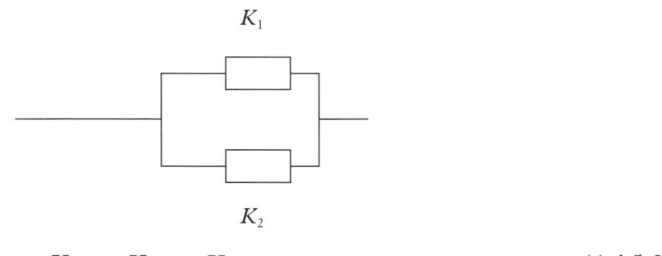

$$K_o = K_1 + K_2 \qquad (4.A5.2)$$

(*b*) In series

For elements of the circuit in series, the overall capacity K_o is given by

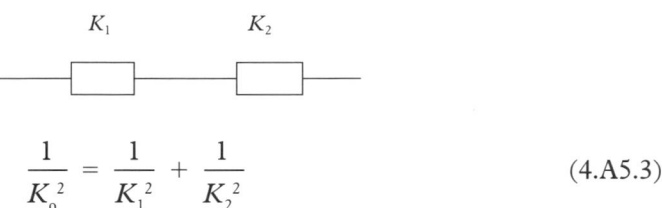

$$\frac{1}{K_o^2} = \frac{1}{K_1^2} + \frac{1}{K_2^2} \qquad (4.A5.3)$$

Thus for a complex circuit, the entire circuit can be simplified bit by bit until the circuit overall, 'total' capacity K_T is obtained. Use of equation (4.A5.1) then permits the calculation of pressure drop for any value of total flow. Thus the pipework or ductwork pressure drop characteristic is obtained.

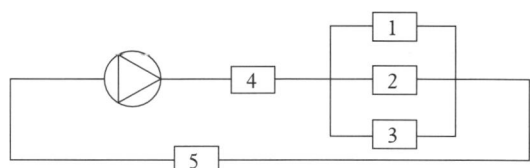

Sub-circuits 1, 2 and 3 having capacities K_1, K_2 and K_3 can be simplified to give K_x using equation (4.A5.2) for circuits in parallel.

Sub-circuits 4, 5 and *x* having capacities K_4, K_5 and K_x can be simplified to give K_T using equation (4.A5.3) for circuits in series.

Equation (4.A5.1) can be used to determine the system characteristic.

5 Fuels and combustion

5.1   Introduction

Major changes have occurred within the supply side of the energy industries since the last revision of this section in 1974. These concern the country wide utilisation of natural gas and the decline of the coal industry; the privatisation and regulation of the energy market; the growing influence of renewable energies and the need to develop environmentally friendly substitutes for fossil fuels. Government obligations to comply with international treaties on greenhouse gas emissions are also changing the mix of fuels available for power generation and place increased emphasis upon energy efficiency. To the year 2025 the supply of gas and oil looks reasonably secure but energy usage options will continue to be driven by market forces with, perhaps, increasing intervention by the Government in the area of energy taxation.

The data presented in this section are limited to those required by a practising building services engineer. They are not intended to represent full property specifications for each class of fuel and reference should be made to suppliers and standard literature for further information. A bibliography is included at the end of the section. The familiar structure of the earlier edition has been retained and, where possible, data assembled according to the previous format.

5.2   Classification of fuels

Fossil fuels may be classified into primary and secondary groups. Primary fuels occur naturally and are formed within a geological time scale from the decay of animal and vegetable matter. In general secondary fuels are prepared from primary fuels, with the intention of modifying the properties to suit particular applications. Electricity is sometimes classed as a secondary fuel.

The carbon to hydrogen ratio of the fuel can also be used as a means of classification. This ratio varies on a mass basis from about 3:1 for natural gas, 7.6:1 for heavy fuel oil up to 30:1 for anthracites. Properties such as gross calorific value and stoichiometric air requirement are seen to vary in a regular manner with carbon to hydrogen ratio. The ratio can also affect flame parameters such as luminosity.

5.3   Primary fuels

5.3.1   Coal

British coals are classified according to caking properties (Gray King Assay) and percentage volatile matter content. These and other properties are related to the rank or age of the coal. The Coal Rank Code (CRC) classification divides coals into groups numbered in hundreds from 100 to 900, the groups being sub-divided into classes and sub-classes for a closer definition of property ranges.

Caking propensity increases from zero at both ends of the scale (groups 100 and 900) to a maximum in groups 300 and 400. Coals in group 100 are the anthracites; those in group 200 comprise the low volatile steam coals (coals in these two groups are natural smokeless fuels); coals in groups 300 to 900 are generally known as bituminous coals of which those in the groups 300 to 600 are generally used as coking coals, with the lower ranking groups (600–900) being general purpose coals for domestic and commercial burning.

5.3.2   Natural gas

Virtually all natural gas distributed in the UK stems from underground deposits beneath the North Sea and Irish

Sea. Gas quality is controlled under the auspices of the Gas Act (1986). Calorific value must be declared as gas is costed on the basis of number of units of energy supplied. Similar restrictions apply to the uniformity of calorific value, supply pressure, Wobbe number, hydrogen sulphide content and smell. The unit of energy 'therm' is 10^5 British Thermal Units (BTU) and this is equivalent to 105.5 MJ.

5.4 Secondary fuels

5.4.1 Smokeless solid fuels

Manufactured smokeless fuels are cokes and bonded briquettes. In the heating field these are usually burned in hand-fired or gravity-fed appliances. The use of briquetted fuels is restricted to domestic appliances.

5.4.2 Petroleum fuel oils

All fuel oils are refinery products. Crude oils, which are obtained from the oil wells as mixtures of hydrocarbons, are processed in the refinery to produce many hundreds of commercial products, including fuel oil. The numerous hydrocarbon compounds have different boiling points and, by controlled heating of the crude oil, the various fractions are distilled and condensed, producing gasolines, kerosenes and gas oils which are termed distillates. Blending of the residual oils with a suitable distillate enables the production of the various grades of residual fuel oils, which are commercially available for application to the larger heating plants. Nitrogen and asphaltene contents of residual fuel oils are significant in determining levels of nitrogen oxide and particulate emissions to the atmosphere. Careful selection of the appropriate equipment, boilers, burners and flues will enable the most effective grade of fuel oil to be applied to the particular plant requirements.

5.4.3 Liquefied petroleum gases (LPG)

Liquefied petroleum gases, commonly known as LPG, are the C3 and C4 members of the hydrocarbon family. They are readily liquefied by the application of moderate pressure at ambient temperature. These are marketed in the UK as two grades, known as 'commercial propane' and 'commercial butane' under various brand names given by the distributing companies. They are transported and stored in the liquid phase but are used and handled as a gas.

5.4.4 Electricity

Electricity is distributed over a high voltage grid system which is reduced at the user to standard conditions of 415 V, three phase, 50 Hz for normal purposes and small demands, but at higher voltages for large power requirements.

5.5 Specification of fuels

5.5.1 Solid fuels

5.5.1.1 Coal

The average properties of coals for the UK are given in Table 5.1 for typical 'as fired' fuel. This is the form in which the data are normally used for calculation. Physical data are given in Tables 5.2 to 5.7.

Coal size is identified according to metric screen sizes appropriate to BS1016, Part 109. Sizes for various graded coals are given in Tables 5.2 to 5.5.

Graded coals are separated into five standard groups for which the upper and lower limits have a permitted range (Table 5.2). Smalls are specified in terms of an upper size only; for stoker firing smalls have a top size of either 25 mm or 50 mm. Treated smalls are washed or dry cleaned. A guide to coal storage and handling is given in reference (18).

Calorific values

The following definitions are adopted in this section for gross and net calorific values:

— *Gross calorific value* — the calorific value of the fuel including the latent heat of condensation of all water vapour in the products of combustion.

— *Net calorific value* — the calorific value of the fuel when no water vapour is condensed from the products of combustion.

The calorific values of the fuels are tabulated for certain moisture contents. When the fuel has a different moisture content from that tabulated, the calorific value can be found from:

$$h_g = \frac{(100 - m)h_{gt}}{(100 - m_t)} \tag{5.1}$$

$$h_n = \frac{(100 - m)h_{nt}}{(100 - m_t)} - 24.5(m_t - m) \tag{5.2}$$

where:

h_g = gross calorific value (mass basis) at moisture content m (kJ.kg^{-1})

h_{gt} = tabulated gross calorific value (mass basis) (kJ.kg^{-1})

m = moisture content of fuel (%)

m_t = tabulated moisture content of fuel (%)

h_n = net calorific value (mass basis) at moisture content m (kJ.kg^{-1})

h_{nt} = tabulated net calorific value (mass basis) (kJ.kg^{-1})

5.5.1.2 Pelletised refuse derived fuel (d-RDF)

Pelletised refuse derived fuel remains a suitable energy source for small boilers despite some problems in manufacture and market viability. Table 5.8 compares the properties

Table 5.1 Average 'as fired' properties of coal

Fuel	British coal rank number (CRC)	Moisture content (%)		Constituent parts by mass (%)							Calorific values	
		Air dried coal	96% rh and 30°C	Moisture	Ash	Carbon	Hydrogen	Nitrogen	Sulphur	Oxygen	Gross MJ.kg^{-1}	Net MJ.kg^{-1}
WASHED SMALLS												
Anthracite	101	2	4	8	8	78.2	2.4	0.9	1.0	1.5	29.65	28.95
	102	1	2	7	8	77.9	3.1	1.1	1.0	1.9	30.35	29.30
Dry steam coals	201	1	1	7	8	77.4	3.4	1.2	1.0	2.0	30.60	29.65
Coking steam coals	202	1	1	7	8	77.1	3.5	1.2	1.0	2.2	30.70	29.75
	204	1	1	7	8	76.8	3.8	1.2	1.0	2.2	30.80	29.80
Medium volatile coking coals	301a	1	1	7	8	75.8	4.1	1.3	1.2	2.6	30.80	29.75
	301b	1	1	7	8	74.8	4.2	1.3	1.2	3.5	30.45	29.40
Heat altered coals	302H	1	2	7	8	74.4	4.2	1.7	1.2	3.5	30.35	29.25
	303H	2	3	8	8	72.7	4.2	1.4	1.2	4.5	29.75	28.60
High volatile coking coals:												
Very strongly caking	401	2	2	9	8	71.6	4.3	1.6	1.7	3.8	29.55	28.40
Strongly caking	501	2	3	9	8	71.0	4.3	1.5	1.7	4.6	29.20	28.05
	502	3	4	10	8	68.8	4.4	1.5	1.7	5.5	28.60	27.40
Medium caking	601	4	6	11	8	67.8	4.3	1.4	1.7	5.9	27.80	26.60
	602	4	5	11	8	67.0	4.4	1.4	1.7	6.1	27.80	26.55
General purpose coals:												
Weakly caking	701	5	6	13	8	65.7	4.0	1.4	1.7	6.2	26.75	25.50
	702	5	7	13	8	65.0	4.2	1.3	1.7	6.8	26.75	25.50
Very weakly caking	802	8	11	16	8	61.3	4.0	1.3	1.7	7.7	25.25	23.95
Non-caking	902	10	13	18	8	59.0	3.7	1.2	1.7	8.4	23.85	22.60
Manufactured fuels:												
Domestic and industrial coke	—	—	—	8–12	7	82.0	0.4	1.7	—	—	27.90	26.30
Low temperature coke	—	—	—	15	6	71.0	2.4	2.4	—	3.2	27.45	25.40
WASHED SINGLES												
Anthracite	101	2	4	4	5	84.7	2.6	1.0	1.1	1.6	32.10	31.45
	102	1	2	3	5	84.3	3.4	1.2	1.1	2.0	32.85	31.80
Dry steam coals	201	1	1	3	5	83.7	3.7	1.3	1.1	2.2	33.10	32.20
Coking steam coals	202	1	1	3	5	83.4	3.8	1.3	1.1	2.4	33.20	32.30
	204	1	1	3	5	83.1	4.1	1.3	1.1	2.4	33.30	32.35
Medium volatile coking coals	301a	1	1	3	5	82.1	4.4	1.4	1.3	2.8	33.30	32.30
	301b	1	1	3	5	81.0	4.5	1.4	1.3	2.8	32.95	31.90
Heat altered coals	302H	1	2	3	5	80.6	4.5	1.8	1.3	3.8	32.85	31.75
	303H	2	3	4	5	78.7	4.6	1.5	1.3	4.9	32.20	31.10
High volatile coking coals:												
Very strongly caking	401	2	2	4	5	78.5	4.7	1.7	1.9	4.2	32.40	31.25
Strongly caking	501	2	3	5	5	77.0	4.7	1.5	1.8	5.0	31.65	30.50
	502	3	4	5	5	75.5	4.8	1.6	1.9	6.2	31.40	30.20
Medium caking	601	4	6	6	5	74.5	4.6	1.5	1.9	6.5	30.60	29.35
	602	4	5	7	5	73.2	4.8	1.5	1.9	6.6	30.15	28.95
General purpose coals:												
Weakly caking	701	5	6	9	5	71.5	4.4	1.5	1.8	6.8	29.15	27.80
	702	5	7	9	5	70.8	4.6	1.4	1.8	7.4	29.15	27.90
Very weakly caking	802	8	11	11	5	67.8	4.4	1.4	1.9	8.5	27.90	26.60
Non-caking	902	10	13	16	5	63.0	3.9	1.3	1.8	9.0	24.45	24.10

Table 5.2 Standard size groups for graded coals

Name of group	Round hole screen size (mm)	
	Upper limit	Lower limit
Large cobbles	>150	75
Cobbles	100–150	50-100
Trebles/large nuts	63–100	38-63
Doubles/nuts	38–63	25-38
Singles	25–38	13-18

Table 5.3 Size limits and bulk density of Welsh anthracite

Name	Size limits (mm)	Bulk density (kg.m^{-3})
Cobbles	80–125	770–800
French nuts	63–80	770–800
Stove nuts	36–63	770–800
Stovesse	20/16–36	750–785
Beans	10–20	750–785
Peas	10–16	750–785
Grains	5–10	750–785
Washed duff	0–5	785–820

of d-RDF with a typical coal used for stoker firing. The calorific value of d-RDF is about two-thirds that of coal and ash yields on combustion significantly higher. Appreciable fuel glass contents may provide low d-RDF ash fusion temperatures and possible clinker formation under adverse combustion conditions.

5.5.1.3　Properties of wood fuels

The sustainable use of wood fuels can provide environmental benefits in terms of reduced carbon dioxide, nitrogen oxide and sulphur oxide emissions in comparision to

Table 5.4 Size limits and bulk density of Welsh dry steam coal

Name	Size limits (mm)	Bulk density (kg.m^{-3})
Cobbles	80–125	720–750
Large nuts	56–80	720–750
Small nuts	18–56	720–750
Beans	16–18	705–735
Peas	10–18	705–735
Washed duff	0–10	705–735

Table 5.5 Size limits and bulk density of hard coke

Name	Size limits (mm)	Bulk density (kg.m^{-3})
Large	Over 90	433
Cobbles	64–90	448
Trebles	40–64	464
Doubles	25–40	464
Singles	16–25	481
Beans	10–16	497

Table 5.6 Natural angle of repose of solid fuels

Approximate size of fuel (mm)	Angle of repose (measured from the horizontal)
20–30	40°
12–20	42°
6–12	52°
0–6	58°

Table 5.7 Bulk density of loosely packed dry coal

Nature and size of coal	Bulk density (kg.m^{-3})
Graded coal	640 ± 60
Small coal	770 ± 60
Coal dust (<3 mm)	530 ± 50
Pulverised fuel (50–90% passing 76 × 76 μm square mesh sieve)	450 ± 50

Note. The bulk density depends on a number of factors and is not reproducible within ±5% except under laboratory conditions. Compaction may increase the density by up to 20%, whereas freshly formed pulverised fuel has a wide variation in range and may be less than 50% of the quoted figure.

coal. Short rotation coppice (SRC) wood fuels, based on fast growing poplar and willow species, are becoming more available in the form of dried chips. For small boiler plant operation it is recommended that fuel moisture content does not exceed 40 per cent. Wood fuels contain high percentage volatile matter and oxygen compositions with low ash. These properties influence smoke emission and the stoichiometric air requirement respectively. Table 5.9 gives an indicative composition of a chipped wood fuel.

5.5.2   Liquid fuels

5.5.2.1   Petroleum oils

British Standard specifications are published for all grades of petroleum oil fuels and are accepted as the basic requirements for the United Kingdom (BS2869, 1998). The five classes shown in Table 5.10 cover the fuels normally used in fixed appliances. Class C1 is a paraffin type fuel for use in free-standing, flueless domestic burners, and is not detailed in this section. Class C2 is a distillate fuel of the kerosene type for vaporising and small atomising burners.

Class D is a distillate grade for larger atomising burners in both domestic and industrial use, generally known as gas oil.

Classes E, F and G are residual or blended fuel oils for atomising burners and generally need preheating before combustion in such burners and normally require storage and handling plant with heating facilities.

Commercial specifications follow the pattern shown in Table 5.11 but usually include additional information so that all points governing fuel performance can be assessed. More detailed property data and updated information may be obtained from the fuel supplier.

BS799, Part 5, 1987, provides specifications for oil storage tank installations.

Viscosity

Kinematic viscosity is a measure of resistance of the liquid to flow and may be defined as the force needed to shear a unit cube at unit speed and is expressed in units of centistoke (cSt) or mm^2.s^{-1}; where 1 cSt = 1 mm^2.s^{-1}.

Figures 5.1 and 5.2 show the viscosity maxima/temperature relationship for the various grades of fuel oil with kinematic viscosity expressed in terms of cSt.

The use of Figures 5.1 and 5.2

The viscosity of the oil, at a particular temperature and in units of cSt, must first be known. Locate this point on the graph and draw a line through it parallel to the sloping lines already shown. Corresponding temperatures and viscosities may then be read off from any point on the line so drawn. The sloping lines shown in Figures 5.1 and 5.2

Table 5.8 Properties of a commercial coal and d-RDF[19]

Fuel	Moisture (%)	Volatile matter (%)	Ash (%)	Calorific value 'as fired' (MJ.kg^{-1})	Bulk density (kg.m^{-3})
Coal	8.4	25.9	10.2	27.2	900
d-RDF	7.3	67.5	15.0	18.7	600

Table 5.9 Indicative composition of a chipped wood fuel

% moisture as fired by mass	Ash as fired by mass (%)	Volatile matter d.a.f.* by mass (%)	O$_2$ d.a.f.* by mass (%)	Gross calorific value d.a.f.* (MJ.kg^{-1})
15	0.6	80	48	19.7

*Dry as fired

Table 5.10 Properties of petroleum burner fuels (BS2869, 1998)

Property		Class C2	Class D	Class E	Class F	Class G
Kinematic viscosity (mm².s⁻¹):	at 40°C					
	min-max	1.00–2.00	1.5–5.5	—	—	—
Kinematic viscosity (mm².s⁻¹)	at 100°C					
	max	—	—	8.20	20.00	40.00
Carbon residue, Ramsbottom on 10% residue, % mass, max		—	0.3	15.0	18.0	20.0
Minimum closed flash point (°C)						
Abel		38	—	—	—	—
Pensky–Martens		—	56	66	66	66
Maximum water content		—	200mg.kg⁻¹	0.5%v/v	0.75%v/v	1.0%v/v
Maximum sediment content by mass (%)		—	0.01	0.10	0.15	0.15
Maximum ash content by mass (%)		—	0.01	0.10	0.10	0.15
Maximum sulphur content by mass (%)		0.20	0.20	3.50	3.50	3.50

Table 5.11 Properties of typical petroleum fuel oils

	Kerosene Class 2	Gas oil Class D	Light fuel oil Class E	Medium fuel oil Class F	Heavy fuel oil Class G
Relative density (specific gravity) at 15°C	0.803	0.850	0.940	0.970	0.980
Minimum closed flash point (°C)	38	60	66	66	66
Kinematic viscosity (cSt)					
at 40°C	—	3.2	—	—	—
at 100°C	—	—	8	16	35
Freezing point (°C)	<-40	—	—	—	—
Maximum pour point (°C)	—	—	−6	24	30
Maximum cloud point (°C)	—	−5 (Mar/Sep) −16 (Oct/Feb)	—	—	—
Gross calorific value (MJ.kg⁻¹)	46.4	45.5	42.5	41.8	42.7
Net calorific value (MJ.kg⁻¹)	43.6	42.7	40.1	39.5	40.3
Maximum sulphur content by mass (%)	0.2	0.2	3.2	3.5	3.5
Maximum water content by volume (%)	negligible	0.05	0.5	0.75	1.0
Maximum sediment content by mass (%)	—	0.01	0.10	0.15	0.15
Maximum ash content by volume (%)	—	0.01	0.05	0.07	0.10
Mean specific heat capacity, 0-100°C (kJ.kg⁻¹.K⁻¹)	2.1	2.06	1.93	1.89	1.89

represent maximum viscosity values for typical fuel oils at various temperatures; viscosity maxima indicated on the plots at 100°C and 40°C correspond to those given in Table 5.10.

Although the temperature scale on Figure 5.2 goes up to 40°C it must be borne in mind that, in many countries, kerosene-type products may have flash points of this order.

Maximum viscosities of 500 cSt for pumping and 12 to 15 cSt for pressure atomisation are normally used. Rotary cup atomisers employ a viscosity range of 50–80 cSt.

Pour point

Pour point (Table 5.11) is a laboratory test by which the lowest temperature at which an oil will flow under careful-ly defined conditions is measured. In order to ensure mobility of the fuel, minimum storage temperatures are required for class E, F and G oil fuels. The distillate grades require no heating.

Heating requirements

Fuel of classes C2 and D may be stored, handled and atom-ised at ambient temperatures, but exposure for a long period to extreme cold should be avoided otherwise restrictions in the flow of oil from the tank may result. The appropriate temperatures for the storage and handling of fuels of classes E, F and G are given in Table 5.12.

Normally oil burners require that residual fuel oils should be presented to the burner at a viscosity between 12 and 15 cSt. The burner manufacturers' actual requirements should be ascertained at the design stage. The appropriate temper-atures can be determined from the viscosity–temperature charts, Figures 5.1 and 5.2.

Table 5.12 Storage and handling temperatures of fuel oils (BS799, Part 5, 1987)

Class of fuel	Minimum storage temperature (°C)	Minimum handling or outflow from storage temperature (°C)
E	10	10
F	25	30
G	40	50

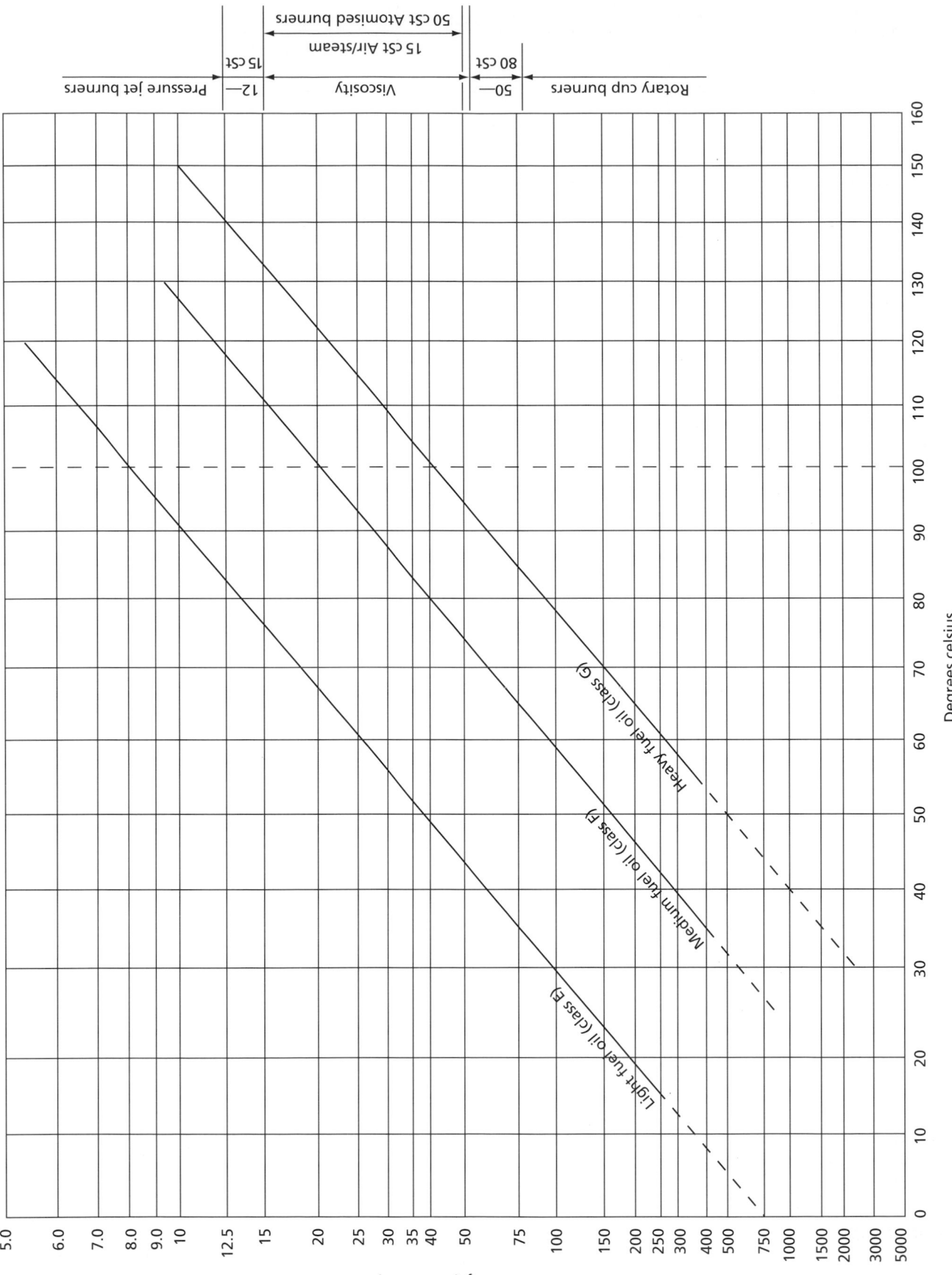

Figure 5.1 Viscosity–temperature chart for class E, F, G fuel oils (courtesy of BP Oil Ltd)

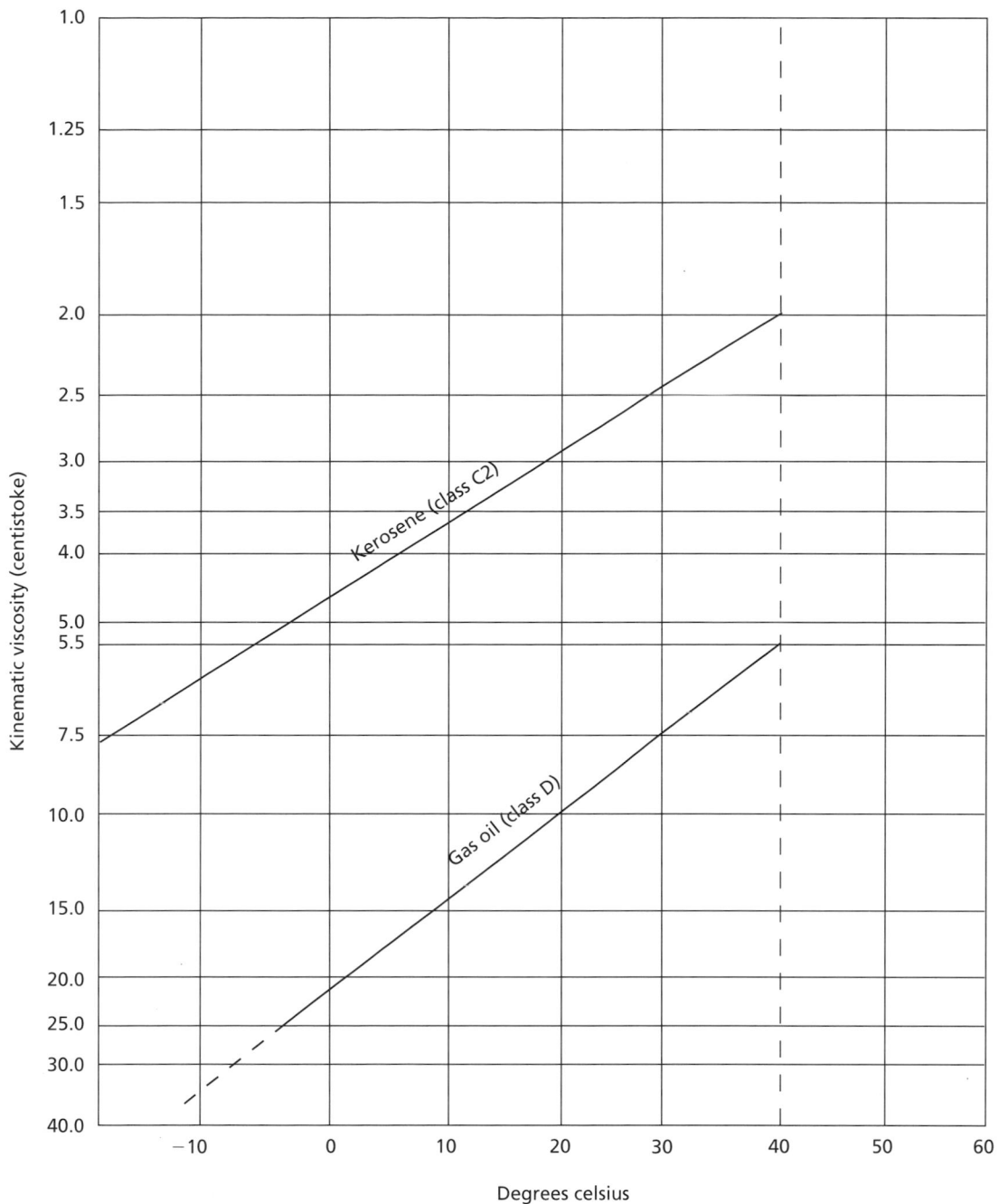

Figure 5.2 Viscosity–temperature chart for distillate fuel oils class C2 and D (courtesy of BP Oil Ltd)

5.5.3   Gaseous fuels

5.5.3.1   Liquefied petroleum gas (LPG)

Typical properties of commerical butane and commercial propane are given in Table 5.13. Limiting requirements for the properties of commercial butane and commercial propane are given in BS4250[6].

Latent heat of vaporisation

Reference to Table 5.13 shows that on a mass basis the latent heats of vaporisation for propane and butane are similar and equivalent to 0.75% of the calorific value of the fuel. Unless the latent heat is supplied artificially with a vaporiser it can only be taken from the atmosphere through the walls of the container; this places a limit on

the rate at which gas can be taken off under natural, ambient temperature conditions.

Specific gravity (relative density)

Both propane and butane are heavier than air and the vapour will therefore tend to collect at a low level in the event of a leak. This point must be borne in mind when designing storage and handling systems. Conversely the liquid density is low and this is important when considering transportation and filling of containers of known internal volume.

Sulphur content

LPG has a low sulphur content controlled by specification limits (BS4250) to 200 mg.kg^{-1} (0.02% by mass).

Table 5.13 Typical properties of commercial butane and commercial propane

Property	Butane (C_4H_{10})	Propane (C_3H_8)
Density at 15°C (kg.m^{-3})	570	500
Specific gravity of liquid at 15°C	0.570–0.580	0.500–0.510
Specific gravity of gas compared with air at stp	1.90–2.10	1.40–1.55
Volume of gas per kg of liquid at stp (m^3)	0.41–0.43	0.53–0.54
Ratio of gas volume to liquid volume at stp	233	274
Boiling point (°C)	0	−45
Absolute vapour pressure (for products of the maximum specified vapour pressure) (kPa)		
Temperature: −40.0°C	—	140
−17.8°C	—	320
0°C	200	560
37.8°C	580	1570
45.0°C	690	1810
Latent heat of vaporisation at 15°C (kJ.kg^{-1})	370	357
Specific heat of liquid (kJ.kg^{-1}.K^{-1})	2.4	2.52
Sulphur content	Negligible to 0.02% by mass	Negligible to 0.02% by mass
Limits of flammability (percentage by volume of gas in gas and air mixture)	Upper, 9.0 Lower, 1.8	Upper, 10.0 Lower, 2.2
Calorific values (dry volumetric basis) (MJ.m^{-3})		
Gross	122	93
Net	113	86
Calorific values (mass basis) (MJ.kg^{-1})		
Gross	49.5	50
Net	46	46.5
Air required for combustion (m^3 per m^3 of gas)	30	24

5.5.3.2 Natural gas

Typical analyses and properties of natural gas are given in Tables 5.14 and 5.15.

Where gas volume data are given they are based on conventional reference conditions of 15°C and 101.3 kPa.

Calorific value

The legal requirement is for the Public Gas Transporter to calculate each day in accordance with the Gas (Calculation of Thermal Energy) Regulations, a 'flow weighted average' calorific value (FWAC) which is passed to all gas shippers for all charging areas.

Sulphur compounds

The contract specification for natural gas limits the total sulphur content to 35 parts per million expressed as hydrogen sulphide. As distributed the sulphur content of natural gas is approximately 0.0011 per cent by volume.

Specific gravity (relative density)

There is no legal requirement for relative density and the average value is given in Tables 5.14 and 5.15.

Wobbe number

The thermal input of an appliance (e.g. central heating boiler) for a given pressure and burner orifice is a function of the Wobbe number. This number is defined as:

Table 5.14 Typical volume analysis and properties of natural gas

	Natural gas
Component parts (%)	
Methane	92.6
Ethane	3.6
Propane	0.8
Butane	0.2
Pentane and above	0.1
Hydrogen	—
Carbon monoxide	—
Carbon dioxide	0.1
Nitrogen	2.6
Properties	
Gross calorific value (MJ.m^{-3})	38.7
Net calorific value (MJ.m^{-3})	34.9
Relative density	0.602
Wobbe No. (dry)	49.9
Air required for combustion (m^3.m^{-3} gas)	9.73

Table 5.15 Operating properties of natural gas

	Natural gas
Declared calorific value	$\approx$38.7 MJ.m^{-3}
Relative density	0.59–0.61
Wobbe No.	Gas Group H 45.7–55.0
Distribution pressure	1750–2750 Pa. Legal requirement: must not fall below 1250 Pa

$$W = \frac{h_g}{d^{0.5}} \qquad (5.3)$$

where:

W = Wobbe number (MJ.m^{-3})

h_g = gross calorific value (volume basis) (MJ.m^{-3})

d = relative density

Natural gases all have Wobbe numbers falling within a narrow range and all appliances are designed to operate on gas corresponding to the mean of that range.

5.5.3.3 Landfill gas

There is significant potential for landfill gas as a fuel for direct boiler firing or in CHP applications. Gas may require cleaning and dewatering before use. The composition is approximately 60 per cent methane and 40 per cent carbon dioxide and also includes many other gases in trace concentrations. The calorific value is in the range 15–25 MJ.m^{-3} depending upon the inert gas content and the extent of raw gas conditioning[19].

5.6 Combustion data

5.6.1 Combustion air and waste gas volume

It is necessary to determine combustion air requirements and waste gas volumes for boiler plant and chimney/flue designs. In order to simplify this assessment, charts are reproduced below for various fuels where the required volume under various CO_2 per cent (excess air) conditions can be read off. Per cent CO_2 values are expressed on a dry gas basis.

Note that all volumes are standardised at 15°C, 101.3 kPa. For volumes at other temperatures use the formula:

$$V_g = \frac{V(t_g + 273)}{288} \qquad (5.4)$$

where:

V_g = volume at temperature required t_g (m^3)

t_g = actual air or flue gas temperature (°C)

V = standard volume read from charts (m^3)

5.6.1.1 Solid fuels

The average figures given in Table 5.16 may be taken for CO_2 per cent, excess air, total air and gas requirements per kilogram of fuel for plant working under normal conditions of combustion efficiency.

Where the anticipated CO_2 per cent (and excess air values) varies considerably from the averaged conditions listed in Table 5.16, refer to the appropriate curves on Figures 5.3 and 5.4, which are accurate for practical design purposes.

Table 5.16 Combustion conditions

Fuel	CO$_2$ (%)	Excess air (%)	Total air volume (m³.kg⁻¹)	Flue gas volume (m³.kg⁻¹)
Solid fuels				
Coke	13.7	50	11.7	11.9
Anthracite	12.7	50	12.3	12.7
Bituminous coal	12.3	50	10.8	11.4
Fuel oils				
Classes C and D	11.6	30	15.4	16.3
Classes E, F and G	12.1	30	14.1	14.7

5.6.1.2 Fuel oils

The average figures in Table 5.16 may be taken for CO_2 per cent, excess air, total air and flue gas volume requirements per kilogram of fuel with plant working under normal conditions of combustion efficiency.

Where the anticipated CO_2 per cent (excess air) varies considerably from the above averaged figures, refer to the appropriate curves on Figures 5.5 and 5.6, which are accurate for practical design purposes.

5.6.1.3 Gaseous fuels

With gas fired boilers the flue gas/air volumes depend on the type of boiler/burner unit installed and the design of the union between the boiler exhaust gas outlet flue and chimney.

The average figures given in Table 5.17 for natural gas have been obtained from site tests. Two types are given:

(a) Boilers fitted with a naturally inspirated burner and a draught diverter.

(b) Boilers fitted with a forced draught burner and a direct flue connection.

Excess air data are given in Figure 5.7. It is important to note that it is possible to have an air deficiency when measuring CO_2 percentage alone. Since natural gas does not produce black smoke when combustion is incomplete it is necessary to measure the CO content in addition to CO_2 to determine if there is sufficient combustion air. For example, from Figure 5.7 it is seen that 10.3% CO_2 could mean either 14% excess air or 15% air deficiency; measuring the CO content will show which air value is correct.

The water dew point of natural gas exhaust gases, appropriate to typical operating conditions of forced draught condensing boilers, is approximately 55°C.

Table 5.17 Percentage volumes of combustion products for boilers burning natural gas

Type of burner	CO$_2$ (%)	CO (%)	Flue gas temperature (°C)
Natural inspirated burner and draught diverter (values measured in primary flue)	7.5–9	0.001–0.008	190–290
Forced draught burner with direct flue connection	8–11	0.001–0.006	55–320

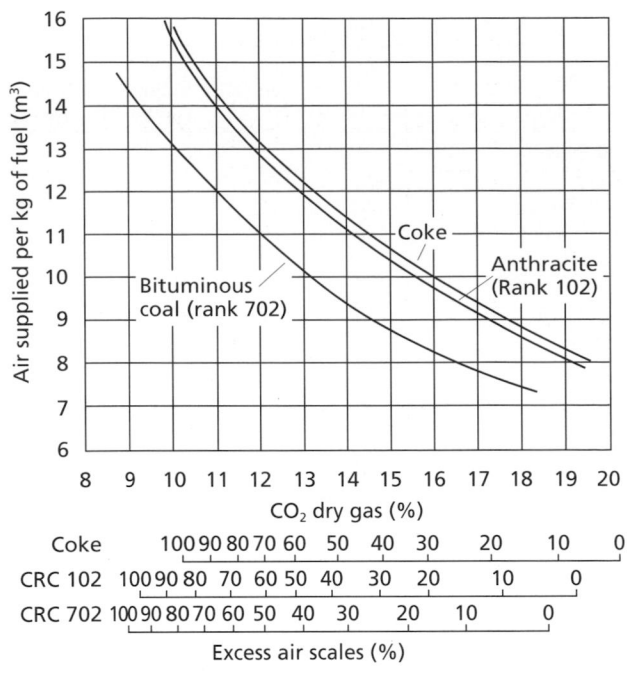

Figure 5.3 Combustion air requirements for solid fuels

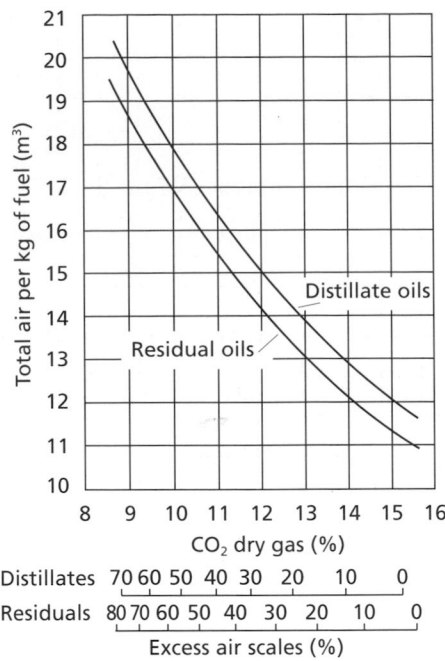

Figure 5.5 Combustion air requirements for petroleum fuel oils

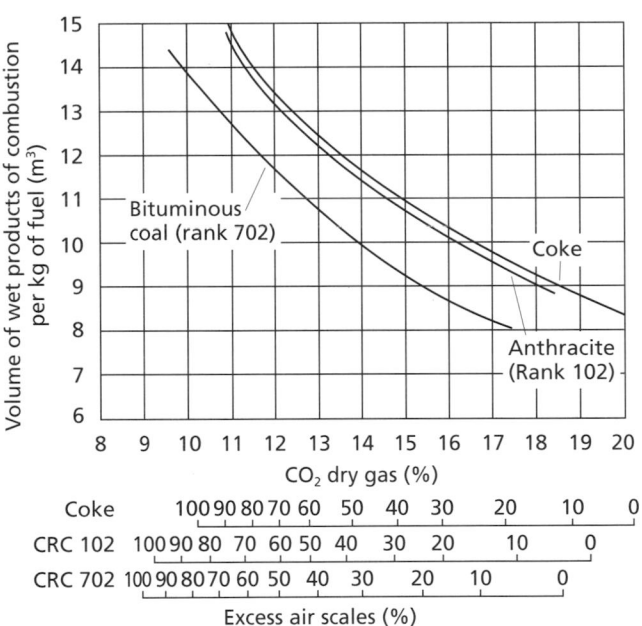

Figure 5.4 Volumes of products of combustion for solid fuels

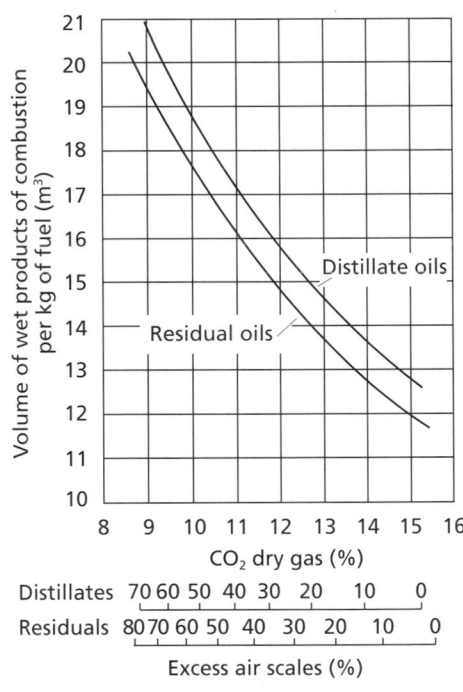

Figure 5.6 Volumes of products of combustion for petroleum fuel oils

5.7   Stack losses

The major heat loss from combustion appliances is the heat carried away in the flue gases. Several formulae have been proposed for assessing these losses based mainly upon variations to the 'Siegert' expression. In order to simplify the assessment for practical requirements, graphs are included in this section which give the stack losses based upon the gross calorific value of the fuel. They do not include any unburned gas loss, i.e. it is assumed that there is no CO in the flue gases.

5.7.1   Solid fuels

The curves given in Figures 5.8, 5.9 and 5.10 show the total flue gas heat loss for various CO_2 per cent values (excess air quantities) expressed as a per cent of gross calorific value. They are provided for general guidance because variables such as rank, grade and moisture content will affect the gross heat loss. CO_2 per cent values are denoted on the basis of dry flue products.

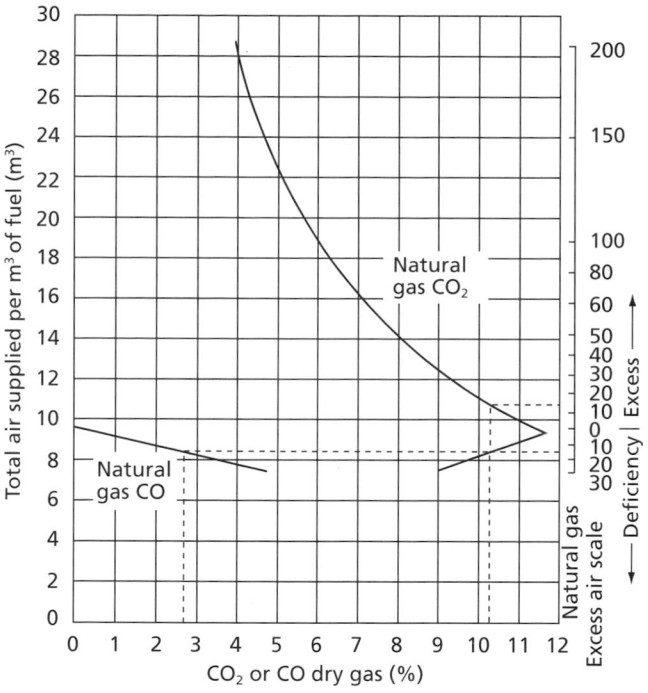

Table 5.7 Combustion air requirements for natural gas

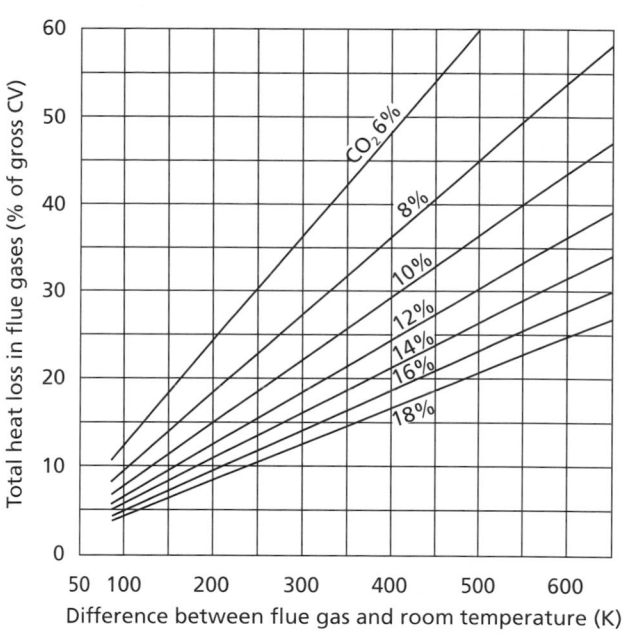

Figure 5.9 Flue gas loss — coke

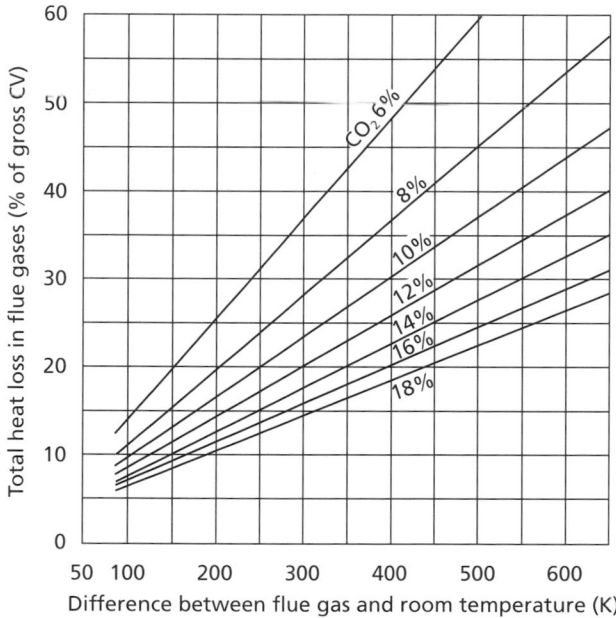

Figure 5.8 Flue gas loss — bituminous coal (CRC 702)

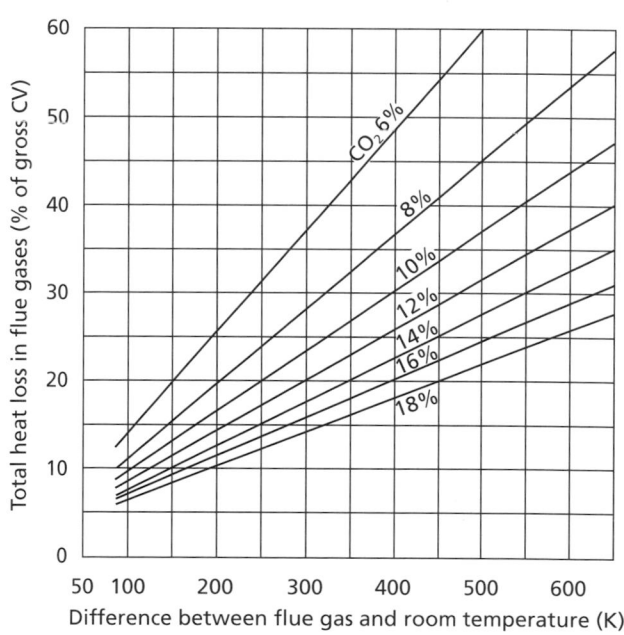

Figure 5.10 Flue gas loss — anthracite (CRC 102)

5.7.2   Liquid fuels

The curves given in Figures 5.11 and 5.12 show the total flue gas loss at various flue gas temperatures for various CO_2 per cent values (excess air quantities) expressed as a per cent of gross calorific value.

5.7.3   Gaseous fuels

The curves given in Figure 5.13 show the total flue gas loss at various flue gas temperatures for various CO_2 per cent values (excess air quantities) expressed as a per cent of gross calorific value.

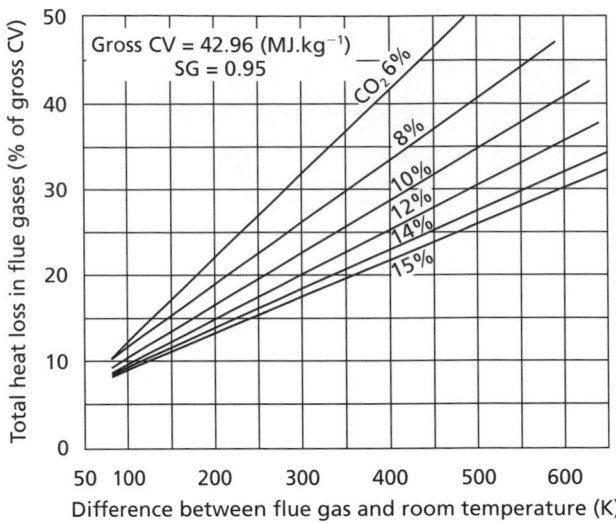

Figure 5.11 Flue gas loss — residual fuel oils

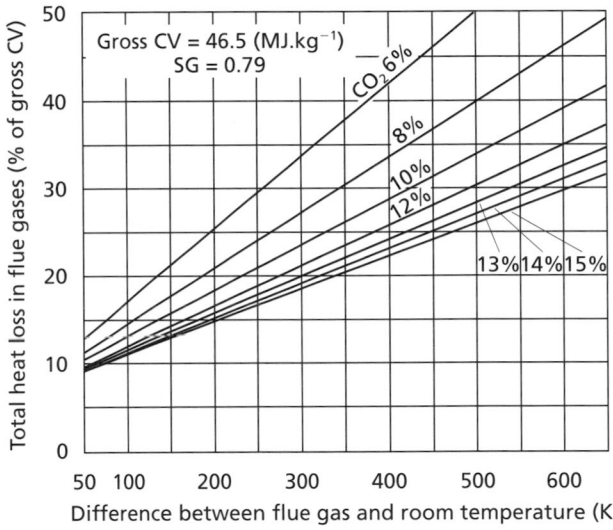

Figure 5.12 Flue gas loss — distillate fuel oils

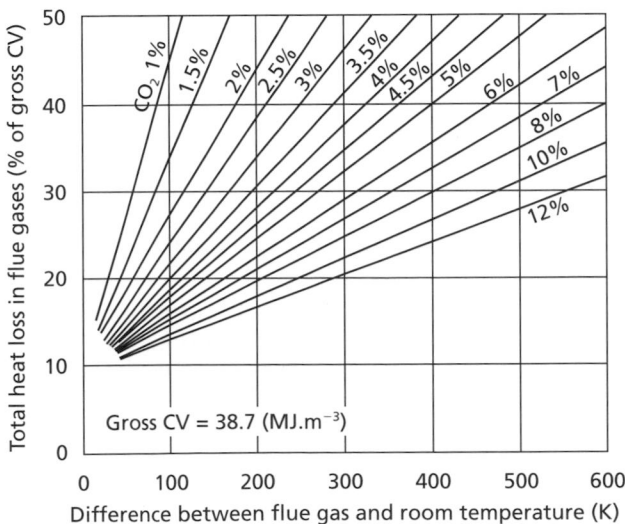

Figure 5.13 Flue gas loss — natural gas

Bibliography

1   Breag G R, Joseph P G and Tariq A S, Biomass Combustion Systems — Flue Gas Losses and Equipment Efficiency: Natural Resources Institute Report 1992. ISBN 085 954 3048

2   BS799, Oil Burning Equipment, Part 3, 1981; Part 4, 1991; Part 5, 1987, BSI (British Standard Institute)

3   BS845, Methods for Assessing Thermal Performance of Boilers for Steam, Hot Water and High Temperature Heat Transfer Fluids, Part 1 (1987) ISBN 0580 15856, Part 2 (1987) ISBN 0580 158578, BSI

4   BS1016, Methods for the Analysis and Testing of Coal and Coke, Parts 1–21 (1970–1998 with earlier parts withdrawn). Also Parts 100–113 (1991–1998). Part 100: Analysis and testing of coal and coke — General introduction and methods for reporting results, ISBN 0580 226 255 (1994)

5   BS2869, Specification for Fuel Oils for Agricultural, Domestic and Industrial Engines and Boilers, ISBN 0580 295 42 (1998)

6   BS4250, Specification for Commercial Butane and Commercial Propane, ISBN 0 580 276821, BSI (1997)

7   *Digest of United Kingdom Energy Statistics* (published annually by the Government Statistical Service), Edition ISBN 011 5154639, The Stationary Office (1999)

8   Dryden I G C (general editor) *The Efficient Use of Energy*, 2nd edition, Butterworth Scientific, ISBN 0 408012501 (1982)

9   Establishing Guidelines for Wood Fuel Standards, Department of Trade and Industry (DTI) Project Summary 369, First Issue (October 1994)

10 *Fuel and Energy Abstracts* (bimonthly journal published on behalf of the Institute of Energy), ISBN 0140–6701

11 Gas Act, Sections 16, 17, 47, ISBN 0 105 555 863, HMSO (1986)

12 Gunn D and Horton R, *Industrial Boilers*, 1st edition, Longman Scientific and Technical, ISBN 0 582 02532 (1989)

13 *Pollution Handbook* (published annually by the National Society for Clean Air and Environmental Protection — NSCA) 2000 Edition, ISBN 0903474484

14 Rose J W and Cooper J R (editors), *Technical Data on Fuel*, 7th edition, Scottish Academic Press, ISBN 0 7073 01 297 (1977)

15 Standard Methods for Analysis and Testing of Petroleum and Related Products, *Institute of Petroleum*, Vols 1 and 2, ISBN 0471 984728 (1999)

16 Environmental Protection Act (UK), ISBN 0 10 544 390 5 (1990)

17 *The Institute of Energy Yearbook* (in association with Energy in Buildings and Industry), available from The Institute of Energy, 18 Devonshire Street, London, W1N 2AU

18 Vesma V (editor), *Industrial Coal Handbook*, 1st edition, Energy Publications, ISBN 0 9055332 40 7 (1985)

19 Williams P T, *Waste Treatment and Disposal*, John Wiley and Sons, ISBN 0471 98166–4 (1998)

6   Units, standard and mathematical data

6.1   Introduction

This section contains information which, although essential to engineers and designers does not on the whole belong exclusively to any other section of the Guide. Inevitably, therefore, a large part of the section is devoted to SI units, which are used throughout the Guide but are not explained elsewhere. The development of SI is traced and the definitions of the base and supplementary units are given. There is also an explanation of how the units adopted by the European Union differ from SI. Comprehensive tables then list the conversion factors for changing from old units to their practical SI equivalents.

Although SI units have been the official standard units in Europe (including the UK) since 1971, conversion from other units is still regularly required for three reasons. Firstly some countries, notably the USA, have not fully adopted SI units, secondly many UK engineers still think in imperial units, and thirdly much of the UK heritage of plant and buildings was built to imperial unit designs.

6.2   The International System of Units (SI)

All international matters concerning the metric system have been the responsibility of the Conférence Générale des Poids et Mesures (CGPM) since the signing of the Metre Convention in 1875. Under the authority of CGPM are the Comité International des Poids et Mesures (CIPM) and the Bureau International des Poids et Mesures (BIPM). UK participation in CGPM work is through the Department of Trade and Industry (DTI). The International Standards Organisation (ISO) provides recommendations and advice and, more recently International Standards on the use and selection of SI units in Industry and Technology. UK participation in ISO work is through the British Standards Institution (BSI). The relevant standards are ISO 31/0 to 31/13 and OSI 2955 and their UK equivalents BS5775, Parts 1 to 13 and BS6430.

At its tenth meeting in 1954, CGPM adopted a coherent system of units based on the metre, kilogram, second, ampere, kelvin and candela. At its 14th meeting in 1971, the mole was added as the seventh base unit. The 11th meeting in 1960 gave the system its formal title 'Le Système International d'Unités – commonly abbreviated to SI. The SI comprises:

> Seven named base units.

> Two named supplementary units.

These units are defined below. Each unit has been allocated an internationally agreed symbol. All other units may be derived from these base and supplementary units. The supplementary units have since 1981 been redefined as derived units.

Certain derived units have been given internationally agreed names and symbols.

The allocation of a name to a derived unit has:

(*a*)   The advantage of simplicity in written and verbal communication.

(*b*)   The disadvantage of masking the derivation of the unit.

6.2.1   Definitions of base units

6.2.1.1   Unit of length

Name: metre   Symbol: m
The metre is the length of the path travelled in a vacuum by light during 1/299 792 458 seconds.
(17th CGPM (1983), Resolution 1.)

6.2.1.2 Unit of mass

Name: kilogram   Symbol: kg
With the object of removing the ambiguity which still occurred in the common use of the word 'weight', the 3rd CGPM (1901) declared: the kilogram is the unit of mass (and not of weight or of force); it is equal to the mass of the international prototype of the kilogram. This international prototype made of platinum–iridium is kept at the BIPM under conditions specified by the 1st CGPM in 1889.

6.2.1.3 Unit of time

Name: second   Symbol: s
The second is the duration of 9 192 631 770 periods of the radiation corresponding to the transition between the two hyperfine levels of the ground state of the caesium-133 atom.
(13th CGPM (1967), Resolution 1.)

6.2.1.4 Unit of electric current

Name: ampere   Symbol: A
The ampere is that constant current which, if maintained in two straight parallel conductors of infinite length, of negligible circular cross-section, and placed 1 metre apart in a vacuum, would produce between these conductors a force equal to 2×10^{-7} newton per metre of length.
(CPIM (1946), Resolution 2 approved by the 9th CGPM 1948.)

6.2.1.5 Unit of thermodynamic temperature

Name: kelvin   Symbol: K
The kelvin, unit of thermodynamic temperature, is the fraction 1/273.16 of the thermodynamic temperature of the triple point of water.
(3th CGPM (1967), Resolution 4.)

Note 1. The 13th CGPM (1967), Resolution 3, also decided that the unit kelvin and its symbol K should be used to express an interval or a difference of temperature.

Note 2. In addition to the thermodynamic temperature expressed in kelvins, use is also made of Celsius temperature defined by:

$$t = T - T_o \qquad (6.1)$$

where

t = Celsius temperature (°C)

T = thermodynamic temperature (K)

T_o = 273.15 K by definition

The Celsius temperature is in general expressed in degree Celsius (symbol °C). The unit 'degree Celsius' is thus equal to the unit 'kelvin' and an interval or a difference of Celsius temperature may also be expressed in degrees Celsius.

6.2.1.6 Unit of luminous intensity

Name: candela   Symbol: cd
The candela is the luminous intensity in a given direction of a source emitting monochromatic radiation at $50 \times 10^{12}\,H_3$ with a radiant intensity that direction of 1/683 Watt per steradian.
(16th CGPM (1979), Resolution 3.)

Table 6.1 SI base units

Quantity	Name	Symbol
Length	metre	m
Mass	kilogram	kg
Time	second	s
Electric current	ampere	A
Thermodynamic temperature	kelvin	K
Luminous intensity	candela	cd
Amount of substance	mole	mol

6.2.1.7 Unit of amount of substance

Name: mole   Symbol: mol
The mole is the amount of substance of a system which contains as many elementary entities as there are atoms in 0.012 kilogram of carbon 12.

Note. When the mole is used, the elementary entities must be specified and may be atoms, molecules, ions, electrons, other particles, or specified groups of such particles.

(14th CGPM (1971), Resolution 3.)

6.2.2 Definitions of supplementary units

6.2.2.1 Unit of plane angle

Name: radian   Symbol: rad
The radian is the angle between two radii of a circle which cuts off on the circumference an arc equal in length to the radius.

6.2.2.2 Unit of solid angle

Name: steradian   Symbol: sr
The steradian is the solid angle which, having its vertex in the centre of a sphere, cuts off an area of the surface of the sphere equal to the square of the radius of the sphere.

6.2.3 Derived units

Derived units are expressed algebraically in terms of base and/or supplementary units, see Table 6.2. Certain of these have been given special names, see Table 6.3.

6.2.4 Prefixes for multiples and submultiples

The magnitude of SI units may be increased or decreased by the use of named prefixes. Each prefix is allocated an internationally agreed symbol which may be added (in front) of the unit symbol.

Table 6.2 Examples of SI derived units expressed in terms of base units

Quantity	SI unit	
	Name	Symbol
Area	square metre	m^2
Volume	cubic metre	m^3
Velocity	metre per second	$m.s^{-1}$
Specific volume	cubic metre per kilogram	$m^3.kg^{-1}$
Thermal conductivity	watt per metre kelvin	$W.m^{-1}.K^{-1}$
Luminance	candela per square metre	$cd.m^{-2}$

Table 6.3 SI derived units with special names

Quantity	Name of SI derived unit	Symbol	Expressed in terms of SI base or supplementary units or in terms of other SI derived units
Plane angle	radian	rad	
Solid angle	steradian	sr	
Frequency	hertz	Hz	$1 \text{ Hz} = 1 \text{ s}^{-1}$
Force	newton	N	$1 \text{ N} = 1 \text{ kg.m.s}^{-2}$
Pressure and stress	pascal	Pa	$1 \text{ Pa} = 1 \text{ N.m}^{-2}$
Work, energy, quantity of heat	joule	J	$1 \text{ J} = 1 \text{ N.m}$
Power	watt	W	$1 \text{ W} = 1 \text{ J.s}^{-1}$
Apparent power	volt ampere	VA	$1 \text{ VA} = 1 \text{ J.s}^{-1}$
Quantity of electricity	coulomb	C	$1 \text{ C} = 1 \text{ A.s}$
Electrical potential, potential difference, electromotive force	volt	V	$1 \text{ V} = 1 \text{ W.A}^{-1} = 1 \text{ J.C}^{-1}$
Electrical capacitance	farad	F	$1 \text{ F} = 1 \text{ A.s.V}^{-1} = 1 \text{ C.V}^{-1}$
Electrical resistance	ohm	Ω	$1 \text{ } \Omega = 1 \text{ V.A}^{-1}$
Electrical conductance	siemens	S	$1 \text{ S} = 1 \text{ } \Omega^{-1}$
Magnetic flux, flux of magnetic induction	weber	Wb	$1 \text{ Wb} = 1 \text{ V.s}$
Magnetic flux density, magnetic induction	tesla	T	$1 \text{ T} = 1 \text{ Wb.m}^{-2}$
Inductance	henry	H	$1 \text{ H} = 1 \text{ V.s.A}^{-1} = 1 \text{ Wb.A}^{-1}$
Luminous flux	lumen	lm	$1 \text{ lm} = 1 \text{ cd.sr}$
Illuminance	lux	lx	$1 \text{ lx} = 1 \text{ lm.m}^{-2}$
Celsius temperature	degree Celsius	°C	$1°C = 1 \text{ K}$
Activity	becquerel	Bq	$1 \text{ Bq} = 1 \text{ s}^{-1}$
Specific energy imparted	gray	Gy	$1 \text{ Gy} = 1 \text{ J.kg}^{-1}$
Dose equivalent	sievert	Sv	$1 \text{ Sv} = 1 \text{ J.kg}^{-1}$

The prefixes given in Table 6.4 may be used to construct decimal multiples of units.

6.2.5 Explanatory notes

The following notes are intended as a guide to some quantities and units.

Pressure and stress

In terms of the SI base units the quantity pressure may be expressed in the units $(\text{kg.m}^{-1}.\text{s}^{-2})$. This derived unit has been given the name pascal (Pa). Pressure could also be expressed as:

$$\text{N.m}^{-2} \quad \text{J.m}^{-3} \quad \text{W.s.m}^{-3} \text{ etc.}$$

A non-SI unit in common use is the bar $(1 \text{ bar} = 10^5 \text{ Pa})$.

Table 6.4 SI prefixes

Multiplying factor	Prefix	Symbol
10^{24}	yotta	Y
10^{21}	zetta	Z
10^{18}	exa	E
10^{15}	peta	P
10^{12}	tera	T
10^9	giga	G
10^6	mega	M
10^3	kilo	k
10^2	hecto	h
$10^1 = 10$	deca	da
$10^{-1} = 0.1$	deci	d
10^{-2}	centi	c
10^{-3}	milli	m
10^{-6}	micro	μ
10^{-9}	nano	n
10^{-12}	pico	p
10^{-15}	femto	f
10^{-18}	atto	a
10^{-21}	zepto	z

Since the bar is a special authorised EU multiple of the pascal it is anticipated that the bar (and mbar) will remain in use indefinitely.

The Institution has adopted the pascal (and the internationally agreed multiples and sub-multiples) for expressing both pressure and stress in documents published after 1973. Operating pressures are expressed in kilopascals followed by the quantity in bars in parentheses; i.e. 60 kPa (0.6 bar). It should be noted that operating pressures should be in terms of 'gauge pressure' and should be unambiguously stated, i.e. gauge pressure 600 kPa (6 bar).

Weight and mass

The term 'weight' has for many years been used in two different senses. In common parlance and in the Weights & Measures Act 1963 it is used to mean 'mass', whereas in some technical work the word 'weight' is used in the sense of 'gravitational force'. There is no explicit SI unit of weight. When 'weight' is used to mean 'mass', then the coherent SI unit is the kilogram. When 'weight' is used to mean 'force', the coherent SI unit is the newton.

Specific heat capacity

Due to the original definitions of British thermal unit and calorie, the quantity 'specific heat capacity' of water approximated to unity when expressed in imperial or technical metric units.

In SI units the specific heat capacity of water is 4.185 5 $\text{kJ.kg}^{-1}.\text{K}^{-1}$ at a reference temperature of 15°C.

Temperature

When adopting the kelvin as the unit of thermodynamic temperature or for expressing temperature difference the 13th CGPM recognised that the term degree Celsius (°C) would continue in everyday use for as long as could be foreseen. The degree Celsius is defined by equation 6.1.

Celsius temperature is in general expressed in degrees Celsius (°C). The unit 'degree Celsius' is equal to the unit 'kelvin' and an interval of, or a difference of, Celsius temperature may also be expressed in degrees Celsius. In the CIBSE Guide temperature differences, particularly in compound units, are expressed in kelvins, i.e. $W.m^{-1}.K^{-1}$.

Sound

The decibel is a unit which compares power and its derivatives. When used in the context of sound, it is known as a measure of sound level defined, in the case of power, as

$$L_w = 10 \log_{10} \frac{W_1}{W_2} \tag{6.2}$$

where

L_w = sound power level (dB)

W_1 = sound power 1 (W)

W_2 = sound power 2 (usually a reference value) (W)

Sound intensity level and sound pressure level are defined by similar equations which take into account their relationship with sound power. Table 6.9 gives standard reference values which are used in sound measurement.

Radioactivity

The activity of a radioactive source is the number of nuclear transformations occurring in a small time interval provided by that time interval. The unit of activity is the Becquerel second.

Concentrations in air are expressed in reciprocal seconds per cubic metre.

The radiation dose absorbed by any material is defined as the mean energy imparted by ionising radiation per unit mass at the point of interest and is expressed in $J.kg^{-1}$. The special unit gray for the quantity absorbed dose is defined as:

$$1 \text{ Gy} = 1 \text{ J.kg}^{-1} \tag{6.3}$$

Doses to people are expressed as dose equivalent, which is obtained by multiplying the absorbed dose at the point of interest in tissue by various modifying factors, which among other things take account of the energy and type of the ionising radiation. Where the radiation dose is measures in grays the dose equivalent is measured in sieverts.

Clothing insulation

The clo is a dimensionless factor expressing the insulation of clothing and is defined as:

$$I_{cl} = \frac{R_{cl}}{R_r} \tag{6.4}$$

where

I_{cl} = clo value

R_{cl} = the total thermal resistance from skin to outer surface of the clothed bodies $(m^2.K.W^{-1})$

R_r = reference thermal resistance $(m^2.K.W^{-1})$

   = $0.155 \text{ m}^2.K.W^{-1}$

A clo value of 1 is the insulation given by a typical business suit with waistcoat.

6.3 Quantities, units and numbers

A physical quantity is an attribute which can be measured. The measurement is described in terms of a number multiplied by a unit. The algebraic relationship:

Physical quantity = Number × Unit

must always be maintained.

Some examples of physical quantities expressed as number of units

Height of Nelson's Column = 43.211 75 metres

Velocity of light in vacuo = 299 792 500 metres per second.

Thermal conductivity of balsa wood = 0.040 watt per metre kelvin.

6.3.1 Some rules for physical quantities

(1) The algebraic symbol should be a single letter of the Latin or Greek alphabet.

(2) When necessary, subcripts, superscripts, or other modifying signs may be attached.

(3) The symbol should, if possible, be printed in sloping (italic) type.

(4) The word 'specific' is restrcted to mean divided by mass.

(5) The word 'molar' is restricted to mean divided by amount of substance.

6.3.2 Some rules for units

(1) There is ONE and ONE ONLY SI unit for each physical quantity.

(2) To find the units of physical quantities other than the BASE physical quantities:

   (a) Define the physical quantity in terms of the base physical quantities.

   (b) Obtain the units of the physical quantity by mutiplying and/or dividing the constituent BASE units.

(3) Only agreed symbols must be used.

(4) The symbol should be printed in upright (Roman) type.

(5) The symbol for a unit derived from a proper name begins with a capital letter.

(6) When written in full, all units are written in lower case letters, e.g. newton, pascal, with the exception of Celsius, which starts with a capital letter.

(7) A product of two units may be represented in a number of ways:

m s   m.s   m·s

but not ms without the space.

(8) A quotient of two units may be represented in a number of ways:

$$m \ s^{-1} \quad m.s^{-1} \quad m/s \quad or \quad \frac{m}{s}$$

but not ms⁻¹, i.e. always a space between the unit symbols.

(9) More than one solidus (/) should not appear in the same expression unless parentheses are used to eliminate ambiguity.

6.3.3   Some rules for numbers

(1) Numbers should be printed in upright type.

(2) The decimal sign can be a point (.) either on the line, i.e. 2.6 (as has been adopted for the Guide and all current British Standards), or half-way up the characters, i.e. 2·6, though in continental countries a comma is normally used, i.e. 2,6.

(3) Digits should be grouped in threes about the decimal sign, separated by a space.

2 576.392 72

(4) When the decimal sign is before the first digit a zero should be placed before the decimal sign.

0.292 *not* .292

(5) The multiplication sign between numbers should be a cross (×) and not the mathematical (.) as may be used with unit symbols.

2.3 × 3.4

(6) Division of one number by another may be indicated in a number of ways.

$$\frac{129}{298} \quad or \quad 129/298 \quad or \quad 129 \times (298)^{-1}$$

(7) More than one solidus should never be used in the same expression unless parentheses are used to eliminate ambiguity.

(129/298)/2.62 or 129/(298 × 2.62)

but never 129/298/2.62.

6.3.4   Some rules for prefixes

A combination of prefix and symbol is regarded as a single symbol:

cm^2 means $(0.01 \ m)^2$ not $0.01 \ m^2$

compound prefixes should not be used.

$$10^{-9} \ m = nm \ not \ m\mu m$$

Note

Decimal multiples of the kilogram are formed by attaching an SI prefix to gram and *not* kilogram:

mg *not* μkg for 10^{-6} kg

Mg *not* kkg for 10^3 kg

6.3.5   Some rules for labelling graphs and tables

When labelling the co-ordinate axes of graphs or the column headings of tables it is essential to distinguish between the quantity itself and the numerical value of the quantity expressed in a particular unit. In graphs or tables it is sets of numbers which are plotted or tabulated. The normal presentation is to enclose the unit in brackets. However, since:

Number = Physical Quantity/Unit

the label may comprise Quantity/Unit. For clarity the unit may be enclosed in brackets.

6.4   Metrication in the European Union

6.4.1   Legislative background

Use of metric and other units within the EU is governed by Council Directive 80/181/EEC of 20th December 1979 as modified 3rd January 1985 and 7th December 1989. This Directive approved the use of metric units as defined in ISO standard 2955 of 1st March 1974, authorising continued use of certain other units in certain circumstances, and repealed Directive 71/354/EEC of 18th October 1971, the original direcive on metric units.

6.4.2   Obligatory units

The obligatory units for use within the EU are the SI base, supplementary and derived units, as given in Tables 6.1 and 6.2. Some special names, given in Table 6.5 are authorised.

6.4.3   Special authorised units

Certain non-SI units are specially authorised for use within member states either for general use or for specific purposes. Each state is entitled to select any or all of these for internal use. These units arc givcn in Table 6.6. Those marked with an asterisk were only authorised until 31st December 1999.

Table 6.5 EU units with special names

Description	Quantity	Unit		
		Name	Symbol	Value
Decimal multiples of SI units	volume	litre*	l	1 l = 1 dm^3
	mass	tonne*	t	1 t = 1 Mg
	pressures and stress	bar*	bar	1 bar = 10^5 Pa
Special field of application	area of farmland and real estate	are*	a	1 a = 100 m^2
	vergency of optical systems	dioptre		1 = 1 m^{-1}
	mass of precious stones	metric carat		1 = 2 × 10^{-4} kg
	mass/unit length, textile yarns and threads	tex	tex	1 tex = 10^{-6} kg.m^{-1}
Defined from SI units but not decimal multiples	plane angle	revolution	r †	1 r = 2π rad
		gon*	gon	1 gon = π/200 rad
		degree	°	1° = π/180 rad
		minute of angle‡	′	1′ = π/10 800 rad
		second of angle‡	″	1″ = π/648 000 rad
	time	minute	min	1 min = 60 s
		hour	h	1 h = 3600 s
		day	d	1 d = 86 400 s
Defined independently of SI units	mass	atomic mass unit*	u	1u ≈ 1.660 565 5 × 10^{-27} kg §
	energy	electronvolt*	eV	1eV ≈ 1.602 189 2 ×10^{-19} J §

* SI prefixes apply to these units
† UK symbol
‡ ISO31: 1992 recommends use of decimal division of the degree, i.e. 30′ = 0.5°
§ Values from CODATA bulletin No. 11, December 1973

Table 6.6 Special EU authorised units

Quantity	Use	Unit		
		Name	Symbol	Value
Length	a	inch	in	2.540 × 10^{-1} mm
	a	foot	ft	3.048 × 10^{-1} m
	a	yard	yd	9.144 × 10^{-1} m
	e	fathom		1.829 m
	a	mile		1.609 km
Area	l	square foot	ft^2*	9.290 × 10^{-2} m^2
	c	acre		4.047 × 10^3 m^2
	l	square yard	yd^2*	8.361 × 10^{-1} m^2
Volume	f	fluid ounce	fl oz	2.841 × 10^{-2} dm^3
	g	gill		1.421 × 10^{-1} dm^3
	b, f	pint	pt	5.683 × 10^{-1} dm^3
	l	quart	qt*	1.137 dm^3
	l	gallon	gal*	4.456 dm^3
Mass	h	ounce	oz	2.835 × 10^1 g
	d	troy ounce	oz tr	3.110 × 10^1 g
	h	pound	lb	4.536 × 10^{-1} kg
Energy	j	therm	*	1.055 × 10^{-1} GJ
Pressure	k	mm mercury	mm hg	1.333 × 10^2 Pa

* authorisation ceased 31st December 1999   f   drinks in returnable containers
a   road traffic signs, distance and speed measures   g   spirits
b   draught beer and cider, milk   h   loose bulk goods
c   land registration   j   gas supply
d   precious metals   k   blood pressure
e   marine navigation   l   any use

6.4.4   Forbidden units

All other units are forbidden although they may be retained for use for products already on the market at the time of decision and for spares for these and obsolete products. They may still be used as supplementary indications but may be no larger than the approved indications, which must predominate.

6.5   Conversion factors

The tables of conversion factors (Tables 6.7 and 6.8) are arranged in five columns: physical quantity, previous unit, factor, SI unit, SI symbol. To convert a quantity in previous units to the equivalent quantity in SI units, multiply by the factor. To convert a quantity in SI units to the equivalent quantity in old units, divide by the factor.

Table 6.7 Conversion factors

Physical quantity	Previous unit	× Factor		= SI unit	SI symbol
SPACE AND TIME					
Length	micron	1	E	micrometre	μm
	thou' (mil)	2.54×10^1	E	micrometre	μm
	inch	2.54×10^1	E	millimetre	mm
	foot	3.048×10^{-1}	E	metre	m
	yard	9.144×10^{-1}	E	metre	m
	mile	1.609		kilometre	km
Area	square inch	6.452×10^2		square millimetre	mm^2
		6.452		square centimetre	cm^2
	square foot	9.290×10^{-2}		square metre	m^2
	square yard	8.361×10^{-1}		square metre	m^2
	are	1×10^2	E	square metre	m^2
	acre	4.047×10^3		square metre	m^2
	hectare	1×10^4	E	square metre	m^2
	square mile	2.590		square kilometre	km^2
Volume	cubic inch	1.639×10^1		cubic centimetre	cm^3
	US pint	4.732×10^{-1}		cubic decimetre	dm^3
	pint	5.683×10^{-1}		cubic decimetre	dm^3
	litre	1	E	cubic decimetre	dm^3
	US gallon	3.785		cubic decimetre	dm^3
	gallon	4.546		cubic decimetre	dm^3
	cubic foot	2.832×10^1		cubic decimetre	dm^3
		2.832×10^{-2}		cubic metre	m^3
	US barrel (petroleum)	1.590×10^{-1}		cubic metre	m^3
	cubic yard	7.646×10^{-1}		cubic metre	m^3
Second moment of area	quartic inch	4.162×10^1		quartic centimetre	cm^4
	quartic foot	8.631×10^{-3}		quartic metre	m^4
Time	minute	6×10^1	E	second	s
	hour	3.6×10^3	E	second	s
	day	8.64×10^4	E	second	s
Angle	second	4.848		microradian	μrad
	minute	2.909×10^{-1}		milliradian	mrad
	grade	1.571×10^{-2}		radian	rad
	gon	1.571×10^{-2}		radian	rad
	degree	1.745×10^{-2}		radian	rad
	right angle	1.571		radian	rad
	revolution	6.283		radian	rad
Velocity	foot/minute	5.080×10^{-3}	E	metre/second	m.s^{-1}
	kilometre/hour	2.778×10^{-1}		metre/second	m.s^{-1}
	foot/second	3.048×10^{-1}	E	metre/second	m.s^{-1}
	mile/hour	4.470×10^{-1}		metre/second	m.s^{-1}
	knot	5.148×10^{-1}		metre/second	m.s^{-1}
Angular velocity	revolution per minute	1.047×10^{-1}		radian/second	rad.s^{-1}
	revolution per second	6.283		radian/second	rad.s^{-1}
Acceleration	foot/square second	3.048×10^{-1}	E	metre/square second	m.s^{-2}
Frequency	cycle/second	1	E	hertz	Hz
MASS AND DENSITY					
Mass	grain	6.480×10^1		milligram	mg
	ounce	2.835×10^1		gram	g
	pound	4.536×10^{-1}		kilogram	kg
	slug	1.459×10^1		kilogram	kg
	hundredweight	5.080×10^1		kilogram	kg
	ton (short)	9.072×10^{-1}		megagram	Mg
	tonne	1	E	megagram	Mg
	ton	1.016		megagram	Mg
Mass per unit length	pound/foot	1.488		kilogram/metre	kg.m^{-1}
	pound/inch	1.786×10^1		kilogram/metre	kg.m^{-1}
Mass per unit area	pound/square foot	4.882		kilogram/square metre	kg.m^{-2}
Concentration	grain/cubic foot	2.288		gram/cubic metre	g.m^{-3}
Density	pound/cubic foot	1.602×10^1		kilogram/cubic metre	kg.m^{-3}
	pound/gallon	9.978×10^1		kilogram/cubic metre	kg.m^{-3}
	pound/cubic inch	2.768×10^1		megagram/cubic metre	Mg.m^{-3}
Specific volume	cubic foot/pound	6.243×10^{-2}		cubic metre/kilogram	m^3.kg^{-1}

E — exact conversion factor　　　　　　　　　　The word 'litre' may be employed as a special name for dm^3

Table 6.7 Conversion factors — *continued*

Physical quantity	Previous unit	× Factor		= SI unit	SI symbol
FLOW RATE					
Mass flow rate	pound/hour	1.260×10^{-1}		gram/second	g.s^{-1}
	kilogram/hour	2.778×10^{-1}		gram/second	g.s^{-1}
	pound/minute	7.560×10^{-3}		kilogram/second	kg.s^{-1}
	kilogram/minute	1.667×10^{-2}		kilogram/second	kg.s^{-1}
Volume flow rate	cubic inch/minute	2.732×10^{-4}		cubic decimetre/second	dm^3.s^{-1}
	litre/hour	2.778×10^{-4}		cubic decimetre/second	dm^3.s^{-1}
	US gallon/hour	1.052×10^{-3}		cubic decimetre/second	dm^3.s^{-1}
	gallon/hour	1.263×10^{-3}		cubic decimetre/second	dm^3.s^{-1}
	cubic foot/hour	7.886×10^{-3}		cubic decimetre/second	dm^3.s^{-1}
	cubic inch/second	1.639×10^{-2}		cubic decimetre/second	dm^3.s^{-1}
	litre/minute	1.667×10^{-2}		cubic decimetre/second	dm^3.s^{-1}
	US gallon/minute	6.309×10^{-2}		cubic decimetre/second	dm^3.s^{-1}
	gallon/minute	7.577×10^{-2}		cubic decimetre/second	dm^3.s^{-1}
	cubic metre/hour	2.778×10^{-1}		cubic decimetre/second	dm^3.s^{-1}
	cubic foot/minute	4.719×10^{-1}		cubic decimetre/second	dm^3.s^{-1}
	cubic metre/minute	1.667×10^{1}		cubic decimetre/second	dm^3.s^{-1}
	cubic foot/second	2.832×10^{-2}		cubic metre/second	m^3.s^{-1}
MOMENTUM					
Momentum	pound foot/second	1.383×10^{-1}		kilogram metre/second	kg.m.s^{-1}
Moment of inertia	pound square foot	4.214×10^{-2}		kilogram square metre	kg.m^2
Moment of momentum	pound square foot/second	4.214×10^{-2}		kilogram square metre/second	kg.m^2.s^{-1}
FORCE AND TORQUE					
Force	dyne	1×10^{1}	E	micronewton	μN
	poundal	1.383×10^{-1}		newton	N
	pound force	4.448		newton	N
	kilogram force	9.807		newton	N
	kilopond	9.807		newton	N
Torque	pound force foot	1.356		newton metre	N.m
PRESSURE AND STRESS					
Pressure	millimetre of water	9.807		pascal	Pa
	pound force/square foot	4.788×10^{1}		pascal	Pa
	millimetre of mercury	1.333×10^{2}		pascal	Pa
	torr	1.333×10^{2}		pascal	Pa
	inch of water	2.491×10^{2}		pascal	Pa
	foot of water	2.989		kilopascal	kPa
	inch of mercury	3.386		kilopascal	kPa
	pound force/square inch	6.895		kilopascal	kPa
	kilogram force/square centimetre	9.807×10^{1}		kilopascal	kPa
	bar	1×10^{2}	E	kilopascal	kPa
		1×10^{-1}	E	megapascal	MPa
	standard atmosphere	1.013×10^{2}		kilopascal	kPa
		1.013×10^{-1}		megapascal	MPa
Pressure drop per unit length	inch of water/hundred feet	8.176		pascal/metre	Pa.m^{-1}
	foot of water/hundred feet	9.810×10^{1}		pascal/metre	Pa.m^{-1}
Stress	pound force/square foot	4.788×10^{1}		pascal	Pa
	pound force/square inch	6.895		kilopascal	kPa
	ton force/square foot	1.073×10^{2}		kilopascal	kPa
	ton force/square inch	1.544×10^{1}		megapascal	MPa
VISCOSITY					
Dynamic viscosity	pound/hour foot	4.134×10^{-1}		millipascal second	mPa.s
	centipoise	1×10^{-3}	E	pascal second	Pa.s
	poise	1×10^{-1}	E	pascal second	Pa.s
	pound force second/square foot	4.788×10^{1}		pascal second	Pa.s
	pound force hour/square foot	1.724×10^{2}		kilopascal second	kPa.s
Kinematic viscosity	stokes	1	E	square centimetre/second	cm^2.s^{-1}
	square metre/hour	2.778		square centimetre/second	cm^2.s^{-1}
	square inch/second	6.452		square centimetre/second	cm^2.s^{-1}
	square foot/minute	1.548×10^{-3}		square metre/second	m^2.s^{-1}
	Redwood No. 1 and No. 2 seconds	No direct conversion			
	SAE grades	No direct conversion			

E — exact conversion factor.

The word 'litre' may be employed as a special name for dm^3

Table 6.7 Conversion factors — *continued*

Physical quantity	Previous unit	× Factor		= SI unit	SI symbol
ENERGY					
Energy, work, quantity of heat	erg	1×10^{-1}	E	microjoule	μJ
	foot pound force	1.356		joule	J
	calorie★	4.187		joule	J
	metre kilogram force	9.807		joule	J
	British thermal unit	1.055		kilojoule	kJ
	frigorie†	4.186		kilojoule	kJ
	kilocalorie★	4.187		kilojoule	kJ
	horsepower hour	2.685		megajoule	MJ
	kilowatt hour	3.6	E	megajoule	MJ
	thermie†	4.186		megajoule	MJ
	therm	1.055×10^{-1}		gigajoule	GJ
POWER					
Power, heat flow rate	British thermal unit/hour	2.931×10^{-1}		watt	W
	kilocalorie/hour	1.163	E	watt	W
	foot pound force/second	1.356		watt	W
	calorie/second	4.187		watt	W
	metric horsepower (cheval vapeur)	7.355×10^{-1}		kilowatt	kW
	horsepower	7.457×10^{-1}		kilowatt	kW
	ton of refrigeration	3.517		kilowatt	kW
	Lloyd's ton of refrigeration	3.884		kilowatt	kW
Intensity of heat flow rate	kilocalorie/hour square metre	1.163	E	watt/square metre	W.m^{-2}
	Btu/hour square foot	3.155		watt/square metre	W.m^{-2}
	watt/square foot	1.076×10^{1}		watt/square metre	W.m^{-2}
Heat emission	Btu/hour cubic foot	1.035×10^{1}		watt/cubic metre	W.m^{-3}
Thermal conductivity	Btu inch/hour square foot degree Fahrenheit	1.442×10^{-1}		watt/metre kelvin	W.m^{-1}.K^{-1}
	kilocalorie/hour metre degree Celsius	1.163	E	watt/metre kelvin	W.m^{-1}.K^{-1}
	Btu/hour foot degree Fahrenheit	1.731		watt/metre kelvin	W.m^{-1}.K^{-1}
	calorie/second centimetre degree Celsius	4.187×10^{2}		watt/metre kelvin	W.m^{-1}.K^{-1}
Thermal conductance	kilocalorie/hour square metre degree Celsius	1.163	E	watt/square metre kelvin	W.m^{-2}.K^{-1}
	Btu/hour square foot degree Fahrenheit	5.678		watt/square metre kelvin	W.m^{-2}.K^{-1}
	calorie/second square centimetre degree Celsius	4.187×10^{1}		kilowatt/square metre kelvin	kW.m^{-2}.K^{-1}
Thermal resistivity	centimetre second degree Celsius/calorie	2.388×10^{-3}		metre kelvin/watt	m.K.W^{-1}
	foot hour degree Fahrenheit/Btu	5.778×10^{-1}		metre kelvin/watt	m.K.W^{-1}
	metre hour degree Celsius/kilocalorie	8.598×10^{-1}		metre kelvin/watt	m.K.W^{-1}
	square foot hour degree Fahrenheit/Btu inch	6.933		metre kelvin/watt	m.K.W^{-1}
Thermal resistance	square centimetre second degree Celsius/calorie	2.388×10^{-5}		square metre kelvin/watt	m^2.K.W^{-1}
	square foot hour degree Fahrenheit/Btu	1.761×10^{-1}		square metre kelvin/watt	m^2.K.W^{-1}
	square metre hour degree Celsius/kilocalorie	8.598×10^{-1}		square metre kelvin/watt	m^2.K.W^{-1}
Thermal diffusivity	square inch/hour	1.792×10^{-1}		square millimetre/second	mm^2.s^{-1}
	square foot/hour	2.581×10^{-5}		square metre/second	m^2.s^{-1}
	square metre/hour	2.778×10^{-4}		square metre/second	m^2.s^{-1}
ENERGY CONTENT					
Heat capacity	Btu/degree Fahrenheit	1.899		kilojoule/kelvin	kJ.K^{-1}
	kilocalorie/degree Celsius	4.187		kilojoule/kelvin	kJ.K^{-1}
Specific enthalpy	Btu/pound	2.326	E	kilojoule/kilogram	kJ.kg^{-1}
	kilocalorie/kilogram	4.187		kilojoule/kilogram	kJ.kg^{-1}
Specific heat capacity	Btu/pound degree Fahrenheit	4.187		kilojoule/kilogram kelvin	kJ.kg^{-1}.K^{-1}
	kilocalorie/kilogram degree Celsius	4.187		kilojoule/kilogram kelvin	kJ.kg^{-1}.K^{-1}
Entropy	Btu/degree Rankine	1.899		kilojoule/kelvin	kJ.K^{-1}
	kilocalorie/kelvin	4.187		kilojoule/kelvin	kJ.K^{-1}
Specific entropy	Btu/pound degree Rankine	4.187		kilojoule/kilogram kelvin	kJ.kg^{-1}.K^{-1}
	kilocalorie/kilogram kelvin	4.187		kilojoule/kilogram kelvin	kJ.kg^{-1}.K^{-1}

E — exact conversion factor
★ Based on the international calorie defined as 4.1868 J
† Based on the 15°C calorie determined as 4.1855 J

The word 'litre' may be employed as a special name for dm^3

Table 6.7 Conversion factors — *continued*

Physical quantity	Previous unit	× Factor		= SI unit	SI symbol
Latent heat	foot pound force/pound	2.989		joule/kilogram	J.kg^{-1}
	Btu/pound	2.326	E	kilojoule/kilogram	kJ.kg^{-1}
	kilocalorie/kilogram	4.187		kilojoule/kilogram	kJ.kg^{-1}
Volumetric calorific value	kilocalorie/cubic metre	4.187		kilojoule/cubic metre	kJ.m^3
	Btu/cubic foot	3.726 × 10^1		kilojoule/cubic metre	kJ.m^3
Specific heat (volume basis)	kilocalorie/cubic metre degree Celsius	4.187		kilojoule/cubic metre kelvin	kJ.m^{-3}.K^{-1}
	Btu/cubic foot degree Farenheit	6.707 × 10^1		kilojoule/cubic metre kelvin	kJ.m^{-3}.K^{-1}

MOISTURE CONTENT

Physical quantity	Previous unit	× Factor		= SI unit	SI symbol
Vapour permeability	grain inch/hour square foot inch of mercury (perminch)	1.45		nanogram metre/newton second	ng.m.N^{-1}.s^{-1}
		1.45		nanogram/second pascal metre	ng.s^{-1}.Pa^{-1}.m^{-1}
	pound foot/hour pound force	8.620		milligram metre/newton second	mg.m.N^{-1}.s^{-1}
		8.620		milligram/second pascal metre	mg.s^{-1}.Pa^{-1}.m^{-1}
Vapour permeance	grain/square foot hour inch of mercury (perm)	5.72 × 10^1		nanogram/newton second	ng.N^{-1}.s^{-1}
	grain/ square foot hour millibar	1.940		microgram/newton second	μg.N^{-1}.s^{-1}
	pound square inch/square foot hour pound force	1.965 × 10^{-1}		milligram/newton second	mg.N^{-1}.s^{-1}
	pound/hour pound force	2.834 × 10^1		milligram/newton second	mg.N^{-1}.s^{-1}
Moisture content	grain/pound	1.428 × 10^{-1}		gram/kilogram	g.kg^{-1}
	pound/pound	1	E	kilogram/kilogram	kg.kg^{-1}
Moisture flow rate	pound/square foot hour	1.357		gram/square metre second	g.m^{-2}.s^{-1}
	grain/square foot hour	1.94 × 10^{-1}		milligram/square metre second	mg.m^{-2}.s^{-1}
Mass transfer coefficient	foot/hour	8.47 × 10^{-2}		millimetre/second	mm.s^{-1}

LIGHT

Physical quantity	Previous unit	× Factor		= SI unit	SI symbol
Luminous intensity	candle	9.810 × 10^{-1}		candela	cd
Illumination	foot candle	1.076 × 10^1		lux	lx
	lumen/square foot	1.076 × 10^1		lux	lx
Luminance	foot lambert	3.426		candela/square metre	cd.m^{-2}
	candela/square inch	1.550 × 10^3		candela/square metre	cd.m^{-2}

ELECTRICTY AND MAGNETISM

Physical quantity	Previous unit	× Factor		= SI unit	SI symbol
Conductance	mho	1	E	siemens	S
Magnetic field strength	oersted	7.958 × 10^1		ampere/metre	A.m^{-1}
Magnetic flux	maxwell	1 × 10^{-2}	E	microweber	μWb
Magnetic flux density	gauss	1 × 10^{-1}	E	millitesla	mT

RADIOACTIVITY

Physical quantity	Previous unit	× Factor		= SI unit	SI symbol
Activity of a radioactive source	curie	3.7 × 10^1		nanosecond^{-1}	ns^{-1}
Absorbed dose	rad	1 × 10^{-2}	E	joule/kilogram	J.kg^{-1}
Equivalent absorbed dose	rem	1 × 10^{-2}	E	joule/kilogram	J.kg^{-1}
Exposure to ionisation	roentgen	2.58 × 10^{-1}		milllicoulomb/kilogram	mC.kg^{-1}

E — exact conversion factor

Example 1

A boiler is rated at 150 000 Btu.h^{-1}. In SI units this is 150 000 × (2.931 × 10^{-1}) = 43 965 W = 43.965 kW.

Example 2

A heating plant has an annual energy consumption of 500 GJ. In imperial units this is 500/(1.055 × 10^{-1}) = 4740 therms. In technical metric units the factor quoted is for converting between thermies and megajoules. Thus, 500 GJ must be transposed to MJ before being divided by the conversion factor, i.e. (500 × 10^3)/4.186 = 119 446 thermies.

Table 6.8 Conversion factors in alphabetical subject order

Physical quantity	Previous unit	$\times$ Factor		= SI unit	SI symbol
Absorbed dose	rad	1×10^{-2}	E	joule/kilogram	J.kg^{-1}
Acceleration	foot/square second	3.048×10^{-1}	E	metre/square second	m.s^{-2}
Angle	second	4.848		microradian	μrad
	minute	2.909×10^{-1}		milliradian	mrad
	grade	1.571×10^{-2}		radian	rad
	gon	1.571×10^{-2}		radian	rad
	degree	1.745×10^{-2}		radian	rad
	right angle	1.571		radian	rad
	revolution	6.283		radian	rad
Angular velocity	revolution per minute	1.047×10^{-1}		radian/second	rad.s^{-1}
	revolution per second	6.283		radian/second	rad.s^{-1}
Area	square inch	6.452×10^{2}		square millimetre	mm^2
		6.452		square centimetre	cm^2
	square foot	9.290×10^{-2}		square metre	m^2
	square yard	8.361×10^{-1}		square metre	m^2
	are	1×10^{2}	E	square metre	m^2
	acre	4.047×10^{3}		square metre	m^2
	hectare	1×10^{4}	E	square metre	m^2
	square mile	2.590		square kilometre	km^2
Concentration	grain/cubic foot	2.288		gram/cubic metre	g.m^{-3}
Conductance, electrical	mho	1	E	siemens	S
Conductance, thermal	kilocalorie/hour square metre degree Celsius	1.163	E	watt/square metre kelvin	W.m^{-2}.K^{-1}
	Btu/hour square foot degree Fahrenheit	5.678		watt/square metre kelvin	W.m^{-2}.K^{-1}
	calorie/second square centimetre degree Celsius	4.187×10^{1}		kilowatt/square metre kelvin	kW.m^{-2}.K^{-1}
Conductivity, thermal	Btu inch/hour square foot degree Fahrenheit	1.442×10^{-1}		watt/metre kelvin	W.m^{-1}.K^{-1}
	kilocalorie/hour metre degree Celsius	1.163	E	watt/metre kelvin	W.m^{-1}.K^{-1}
	Btu/hour foot degree Fahrenheit	1.731		watt/metre kelvin	W.m^{-1}.K^{-1}
	calorie/second centimetre degree Celsius	4.187×10^{2}		watt/metre kelvin	W.m^{-1}.K^{-1}
Density	pound/cubic foot	1.602×10^{1}		kilogram/cubic metre	kg.m^{-3}
	pound/gallon	9.978×10^{1}		kilogram/cubic metre	kg.m^{-3}
	pound/cubic inch	2.768×10^{1}		megagram/cubic metre	Mg.m^{-3}
Diffusivity, thermal	square inch/hour	1.792×10^{-1}		square millimetre/second	mm^2.s^{-1}
	square foot/hour	2.581×10^{-1}		square centimetre/second	cm^2.s^{-1}
	square metre/hour	2.778		square centimetre/second	cm^2.s^{-1}
Energy, work, quantity of heat	erg	1×10^{-1}	E	microjoule	μJ
	foot pound force	1.356		joule	J
	calorie*	4.187		joule	J
	metre kilogram force	9.807		joule	J
	British thermal unit	1.055		kilojoule	kJ
	frigorie†	4.186		kilojoule	kJ
	kilocalorie*	4.187		kilojoule	kJ
	horsepower hour	2.685		megajoule	MJ
	kilowatt hour	3.6	E	megajoule	MJ
	thermie†	4.186		megajoule	MJ
	therm	1.055×10^{-1}		gigajoule	GJ
Enthalpy, specific	Btu/pound	2.326	E	kilojoule/kilogram	kJ.kg^{-1}
	kilocalorie/kilogram	4.187		kilojoule/kilogram	kJ.kg^{-1}
Entropy	Btu/degree Rankine	1.899		kilojoule/kelvin	kJ.K^{-1}
	kilocalorie/kelvin	4.187		kilojoule/kelvin	kJ.K^{-1}
Entropy, specific	Btu/pound degree Rankine	4.187		kilojoule/kilogram kelvin	kJ.kg^{-1}.K^{-1}
	kilocalorie/kilogram kelvin	4.187		kilojoule/kilogram kelvin	kJ.kg^{-1}.K^{-1}
Equivalent absorbed dose	rem	1×10^{-2}	E	joule/kilogram	J.kg^{-1}
Exposure to ionisation	roentgen	2.58×10^{-1}		millicoulomb/kilogram	mC.kg^{-1}
Flow rate, mass	pound/hour	1.260×10^{-1}		gram/second	g.s^{-1}
	kilogram.hour	2.778×10^{-1}		gram/second	g.s^{-1}
	pound/minute	7.560×10^{-3}		kilogram/second	kg.s^{-1}
	kilogram/minute	1.667×10^{-2}		kilogram/second	kg.s^{-1}

E — exact conversion factor
* Based on the international calorie defined as 4.186 8 J
† Based on the 15°C calorie determined as 4.185 5 J

Table 6.8 Conversion factors in alphabetical subject order — *continued*

Physical quantity	Previous unit	× Factor		= SI unit	SI symbol
Flow rate, volume	cubic inch/minute	2.3732×10^{-4}		cubic decimetre/second	dm³.s⁻¹
	litre/hour	2.778×10^{-4}		cubic decimetre/second	dm³.s⁻¹
	US gallon/hour	1.052×10^{-3}		cubic decimetre/second	dm³.s⁻¹
	gallon/hour	1.263×10^{-3}		cubic decimetre/second	dm³.s⁻¹
	cubic foot/hour	7.866×10^{-3}		cubic decimetre/second	dm³.s⁻¹
	cubic inch/second	1.639×10^{-2}		cubic decimetre/second	dm³.s⁻¹
	litre/minute	1.667×10^{-2}		cubic decimetre/second	dm³.s⁻¹
	US gallon/minute	6.309×10^{-2}		cubic decimetre/second	dm³.s⁻¹
	gallon/minute	7.577×10^{-2}		cubic decimetre/second	dm³.s⁻¹
	cubic metre/hour	2.778×10^{-1}		cubic decimetre/second	dm³.s⁻¹
	cubic foot/minute	4.719×10^{-1}		cubic decimetre/second	dm³.s⁻¹
	cubic metre/minute	1.667×10^{1}		cubic decimetre/second	dm³.s⁻¹
	cubic foot/second	2.832×10^{-2}		cubic metre/second	m³.s⁻¹
Force	dyne	1×10^{1}	E	micronewton	μN
	poundal	1.383×10^{-1}		newton	N
	pound force	4.448		newton	N
	kilogram force	9.807		newton	N
	kilopond	9.807		newton	N
Frequency	cycle/second	1	E	hertz	Hz
Heat capacity	Btu/degree Fahrenheit	1.899		kilojoule/kelvin	kJ.K⁻¹
	kilocalorie/degree Celsius	4.187		kilojoule/kelvin	kJ.K⁻¹
Heat capacity, specific	Btu/pound degree Fahrenheit	4.187		kilojoule/kilogram kelvin	kJ.kg⁻¹.K⁻¹
	kilocalorie/kilogram degree Celsius	4.187		kilojoule/kilogram kelvin	kJ.kg⁻¹.K⁻¹
Heat emission	Btu/hour cubic foot	1.035×10^{1}		watt/cubic metre	W.m⁻³
Illumination	foot candle	1.076×10^{1}		lux	lx
	lumen/square foot	1.076×10^{1}		lux	lx
Intensity of heat flow rate	kilocalorie/hour square metre	1.163	E	watt/square metre	W.m⁻²
	Btu/hour square foot	3.155		watt/square metre	W.m⁻²
	watt/square foot	1.076×10^{1}		watt/square metre	W.m⁻²
Latent heat	foot pound force/pound	2.989		joule/kilogram	J.kg⁻¹
	Btu/pound	2.326	E	kilojoule/kilogram	kJ.kg⁻¹
	kilocalorie/kilogram	4.187		kilojoule/kilogram	kJ.kg⁻¹
Length	micron	1	E	micrometre	μm
	thou' (mil)	2.54×10^{1}	E	micrometre	μm
	inch	2.54×10^{1}	E	millimetre	mm
	foot	3.048×10^{-1}	E	metre	m
	yard	9.144×10^{-1}	E	metre	m
	mile	1.609		kilometre	km
Luminance	foot lambert	3.426		candela/square metre	cd.m⁻²
	candela/square inch	1.550×10^{3}		candela/square metre	cd.m⁻²
Luminous intensity	candle	9.810×10^{-1}		candela	cd
Magnetic field strength	oersted	7.958×10^{1}		ampere/metre	A.m⁻¹
Magnetic flux	maxwell	1×10^{-2}	E	microweber	μWb
Magnetic flux density	gauss	1×10^{-1}	E	millitesla	mT
Mass	grain	6.480×10^{1}		milligram	mg
	ounce	2.835×10^{1}		gram	g
	pound	4.536×10^{-1}		kilogram	kg
	slug	1.459×10^{1}		kilogram	kg
	hundredweight	5.080×10^{1}		kilogram	kg
	ton (short)	9.072×10^{-1}		megagram	Mg
	tonne	1	E	megagram	Mg
	ton	1.016		megagram	Mg
Mass per unit area	pound/square foot	4.882		kilogram/square metre	kg.m⁻²
Mass per unit length	pound/foot	1.488		kilogram/metre	kg.m⁻¹
	pound/inch	1.786×10^{1}		kilogram/metre	kg.m⁻¹
Mass transfer coefficient	foot/hour	8.47×10^{-2}		millimetre/second	mm.s⁻¹
Moisture content	grain/pound	1.428×10^{-1}		gram/kilogram	g.kg⁻¹
	pound/pound	1	E	kilogram/kilogram	kg.kg⁻¹
Moisture flow rate	pound/square foot hour	1.357		gram/square metre second	g.m⁻².s⁻¹
	grain/square foot hour	1.94×10^{-1}		milligram/square metre second	mg.m⁻².s⁻¹
Moment of inertia	pound square foot	4.214×10^{-2}		kilogram square metre	kg.m²

E — exact conversion factor　　　　　　　　　　　　　　The word 'litre' may be employed as a special name for dm³

Table 6.8 Conversion factors in alphabetical subject order — *continued*

Physical quantity	Previous unit	× Factor		= SI unit	SI symbol
Moment of momentum	pound square foot/second	4.214×10^{-2}		kilogram square metre/second	$kg.m^2.s^{-1}$
Momentum	pound foot/second	1.383×10^{-1}		kilogram metre/second	$kg.m.s^{-1}$
Permeability, vapour	grain inch/hour square foot inch of mercury (perminch)	1.45		nanogram metre/newton second	$ng.m.N^{-1}.s^{-1}$
		1.45		nanogram/second pascal metre	$ng.s^{-1}.Pa^{-1}.m^{-1}$
	pound foot/hour pound force	8.620		milligram metre/newton second	$mg.m.N^{-1}.s^{-1}$
		8.620		milligram/second pascal metre	$mg.s^{-1}.Pa^{-1}.m^{-1}$
Permeance, vapour	grain/square foot hour inch of mercury (perm)	5.72×10^1		nanogram/newton second	$ng.N^{-1}.s^{-1}$
	grain/square foot hour millibar	1.940		microgram/newton second	$\mu g.N^{-1}.s^{-1}$
	pound square inch/square foot hour pound force	1.965×10^{-1}		milligram/newton second	$mg.N^{-1}.s^{-1}$
	pound/hour pound force	2.834×10^1		milligram/newton second	$mg.N^{-1}.s^{-1}$
Power, heat flow rate	British thermal unit/hour	2.931×10^{-1}		watt	W
	kilocalorie/hour	1.163	E	watt	W
	foot pound force/second	1.356		watt	W
	calorie/second	4.187		watt	W
	metric horsepower (cheval vapeur)	7.355×10^{-1}		kilowatt	kW
	horsepower	7.457×10^{-1}		kilowatt	kW
	ton of refrigeration	3.517		kilowatt	kW
	Lloyd's ton of refrigeration	3.884		kilowatt	kW
Pressure	millimetre of water	9.807		pascal	Pa
	pound force/square foot	4.788×10^1		pascal	Pa
	millimetre of mercury	1.333×10^2		pascal	Pa
	torr	1.333×10^2		pascal	Pa
	inch of water	2.491×10^2		pascal	Pa
	foot of water	2.989		kilopascal	kPa
	inch of mercury	3.386		kilopascal	kPa
	pound force/square inch	6.895		kilopascal	kPa
	kilogram force/square centimetre	9.807×10^1		kilopascal	kPa
	bar	1×10^2	E	kilopascal	kPa
		1×10^{-1}	E	megapascal	MPa
	standard atmosphere	1.013×10^2		kilopascal	kPa
		1.013×10^{-1}		megapascal	MPa
Pressure drop per unit length	inch of water/hundred feet	8.176		pascal/metre	$Pa.m^{-1}$
	foot of water/hundred feet	9.810×10^1		pascal/metre	$Pa.m^{-1}$
Radioactivity	curie	3.7×10^1		nanosecond^{-1}	ns^{-1}
Resistance, thermal	square centimetre second degree Celsius/calorie	2.388×10^1		square decimetre kelvin/watt	$dm^2.K.W^{-1}$
	square foot hour degree Fahrenheit/Btu	1.761×10^{-1}		square metre kelvin/watt	$m^2.K.W^{-1}$
	square metre hour degree Celsius/kilocalorie	8.598×10^{-1}		square metre kelvin/watt	$m^2.K.W^{-1}$
Resistivity, thermal	centimetre second degree Celsius/calorie	2.388×10^{-3}		metre kelvin/watt	$m.K.W^{-1}$
	foot hour degree Fahrenheit/Btu	5.778×10^{-1}		metre kelvin/watt	$m.K.W^{-1}$
	metre hour degree Celsius/kilocalorie	8.598×10^{-1}		metre kelvin/watt	$m.K.W^{-1}$
	square foot hour degree Fahrenheit/Btu inch	6.933		metre kelvin/watt	$m.K.W^{-1}$
Second moment of area	quartic inch	4.162×10^5		quartic decimetre	dm^4
	quartic foot	8.631×10^{-3}		quartic metre	m^4
Specific heat (volume basis)	kilocalorie/cubic metre degree Celsius	4.187		kilojoule/cubic metre kelvin	$kJ.m^{-3}.K^{-1}$
	Btu/cubic foot degree Fahrenheit	6.707×10^1		kilojoule/cubic metre kelvin	$kJ.m^{-3}.K^{-1}$
Specific volume	cubic foot/pound	6.243×10^{-2}		cubic metre/kilogram	$m^3.kg^{-1}$
Stress	pound force/square foot	4.788×10^1		pascal	Pa
	pound force/square inch	6.895		kilopascal	kPa
	ton force/square foot	1.073×10^2		kilopascal	kPa
	ton force/square inch	1.544×10^1		megapascal	MPa
Time	minute	6×10^1	E	second	s
	hour	3.6×10^3	E	second	s
	day	8.64×10^4	E	second	s

E — exact conversion factor

Table 6.8 Conversion factors in alphabetical subject order — *continued*

Physical quantity	Previous unit	× Factor		= SI unit	SI symbol
Torque	pound force foot	1.356		newton metre	N.m
Velocity	foot/minute	5.080×10^{-3}	E	metre/second	m.s^{-1}
	kilometre/hour	2.778×10^{-1}		metre/second	m.s^{-1}
	foot/second	3.048×10^{-1}	E	metre/second	m.s^{-1}
	mile/hour	4.470×10^{-1}		metre/second	m.s^{-1}
	knot	5.148×10^{-1}		metre/second	m.s^{-1}
Viscosity, dynamic	pound/hour foot	4.134×10^{-1}		millipascal second	mPa.s
	centipoise	1×10^{-3}	E	pascal second	Pa.s
	poise	1×10^{-1}	E	pascal second	Pa.s
	pound force second/square foot	4.788×10^{1}		pascal second	Pa.s
	pound force hour/square foot	1.724×10^{2}		kilopascal second	kPa.s
Viscosity, kinematic	stokes	1×10^{-2}	E	square decimetre/second	dm^2.s^{-1}
	square metre/hour	2.778×10^{-2}		square decimetre/second	dm^2.s^{-1}
	square inch/second	6.452		square decimetre/second	dm^2.s^{-1}
	square foot/minute	1.548×10^{-3}		square metre/second	m^2.s^{-1}
	Redwood No. 1 and No. 2 seconds	No direct conversion			
	SAE grades	No direct conversion			
Volume	cubic inch	1.639×10^{-2}		cubic decimetre	dm^3
	US pint	4.732×10^{-1}		cubic decimetre	dm^3
	pint	5.683×10^{-1}		cubic decimetre	dm^3
	litre	1	E	cubic decimetre	dm^3
	US gallon	3.785		cubic decimetre	dm^3
	gallon	4.546		cubic decimetre	dm^3
	cubic foot	2.832×10^{1}		cubic decimetre	dm^3
		2.832×10^{-2}		cubic metre	m^3
	US barrel (petroleum)	1.590×10^{-1}		cubic metre	m^3
	cubic yard	7.646×10^{-1}		cubic metre	m^3
Volumetric calorific value	kilocalorie/cubic metre	4.187		kilojoule/cubic metre	kJ.m^{-3}
	Btu/cubic foot	3.726×10^{1}		kilojoule/cubic metre	kJ.m^{-3}

E — exact conversion factor

The word 'litre' may be employed as a special name for dm^3

Bibliography

1   BS350 Conversion factors and tables: Part 1: 1974 (1983) Basis of tables. Conversion factors; Part 2: Supplement No. 1: 1967 (1982) Additional tables for SI conversion (London: British Standards Institution) (dates as shown)

2   BS5775 Specifications for quantities, units and symbols: Part 0: 1993 General principles; Part 1: 1993 Space and time; Part 2: 1993 Periodic and related phenomena; Part 3: 1993 Mechanics; Part 4: 1993 Heat; Part 5: 1993 Electricity and magnetism; Part 6: 1993 Light and related electromagnetic radiations; Part 7: 1993 Acoustics; Part 8: 1993 Physical chemistry and molecular physics; Part 9: 1993 Atomic and nuclear physics; Part 10: 1993 Nuclear reactions and ionizing radiations; Part 11: 1993 Mathematical signs and symbols for use in physical sciences and technology; Part 12: 1993 Characteristic numbers; Part 13: 1993 Solid state physics (London: British Standards Institution) (1993) (equivalent to ISO 31-0 to ISO 31-13)

3   BS6430: 1983 (1988) Method for representing SI and other units in information processing systems with limited character sets (London: British Standards Institution) (1988) (equivalent to ISO 2955)

4   BS6430: 1983 (1988) Method for representing SI and other units in information processing systems with limited character sets (equivalent to ISO 2955)

5   BS5555: 1993 Specification for SI units and recommendations for the use of their multiples and of certain other units (equivalent to ISO 1000: 1992)

6   Council Directive 80/181/EEC of 20 December 1979 on the aproximation for the laws of the Member States relating to units of measurement and on the repeal of Directive 71/354/EEC (Brussels: Commission for the European Communities) (1979)

7   Official Journal 15.02.80 p. 40

8   Official Journal 03.01.85 p. 11

9   Official Journal 07.12.89 p. 28

Table 6.9 Degrees Fahrenheit to degrees Celsius

°F	0	−1	−2	−3	−4	−5	−6	−7	−8	−9
−40	−40.0	−40.6	−41.1	−41.7	−42.2	−42.8	−43.3	−43.9	−44.4	−45.0
−30	−34.4	−35.0	−35.6	−36.1	−36.7	−37.2	−37.8	−38.3	−38.9	−39.4
−20	−28.9	−29.4	−30.0	−30.6	−31.1	−31.7	−32.2	−32.8	−33.3	−33.9
−10	−23.3	−23.9	−24.4	−25.0	−25.6	−26.1	−26.7	−27.2	−27.8	−28.3
0	−17.8	−18.3	−18.9	−19.4	−20.0	−20.6	−21.1	−21.7	−22.2	−22.8

°F	0	+1	+2	+3	+4	+5	+6	+7	+8	+9
0	−17.8	−17.2	−16.7	−16.1	−15.6	−15.0	−14.4	−13.9	−13.3	−12.8
10	−12.2	−11.7	−11.1	−10.6	−10.0	−9.4	−8.9	−8.3	−7.8	−7.2
20	−6.7	−6.1	−5.6	−5.0	−4.4	−3.9	−3.3	−2.8	−2.2	−1.7
30	−1.1	−0.6	0	0.6	1.1	1.7	2.2	2.8	3.3	3.9
40	4.4	5.0	5.6	6.1	6.7	7.2	7.8	8.3	8.9	9.4
50	10.0	10.6	11.1	11.7	12.2	12.8	13.3	13.9	14.4	15.0
60	15.6	16.1	16.7	17.2	17.8	18.3	18.9	19.4	20.0	20.6
70	21.1	21.7	22.2	22.8	23.3	23.9	24.4	25.0	25.6	26.1
80	26.7	27.2	27.8	28.3	28.9	29.4	30.0	30.6	31.1	31.7
90	32.2	32.8	33.3	33.9	34.4	35.0	35.6	36.1	36.7	37.2
100	37.8	38.3	38.9	39.4	40.0	40.6	41.1	41.7	42.2	42.8
110	43.3	43.9	44.4	45.0	45.6	46.1	46.7	47.2	47.8	48.3
120	48.9	49.4	50.0	50.6	51.1	51.7	52.2	52.8	53.3	53.9
130	54.4	55.0	55.6	56.1	56.7	57.2	57.8	58.3	58.9	59.4
140	60.0	60.6	61.1	61.7	62.2	62.8	63.3	63.9	64.4	65.0
150	65.6	66.1	66.7	67.2	67.8	68.3	68.9	69.4	70.0	70.6
160	71.1	71.7	72.2	72.8	73.3	73.9	74.4	75.0	75.6	76.1
170	76.7	77.2	77.8	78.3	78.9	79.4	80.0	80.6	81.1	81.7
180	82.2	82.8	83.3	83.9	84.4	85.0	85.6	86.1	86.7	87.2
190	87.8	88.3	88.9	89.4	90.0	90.6	91.1	91.7	92.2	92.8
200	93.3	93.9	94.4	95.0	95.6	96.1	96.7	97.2	97.8	98.3
210	98.9	99.4	100.0	100.6	101.1	101.7	102.2	102.8	103.3	103.9
220	104.4	105.0	105.6	106.1	106.7	107.2	107.8	108.3	108.9	109.4
230	110.0	110.6	111.1	111.7	112.2	112.8	113.3	113.9	114.4	115.0
240	115.6	116.1	116.7	117.2	117.8	118.3	118.9	119.4	120.0	120.6
250	121.1	121.7	122.2	122.8	123.3	123.9	124.4	125.0	125.6	126.1
260	126.7	127.2	127.8	128.3	128.9	129.4	130.0	130.6	131.1	131.7
270	132.2	132.8	133.3	133.9	134.4	135.0	135.6	136.1	136.7	137.2
280	137.8	138.3	138.9	139.4	140.0	140.6	141.1	141.7	142.2	142.8
290	143.3	143.9	144.4	145.0	145.6	146.1	146.7	147.2	147.8	148.3
300	148.9	149.4	150.0	150.6	151.1	151.7	152.2	152.8	153.3	153.9
310	154.4	155.0	155.6	156.1	156.7	157.2	157.8	158.3	158.9	159.4
320	160.0	160.6	161.1	161.7	162.2	162.8	163.3	163.9	164.4	165.0
330	165.6	166.1	166.7	167.2	167.8	168.3	168.9	169.4	170.0	170.6
340	171.1	171.7	172.2	172.8	173.3	173.9	174.4	175.0	175.6	176.1
350	176.7	177.2	177.8	178.3	178.9	179.4	180.0	180.6	181.1	181.7
360	182.2	182.8	183.3	183.9	184.4	185.0	185.6	186.1	186.7	187.2
370	187.8	188.3	188.9	189.4	190.0	190.6	191.1	191.7	192.2	192.8
380	193.3	193.9	194.4	195.0	195.6	196.1	196.7	197.2	197.8	198.3
390	198.9	199.4	200.0	200.6	201.1	201.7	202.2	202.8	203.3	203.9
400	204.4	205.0	205.6	206.1	206.7	207.2	207.8	208.3	208.9	209.4
410	210.0	210.6	211.1	211.7	212.2	212.8	213.3	213.9	214.4	215.0
420	215.6	216.1	216.7	217.2	217.8	218.3	218.9	219.4	220.0	220.6
430	221.1	221.7	222.2	222.8	223.3	223.9	224.4	225.0	225.6	226.1
440	226.7	227.2	227.8	228.3	228.9	229.4	230.0	230.6	231.1	231.7
450	232.2	232.8	233.3	233.9	234.4	235.0	235.6	236.1	236.7	237.2
460	237.8	238.3	238.9	239.4	240.0	240.6	241.1	241.7	242.2	242.8
470	243.3	243.9	244.4	245.0	245.6	246.1	246.7	247.2	247.8	248.3
480	248.9	249.4	250.0	250.6	251.1	251.7	252.2	252.8	253.3	253.9
490	254.4	255.0	255.6	256.1	256.7	257.2	257.8	258.3	258.9	259.4
500	260.0	260.6	261.1	261.7	262.2	262.8	263.3	263.9	264.4	265.0
510	265.6	266.1	266.7	267.2	267.8	268.3	268.9	269.4	270.0	270.6
520	271.1	271.7	272.2	272.8	273.3	273.9	274.4	275.0	275.6	276.1
530	276.7	277.2	277.8	278.3	278.9	279.4	280.0	280.6	281.1	281.7
540	282.2	282.8	283.3	283.9	284.4	285.0	285.6	286.1	286.7	287.2
550	287.8	288.3	288.9	289.4	290.0	290.6	291.1	291.7	292.2	292.8
560	293.3	293.9	294.4	295.0	295.6	296.1	296.7	297.2	297.8	298.3
570	298.9	299.4	300.0	300.6	301.1	301.7	302.2	302.8	303.3	303.9
580	304.4	305.0	305.6	306.1	306.7	307.2	307.8	308.3	308.9	309.4
590	310.0	310.6	311.1	311.7	312.2	312.8	313.3	313.9	314.4	315.0

Table 6.10 SI units for catalogues

Quantity	Unit	Quantity	Unit
BOILERS		FUELS	
Heat output	kW	Calorific value:	
Heat input	kW	Solid	MJ.kg^{-1}
Steam generation rate	kg.s^{-1}	gaseous	MJ.m^{-3}
Fuel firing rate:		liquid	MJ.kg^{-1}
solid	kg.s^{-1}		
gaseous	dm^3.s^{-1}★	HEAT EXCHANGERS	
liquid	kg.s^{-1}	Heat output	kW
Volume flow rate (combustion products)	m^3.s^{-1}	Mass flow rate	kg.s^{-1}
Power to input (to drives)	kW	Hydraulic resistance	Pa
Operating pressure	kPa (bar)	Operating pressure	kPa (bar)
Hydraulic resistance	Pa	Flow velocity	m.s^{-1}
Draft conditions	Pa	Heat exchange surface	m^2
COIL, COOLING AND HEATING		INDUCTION TERMINALS	
Heat, exchange rate	kW	Heating or cooling output	kW
Primary medium:		Primary air volume flow rate	m^3.s^{-1}
mass flow rate	kg.s^{-1}	Primary air static pressure loss	Pa
hydraulic resistance	Pa	Secondary water mass flow rate	kg.s^{-1}
Air volume flow rate	m^3.s^{-1}	Secondary water hydraulic resistance	Pa
Air flow static pressure loss	Pa		
		PUMPS	
CONTROLS AND INSTRUMENTS		Mass flow rate	kg.s^{-1}
Flow rate:		Volume flow rate	dm^3.s^{-1}★
mass	kg.s^{-1}	Power input (to drive)	kW
volume	m^3.s^{-1}	Developed pressure	Pa
Operating pressure	kPa (bar)	Operating pressure	kPa (bar)
Hydraulic resistance	Pa	Rotational frequency	rev.s^{-1}
Rotational frequency	rev.s^{-1}		
		SPACE HEATING APPARATUS	
COOLING TOWERS		Heat output	kW
Heat extraction rate	kW	Air flow volume flow rate	m^3.s^{-1}
Volume flow rate:		Power input (to drive)	kW
air	m^3.s^{-1}	Primary medium mass flow rate	kg.s^{-1}
water	dm^3.s^{-1}★	Hydraulic resistance	Pa
Power input (to drive)	kW	Operating pressure	kPa (bar)
		Air flow static pressure loss	Pa
DIFFUSERS AND GRILLES			
Air volume flow rate	m^3.s^{-1}	VESSELS	
Air flow pressure loss	Pa	Operating pressure	kPa (bar)
Specific velocity	m.s^{-1}	Volumetric capacity	dm^3★ or m^3
FANS		WASHERS (AIR)	
Air volume flow rate	m^3.s^{-1}	Volume flow rate:	
Power input (to drive)	kW	air	m^3.s^{-1}
Fan static pressure	Pa	water	dm^3.s^{-1}★
Fan total pressure	Pa	Mass flow rate, water	kg.s^{-1}
Rotational frequency	rev.s^{-1}	Power input (to drive)	kW
Outlet velocity	m.s^{-1}	Air flow static pressure loss	Pa
		Hydraulic resistance	Pa
FILTERS			
Air volume flow rate	m^3.s^{-1}	WATER CHILLERS	
Liquid volume flow rate	dm^3.s^{-1}★	Cooling capacity	kW
Static pressure loss	Pa	Mass flow rate, water	kg.s^{-1}
		Power input (to drive)	kW
		Refigerant pressure	kPa (bar)
		Hydraulic resistance	Pa

★ The word 'litre' may be employed as a special name for dm^3

Table 6.11 Birmingham gauge and standard wire gauge thickness

BG	SWG	Thickness/mm
52	—	0.024
—	50	0.025
50	—	0.030
48	—	0.039
—	48	0.041
46	—	0.049
44	46	0.061
42	—	0.078
—	44	0.081
40	—	0.098
—	42	0.102
38	40	0.122
—	38	0.152
36	—	0.155
—	36	0.193
34	—	0.196
—	34	0.234
32	—	0.249
—	32	0.274
30	—	0.312
—	30	0.315
—	28	0.376
28	—	0.397
—	26	0.457
26	—	0.498
—	24	0.559
24	—	0.629
—	22	0.711
22	—	0.794
—	20	0.914
20	—	0.996
—	18	1.219
18	—	1.257
16	—	1.588
—	16	1.626
14	—	1.994
—	14	2.032
12	—	2.517
—	12	2.642
10	—	3.175
—	10	3.251
8	—	3.988
—	8	4.064
—	6	4.877
6	—	5.032
—	4	5.893
4	—	6.350
—	2	7.010
2	—	7.993
—	0	8.230
—	2/0	8.839
—	3/0	9.449
0	—	10.07
—	4/0	10.16
—	5/0	10.97
2/0	—	11.31
—	6/0	11.79
3/0	7/0	12.70
4/0	—	13.76
5/0	—	14.94

Table 6.12 Preferred iso metric screw thread sizes

Nominal diameter/mm	Pitch/mm Coarse	Fine
1.0	0.25	0.20
1.2	0.25	0.20
1.6	0.35	0.20
2.0	0.40	0.25
2.5	0.45	0.35
3.0	0.50	0.35
4.0	0.70	0.50
5.0	0.80	0.50
6.0	1.0	0.75
8.0	1.25	0.75
10	1.5	0.75; 1.0; 1.25
12	1.75	1.0; 1.25; 1.5
16	2.0	1.0; 1.5
20	2.5	1.0; 1.5; 2.0
24	3.0	1.0; 1.5; 2.0
30	3.5	1.0; 1.5; 2.0; 3.0
36	4.0	1.5; 2.0; 3.0
42	4.5	1.5; 2.0; 3.0; 4.0
48	5.0	1.5; 2.0; 3.0; 4.0
56	5.5	1.5; 2.0; 3.0; 4.0
64	6.0	1.5; 2.0; 3.0; 4.0

Note: Hexagon head bolts and screws are classified as M followed by the nominal diameter, e.g. M10 is a 10 mm diameter bolt.

Table 6.13 Standard values and reference values

Physical quantity	Number	Unit
Avogadro's number	$6.022\ 17 \times 10^{23}$	mol^{-1}
e	2.718 28	★
Gas constants:		
Universal	8.314	$J.mol^{-1}.K^{-1}$
Dry air	287	$J.kg^{-1}.K^{-1}$
Steam	461	$J.kg^{-1}.K^{-1}$
Gravitational acceleration	9.806 65	$m.s^{-2}$
Gravitational constant	66.7	$pN.m^2.kg^{-2}$
π	3.141 59	★
Planck's constant	6.626×10^{-34}	J.s
Permeability of free space	1.257	$\mu H.m^{-1}$
Permittivity of free space	8.854	$pF.m^{-1}$
Sound reference intensity	1	$pW.m^{-2}$
Sound reference power	1	pW
Sound reference pressure	20	μPa
Speed of light in vacuo	299.792	$Mm.s^{-1}$
Speed of sound:		
In dry air at 20°C	343.6	$m.s^{-1}$
In water at 20°C	1 497	$m.s^{-1}$
In copper	4 760	$m.s^{-1}$
In mild steel	5 960	$m.s^{-1}$
Stefan–Boltzmann constant	56.696	$nW.m^{-2}.K^{-4}$
CIBSE reference conditions for air:		
Density	1.200	$kg.m^{-3}$
Pressure	101.325	kPa
Relative humidity	43	%
Specific heat capacity	1.02	$kJ.kg^{-1}.K^{-1}$
Temperature (dry bulb)	20	°C

★ dimensionless quantity

Table 6.14 Dimensionless constants

	Name	Symbol	Definition
Momentum transport	Reynolds number	Re	$Re = \dfrac{\rho v l}{\eta} = \dfrac{vl}{v}$
	Euler number	Eu	$Eu = \dfrac{\Delta p}{\rho v^2}$
	Froude number	Fr	$Fr = \dfrac{v}{\sqrt{lg}}$
	Grashof number	Gr	$Gr = \dfrac{l^3 g \gamma \Delta \theta}{v^2}$
	Weber number	We	$We = \dfrac{\rho v^2 l}{\sigma}$
	Mach number	Ma	$Ma = \dfrac{v}{c}$
	Knudsen number	Kn	$Kn = \dfrac{\lambda^{\star}}{l}$
	Strouhal number	Sr	$Sr = \dfrac{lf}{v}$
Transport of heat	Fourier number	Fo	$Fo = \dfrac{\lambda t}{c_{\mathrm{p}} \rho l^2} = \dfrac{at}{l^2}$
	Péclet number	Pe	$Pe = \dfrac{\rho c_{\mathrm{p}} v l}{\lambda} = \dfrac{vl}{e} = Re.Pr$
	Rayleigh number	Ra	$Ra = \dfrac{l^3 \rho^2 c_{\mathrm{p}} g \gamma \Delta \theta}{\eta \lambda} = \dfrac{l^3 g \gamma \Delta \theta}{va} = Gr.Pr$
	Nusselt number	Nu	$Nu = \dfrac{hl}{\lambda}$
	Stanton number	St	$St = \dfrac{h}{\rho v c_{\mathrm{p}}} = Nu/Pe$
Transport of matter in binary mixture	Fourier number for mass transfer	$Fo^{\star}$	$Fo^{\star} = \dfrac{Dt}{l^2} = Fo/Le$
	Péclet number for mass transfer	$Pe^{\star}$	$Pe^{\star} = \dfrac{vl}{D} = Pe.Le$
	Grashof number for mass transfer	$Gr^{\star}$	$Gr^{\star} = \dfrac{l^3 g \beta \Delta x}{v^2}$
	Nusselt number for mass transfer	$Nu^{\star}$	$Nu^{\star} = \dfrac{kl}{\rho D}$
	Stanton number for mass transfer	$St^{\star}$	$St^{\star} = \dfrac{k}{\rho v} = \dfrac{Nu^{\star}}{Pe^{\star}}$
Constants of matter	Prandtl number	Pr	$Pr = \dfrac{\eta c_{\mathrm{p}}}{\lambda} = \dfrac{v}{a}$
	Schmidt number	Sc	$Sc = \dfrac{\eta}{\rho D} = \dfrac{v}{D}$
	Lewis number	Le	$Le = \dfrac{\lambda}{\rho c_{\mathrm{p}} D} = \dfrac{a}{D} = \dfrac{Sc}{Pr}$

Table 6.15 Symbols used in Table 6.14

Symbol	Name of quantity
l	characteristic length
v	characteristic velocity
$\Delta\theta$	characteristic temperature difference
Δp	pressure difference
θ	temperature
ρ	density (mass density)
η	viscosity (dynamic viscosity)
v	kinematic viscosity : η/ρ
σ	surface tension
g	acceleration due to gravity
γ	cubic expansion coefficient: $-\dfrac{1}{\rho}\left(\dfrac{\partial\rho}{\partial\theta}\right)_{\mathrm{p}}$
$\lambda\star$	mean free path
f	characteristic frequency
c	velocity of sound
h	coefficient of heat transfer: heat/(time $\times$ cross-sectional area $\times$ temperature difference)
t	characteristic time interval
Δx	characteristic difference of mole fraction
β	$\beta = -\dfrac{1}{\rho}\left(\dfrac{\partial\rho}{\partial x}\right)_{\theta,\mathrm{p}}$
D	diffusion coefficient
k	mass transfer coefficient: mass /(time $\times$ cross-sectional area $\times$ mole fraction difference)
c_{p}	specific heat capacity at constant pressure
λ	thermal conductivity
a	thermal diffusivity: $\lambda/\rho c_{\mathrm{p}}$

Table 6.16 The Beaufort scale

Beaufort number	Description of wind	Observations	Limit of wind speed /m.s^{-1}
0	Calm	Smoke rises vertically	Less than 0.5
1	Light air	Direction of wind shown by smoke drift but not by wind vanes	0.5 to 1.5
2	Light breeze	Wind felt on face; leaves rustle; ordinary vane moved by wind	1.5 to 3.0
3	Gentle breeze	Leaves and small twigs in constant motion; wind extends light flag	3 to 6
4	Moderate breeze	Raises dust and loose paper; small branches are moved	6 to 8
5	Fresh breeze	Small trees in leaf begin to sway	8 to 11
6	Strong breeze	Large branches in motion; umbrellas used with difficulty	11 to 14
7	Moderate gale	Whole trees in motion; inconvenience felt when walking into wind	14 to 17
8	Fresh gale	Twigs broken off trees; generally impedes progress	17 to 21
9	Strong gale	Slight structural damage occurs (slates and chimney pots removed from roofs)	21 to 24
10	Whole gale	Seldom experienced inland; trees uprooted; considerable structural damage occurs	24 to 28
11	Storm	Very rarely experienced; accompanied by widespread damage	28 to 32
12	Hurricane	(Yacht crews take up golf)	32 to 36

With acknowledgement to P. Heaton

Table 6.17 Velocity pressure of wind

Velocity/m.s^{-1}	Pressure/Pa	Velocity/m.s^{-1}	Pressure/Pa
0.5	1.56×10^{-1}	10	6.25×10
1	6.25×10^{-1}	11	7.55×10
2	2.5	15	1.39×10^2
3	5.6	20	2.5×10^2
4	10	25	3.9×10^2
5	1.56×10	30	5.63×10^2
6	2.25×10	35	7.61×10^2
7	3.05×10	40	1.0×10^3
8	4.0×10	45	1.27×10^3
9	5.05×10	50	1.56×10^3

Table 6.18 Area and circumference of circles

Diameter/m	Area/m^2	Circumference/m	Diameter/m	Area/m^2	Circumference/m
0.025(25 mm)	0.000 491(491 mm^2)	0.078 5 (78.5 mm)	1.725	2.338	5.419
0.050	0.001 963	0.157 0	1.750	2.406	5.498
0.075	0.004 419	0.235 5	1.775	2.475	5.576
0.100	0.007 854	0.314 2	1.800	2.545	5.652
0.125	0.012 27	0.392 5	1.825	2.616	5.733
0.150	0.017 68	0.471 0	1.850	2.688	5.812
0.175	0.024 06	0.549 5	1.875	2.761	5.891
0.200	0.031 43	0.628 5	1.900	2.836	5.969
0.225	0.039 75	0.707 0	1.925	2.912	6.048
0.250	0.049 10	0.785 5	1.950	2.987	6.126
0.275	0.059 41	0.864 0	1.975	3.064	6.205
0.300	0.070 71	0.942 0	2.000	3.143	6.285
0.325	0.082 96	1.020	2.025	3.219	6.362
0.350	0.096 23	1.099	2.050	3.301	6.440
0.375	0.110 4	1.177	2.075	3.382	6.519
0.400	0.125 7	1.256	2.100	3.465	6.597
0.425	0.141 9	1.334	2.125	3.544	6.676
0.450	0.159 2	1.413	2.150	3.632	6.754
0.475	0.177 2	1.491	2.175	3.715	6.833
0.500	0.196 3	1.570	2.200	3.803	6.911
0.525	0.216 5	1.648	2.225	3.886	6.990
0.550	0.237 5	1.727	2.250	3.975	7.070
0.575	0.259 8	1.805	2.275	4.065	7.147
0.600	0.282 8	1.884	2.300	4.157	7.226
0.625	0.307 0	1.962	2.325	4.246	7.304
0.650	0.332 0	2.041	2.350	4.338	7.383
0.675	0.357 9	2.119	2.375	4.430	7.461
0.700	0.385 0	2.198	2.400	4.526	7.540
0.725	0.413 0	2.276	2.425	4.619	7.618
0.750	0.441 9	2.355	2.450	4.714	7.697
0.775	0.472 0	2.433	2.475	4.811	7.775
0.800	0.502 9	2.512	2.500	4.910	7.855
0.825	0.534 7	2.590	2.525	5.006	7.932
0.850	0.567 6	2.669	2.550	5.109	8.011
0.875	0.601 3	2.747	2.575	5.208	8.090
0.900	0.636 3	2.826	2.600	5.310	8.168
0.925	0.672 2	2.904	2.625	5.412	8.247
0.950	0.708 8	2.983	2.650	5.517	8.325
0.975	0.746 6	3.063	2.675	5.620	8.404
1.000	0.785 4	3.140	2.700	5.727	8.482
1.025	0.825 4	3.218	2.725	5.830	8.561
1.050	0.866 0	3.297	2.750	5.941	8.640
1.075	0.908 0	3.375	2.775	6.048	8.718
1.100	0.950 3	3.454	2.800	6.159	8.796
1.125	0.994 2	3.532	2.825	6.268	8.875
1.150	1.040	3.611	2.850	6.381	8.953
1.175	1.085	3.689	2.875	6.492	9.032
1.200	1.131	3.768	2.900	6.605	9.111
1.225	1.180	3.848	2.925	6.719	9.189
1.250	1.227	3.925	2.950	6.835	9.268
1.275	1.277	4.003	2.975	6.951	9.346
1.300	1.327	4.082	3.000	7.071	9.425
1.325	1.379	4.163	3.025	7.186	9.503
1.350	1.432	4.239	3.050	7.307	9.582
1.375	1.486	4.320	3.075	7.426	9.660
1.400	1.540	4.396	3.100	7.550	9.739
1.425	1.595	4.477	3.125	7.670	9.816
1.450	1.651	4.553	3.150	7.793	9.896
1.475	1.710	4.634	3.175	7.917	9.975
1.500	1.768	4.710	3.200	8.044	10.05
1.525	1.828	4.791	3.225	8.168	10.13
1.550	1.888	4.867	3.250	8.296	10.20
1.575	1.949	4.948	3.275	8.424	10.29
1.600	2.011	5.027	3.300	8.553	10.37
1.625	2.074	5.105	3.325	8.683	10.44
1.650	2.139	5.184	3.350	8.816	10.52
1.675	2.205	5.262	3.375	8.946	10.60
1.700	2.271	5.341	3.400	9.080	10.68

Table 6.19 Geometric formulae

Shape	Plan area A	Centre of gravity G	Radius of gyration k ($I = Ak^2$)
Triangle	$\dfrac{ah}{2}$	$\dfrac{h}{3}$ up $\dfrac{2a_1 + a_2}{3}$ across	about G. x plane $k^2 = \dfrac{h^2}{18}$ about G. y plane $k^2 = \dfrac{(a_1^2 + a_1 a_2 + a_2^2)}{18}$
Square	a^2	at centre	about base $k^2 = \dfrac{a^2}{3}$ about G $k^2 = \dfrac{a^2}{12}$
Rectangle	ab	at centre	about base $k^2 = \dfrac{b^2}{3}$ about G. x plane $k^2 = \dfrac{b^2}{12}$
Hollow square tube	$A^2 - a^2$	at centre	about G. $k^2 = \dfrac{(A^4 - a^4)}{12(A^2 - a^2)}$
Hollow rectangular tube	$AB - ab$	at centre	about G. x plane $k^2 = \dfrac{(AB^3 - ab^3)}{12(AB - ab)}$ about G. y plane $k^2 = \dfrac{(A^3B - a^3b)}{12(AB - ab)}$
I	$AB - 2ab$ $= at + 2BT$	at centre	about G. x plane $k^2 = \dfrac{(B^3(A - a) + a(B - 2b)^3)}{12(AB - 2ab)}$ about G. y plane $k^2 = \dfrac{(A^3B - 2a^3b)}{12(AB - 2ab)}$
Circle	$\dfrac{\pi d^2}{4}$	at centre	about G $k^2 = \dfrac{d^2}{16}$

Table 6.19 — *continued*

Shape	Plan area A	Centre of gravity G	Radius of gyration k $(I = Ak^2)$
Sector of circle θ in radians	$\dfrac{d^2\theta}{8}$	$\dfrac{2d \sin \theta/2}{3\theta}$	about G. x plane $k^2 = \dfrac{d^2}{4}\left(\dfrac{\theta - \sin\theta}{4\theta}\right)$ about G. y plane $k^2 = \dfrac{d^2}{4}\left(\dfrac{\theta + \sin\theta}{4\theta} - \dfrac{16\sin^2\theta/2}{9\theta^2}\right)$
Cone	Volume V $\dfrac{\pi d^2 h}{12}$	$\dfrac{dh}{2}$ $\dfrac{h}{4}$ up at centre across	about G. x plane $k^2 = \dfrac{3(h^2 + d^2)}{80}$ about tip $k^2 = \dfrac{3d^2}{40}$
Cylinder	$\dfrac{\pi h d^2}{4}$	x plane $\dfrac{\pi d^2}{4}$ z plane dh at centre	about G. x plane $k^2 = \dfrac{(4h^2 + 3d^2)}{48}$ about G. z plane $k^2 = \dfrac{d^2}{8}$
Hollow cylinder	$\dfrac{\pi h(D^2 - d^2)}{4}$	x plane $\dfrac{\pi(D^2 - d^2)}{4}$ z plane $h(D - d)$ at centre	about G. x plane $k^2 = \dfrac{(4h^2 + 3(D^2 + d^2))}{48}$ about G. z plane $k^2 = \dfrac{D^2 + d^2}{8}$
Sphere	$\dfrac{\pi d^3}{6}$	$\dfrac{\pi d^2}{4}$ at centre	about G $k^2 = \dfrac{d^2}{10}$
Rectangular block	abc	x plane bc at centre	about G. x plane $k^2 = \dfrac{a^2 + c^2}{12}$

Index